Logic: A History of its Central Concepts

Handbook of the History of Logic

General Editors

Dov Gabbay
John Woods

Handbook of the History of Logic

Volume 11

Logic: A History of its Central Concepts

Edited by

Dov M. Gabbay
Department of Informatics
King's College London, UK

Francis Jeffry Pelletier
Department of Philosophy
University of Alberta, Canada
and
Departments of Philosophy and Linguistics
Simon Fraser University, Canada

John Woods
Department of Philosophy
University of British Columbia, Canada
and
Department of Philosophy
University of Lethbridge, Canada
and
Group on Logic, Information and Computation
King's College London, UK

ELSEVIER

AMSTERDAM • BOSTON • HEIDELBERG • LONDON • NEW YORK • OXFORD
PARIS • SAN DIEGO • SAN FRANCISCO • SINGAPORE • SYDNEY • TOKYO
North Holland is an imprint of Elsevier

N·H

North Holland is an imprint of Elsevier
The Boulevard, Langford lane, Kidlington, Oxford, OX5 1GB, UK
Radarweg 29, PO Box 211, 1000 AE Amsterdam, The Netherlands
225 Wyman Street, Waltham, MA 02451, USA

First edition 2012

British Library Cataloguing in Publication Data
A catalogue record for this book is available from the British Library

Library of Congress Cataloging-in-Publication Data
A catalog record for this book is available from the Library of Congress

ISBN: 978-0-444-52937-4
ISSN: 1874-5857

For information on all North Holland publications
visit our web site at store.elsevier.com

Printed and bound by CPI Group (UK) Ltd, Croydon, CR0 4YY
Transferred to digital print 2012

This book is dedicated with gratitude to
Jane Spurr and Carol Woods.

CONTENTS

PREFACE

The present volume marks the conclusion of the *Handbook of the History of Logic* series. This capstone volume addresses central topics in the history of logic, showing how logicians, philosophers, mathematicians and others understood these topics over the years and how they guided their development down to the present century.

Certainly the most central topic in logic is the notion of *logical consequence.* Asmus and Restall start with Aristotle's definition of a syllogism as "an argument in which, certain things having been assumed, something other than these follows of necessity from their truth, without needing any term from outside" and carry the explanation of this conception through the middle ages and into the twenty-first century. Any account of logical consequence must determine the type of entities that can be premises and conclusion, must explain what ways premises can combine, and crucially, must explain the types of connection that that are allowed to hold between premises and conclusion in order for it to really be a consequence.

A part of this explanation will involve certain connected concepts: the quantifiers and the connectives. Bonevac traces the notion of quantification from Aristotle through modern generalized quantifiers. It is important to note, he says, that there is no theory-neutral way of defining quantification or even of delineating the class of quantifiers, and so a history of quantification has to trace the development of both *what* is to be explained along with *how* it is to be explained. Alongside the account of quantifiers and quantification needs to be an account of the logical particles — the connectives. Bonevac and Dever discuss the implicit treatment of propositional connectives in Aristotle before moving on to the explicit theory of them developed by the Stoics. The development of an understanding of the connectives took a winding path from the Stoics through the medieval *logica vetus* (Old Logic) and the revolutionary *logica nova* (New Logic) of the 13th century, through the under-appreciated algebraic understanding of Leibniz, and to the "Modern-Era logicians" of the 19th and 20th century.

As Bonevac and Dever remark, the history of the connectives is marked by an ambivalence between the attitude that the connectives are operators on the *content* of the items being connected and an attitude that they are operators on the *speech act force* of the items (say, in a *presentation* of an argument). And again, there is the ambivalence between the view that negation is a *propositional* operator ("It is not the case that –") and that it is a *term* operator ("– is not-pale"). The 20th century saw the latter issue decided in favor of the propositional approach for negation. However, the question of whether negation is a content or a speech-act operator (for instance, denial) is disputed. Speranza and Horn start with Paul Grice's account of negation, using it as a springboard to discuss the ways this difference has been in effect over the history of logic.

Aristotle's notion of "following of necessity from the truth of the premises" is often described in terms of the truth-values of the premises and the conclusion: It is not possible for the truth-value of every premise to be true while the truth-value of the conclusion is false. But of course, the history of logic has seen accounts where there are more than two truth-values, and furthermore where there are "gaps" and "gluts" of truth-value. Indeed, the notion of truth-value permeates much of the broader realms of philosophy, linguistics, mathematics and computer science, and Béziau undertakes a very broad-ranging discussion of the "mathematical conception of truth-value" to show how it underlies many of the more familiar conceptions that are associated with that concept.

Modality is yet another central concept in logic. Not only is it employed in Aristotle's definition of a correct argument, but also it features in the characterization of modalized sentences, and thereby into metaphysics and language (*de re* and *de dicto* modalities). Knuuttila explores the ancient and medieval traditions in modality — distinguishing modality as "extensional" (all possibilities will be actualized) from modality as "alternativity" — and showing where these two conceptions emerge in more recent accounts of modality. These differing accounts also are manifested in the modal syllogistic and logics that were developed in ancient and medieval times, Knuuttila shows, as were interpretations of the modalities in terms of epistemic operators like *knows* and *believes*.

One version of logic employs no independently-claimed-to-be-necessarily-true statements (axioms) but instead employs only rules. Although some have claimed that Aristotle's syllogistic is such a system, the more modern version traces its history to 1934. Despite this very recent invention, most logic that is currently taught in philosophy is of this nature — "natural deduction". Pelletier and Hazen discuss the history (since 1934) of this development, its relationship with other conceptions of logic, and the metatheoretic facts that allow it to have such a prominent position in modern logic.

A rather different conception of logic arises when one thinks of sentential implication as the basic operation, and thinks of that operation as asserting some sort of "connection" between the antecedent and consequent. McCall traces this conception from the Stoics (and also Aristotle) through its development in the middle ages, to Ramsey, Nelson and Angell in the 20th century and the interpretation of connexive implication within relevant logics. McCall also displays the results of some empirical studies that seem to favor a "connexive interpretation" of *if–then*.

The notion of logical type was brought into logical prominence with the publication of *Principia Mathematica* in 1910, although as Kamareddine, Laan, and Nederpelt show, the notion was always present in mathematics before then. They trace the development of the theory of types from Russell-Whitehead to Church's simply-typed λ-calculus of 1940, and they show how the logical paradoxes that entered into the formal systems of Frege, Cantor and Peano brought forth the first explicit theory of types in Russell's *Principles of Mathematics*.

A wider notion of logic involves the contrast between good arguments and merely good-seeming arguments. A good argument need not be deductively valid, as we all learn in elementary practical logic; and furthermore, some deductively valid arguments are not good arguments in this sense (e.g., the ones that are obviously circular). So another part of logic that has been passed down to us over the ages is the study of the distinction between good and merely good-seeming arguments. Aristotle was the first to codify this (in the *Topics* and *Sophistical Refutations*), just as he was the first to codify the notion of deductive correctness (in the *Prior Analytics*). Woods traces the evolving notion of a fallacy in argumentation from its Aristotelian beginnings through the late 20th century.

The final topic in this survey of the central topics in the history of logic is the use of diagrams in logical reasoning. Moktefi and Shin survey the very well-known Euler diagrams, showing their modification by Venn and Peirce. Additionally, other diagrammatic traditions — e.g., the use of tables and linear diagrams — are surveyed. With regard to the full predicate logic, Moktefi and Shin explore Frege's two-dimensional graphical notation and Peirce's Existential Graphs. They also address the important question of the place of diagrams as representational systems in their own right, and the possibility of having rules of inference that directly characterize logical consequence in such a visual representation scheme.

The Editors are in the debt of the volume's superb authors. For support and encouragement thanks are also due Paul Bartha, Head of Philosophy at UBC, and Christopher Nicol, Dean of Arts and Science, and Kent Peacock, Chair of Philosophy, both at the University of Lethbridge.

The entire eleven-volume series of *The Handbook of the History of Logic* owes a very special thanks to Jane Spurr, Publications Administrator in London; to Carol Woods, Production Associate in Vancouver; and to our colleagues at Elsevier, Associate Acquisitions Editor Susan Dennis and Senior Developmental Editor (Physical Sciences Books) Derek Coleman. The series really is something very special, and all those people have played a central role in bringing the project to fruition.

The *Handbook* owes its existence to former sponsoring editor Arjen Sevenster, whose support, guidance and frendship the Editors will always remember, with gratitude and affection. The Editors also wish to record their indebtedness to Drs. Sevenster's very able associate, Andy Deelen.

Dov M. Gabbay

Francis Jeffry Pelletier

John Woods

LIST OF AUTHORS

Conrad Asmus
Japan Advanced Institute of Science and Technology, Japan;
conrad.asmus@gmail.com

Jean-Yves Béziau
University of Brazil, Rio de Janeiro (UFRJ), and Brazil Research Council (CNPq), Brazil; jyb@ifcs.ufrj.br

Daniel Bonevac
University of Texas at Austin, USA; bonevac@austin.utexas.edu

Josh Dever
University of Texas at Austin, USA; dever@mail.utexas.edu

Allen P. Hazen
University of Alberta, Canada; aphazen@ualberta.ca

Laurence R. Horn
Yale University, USA; laurence.horn@yale.edu

Fairouz Kamareddine
Heriot Watt University, Scotland; fairouz@macs.hw.ac.uk

Simo Knuuttila
University of Helsinki, Finland; simo.knuuttila@helsinki.fi

Twan Laan
Olten, Switzerland; twanlaan@gmail.com

Storrs McCall
McGill University, Canada; storrs.mccall@gmail.ca

Amirouche Moktefi
IRIST, University of Strasbourg, France; moktefi@unistra.fr

Rob Nederpelt
Technische Universiteit Eindhoven, The Netherlands; r.p.nederpelt@tue.nl

Francis Jeffry Pelletier
University of Alberta, Canada; jeff.pelletier@ualberta.ca

Greg Restall
University of Melbourne, Australia; restall@unimelb.edu.au

Sun-Joo Shin
Yale University, USA; sun-joo.shin@yale.edu

J. L. Speranza
Argentine Society for Philosophical Analysis, Buenos Aires, Argentina;
jlsperanza@aol.com

John Woods
University of British Columbia, Canada; john.woods@ubc.ca

A HISTORY OF THE CONSEQUENCE RELATIONS

Conrad Asmus and Greg Restall

1 INTRODUCTION

Consequence is a, if not *the*, core subject matter of logic. Aristotle's study of the syllogism instigated the task of categorising arguments into the logically good and the logically bad; the task remains an essential element of the study of logic. In a logically good argument, the conclusion follows validly from the premises; thus, the study of consequence and the study of validity are the same.

In what follows, we will engage with a variety of approaches to consequence. The following neutral framework will enhance the discussion of this wide range of approaches. Consequences are conclusions of valid arguments. Arguments have two parts: a conclusion and a collection of premises. The conclusion and the premises are all entities of the same sort. We will call the conclusion and premises of an argument the *argument's components* and will refer to anything that can be an argument component as a *proposition*. The class of propositions is defined functionally (they are the entities which play the functional role of argument components); thus, the label should be interpreted as metaphysically neutral. Given the platonistic baggage often associated with the label "proposition", this may seem a strange choice but the label is already used for the argument components of many of the approaches below (discussions of Aristotlean and Medieval logic are two examples). A consequence relation is a relation between collections of premises and conclusions; a collection of premises is related to a conclusion if and only if the latter is a consequence of the former.

Aristotle's and the Stoics' classes of arguments were different, in part, because their classes of propositions differed. They thought that arguments were structures with a single conclusion and two or more premises;[1] conclusions and premises (that is, propositions) were the category of things that could be true or false. In Aristotlean propositions, a predicate is applied to a subject; the Stoics allowed for the *recombination* of propositions with connectives. Later on, some medieval logicians restricted propositions to particular concrete tokens (in the mind, or spoken, or written).

[1]This is until Antipater, head of the Stoic school around 159 – 130 BCE, who "recognized inference from one premise, his usage was regarded as an innovation" [Kneale and Kneale, 1962, p 163].

Handbook of the History of Logic. Volume 11: Logic: A History of its Central Concepts.
Volume editors: Dov M. Gabbay, Francis Jeffry Pelletier and John Woods.
General editors: Dov M. Gabbay and John Woods.

Changing the class of realisers of the propositional functional role affects the consequence relation. A relation involving only abstract propositions must differ from a relation which involves some of concrete realisers. Not every change in the composition of propositions, however, is equal. If there is a mapping that connects the abstract propositions with the concrete sentences, and the consequence relation on these collections respects this mapping, then the differences are more metaphysical than they are logical. If there is no such mapping, then the choice between these implementations is of serious logical importance.

Aristotle and the Stoics dealt with arguments with two or more premises. Without further investigation of historical details, this can be interpreted in two ways: (1) any argument with fewer than two premises is invalid, or (2) arguments cannot have fewer than two premises. On the first interpretation, there is some necessary requirement for validity that zero and one premise arguments always fail to satisfy. According to some schools: for a conclusion to be a consequence of the premises, it must be genuinely *new*. This makes all single premise arguments invalid. Similarly, a zero premise argument is not one where the conclusion *results from* the premises. This is a choice about what the consequence relation is: whether a consequence has to be new, whether it must result from the premises, and so on. Different approaches to this issue have be taken through the history of logical consequence. Sometimes a rigid adherence to the motivations of a consequence being *new* and *resulting from* premises is maintained; at other times, this is sacrificed for the sake of simplicity and uniformity.

The second interpretation limits how a collection of premises can be structured in an argument. The combination of two propositions (one as a premise and the other as a conclusion) isn't a good argument because it isn't an argument. Premise combination has often been treated rather naively. Recently, careful discussions of premise combination have come out of Gentzen's proof systems and substructural logic. In substructural logics, premises are not combined as unordered sets. Different structural restrictions on the combination of premises, and the ways one is able to manipulate them (structural rules), result in different consequence relations. There has also been a loosening in the forms that conclusions take. Typical arguments *seem* to have exactly one conclusion (see [Restall, 2005] for an argument against this). This led to a focus on single conclusions as consequences of premises. More generally, however, we can investigate whether a *collection* of conclusions is a consequence of a collection of premises.

Any theorist of consequence needs to answer the following questions:

1. What sort of entity can play the role of a premise or of a conclusion? That is, what are propositions?

2. In what ways can premises combine in an argument? In what ways can conclusions combine in an argument?

3. What connection must hold between the premises and the conclusion(s) for the conclusion(s) be a consequence of the premises?

An answer to the first question has two main parts. There is the form of propositions (for example, on Aristotle's view propositions always predicate something of a subject) and the composition of propositions (for example, on a medieval nominalist's theory of propositions they are concrete singulars).

There are two broad approaches to the third question. Some theorists focus on a property of propositions; some theorists focus on connections between conclusions and premises. In both cases, consequence is explicated in terms of something else. In the first approach, the conclusion is a consequence of the premises if and only if, whenever the premises have some specified property, so does the conclusion. This approach focusses on whether the premises and conclusion have the designated property or not, it doesn't *rely* on a strong connection between premises and conclusion. In the paradigmatic example, this property is truth. The second approach is more concerned with the relation between the premises and conclusion. The consequence relation is build on top of another relation between premises and conclusions. If the premises and conclusion of an argument are connected by any number of steps by the basic relation, then the conclusion is a consequence of the premises. Paradigmatic examples are based on proof theories. We will refer to the first type of approaches as property based approaches, and the second as transference based approaches. There are many hybrids of the two approaches. A truth preservation approach sounds like a property based approach, but this depends on what we make of *preservation*. If it is important that the truth of the conclusion is connected via a processes of transference to the truth of the premises, then the approach has both property and transference features.

Different answers to these three questions originate from a variety of sources. Sometimes answers (especially to the first question) come from metaphysics; sometimes answers (especially to the third question) come from epistemology. Importantly, different answers are connected to different properties that consequence relations are expected to have. In the next three sections, we will look at some features that have struck theorists as important properties for consequence relations. Different answers to the three questions often correspond to different emphases on these properties.

Theorists, like Tarski in the quote below, have been aware that there are many tensions in developing an account of consequence. There is usually a trade off between precision, adherence to everyday usage of the concept, and with adherence to past accounts. Any precise account will be, to some extent, revisionary. In [Tarski, 1956b, p 409] Tarski says,

> The concept of *logical consequence* is one of those whose introduction into the field of strict formal investigation was not a matter of arbitrary decision on the part of this or that investigator; in defining this concept, efforts were made to adhere to the common usage of the language of everyday life. But these efforts have been confronted with the difficulties which usually present themselves in such cases. With respect to the clarity of its content the common concept of consequence is in no way superior to other concepts of everyday language. Its extension

> is not sharply bounded and its usage fluctuates. Any attempt to bring into harmony all possible vague, sometimes contradictory, tendencies which are connected with the use of this concept, is certainly doomed to failure. We must reconcile ourselves from the start to the fact that every precise definition of this concept will show arbitrary features to a greater or less degree.

This leaves the theorist with a final question to answer: What is the point of the theory? Precision, accord with everyday usage, accord with the normative constraints on reasoning, and many other answers have been forthcoming in the history of logical consequence.

1.1 Necessity and Counterexamples

Aristotle categorised syllogisms into those that are deductions and those that are not. The distinguishing feature of a deduction is that the conclusion *necessarily* results from the premises. That consequences follow of necessity from premises was one of the earliest characteristic features of consequence to be emphasised. It is not always easy to determine, however, what substance theorists impart into this necessity. The way in which theorists categorise arguments provides insight into how they understand necessity.

Aristotle, the Stoics, the medievals, Leibniz, Kant, and many more of the logicians and philosophers dealt with in this entry discuss necessity and modal logic. Of particular importance is Leibniz's account of necessity. A proposition is necessary if it is true in all possible worlds. There are two important parts of this move. Firstly, the notion of *possible world* is introduced. Possible worlds can serve as a type of *counterexample*. If it is possible for the premises of an argument to be true, and the conclusion false, then this is taken to demonstrate that the argument is invalid, and thus that the conclusion is not a consequence of the premises. Secondly, necessity is fixed as truth in *every* possible world. Universal quantification over possible worlds is a genuine advancement: for example, consider the equivalence of $\Box(A \wedge B)$ with $\Box A \wedge \Box B$.

A conclusion that is a consequence of a collection of premises should hold in *any* situation in which the premises do. Logical consequence can be used to reason about hypothetical cases as well as the actual case; the conclusion of a good argument doesn't merely follow *given the way things are* but will follow *no matter how things are*.

A characterisation of logical consequence in terms of necessity can lead away from the transference approach to consequence. A demonstration that there are no counterexamples to an argument needn't result in a recipe for the connecting the premises and conclusion in any robust sense. Necessity is not, however, anathema to the transference approach. If the appropriate emphasis is placed in "necessarily *results from*" and "consequences *follow* of necessity", and this is appropriately implemented, then transference can still be respected.

1.2 *Formality and Structure*

Necessity is not sufficient for *logical* consequence. Consider the argument:

All logicians are blue.

Some blue objects are coloured.

Therefore, all logicians are coloured.

It seems that, if the premises of the argument are true, the conclusion *must* also be; the conclusion seems to follow of necessity from the first premise. This is not a *formally* valid argument. That the conclusion is necessitated *relies* on all blue objects being coloured. This reliance disqualifies it as a logical consequence. A conclusion is a formal consequence of a collection of premise not when there is merely no possibility of the premises being true and conclusion false, but when it has an *argument form* where there is no possibility of *any* instance of the form having true premises and a false conclusion. Counterexamples are not only counterexamples to arguments but to argument forms.

In this example, there are counterexamples to the argument form:

All αs are βs.

Some βs are γs.

Therefore, all αs are γs.

If the argument is not an instance of any other valid argument form, it is not valid and the conclusion is not a formal consequence of the premises. Argument forms and instances of argument forms play a crucial role in logical consequence; in some ways they are more central than arguments. Logical consequence is *formal* in at least this respect.

Formal consequence is not the only relation of consequence that logicians have studied. Some logicians have placed a high level of importance on *material* consequence. A conclusion is a material consequence of a collection of premises if it follows either given the way things are (so not of necessity) or follows of necessity but not simply because of the form of the argument. In order to properly distinguish material and formal consequence we require a better characterisation of the forms of propositions and of arguments.

That logical consequence is schematic, and in this sense formal, is a traditional tenet of logical theory. There is far more controversy over other ways in which consequence may be formal. The use of schemata is not sufficient for ruling out the sample argument about blue logicians. The argument appears to be of the following form:

$(\forall x)(Lx \rightarrow x \text{ is blue})$

$(\exists x)(x \text{ is blue} \wedge x \text{ is coloured})$

Therefore, $(\forall x)(Lx \to x \text{ is coloured})$,

where L is the only schematic letter. There are no instances of this schema where it is possible for the premises of the argument to be true and the conclusion false. Whether this counts as a legitimate argument form depends on what must be, and what may be, treated schematically. This choice, in turn, rests on the other ways in which consequence is formal.

Sometimes logic is taken to be "concerned merely with the form of thought" [Keynes, 1906, p 2]. This can be understood in a number of ways. Importantly, it can be understood as focussing on the general structure of propositions. If propositions have some general form (a predicate applied to a subject, has some recursive propositional structure, and so on) then consequence is formal in that it results from the logical connections between these forms. In MacFarlane's discussion of the formality of logic, this is described as (1) "logic provides constitutive norms for thought as such" [MacFarlane, 2000, p ii]. The other two ways in which logic can be formal what MacFarlane points out are:

(2) logic is "indifferent to the particular identities of objects."

(3) logic "abstracts entirely from the *semantic content* of thought."

He argues, convincingly, that Kant's logic was formal in all three senses, but that later theorists found themselves pressured into choosing between them.

1.3 A Priori *and Giving Reasons*

Logical consequence is often connected to the practice of reason giving. The premises of a valid argument are reasons for the conclusion. Some transference approaches take logical consequence to rest on the giving of reasons: C is a consequence of the premises Δ if and only if a justification for C can be constructed out of justifications for the premises in Δ. Logical consequence, on this view, is about the transformation of reasons for premises into reasons for conclusions.

Most reason giving doesn't rely entirely on logical consequence. Lots of reasoning is *ampliative*; the conclusion genuinely says more than the combination of the premises. The common example is that *there is smoke* is a reason for that *there is fire*. The argument:

There is smoke.

Therefore, there is fire.

is invalid — the conclusion is not a logical consequence of the premise. It is a *material consequence* of the premise. In this entry, we ill focus on *logical* consequence. In logical reason giving, the reasons are *a priori* reasons for the conclusion. That the premises of a valid argument are reasons for the conclusion does not rely on any further evidence (in this example, regarding the connections between smoke and fire).

Some rationalists, the rationalistic pragmatists, hold that material consequence is also, in some sense, *a priori* (e.g. Sellars [Sellars, 2007, especially p 26]). Material consequences are, however, not *necessary* in the same way. A counterexample to a material consequence does not immediately force a revision of our conceptual scheme on us. This is not true with logical consequence: either the purported counterexample must be rejected, or the purported logical consequence must be. This necessity is closely connected to the normativity of logical and material consequence. I can believe that there is smoke and that there isn't fire, *so long as* I also believe that this is an exceptional situation. There is no similar exception clause when I accept the premises of an instance of *modus ponens* and reject its conclusion.

The connection between logical consequence and the giving of reasons highlights the normative nature of consequence. If an argument is valid and I am permitted to accept the premises, then I am permitted to accept the conclusion. Some theoriests make the stronger claim that if one accepts the premises of a valid argument, then one ought to accept the conclusion. One of the many positions between these positions is that if one accepts the premises of a valid argument, then one ought not reject the conclusion.

A focus on the giving of reasons and the normativity of logical consequence is often the result of an aim to connect logical consequence to human activity — to concrete cases of reasoning. Logical consequence, from this perspective, is the study of a particular way in which we are obligated and entitled to believe, accept, reject and deny.

2 ARISTOTLE [384 BCE–322 BCE]

Aristotle's works on logic are the proper place to begin any history of consequence. They are the earliest formal logic that we have and have been immensely influential. Kant is merely one example of someone who thought that Aristotle's logic required no improvement.

> It is remarkable also that to the present day this logic has not been able to advance a single step, and is thus to all appearance a closed and completed body of doctrine. If some of the moderns have thought to enlarge it . . . , this could only arise from their ignorance of the peculiar nature of logical science.
>
> [Kant, 1929, Bviii–ix]

Aristotle categorised syllogisms based on whether they were *deductions*, where the conclusion is a consequence of the premises.

According to Aristotle, propositions are either simple — predicating a property of a subject in some manner — or can be analysed into a collection of simple propositions. There are three parts to any simple proposition: subject, predicate and kind. In non-modal propositions predicates are either affirmed or denied of the subject, and are affirmed or denied either in part or universally (almost everything is controversial in the modal cases).

Subjects and predicates are *terms.* Terms come in two kinds: universal and individual. Universal terms can be predicates and subjects (for example: children, parent, cat, weekend). Individual terms can only be the subject of a proposition (for example: Plato, Socrates, Aristotle). A proposition which seems to have a individual term in the predicate position is, according to Aristotle, not a genuine proposition but merely an accidental predication that depends on a genuine predication for its truth (for example, "The cat on the mat is Tully" depends on "Tully is on the mat").

A proposition can be specified by nominating a subject, a predicate and a kind. Here are some examples with universal terms and the four non-modal kinds: universal affirmation, partial affirmation, universal denial and partial denial:

Example	Kind	Code
All *children* are *happy.*	Universal Affirmative	A
No *weekends* are *relaxing.*	Universal Negative	E
Some *parents* are *tired.*	Particular Affirmative	I
Some *cats* are not *friendly.*	Particular Negative	O

Any collection of propositions is a syllogism; one proposition is the conclusion and the rest are premises. Aristotle gives a well worked out categorisation of a subclass of syllogisms: the *categorical* syllogisms. A categorical syllogism has exactly two premises. The two premises share a term (the middle term); the conclusion contains the other two terms from the premises (the extremes). There are three resulting figures of syllogism, depending on where each term appears in each premise and conclusion. Each premise and conclusion (in the non-modal syllogisms) can be one of the four kinds in the table above.

The syllogisms are categorised by whether or not they are deductions.

> A deduction is a discourse in which, certain things having been supposed, something different from the things supposed results of necessity because these things are so. By 'because these things are so', I mean 'resulting through them,' and by 'resulting through them' I mean 'needing no further term from outside in order for the necessity to come about.' [Smith, 1989, Prior Analytics A1:24b]

The following example is a valid syllogism in the second figure with E and I premises and an O conclusion (it has come to be called "Fest$i$no").

No weekends are relaxing.

Some holidays are relaxing.

Therefore, some holidays are not weekends.

Aristotle categorises syllogisms based on their form. He justifies this particular argument's form:

No Bs are As.

Some Cs are As.

Therefore, some Cs are not Bs.

in a two step procedure. Aristotle transforms the argument form by converting the premise "No Bs are As" into the premise "No As are Bs". This transforms the second figure Festino into the first figure Ferio. The justification of Festino rests on the justification of the conversion and the justification of Ferio. Here is Aristotle's justification of the former:

> Now, if A belongs to none of the Bs, then neither will B belong to any of the As. For if it does belong to some (for instance to C), it will not be true that A belongs to none of the Bs, since C is one of the Bs. [Smith, 1989, Prior Analytics A2:25a]

There is no justification for the latter: merely an assertion that the conclusion follows of necessity.

Aristotle uses a counterexample to show that the syllogistic form:

All Bs are As

No Cs are Bs

Therefore, All Cs are As

is invalid. He reasons in the following way:

> However, if the first extreme [A] follows all the middle [B] and the middle [B] belongs to none of the last [C], there will not be a deduction of the extremes, for nothing necessary results in virtue of these things being so. For it is possible for [A] to belong to all as well as to none of the last [C]. Consequently, neither a particular nor a universal conclusion becomes necessary; and, since nothing is necessary because of these, there will not be a deduction. Terms for belonging to every are animal, man, horse; for belonging to none, animal, man, stone. [Smith, 1989, Prior Analytics A4:26a]

Aristotle concludes that the argument form is not a deduction as the syllogism:

All men are animals

No stones are men

Therefore, All stones are animals

is of the same form but one has true premises and a false conclusion, so the conclusion of the other syllogism cannot follow of necessity.

3 STOICS [300 BCE–200 CE]

The Stoic school of logicians provided an alternative to Aristotle's logic. The Stoic school grew out of the Megarian and Dialectical schools.[2] The Megarians and the members of the Dialectical school contributed to the development of logic by their attention to paradoxes, a careful examination of modal logic and by debating the nature of the conditional (notably by Philo of Megara). Eubulides was particularly noted among the Megarians for inventing paradoxes, including the liar paradox, the hooded man (or the Electra), the sorites paradox and the horned man. As we will return to the lair paradox when discussing the medieval logicians, we will formulate it here. "A man says that he is lying. Is what he says true or false?" [Kneale and Kneale, 1962, p 114]. If the man says something true, then it seems that he is indeed lying — but if he is lying he is not saying something true. Similarly, if what the man says is false, then what he says is not true and, thus, he must be lying — but he says that he is lying and we have determined that he is lying, so what he says is true. Diodorus Cronus is well know for his *master argument*. Diodorus' argument is, plausibly, an attempt to establish his definition of modal notions.

According to Epictetus:

> The Master Argument seems to have been formulated with some such starting points as these. There is an incompatibility between the three following propositions, "Everything that is past and true is necessary", "The impossible docs not follow from the possible", and "What neither is nor will be is possible". Seeing this incompatibility, Diodorus used the convincingness of the first two propositions to establish the thesis that nothing is possible which neither is nor will be true. [Kneale and Kneale, 1962, p 119]

The reasoning involved in the argument is clearly non-syllogistic and the modal notions involved are complex.

The Stoic school was founded by Zeno of Citium, succeeded in turn by Cleanthes and Chrysippus. The third of these was particularly important for the development on Stoic logic. Chrysippus produced a great many works on logic; we encourage the reader to look at the list of works that Diogenes Laertius attributes to him [Hicks, 1925, pp 299 – 319].

A crucial difference between the Stoic and Aristotelean schools is the sorts of propositional forms they allowed. In Aristotle's propositions, a predicate is affirmed or denied of a subject. The Stoics allowed for complex propositions with a recursive structure. A proposition could be basic or could contain other propositions put together with propositional connectives, like the familiar *negation*,

[2]There is some controversy regarding who belongs to which school. The members of the Dialectical group were traditionally thought of as Megarians. For our discussion, the most noticeable of these are Philo of Megara and Diodorus Cronus. In our limited discussion it will not hurt to consider the groups as one.

conditional, *conjunction* and *disjunction*, but also the connectives *Not both ... and ...*; *... because ...*; *... rather than ...* and others. The Stoics had accounts of the meaning and truth conditions of complex propositions. This came close to modern truth table accounts of validity but, while meaning and truth were sometimes dealt with in a truth-table-like manner, validity was not.

Chrysippus recognised the following five indemonstrable moods of inference, [Kneale and Kneale, 1962, p 163] [Bury, 1933, *Outlines of Pyrrhonism* II. 157f]:

1. If the first, then the second; but the first; therefore the second.
2. If the first, then the second; but not the second; therefore not the first.
3. Not both the first and the second; but the first; therefore not the second.
4. Either the first or the second; but the first; therefore not the second.
5. Either the first or the second; but not the second; therefore the first.

These indemonstrable moods could be used to justify further arguments. The arguments, like Aristotle's categorical syllogisms, have two premises. The first premise is always complex. Notice that, even though the Stoics had a wide range of propositional connectives, only the conditional, disjunction and negation conjunction and (possibly) negation appear in these indemonstrables. This is an example of a transference style approach to logical consequence.

4 MEDIEVALS [476 CE–1453 CE]

Logic was a foundational discipline during the medieval period. It was considered to have intrinsic value and was also regarded as an important groundwork for other academic study. Medieval logic is often divided into two parts: the old and the new logic. The demarcation is based on which Aristotelian texts were available. The old logic is primarily based on Aristotle's *Categories* and *De interpretatione* (this includes discussions on propositions and the square of opposition, but importantly lacks the *prior analytics*, which deals with the syllogism) while the new logic had the benefit of the rest of Aristotle's *Organon* (in the second half of the 12th century). Many medieval logicians refined Aristotle's theory of the syllogism, with particular attention to his theory of modal logic. The medieval period, however, was not confined to reworking ancient theories. In particular, the *terminist* tradition produced novel and interesting directions of research. In the later medieval period, great logicians such as Abelard, Walter Burley, William of Ockham, the Pseudo Scotus, John Buridan, John Bradwardine and Albert of Saxony made significant conceptual advances to a range of logical subjects.

It is not always clear *what* the medieval logicians were doing, nor *why* they were doing it [Spade, 2000]. Nevertheless, it is clear that consequence held an important place in the medieval view of logic, both as a topic of invesitagation and as a tool to use in other areas. Some current accounts of logical consequence have remarkable similarities to positions from the medieval era. It is particularly interesting that early versions of standard accounts of logical consequence were considered *and rejected* by thinkers of this period (in particular, see Pseudo-Scotus and Buridan below).

The medievals carried out extensive logical investigations in a broad range of areas (including: inference and consequence, grammar, semantics, and a number of disciples the purpose of which we are still unsure). This section will only touch on three of these topics. We will discuss theories of consequentiæ, the medieval theories of consequence. We will describe how some medievals made use of consequence in solutions to insolubilia. Lastly, we'll discuss the role of consequence in the medieval area of obligationes. This third topic is particularly obscure; it will serve as an example of where consequence plays an important role but is not the focus of attention.

4.1 Consequentiæ

The category of consequentiæ was of fluctuating type. It is clear that in Abelard's work a consequentiæ was a true conditional but that in later thinkers there was equivocation between true conditionals, valid one premise arguments, and valid multiple premise arguments. This caused difficulties at times but what is said about consequentiæ is clearly part of the history of logical consequence.

The medievals broadened the range of inferences dealt with by accounts of consequentiæ from the range of consequences that Aristotle and the Stoics considered. In the following list, from [Kneale and Kneale, 1962, pp 294 – 295], items (3), (4), (9) and (10) are particularly worth noting:

1. From a conjunctive proposition to either of its parts.
2. From either part of a disjunctive proposition to the whole of which it is a part.
3. From the negation of a conjunctive proposition to the disjunction of the negations of its parts, and conversely.
4. From the negation of a disjunctive proposition to the conjunction of the negations of its parts, and conversely.
5. From a disjunctive proposition and the negation of one of its parts to the other part.
6. From a conditional proposition and its antecedent to its consequent.
7. From a conditional proposition and the negation of its consequent to the negation of its antecedent.
8. From a conditional proposition to the conditional proposition which has for antecedent the negation of the original consequent and for consequent the negation of the original antecedent.
9. From a singular proposition to the corresponding indefinite proposition.

10. From any proposition with an added determinant to the same without the added determinant.

Like Aristotle and the Stoics, the medievals investigated the logic of modalities. The connections they drew between modalities, consequentiæ and the "follows from" relation are interesting. Ockham gives us the rules [Kneale and Kneale, 1962, p 291]:

1. The false never follows from the true.
2. The true may follow from the false.
3. If a consequentiæis valid, the negative of its antecedent follows from the negative of its consequent.
4. Whatever follows from the consequent follows from the antecedent.
5. If the antecedent follows from any proposition, the consequent follows from the same.
6. Whatever is consistent with the antecedent is consistent with the consequent.
7. Whatever is inconsistent with the consequent is inconsistent with the antecedent.
8. The contingent does not follow from the necessary.
9. The impossible does not follow from the possible.
10. Anything whatsoever follows from the impossible.
11. The necessary follows from anything whatsoever.

Unlike the Stoics approach to "indemonstrables", the medievals provided analyses of consequentiæ. According to Abelard, consequentiæ form a sub-species of *inferentia*. An inferentia holds when the premises (or, in Abelard's case the antecendent) necessitate the conclusion (consequence) in virtue of their meaning (in modern parlance, an inferentia is an entailment, and the "in virtue of" condition makes the relation *relevant*). The inferentia are divided into the perfect and the imperfect. In perfect inferentia, the necessity of the connection is based on the structure of the antecendent — "if the necessity of the consecution is based on the arrangement of terms regardless of their meaning" [Boh, 1982, pp 306]. The characteristic features of perfect inferentia are remarkably close to Balzano's analysis of logical consequence.

Buridan and Pseudo-Scotus

Buridan, Pseudo-Scotus and other medieval logicians argued against accounts of consequence that were based on necessary connections. Pseudo-Scotus and Buridan provide apparent counterexamples to a range of definitions of consequence. In this section, we look at three accounts of consequence and corresponding purported counterexamples. (We rely heavily on [Boh, 1982] and [Klima, 2004].)

The first analysis we consider is:

(A) A proposition is a consequence of another if it is impossible for the premise (antecedent) to be true and conclusion (consequent) not to be true.

Buridan offers a counterexample:

> the following is a valid consequence: 'every man is running; therefore, some man is running' still, it is possible for the first proposition to be true and for the second not to be true, indeed, for the second not to be.[3] [Klima, 2004, p 95 – 96]

Buridan's argument is a counterexample because his propositions are contingent objects; the proposition "Some man is running" could fail to exist. What doesn't exist, can't be true; so, it is possible for the premise to be true and not the conclusion. The ontological status of propositions can be as important to an account of consequence as the forms of propositions. Whereas our introduction to this entry argued that necessity is not a *sufficient* condition for consequence, Buridan's example is purported to show that it is not even a *necessary* condition.

Pseudo-Scotus provides a counterexample in a similar style: "Every proposition is affirmative, therefore no proposition is negative" [Boh, 1982, p 308]. In this case, the premise is true in situations where all negative propositions (including the conclusion) are destroyed.

In order to avoid these counterexamples, a definition of consequence has to take the contingency of propositions into account. Pseudo-Scotus responds to the definition

(B) For the validity of a consequence it is necessary and sufficient that it be impossible for things to be as signified by the antecedent without also being as signified by the consequent. [Boh, 1982, p 308]

with the argument *No chimaera is a goat-stag; therefore a man is a jack-ass.* The argument is thought to be invalid (having a true premise and false conclusion), but not ruled out by the proposed definition.

Both Pseudo-Scotus and Buradin consider the definition

(C) For a consequence to be valid it is necessary and sufficient that it be impossible that if the antecedent and the consequent are formed at the same time, the antecedent be true and the consequent false.[4] [Boh, 1982, p 308]

Pseudo-Scotus gives the example: *God exists; therefore this consequence is not valid* [Boh, 1982, p 308], which is meant to be invalid but satisfies the definition. The argument cannot be valid — assuming that the argument is valid is self defeating. The premise and the conclusion are both, apparently, necessary propositions

[3]The validity of the argument depends on existential import for universal quantifiers, a topic which we need not go into here.

[4]Alternatively: "that proposition is the antecedent with respect to another proposition which cannot be true while the other is not true, when they are formed together." [Klima, 2004, p 96]

and so it is impossible that they are formed at the same time when the former is true and the latter false. Pseudo-Scotus ultimately accepts this definition, but allows for exceptions, in light of this example, and calls for further investigation.

Buridan's counterexample is "No proposition is negative; therefore, no donkey is running" [Klima, 2004, p 96]. In this example, the premise cannot be formed without falsifying itself. It is impossible for the premise to be formed and true, so it is impossible for premise and conclusion to be formed with the premise true and the conclusion false. The argument meets the definition but isn't valid. The invalidity of the argument is justified on the assumption that logical consequence supports contraposition (If the argument A therefore B is valid, so is the argument Not A therefore Not B), and the contrapositive of this argument is clearly invalid.

Buridan favours a blending of (B) and (C):

> Therefore, others define [antecedent] differently, [by saying that] that a proposition is antecedent to another which is related to it in such a way that it is impossible for things to be in whatever way the first signifies them to be without their being in whatever way the other signifies them to be, when these propositions are formed at the same time. [Dutilh Novaes, 2007, p 103]

His approach requires using both the notions of what a proposition signifies and a proposition being formed; for more details see [Dutilh Novaes, 2007].

4.2 Self-Reference and Insolubilia

The medieval logicians devoted considerable effort to paradoxes and *insolubilia*. There was no sense of danger in discussions of insolubilia [Spade, 2002]. There was no fear that logic might be unable to solve insolubilia nor any fear that insolubilia were signs of incoherences in logic. Discussions of insolubilia were aimed at discovering the right way to deal with these examples, not at vindicating logic. The medieval's primary example was the lair paradox (introduced in the Stoics section above); we also will focus on this insolubilia.

There was a range of approaches to insolubilia in the medieval period.[5] There are many similarities between the types of medieval and modern day responses to the lair. For example, some theories dealt with the paradox by restricting self-reference (e.g. Burley and Ockham). The approaches of Roger Swyneshead, and William Heytesbury, Gregory of Rimini, John Buridan and Albert of Saxony are all worth further discussion, but we will restrict ourselves to Bradwardine and Buridan (with a brief mention of Swyneshead).

Bradwardine's solution relies on the connection between signification and truth. A proposition is true if it only signifies things which are the case, a proposition is false it signifies anything which is not the case. Bradwardine held that

[5]For a survey of these types, and their connections to modern approaches, see [Dutilh Novaes, 2008a].

a proposition signifies everything which, in some sense, *follows* from it (the closure principle); the details, however, of the closure principle are currently contested. What is crucial is that this is meant to justify the thesis: Any proposition which signifies that it is false, signifies that it is true and false. From this it follows that the liar is false, but not that it is true. Spade's [Spade, 1981; Spade and Read, 2009] interpretation of the closure principle is that any consequence of s is signified by s (where $s : P$ is that s signifies P and $\Rightarrow$ is "follows from", this is $(\forall s)(\forall P)((s \Rightarrow P) \rightarrow s : P)$). Read's version [Read, 2002], based on making sure that Bradwardine's proof of the thesis works, is that s signifies anything which is a consequence of something it signifies $((\forall s)(\forall P)(\forall Q)((s : P \wedge (P \Rightarrow Q)) \rightarrow s : Q))$. In both versions, a type of consequence is an important element of signification.[6] In the previous section, we saw that Buridan's account of consequentiæ relied on signification; this section shows that some theories of signification involve some sort of consequence.

Buridan, like may other medieval logicians, blocked the liar paradox in a similar way to Bradwardine. In these approaches, the liar is false but it doesn't follow from this that it is true. If the liar is true, then everything that it signifies follows, including that it is false. If the liar is false, it *doesn't* follow that it is true. Both Bradwardine's and Buradin's solutions have additional requirements for truth, and thus the paradox is blocked. In Bradwardine's case, the liar signifies its truth and its falsity. To show that it is true, one has to demonstrate that it is true and that it is false. Buridan's early solution was similar; he thought that *every* proposition signified its own truth. He rejected this solution on metaphysical grounds; his nominalism and propositional signification were irreconcilable. He replaced this with the principle that every proposition entails another that claims that the original proposition is true. The liar is simply false because, while what it claims is the case (it claims that it is false, and is false), no entailed proposition which claims the lair's truth can be supplied. A form of consequence is again crucial to the solution. Buridan's account of the lair is connected to complex metaphysical and semantic theories, and has we have already seen, this drives him to define logical consequence in a very particular way.

Medieval logicans were aware that their solutions to the lair paradox via theories of truth were connected to theories of consequence. Klima [Klima, 2004, Section 4] argues that, as Buridan's theory of consequence doesn't require an account of truth, his solution to the liar is not bound by the same requirements as Bradwardine's. One of the consequences of Roger Swineshead's solution to the liar is that the argument:

> The consequent of this consequence is false; therefore, the consequent of this consequence is false.
>
> [Kretzmann *et al.*, 1982, p 251]

is a valid argument which doesn't preserve truth (the premise is true, and the conclusion is false)! Swineshead's position is that while not all consequences preserve

[6] See [Dutilh Novaes, 2009] for further discussion.

truth, they all preserve correspondence to reality.

4.3 *Obligationes*

Obligationes are among the more obscure areas in which medieval logicians worked [Spade, 2000; Stump and Spade, 1982]. An obligationes is a stylised dispute between two parties: the *opponent* and the *respondent*. The name 'obligationes' seems to be drawn from the manner in which the parties are obligated to respond within the dispute.

There are a variety of types of obligationes, the most common of which are called *positio*. In this form of obligatione, the opponent begins by making a posit — the *positum*. The *positum* is either admitted or denied by the respondent. If the respondent admits the proposition, then the opponent continues with further propositions. This time the respondent has three options: they can concede, deny or doubt. Their responses have to be in accord with their obligations. How this is dealt with varies between authors.

The rules of Walter Burley's were standard among earlier authors:

> For positio, Burley gives three fundamental rules of obligations:
>
> (1) Everything which follows from (a) the positum, with (b) a granted proposition or propositions, or with (c) the opposite(s) of a correctly denied proposition or propositions, known to be such, must be granted.
>
> (2) Everything which is incompatible with (a) the positum, with (b) a granted proposition or propositions, or with (c) the opposite(s) of a correctly denied proposition or propositions, known to be such, must be denied.
>
> (3) Everything which is irrelevant (impertinens) [that is, every proposition to which neither rule (1) nor rule (2) applies] must be granted or denied or doubted according to its own quality, that is, according to the quality it has in relation to us [i.e., if we know it to be true, we grant it; if we know it to be false, we deny it; if we do not know it to be true and do not know it to be false, we doubt it].
>
> [Stump and Spade, 1982, p 322]

The positio ends in one of two ways. If the respondent is forced to a contradictory position, they lose. If the respondent doesn't lose in a predetermined amount of time, the opponent loses.

There are numerous suggestions as to what purpose obligationes served. One suggestion is that they were logic exercises for students. This is often dismissed on the grounds that some of highly respected logicians (for example, Burley and Ockham) seem to have put more effort into them than mere exercises deserve

[Dutilh Novaes, 2007, p 147]. Spade [Spade, 1982] suggested that obligationes provided a framework for exploring counterfactuals (but later withdrew this suggestion). Stump has [Stump and Spade, 1982; Stump, 1985] suggested that this was a framework for dealing with *insolubilia*. Catarina Dutilh Novaes [2005b; 2006; 2007] has suggested that obligationes correspond to game theoretic consistency maintenance.

Consequence plays an important role within obligationes. The parties must proceed without violating the rules, and the rules often depend on consequence and the closely related notions of incompatibility and relevance. If Dutilh Novaes' suggestion is correct, then obligationes are a different mechanism for determining consequences of some type; the disputes tell us what we ought to and may accept or reject, based on our prior commitments.

5 LEIBNIZ [1646–1716]

Gottfried Leibniz's contributions to philosophy, mathematics, science and other areas of knowledge are astonishing. He was, quite simply, a genius. He is, perhaps, best known for discovering the calculus (at roughly the same time as Newton), but his contributions to philosophy (metaphysics, epistemology, philosophy of religion), to physics and other mathematical achievements should not be ignored. His work in logical theory foreshadowed many later advancements [Leibniz, 1966; Peckhaus, 2009]. Here, we note two directions that his work took. First, Leibniz pushed for a mathematization of logic.

> ... it seems clear that Leibniz had conceived the possibility of elaborating a basic science which would be like mathematics in some respects, but would include also traditional logic and some studies as yet undeveloped. Having noticed that logic, with its terms, propositions, and syllogisms, bore a certain formal resemblance to algebra, with its letters, equations and transformations, he tried to present logic as a calculus, and he sometimes called his new science universal mathematics ... There might be calculi concerned with abstract or formal relations of a non-quantitative kind, e.g. similarity and dissimilarity, congruence, inclusion ... It would cover the theory of series and tables and all forms of order, and be the foundation of other branches of mathematics such as geometry, algebra, and the calculus of chances. But most important of all it would be an instrument of discovery. For according to his own statement it was the *ars combinatoria* which made possible his own achievements in mathematics ... [Kneale and Kneale, 1962, pp 336–337]

Leibniz's approach to logic and philosophy required representing language as a calculus. Complex terms (subjects and predicates) were analysed into parts and given numerical representations. The numerical representations could be used to determine the truth of certain propositions. Surprisingly, Leibniz thought that the calculus, if developed correctly, would not only determine logical or analytic truths but all universal affirmative propositions. According to Leibniz, a universal affirmative proposition was true only if the representation of the subject *contained* the representation of the predicate.

> From this, therefore, we can know whether some universal affirmative proposition is true. For in this proposition the concept of the subject ... always contains the concept of the predicate. ... [I]f we want to know whether all gold is metal (for it can be doubted whether, for example fulminating gold is still a metal, since it is in the form of powder and explodes rather than liquefies when fire is applied to it in a certain degree) we shall only investigate whether the definition of metal is in it. That is, by a very simple procedure ... we shall investigate whether the symbolic number of gold can be divided by the symbolic number of metal. [Leibniz, 1966, pp 22]

Leibniz's very strong notion of truth is closely connected with Kant's later notion of analytical truth.

Secondly, Leibniz provided a detailed account of necessity based on possible worlds. He had a theory of possible worlds as collections of individuals (or, more precisely, of individual concepts), the actual world being the only world with all and only the actual individuals. This possible world based approach to necessity is highly influential in current approaches to necessity. As necessity is a core feature of consequence, this was a remarkable advancement in understanding logical consequence. Leibniz recognized that possible worlds could be used in understanding consequence, as well as connecting it to probability theory and other areas.

6 KANT [1724–1804]

Kant's characterisation of logic was immensely important for the later development of philosophy of logic. Kant seems to have taken Aristotle's work to provide a "completed body of doctrine" to which nothing sensible could be added.

> That logic has already, from the earliest times, proceeded upon this sure path is evidenced by the fact that since Aristotle it has not required to retrace a single step, unless, indeed, we care to count as improvements the removal of certain needless subtleties or the clearer exposition of its recognised teaching, features which concern the elegance rather than the certainty of the science. It is remarkable also that to the present day this logic has not been able to advance a single step, and is thus to all appearance a closed and completed body of doctrine. If some of the moderns have thought to enlarge it by introducing *psychological* chapters on the different faculties of knowledge

> (imagination, wit, etc.), *metaphysical* chapters on the origin of knowledge or on the different kinds of certainty according to difference in the objects (idealism, scepticism, etc.), or *anthropological* chapters on prejudices, their causes and remedies, this could only arise from their ignorance of the peculiar nature of logical science. We do not enlarge but disfigure sciences, if we allow them to trespass upon one another's territory. The sphere of logic is quite precisely delimited; its sole concern is to give an exhaustive exposition and strict proof of the formal rules of all thought, whether it be *a priori* or empirical, whatever be its origin or its object, and whatever hindrances, accidental or natural, it may encounter in our minds.
>
> [Kant, 1929, Bviii–ix]

Kant's main importance is what he took logic to be, rather than making changes within it. Logic, according to Kant, is the study of the forms of judgements and, as such, the study of the formal rules of all thought. This extended the characterisation of *formal* consequence beyond the study of schemata. The kantian forms are characterised by the *table of judgements.* The table is similar to Aristotle's characterisation of propositions. Propositions and judgements vary in their quantity (whether they are universal, particular or singular judgements), in their quality (whether they are affirmative, negative or infinite), in their relation (whether they are categorical, hypothetical or disjunctive) and in their modality.

> The table of judgments, in turn, captures a fundamental part of the science of pure general logic: *pure*, because it is *a priori*, necessary, and without any associated sensory content; *general*, because it is both universal and essentially formal, and thereby abstracts away from all specific objective representational contents and from the differences between particular represented objects; and *logic* because, in addition to the table of judgments, it also systematically provides normative cognitive rules for the truth of judgments (i.e., the law of non-contradiction or logical consistency) and for valid inference (i.e., the law of logical consequence)
>
> [Hanna, 2011, (A52-55/B76-79) (9: 11-16)]

Logical consequence, as the relations which arises from the forms of judgements, inherits these characteristic properties.

Kant had a clear idea of what logic is and what logic isn't. Logic is the study of the formal rules of thought and the form of judgements. Logic is general: it is not concerned with whether a thought is "*a priori* or empirical", nor with what its origins or objects are, nor how it is processed in my minds. Psychological, metaphysical and anthropological considerations are not part of a pure general logic.

Kant famously drew two dichotomies on judgements: the *a priori* and the *a posteriori*; and the analytic and the synthetic. Pure general logic, according to

Kant is *a priori* analytic while arithmetic and geometry are *a priori* synthetic. In *Critique of Pure Reason*, Kant sets out to provide grounds for synthetic *a priori* judgements. Later, Frege's intent to show that arithmetic is in fact analytic *a priori* would result in a revolution in logic and the study of consequence.

7 BOLZANO [1781–1848]

In the 19th Century, mathematicians carefully examined the reasoning involved in infinite and infinitesimal numbers. Bolzano played an important role in clarifying these mathematical concepts, which were fraught with paradox and confusion. The Kantian account of intuition as the grounds for *a priori* synthetic judgements was coming into question. Bolzano managed to give a definition of continuity for real-valued functions. In the course of this, he made important philosophical and logical contributions. Part of Bolzano's project for clarity included an analysis of propositions and of consequence.

> We often take certain representations in a given proposition to be variable and, without being clearly aware of it, replace these variable parts by certain other representations and observe the truth values which these propositions take on ... Given a proposition, we could merely inquire whether it is true or false. But some very remarkable properties of propositions can be discovered if, in addition, we consider the truth values of all those propositions which can be generated from it, if we take some of its constituent representations as variable and replace them with any other representations whatever. [Bolzano, 1972, p 194]

Bolzano uses this account of propositions, where variable components can be replaced by other representations, to give an analysis of logical consequence.

> The 'follows of necessity' can hardly be interpreted in any other way than this: that the conclusion becomes true *whenever* the premises are true. Now it is obvious that we cannot say of one and the same class of propositions that one of them becomes true *whenever* the others are true, unless we envisage some of their parts as variable ... The desired formulation was this: as soon as the exchange of certain representations makes the premises true, the conclusion must also become true. [Bolzano, 1972, p 220]

Consider the following three premise argument:

- If Fred lives in New Zealand, then he is further away from Sally than if he lived in Australia.
- If Fred is further away from Sally than if he lived in Australia, then Fred is very sad.

- Fred isn't very sad
- Therefore, Fred doesn't live in New Zealand.

The argument is valid; the conclusion follows necessarily from the premises. Suppose that "Fred lives in New Zealand", "Fred is further away from Sally than if he lived in Australia", and "Fred is very sad" are the *variable* parts of the premises and conclusion. In this case, the argument is valid according to Bolzano's analysis of consequence if and only if, for any variation of these variable parts, the conclusion is true whenever all the premises are. That is, whenever p, q and r are uniformly replaced by propositions in the following form: either a premise is false or the conclusion is true.

- If p, then q.
- If q, then r.
- Not r
- Therefore, not p.

This is remarkably similar to current definitions of logical consequence and to accounts rejected by medieval logicians. Bolzano's account relies on substitution of representations into the variable parts of propositions. This differs from the later Tarskian definition of consequence where open sentences are satisfied, or not, by objects. Bolzano's account works on a single level, where Tarski's involves two levels. A single level definition can either be given at the level of the language (e.g. truth preservation from *sentences* to *sentences* across substitution of *names* for *names* and *predicates* for *predicates*) or at the level of the semantic values of sentences (truth preservation from *semantic values of sentences* to *semantic values* across substitution of *objects* for *objects*, *properties* for *properties*). With Bolzano's talk of replacement of representations for representations, his account seems to be of the latter sort (*pace* [Etchemendy, 1990]).

Bolzano's analysis of consequence depends on an account of the variable components of propositions. The logical components must remain fixed while the other components must be suitably varied. This requires that a line is drawn between the logical and non-logical vocabulary. It is still a difficult task (some argue that it is impossible, [Etchemendy, 1990]) to provide such a line. Bolzano provided examples of logical components (including ... *has* ..., *non-*... and *something*) but confessed that he had no sharp distinction to provide. This is a recurring theme in the history of logical consequence. Aristotle, the Stoics, Kant and others had theories on what the forms of propositions were. Providing reasons for why these forms exhausts the logical forms is a difficult endeavour. Bolzano is interesting here, as he says that the matter is entirely conventional, this is a foreshadow of Carnap's philosophy of consequence.

8 BOOLE [1815–1864]

In *The Laws of Thought* [Boole, 1951], Boole advanced the mathematization of logic. Boole's work applied the current accounts of algebra to thought and logic, resulting in algebraic techniques for determining formal logical consequences. Boole thought that an argument held in logic if and only if, once translated in the appropriate way into equations, it was true in the common algebra of 0 and 1 [Boole, 1951, pp 37 – 38] [Burris, 2010, section 4].

Before explaining what this *appropriate translation* is and why Boole choose 0 and 1 we need his account of the forms of propositions. Every proposition has a negation or denial (the negation of p is $\neg p$). Every pair of propositions has a disjunction and a conjunction ($p \vee q$ and $p \wedge q$).

A simple proposition is translated as, what we now call, a variable. The negation of the proposition p is translated as $1 - p$ (that is, the translation of p subtracted from one). From this it follows that $\neg\neg p = p$, as $1 - (1 - p) = p$. The disjunction of two propositions corresponds to the multiplication of their translation; for this reason Boole notated $p \vee q$ as pq. The important features of 1 and 0 are that they are idempotent: their squares are equal to themselves. This is required if a disjunction of a proposition with itself is equivalent to itself. The conjunction of two propositions is similar to addition (and Boole represents it with an addition sign $+$) but with the difference that $1 + 1 = 1$.

Boole provides a structure that is meant to characterise all systems of propositions. He gives their form in terms of negation, conjunction and disjunction, and he gives algebraic laws which characterise how the collection of these propositions is structured. The result is that we have algebraic devices for determining consequences. The resulting structures are called *Boolean Algebras* in his honour. The study of Boolean Algebras, related structures and algebraic approaches to logic now is a core tradition in the study of logic.

Boole's advancement of the mathematization of logic coincides with a waning interest in logical consequence. Emphasis is placed on axioms, tautologies and logical truths. This continues through the sections on Frege and Russell below. The mathematical techniques that were developed, and the philosophical insights gained, in this period were important for later studies of consequence but the focus on consequence that runs from the Greeks and medievals to Bolzano is diffused until Tarski and Carnap.

9 FREGE [1848–1925]

Gottlob Frege is one of the fathers of modern logic. He profoundly influenced the disciplines of logic, the philosophy of mathematics and the philosophy of language. Frege developed a logical notation which was meant to clarify and improve on natural languages. The *begrifftschrift*, or concept script, is a precise regimentation of Frege's own natural language, German. His intention was to remove the ambiguities, inconsistencies and misleading aspects of natural language. For the project to succeed, Frege's logic had to be much more than a *mere* calculating device; thus, he rejected the boolean algebraic tradition.

> Frege devoted considerable effort to separating his own conceptions of "logic" from that of the mere computational logicians such as Jevons, Boole and Schroeder. Whereas these people, he explained, were engaged in the Leibnizian project of developing a calculus ratiocinator, his own goal was the much more ambitious one of designing a *lingua characteristica*. Traditional logicians were concerned basically with the problem of identifying mathematical algorithms aimed at solving traditional logical problems—what follows from what, what is valid, and so on. Frege's goal went far beyond what we now call formal logic and into semantics, meanings, and contents, where he found the ultimate foundation of inference, validity, and much more.
>
> [Coffa, 1993, p 65]

Frege's intention was to show, in opposition to Kant, that arithmetic is analytic. According to both Kant and Frege, geometry is *a priori* synthetic, but Kant and Frege differed on the status of arithmetic. Frege's logicism aimed at a reduction of arithemtic to logic; Kant thought that arithmetic was synthetic. There is no direct opposition between Frege and Kant here. Kant and Frege's categories of *analytic* are different because they are based on different accounts of the forms of propositions; it is here that they are in opposition. The purely logical forms of propositions, according to Kant, are quite limited. Frege abandoned the Aristotlean forms of propositions. In Aristotle's and Kant's categorisations, the following are all propositions where a term is predicated of a subject.

- Socrates is mortal.
- Every human is mortal.
- No-one is mortal.

The forms of these propositions still differ, according to Aristotle and Kant (for example, according to Kant's table of judgements: the first is singular, the second is universal; the first two are positive, the third is negative). Frege, however, is quite clear that "a distinction of *subject* and *predicate* finds *no place* in [his] way of representing a judgement"[Geach and Black, 1952, p 2]. These three proposition have very different fregean structures. The first does predicate mortality of a person, Socrates, but neither of the other statements have a subject in the same way. The second statement is understood as the universal quantification of the incomplete statement *if x is human, x is mortal.* The statement is true if every object satisfies the incomplete statement. Frege changed what we mean by the word "predicate". Aristotlean predicates are terms, which are predicated of subjects in some manner, but can also be subjects. Fregean names and predicates are not of the same type. Names are complete expressions; fregean predicates are incomplete sentences.

Frege's *begrifftschrift* is far more powerful than Kant's logic. Indeed, at some points Frege's logic was inconsistent. Russell showed that Frege's infamous law five results in a contradiction. Inconsistency aside, Frege's *begrifftschrift* outstretches what is commonly used today. Frege makes use of second order quantification and no level of higher order quantification is ruled out. This was the result of his new approach to the pure logical forms of propositions.

Frege's *begrifftschrift* is the foundation on which modern logic is based. Its use of predicates, names, variables and quantifiers gives the structure that most logical systems use.

10 RUSSELL [1872–1970]

At the beginning of the 20th century, Russell was involved in a reductive project similar to Frege's. Russell employed the methods of Peano to give derivations of mathematical results based on logic alone. When he applied this approach to Cantor's proof that there is no greatest largest cardinal number he stumbled on paradox. The result appeared to conflict with Russell's assumption that there is a class which contains all other objects; this class would be the greatest cardinal. Running through Cantor's argument with this supposed universal class leads to Russell's class: the class of all classes which do not contain themselves. This class, were it to exist, would and would not contain itself. Russell showed that this conflicts with Frege's law five (roughly, that there is a class for any concept). Russell's paradox, along with others, became very important for modern logic. Unlike in the medieval era, paradoxes triggered serious doubts about core logical notions in many of the best logicians to follow Russell.

In *Principia Mathematica* [Whitehead and Russell, 1925–1927], Russell and Alfred North Whitehead aimed at producing a paradox free reduction of mathematics to logic. In order to achieve this, they had to steer between the weak logic of Kant and Aristotle, and the inconsistent strength of Frege's logic. Russell and Whitehead certainly did not take a predefined notion of "logic" and reduce mathematics to it. In the preface to *Principia Mathematica*, Russell and Whitehead say:

> In constructing a deductive system such as that contained in the present work, there are two opposite tasks which have to be concurrently performed. On the one hand, we have to analyse existing mathematics, with a view to discovering what premisses are employed, whether these premisses are mutually consistent, and whether they are capable of reduction to more fundamental premisses. On the other hand, when we have decided upon our premisses, we have to build up again as much as may seem necessary of the data previously analysed, and as many other consequences of our premisses as are of sufficient general interest to deserve statement. The preliminary labour of analysis does not appear in the final presentation, which merely sets forth the outcome of the analysis in certain undefined ideas and undemonstrated propo-

> sitions. It is not claimed that the analysis could not have been carried farther: we have no reason to suppose that it is impossible to find simpler ideas and axioms by means of which those with which we start could be defined and demonstrated. All that is affirmed is that the ideas and axioms with which we start are sufficient, not that they are necessary.

There is a sense in which the the reduction of mathematics is an uncovering of what logic must be. As a result, Russell's logic is much more intricate than Kant's. Nonetheless, russellian logic retained some of the *form of thought* characteristics it had for Kant and Frege. Russell says:

> The word symbolic designates the subject by an accidental characteristic, for the employment of mathematical symbols, here as elsewhere, is merely a theoretically irrelevant convenience. ...
>
> Symbolic logic is essentially concerned with inference in general, and is distinguished from various special branches of mathematics mainly by its generality. [Russell, 1937, section 11 and 12]

The (ramified) theory of types is central to Russell's approaches to logicism. We can think of the theories of types as theories of the forms propositions. The pure forms of propositions are more complex than in Aristotelian logic, but they are more restricted than in Frege's logic.

The ramified theory of types (we follow Church's [Church, 1976] rather than [Whitehead and Russell, 1925–1927] in our exposition) begins with *types*, or different domains of quantification. Domains of quantification are associated with variables which range over them; so, variables have specified types. Individual variables are of some basic type i. If $\beta_1, \ldots, \beta_m$ are types of variables, then there is a further type $(\beta_1, \ldots, \beta_m)/n$ of variables which contains m-place functional variables of level n. A type $(\alpha_1, \ldots, \alpha_m)/k$ is *directly lower* than the $(\beta_1, \ldots, \beta_m)/n$ if $\alpha_i = \beta_i$ and $k < n$.

0-place functional variables of level n are propositional variables of level n. Functional variables of 1 or more places are *propositional functions*. A 1-place functional variable of level n is a one place predicate of that level.

Formulas are restricted to the following forms. Propositional variables are well formed formulas. If f is a variable of a type $(\beta_1, \ldots, \beta_m)/n$ and x_i is a variable of, or directly lower than, type β_i, then $f(x_1, \ldots, x_m)$ is a well formed formula. On this base we add a recursive clause for negation, disjunction and universal quantification.

The forms of propositions are restricted by the requirement that variables are only applied to others of lower levels. The levels are cumulative in that the range of a variable of one type includes all the range of the variables of directly lower types.

Each variable type has an associated *order*. The order of a type is defined recursively. Individual variables (variables of type i) have order 0. A type $(\beta_1, \ldots, \beta_m)/n$

has order $N + n$ where N is the greatest order of the types $\beta_1, \ldots, \beta_m$. Orders of types are used to allow for controlled versions of Frege's law five.

- $(\exists p)(p \leftrightarrow P)$
- $(\exists f)(f(x_1, \ldots, x_m) \leftrightarrow P(x_1, \ldots, x_m))$

The crucial feature of these axioms is that the variables in the well formed formula P on the right hand side of the biconditionals are restricted by the order of the propositional variable or functional varible on the left hand side.

In order to achieve the results they wanted, Russell and Whitehead needed to introduce axioms of reducibility. The axioms ensure that every propositional function has a logically equivalent propositional function of level 1.

The logicist project of *Principia Mathematica* was intended to reduce mathematics by providing an account of what is *said* by mathematical claims. The account had trouble because of what the "logical" base was committed to. The base required axioms like *the axiom of infinity* and *the axiom of reducibility* to be able to provide deductions of mathematics that Russell and Whitehead aimed at recapturing. Neither axiom seems *obviously* true. This leads to careful discussion of the epistemology of logic.

> In fact self-evidence is never more than a part of the reason for accepting an axiom, and is never indispensable. The reason for accepting an axiom, as for accepting any other proposition, is always largely inductive, namely that many propositions which are nearly indubitable can be produced from it, and that no equally plausible way is known by which these propositions could be true if the axiom were false, and nothing which is probably false can be deduced from it. [Whitehead and Russell, 1962, p 59]

Russell has some clear ideas about what logic is, but he is also clear that he had no adequate definition. In some places he rejects the axiom of infinity as logical, because it is not adequately *tautological.*

> It is clear that the definition of "logic" or "mathematics" must be sought by trying to give a new definition of the old notion of "analytic" propositions. ... They all have the characteristic which, a moment ago, we agreed to call "tautology". This, combined with the fact that they can be expressed wholly in terms of variables and logical constants (a logical constant being something which remains constant in a proposition even when all its constituents are changed) will give the definition of logic or pure mathematics. For the moment, I do not know how to define "tautology". It would be easy to offer a definition which might seem satisfactory for a while; but I know of none that I feel to be satisfactory, in spite of feeling thoroughly familiar with the characteristic of which a definition is wanted. At this point, therefore, for the moment,

> we reach the frontier of knowledge on our backward journey into the logical foundations of mathematics. [Russell, 1919, pp 204 – 205]

Later he says:

> It seems clear that there must be some way of defining logic otherwise than in relation to a particular logical language. The fundamental characteristic of logic, obviously, is that which is indicated when we say that logical propositions are true in virtue of their form. The question of demonstrability cannot enter in, since every proposition which, in one system, is deduced from the premises, might, in another system, be itself taken as a premise. If the proposition is complicated, this is inconvenient, but it cannot be impossible. All the propositions that are demonstrable in any admissible logical system must share with the premises the property of being true in virtue of their form; and all propositions which are true in virtue of their form ought to be included in any adequate logic.
>
> Some writers, for example Carnap in his "Logical Syntax of Language," treat the whole matter as being more a matter of linguistic choice than I can believe it to be. In the above mentioned work, Carnap has two logical languages, one of which admits the multiplicative axiom and the axiom of infinity, while the other does not. I cannot myself regard such a matter as one to be decided by our arbitrary choice. It seems to me that these axioms either do, or do not, have the characteristic of formal truth which characterises logic, and that in the former event every logic must include them, while in the latter every logic must exclude them. I confess, however, that I am unable to give any clear account of what is meant by saying that a proposition is "true in virtue of its form." But this phrase, inadequate as it is, points, I think, to the problem which must be solved if an adequate definition of logic is to be found.
>
> [Russell, 1937, Introduction]

11 CARNAP [1891–1970]

Rudolf Carnap's intellectual development began within a dominantly Kantian tradition. He had the benefit of attending logic lectures by Frege in Jena (in the early 1910's), but this exposure to a father of modern logic only had significant philosophical impact on Carnap after a "conversion experience" through reading Bertrand Russell. Carnap was particularly struck by Russell's insistence that "[t]he study of logic becomes the central study in philosophy" [Carus, 2007, p 25]. Carnap was won over by the combination of rigour and philosophical applicability of Russell's work [Coffa, 1993].

Carnap began using logical methods in all of his work, following Russell but also heavily influenced by Frege's classes and also (unlike Russell and Frege) axiomatic theories in mathematics (especially Hilbert's program). Carnap was a great populariser of modern logic. He lead the way in producing textbooks and introductions to the area. He also did cutting edge work on the relations between completeness, categoricity and decidability. The core conjectures that he focussed on (formulated for the modern reader) are:

1. An axiomatic system S is consistent (no contradiction is deducible from it) if and only if it is satisfiable, i.e., has a model.
2. An axiomatic system S is semantically complete (non-forkable) if and only if it is categorical (monomorphic).
3. An axiomatic system S is deductively complete if and only if it is semantically complete (non-forkable).

[Reck, 2007, p 187]

This work was closely connected to important results of Gödel and Tarski (see [Reck, 2007] for further details).

As we saw in an earlier quote from Russell, Carnap's philosophy and use of these tools were very different to Russell's. Carnap's works the *Der Logische Aufbau der Welt* [Carnap, 1928; Carnap, 1967] and the *Logical Syntax of Language* [Carnap, 1959] use techniques inspired by Russell and Frege, but the resulting philosophical picture is very different. Russell's theory of types provided one logical base for the reduction of mathematics but there were alternatives. In *Logical Syntax of Language*, Carnap aimed to show that there is no need to *justify* deviation from Russell's theory of types. His position was that no "new language-form must be proved to be 'correct' [nor] to constitute a faithful rendering of 'the true logic"' [Carnap, 1959, p xiv]. Carnap's *Principle of Tolerence* gives us "complete liberty with regard to the forms of a language; that both the forms of construction for sentences and the rules of transformation ... may be chosen quite arbitrarily." [Carnap, 1959, p xv] *Correctness* can only be determined within a system of rules; the adoption of a logical system (or language-form) is not done in this way. There may be reasons for or against adopting a particular system, but these are pragmatic choices — choices about which system will be more useful, rather than which system is correct.

Carnap argued for a viewpoint in which *philosophy was syntax*. The rules of formation and transformation of a language are conventional and we are at liberty to choose between systems. The rules of formation govern the forms of formulas (and propositions), and the transformation rules are the basis for logical consequence. The transformation rules determine a collection of legitimate transformations of propositions. If a proposition is the result of legitimate transformations of a collection of assumptions, then it is a consequence of these assumptions. This bases consequence on "syntactic" manipulation rather than semantic notions like truth or meaning. This was important for earlier followers of Carnap, like Quine [Carnap *et al.*, 1990], but later (inspired by the work of Tarski) Carnap came to change his approach. In the end, Carnap offered both transference and property style accounts of logical consequence.

Carnap was a logical empiricist (logical positivist). One facet of logical empiricism was the result of combining Russell's logicism with Wittgenstein's account of tautologies. The logical empiricists wanted to account for the necessity of mathematics without granting it any empirical substance. If *Principia Mathematica* [Whitehead and Russell, 1925–1927] was successful in reducing mathematics to logic (and the logical empiricists thought this plausible) then it was only logic that needed accounting for. For this, they turned to Wittgenstein's *Tractatus* [Wittgenstein, 1922]. In the *Tractatus*, logical truths were true in virtue of language, not in virtue of how they represent the world; they were *empty* tautologies. The necessities of mathematics (via their reduction to logic) were necessary but this necessity only stems from the logical system one has adopted. How do we determine what logical system should be adopted? This is the point at which the Principle of Tolerance came in.

> In the foregoing we have discussed several examples of negative requirements (especially those of Brouwer, Kaufmann, and Wittgenstein) by which certain common forms of language — methods of expression and of inference — would be excluded. Our attitude to requirements of this kind is given a general formulation in the *Principle of Tolerance: It is not our business to set up prohibitions, but to arrive at conventions.* ... *In logic, there are no morals.* Everyone is at liberty to build up his own logic, i.e. his own form of language, as he wishes. All that is required of him is that, if he wishes to discuss it, he must state his methods clearly, and give syntactical rules instead of philosophical arguments.
>
> [Carnap, 1959, pp 51 – 52, emphasis in original]

Carnap made suggestions regarding what choices may be better for certain purposes (in particular for unified empirical science) but this doesn't make these choices *correct.* Carnap developed two different systems in *Logical Syntax.* Logical consequence is an important feature of both. *Language I* is a weaker system which implements some constructivist constraints on consequence. This system is safer; it is less likely to produce inconsistency. *Language II* is a richer system based on the theory of types in Russell's work. If we prefer the safety of Brouwer's intuitionstism, we can adopt the former system. If we prefer the strength of the classical theory of types, we can adopt the latter. Neither give the *correct* consequence relation. Rightness or wrongness of arguments can be determined with respect to the logical system they are framed in, but a consequence relation is something we adopt for a purpose. The choice of logical system and consequence relation is not arbitrary; pragmatic considerations (like simplicity, fruitfullness, safety from inconsistency and the like) determine that some choices may be *better* than others.

12 GENTZEN [1909–1945]

Gerhard Gentzen's work in logic, particularly his presentation of proof systems, has been highly influential. His natural deduction systems and sequent calculi were tremendous advances in proof theory. The ideas underlying these systems, and their connections to logical consequence, are important for understanding different ways of reasoning.

Gentzen's first publicly submitted work, in which he demonstrated significant technical ability, was a paper *On the Existence of Independent Axiom Systems for Infinite Sentence Systems* [Gentzen, 1932]. Importantly for us, he gave a clear description of the connection between proof (and inference) in a logical system and an intuitive notion of logical consequence. With respect to the notion of proof used in the paper, he says:

> Our formal definition of provability and, more generally, our choice of the forms of inference will seem appropriate only if it is certain that a sentence $\mathfrak{q}$ is 'provable' from the sentences $\mathfrak{p}_1 \ldots \mathfrak{p}_v$ if and only if it represents informally a consequence of the $\mathfrak{p}$'s. We shall be able to show that this is indeed so as soon as we have fixed the meaning of the still somewhat vague notion of 'consequence'.
>
> [Gentzen, 1969, p 33]

Gentzen gave a regimented version of the "somewhat vague" notion and then proved the equivalence (in separate soundness and completeness stages) between proof and consequence for his system. A complication worth mentioning is that a "sentence" in this system is more like a sequent (or argument) than a formula.

Gentzen's natural deduction and sequent calculi, and the associated notions of normal proof and cut elimination, have become central to proof theory. In his *Investigations into Logical Deduction* [Gentzen, 1935a; Gentzen, 1935b] Gentzen presents systems of proof which are intended to "[come] as close as possible to actual reasoning" [Gentzen, 1969, p 68]. In calculi of *natural deduction* one demonstrates that a formula follows from others by means of inference figures which license derivations. Each connective has both an introduction and an elimination rule.

The following example is a natural deduction style derivation of $(A \to C) \land (B \to C)$ from the premise $(A \lor B) \to C$. The derivation introduces disjunctions and conjunctions, and introduces and eliminates conditionals. Notice that the temporary premise A is discharged when the conditional $A \to C$ is introduced. (Some medieval discussions of inference and consequence come very close to natural deduction; one point of difference is that, even though the medievals performed hypothetical reasoning, it seems that they had no way to understand how one might discharge assumptions in order to introduce conditionals [Hodges, 2009]. Consequence is always a local matter, and rules which allow 'action at a distance' are absent.)

$$
\dfrac{
\dfrac{
\dfrac{(A \vee B) \rightarrow C \quad \dfrac{[A]_a}{A \vee B}\vee I}{C} \rightarrow E
}{(A \rightarrow C)} \rightarrow I, a
\qquad
\dfrac{
\dfrac{(A \vee B) \rightarrow C \quad \dfrac{[B]_b}{A \vee B}\vee I}{C} \rightarrow E
}{(B \rightarrow C)} \rightarrow I, b
}{(A \rightarrow C) \wedge (B \rightarrow C)} \wedge I
$$

A natural deduction derivation proves that a conclusion *follows from* a collection of assumptions. Derivations show that formulas are consequences of premises. Inference rules show that consequences hold, if other consequences hold. This became clearer when Gentzen introduced his *sequent calculi.*

Sequent calculi make explicit the reasoning involved in inference steps. The inferences of sequent calculi operate explicitly on sequents (or argument forms), which have a sequence of antecedent formulas (or premises) and either a single succedent formula (or conclusion) in the case of intuitionistic logic or a sequence of succedent formulas in the case of classical logic. The inferences of the system can be divided into two kinds. The structural inferences *thinning*, *contraction*, *interchange* and *cut* are concerned with the structure of the antecedent and succedent formulas. The operational inferences introduce logical connectives into either the antecedent or the succeedent. A sequent calculus style proof of the argument $(A \vee B) \rightarrow C \therefore (A \rightarrow C) \wedge (B \rightarrow C)$ is in the substructural logic section below. Note the correspondence between natural deduction introduction rules and sequent right hand side rules, elimination rules and sequent left rules.

A derivation of a sequent in in a sequent calculi proves that it corresponds to a valid argument. As classical sequents have sequences of conclusions, this broadens the category of arguments to include structures with multiple conclusions.

13 TARSKI [1902–1983]

Alfred Tarski is a foundational figure for modern logic. Tarski's contributions to logic and mathematics guided much of the development of logic in the 20th and early 21st centuries. Tarski is particularly well known for the new degree of clarity that he brought to the study of logic. He advocated a *metamathematical* approach to logic. This has been very important for clarifying the central notions of logic. Tarski's metamathematical approach allowed him to study many semantic notions, like consequence and truth, at a time when they were thought troublesome at best and incoherent at worst.

Metamathematics is the branch of mathematics dedicated to the study of formalised deductive disciples: that is, "formalized deductive disciples form the field of research of metamathematics roughly in the same sense in which spactial entities form the field of research in geometry" [Tarski, 1956a, p 30]. A deductive discipline is composed of a collection of *meaningful sentences*. From a collection of sentences one can make various inferences; the resulting sentences are *consequences*. Every deductive science has associated rules of inference; "To establish these rules of inference, and with their help to define exactly the concept of consequence, is again a, task of special metadisciplines" [Tarski, 1956a, p 63]. Any sentence A which is derived from the collection of sentences Γ by means of these rules is a consequence of Γ. The consequences of Γ can be defined as the intersection of all the sets which contain the set A and are closed under the given rules of inference. This is a transference style approach to consequence; Tarski later gave a property based conception of consequence.

In *Fundamental Concepts of the Methodology of the Deductive Sciences* [Tarski, 1956a, chapter 5], Tarski studies deductive disciplines and their consequence relations at a high level of abstraction. On the basis of a small number of axioms regarding sentences and consequence, Tarski defines a number of important concepts for deductive disciplines. In the following axioms, S is a collection of sentences, $\subseteq$ is the standard sub-set or equals relation between sets, and Cn is an operation on sets of sentences. The set $Cn(X)$ is the set of consequences of the sentences in the set X. Sentences and consequence in a deductive discipline is subject to the following axioms:

Axiom 1 The collection of sentences, S, is denumerable.

Axiom 2 If $X \subseteq S$ (that is, X is a set of sentences), then $X \subseteq Cn(X) \subseteq S$.

Axiom 3 If $X \subseteq S$, then $Cn(Cn(X)) = Cn(X)$

Axiom 4 If $X \subseteq S$, then $Cn(X) = \bigcup\{Cn(A) : A \text{ is a finite subset of } X\}$

[Tarski, 1956a, pp 63 – 64]

On the basis of these axioms, Tarski proves a number of general theorems about sentences and consequence. He then focusses on *closed deductive systems* which contain all the consequences of their subsets. In *Fundamental Concepts of the Metamathematics* [Tarski, 1956a, chapter 3], Tarski uses a similar approach to investigate a narrowed field of deductive systems. In this approach, additional axioms are imposed on the sentences in the collection S. The first restriction is that S has at least one member which has every element of S as a consequence — an absurd sentence. The collection of sentences is required to contain the conjunction of any two of its members, and the negation of any of its members. The logical properties of these sentences are characterised in terms of the consequence operator. For example,

Axiom 9 If $x \in S$, then $Cn(\{x, n(x)\}) = S$ [Tarski, 1956a, p 32]

where $n(x)$ is the negation of x, ensures that every sentence follows from a set which contains a sentence and its negation.

It is now common to refer to relations, R, between sets of sentences (X and Y in the properties below) and sentences (a and c) with the properties:

Reflexivity If $a \in X$ then RaX

Transitivity If RcY and $Ra(X \cup \{c\})$, then $Ra(X \cup Y)$

Monotonicity If RaX and $X \subseteq Y$ then RaY

as a *Tarski consequence relation.* If the relation is between sets of sentences, that is a set of sentences can have a set of sentences as a consequence, and if the properties are appropriately altered, it is a *Tarski-Scott consequence relation.*

Tarski is famous for his definition of truth in formal languages, but his analysis of logical consequence for predicate logic has become a central part of orthodox logic. In *On the concept of logical consequence*, Tarski used *models* and *satisfaction* to give a theory of logical consequence for predicate logic. Sentences of predicate logic have a recursive structure. They are either atomic or are formed using clauses for conjunction, disjunction, negation, universal quantification, existential quantification, etc. Tarski shows how the truth of each type of sentence is to be determined relative to a model. The propositional part of this is relatively straight forward. The quantifiers, however, required particular care. The truth of the formula $(\forall x)(Fx \supset Gx)$ depends on the properties of $Fx \supset Gx$. It cannot depend on the *truth* of $Fx \supset Gx$ as it contains the unbound variable x. Tarski used *satisfaction in a model relative to an assignment of values to variables* to define truth in a model. The formula $Fx \supset Gx$ may be satisfied in a model relative to some assignments but not others. For $(\forall x)(Fx \supset Gx)$ to be satisfied relative to an assignment requires that $Fx \supset Gx$ is satisfied in the model relative to *all* assignments of values to variables. A model is a *model of a sentence* if the sentence is satisfied by an assignment of variables in the model. This leads to the definition of consequence:

> The sentence X *follows logically* from the sentences of the class K if and only if every model of the class K is also a model of the sentence X.
>
> [Tarski, 1956b, p 417, emphasis in original]

This gives a very restricted version of the intuitive notion of logical consequence. Tarski was well aware that some semantic notions cannot be defined in full generality. As another example, his definition of truth is for a carefully regimented language. He says that the intuitive concept as found in natural languages is inconsistent, and he demonstrates that there are a number of contexts in which truth cannot be defined. He is similarly aware that there are restrictions on definitions of logical consequence.

> Any attempt to bring into harmony all possible vague, sometimes contradictory, tendencies which connected with the use of this concept, is certainly doomed to failure. We must reconcile ourselves from the start to the fact that every precise definition of this concept will show arbitrary features to a greater or less degree
>
> [Tarski, 1956a, p 411]

He steered the investigation in a meta-logical or metamathematical direction in recognition that no attempt to supplement rules of inference in deductive systems can capture the intuitive notion of logical consequence.

> By making use of the results of K. Gödel we can show that this conjecture is untenable. In every deductive theory (apart from certain theories of a particularly elementary nature), however much we supplement the ordinary rules of inference by new purely structural rules, it is possible to construct sentences which follow, in the usual sense, from the theorems of this theory, but which nevertheless cannot be proved in this theory on the basis of the accepted rules of inference.
>
> [Tarski, 1956a, p 413]

14 GÖDEL [1906–1978]

Kurt Gödel and Alfred Tarski were, arguably, the greatest logicians of the 20th Century. Gödel's results that are most relevant to our discussion are the completeness of predicate logic and the incompleteness of arithmetic.

In his thesis [Godel *et al.*, 1986, see 1929], Gödel proved the *completeness* of predicate logic. He proved that a proof calculus of Hilbert and Ackerman proves all the formulas which are correct for any domain of individuals. Gödel proved that any formula is either refutable in the proof system (meaning that its negation can be proved from the axioms and proof rules) or is satisfiable (meaning that it is true in some domain of individuals). This showed that the proof system is complete: any unsatisfiable formula can be refuted by the proof system. In conjunction with Tarski's model theoretic analysis of consequence, this gives the result that logical consequence for predicate (classical) logic is captured by Hilbert and Ackerman's proof system. This shows that, in at least this case, there are transference and property style approaches to consequence that agree.

Gödel's completeness results were quickly followed by his *incompleteness* results. These were the results that Tarski referred to (in the above quote) in dismissing a derivation based approach to defining logical consequence. Gödel's incompleteness theorems were two of the most unexpected and outstanding results of the last century. The first theorem is given by the following quote:

> In the proof of Theorem VI no properties of the system P were used besides the following:

> 1. The class of axioms and the rules of inference (that is, the relation "immediate consequence") are recursively definable (as soon as we replace the signs in some way by natural numbers).
> 2. Every recursive relation is definable ... in the system P.
>
> Therefore, in every formal system that satisfies the assumptions 1 and 2 and is ω-consistent, there are undecidable propositions of the form $(x)Fx$, where F is a recursively defined property of natural numbers, and likewise in every extension of such a system by a recursively definable ω-consistent class of axioms. As can easily be verified, included among the systems satisfying the assumptions 1 and 2 are the Zermelo-Fraenkel and the von Neumann axiom systems of set theory, as well as the axiom systems of number theory consisting of the Peano axioms, recursive definition, and the rules of logic. [Godel *et al.*, 1986, p 181]

The second incompleteness theorem is that in every *consistent* (rather than the stronger ω-consistent) formal system that satisfies assumptions 1 and 2 there are undecidable propositions. In particular, a system of this type cannot prove the coded sentence which expresses the systems consistency.

> [Gödel's theorems] sabotage, once and for all, any project of trying to show that all basic arithmetical truths can be thought of as deducible, in some standard deductive system from one unified set of fundamental axioms which can be specified in a tidy, [primitively recursive], way. In short, *arithmetical truth isn't provability in some single axiomatizable system.* [Smith, 2008, p 161]

Gödel's incompleteness theorems show that certain types of transference/deductive accounts of consequence and property/truth accounts come apart for languages with strong enough expressive resources. Any formal system that is as strong as Peano arithmetic (for example, second order logic) cannot have a complete primitive recursive proof theory.

It doesn't follow that transference style approaches to consequence must fail. Carnap provided transference style approaches for arithmetic consequence while fully aware of Gödel's results (see [Ricketts, 2007, section 3] for a good discussion). Gentzen gave proof theoretic proofs of the consistency of arithmetic [Gentzen, 1969, # 4]. In both cases, infinitary (in some sense) techniques must be admitted. Nonetheless, Gödel's results show that no single member of the class of primitive recursive transference approaches to consequence can capture all of classical arithmetic (while remaining consistent). This leaves the consequence theorist with a choice: abandon primitive recursive transformations (whether by adopting an appropriate property based account or by incorperating some infinitary transformations) or abandon the consequence relation of classical arithmetic.

15 MODAL LOGICS

The history of modal logic is woven throughout the history of logic and logical consequence. Modal logic began with Aristotle's syllogisms and was further developed by the Stoics and medievals. In the medieval period, necessity was connected via modal logic to logical consequence. MacColl investigated modal logic in the algebraic tradition of Boole. Many of the logicians focussed on by this entry worked on modal logic (for example: Carnap, Gödel, and Tarski). Modern modal logic began with C. I. Lewis' response to Whitehead and Russell's *Principia Mathematica.* The controversy was over the material conditional of the *Principia.* Lewis argued that Whitehead and Russell's conditional doesn't adequately capture implication. There was still confusion over to the connections between conditionals, implication, entailment and logical consequence (Peano used "C" for consequence and a backwards "C" for ... *is deducible from* ..., the backwards 'C' became our hook symbol $\subset$ for the material conditional [van Heijenoort, 1967, p 84]). Lewis drew attention to the *paradoxes of material implication.* His various attempted improvements make use of modal notions, for example: necessity, possibility and compatibility.

The modern semantic period of modal logic started with the work of Krikpe, Prior, Hintikka, Montague and others (see [Goldblatt, 2006, section 4] for some discussion of the controversy involved). The model theoretic approach to modal logic draws on the ideas of Leibniz: truth is relativised to possible worlds. In a modal propositional language with the operators $\Diamond$ (for possibility) and $\Box$ (for necessity) the formulas $\Diamond A$ and $\Box A$ are assigned truth values relative to each world in the model (which is composed of a collection of worlds, a binary *accessibility* relation, and a valuation):

Diamond $\Diamond A$ is true at world w in the model $\mathfrak{M} = \langle W, R, v\rangle$ (in symbols: $\mathfrak{M}, w \Vdash \Diamond A$) if and only if there is some world u related to w by R (that is, Rwu) and A is true at u in $\mathfrak{M}$.

Box $\mathfrak{M}, w \Vdash \Box A$ if and only if, for all u such that Rwu, $\mathfrak{M}, u \Vdash A$

There are two notions of logical consequence which can be defined in terms of truth in modal, or Krikpe, models. *Local* consequence is based on preservation of truth at each individual world in a model, while *global* consequence is based on preservation of truth-at-all-worlds in each model. Here is the formal characterisation of these distinct notions

- The formula C is a *local consequence* of the formulas Γ relative to the models $\mathbf{M}$ if and only if, for any model $\mathfrak{M}$ in $\mathbf{M}$, and any world w in $\mathfrak{M}$ if, for each $A \in \Gamma$, $\mathfrak{M}, w \Vdash A$, then $\mathfrak{M}, w \Vdash C$.
- The formula C is a *global consequence* of the formulas Γ relative to the models $\mathbf{M}$ if and only if, for any model $\mathfrak{M}$ in $\mathbf{M}$, if for any $A \in \Gamma$ and any world w in $\mathfrak{M}$, $\mathfrak{M}, w \Vdash A$, then for any world w in $\mathfrak{M}$, we have $\mathfrak{M}, w \Vdash C$.

The two notions are both consequence relations in at least the sense that they are Tarski consequence relations (both relations are reflexive, transitive and monotonic). They also differ in important respects. The $\Box p$ is a global consequence of p in all classes of Kripke models (if p is true at every world in a model, then so is $\Box p$) but $\Box p$ is rarely a local consequence of p. There are many models in which we have worlds w at which p holds but $\Box p$ does not.

Various logicians and philosophers of logic have thought that consequence should be a necessary, *a priori* relation based on the meaning of logical vocabulary. The categories of the necessary, the *a priori* and the analytic have all received criticism during the history we have discussed. The *a priori* and necessity were attacked by empiricists, including Carnap and the logical empiricists. The logical empiricists allowed a remnant of necessity to remain under the guise of the analytic but insisted, inspired by Wittgenstein, that analytical truths said nothing. This was not far enough for Quine, who continued the attack on necessity and analyticity. The recent approach of two dimensional semantics attempts to capture all three notions in a single framework.

> First, Kant linked reason and modality, by suggesting that what is necessary is knowable a priori, and vice versa. Second, Frege linked reason and meaning, by proposing an aspect of meaning (sense) that is constitutively tied to cognitive significance. Third, Carnap linked meaning and modality, by proposing an aspect of meaning (intension) that is constitutively tied to possibility and necessity. ...The result was a golden triangle of constitutive connections between meaning, reason, and modality.
>
> Some years later, Kripke severed the Kantian link between apriority and necessity, thus severing the link between reason and modality. Carnaps link between meaning and modality was left intact, but it no longer grounded a Fregean link between meaning and reason. In this way the golden triangle was broken: meaning and modality were dissociated from reason. Two-dimensional semantics promises to restore the golden triangle. [Chalmers, 2006, p 55]

We will focus on necessity and *a priori* knowable and a two dimensional system based on Davies and Humberstone's [Davies and Humberstone, 1980].

In a two dimensional model, truth is relativised to pairs of worlds, rather than single worlds. A proposition A is true at the pair $\langle u, v\rangle$ if and only if, were u the actual world, then A would have been true in the possible world v. There are typically three important modal operator:

Box $\mathfrak{M}, \langle u, v\rangle \Vdash \Box A$ if and only if, for all w, $\mathfrak{M}, \langle u, w\rangle \Vdash A$

Fixedly $\mathfrak{M}, \langle u, v\rangle \Vdash FA$ if and only if, for all w, $\mathfrak{M}, \langle w, v\rangle \Vdash A$

At $\mathfrak{M}, \langle u, v\rangle \Vdash @A$ if and only if $\mathfrak{M}, \langle u, u\rangle \Vdash A$

Intuitively: $\Box A$ is true if A is true in all the ways the world could be; FA is true if A is true in all the epistemic alternatives (the ways the world could, for all we know, turn out to be); and $@A$ is true if A is true the actual world as it actually will turn out to be. In this two dimensional analysis $\Box$ plays the role of necessity and $F@$ plays the role of knowable *a priori*.

The logical and philosophical details of two dimensional approaches are varied and often far more complex than this short exposition. There are two points which we want to draw out of this short description. First, the different properties that consequence is sometimes thought to have are brought together in this single framework. Secondly, as with other modal logics, there are a number of different consequence relations for two dimensional modal logic, and they connect to necessity and *a priori* knowledge in different ways. Are the arguments "A therefore $@A$" and "$@A$ therefore A" logically good ones? Are the conclusions consequences of the premises? If consequence is defined as truth preservation at *diagonal* pairs (where the first and second element are the same world), these are valid arguments. This notion of consequence (according to this two dimensional analysis), is one where the premises are *a priori* reasons for the conclusion. The premises, however, do not *necessitate* (according to this two dimensional analysis) the conclusion. If the premises necessitate the conclusion, then the truth of the premise is sufficent for the truth of the conclusion in *any* hypothetical situation. Neither the truth of $@A$ nor of A is sufficient for the other in arbitrary hypotheticals. For this consequence relation we require truth preservation over all the pairs in the model.

16 NONMONOTONIC OPTIONS

Tarski's conditions of reflexivity, transitivity and monotonicity have been central in the study of all sorts of consequence relations. However, they are not necessarily all that we want in any notion of consequence, broadly understood. Let's remind ourselves of Tarski's three conditions, for a consequence relation $\vdash$.

Reflexivity If $A \in X$ then $X \vdash A$

Transitivity If $Y \vdash C$ and $X \cup \{C\} \vdash A$, then $X \cup Y \vdash C$

Monotonicity If $X \vdash A$ and $X \subseteq Y$ then $Y \vdash A$

Not everything that we might want to call a consequence relation satisfies these three conditions. In particular, the monotonicity condition rules out a lot of what we might broadly consider 'consequence.' Sometimes we wish to conclude something beyond what is deductively entailed. If I learn that someone is a Quaker, I might quite reasonably conclude that they are a pacifist. To be sure, this conclusion may be tentative, and it may be be undercut by later evidence, but it may still be a genuine *conclusion* rather than an hypothesis or a conjecture. Work in the second half of the 20th Century and beyond has drawn out some of the structure of these more general kinds of consequence relations. There is a structure here, to be examined, and to be considered under the broader banner of 'logic.'

The first thing to notice is that these relations do not satisfy monotonicity. Using '$\mid\!\sim$' to symbolise this sort of consequence relation, we may agree that

$$x \text{ is a Quaker} \mid\!\sim x \text{ is a Pacifist}$$

while we deny that

$$x \text{ is a Quaker}, x \text{ is a Republican} \mid\!\sim x \text{ is a Pacifist}$$

We may have $Q \mid\!\sim P$ and $Q, R \not\mid\!\sim P$, and so, the monotonicity condition fails for $\mid\!\sim$. However, work on such 'nonmonotonic' consequence relations has revealed a stricter monotonicity condition which plausibly does hold.

Cautious Monotonicity If $X \mid\!\sim A$ and $X \mid\!\sim B$ then $X, B \mid\!\sim A$.

Adding *arbitrary* extra premises may defeat a conclusion, but we add as a premise something which may be concluded from premises this does not undercut the original conclusion [Gabbay, 1985]. This seems to be a principle that is satisfied by a number of different ways to understand nonmonotonic consequence relation and to distinguish it from broader notions, such as compatibility.[7]

Much work has been done since the 1970s on nonmonotonic consequence relations and ways that they might come about. John McCarthy's work on circumscription analyses nonmonotonic consequence relations in terms of the minimality of the extensions of certain predicates. The core idea is that, *ceteris paribus*, some predicates are to have as small extensions as possible. If we take all *normal* or *typical* Quakers to be Pacifists, then the idea is that we want there to be as few *abnormal* or *atypical* items as possible, unless we have positive information to the effec that there are abnormal or atypical items. McCarthy represents this constraint formally, using second order quantification to express the condition [McCarthy, 1980].

Other accounts of nonmonotonic logic—such as Reiter's *Default Logic* [Reiter, 1980] characterise the relation $\mid\!\sim$ in terms of primitive default rules. We can specify models in terms of default rules (stating that Quakers are generally Pacifists and Republicans are generally not Pacifists). Networks of default rules like these can be used to define a nonmonotonic consequence relation over a whole language, interpreting default rules as defeasible *inheritance* rules, where properties flow through the network as far as possible. Different ways to treat conflicts in inheritance graphs (e.g. the conflict in the case of Republican Quakers) have been characterised as *skeptical* and *credulous* approaches to conflict. A useful and straightforward introduction to this and many other issues in the interpretation and application of nonmonotonic consequence relations is given by Antonelli [Antonelli, 2010].

---

[7]If we think of $X \mid\!\approx A$ as 'A is compatible with X' then cautious monotonicity fails for $\mid\!\approx$. Here $p \mid\!\approx q$ and $p \mid\!\approx \neg q$ but we do not have $p, \neg q \mid\!\approx q$. Whatever satisfies cautious monotonicity, it is a more discriminating relation than mere compatibility.

17 THE SUBSTRUCTURAL LANDSCAPE

Tarski-Scott consequence relations are relations between sets. Gentzen's sequent calculi are proof systems with sequents of sequences but, with their structural rules, they function very much like the sets of Tarski-Scott consequence relations. There are, however, consequence relations between other structures. There are numerous ways of refining the structure of premises and conclusions in arguments and of restricting the available structural rules. This tends to result in weaker consequence relations. A consequence relation where the structure of premises and conclusions is increased and structural rules are restricted is a *substructural* consequence relation.

The following example shows a Gentzen style sequent calculi derivation (it is a proof for the same argument as in the natural deduction from the Gentzen section). The structures on the left and right of the symbol $\Rightarrow$ are *multi-sets* (the order of presentation is irrelevant but the number of appearances of a formula in the presentation is). Disjunctions and a conjunction are introduced on the right of the double arrow. Conditionals are introduced on both the right and the left. Note that the disjunctions are introduced from multiple conclusions. This proof looks much bigger than the natural deduction above. The reason is that more attention has been paid to the structure of the premises and conclusions. The natural deduction example took no notice that there were two instances of the premise at the end of the deduction (one on each branch of the tree), this deduction uses the structural rule of contraction on the left (LW) to remove the duplication. The structural rule of weakening on the right (RK) is used to introduce the additional disjunct before the disjunction is introduced.

$$\dfrac{\dfrac{\dfrac{\dfrac{\dfrac{\dfrac{\dfrac{A \Rightarrow A}{A \Rightarrow A, B}\,RK}{A \Rightarrow A \vee B}\,R\vee \qquad C \Rightarrow C}{(A \vee B) \to C, A \Rightarrow C}\,R\to}{(A \vee B) \to C \Rightarrow (A \to C)}\,L\to \qquad \dfrac{\dfrac{\dfrac{\dfrac{B \Rightarrow B}{B \Rightarrow A, B}\,RK}{B \Rightarrow A \vee B}\,R\vee \qquad C \Rightarrow C}{(A \vee B) \to C, B \Rightarrow C}\,R\to}{(A \vee B) \to C \Rightarrow (B \to C)}\,L\to}{(A \vee B) \to C, (A \vee B) \to C \Rightarrow (A \to C) \wedge (B \to C)}\,R\wedge}{(A \vee B) \to C \Rightarrow (A \to C) \wedge (B \to C)}\,LW}{}$$

As can be seen from the above history of consequence, there are close connections between logical consequence and conditionals. The distinctions between these two have not always been very clear. The *Residuation condition* is one way of expressing the connection.

$$A, B \Rightarrow C \text{ if and only if } A \Rightarrow B \to C$$

The left to right direction of this law can be seen in the above example derivation in the application of the conditional on the right rule. The law of risiduation connects three notions together: logical consequence (indicated by $\Rightarrow$), conditionals (by $\rightarrow$) and *premise combination* (indicated by the comma). Structural rules (excluding identity and cut) operate on the comma in premise combination and, thus by the law of residuation, impact the properties of the conditional.

In the example above, conclusions are *weakened* into the proof. Logics which permit weakening on both the left and right are monotonic. The nonmonotoic logics from the previous section are all substructural in the sense that they do not respect the substructural rule of weakening. It is particularly important for relevant/relevance logics that weakening is abandoned. It doesn't follow from the consequence $X \Rightarrow A$ where the conclusion follows relevantly from the premises (each premise is important in the *following from* relation), that in the argument $X, B \Rightarrow A$ the conclusion *follows from* the premises in the same relevant way. Relevance in deduction and consequence was studied by Moh [1950], Church [1951] and Ackermann [1956]. The canonical text on later developments is Anderson and Belnap's *Entailment* [Anderson and Belnap, 1975; Anderson *et al.*, 1992].

While weakening allows for the addition of premises and conclusions, contraction removes duplicated premises and conclusions. In resource conscious logics (like Girard's *linear logic* [Girard, 1987]) contraction is dropped. Resource conscious logics pay attention to the number of times a formula is required in deriving a conclusion. So, a conclusion may follow from n uses of A, but not from any fewer. A number of contraction free logics are also weakening free.

Where the weakening rule has been connected to the paradoxes of material implication, the contraction rule is connected to Curry's paradox. Logics without contraction, like Łukasiewicz many-valued logics, have sustained interest because of this connection.

The example above uses multi-sets in the left and right of the sequents. Gentzen's original formulation used sequence, rather than multi-sets. In this case rules like *exchange*

$$\frac{\Gamma, A, B, \Gamma' \Rightarrow \Delta}{\Gamma, B, A, \Gamma' \Rightarrow \Delta} \qquad \frac{\Gamma \Rightarrow \Delta, A, B, \Delta'}{\Gamma \Rightarrow \Delta, B, A, \Delta'}$$

need to be included for classical and intuitionistic logic. Substructural logics like Lambek's calculus [Lambek, 1958; Lambek, 1961] drop these rules. Lambek used mathematical techniques to model language and syntax. "Premise" combination is used to model composition of linguistic units. In these cases, it is inappropriate to change the order of the "premises".

In the sequent derivation above, the sequents have finite sets of formulas in the premise positions and in the conclusion positions. By restricting the maximum number of formulas which can appear in the antecedent or consequent sequence, you restrict the consequence relation. Sequent system for classical logic have multi-sets or sequences of any (finite) number of formulas on either consequent or antecedent position. Gentzen's system for intuitionist logic, however, is restricted to no more than one formula on the right hand side. This restriction on the structure on the right hand side makes the difference between intuitionist and classical logic. This restriction is a restriction on the structural rules of Intuitionist logic, the identity and cut rules are suitably restricted to single formulas on the right of sequents and weakening on the right is restricted.indexOne True Logic

These structural rules can be included, dropped entirely, or restricted in some way. The structural rule *mingle*:

$$\frac{\Gamma, A \Rightarrow \Delta}{\Gamma, A, A \Rightarrow \Delta} \qquad \frac{\Gamma \Rightarrow \Delta, A}{\Gamma \Rightarrow \Delta, A, A}$$

is a restricted version of weakening. The rule doesn't allow for arbitrary additions of premises, but if a premise has already been included in a relevant deduction of the conclusion, it can be weakened in any number of times.

We began this entry by claiming that theorists of logical consequence have to answer the question "In what ways can premises combine in an argument?". The substructural landscape shows that this is a genuine requirement on theorists. Different answers, in combination with different structural rules, produce very different consequence relations.

18 MONISM OR PLURALISM

Given such a variety of different accounts of logical consequence, what are we to say about them? Are some of these accounts correct and others incorrect? Is there One True Logic, which gives the single correct answer to the question "is this valid?" when presented with an argument, or is there no such logic but rather, a Plurality of consequence relations? This is the issue between *Monists* [Priest, 2001; Read, 2006] and *Pluralists* [Beall and Restall, 2000; Beall and Restall, 2006] about logical consequence.

In this debate we can at first set aside some consequence relations. Some formal consequence relations we call 'logics' are clearly logical consequence relations by courtesy only. They are not intended as giving an account of a consequence relations between statements or propositions. They are logical consequence relations as abstract formal structures designed to model something *quite like* a traditional consequence relation. The substructural logic of the Lambek calculus, when interpreted as giving us information about syntactic types is a good example of this. It is a 'logic' because it has a formal structure like other logics, but we do not think of the consequence relation between statements as anything like *truth* preservation.

To know that A follows from B in the Lambek calculus interpreted in this way means that any syntactic string of type A is also a string of type B.[8] So, we may set these logics (or logical consequence relations interpreted in these ways) aside from our consideration, and consider consequence relations where the verdict that A logically entails B is to tell us something about the relationship between the *statements* A and B.

Consider four examples where there is thought to be a genuine disagreement about consequence between given statements, in debates over second order consequence; intuitionist logic; classical logic; and indexicality.

SECOND ORDER LOGIC: Does $a \neq b$ entail $\exists X(Xa \wedge \neg Xb)$? According to standard accounts of second order logic, it does.[9] If a and b are distinct, there is some way to assign a value to the second order variable X to ensure that (relative to this assignment), $Xa \wedge \neg Xb$ is satisfied.[10] If second order quantification is not properly logical, then perhaps the entailment fails. What are we to say? Perhaps there is to be addressed by singling out some class of vocabulary as the *logical constants*, and then we are to decide which side of the boundary the second order quantifiers are to fall. Perhaps, on the other hand, we are to address the question of the validity of this argument by other means. Perhaps we are to ask what are the admissible interpretatons of the statment $(\exists X)(Xa \wedge \neg Xb)$, or the circumstances in which its truth or falsity may be determined.

Monists take there to be a single answer to this question. If they accept the distinction between logical constants and non-logical vocabulary, then second order logic must fall on one side or other of that boundary. If they do not accept such a distinction, then if consequence is understood in terms of truth preservation across a range of cases, there is a definitive class of cases in which one is to interpret the second order vocabulary.

Pluralists, on the other hand, can either say that the distinction between logical and non-logical vocabulary admits of more than one good (and equally correct) answer, or that if we do not accept such a distinction, we can say that there is a range of circumstances appropriate for interpreting second order vocabulary. For example, in this case we could say that in standard models of the second order vocabulary, $a \neq b$ does entail $\exists X(Xa \wedge \neg Xb)$, but in the more generous class of Henkin models, we can find interpretations in which $a \neq b$ holds yet $\exists X(Xa \wedge \neg Xb)$ fails, because we have fewer possible assignments for the second order variable X. Pluralists in this case go on to say that the choice between the wider and narrower class of models need not be an all-or-nothing decision. We can

[8] So, we could interpret the consequence relation between A and B as telling us that the typing judgement 'x is of type B' somehow follows from the typing judgement 'x is of type A,' but these typing judgements are not the relata of the 'consequence relation' in question. The types A and B are those relata.

[9] This is a relatively uncontroversial example. For a more difficult case, consider the statement $\forall R(\forall x \exists y Rxy \rightarrow \exists f \forall x Rxfx)$, which is a statement of the axiom of choice in the vocabulary of second order logic. Is this a tautology (a consequence of the empty set of premises)?

[10] Perhaps we can assign X the extension $\{d\}$ where d is the object in the domain assigned as the denotation of the name a.

have two consequence relations: according to one, the argument is valid (and has no—standard model—counterexamples), and according to the other it is invalid (and has a—Henkin model—counterexample).

Is this sort of answer enough to satisfy? For the pluralist, it may well. For the monist, it does not. The monist would like to know whether the putative counterexample to the argument is a *genuine* counterexample or not, for then and only then will we know whether the argument is valid or not.

Not all disagreements between pluralists and monists emerge on the ground where we may dispute over whether or not we have a logical constant. The next two disagreements over logical consequence are on the battleground of the standard propositional connectives, and the consensus is that these are logical constants if anything deserves the name.

CONSTRUCTIVE LOGIC: Is $p \vee \neg p$ a tautology? Is the argument from $\neg\neg p$ to p valid? The intuitionist says that they are not—that we may have a construction that shows $\neg\neg p$ (by reducing $\neg p$ to absurdity) while not also constructing the truth of p. Constructions may be incomplete. Similarly, not all constructions will verify $p \vee \neg p$—not through *refuting* it by verifying $\neg p \wedge \neg\neg p$ but by neither verifying p nor $\neg p$. In these cases, the genuine intuitionists claim to have counterexamples to these classically valid arguments, while the orthodox classical logicians take these arguments to have no counterexamples. The debate here is *not* over whether negation or disjunction are logical constants. The monist (whether accepting classical or intiuitionist logic) holds there to be a definitive answer to the question of whether these constructive counterexamples are worth the name. If they are, then the arguments are *invalid* (and classical logic is not correct) and if they are not, then classical logic judges some invalid arguments to be invalid, and is hence to be rejected. This is the standard intuitioinst response to classical logic.[11]

The pluralist, on the other hand, cannot say this. The pluralist can say, on the other hand, that the argument from $\neg\neg p$ to p is valid in one sense (it is classically valid: it has no counterexample in consistent and complete *worlds*) and invalid in another (it is constructively invalid: there are *constructions* for $\neg\neg p$ that are not constructions for p). The pluralist goes on to say that these two verdicts do not conflict with one another. The one person can agree with both verdicts. It is just that the notion of a counterexample expands from the traditional view according to which an argument either has counterexamples or it does not, and that is the end of the matter. Instead, we have narrower (classical, worlds) and wider (intuitionistic, constructions) classes of circumstances, and for some theoretical purposes we want the narrower class, and for others the wider. There are two consequence relations here, not one [Restall, 2001].

RELEVANCE: The same sort of consideration holds over debates of relevance. For

[11]It may be augmented, of course, by saying that classical logic has its place as a logic of decidable situations, which is to say that in *some* circumstances, $p \vee \neg p$ is true, for some statements p. It is not necessarily to say that classical logic is the right account of *validity* for any class of arguments.

relevantists, the argument form of explosion, from $p \wedge \neg p$ to q and disjunctive syllogism, from $p \wedge (\neg p \vee q)$ to q are invalid, while for classical and constructive accounts of validity, they are valid. The monist holds either that classical logic is correct in this case, or that it is incorrect and a relevant account of consequence is to be preferred. This has proved to be a difficult position to take for the relevantist: for there seems to be a clear sense in which disjunctive syllogism is *valid*—there are no consistent circumstances in which $p \wedge (\neg p \vee q)$ is true and q is not. There are, for the relevantist, counterexamples to this argument, but they involve inconsistent circumstances, which make p and $\neg p$ both true. Now, it may be the case that inconsistent circumstances like these are suitable for individuating content in a finely-grained way—a situation inconsistent about p need not be inconsistent about q so we can distinguish them as having different subject matter, even in the case where the premise $p \wedge \neg p$ is inconsistent. However, for many purposes, we would like to ignore these inconsistent circumstances. In many practical reasoning situations, we would like very much to deduce q from $p \wedge (\neg p \vee q)$, and inconsistent situations where $p \wedge (\neg p \vee q)$ holds and q doesn't seem neither here nor there.

In this case the pluralist is able to say that there is more than one consequence relation at work, and that we need not be forced to choose. If we wish to include incosistent situations as 'counterexamples' then we have a strong—relevant—consequence relation. Without them, we do not. The monist has no such response, and this has caused some consternation for the monist who wishes to find a place for relevant consequence [Belnap and Dunn, 1981].

INDEXICALITY: Is the argument from 'It is raining' to 'It is raining *here*' a valid one? Is the argument form from p to $@p$ valid? Again, it depends on the admissible ways to interpret the actuality operator @ or the indexical 'here.' In one sense it is clearly valid, and in another sense, it is cleary not valid. If we take the constituents of arguments here to be *sentences*, then there is a sense in which the argument from 'It is raining' to 'It is raining *here*' is valid, for any circumstance in which the sentence 'It is raining' is expressed to state something true, the sentence 'It is raining here' expressed in that circumstance is also true. If, on the other hand, we take the constituents of the argument to be the *contents* of the sentences, then the argument can be seen to be invalid. For the claim that it is raining can be true without it being true that it is raining *here*. Had it been raining over there, then relative to that circumstance, it would be raining, but it wouldn't be raining *here*.[12] Another way to be a pluralist, then, is to say that sometimes it is appropriate to evaluate consequence as a relation between sentences, and sometimes it is appropriate to think of it as a relation between contents of those sentences. In any case, we must pay attention to the kinds of items *related* by our consequence relations, as our choices here will also play a part in determining what kind of relation we have [Russell, 2008], and perhaps, how many of those relations there might be.

[12] This is equivalent, formally, to evaluating the sentences 'off the diagonal' in a two-dimensional modal model.

BIBLIOGRAPHY

[Anderson and Belnap, 1975] Alan R. Anderson and Nuel D. Belnap. *Entailment: The Logic of Relevance and Necessity*, volume 1. Princeton University Press, Princeton, 1975.

[Anderson *et al.*, 1992] Alan Ross Anderson, Nuel D. Belnap, and J. Michael Dunn. *Entailment: The Logic of Relevance and Necessity*, volume 2. Princeton University Press, Princeton, 1992.

[Antonelli, 2005] Aldo Antonelli. *Grounded Consequence for Defeasible Logic*. Cambridge University Press, Cambridge, 2005.

[Antonelli, 2010] Aldo Antonelli. Non-monotonic logic. In Edward N. Zalta, editor, *The Stanford Encyclopedia of Philosophy*. Fall 2011 edition, 2010.

[Aristotle and Ackrill, 1963] Aristotle and J.L. Ackrill. *Aristotle's Categories and De interpretatione*. Clarendon Aristotle series. Clarendon Press, 1963.

[Armstrong, 1967] A. H. Armstrong, editor. *The Cambridge History of Later Greek and Early Medieval Philosophy*. Cambridge University Press, 1967.

[Barnes, 1995] J. Barnes. *The Cambridge Companion to Aristotle*. Cambridge Companions to Philosophy. Cambridge University Press, 1995.

[Beall and Restall, 2000] JC Beall and Greg Restall. Logical pluralism. *Australasian Journal of Philosophy*, 78:475–493, 2000.

[Beall and Restall, 2006] JC Beall and Greg Restall. *Logical Pluralism*. Oxford University Press, Oxford, 2006.

[Belnap and Dunn, 1981] Nuel D. Belnap and J. Michael Dunn. Entailment and the disjunctive syllogism. In F. Fløistad and G. H. von Wright, editors, *Philosophy of Language / Philosophical Logic*, pages 337–366. Martinus Nijhoff, The Hague, 1981.

[Blackburn *et al.*, 2001] Patrick Blackburn, Maarten de Rijke, and Yde Venema. *Modal Logic*. Cambridge University Press, 2001.

[Bobzien, 2011] Susanne Bobzien. Dialectical school. In Edward N. Zalta, editor, *The Stanford Encyclopedia of Philosophy*. Fall 2011 edition, 2011.

[Boh, 1982] Ivan Boh. *The Cambridge History of Later Medieval Philosophy.*, chapter 15 Consequences, pages 300–314. Cambridge University Press, 1982.

[Bolzano, 1972] B. Bolzano. *Theory of Science*. University of California Press, Berkeley and Los Angeles, 1972.

[Boole, 1951] George Boole. *An Investigation of the Laws of Thought: on which are founded the mathematical theories of logic and probabilities*. Dover Press, 1951.

[Boole, 2009] G. Boole. *The Mathematical Analysis of Logic: Being an Essay Towards a Calculus of Deductive Reasoning*. Cambridge Library Collection - Mathematics. Cambridge University Press, 2009.

[Brandom, 1994] Robert B. Brandom. *Making It Explicit*. Harvard University Press, 1994.

[Buridan, 1985] Jean Buridan. *Jean Buridan's Logic: the Treatise on supposition, the Treatise on consequences; translated, with a philosophical introduction by Peter King*. D. Reidel Publishing Company, 1985.

[Burris, 2010] Stanley Burris. George Boole. In Edward N. Zalta, editor, *The Stanford Encyclopedia of Philosophy*. Summer 2010 edition, 2010.

[Bury, 1933] R. G. Bury. *Sextus Empiricus*. William Heinemann, 1933.

[Carnap *et al.*, 1990] R. Carnap, W.V.O. Quine, and R. Creath. *Dear Carnap, dear Van: the Quine Carnap correspondence and related work*. Centennial Books. University of California Press, 1990.

[Carnap, 1928] Rudolf Carnap. *Der Logische Aufbau der Welt*. Felix Meiner, 1928.

[Carnap, 1959] Rudolf Carnap. *The Logical Syntax of Language*. Littlefield, Adams and Co., 1959.

[Carnap, 1967] Rudolf Carnap. *The Logical Structure of the World and Pseudoproblems in Philosophy*. Routledge & Kegan Paul, 1967.

[Carus, 2007] A. W. Carus. Carnap's intellectual development. In *The Cambridge Companion to Carnap*, pages 19–42. Cambridge University Press, 2007.

[Chalmers, 2006] David Chalmers. The foundations of two-dimensional semantics. In Manuel Garcia-Carpintero and Josep Maciá, editors, *Two-Dimensional Semantics*, pages 55–140. Clarendon Press, 2006.

[Chihara, 1979] Charles Chihara. The semantic paradoxes: A diagnostic investigation. *The Philosophical Review*, 88:590–618, 1979.

[Church, 1976] Alonzo Church. Comparison of Russell's resolution of the semantical antinomies with that of Tarski. *The Journal of Symbolic Logic*, 41:747–760, 1976.

[Coffa, 1993] J. Alberto Coffa. *The Semantic Tradition from Kant to Carnap.* Cambridge University Press, 1993.

[Czelakowski and Malinowski, 1985] Janusz Czelakowski and Grzegorz Malinowski. Key notions of Tarski's methodology of deductive systems. *Studia Logica*, 44:321–351, 1985.

[Davies and Humberstone, 1980] Martin Davies and Lloyd Humberstone. Two notions of necessity. *Philosophical Studies*, 38(1):1–30, 1980.

[Dutilh Novaes, 2005a] Catarina Dutilh Novaes. Buridan's consequentia: Consequence and inference within a token-based semantics. *History and Philosophy of Logic*, 26(4):277–297, 2005.

[Dutilh Novaes, 2005b] Catarina Dutilh Novaes. Medieval obligationes as logical games of consistency maintenance. *Synthese*, 145(3):371–395, 2005.

[Dutilh Novaes, 2006] Catarina Dutilh Novaes. Roger Swyneshed's obligationes: A logical game of inference recognition? *Synthese*, 151:125–153, 2006.

[Dutilh Novaes, 2007] Catarina Dutilh Novaes. *Formalizing Medieval Logical Theories: Suppositio, Consequentiae and Obligationes.* Number 7 in Logic, Epistemology, and the Unity of Science. Springer, 2007.

[Dutilh Novaes, 2008a] Catarina Dutilh Novaes. A comparative taxonomy of medieval and modern approaches to liar sentences. *History and Philosophy of Logic*, 29:227–261, 2008.

[Dutilh Novaes, 2008b] Catarina Dutilh Novaes. An intensional interpretation of Ockham's theory of supposition. *Journal of the History of Philosophy*, 46(3):365–394, 2008.

[Dutilh Novaes, 2009] Catarina Dutilh Novaes. Lessons on sentential meaning from medieaeval solutions to the liar paradox. *The Philosophical Quarterly*, 59:682–704, 2009.

[Epictetus, 1925] Epictetus. *Epictetus: The Discourses as reported by Arrian, the Manual, and Fragments.* Harvard University Press, 1925.

[Etchemendy, 1990] John Etchemendy. *The Concept of Logical Consequence.* Harvard University Press, Cambridge, Mass., 1990.

[Ewald, 1996a] William Ewald. *From Kant to Hilbert*, volume 1. Oxford University Press, 1996.

[Ewald, 1996b] William Ewald. *From Kant to Hilbert*, volume 2. Oxford University Press, 1996.

[Field, 2008] Hartry Field. *Saving Truth From Paradox.* Oxford University Press, 2008.

[Frege *et al.*, 2004] G. Frege, R. Carnap, E.H. Reck, S. Awodey, and G. Gabriel. *Frege's lectures on logic: Carnap's student notes, 1910-1914.* Full circle. Open Court, 2004.

[Frege, 1953] Gottlob Frege. *The Foundations of Arithmetic.* Harper & Bros., New York, 1953.

[Friedman and Creath, 2007] Michael Friedman and Richard Creath, editors. *The Cambridge Companion to Carnap.* Cambridge University Press, 2007.

[Gabbay and Woods, 2004a] D.M. Gabbay and J. Woods. *Handbook of the History of Logic: Greek, Indian, and Arabic logic.* Handbook of the History of Logic: Greek, Indian and Arabic Logic. Elsevier, 2004.

[Gabbay and Woods, 2004b] D.M. Gabbay and J. Woods. *Handbook of the History of Logic: The rise of modern logic: from Leibniz to Frege.* Handbook of the History of Logic. Elsevier, 2004.

[Gabbay and Woods, 2009] D.M. Gabbay and J. Woods. *Logic from Russell to Church.* Handbook of the History of Logic. Elsevier, 2009.

[Gabbay *et al.*, 2006] D.M. Gabbay, J. Woods, and J. Woods. *Logic and the Modalities in the Twentieth Century*, volume 7 of *Handbook of the History of Logic.* Elsevier North Holland, 2006.

[Gabbay, 1985] D. M. Gabbay. Theoretical foundations for nonmonotonic reasoning in expert systems. In K. Apt, editor, *Logics and Models of Concurrent Systems*, pages 439–459. Springer Verlag, Berlin and New York, 1985.

[Garcia-Carpintero and Maciá, 2006] Manuel Garcia-Carpintero and Josep Maciá, editors. *Two-Dimensional Semantics.* Clarendon Press, 2006.

[Geach and Black, 1952] Peter Geach and Max Black. *Translations from the Philosophical Writings of Gottlob Frege.* Oxford University Press, 1952.

[Gentzen, 1932] Gerhard Gentzen. Über die Existenz unabhängiger Axiomensysteme zu unendlichen Satz-systemen. *Mathematische Annalen*, 107:329–350, 1932.

[Gentzen, 1935a] Gerhard Gentzen. Untersuchungen über das logische Schließen. I. *Mathematische Zeitschrift*, 39(1):176–210, 1935.

[Gentzen, 1935b] Gerhard Gentzen. Untersuchungen über das logische Schließen. II. *Mathematische Zeitschrift*, 39(1):405–431, 1935.

[Gentzen, 1969] Gerhard Gentzen. *The Collected Papers of Gerhard Gentzen*. North Holland, Amsterdam, 1969.

[Girard, 1987] Jean-Yves Girard. Linear logic. *Theoretical Computer Science*, 50:1–101, 1987.

[Godel *et al.*, 1986] K. Godel, S. Feferman, J.W. Dawson, S.C. Kleene, G.H. Moore, R.M. Solovay, and J. Heijenoort. *Collected Works: Volume I: Publications 1929-1936*. Oxford University Press, USA, 1986.

[Goldblatt, 2006] Robert Goldblatt. Mathematical modal logic: A view of its evolution. In D.M. Gabbay, J. Woods, and J. Woods, editors, *Logic and the Modalities in the Twentieth Century*, volume 7 of *Handbook of the History of Logic*, pages 1–98. Elsevier North Holland, 2006.

[Hailperin, 1986] Theodore Hailperin. *Boole's Logic and Probability: A Critical Exposition from the Standpoint of Contemporary Algebra, Logic, and Probability Theory*. North-Holland Pub. Co., Amsterdam, Netherlands; New York; New York, N.Y., U.S.A, 1986.

[Hanna, 2011] Robert Hanna. Kant's theory of judgment. In Edward N. Zalta, editor, *The Stanford Encyclopedia of Philosophy*. Summer 2011 edition, 2011.

[Hicks, 1925] R. D. Hicks. *Diogenes Laertius: Lives of Eminent Philosophers*, volume 2. William Heinemann, 1925.

[Hodges, 2009] Wilfrid Hodges. Traditional logic, modern logic and natural language. *Journal of Philosophical Logic*, 38(589-606), 2009.

[Irvine, 2009] Andrew D. Irvine. Bertrand Russell's logic. In Dov M. Gabbay and John Woods, editors, *Logic From Russell to Church*, volume 5 of *Handbook of the History of Logic*, pages 1–28. Elsevier, 2009.

[Kant, 1929] Immanuel Kant. *Critique of Pure Reason*. Macmillan, 1929.

[Kennedy, 2011] Juliette Kennedy. Kurt Gödel. In Edward N. Zalta, editor, *The Stanford Encyclopedia of Philosophy*. Fall 2011 edition, 2011.

[Keynes, 1906] J. N. Keynes. *Studies and Exercises in Formal Logic*. Macmillian & co., London, 4th edition, 1906.

[King, 1985] Peter King. *John Buridan's Logic: The Treatise on Supposition; The Treatise on Consequences, Translation from the Latin with a Philosophical Introduction*. Reidel, Dordrecht, 1985.

[King, 2010] Peter King. Peter Abelard. In Edward N. Zalta, editor, *The Stanford Encyclopedia of Philosophy*. Winter 2010 edition, 2010.

[Klima, 2003] Gyula Klima. *Companion to Philosophy in the Middle Ages*, chapter John Buridan, pages 340–348. Blackwell, 2003.

[Klima, 2004] Gyula Klima. Consequences of a closed, token-based semantics: the case of John Buridan. *History and Philosophy of Logic*, 25:95–110, 2004.

[Kneale and Kneale, 1962] William Kneale and Martha Kneale. *The Development of Logic*. Oxford University Press, 1962.

[Kretzmann *et al.*, 1982] Norman Kretzmann, Anthony Kenny, and Jan Pinborg, editors. *The Cambridge History of Later Medieval Philosophy*. Cambridge University Press, 1982.

[Lambek, 1958] Joachim Lambek. The mathematics of sentence structure. *American Mathematical Monthly*, 65(3):154–170, 1958.

[Lambek, 1961] Joachim Lambek. On the calculus of syntactic types. In R. Jacobsen, editor, *Structure of Language and its Mathematical Aspects*, Proceedings of Symposia in Applied Mathematics, XII. American Mathematical Society, 1961.

[Leibniz, 1966] Gottfried Leibniz. *Logical Papers*. Clarendon Press, Oxford, 1966.

[Lewis, 1912] C. I. Lewis. Implication and the algebra of logic. *Mind*, 21:522–531, 1912.

[MacFarlane, 2000] John MacFarlane. *What Does it Mean to Say that Logic is Formal?* PhD thesis, University of Pittsburgh, 2000.

[Marenbon, 2010] John Marenbon. Anicius Manlius Severinus Boethius. In Edward N. Zalta, editor, *The Stanford Encyclopedia of Philosophy*. Fall 2010 edition, 2010.

[Mates, 1953] Benson Mates. *Stoic Logic*. University of California Publications in Philosophy. University of California Press, 1953.

[McCarthy, 1980] J. McCarthy. Circumscription—a form of non-monotonic reasoning. *Artificial Inteligence*, 13:27–39, 1980.

[Peckhaus, 2009] Volker Peckhaus. Leibniz's influence on 19th century logic. In Edward N. Zalta, editor, *The Stanford Encyclopedia of Philosophy*. Fall 2009 edition, 2009.

[Peirce, 1870] Charles S. Peirce. Description of a notation for the logic of relatives, resulting from an amplification of the conceptions of Boole's calculus of logic. *Memoirs of the American Academy of Science*, 9:317–378, 1870.

[Priest, 2001] G. Priest. Logic: One or many? In Bryson Brown and John Woods, editors, *Logical Consequence: Rival Approaches. Proceedings of the 1999 Conference of the Society of Exact Philosophy*, pages 23–38. Hermes, Stanmore, 2001.

[Read, 2002] Stephen Read. The liar paradox from John Buridan back to Thomas Bradwardine. *Vivarium*, 40(2):189–218, 2002.

[Read, 2006] Stephen Read. Monism: The one true logic. In David Devidi and Tim Kenyon, editors, *A Logical Approach to Philosophy: Essays in Honour of Graham Solomon*, volume 69 of *The Western Ontario Series in Philosophy of Science*, pages 193–209. Springer, 2006.

[Reck, 2007] Erich H. Reck. Carnap and modern logic. In *The Cambridge Companion to Carnap*, pages 176–199. Cambridge University Press, 2007.

[Reiter, 1980] R. Reiter. A logic for default reasoning. *Artificial Intelligence*, 13(1-2):81–132, 1980.

[Restall, 2001] Greg Restall. Constructive logic, truth and warranted assertibility. *Philosophical Quarterly*, 51:474–483, 2001.

[Restall, 2005] Greg Restall. Multiple conclusions. In Petr Hájek, Luis Valdés-Villanueva, and Dag Westerståhl, editors, *Logic, Methodology and Philosophy of Science: Proceedings of the Twelfth International Congress*, pages 189–205. KCL Publications, 2005.

[Restall, to appear] Greg Restall. A cut-free sequent system for two-dimensional modal logic, and why it matters. *Annals of Pure and Applied Logic*, to appear.

[Ricketts, 2007] Thomas Ricketts. Tolerance and logicism: logical syntax and the philosophy of mathematics. In *The Cambridge Companion to Carnap*, pages 200–225. Cambridge University Press, 2007.

[Russell, 1919] Bertrand Russell. *Introduction to Mathematical Philosophy.* Allen and Unwin, 1919.

[Russell, 1937] Bertrand Russell. *The Principles of Mathematics.* George Allen & Unwin, second edition, 1937.

[Russell, 2008] Gillian Russell. One true logic? *Journal of Philosophical Logic*, 37(6):593–611, 2008.

[Scott, 1971] Dana Scott. On engendering an illusion of understanding. *The Journal of Philosophy*, 68:787–807, 1971.

[Sebestik, 2011] Jan Sebestik. Bolzano's logic. In Edward N. Zalta, editor, *The Stanford Encyclopedia of Philosophy.* Winter 2011 edition, 2011.

[Sedley, 2009] David Sedley. Diodorus Cronus. In Edward N. Zalta, editor, *The Stanford Encyclopedia of Philosophy.* Fall 2009 edition, 2009.

[Sellars, 2007] W. Sellars. Inference and meaning. In Kevin Scharp and Robert Brandom, editors, *In the Space of Reasons*, pages 3–27. Harvard University Press, 2007.

[Shoesmith and Smiley, 1978] D. J. Shoesmith and T. J. Smiley. *Multiple Conclusion Logic.* Cambridge University Press, Cambridge, 1978.

[Siebel, 2002] Mark Siebel. Bolzano's concept of consequence. *Monist*, 85(4):580–599, 2002.

[Simmons, 2009] Keith Simmons. Tarski's logic. In Dov M. Gabbay and John Woods, editors, *Logic From Russell to Church*, volume 5 of *Handbook of the History of Logic*, pages 511–616. Elsevier, 2009.

[Smith, 2008] Peter Smith. *An Introduction to Gödel's theorems.* Cambridge University Press, Cambridge, 2008.

[Smith, 1989] Robin Smith (tr. and comm.). *Aristotle: Prior Analytics.* Hackett Publishing Company, 1989.

[Smith, 2011] Robin Smith. Aristotle's logic. In Edward N. Zalta, editor, *The Stanford Encyclopedia of Philosophy.* Fall 2011 edition, 2011.

[Spade and Read, 2009] Paul Vincent Spade and Stephen Read. Insolubles. In Edward N. Zalta, editor, *The Stanford Encyclopedia of Philosophy.* Winter 2009 edition, 2009.

[Spade, 1981] Paul Vincent Spade. *Insolubilia* and Bradwardine's theory of signification. *Medioevo: Revista di storia della filosofia medievale*, 7:115–34, 1981.

[Spade, 1982] Paul Vincent Spade. Three theories of *obligationes*: Burley, Kilvington and Swyneshed on counterfactual reasoning. *History and Philosophy of Logic*, 3:1–32, 1982.

[Spade, 2000] Paul Vincent Spade. Why don't mediaeval logicians ever tell us what they're doing? Or, what is this, a conspiracy? 2000.

[Spade, 2002] Paul Vincent Spade. Thoughts, words and things: An introduction to late mediaeval logic and semantic theory. Version 1.1, 2002.

[Spade, 2008] Paul Vincent Spade. Medieval theories of obligationes. In Edward N. Zalta, editor, *The Stanford Encyclopedia of Philosophy*. Fall 2008 edition, 2008.

[Stump and Spade, 1982] Eleonore Stump and Paul Vincent Spade. *The Cambridge History of Later Medieval Philosophy*, chapter 16 Obligations, pages 315–341. Cambridge University Press, 1982.

[Stump, 1985] E. Stump. The logic of disputation in Walter Burley's treatise on obligations. *Synthese*, 63:355–374, 1985.

[Swyneshed, 2006] Roger Swyneshed. Obligationes: A logical game of inference recognition? *Synthese*, 151(1):125–153, 2006.

[Tarski, 1956a] Alfred Tarski. *Logic, Semantics, Metamathematics: papers from 1923 to 1938.* Clarendon Press, Oxford, 1956.

[Tarski, 1956b] Alfred Tarski. On the concept of logical consequence. In *Logic, Semantics, Metamathematics: papers from 1923 to 1938*, chapter 16, pages 409–420. Clarendon Press, Oxford, 1956.

[van Atten and Kennedy, 2009] Mark van Atten and Juliette Kennedy. Gödel's logic. In Dov M. Gabbay and John Woods, editors, *Logic From Russell to Church*, volume 5 of *Handbook of the History of Logic*, pages 449–510. Elsevier, 2009.

[van Benthem, 1985] Johan van Benthem. The variety of consequence, according to bolzano. *Studia Logica*, 44:389–403, 1985.

[van Heijenoort, 1967] Jean van Heijenoort. *From Frege to Gödel: a a source book in mathematical logic, 1879–1931.* Harvard University Press, Cambridge, Mass., 1967.

[von Plato, 2009] Jan von Plato. Gentzen's logic. In Dov M. Gabbay and John Woods, editors, *Logic From Russell to Church*, volume 5 of *Handbook of the History of Logic*, pages 667–722. Elsevier, 2009.

[Whitehead and Russell, 1925–1927] Alfred North Whitehead and Bertrand Russell. *Principia Mathematica*. Cambridge University Press, 1925–1927.

[Whitehead and Russell, 1962] A. N. Whitehead and B. Russell. *Principa Mathematica to *56.* Cambridge University Press, Cambridge, 1962.

[Wittgenstein, 1922] Ludwig Wittgenstein. *Tractatus Logico-Philosophicus.* Routledge, 1922.

[Zupko, 2011] Jack Zupko. John Buridan. In Edward N. Zalta, editor, *The Stanford Encyclopedia of Philosophy*. Fall 2011 edition, 2011.

A HISTORY OF QUANTIFICATION

Daniel Bonevac

Aristotle (384–322 BC), the founder of the discipline of logic, also founded the study of quantification. Normally, Aristotle begins a topic by reviewing the common opinions, including the opinions of his chief predecessors. In logic, however, he could not adopt the same strategy; before him, he reports, "there was nothing at all" (*Sophistical Refutations* 183b34–36). Aristotle's theory dominated logical approaches to quantification until the nineteenth century.

That is not to say that others did not make important contributions. Medieval logicians elaborated Aristotle's theory, structuring it in the form familiar to us today. They also contemplated a series of problems the theory generated, devising increasingly complex theories of semantic relations to account for them. Textbook treatments of quantification in the seventeenth and nineteenth centuries made important contributions while also advancing some peculiar theories based on medieval contributions.

Modern quantification theory emerged from mathematical insights in the middle and late nineteenth century, displacing Aristotelian logic as the dominant theory of quantifiers for roughly a century. It has become common to see the history of logic as little more than a prelude to what we now call classical first-order logic, the logic of Frege, Peirce, and their successors. Aristotle's theory of quantification is nevertheless in some respects more powerful than its modern replacement. Aristotle's theory combines a relational conception of quantifiers with a monadic conception of terms. The modern theory combines a monadic conception of quantifiers with a relational theory of terms. Only recently have logicians combined relational conceptions of quantifiers and terms to devise a theory of generalized quantifiers capable of combining the strengths of the Aristotelian and modern approaches.

There is no theory-neutral way of defining quantification, or even of delineating the class of quantifiers. Some logicians treat determiners such as 'all,' 'every,' 'most,' 'no,' 'some,' and the like as quantifiers; others think of them as denoting quantifiers. Still others think of quantifiers as noun phrases containing such determiners ('all men,' 'every book,' etc.). Some include other noun phrases ('Aristotle,' 'Peter, Paul, and John,' etc.). Some define quantifiers as variable-binding expressions; others lack the concept of a variable. My sketch of the history of our understanding of quantification thus traces the development of understandings of *what* is to be explained as much as *how* it is to be explained.

1 ARISTOTLE'S QUANTIFICATION THEORY

Aristotle first developed a theory of quantification in the form of his well-known theory of syllogisms. The theory's familiarity, not only from ubiquitous textbook treatments but

Handbook of the History of Logic. Volume 11: Logic: A History of its Central Concepts.
Volume editors: Dov M. Gabbay, Francis Jeffry Pelletier and John Woods.
General editors: Dov M. Gabbay and John Woods.

also from important scholarly studies, should not blind us to some of its less-remarked but critically important features.[1] Aristotle recognizes that validity is a matter of *form*. He aspires to *completeness*; he characterizes a realm of inquiry and seeks to identify all valid argument forms within it. He develops the first theory of *deduction*, and offers the first *completeness proof*, showing by means of his method of deduction that all the valid argument forms within that realm can be shown to be valid on the basis of two basic argument forms. He proceeds to prove several *metatheorems*, which taken together constitute an alternative decision procedure for arguments. More importantly for our purposes, Aristotle develops an understanding of quantifiers that is in some ways more powerful than that of modern logic, and was not superceded until the development of the theory of generalized quantifiers.

1.1 Validity as a Matter of Form

Aristotle restricts his attention to *statements*, *assertions*, or *propositions* (*apophanseis*), sentences that are (or perhaps: can be) true or false.[2] Like most subsequent logicians, he focuses on a limited set of quantifiers. Every premise, he says, is universal, particular, or indefinite:

> A proposition (*protasis*), then, is a sentence affirming or denying something of something; and this is either universal or particular or indefinite. By universal I mean a statement that something belongs to all or none of something; by particular that it belongs to some or not to some or not to all; by indefinite that it does or does not belong, without any mark of being universal or particular, e.g. 'contraries are subjects of the same science', or 'pleasure is not good'. (*Prior Analytics* I, 1, 24a16–21.)[3]

Aristotle is here defining the subject matter of his theory; interpreted otherwise, his claim is obviously false. Not only are some statements singular, e.g., 'Socrates is running,' as Aristotle recognizes at *De Interpretatione* I, 7, while others are compound, but there are also many other quantifiers: 'many,' ''most,' 'few,' 'exactly one,' 'almost all,' 'finitely many,' and so on. The theory he presents, in fact, focuses solely on universal and particular quantifiers; he has almost nothing to say about the third sort he mentions, indefinites (or bare plurals, as they are known today), and the little he says holds only in limited linguistic contexts.[4]

---

[1]Important scholarly studies include [Kneale and Kneale, 1962; Patzig, 1969; Corcoran, 1972; 1973; 1974; Smiley, 1974; 1994; Lear, 1980; Barnes, 1981; Irwin, 1988; Wedin, 1990; McKirahan, 1992; Johnson, 1994; Boger, 2004; Woods and Irvine, 2004].

[2]See *De Interpretatione* 4, 17a1: "Every sentence is significant... but not every sentence is a statement-making sentence, but only those in which there is truth and falsity."

[3]Hereafter all references to Aristotle will be to the *Prior Analytics* unless otherwise noted.

[4]"We shall have the same deduction (*syllogismos*) whether it is indefinite or particular" (1, 4, 26a30; see also I, 7, 29a27–29: "It is evident also that the substitution of an indefinite for a particular affirmative will effect the same deduction in all the figures."). Aristotle's commentators follow him in this. Alexander of Aphrodisias, for example, writes: "He [Aristotle] doesn't speak about [converting] indefinites, because they are of no use for syllogisms and because they can be [regarded as] equal to particulars" [Alexander of Aphrodisias, 1883, 30].

The sentences within the scope of Aristotle's theory, then, are either universal or particular. They are also either affirmative or negative (I, 2, 25a1). (Medieval logicians refer to these as the *quantity* and *quality* of the statements, respectively. Peter of Spain may have been the first to use those terms. C. S. Peirce credits their introduction to Apuleius (125?–180?), remarking that they are "more assified than golden" [1893, 279n], but this rests on a misattribution.) Aristotle thus concerns himself with four kinds of sentences, known as *categorical propositions*. These have the forms:

Universal affirmative (A): Every S is P (SaP)

Universal negative (E): No S is P (SeP)

Particular affirmative (I): Some S is P (SiP)

Particular Negative (O): Some S is not P (SoP)

I shall call these *categorical statement forms*. The abbreviations A, E, I, and O are early medieval inventions; the symbolic forms SaP, etc., are not in Aristotle or the medievals, though a number of modern commentators employ them. They point to two central features of Aristotle's theory. First, Aristotle analyzes inference in terms of logical form. He does not turn to a discussion of the content of the terms that appear in the arguments he uses as illustrations (e.g., 'contraries,' 'pleasure,' 'good,' and the like, from the arguments in the very first section). He sees logical validity as structural, as a matter of form. He introduces the use of variables to represent the forms of the sentences under consideration, leaving only the determiners, the copula, and negation as logical constants. In medieval language, he treats them as *syncategorematic*. He proceeds to develop a theory of validity, assuming that telling good arguments from bad is solely a matter of identifying and classifying logical forms.

Second, the notation SaP, etc., makes explicit Aristotle's understanding of quantifiers as relations. It is not so easy to say what they relate; given Aristotle's lack of any theory of the semantic value of terms (*horoi*), it is probably most accurate to say that, for him, quantifiers are relations between terms.[5] This marks a critical difference between Aristotle's theory and modern quantification theory, which effectively takes a quantifier as a monadic predicate.

Aristotle uses the singular (e.g., 'Every S is P') rather than the plural, which sounds more natural in English, to stress that terms such as S and P are affirmed or denied of objects taken one by one. His definition of 'term' is obscure — "I call that a term into which the proposition is resolved, i.e. both the predicate and that of which it is predicated, 'is' or 'is not' being added" (I, 1, 24b17–18) — but it is plainly in the general spirit of his theory to define a term as a linguistic expression that can be true or false of individual objects. He does not put it in quite that way, for he thinks of truth and falsehood solely as

'Dogs are barking' is equivalent to 'Some dogs are barking,' but 'Dogs bark' is plainly not equivalent to 'Some dogs bark.' The position thus has some peculiar consequences. There are no valid syllogisms with two indefinite premises, for example; hence, 'Dogs are mammals; mammals are animals; so, dogs are animals' and 'Dogs are mammals; dogs bark; so, some mammals bark' both fail.

[5]For a discussion of extensional and intensional interpretations of the role of quantifiers, see Boger [2004, 164–165].

properties of statements; he has no concept of an expression being true or false *of* something, though he does have the closely related concept of an expression being affirmed or denied of something.[6]

The determiner-first phrasing ('Every S is P,' 'No S is P,' etc.) is from Boethius (480–525). Aristotle's usual phrasing places the determiner near the end: 'P belongs to every S,' 'P belongs to no S,' etc.), which perhaps makes it easier to see the validity of some syllogistic forms, as certain commentators have suggested, but also makes it harder to see the determiner as a relation between terms. To a modern ear, it furthermore suggests a misleading analogy with set theory. I will consequently continue to use the Boethian representations, even though they post-date Aristotle by some 800 years.

1.2 Aristotle's Completeness Proof

Aristotle has an informal definition of validity:

> A syllogism is discourse in which, certain things being stated, something other than what is stated follows of necessity from their being so. I mean by the last phrase that they produce the consequence, and by this, that no further term is required from without in order to make the consequence necessary. (I, 1)

Notice that syllogisms are thus valid by definition; most arguments of the general sort addressed by the theory are not syllogisms. Notice also that the definition has a circular feel to it, for the definiens contains 'follows,' and its explication contains 'consequence.' Aristotle thus says that a syllogism is one whose premises entail its conclusion. But he does not give even an informal characterization of entailment. Some commentators read the contemporary conception of deductive validity as truth-preservation — if the premises are true, the conclusion must be true as well — into Aristotle's definition on the ground that he speaks of the consequence following by necessity (*ex anankes sumbainein*). It seems fairer to treat the modern conception as compatible with Aristotle's definition, but as not itself a part of the theory, since Aristotle's definition makes no reference to truth. He seems to take the concept of necessity as primitive [Lear, 1980]. Aristotle also insists that the conclusion must be a proposition that does not appear among the premises. Some ancient commentators, taking Aristotle's plural form seriously — in my view, too seriously — furthermore see the definition as ruling out arguments with a single premise. In any case Aristotle does concern himself with such inferences under the heading of conversion.

It has become common to see in Aristotle not only the first development of a logical system but also the first completeness proof. In one sense, this is correct, as we shall see in a moment. But it is also misleading, for Aristotle has nothing more than an intuitive conception of validity. He restricts himself to arguments with two premises, one conclusion, and three terms, each appearing in two propositions. All the propositions are of one of four categorical statement forms. That makes his task finite. There are 256 possible arguments of that general form, 24 of which he considers valid. He then shows

[6]Compare *De Interpretatione* 10, 19b11–12: "Without a verb there will be no affirmation or negation."

that all the valid forms reduce to some basic forms in the sense that their validity can be demonstrated from the validity of the basic forms by applying certain rules. From a modern perspective, the finite character of the deduction system makes it something of a toy. One could simply take all twenty-four valid forms as basic and leave it at that. The real point of axiomatization is, after all, to give a finite characterization of an infinite set, something Aristotle has no need to do within the confines of the theory of categorical syllogisms. The interesting completeness claim lies not in the system of the *Prior Analytics* but instead in the *Posterior Analytics*, where Aristotle argues that all deductively valid arguments can be shown to be valid by combinations of syllogistic techniques.[7]

Aristotle's system is nevertheless an impressive achievement. He effectively presents something akin to a natural deduction system for syllogistic forms. He presents the rules as licensing moves from one categorical sentence form to another, and most commentators follow him in thinking of the deduction system as licensing inferences to categorical statement forms given other categorical statement forms. Lukasiewicz [1951], noting that Aristotle always phrases syllogisms as conditionals rather than inference patterns, constructs an Aristotelian calculus with some interesting features. Given that Aristotle's conditionals are logical truths, however, and given that Lukasiewicz construes them as material conditionals, the deduction theorem makes a conditional interpretation equivalent to an inference rule interpretation.[8] It would also be possible to interpret the derivations Aristotle provides in the style of a Gentzen consecution calculus, as licensing moves from one syllogism to another. I shall present his system in both forms.

Aristotle distinguishes three syllogistic *figures*, or configurations of terms. Syllogisms contain three categorical propositions and three terms, each of which appears in two different propositions. The *middle* term occurs in both premises; the other terms are *extremes*. In the *first* figure, as Aristotle understands it, the middle term is subject of one premise and predicate of the other. (This, at any rate, is the definition offered by Theophrastus, Aristotle's successor as head of the Lyceum; Aristotle himself gives no definition.[9]) In the *second* figure, the middle term appears as predicate in both premises (26b34–35). In the *third* figure, the middle term appears as subject in both (28a10–11).

Aristotle's commentators, as well as many medieval logicians, debate the need for a fourth figure. The issue comes down to distinguishing the extremes as major and minor terms. Aristotle gives no general definition; instead, he distinguishes major from minor terms relative to a figure. Here are his definitions:

Major term

First figure: "in which the middle is contained" (26a22)
Second figure: "that which lies near the middle" (26b37)

[7]For an extended discussion, see Lear [1980].

[8]There are reasons other than tradition and this equivalence for taking syllogisms as arguments rather than conditionals. Lukasiewicz has to attribute to Aristotle a modern propositional logic; he also has to reinterpret talk of premises and conclusions as talk of antecedents and consequents.

[9]See Lukasiewicz [1951, 97]. Some commentators see a definition in Aristotle, at 25b32–33 or 40b30–41a20. I find it hard to interpret these remarks as definitions, much less as a definition specifying that the middle term is subject of the major premise and predicate of the minor premise.

Third figure: "that which is further from the middle" (28a13)

Minor term

First figure: "which comes under the middle" (26a23)
Second figure: "that which is further away from the middle" (26b38)
Third figure: "that which is nearer to [the middle]" (28a14)

These are unsatisfactory in several respects. They suggest no general meaning of 'major' and 'minor'; the two switch roles completely between the second and third figures. The definition for first figure works well for the syllogism known to the medievals as *Barbara*:

Every M is P
Every S is M
$\therefore$ Every S is P

But it fails for others. Consider *Ferio*:

No M is P
Some S is M
$\therefore$ Some S is not P

Here the middle term, M, is not contained in P; in fact, the two are completely disjoint.

John Philoponus (490–570), a sixth-century Alexandrian commentator, first defines major and minor terms in the way that is now standard: The major term is the predicate of the conclusion; the minor term is the subject of the conclusion.[10] That definition, of course, makes it easy to distinguish two figures within what Aristotle considered the first: those in which the middle term is the subject of the major premise (which remain first figure) and those in which it is subject of the minor premise (fourth figure). It is then possible to specify a syllogism completely by indicating its figure (first, second, third, or fourth) and its mood (the categorical statement forms in the order ⟨major premise, minor premise, conclusion⟩): thus, 1AAA, 2EIO, 3IAI, etc.

Aristotle's deduction system proceeds by listing acceptable immediate inferences, inferences from one categorical statement form to another, that act as rules of inference. The first and most important is *conversion*. Particular affirmative and universal negative statement forms convert:

Conversion

Some S is $P \Leftrightarrow$ Some P is S
No S is $P \Leftrightarrow$ No P is S

[10] *In Aristotelis Analytica Priora commentaria*, Wallies, ed., 67.27–29, quoted in Spade [2002, 20]: "So we should use the following rule for the three figures, that the major is the term in predicate position in the conclusion, and the minor [is the term] in subject position in the conclusion."

A second inference rule is *conversion per accidens*, and raises issues of existential import, to which we will turn in the next section. Universal affirmatives do not convert *simply*, as in the above rule; 'every S is P' and 'every P is S' are not equivalent. But Aristotle assumes that all terms have nonempty extensions, and so allows the move from A forms to I forms:

Conversion per accidens

Every S is $P \Rightarrow$ Some P is S

Aristotle's third inference rule is *reductio ad absurdum*, or indirect proof, together with rules about which categorical statement forms contradict which:

Contradictories

NOT Every S is $P \Leftrightarrow$ Some S is not P
NOT Some S is not $P \Leftrightarrow$ Every S is P
NOT Some S is $P \Leftrightarrow$ No S is P
NOT No S is $P \Leftrightarrow$ Some S is P

Thus, A and O forms are contradictories, as are I and E forms; they always have opposite truth values.

Armed with these rules, we can reduce some syllogisms to others. Aristotle first shows that we can reduce all syllogisms to first-figure patterns. The first-figure syllogisms serve as axioms, if we think of the deduction system as moving us from syllogisms to syllogisms in the style of a consecution calculus, or as inference rules, if we think of it as moving us from statement forms to statement forms in a more typical natural deduction system. From first-figure syllogisms we can deduce all syllogisms by means of the other rules of inference.

Most deductions are direct. Consider, for example, the second figure pattern known as *Cesare* (2EAE):

No P is M
Every S is M
$\therefore$ No S is P

We can convert the first premise to obtain the first-figure syllogism *Celarent* (1EAE):

No M is P
Every S is M
$\therefore$ No S is P

We can think of this deduction in two ways. The first is a series of steps, each of which is a categorical statement form:

1. No P is M (Assumption)
2. Every S is M (Assumption)

3. No M is P (Conversion, 1)

4. No S is P (1EAE, 3, 2)

Note that we here use a first-figure syllogism as a rule of inference.

The second, in the style of the Gentzen consecution calculus, is a quick deduction from one syllogism to another:

1. No M is P, Every S is $M \vdash$ No S is P (1EAE)

2. No P is M, Every S is $M \vdash$ No S is P (Conversion, 1)

Which style of deduction most closely matches Aristotle's conception? Probably the former. Here is his deduction of *Cesare* (substituting variables appropriately):

> Let M be predicated of no P, but of every S. Since, then, the negative is convertible, P will belong to no M: but M was assumed to belong to every S: consequently P will belong to no S. This has already been proved.

This seems to take the lines of the proof as categorical statement forms rather than syllogisms.

Two syllogisms cannot be reduced to first-figure syllogisms directly. Consider this second-figure syllogism, *Baroco* (2AOO):

> Every P is M
> Some S is not M
> ∴ Some S is not P

We can pursue an indirect proof:

1. Every P is M (Assumption)

2. Some S is not M (Assumption)

3. NOT Some S is not P (Assumption for reductio)

4. Every S is P (Contradictories, 3)

5. Every S is M (1AAA, 1, 4)

6. NOT Every S is M (Contradictories, 2)

7. Some S is not P (Reductio, 3–6)

If we think of the system as moving from syllogisms to syllogisms, we can frame the argument in this way, where we view reductio as a rule allowing us to move from a syllogism $p, q \vdash r$ to p, NOT $r \vdash$ NOT q (or q, NOT $r \vdash$ NOT p), and then switch the order of the premises:

1. Every P is M, Every S is $P \vdash$ Every S is M (1AAA)

2. NOT Every S is M, Every P is $M \vdash$ NOT Every S is P (Reductio, 1)

3. Some S is not M, Every P is $M \vdash$ Some S is not P (Contradictories, 2)

4. Every P is M, Some S is not $M \vdash$ Some S is not P (Order, 3)

Aristotle can show, in either fashion, that all syllogisms reduce to first-figure syllogisms.

In fact, they all reduce to just two patterns, 1AAA (*Barbara*), which we have already met, and 1AII (*Darii*):

Every M is P
Some S is M
∴ Some S is P

For the moment, consider Aristotle's more restricted claim. All we need to do to show this is reduce the other first-figure patterns to those two. To take an example, consider the first-figure syllogism *Ferio* (1EIO):

No M is P
Some S is M
∴ Some S is not P

We can reduce this as follows:

1. No M is P (Assumption)

2. Some S is M (Assumption)

3. NOT Some S is not P (Assumption for reductio)

4. Every S is P (Contradictories, 3)

5. Some M is S (Conversion, 2)

6. Some M is P (1AII, 4, 5)

7. NOT Some M is P (Contradictories, 1)

8. Some S is not P (Reductio, 3–7)

Or, in consecution form,

1. Every S is P, Some M is $S \vdash$ Some M is P (1AII)

2. Every S is P, Some S is $M \vdash$ Some M is P (Conversion, 1)

3. NOT Some M is P, Some S is $M \vdash$ NOT Every S is P (Reductio, 2)

4. No M is P, Some S is $M \vdash$ Some S is not P (Contradictories, 3)

Aristotle does not take *Darii* as fundamental. Instead, he argues that all syllogisms reduce to universal first-figure syllogisms. He shows, for example, how *Darii* reduces by reducing it to a second-figure syllogism, which in turn reduces to first-figure. His argument (again, replacing variables):

> ... if *P* belongs to every *M*, and *M* to some *S*, it follows that *P* belongs to some *S*. For if it belonged to no *S*, and belongs to every *M*, then *M* will belong to no *S*: this we know by means of the second figure. (29b8–10)

We can represent this as follows:

1. Every *M* is *P* (Assumption)
2. Some *S* is *M* (Assumption)
3. NOT Some *S* is *P* (Assumption for reductio)
4. No *S* is *P* (Contradictories, 3)
5. No *S* is *M* (2AEE, 1, 4)
6. NOT No *S* is *M* (Contradictories, 2)

This reduces *Darii* to the second-figure syllogism *Camestres*, which in turn reduces to the universal first-figure syllogism *Celarent*:

1. Every *M* is *P* (Assumption)
2. No *S* is *P*
3. No *P* is *S* (Conversion, 2)
4. No *M* is *S* (1EAE, 3, 1)
5. No *S* is *M* (Conversion, 4)

But we could also reduce universal negative first-figure syllogisms to *Darii*. So, it is possible to see every syllogism as derived rule of inference if we take *Barbara* and either *Celarent* or *Darii* as basic rules.

1.3 The Square of Opposition

In *De Interpretatione* 7 Aristotle discusses logical relations between categorical propositions, formulating what has become known as the square of opposition. As early as the second century CE, diagrams of the square began to appear in logic texts; they became commonplace by the twelfth century. As we have seen, Aristotle takes universal affirmatives and particular negatives as contradictories, propositions having opposite truth values. He similarly takes particular affirmatives to contradict universal negatives. Aristotle adds to these relations several others.

First, and explicitly, universal affirmatives and universal negatives are *contraries*: they cannot both be true. 'Every S is P' and 'No S is P' may both be false, but cannot both be true (17b3–4, 21).

Second, in consequence, particular affirmatives and negatives are what later came to be known as *subcontraries*: they can both be true, but cannot both be false. Suppose 'Some S is P' and 'Some S is not P' were both false. Then their contradictories, 'Every S is P' and 'No S is P,' would both be true. As contraries, however, they cannot be.

Third, universals imply their corresponding particulars. Suppose 'Every S is P' is true. Then its contrary 'No S is P' is false, so its contradictory 'Some S is P' must be true. A similar argument shows that 'No S is P' implies 'Some S is not P.'

Aristotle's theses about contraries and contradictories thus generate a problem. 'Some S is P' and 'Some S is not P' cannot both be false, so 'Some S is P or some S is not P' is evidently a logical truth. But it appears to be equivalent to 'Some S is either P or not P,' and that appears to be equivalent to 'There is an S.' So, Aristotle's theses imply that certain existence claims are logical truths.

We can put the point simply. Suppose there are no Ss. Then evidently 'Some S is P' and 'Some S is not P' are both false, in which case their contradictories, 'Every S is P' and 'No S is P,' must both be true. But then they cannot be contraries.

If we are to maintain the core Aristotelian claims that generate the square of opposition, there are only two options. One is to restrict the logic to nonempty terms.[11] The other is to insist that 'Some S is not P' — or 'Not every S is P,' as Aristotle phrases the particular negative form in this context — can be true even though there are no Ss. That strategy in turn divides into two: to treat the two as equivalent, and use the latter to motivate the thought that 'Some S is not P' does not imply that there are Ss, or to deny the equivalence of 'Some S is not P' and 'Not every S is P' as Abelard (1079–1142) does, holding that only the former has existential import and that the latter is the proper inhabitant of the square of opposition. It is interesting that Aristotle used that form when articulating the square. Boethius substitutes the usual particular negative form, however, and most logicians followed his lead, even once Abelard pointed out the possibility of drawing a distinction.

2 QUANTIFIERS IN MEDIEVAL LOGIC

2.1 *The Old Logic*

The disappearance of Aristotle's logical works after the sixth century, when John Philoponus (490–570) and Simplicius (490–560) had access to them and wrote their important commentaries, placed medieval logicians at a great disadvantage. They knew logic chiefly

[11] Thus, Buridan [2001]: "Aristotle did not intend to speak about fictive terms, namely, ones that supposit for nothing, such as 'chimera' or 'goatstag,' but about terms each of which supposit for something" (382). Inspired by the square of opposition for modal propositions, which modern logic generally still endorses ('It is necessary that p' and 'It is necessary that not p' are contraries, 'It is necessary that p' and 'It is possible that not p' are contradictories, etc.), one might think of construing Aristotle's quantifiers as ranging over possible objects. The problem of empty terms then becomes a problem of impossible terms; what is one to say about 'Some round squares are not square'?

through the works of Porphyry and Boethius. The logical tradition that grew in that somewhat thin soil became known as the *logica vetus* — the Old Logic.

Porphyry (234–305), a Phoenician neo-Platonist, wrote commentaries on Aristotle's *Categories* that are now mostly lost. But he also wrote the *Isagoge* (Introduction), which became hugely influential, largely by way of Boethius's commentary on it, throughout the following millennium. Porphyry sought to avoid deep philosophical questions, but his formulation of them influenced medieval thinkers for centuries.

Boethius (475?–526?), son of a Roman consul and father of two others, translated and commented on Porphyry's *Isagoge* as well as Aristotle's *Prior Analytics*. He wrote two works on categorical syllogisms. Though most scholars see him as making few original contributions to logic, and in particular to the theory of quantification, we have already seen one vital contribution, namely, the reformulation of categorical statement forms to put determiners in initial position, e.g., of 'P belongs to every S' as 'Every S is P.' That reformulation made it possible for medieval thinkers to see quantifiers straightforwardly as relations between terms, enabling them to develop the doctrine of distribution, the *dictum de omni et nullo*, and the general theory of terms that transformed the Old Logic first into the *logica nova* and then into the late medieval theory of terms.[12] Boethius also discusses conversion at length, adding to Aristotle's theory in several respects. He has a great interest in the infinite (*infinitum*) terms discussed briefly by Aristotle in *De Interpretatione* 10, and in particular in the equivalence of 'No S is P' and 'Every S is *nonP*' (20a20–21), called *equipollence* throughout the medieval period and later known as *obversion*. This leads him to supplement Aristotle's theory of categorical syllogisms in various ways. First, he adds a new form of conversion *per accidens*, allowing the transition from universal to particular negatives:

Boethian conversion per accidens

No S is $P \Rightarrow$ Some *nonP* is S

This follows directly from Aristotelian conversion *per accidens* given the above equivalence. It also follows, of course, from subalternation and simple conversion. Second, Boethius adds a thoroughly new form of conversion, *contraposition*.

Discussions of contraposition are common in Old Logic texts. Illustrative is the *Abbreviatio Montana*, a short summary of logic written in the mid-twelfth century by the monks of Mont Ste. Geneviève. It provides an overview of the Old Logic, the theory that had developed from Porphyry and Boethius without the direct influence of the *Prior Analytics*. Though it follows Aristotle in most respects, it diverges from the *Prior Analytics* in some striking ways.

The first divergence concerns the nature of logic itself. Aristotle's subject is demonstration: what follows from what. The *Abbreviatio Montana*, in contrast, locates logic within

[12]See [Martin, 2009]. I am setting aside here another important contribution with some connection to our subject, namely, Boethius's contribution to the theory of topics, which eventually blended into the theory of quantification in the *logica nova*. Boethius took two very different approaches to the theory of topics — those of Aristotle and Cicero (106–43 BC) — and unified them within a generally Aristotelian framework, organizing what appears in the *Topics* of both Aristotle and Cicero to be a disorganized collection of insights and putting the theory into a form in which it remained influential until the Renaissance.

the sphere of rhetoric. The art with which it is concerned is dialectic, whose purpose is "to prove in the basis of readily believable arguments a question that has been proposed" (77). Its goal is persuasion, "to produce belief regarding the proposed question."

A second divergence concerns the nature of quantifiers. The *Abbreviatio Montana* includes singular propositions along with universals, particulars, and indefinites in its catalogue of categorical propositions, and defines them in an interesting way, distinguishing the determiner from the noun phrase. A universal proposition is one that "has a universal subject, stated with a universal sign" (80). A particular proposition, similarly, has a particular subject stated with a particular sign. An indefinite has no sign at all. A singular proposition "has a singular subject stated with a singular sign, as in 'Socrates is reading.'" A proper name is thus both a subject and a sign.

The *Abbreviatio Montana* follows Boethius, distinguishing *finite* terms such as 'man,' 'stone,' and so on from *infinite* terms such as 'non-man,' 'non-stone,' and so on. In contraposition,

> The predicate is turned into a subject, and the subject is turned into a predicate; the finite terms are made infinite, while the signs remain the same. Universal affirmatives and particular negatives are converted with this sort of conversion. How? In this way: 'Every man is an animal': 'Every non-animal is a non-man'; 'Some man is not an animal': 'Some non-animal is [not] a non-man'; 'No man is an animal': 'No non-animal is a non-man'; 'Some man is an animal'; 'Some non-animal is a non-man.' (*Abbreviatio Montana* 83)

Following Boethius, the Old Logic texts define contraposition accordingly:

Contraposition

Every S is $P \Leftrightarrow$ Every *nonP* is *nonS*
Some S is not $P \Leftrightarrow$ Some *nonP* is not *nonS*

The *Abbreviatio Montana* lists contraposition among the conversions applicable to propositions sharing both terms. But Aristotle would hardly count 'Every man is an animal' and 'Every non-animal is a non-man' as sharing terms. Allowing infinite terms into Aristotle's framework, moreover, threatens to break down the entire system. It allows the expression of categorical propositions such as 'No non-animal is a non-man' and 'Some non-animal is a non-man' which have no categorical equivalents expressible in finite terms.[13] The *Introductiones Norimbergenses* provides a table including such propositions (309, f. 54r):

Every man is an animal $\Leftrightarrow$ No man is a non-animal
Every man is a non-animal $\Leftrightarrow$ No man is an animal
Every non-man is an animal $\Leftrightarrow$ No non-man is a non-animal
Every non-man is a non-animal $\Leftrightarrow$ No non-man is an animal

The presence of infinite terms also allows the formulation of valid inference patterns that have no place in Aristotle's system, such as

[13] The former is equivalent to 'Everything is an animal or a man,' the latter to its negation.

No *nonM* is *nonP*
No S is M
∴ Every S is P

Such inference patterns went unnoticed in the Old Logic, and even in the New. It would be two hundred years before John Buridan would expand the theory of the syllogism to include them, and more than seven hundred years before Lewis Carroll [1896; 1977] would write about them and use them to mock the syllogistic rules of the New Logicians, which nineteenth-century textbook writers would regurgitate — Buridan's contributions long having been forgotten. The syllogism above, for example, has several shocking features. It appears to equivocate on its middle term. It derives an affirmative conclusion from two negative premises. It is first-figure, with a universal affirmative conclusion, but it certainly is not *Barbara*.

It does, however, reduce readily to *Barbara*, given a rule of Obversion recognizing the equivalences:

Some S is *nonP* ⇔ Some S is not P
Every S is *nonP* ⇔ No S is P
No S is *nonP* ⇔ Every S is P
Some S is not *nonP* ⇔ Some S is P

The deduction goes as follows:

1. No *nonM* is *nonP* (Assumption)
2. No S is M (Assumption)
3. Every *nonM* is P (Obversion, 1)
4. Every S is *nonM* (Obversion, 2)
5. Every S is P (Barbara, 3, 4)

It is somewhat surprising, then, that no one noticed the additional power that recognition of infinite terms and immediate inferences of contraposition and obversion grant. A defender of the Old Logic who realized its power might have been in an excellent position to combat the New Logic's shift in focus.

Another manifestation of the added power of the Old Logic lies in its effect on Aristotle's reduction of syllogisms to first-figure forms. Aristotle shows that all valid syllogisms reduce to *Barbara* and either *Darii* or a negative form such as *Celarent*. We could go even further with obversion, reducing *Celarent* and *Darii* to *Barbara*, thus deriving every syllogism from the single form 1AAA.[14]

The additional power of the Old Logic, however, creates additional problems with the square of opposition. We saw earlier that if S is empty, 'Some S is P' and 'Some S

[14]Surprisingly, no Old Logician seems to recognize this. But the deduction is not difficult. Assume *Celarent*: No M is P, Every S is $M \vdash$ No S is P. Two applications of Obversion reduce this to *Barbara*: Every M is *nonP*, Every S is $M \vdash$ Every S is *nonP*.

is not *P*' are both false. By contradictories, then, 'Every *S* is *P*' and 'No *S* is *P*' must both be true. But then they cannot be contraries. That is bad enough, perhaps, though we might reassure ourselves that Aristotle intended his theory as a tool for demonstrative science, the terms of which would not be empty. Adding obversion and contraposition, however, undermine any such reassurance. The *Abbreviatio Montana* passes over the difficulties contraposition introduces, but the monks of Mount Ste. Geneviève discuss them at length in their accompanying twelfth-century manuscript *Introductiones Montana Minores*. They worry that contraposition entails not only that the subject term is nonempty, but that *no term in the language* is empty. Here, for example, is a simple argument from a completely unrelated categorical proposition to the existence of stones:

> If 'every man is an animal' is true, so is 'every non-animal is a non-man,' by the same reasoning [i.e., by contraposition]. And if every non-animal is a non-man, every stone is a non-man, by the topic "from the whole" [i.e., since every stone is a non-animal]. And if every stone is a non-man, every stone exists ("from the part"). (34, f. 41rb, my translation)

Not only may there be no empty terms, whether involved in the reasoning under consideration or not; there can be no universally applicable terms.

> 'Every man is a thing.' This proposition is true; therefore its converse by contraposition, 'every non-thing is not a man,' is too. But if that is so, then every non-thing exists, because whatever is a non-man exists.
>
> This shows that it is not possible to convert propositions having terms applying to everything. (34–35, f. 160r, my translation)

The monks, following Abelard, conclude that the particle 'not' should not be used to form infinite terms, but instead be restricted to negating propositions.

The monks also consider objections to simple conversion, but reject them as relying on misleading linguistic forms. 'Some old man was once a boy' might appear, outrageously, to imply that some boy was once an old man, but this is a fallacy based on tense.

2.2 *The New Logic*

No one knows how the logical works of Aristotle that had been missing for centuries were once again found, but increased travel to Constantinople, due in part to economic growth and in part to the Crusades, seems to have led a number of scholars to investigate Byzantine libraries and begin translation projects that included the works of Aristotle. The discovery of Aristotle's logical works in the late twelfth century changed logic significantly, leading to the development of the *logica nova* — the New Logic.

The texts of the New Logic begin just as those of the Old Logic do, with an account of sounds, words, nouns, verbs, and the standard Aristotelian definition of a statement or proposition as an expression signifying what is true or false (*Propositio est oratio verum vel falsum significans*). But the remainder of the texts diverge significantly from those of the Old Logic in emphasis. They also include a number of new theoretical developments, including the doctrine of distribution, the development of rules for determining the validity of syllogisms, and the theory of supposition.

One sign of the shift in emphasis is the emergence of mnemonics for syllogisms and deductions. The *Ars Emmerana* and *Ars Burana*, twelfth-century Old Logic texts, end their treatments of syllogisms with some syllables that stand for the valid moods in various syllogistic figures.[15] Those syllables did not catch on. But by the time of Peter of Spain's *Summulae Logicales* a century later, a verse encoding such information had caught on and become standard. As Peter has it:

> Barbara Celarent Darii Ferio Baralipton
> Celantes Dabitis Fapesmo Frisesomorum
> Cesare Cambestres Festino Barocho Darapti
> Felapto Disamis Datisi Bocardo Ferison.[16]

These expressions, seemingly nonsensical, encode a great deal of information. The first two lines are the first-figure syllogisms in Aristotle's scheme. (The second line would later be reckoned as fourth figure.) The third line, except for the last word, lists the second-figure syllogisms; the final word of that line and the final line list the third-figure syllogisms. The vowels represent the categorical statement forms that make up the syllogism. Thus, *Barbara* consists of three universal affirmatives; *Celarent*, a universal negative and a universal affirmative, with a universal negative conclusion; and so on. The initial consonant specifies the first-figure syllogism to which the syllogism in question reduces. Thus, *Festino* reduces to *Ferio*; *Datisi* reduces to *Darii*. An 's' indicates simple conversion; a 'p,' conversion *per accidens*; an 'm,' a transposition of premises; a 'c,' a *reductio ad absurdum* (that is, by contradiction).

The emergence of such verses might be taken as indicating the importance of Aristotle's deductive method in the theory of syllogisms from a thirteenth-century point of view. But it is more likely just the reverse. Logic students memorized the verse; that removed the need to become skilled at the deductive technique. As Lukasiewicz has shown, Aristotle's deductive method allows one to establish the validity of a wide variety of complex arguments that go far beyond Aristotle's simple forms. Reducing the method to a memorized verse, however, removed it as a living logical technique. It also removed the need for the extensive discussions of metaprinciples governing each figure that occupied most Old Logic texts.

Peter of Spain's *Summulae Logicales*, also known as the *Tractatus*, is perhaps the most influential logic textbook in history.[17] It became the standard textbook in the universities

[15]See *Ars Emmerana* 173, 49ra; *Ars Burana* 200, f. 111r, 203, 112r, and 205, 112v. The syllables, VIO NON EST LAC VIA MEL VAS ERB ARC/ REN ERM RAC OBD/ EVA NEC AVT ESA DVC NAC, do not follow an obvious code and convey much less information than the more complex expressions that succeeded them.

[16]Peter of Spain [1972, 52]. This appears unaltered in Buridan [2001, 320], except that he adds punctuation after 'Frisesomorum' and 'Barocho' to signal the shift from one figure to another. Kretzmann [1966, 66n], guesses that William of Sherwood may have invented this verse, though some of the terms appear to have been in use earlier.

[17]Historians disagree about who Peter of Spain was; a leading theory is that he became Pope John XXI. For issues surrounding the text, see [de Rijk, 1968; 1972; 1982]. Peter probably wrote the *Tractatus* around 1240; see de Rijk [1972, lvii].

for at least four centuries. It makes or at least marks the theoretical innovations of the New Logicians — a group, often known as *terminists*, that includes William of Sherwood (1190–1249) and Lambert of Auxerre.

Peter's *Tractatus* reviews the rules governing syllogisms in each figure, but, beforehand, does something quite new by proposing rules that apply to syllogisms of any figure. His rules:

1. No syllogism can be made of propositions that are entirely particular, indefinite, or singular.
2. No syllogism in any figure can be made of propositions that are entirely negative.
3. If one of the premises is particular, the conclusion must be particular; but not conversely.[18]
4. If one if the premises is negative, the conclusion is negative; and conversely.
5. The middle term must never be placed in the conclusion.

These rules plainly do not suffice to determine the validity or invalidity of any argument of generally syllogistic form. They narrow the possible syllogistic moods to AAA, EAE, AEE, AII, IAI, AOO, OAO, EIO, IEO, and the subaltern moods AAI, AEO, and EAO. But they say nothing about figure; they do not distinguish major terms and premises from minor terms and premises. 1AEE, for example, satisfies all these rules but fails to be valid. ("Every M is P; No S is M; ∴ No S is P.") The rules are nevertheless important, for they mark the beginning of a quest for a complete set of rules capable of serving as a decision procedure for syllogisms.

Peter's *Tractatus* outlines the doctrine of distribution. His definition of distribution is somewhat obscure: "Distribution is the multiplication of a common term effected by a universal sign" ("*Distributio est multiplicatio termini communis per signum universale facta*" (209)). Lambert of Auxerre reverses the metaphor: "Distribution is the division of one thing into divided [parts]" (139). Both, however, see distribution as something a determiner such as 'all' does to a common noun. They distinguish collective from distributive readings, and observe that universal affirmative determiners ('all,' 'every,' 'each,' etc.) distribute the subject term but not the predicate term, while universal negative determiners distribute both terms. Particular determiners do not distribute their subject or predicate terms. Negated terms reverse in distribution, however, so the predicate of a particular negative proposition is distributed. The intuitive significance of distribution is that distributed terms say something about everything in their extensions. To say that every man is an animal is to say something about each and every man. To say that no animal is a stone is to say something about every animal and also about every stone.

Both Peter and Lambert recognize that distribution plays a role in syllogistic reasoning. There can be no syllogism without a universal premise, they note. But not until the fourteenth century does anyone see how to use distribution to extend Peter's rules to a decision procedure.

[18]Oddly, Peter lists this rule twice.

2.3 *The Mature Logic of Terms*

Walter Burley (also Burleigh; 1275–1344) develops a theory of consequences relevant mostly to the history of the logical connectives.[19] But he links his theory to distribution in a way important to the development of the doctrine, for he sees how distribution matters to syllogistic reasoning. Burley thinks first about conversion. Universal negatives convert simply; universal affirmatives do not. Similarly, particular affirmatives convert simply; particular negatives do not. Universal negatives and particular affirmatives are contradictories, but have something in common: their subject and predicate terms agree in distribution. In 'No S is P,' both S and P are distributed. In 'Some S is P,' both S and P are undistributed. Universal affirmatives convert *per accidens*; from 'Every S is P' we can infer 'Some P is S.' This suggests to Burley a pattern. When terms remain unchanged in distribution, or go from distributed to undistributed, the inference works; when they go from undistributed to distributed, it fails. There is good reason for this: one cannot conclude something about the entire extension of a term on the basis of something pertaining to only part of its extension.

He captures this in a general rule.

> Whenever a consequent follows from an antecedent, the distribution of the antecedent follows from the distribution of the consequent.

To put this another way, any term distributed in the conclusion must be distributed in the premises. If we add this rule to those of Peter of Spain (omitting his last as implied by an argument's being of proper syllogistic form), we get

1. No syllogism can be made of propositions that are entirely particular, indefinite, or singular.
2. No syllogism in any figure can be made of propositions that are entirely negative.
3. If one of the premises is particular, the conclusion must be particular.
4. If one if the premises is negative, the conclusion is negative; and conversely.
5. Any term distributed in the conclusion must be distributed in the premises.

This is still not a complete set; it allows for the fallacy of the undistributed middle. To complete the rules, one needs in addition to require that the middle term is distributed at least once. This makes good sense, given the thought that distributed terms say something of everything falling under them, while undistributed terms do not. If neither premise says something about everything that falls under the middle term, the argument fails to relate the parts of the middle term's extension relevant to the premises.

John Buridan (1300?–1358?) adds precisely that requirement. In the *Summulae de Dialectica* he merely reviews Peter of Spain's rules, explaining why each should be true. In the *Treatise on Consequence* [Buridan, 1976; King, 1985], however, he brings the doctrine of distribution to bear on syllogisms, specifying the following rules:

[19]For more on Burley and fourteenth-century logical developments, see [Geach, 1962; Ockham, 1974; 1980; King, 1985; Normore, 1999; Spade, 2002; Dutilh-Novaes, 2008].

1. No syllogism can be made of propositions that are entirely particular, indefinite, or singular.
2. No syllogism in any figure can be made of propositions that are entirely negative.
3. If one of the premises is particular, the conclusion must be particular.
4. If one if the premises is negative, the conclusion is negative; and conversely.
5. Any term distributed in the conclusion must be distributed in the premises.
6. The middle term must be distributed at least once.

This is a complete set of rules; it approves all and only syllogisms, providing a simple decision procedure.

Buridan does not rest content with devising a decision procedure for syllogistic reasoning. He develops a general theory of infinite terms — that is, terms such as "non-animal," formed by negating other terms — and extends his theory to syllogistic-like reasoning involving such terms. As we have seen, practitioners of the Old Logic were in a position to do just that, but failed to accomplish it. Buridan does, writing an entire chapter (5.9, "About Syllogisms with Infinite Terms") on the subject.

Fourteenth-century logicians made other significant advances in our understanding of quantification. They began to recognize inferential properties of determiners that were to become central to the contemporary theory of generalized quantifiers. Burley, for example, states rules for consequences that correspond to characteristic properties of certain kinds of determiners. For example:

> A consequence from a distributed superior to its inferior taken with distribution and without distribution holds good, but a consequence taken from an inferior to its superior with distribution does not hold good. (300)

His example: 'Every animal is running' implies 'Every man is running,' but not conversely. Burley is here noticing that universal affirmative determiners are *antipersistent*: If every S is P, then every T is P, for any T whose extension is a subset of the extension of S.

Burley and Buridan link this to distribution. If something is said of everything that falls under a term, then it is said of each object individually — that, in fact, is Buridan's definition of distribution[20] — and of everything that falls under any subset of the term's extension. They recognized the connection to inference: distributed terms are, in modern parlance, *monotonic decreasing*. Undistributed terms are monotonic increasing: what is said of a part of a term's extension is also said of a part of any superset. If we use the convenient notation using arrows to indicate monotonic decreasing or increasing effects on terms — the first arrow indicating the subject term, and the second the predicate — then we can summarize the properties of determiners and correlatively of categorical propositions very simply:

[20]"Distributive supposition is that in accordance with which from a common term any of its supposita can be inferred separately, or all of then at once conjunctively, in terms of a conjunctive proposition" (264). Buridan thus expresses the idea that universal quantification is a generalized conjunction.

Universal affirmative: 'every,' 'all,' 'each' — ↓mon↑
Universal negative: 'no' — ↓mon↓
Particular affirmative: 'some' — ↑mon↑
Particular negative: 'some... not" — ↑mon↓

This is enough information to deduce all the inferential properties of categorical propositions, provided that terms are nonempty. Otherwise, the distribution of the universal subject terms, which makes them antipersistent (in this notation, ↓mon), suggests that they should hold when their subject terms are empty. Admittedly, a null "part" of an extension is an odd sort of part. But taking this understanding of determiners to its conclusion leads to modern logic's treatment of universal affirmatives and negatives as vacuously true when nothing falls under their subject terms. This breaks the square of opposition, of course, for then such propositions are not contraries; they can be true at the same time.

Burley and Buridan raise a number of issues of fundamental importance for an adequate theory of quantification. They notice, for example, the plethora of determiners in natural language. Among universal affirmatives, Buridan lists 'every,' 'whichever,' 'whoever,' 'whosoever,' 'both,' 'however much,' 'however many,' 'however many times,' 'whatever... like,' 'whenever,' 'wherever,' 'always,' perpetually,' 'eternally,' 'howsoever much,' 'anyhow,' 'anywhere.' (265), and proceeds to discuss the distinctions between them. This may seem to be a minor point. But Buridan begins the process of thinking not just about certain words ('every,' 'no,' 'some') but about a general category of expressions that he recognizes as playing the same logical and grammatical role.[21]

Burley and Buridan also spend considerable time discussing relational predicates. Aristotle was aware of such predicates; he mentions oblique contexts and gives as an example "there is a genus of that of which there is a science, and if there is a science of the good, we conclude that there is a genus of the good" (I, 36). But no one before the fourteenth century paid such inferences much attention. Armed with a theory of supposition, Burley and Buridan address such inferences in detail. For example, Buridan discusses the inferential properties of these in the *Summulae de Dialectica*:

If someone is father of a daughter, then someone is daughter of a father. (179)
One seeing every donkey is an animal. (273)
Every man's donkey is running. (274, 366)
Every man is seen by some donkey. (274)
Any animal of a king is a horse. (274)
No man sees every donkey. (277)
Every man sees every donkey. (277)
A horse is greater than a man. (277)
A horse is greater than every man. (277)

[21]One earlier (twelfth-century) text that pays similar attention to the variety of determiners is the *Ars Emmerana*, which lists, among universal determiners, 'every,' 'whichever,' 'whoever,' 'whosoever,' 'both,' 'always,' 'anytime' [literally, 'every day'], 'everywhere,' 'no,' 'nobody,' 'nothing,' 'never,' 'nowhere,' and 'neither.' Among particular determiners it lists 'some,' 'somebody,' 'sometimes,' 'somewhere,' 'one,' 'another,' and 'someday.' See 154, f. 46rb.

A horse is greater than the smallest man. (277)
Socrates loves himself. (283)
Socrates acquires something for himself. (283)
Every man likes himself. (286)
Every man sees his own horse. (287)
Socrates rides his own horse. (287)
Socrates is the same height as Plato. (288)
When Socrates arrives, then Plato greets him. (288)
Socrates sees Plato. (298)

This long list of examples may give some sense of how extensive Buridan's discussion of relations is. He also gives many examples of oblique inferences, including:

Every man's donkey is running; every king is a man; therefore, every king's donkey is running. (366)
Every man sees every man; a king is a man; therefore a king sees a king. (367)
No man's donkey is running; a man is an animal; therefore, some animal's donkey is not running. (369)
Any man's donkey is running and any man's horse is running; therefore, [he] whose horse is running [is such that] his donkey is running. (369–70)
Of whatever quality the thing that Socrates bought was, such was the thing he ate; it was a raw thing that Socrates bought; therefore, it is a raw thing that Socrates ate. (370; see 877)

Buridan thinks seriously about issues we would now consider matters of scope in the context of his theory of supposition:

A white man is going to dispute. (293)
Every white man will be good. (301)
A horse was white. (301)

Many examples of scope ambiguities involve intensional contexts. Buridan, foreshadowing Quine [1956; 1960] and Montague [1973; 1974], notes that 'I think of a rose' may be true even if there are no roses.

I recognize a triangle. (279)
I owe you a horse. (279)
I know the one approaching. (279, 294)
The one approaching, I know. (279, 294)
Socrates sees a man. (282, 296)
How Socrates looks, Plato wants to look. (288)
The First Principle Averroes did not believe to be triune. (295)
The First Principle Averroes did believe to be God. (295)
The Triune Averroes believed to be God. (295)

I can see every star. (296)
I think of a rose. (299)
A golden mountain can be as large as Mount Ventoux. (299)
The one creating is of necessity God. (300)
A man able to neigh runs. (302)
If something is sensible, something is sensitive. (182)
If something is knowable, something is cognitive. (182)
A sense is a sense of what is sensed. (179)
What is sensed is sensed by a sense. (179)

Burley and Buridan make another remarkable advance over earlier discussions. They recognize that quantified expressions not only relate terms but introduce "discourse referents," enabling further anaphora. They are not the first to realize this — there are a few examples in the *Tractatus Anagnini* two centuries earlier — but the attention they pay anaphoric issues is unprecedented.[22] They moreover give examples of expressions that are *not* well-formed to show what role certain expressions can and cannot play. Burley, of course, is the source of the famous "donkey" sentence,

Every farmer who owns a donkey beats it.

But this is only one of many. Consider these examples from Buridan:

An animal is a man and he is a donkey. (283)
A man runs and he disputed. (283)
A man is a stone and he runs. (284)
'Mirror' is a noun and it has two syllables. (284)
* A man runs and every he is white. (284)
* an animal is running and no man is that. (285)
A man is running and that man earlier was disputing. (285)
an animal is running and it is a man. (286)
A man runs and another thing is white. (289)

The puzzle these present is that the first conjunct appears to be a categorical proposition, something of the form 'Some *S* is *P*.' The proposition conjoined to it, however, cannot be interpreted as another categorical proposition, but contains an anaphor making reference back to something in the first conjunct. Trying to construe the second conjunct as a categorical proposition yields either something with the wrong truth conditions — 'an animal is a man and he is a donkey,' which is false, is not equivalent to 'an animal is a man and an animal is a donkey,' which is true — or something ungrammatical.

[22] *Tractatus Anagnini* (240, 69v) mentions several under the heading *secundum relationem*: 'Every man is an animal who is capable of laughter; therefore, every man is an animal, and he is capable of laughter'; 'some animal lives that neither lives nor moves; therefore, some animal lives, though it neither lives nor moves'; 'something is not a animal that is a man; therefore, something is not an animal, and it is a man'; and 'Only Socrates is an animal who is Socrates; therefore, only Socrates is an animal, and he is Socrates.'

3 THE TEXTBOOK THEORIES OF QUANTIFICATION

The fourteenth century was unquestionably a high point in the history of logic. The sophistication and subtlety of Burley, Buridan, and Ockham would not be equalled for five hundred years. In fact, logic fell into a rapid and steep decline.

Some of the reasons for the decline were intellectual. The humanism of the Renaissance led writers to disparage theology and everything associated with it, including philosophy and, in particular, logic. The rise of vernacular languages and the decline of Latin, manifested in the works of Dante, Petrarch, Boccaccio, and Machiavelli in Italian, Rabelais in French, Cervantes in Spanish, and Chaucer, Mallory, Marlowe, and Shakespeare in English, led to a loss of interest in a tradition that had, since the time of Boethius, found expression almost entirely in Latin. The rise of rhetoric pushed logic aside as largely beside the point of persuasion.

The chief reasons, however, were material. The end of the medieval warm period led to a series of cold winters and cool and rainy summers that devastated farming throughout Europe, choked economic growth, and led to repeated famines. The Black Death first struck Europe in 1348; it killed roughly a third of the population of Europe. University communities in Oxford, Paris, and other areas were especially hard hit. More than half the faculty died. Latin became unpopular, in fact, largely because the plague killed most of the educated people who were fluent in it. The effects on logic were devastating. Serious attention would not be paid to it until the seventeenth century, and the problems occupying Burley, Buridan, and Ockham were forgotten until well into the twentieth.

3.1 *The Port-Royal Logic*

Antoine Arnauld (1612–1694) and Pierre Nicole (1625–1695) wrote *Logic, or, The Art of Thinking* [Arnauld and Nicole, 1662; 1861] at the Port-Royal convent outside Paris, perhaps with some contributions by Blaise Pascal (1623–1662). The book, written in French rather than Latin, revived logic after centuries of neglect.[23] The book expresses traditional Aristotelian logic in the language of Descartes's new way of ideas. From a strictly logical point of view, it contributes few innovations. It may be worth remarking that the *Port-Royal Logic* is probably the first significant logical work in centuries not to contain the word 'donkey.' Its examples are not artificial and restricted to standard logical illustrations, but come from actual texts, chiefly, from the Bible and classical literature. It is rich in linguistic insight, even if somewhat sparse from a logical point of view.

One logical innovation, however, is important: the distinction between comprehension, or, as logicians writing in English have generally called it, *intension*, and *extension*.

> Now, in these universal ideas there are two things, which it is very important accurately to distinguish: COMPREHENSION and EXTENSION. I call the COMPREHENSION of an idea, those attributes which it involves in itself,

[23]Between 1676 and 1686, Leibniz too engaged in highly original logical work, influenced not by Arnauld and Nicole but by Joachim Jungius, whose *Logica Hamburgensis* [1638; 1977] combined Aristotelian and Ramist approaches. However, the distribution of Leibniz's work was so limited that it had little immediate impact.

> and which cannot be taken away from it without destroying it; as the comprehension of the idea triangle includes extension, figure, three lines, three angles, and the equality of these three angles to two rigid angles, I call the EXTENSION of an idea those subjects to which that idea applies, which are also called the inferiors of a general term, which, in relation to them, is called superior, as the idea of triangle in general extends to all the different sorts of triangles. [1861, 49]

It is natural to think of this as the origin of our idea of a term's extension as the set of things of which it is true. Arnauld and Nicole, however, lack the concept of a set; they use the plural ("those subjects to which the idea applies"). Arguably, therefore, they make only a small advance over the fourteenth-century logicians talk of *supposita*. Their talk of "inferiors" and of "sorts of triangles" moreover raises the possibility that they have in mind not, or not only, the objects to which the term applies but the subsets of what we would today call the extension. The language of inferiors and superiors stems ultimately from Porphyry — indeed, Peirce ridicules Baines's position that Arnauld and Nicole invented the distinction between intension and extension, crediting it to Porphyry ([1893, 237–238] and finding it in Ockham (241–242) — and has the disadvantage of blending these ideas together, obscuring the distinction between members and subsets. That said, first drawing the distinction between intension and extension, however fuzzily, and providing labels for them constitutes an important achievement.

Arnauld and Nicole avoid speaking of distribution, but the idea clearly motivates an argument that all propositions must be either particular or universal:

> But there is another difference of propositions which arises from their subject, which is according as this is universal, particular, or singular. For terms, as we have already said in the First Part, are either singular, or common, or universal. And universal terms may be taken according to their whole extension, by joining them to universal signs, expressed or understood: as, *omnis*, all, for affirmation; *nullus*, none, for negation; all men, no man.
>
> Or according to an indeterminate part of their extension, which is, when there is joined to them *aliquis*, some, as some man, some men; or others, according to the custom of languages. Whence arises a remarkable difference of propositions; for when the subject of a proposition is a common term, which is taken in all its extension, propositions are called universal, whether affirmative, as, 'Every impious man is a fool,' or negative, as, 'No vicious man is happy.'
>
> And when the common term is taken according to an indeterminate part only of its extension, since it is then restricted by the indeterminate word 'some,' the proposition is called particular, whether it affirms, as, 'some cruel men are cowards,' or whether it denies, as, 'some poor men are not unhappy.' (110)

Medieval discussions of distribution do not mark the alternatives so starkly or explicitly, though the terms 'distributed' and 'undistributed' do sound exhaustive.[24] As Arnauld and

[24]Fourteenth-century logicians may have assumed that all terms, in the context of a proposition, are taken with

Nicole see it, a term must be either "taken according to [its] whole extension" or not, and thus the subject term must be distributed, in which case the proposition is universal, or undistributed, in which case the proposition is particular.

This seems to leave no place for singular propositions. But Arnauld and Nicole see singulars as reducing to universals or particulars. They continue, arguing that singular terms are to be taken as universals:

> And if the subject of a proposition is singular, as when I say, 'Louis XIII took Rochette,' it is called singular. But though this singular proposition may be different from the universal, in that its subject is not common, it ought, nevertheless, to be referred to it, rather than to the particular; for this very reason, that it is singular, since it is necessarily taken in all its extension, which constitutes the essence of a universal proposition, and which distinguishes it from the particular. For it matters little, so far as the universality of a proposition is concerned, whether its subject be great or small, provided that, whatever it may be, the whole is taken entire. (110)

More seriously, the *Port-Royal Logic* has no place for definite articles and numerical quantifiers.

Arnauld and Nicole have an extended and sophisticated discussion of indefinites (or bare plurals). Far from reducing them to particulars, as Aristotle and his commentators generally do, or to universals, they point out that neither will do. The existence of white bears and English Quakers is not enough, they observe, to make 'Bears are white' and 'Englishmen are Quakers' true. They distinguish moral from metaphysical universality:

> We must distinguish between two kinds of universality, the one, which may be called *metaphysical*, the other *moral*.
>
> We call universality, *metaphysical*, when it is perfect without exception, as, *every man is living*, which admits of no exception.
>
> And universality, *moral*, when it admits of some exception, since in moral things it is sufficient that things are generally such, *ut plurimum*, as, that which St. Paul quotes and approves of: "*Cretenses semper mendaces, malae bestiae, centres pigri*" [Cretans are always liars, evil beasts, great gluttons (Titus 1:12)]. Or, what the same apostle says : "*Omnes quae sua sunt quaerunt, non quae Jesu-Christi*" [Everyone seeks his own [good], not that of Jesus Christ (Philippians 2:21)]; Or, as Horace says : "*Omnibus hoc vitium est cantoribus, inter amicos Ut nunquam inducant animum cantare rogati; Injussi nunquam desistant*" [It is a flaw of every singer, among friends, that, if asked to sing, they will never start; if not asked, will never stop (Satire III)]; Or, the common aphorisms:

respect to all their supposita or only some. Thirteenth-century logicians such as Peter of Spain, however, made no such assumption. Distribution, for Peter, is the multiplication of a common term. There is no direct path from a failure of multiplication to the "involves a part" idea that implies the inferential properties of particulars. This is to Peter's credit, for, as we have seen, "involves the whole" suggests monotonic decreasing behavior, and "involves a part" implies monotonic increasing behavior. But some determiners are neither: 'the,' 'exactly one,' 'exactly two,' 'most,' etc.

> That all women love to talk.
> That all young people are inconstant.
> That all old people praise past times.
>
> It is enough, in all such propositions, that the thing be commonly so, and we ought not to conclude anything strictly from them. (147)

Moral universality differs from universality as usually understood, for it does not license universal instantiation:

> For, as these propositions are not so general as to admit of no exceptions, the conclusion may be false, as it could not be inferred of each Cretan in particular, that he was a liar and an evil beast....
>
> Thus the moderation which ought to be observed in these propositions, which are only morally universal, is, on the one hand, to draw particular conclusions only with great judgment, and, on the other, not to contradict them, or reject them as false, although instances may be adduced in which they do not hold.... (147–148)

Arnauld and Nicole are content with "moral universality" for contingent (and politically incorrect) generalizations:

> Whence we may very well say, *The French are brave; the Italians suspicious; the Germans heavy; the Orientals voluptuous*; although this may not be true of every individual, because we are satisfied that it is true of the majority. (152)

That last remark suggests that bare plurals are equivalent to noun phrases with the determiner 'most.' But that it is not quite right. Arnauld and Nicole point out some sentences that are difficult to understand even by such a relaxed criterion, and have become known in the linguistics literature as Port-Royal sentences:

> There are also many propositions which are morally universal in this way only, as when we say, *The French are good soldiers, The Dutch are good sailors, The Flemish are good painters, The Italians are good comedians*; we mean to say that the French who are soldiers, are commonly good soldiers, and so of the rest. (150)

Arnauld and Nicole present the square of opposition uncritically, with no consideration of empty terms (112–113). They decline to present rules for the Aristotelian reduction of syllogisms, "because this is altogether useless, and because the rules which are given for it are, for the most part, true only in Latin" (113)! They do, however, give rules derived from those of Buridan for syllogisms:

1. The middle term cannot be taken twice particularly, but it ought to be taken, once at, least, universally.

2. The terms of the conclusion cannot be taken more universally in the conclusion than they are in the premises.

3. No conclusion can be drawn from two negative propositions.

4. A negative conclusion cannot be proved by two affirmative propositions.

5. The conclusion always follows the weaker part, that is to say, if two propositions be negative, it ought to be negative, and if one of them be particular, it ought to be particular.

6. From two particular propositions nothing follows.

Their only innovation is the rigorous demonstration of corollaries from these rules:

- There must always be in the premises one universal term more than in the conclusion.
- When the conclusion is negative, the greater term must necessarily be taken generally in the major.
- The major (proposition) of an argument whose conclusion is negative, can never be a particular affirmative.
- The lesser term is always in the conclusion as in the premises.
- When the minor is a universal negative, if we wish to obtain a legitimate conclusion, it must always be general.
- The particular is inferred from the general.

Arnauld and Nicole also recognize a fourth figure and set out the valid moods in each figure on the basis of the rules and corollaries. They use the medieval names for the moods, but mention the significance only of the vowels, since they reject the project of reducing them to the first figure.

3.2 *The English Textbook Tradition*

The eighteenth century — the century of Bach, Handel, Haydn, and Mozart, of Burns, Sterne, Fielding, and Goethe, of Berkeley, Rousseau, Hume, and Kant — yielded almost no significant work in logic. Perhaps the most important work, Isaac Watts's *Logic, or, The Right Use of Reason* (1725), is a work of metaphysics and epistemology modeled on Locke's *Essay* more than a work of logic in any traditional sense. Watts follows Arnauld and Nicole in his rather brief treatment of inference, as well as in his discussion of moral universality and Port-Royal sentences (120–21). He discusses categorical propositions and conversion in a page or two, and offers the Port-Royal rules for syllogisms, but with a mistake, stipulating that "if either of the premises be negative, the conclusion must be particular." Perhaps the most interesting aspect of Watts's text is his discussion of relative and connective syllogisms, such as

As is the captain so are his soldiers.
The captain is a coward.
Therefore, his soldiers are so too. (233)

Meekness and humility always go together.
Moses was a man of meekness.
Therefore, Moses was a man of humility. (234)

London and Paris are in different latitudes.
The latitude of London is 51.5 degrees.
Therefore, this cannot be the latitude of Paris. (234)

The first of these is interesting not only in being relational but also in being higher-order.

In the nineteenth century, British textbooks revived an interest in logic. Initially, they did little more than review the traditional theory of the syllogism, at a level of detail and sophistication comparable to that of the early twelfth century or the *Port-Royal Logic*. But Richard Whately (1787–1863) wrote *Elements of Logic* (1826), the most extensive discussion of logical subjects since the fourteenth century. The book sets an organizational pattern — language, deductive logic (categorical propositions, conversion, syllogisms, hypotheticals and other connectives), fallacies, inductive reasoning, questions of method — that continues in most contemporary logic textbooks. It moreover contains some significant innovations. Whately, "the restorer of logical study in England," in the words of DeMorgan (quoted in [Valencia 2004, 404]), defines a proposition simply as an indicative sentence. In sharp contrast to the *Port-Royal Logic*, he discusses distribution, and in fact uses it to define universal and particular "signs," i.e., determiners. The subjects of universal sentences are distributed in the sense that they "stand, each, for the whole of its Significates" while particular subjects "stand for a part only of their Significates" (76). This is the doctrine of distribution that Peter Geach famously attacks in *Reference and Generality* (1962), that distributed terms refer to all, and undistributed terms some, of the objects falling under them.[25]

It is worth noting, however, that this way of putting the doctrine is not inevitable. It is not quite a nineteenth-century invention; Ockham treats universal signs as "mak[ing] the term to which it is added stand for all its significata and not just for some" [1980, 96]. The dominant fourteenth-century conception, however, is inferential: a term is distributed in a proposition if and only if that proposition implies all its substitution instances with respect to that term. Thus, in 'every man is mortal,' 'man' is distributed, for the sentence implies, for each individual man, that he is mortal. The idea is that 'Every man is mortal' and 'Socrates is a man' imply 'Socrates is mortal.' For particular terms, the pattern reverses; from 'Socrates is a man' and 'Socrates is mortal' we can infer 'Some man is mortal.' There is no temptation, on that inferential conception, to think that every term must be universal or particular, or that one is forced to any specific thesis about the semantic values of terms.

[25] See [Geach, 1962], and, for a response, [King, 1985].

Whately discusses bare plurals only briefly. They must, he reasons, be either universal or particular; he sees no other option. But they differ from noun phrases with a universal or particular determiner in that their significance depends on the context and content of the sentence in which they appear. That is an important insight.[26] But his illustrations are not very helpful. 'Birds have wings' and 'birds are not quadrupeds' he takes to be universal; 'food is necessary to life,' 'birds sing,' and 'birds are not carnivorous,' as particular.

Whately uncritically describes the square of opposition, with the standard diagram, but his discussion of conversion links it to distribution. His rule is, "No term is distributed in the Converse, which was not distributed in the Exposita" (82). Since conversion switches terms, this means that categorical propositions in which the subject and predicate terms agree in distribution — that is, particular affirmatives and universal negatives — convert simply, but universal affirmatives convert only *per accidens*, and particular negatives convert only by contraposition. His rule also explains why universal negatives also convert *per accidens*, and universal affirmatives convert by contraposition.

Whately is perhaps the first logician to think of categorical propositions as concerned with classes:

> ...in the first Premiss ("*X* is *Y*") it is assumed universally of the *Class* of things (whatever it may be) which "*X*" denotes, that "*Y*" may be affirmed of them; and in the other Premiss ("*Z* is *X*") that "*Z*" (whatever it may stand for) is referred to that Class, as comprehended in it. Now it is evident that whatever is said of the whole of a Class, may be said of anything that is comprehended (or "included," or "contained,") in that Class: so that we are thus authorized to say (in the conclusion) that "*Z*" is "*Y*". (86; quoted in [Hailperin, 2004, 343])

The psychologistic talk of ideas introduced by Arnauld and Nicole thus yielded to talk of classes, which made possible the advances of the nineteenth and twentieth centuries. Augustus DeMorgan and George Boole both read Whately; whether the idea that logic concerns relations of classes was Whately's insight or something "in the air" in nineteenth-century Britain is hard to say.

Perhaps Whately's most important contribution, however, is the modern conception of deductive validity. Aristotle, recall, defines validity in terms of the conclusion following necessarily from the premises. Medieval discussions of conversion and other forms of inference employ a notion of truth-preservation, but do not invoke it explicitly to characterize argumentative success. Whately does:

> ... an argument is an expression in which *from something laid down and granted as true (i.e. the premises) something else (i.e. the Conclusion,) beyond this must be admitted as true, as following necessarily (or resulting) from the other...* (86)

Because logic is concerned with language, moreover, Whately observes that logical validity is a matter of form rather than content, in which we consider "the mere *form of the expression*" rather than "the meaning of the terms" (86).

[26]See, for example, [Carlson, 1977].

Whately cites Aristotle for the *dictum de omni et nullo*, which he phrases in terms of distribution: "whatever is predicated of a term distributed, whether affirmatively or negatively, may be predicated in like manner of everything contained under it" (87). He takes this, immediately, as the principle underlying first-figure syllogisms, and only mediately as the principle of all syllogisms. He speaks of the danger of equivocation on the middle term, and uses it as an argument for the rule that the middle must be distributed at least once. By a similar argument, he justifies the rule that no term may be distributed in the conclusion that was not distributed in the premises. He gives two other rules: from negative premises, nothing follows, and if a premise is negative, the conclusion must be negative, and vice versa. Whately is thus the source of the rules for syllogisms as they are sometimes given in contemporary textbooks:

1. The middle term must be distributed at least once.

2. No term may be distributed in the conclusion without being distributed in the premises.

3. Nothing follows from two negative premises.

4. If a premise is negative, the conclusion must be negative, and vice versa.

He observes that the other standard rules, namely, that from two particular premises nothing follows, and if a premise is particular, the conclusion must be particular, follow from these rules. Whately gives illustrations rather than arguments for these observations. But he is right.[27] Unlike previous logicians, including Arnauld and Nicole, he does not neglect subaltern moods, but notes that "when we can infer a universal, we are always at liberty to infer a particular" (91). Whately goes through the valid syllogisms, showing that each satisfies the rules and moreover showing, in detail, how to reduce each to first-figure syllogisms.

[27]The omitted arguments are somewhat interesting, and use all four rules. To show that nothing follows from two particular premises: Suppose that an argument of syllogistic form has two particular premises. Then both subject terms must be undistributed. Since the middle term must be distributed at least once (Rule 1), at least one of the predicate terms must be distributed. At least one of the premises, then, must be negative, so the conclusion must be negative as well (Rule 4). But then the major term is distributed in the conclusion, so it must be distributed in the major premise (Rule 2). Since the subject term of the major is undistributed, the major term must be the predicate of the major premise, making it negative. But then both premises are negative, so nothing follows (Rule 3).

To show that if a premise is particular, the conclusion must be particular, assume that a premise is particular. By the previous result, the other must be universal. So, one subject term is distributed, while the other is undistributed. Suppose that the conclusion is universal, so that its subject term is distributed. Then the minor term must be distributed in the premises (Rule 2). The middle must be distributed at least once (Rule 1), so the premises must contain at least two distributed terms, only one of which may be in the subject position. So, one of the premises must be negative. It follows that the conclusion must also be negative (Rule 4). So, the major, being distributed in the conclusion, must be distributed in the premises (Rule 2). But that means that there must be three distributed terms in the premises, only one of which appears in subject position. But that means both premises must be negative, and nothing follows from two negative premises (Rule 3).

3.3 *Quantifiers in the Predicate*

In the middle of the nineteenth century, Augustus de Morgan and Sir William Hamilton argued publicly about one of the most peculiar ideas in the history of logic — the idea that the logical form of categorical propositions includes quantification governing the predicate as well as the subject. Since Aristotle, quantifiers had been conceived as relations between terms. The doctrine of distribution, however, suggests another way of thinking about quantifiers.

The idea of quantification of the predicate occurs in William of Ockham (1287–1347), one of the greatest medieval philosophers, who shares with Buridan the thought that universality is a generalization of conjunction and particularity a generalization of disjunction.[28] Suppose that we can enumerate the Ss as $s_1, ..., s_n$ and the Ps as $p_1, ..., p_m$. Then 'Every S is P' is equivalent to 's_1 is P and ... and s_n is P.' But we can go further. If s_1 is P, it must be one of the Ps; so, $s_1 = p_1$ or ... or $s_1 = p_m$. We can do this for each of the Ss. So, 'Every S is P' is equivalent to

$(s_1 = p_1$ or ... or $s_1 = p_m)$ and $(s_2 = p_1$ or ... or $s_2 = p_m)$ and ... and $(s_n = p_1$ or ... or $s_n = p_m)$.

This, in turn, is equivalent to 'Every S is some P.' That suggests the possibility of quantifying the predicate term explicitly, generating additional basic forms [1980, 101], only some of which Ockham discusses:

Every S is every P
Every S is some P
Some S is every P
Some S is some P
No S is every P
No S is some P
Some S is not every P
Some S is not some P

To understand the meanings of these forms, simply place conjunctions (for 'every') or disjunctions (for 'some') in the above schema. The first thus becomes '$(s_1 = p_1$ and ... and $s_1 = p_m)$ and $(s_2 = p_1$ and ... and $s_2 = p_m)$ and ... and $(s_n = p_1$ and ... and $s_n = p_m)$,' which is true iff there is exactly one S, and exactly one P, and those things are identical. The second is a universal affirmative; the third asserts that there is exactly one P, and it is S. The fourth is a particular affirmative, 'Some S is P.'

To understand the negatives, treat them as contradictories of corresponding affirmatives. The sixth and eighth are thus equivalent to the familiar universal and particular negatives. The fifth denies 'Some S is every P,' and the seventh, 'Every S is every P.'

Hamilton's idea of quantification in the predicate is not Ockham's. What distinguishes the four kinds of categorical propositions, Hamilton thinks, is the distribution of subject

[28]Ockham may have been inspired by reflecting on a fourteenth-century sophism. Here, for example, is William Heytesbury (1335): "Every man is every man. Proof: This man is this man, and that man is that man, and thus for each one. Therefore every man is every man" ([Wilson, 1960, 154]; my translation).

and predicate terms. Think of a determiner as a *predicate* of a term, indicating its distribution or lack thereof. Then we would need to mark the distribution (or not) of the subject as well as the distribution (or not) of the predicate. Hamilton sees conversion as forcing recognition of quantification in the predicate. He writes,

> The second cardinal error of the logicians is the not considering that the predicate has always a quantity in thought, as much as the subject, though this quantity frequently be not explicitly enounced.... But this necessity recurs, the moment that, by conversion, the predicate becomes the subject of the proposition; and to omit its formal statement is to degrade Logic from the science of the necessities of thought, to an idle subsidiary of the ambiguities of speech. (1860, quoted in [Bochenski, 1961, 263–64])

This is puzzling, since, traditionally, the quality of the proposition as affirmative or negative, not the quantity (as universal or particular) specified by the determiner, fixes the distribution of the predicate term. In any case, Hamilton thinks of the basic statement forms as

All S is all P
All S is some P
Some S is all P
Some S is some P
Any S is not any P
Any S is not some P
Some S is not any P
Some S is not some P

These make little grammatical or logical sense. (Why the singular copula 'is' with the plural subject 'all S'? Why the switch from 'all' to 'any'?) Evidently the usual understanding of the universal affirmative is 'All S is some P'; of the particular affirmative, 'Some S is some P.' But what do the other forms mean? What are their negations? Which contradict which? The generally distribution-driven motivation suggests that there are only four statement forms to be recognized, for each of the two terms may be distributed or undistributed. What is left for the other forms to do? Despite that rather obvious puzzle, logicians of the stature of Boole [1848] and De Morgan [1860] adopted Hamilton's idea.

Biologist George Bentham had developed this approach two decades before Hamilton, and somewhat more clearly.[29] He evidently thought of terms collectively, and thought of categorical propositions as relating all or part of one collective to all or part of another. The only relations he recognized, however, were identity and diversity. So, to simplify his notation somewhat, letting p represent part, the forms become

$S = P$
$S = pP$
$pS = P$
$pS = pP$

[29]George Bentham, *Outline of a New System of Logic*, 1827.

$S \neq P$
$S \neq pP$
$pS \neq P$
$pS \neq pP$

Imagine diagramming these forms as relations between sets, with each set represented by a circle. We might interpret the first as saying that the circle representing the Ss just is the circle representing the Ps — that is, that S and P have the same extension. We might interpret the second as holding that the Ss are a subset of the Ps, and the third as holding the converse. The fourth we might interpret as holding that the extensions of S and P overlap. The next four would be the negations of these. Bentham's scheme thus has the advantage of making it clear what these forms mean. Unfortunately, it also makes clear its pointlessness. All these relations are easily expressible in traditional terms:

Every S is P and every P is S
Every S is P
Every P is S
Some S is P
Some S is not P or some P is not S
Some S is not P
Some P is not S
No S is P

Bentham's notation might nevertheless be justified if it facilitated a simple method for determining validity. Neither Bentham nor Hamilton developed any such method. Still, given a logic of identity, and the principle that a part of a part is a part ($ppX = pX$), it is not hard to do. This, for example, is *Barbara*:

Every M is P	$M = pP$
Every S is M	$S = pM$
∴ Every S is P	$S = pP$

The reasoning: $S = pM = ppP = pP$. *Celarent*:

No M is P	$pM \neq pP$
Every S is M	$S = pM$
∴ No S is P	$pS \neq pP$

This too admits of easy proof: $pS = ppM = pM \neq pP$. *Darii*:

Every M is P	$M = pP$
Some S is M	$pS = pM$
∴ Some S is P	$pS = pP$

The proof: $pS = pM = ppP = pP$. Finally, *Ferio*:

No M is P	$pM \neq pP$
Some S is M	$pS = pM$
∴ Some S is not P	$S \neq pP$

This proof is only slightly trickier: say $S = pP$. Then $pS = ppP = pP$ and $pS = pM$, so $pM = pP$, contrary to the first premise. So, $S \neq pP$.

Immediate inferences, too, are straightforward. The notation makes it obvious that A and O forms, like I and E forms, are contradictories. A and E forms are contraries, since A forms imply I forms; if $S = pP$, $pS = ppP = pP$. I and O forms are subcontraries: one or the other must both be true. Say the O form is false, so that $S = pP$. Then $pS = ppP = pP$, and the I form is true. Simple conversion rules follow immediately from the symmetry of identity.

Though the position of Bentham and Hamilton seems ill-motivated and even confused from a semantic point of view, therefore, it does yield a strikingly simple method for demonstrating the validity of syllogisms. It requires some elaboration, however, to become a way of showing invalidity — as it stands, the method is unsound — or even establishing validity within the formal language. One supplement is a principle of extensionality: $S = pP \wedge P = pS \Rightarrow S = P$. Another is to elaborate the part function into a family of functions, stipulating that if a premise would employ an *identity* statement with a term pX, or the conclusion would employ a *diversity* statement with a term pX, that has already appeared in an identity statement in the premises, write instead $p'X$, to indicate a *different* part of X. To see why this matters, consider a simple case of undistributed middle, 2AAI. In the unelaborated notation, it becomes $P = pM, S = pM; \therefore\ pS = pP$. This is valid; in fact, the premises imply $S = P$. As elaborated, however, it becomes $P = pM, S = p'M; \therefore\ pS = pP$, which fails. Similarly, consider an invalid argument cited above: 'Every M is P; No S is M; ∴ No S is P.' The premises become $M = pP$ and $pS \neq pM$; we must represent the conclusion as $pS \neq p'P$, lest we get a form easily shown to be valid.

Though the Bentham/Hamilton doctrine of quantification in the predicate today strikes us as strange, it did contribute two ideas that influenced subsequent algebraic logicians. The first is that a proposition is an equation; the second, that a symbol for a part of a class — something Boole would call an *elective* symbol — could represent existence. Few logicians today would find either idea meritorious. But they played an important role in the development of nineteenth-century logic.

4 THE RISE OF MODERN LOGIC

Most turning points in the history of ideas are less sharply defined than they first appear. Those who make significant advances frequently worked in a setting in which a variety of other thinkers were working on similar problems and might easily have made the discovery first. Not so with modern logic, which has a well-defined starting point. Modern logic began in 1847, when George Boole (1815–1864) published his short book, *The Mathematical Analysis of Logic, being as Essay towards a Calculus of Deductive Reasoning*. He followed it with *An Investigation of the Laws of Thought, on Which are Founded the Mathematical Theories of Logic and Probabilities* in 1854. Boole's work displays several advances of vital significance in our understanding of logic and, specifically, of quantification. Boole himself lists them in his 1848 paper, "The Calculus of Logic":

1. *That the business of Logic is with the relations of classes, and with the modes in which the mind contemplates those relations.* Seeing logic as resting on a theory of classes not only makes mathematical logic possible but also gives logicians tools to create formal theories of syntax, semantics, and pragmatics. Medieval logicians spoke of a term's *supposita*; the Port-Royal Logic, of a term's extension. But both of these were conceived in the plural, even if the grammatical singular 'extension' suggested otherwise. Thinking of a term's *supposita* or extension as an entity, a class, governed by mathematical laws was something altogether new when Whately introduced it about two decades before Boole. Boole was the first to use the insight as the basis of a logical theory. That brings us to Boole's second point:

2. *That antecedently to our recognition of the existence of propositions, there are laws to which the conception of a class is subject, — laws which are dependent upon the constitution of the intellect, and which determine the character and form of the reasoning process.* Georg Cantor is rightly viewed as the founder of set theory. But the *idea* of a theory of classes is already in Boole, and he goes some way toward developing it.

3. *That those laws are capable of mathematical expression, and that they thus constitute the basis of an interpretable calculus.* Boole not only conceives of a calculus of classes — something that might, arguably, be attributed to Leibniz almost two centuries earlier — but constructs one that forms a part of Cantor's set theory and continues to be used today.

4. *That those laws are, furthermore, such, that all equations which are formed in subjection to them, even though expressed under functional signs, admit of perfect solution, so that every problem in logic can be solved by reference to a general theorem.* Boole achieves a higher degree of abstraction in his calculus than contemporary logical theories do, for he brings propositional logic, quantification theory, and abstract algebra together in a single theory. Thus:

5. *That the forms under which propositions are actually exhibited, in accordance with the principles of this calculus, are analogous with those of a philosophical language.* Boolean algebra continues to be an important field of investigation, underlying not only propositional logic but the theory of computation. The idea of a proposition as a class — of "cases," as Boole puts it, or possible worlds, or indices — continues to be influential in formal semantics.

6. *That although the symbols of the calculus do not depend for their interpretation upon the idea of quantity, they nevertheless, in their particular application to syllogism, conduct us to the quantitative conditions of inference.* Boole's calculus is not only a propositional logic but a theory of quantification. The classes it deals with may be conceived as classes of cases in which a proposition is true or as classes of objects of which a term is true. The thought that the problems of the relations between propositions and the relations between terms are essentially the same problem is a profound insight that underlies the unification of the two in modern

predicate logic, though it also lies deeper, and remains independent of the particular fashion in which that theory combines them.

4.1 Algebraic Approaches to Quantification

Since the systems of Boole [1847] and [1854] have been subjects of extensive discussion — see, for example, [Hailperin, 2004; Valencia, 2004] — I shall focus here on Boole [1848], an intermediate approach that has received less attention. Boole's calculus starts with a universe of discourse, represented by the numeral 1. We might think of that as a class of objects, or of cases, or of possible worlds; the theory is independent of any particular conception of what the universe of discourse contains. Everything else is in effect a subclass, "formed by limitation" from that universe.

Concatenating two class symbols denotes the intersection of the classes: xy is the class of things in both x and y.[30] Thus, $x1 = x$. $1 - x$ is the class of all things in the universe of discourse that are not in x. So, $x(1 - y)$ is the class of things in x but not y; $(1 - x)(1 - y)$ is the class of thing in neither x nor y. Boole uses addition to represent disjunction or union.[31]

Boole endorses these algebraic laws governing his operations:

$$x(y + z) = xy + xz$$

$$xy = yx$$

$$x^n = x$$

The last is the most striking, and differentiates his calculus from usual algebraic theories.[32]

Boole turns to the analysis of categorical statement forms. The contemporary reader might guess at what is coming next. If we were to define $0 = 1 - 1$:

- All Ys are Xs: $y = yx$
- No Ys are Xs: $yx = 0$
- Some Ys are Xs: $yx \neq 0$
- Some Ys are not Xs: $y \neq yx$

But that is *not* what Boole does.[33] Instead, influenced by the "quantification in the predicate" idea of Bentham and Hamilton, he writes,

[30] Boole equivocates on what variables such as x and y denote. Most of the time, he thinks of them as class symbols, but he also refers to them as operations.

[31] Boole introduces this operation with the equation $x(y + z) = xy + xz$; he never defines the operation itself. He seems to think of disjunction as exclusive, though most of what he says applies as well if we understand disjunction as inclusive. His equation holds for both interpretations.

[32] Boole uses, but never states explicitly, laws of associativity, even though they had already been stated and named by Hamilton, among others.

[33] He does represent universal affirmatives as $y = yx$ in Boole [1847], but immediately moves to an alternative representation $y(1 - x) = 0$. In [1847] he represents particular affirmatives as $xy = v$, where v is an elective symbol representing, evidently, some nonempty class. In [1848], however, he represents universal affirmatives as $y = vx$. Since he denies that universals have existential import, denying the traditional thesis of subalternation, he evidently intends to allow for the possibility that v is empty. But that makes nonsense of his representing particular affirmatives as $vy = v'x$.

> The expression All Ys represents the class Y and will therefore be expressed by y, the copula are by the sign $=$, the indefinite term, Xs, is equivalent to Some Xs. It is a convention of language, that the word Some is expressed in the subject, but not in the predicate of a proposition. The term Some Xs will be expressed by vx, in which v is an elective symbol appropriate to a class V, some members of which are Xs, but which is in other respects arbitrary. Thus the proposition A will be expressed by the equation $y = vx$.

Note the clear implication: universal affirmatives have existential import. What Boole actually offers, therefore, is this:

- All Ys are Xs: $y = vx$
- No Ys are Xs: $y = v(1 - x)$
- Some Ys are Xs: $vy = v'x$
- Some Ys are not Xs: $vy = v'(1 - x)$

where v' is another arbitrary class. This should look familiar; v, v', etc., are functioning much as p, p', etc., function in the Bentham-inspired system described above. He sees that he has the tools to handle everything in the expanded syllogistic language with infinite terms, and gives these equivalences:

- All not-Ys are Xs: $1 - y = vx$
- All not-Ys are not-Xs: $1 - y = v(1 - x)$
- Some not-Ys are Xs: $v(1 - y) = v'x$
- Some not-Ys are not-Xs: $v(1 - y) = v'(1 - x)$

Boole demonstrates the validity of contraposition, arguing for the equivalence of $y = vx$ and $1 - x = v(1 - y)$. His method of development for elective equations, as he calls it, is highly general but quite complex, amounting to a way of expressing equations in what we now call disjunctive normal form. The method consists in expressing equations as sums of products of coefficients and terms. For a single variable x, the equation has the form

$$\phi(x) = \phi(1)x + \phi(0)(1 - x)$$

For two variables x and y, the equation has the form

$$\phi(xy) = \phi(11)xy + \phi(10)x(1 - y) + \phi(01)(1 - x)y + \phi(00)(1 - x)(1 - y)$$

For three, it has the form

$$\phi(xyz) = \phi(111)xyz + \phi(110)xy(1 - z) + \phi(101)x(1 - y)z + \phi(100)x(1 - y)(1 - z) + \phi(011)(1 - x)yz + \phi(010)(1 - x)y(1 - z) + \phi(001)xy(1 - z) + \phi(000)(1 - x)(1 - y)(1 - z).$$

The general solution to an equation in three variables is

$$z = \frac{\phi(110)}{\phi(110)-\phi(111)}xy + \frac{\phi(100)}{\phi(100)-\phi(101)}x(1-y) + \frac{\phi(010)}{\phi(010)-\phi(011)}(1-x)y + \frac{\phi(000)}{\phi(000)-\phi(001)}(1-x)(1-y).$$

Notice that the key, in every case, is to substitute 0 and 1 in for the variables to discover the appropriate coefficients.

We can use this last solution to demonstrate the validity of contraposition. Assume $y = vx$. Let $z = 1 - x$; then $y = v(1 - z)$, so $y - v(1 - z) = 0$. Think of this as $\phi(vyz) = 0$. We can compute the value of this function for each combination of 1s and 0s to derive the equation

$$1 - x = v(1 - y) + \frac{1}{0}(1 - v)y + \frac{0}{0}(1 - v)(1 - y)$$

The second and third terms appear difficult to interpret. Say, however, that $\frac{1}{0}(1 - v)y = t$; then $(1 - v)y = 0$. So, the second term simply drops out. The third term cannot be handled so easily; Boole replaces $\frac{0}{0}$ with an arbitrary elective symbol w. The equation thus becomes

$$1 - x = v(1 - y) + w(1 - v)(1 - y) = (v + w(1 - v))(1 - y)$$

Replacing $(v + w(1 - v))$ with another elective symbol u, we obtain

$$1 - x = u(1 - y)$$

which represents 'All not-*X*s are not-*Y*s.'

Strangely, perhaps, syllogistic inferences are less complex than immediate inferences in Boole's system. Consider *Barbara*:

Every *M* is *P*	$m = vp$
Every *S* is *M*	$s = um$
∴ Every *S* is *P*	$s = wp$

This is very simple; just let $w = uv$. *Darii* is similarly straightforward:

Every *M* is *P*	$m = vp$
Some *S* is *M*	$u's = um$
∴ Some *S* is *P*	$u''s = wp$

Let $u'' = u'$ and $w = uv$. Consider *Celarent*:

No *M* is *P*	$m = v(1 - p)$
Every *S* is *M*	$s = um$
∴ No *S* is *P*	$s = w(1 - p)$

Simply let $w = uv$. Finally, consider *Ferio*:

No *M* is *P*	$m = v(1 - p)$
Some *S* is *M*	$us = u'm$
∴ Some *S* is not *P*	$ws = w'(1 - p)$

Let $w = u$, $w' = u'v$.

Boole's [1848] technique thus yields a simple way to demonstrate the validity of syllogisms. It evidently assigns universal affirmatives existential import, validating subalternation:

$$\begin{array}{ll} \text{Every } S \text{ is } P & s = vp \\ \therefore \text{Some } S \text{ is } P & us = u'p \end{array}$$

If $s = vp$, $us = uvp$, and, letting $u' = uv$, $us = u'p$. Universal affirmatives and negatives end up as contraries, for, if $s = 0$, $s = vp$ and $s = v(1-p)$ are both false.

Boole's systems of [1847; 1848], and [1854] differ in details, but share many of the same strengths and weaknesses. They unite logic and algebra, addressing the problems of the former using the techniques of the latter. They are highly abstract, applying to both predicate and propositional logic. They are in some ways quite elegant. Their handling of elective symbols, however, is far from rigorous. They leave certain key concepts, including addition (disjunction) and subtraction (negation), undefined, while leaving many of the rules governing them implicit. Most seriously, perhaps, they are strictly monadic, finding no place for relations.

4.2 *Peirce's Quantification Theory*

American philosopher Charles Sanders Peirce (1839–1914) developed a comprehensive theory of quantification that had far less impact than it might have had, due to the relative inaccessibility of much of Peirce's writings. Even within the field of logic, he developed not one logical system but many, making it difficult to assess his overall contributions. One of his systems, however, is essentially modern quantification theory, which he presents in a form almost identical to that used by most twentieth-century logicians. This is no accident; Peirce influenced Schröder, Löwenheim, and Skolem, who took over many aspects of his notation. Indeed, the word 'quantifier' is Peirce's. (See [Hilpinen, 2004, 615–616].)

Perhaps the best way to begin is with Peirce's concept of a predicate. A proposition, he says, is "a statement that must be true or false" [1893, 208]. A predicate is a residue of a proposition: "let some parts be struck out so that the remnant is not a proposition, but is such that it becomes a proposition when each blank is filled by a proper name" (203). So far, this is traditional, though expressed in language very closely to Russell's talk of propositional functions. Peirce's examples would have shocked Aristotle or Arnauld:

> Thus, take the proposition "Every man reveres some woman." This contains the following predicates, among others:
> "'...reveres some woman."
> "...is either not a man or reveres some woman."
> "Any previously selected man reveres ..."
> "Any previously selected man is..."

Predication is "the joining of a predicate to a subject of a proposition so as to increase the logical breadth without diminishing the logical depth" (209). He intends this to be vague, admitting different interpretations. But Peirce is concerned to argue that predicate logic encompasses all of logic, subsuming propositional logic under the general heading of quantification theory. Predication is involved in every proposition, even 'It is raining,' which, he argues, has an indexical subject, and in conditionals, "in the same sense that some recognized range of experience or thought is referred to" (209).

Peirce introduces quantification in terms of the traditional syllogistic conception of quantity. "The quantity of a proposition is that respect in which a universal proposition is regarded as asserting more than the corresponding particular proposition: the recognized quantities are Universal, Particular, Singular and — opposed to these as 'definite' — Indefinite" (214). He mentions Hamilton's idea of quantification of the predicate, but dismisses it "as an instructive example of how not to do it" [1893, 323].[34] He also dismisses DeMorgan's system of eight quantified propositions. If a proposition's quantity is that by virtue of which a universal proposition asserts more than a particular proposition, the real question is what that is.

Peirce begins by concurring with Leibniz that a universal proposition neither asserts nor implies the existence of its subject. In fact, his initial example of a universal proposition is 'Any phoenix rises from its ashes' (216). He gives two arguments. The first is the *dictum de omni*: what is asserted universally of a subject is "predicable of whatever that subject may be predicable' (217). That makes no commitment to the existence of anything. The second is that formal logic should have as wide application as possible; a theory allowing empty terms has broader scope than one that restricts itself to terms with nonempty extensions. The third is the square of opposition. Formal logic, Peirce says, needs to specify a negation for each simple proposition. If 'All inhabitants of Mars have red hair' makes no existential commitments, then its negation is simply 'Some inhabitants of Mars do not have red hair.' If it means 'There are inhabitants of Mars and every one of them without exception has red hair,' then its negation must be 'Either there is no inhabitant of Mars, or if there be, there is one at least who has not red hair' (217), which is not a categorical proposition. Peirce sees clearly that, if the negation of a categorical proposition is to be another categorical proposition — something maintained by the traditional square of opposition — then there is a straightforward choice: either universals carry existential import, or particulars do, but not both. Pick one. It is natural to pick particulars, since 'Some inhabitants of Mars have red hair' and 'There are inhabitants of Mars with red hair' seem obviously equivalent. The consequence is that A and O propositions contradict each other, as do E and I propositions. But, Pierce notes [1893, 283], those are the only relations in the square of opposition that actually hold. A and E propositions can both be true if their subject terms are empty; I and O propositions can both be false in the same circumstance. Universal propositions do not entail the corresponding particular propositions, for the latter have an existential import the former lack.

Peirce cautions that we should not take existence as a predicate, citing Kant's critique

[34]This is only the beginning of his invective: "The reckless Hamilton flew like a dor-bug into the brilliant light of DeMorgan's mind in a way which compelled the greatest formal logician who ever lived to examine and report upon the system" [1893, 324]. He calls Hamilton's followers "perfect know-nothings" (324–325), and refers to Hamilton himself as "utterly incapable of doing the simplest logical thinking," "an exceptionally weak reasoner" (326); his doctrines are "of no merit," "glaringly faulty" (345), "all wrong," "ridiculous" (326).

of the ontological argument. He says some intriguing things in light of later discussions, developing the concept of a universe of discourse [1893, 281], now often called a contextual domain: "Every proposition refers to some index: universal propositions to the universe, through an environment common to speaker and auditor, which is an index of what the speaker is talking about" [1893, 218].[35] He also speaks in a language friendly to game-theoretic semantics and even to intuitionism: "But the particular proposition asserts that, with sufficient means, in that universe would be found an object to which the subject term would be applicable, and to which further examination would prove that the image called up by the predicate was also applicable" (218).

Peirce renders particular and universal propositions in forms now familiar, linking categorical propositions to propositions with quantifiers and sentential connectives:

> Some inhabitants of Mars have red hair ⇔ Something is, at once, an inhabitant of Mars and is red haired
> All inhabitants of Mars have red hair ⇔ Everything that exists in the universe is, if an inhabitant of Mars, then also red haired

The quantifier is independent of the proposition's terms, asserting "the existence of a vague something to which it pronounces 'inhabitant of Mars' and 'red haired' to be applicable" (218). Thus,

> A particular proposition is one which gives a general description of an object and asserts that an object to which that description applies occurs in the universe of discourse.... (221)

Peirce commits himself to what would later be called Quine's dictum — that to be is to be a value of a variable [Quine, 1939] — when he declares that 'Something is non-existent' "is an absurdity, and ought not to be considered as a proposition at all" (222). The quantifier asserts existence, so 'Something is non-existent' is a contradiction. Peirce seems to think that contradictory propositions are not propositions at all. At other places, however, he recognizes that the universe of discourse of a proposition is defined by "the circumstances of its enunciation" (326), and consists of "some collection of individuals or of possibilities": "At one time it may be the physical universe, at another it may be the imaginary 'world' of some play or novel, at another a range of possibilities" (326–327).

Peirce's [1885] formalism for quantification looks remarkably modern. He uses the material conditional, and adopts as axioms and rules, which he calls, indiscriminately, *icons*:

$$x \rightarrow x$$

$$x \rightarrow (y \rightarrow z) \vdash y \rightarrow (x \rightarrow z)$$

$$(x \rightarrow y) \rightarrow ((y \rightarrow z) \rightarrow (x \rightarrow z))$$

$$b \rightarrow x \vdash \bar{b}$$

$$((x \rightarrow y) \rightarrow x) \rightarrow x$$

[35] Peirce credits this to DeMorgan [1847, 380].

Note that the second and fourth of these are rules of inference; the fourth is stated imprecisely, meaning that the negation of a formula b follows from the claim that, for any x, if b, then x. Peirce leaves the propositional quantification here implicit, stating this rule only in English paraphrase. We could express his intentions more precisely by introducing a propositional constant for the false, and writing the axiom as $b \to \bot \vdash \bar{b}$. Peirce's system is then a sound and complete axiomatization of a pure implicational propositional logic.

Peirce then introduces disjunction and conjunction, defining $x + y$ as $\bar{x} \to y$ and xy as $(\overline{\bar{x} + \bar{y}})$. He introduces quantifiers by distinguishing "a pure Boolian expression referring to an individual and a Quantifying part saying what individual this is" [1885, 194]. As a first approximation, he writes Any $(\bar{k} + h)$ to mean that any king in the universe of discourse is happy, and Some kh to mean that some king in that universe is happy. To handle relational terms, he introduces subscripts for what would now be considered individual variables, and uses Π and Σ to represent universal and existential quantification, respectively. So, the above would more precisely be written as $\Pi_i(\bar{k}_i + h_i)$ and $\Sigma_i k_i h_i$. He defines the quantifiers in terms of (potentially) infinite conjunctions and disjunctions, but notes that they are "only similar" to them, because the universe of discourse may be uncountable (195). Peirce quickly shows off the power his new notation, writing $\Pi_i \Sigma_j l_{ij} b_{ij}$ to represent "everything is a lover and benefactor of something," $\Pi_i \Sigma_j l_{ij} b_{ji}$ to represent "everything is a lover of a benefactor of itself," $\Sigma_i \Pi_j (g_i l_{ij} + \bar{c}_j)$ to represent "if there be any chimeras there is some griffin that loves them all," and $\Pi_j \Sigma_i (g_i l_{ij} + \bar{c}_j)$ to represent "each chimera (if there is any) is loved by some griffin or other" (195).

Peirce's proof method requires converting formulas to prenex normal form, something that, he shows, preserves equivalence, and then using a technique similar to semantic tableaux. I will not go into details here. But note the dramatic shift from earlier conceptions of logic. Peirce formulates both axioms and rules — though not drawing attention to the difference between them — develops a logic of quantification, and then formulates a proof system. All of this advances significantly beyond the earlier algebraic approaches to logic, not to mention the rest of the logical tradition.

4.3 Frege's Begriffschrift

Gottlob Frege (1848–1925) published his *Begriffschrift*, "ideography" or "concept writing," in 1879. It has been called "perhaps the most important single work ever written in logic" [van Heijenoort, 1967, 1]. Frege's system forms the basis for modern predicate logic. The twentieth-century achievements of Peano, Russell, Whitehead, Gödel, Tarski, Church, Carnap, and many others would have been impossible, or at least would have had a rather different character, without him. Frege's goal, he wrote [1882], was to devise "a *lingua characterica* in Leibniz's sense" — a universal language that, in Leibniz's words, "depicts not words, but thoughts" (PW 13, NS 14) — the structure of which reflects not surface forms but underlying semantic structure, which Frege refers to as "conceptual content" (12).

Frege's system parallels Peirce's, not only in introducing quantifiers but also in employing, explicitly, the material conditional. Like Peirce's system, it rests on the shoulders of Boole, DeMorgan, and others. But he makes a number of vital contributions. First, he rests his logic not on the distinction between propositions and terms or between subject and predicate but instead on the distinction between function and argument.

> If in an expression, whose content need not be capable of becoming a judgment, a simple or a compound sign has one or more occurrences and if we regard that sign as replaceable in all or some of these occurrences by something else (but everywhere by the same thing), then we call that part that remains invariant in the expression a function, and the replaceable part the argument of the function. [1879, 22]

Second, he presents his system purely formally; for the first time, logic concerns formulas and their manipulation according to strictly syntactic operations. Third, he correspondingly draws a sharp distinction between axioms and rules of inference, with *modus ponens* and substitution as the only inference rules. Finally, he presents for the first time a complete set of rules for quantifiers, formulating universal instantiation and universal generalization as well as rules for identity. Frege is the first to see a quantifier as a variable-binding operator. He offers the first system in the history of logic that unites propositional and predicate logic into a single unified system capable of handling nested quantifiers. He also presents the first higher-order logic. If modern logic begins in 1847, modern quantification theory begins in 1879.

Frege employs a notation that is difficult to reproduce, using strokes with concavities filled by German letters as universal quantifiers. Giuseppe Peano (1858–1932), an Italian mathematician, deserves credit for putting quantification theory into something close to its modern notation [Peano, 1889], and I will use the Peano-inspired notation here. Frege's account of universal quantification is this:

> In the expression of a judgment we can always regard the combination of signs to the right of $\vdash$ [the assertion stroke] as a function of one of the signs occurring in it. *If we replace this argument by a German letter and if in the content stroke we introduce a concavity with this German letter in it, ... this stands for the judgment that, whatever we may take for its argument, the function is a fact.* [1879, 24]

Suppose, in other words, we have the assertion that p: $\vdash p$. We can analyze p into function and argument: $\vdash \Phi(a)$. We can then replace the argument sign with a variable (German letters in Frege, Roman letters from the end of the alphabet, here), and prefix a quantifier on that variable, to obtain $\vdash \forall x \Phi(x)$. Frege explains the meaning of this kind of expression in terms of a rule of universal instantiation:

> From such a judgment, therefore, we can always derive an arbitrary number of *judgments of less general content* by substituting each time something else for the German letter and then removing the concavity in the content stroke. [1879, 24][36]

[36]Frege never gives an independent semantic characterization of what universal quantification means. Taking this rule as giving such a characterization yields a substitutional conception of quantification. Frege, unlike Peirce, does not discuss the universe of discourse, and, to the extent that he has the concept of a universe at all, treats it as something given and static, not as something capable of varying from context to context, as in Peirce, or from model to model, as in contemporary model theory.

From $\vdash \forall x\Phi(x)$, that is, we can derive $\vdash \Phi(a)$, $\vdash \Phi(b)$, and so on. Frege discusses scope explicitly (though he does not give a formal definition of it), and cautions that we cannot apply this rule when a universal quantifier occurs within the scope of an arbitrary operator. He thus restricts the rule to universal quantifiers that have as their scope the entire formula.

Frege formulates the rule of universal generalization: "*A Latin letter may always be replaced by a German one that does not yet occur in the judgment*" [1879, 25]. Thus, given $\vdash \Phi(a)$, we may infer $\vdash \forall\Phi(x)$, provided that x does not occur in Φ. This, together with his propositional rules and universal instantiation, gives him a complete set of rules for predicate logic.

Frege also introduces a derivable rule allowing the deduction from $\vdash \Phi(a) \to A$ to $\vdash \forall x\Phi(x) \to A$, provided that a does not occur in A and is replaced in every occurrence by x. He distinguishes $\neg\forall x\Phi(x)$ from $\forall x\neg\Phi(x)$, and defines "There are Φ" as $\neg\forall x\neg\Phi(x)$. Frege shows how to express categorical propositions in his system, and constructs the square of opposition, without remarking that he has actually given up all relations except contradictories. His representations:

All S are P: $\forall x(S(x) \to P(x))$
No S are P: $\forall x(S(x) \to \neg P(x))$
Some S are P: $\neg\forall x(S(x) \to \neg P(x))$
Some S are not P: $\neg\forall x(S(x) \to P(x))$

The notation makes it obvious that A and O propositions are contradictories, as are E and I propositions. A and E propositions, however, are not truly contraries, for they can be true together if there are no Ss. Under just those circumstances, I and O propositions can also be false together, so they are not true subcontraries. Finally, A and E propositions do not entail I and O propositions, respectively, for the former lack the existential import of the latter.

Frege develops his system from nine axioms, together with rules of *modus ponens* and substitution. The axioms:

1. $a \to (b \to a)$
2. $(c \to (b \to a)) \to ((c \to b) \to (c \to a))$
3. $(d \to (b \to a)) \to (b \to (d \to a))$
4. $(b \to a) \to (\neg a \to \neg b)$
5. $\neg\neg a \to a$
6. $a \to \neg\neg a$
7. $c \equiv d \to (f(c) \to f(d))$
8. $c \equiv c$
9. $\forall x f(x) \to f(c)$

Frege claims completeness, but does not formulate the notion precisely, let alone offer an argument. Indeed, he cannot formulate it precisely, for he lacks the concept of an interpretation, and so has no precise concept of validity [Goldfarb, 1979]. Lukasiewicz [1934] shows that the first six are nevertheless a complete axiom system for propositional logic, and, indeed, that (3) is superfluous.

First-order logic makes a huge advance over Aristotelian logic in thinking about quantifiers as variable-binding expressions and thus in having the capacity to handle relational predicates. Aristotle himself realized that his theory did not incorporate inferences such as

> Every perceiving is a thinking
> ∴ Every object of perception is an object of thought.

In the fourteenth century, and again in the seventeenth and nineteenth centuries, logicians returned to those worries, considering such inferences as

> Every man's donkey is running.
> Every king is a man
> ∴ Every king's donkey is running. [Buridan, 2001]
>
> Every circle is a figure.
> ∴ Whoever draws a circle draws a figure. [Junge, 1638]
>
> All horses are animals
> ∴ Every horse's tail is an animal's tail. [DeMorgan, 1847]

First-order logic, by incorporating relational predicates, solves these problems, unifying propositional and quantificational logic in a system powerful enough to express propositions and inferences that had escaped each on its own.

Frege's system is very powerful — in fact, as Russell later observed, *too* powerful — for it permits quantification over predicates as well as individual argument positions. It is thus tantamount to a full, impredicative second-order logic. Russell's (1902) letter to Frege praises the *Begriffsschrift* and the *Grundgesetze der Arithmetik*, which develops the system further, reducing the set of basic rules and axioms. But then he drops the bomb, known as Russell's paradox:

> There is just one point where I have encountered a difficulty. You state [23] that a function, too, can act as the indeterminate element. This I formerly believed, but now this view seems doubtful to me because of the following contradiction. Let *w* be the predicate: to be a predicate that cannot be predicated of itself. Can *w* be predicated by itself? From each answer its opposite follows. Therefore we must conclude that *w* is not a predicate. Likewise there is no class (as a totality) of those classes which, each taken as a totality, do not belong to themselves. From this I conclude that under certain circumstances a definable collection [*Menge*] does not form a totality. [1902, 124–125]

Frege's response expresses consternation, but also the hope that Russell's discovery "will perhaps result in a great advance in logic, unwelcome as it may seem at first glance" [1902, 128]. He nods in the direction of type theory:

> Incidentally, it seems to me that the expression "a predicate is predicated of itself" is not exact. A predicate is as a rule a first-level function, and this function requires an object as argument and cannot have itself as argument (subject). Therefore I would prefer to say "a concept is predicated of its own extension". [1902, 128]

In itself, however, that provides no solution. Russell's paradox cuts at the heart of Frege's reconstruction of arithmetic from logic. It does not, however, strike at Frege's system viewed as a system of first-order predicate logic. Peano, Russell, Whitehead, and others would simplify the unwieldy notation of the *Begriffsschrift*, turning first-order predicate logic into the *lingua franca* of twentieth-century philosophy. Indeed, so influential would Frege's system become that philosophers as prominent as W. V. O. Quine [1960], Donald Davidson [1967], and John Wallace [1970] would write of it as "*the* frame of reference," the logic into which all assertions must be translated if they are to be considered meaningful at all.

5 CONTEMPORARY QUANTIFICATION THEORY

In later work, especially Frege [1892], Frege offers two ideas that have led to contemporary approaches to quantification. First, he makes explicit the Aristotelian thought that determiners such as 'some' and 'all' are relations — in Frege's language, relations between concepts:

> We may say in brief, taking 'subject' and 'predicate' in the linguistic sense: A concept is the reference of a predicate; An object is something that can never be the whole reference of a predicate, but can be the reference of a subject. It must here be remarked that the words 'all', 'any', 'no'', 'some', are prefixed to concept-words. In universal and particular affirmative and negative sentences, we are expressing relations between concepts; we use these words to indicate the special kind of relation. They are thus, logically speaking, not to be more closely associated with the concept-words that follow them, but are to be related to the sentence as a whole. [Frege, 1892; 1951, 173]

Second, Frege treats quantifiers as second-level concepts. "I have called existence a property of a concept," he writes [1892, 174]; he uses the same analysis for quantifiers. These two ideas have led to the contemporary theory of generalized quantifiers. It took more than sixty years, however, for that generalization to be developed, and almost a century before its significance began to be appreciated.

5.1 *Limitations of Classical First-Order Logic*

The theory of quantification that issued from the work of Peirce and Frege became known as classical first-order logic. The early part of the twentieth century witnessed the discovery of many of its deepest properties through the development of metalogic, especially in the form of model theory and recursion theory. Some highlights:

1. Leopold Löwenheim [1915] and Thoralf Skolem [1920], starting from the work of Peirce and Schröder, show that, in Skolem's formulation, "Every proposition in normal form either is a contradiction or is already satisfiable in a finite or countably infinite domain" [Skolem, 1920, 256]. Since every proposition is equivalent to one in normal form, this shows that every satisfiable proposition is satisfiable in a countable domain. He extends this to infinite sets of formulas, and shows that every uncountable domain satisfying a set of formulas has a countable subset also satisfying that set. Alfred Tarski and Carl Vaught [1956] later prove an upward form of this theorem, showing that any set satisfiable in an infinite domain is satisfiable in an uncountable domain.

2. Skolem [1928] shows that every formula of first-order logic can be "Skolemized" by finding an equivalent in prenex normal form, dropping all quantifiers, and then replacing existentially quantified variables with function terms containing variables bound universally to the left of that existential quantifier. Thus, a Skolemized version of $\forall x \forall y \exists z(Fxy \rightarrow (Fxz \wedge Fyz))$ would be $Fxy \rightarrow (Fxf(xy) \wedge Fyf(xy))$. He uses this to develop a non-axiomatic proof method for first-order logic that amounts to a decision procedure for determining, for a given formula A and natural number n, whether A is satisfiable in a domain of cardinality n.

3. Kurt Gödel [1930] proves the completeness theorem, showing that every valid formula is provable. Specifically, Gödel shows that every formula "*is either refutable or satisfiable* (and, moreover, satisfiable in the denumerable domain of individuals)" [1930, 585]. He deduces from it the compactness theorem, showing that a set of formulas is satisfiable if and only if every finite subset of it is satsfiable. The Löwenheim-Skolem theorem is of course also a consequence.

4. Alfred Tarski [1936], in his investigations of the concept of truth, constructs a formal semantics giving truth conditions for quantified sentences and thus lays the foundation for model theory. Earlier thinkers had given informal characterizations of the truth conditions for quantified sentences, linking them to generalized conjunction or disjunction and perhaps to a universe of discourse. Tarski gives the first formal characterization of the semantics of quantified sentences, counting $\forall x\Phi(x)$ true on a model $\mathfrak{M}$, relative to an assignment g of objects from $\mathfrak{M}$'s domain to variables, if and only if $\Phi(x)$ is true on every assignment g' differing from g solely in its assignment to x. The relativization to an assignment function disappears for closed sentences, i.e., those without free variables. Tarski also shows that, although the concept of truth can be given a consistent and precise characterization in the metalanguage, it cannot be given a similar characterization in the object language.

5. Alonzo Church [1936] proves that first-order logic is undecidable. Since propositional logic is decidable, this shows that Boole's idea that a single abstract theory might be at once a theory of propositional relations (on one interpretation) and a theory of quantification (on another interpretation) is hopeless.

6. Per Lindström [1969, 1974] proves that classical first-order logic is the strongest compact logic having the downward Löwenheim-Skolem property, developing the foundations of abstract model theory.

These results represent impressive intellectual achievements. They also point to some limitations of classical first-order logic. Löwenheim and Skolem's result, for example, shows that first-order logic cannot distinguish countable from uncountable domains. 'There are countably many' and 'there are uncountably many' thus have no first-order representations. It is not hard to show that 'there are finitely many' and 'there are infinitely many' similarly have no first-order representations.

Subsequent developments point to additional limitations. Peter Geach, by way of his discussions of medieval (specifically, fourteenth-century) logic, points out many of the most important. First, he draws attention to determiners beyond those analyzed by earlier logical systems. Remarkably, for more than two thousand years, the study of quantification was essentially the study of 'every,' 'no,' and 'some,' together with determiners taken as equivalent or reducible to them ('all,' 'never,' etc.). De Morgan [1847; 1860] broke new ground in thinking about determiners such as 'most,' noting the validity of 'Most Ss are Ps; most Ss are P's; $\therefore$ some Ps are P's.' De Morgan also thought about numerical quantifiers, noting the validity, where s is the number of Ys and $n + m > s$, of 'n Ys are Xs; m Ys are Zs; $\therefore$ $n + m - s$ Xs are Zs.' Subsequent mathematical logicians nevertheless ignored such quantifiers, and indeed developed systems in which 'most' could not receive a treatment analogous to that accorded 'some' and 'all.' Nicholas Rescher [1964] showed that 'most,' like 'there are finitely many,' 'there are infinitely many,' 'there are uncountably many,' and the like, has no first-order representation. 'Most' poses a puzzle for Aristotelian logic in its nineteenth-century formulation, according to which every term must be distributed or undistributed, for 'most' fits neither pattern. But it also poses a puzzle for first-order logic, which lacks the resources to define it or even introduce it on analogy with $\forall$ or $\exists$. David Lewis [1975] similarly observes the wide range of quantifiers available in English, few of which find representation in first-order logic.

Second, Geach draws attention to sentences such as this, known as the Geach-Kaplan sentence: 'Some critics admire only one another.' 'Some critics admire only each other' has a reading expressible in first-order logic — as $\exists x \exists y((Cx \wedge Cy) \wedge \forall z(Axz \rightarrow z = y) \wedge \forall w(Ayw \rightarrow w = x))$ — but the intended reading of the 'one another' version, which encompasses sets of mutually admiring critics of any cardinality, does not. It requires quantification over sets: $\exists X \forall x(x \in X \rightarrow (Cx \wedge \forall y(Axy \rightarrow y \in X)))$.

Third, Geach draws attention to so-called "donkey sentences," such as 'Every farmer who owns a donkey beats it,' in which there is an anaphoric connection between subject and predicate. The difficulty is not in finding a first-order expression with the appropriate truth conditions — $\forall x \forall y((Fx \wedge Dy \wedge Oxy) \rightarrow Bxy)$ serves adequately — but in reaching that representation in a compositional fashion. Symbolizing 'Every farmer owns a donkey' requires an existential quantifier: $\forall x(Fx \rightarrow \exists y(Dy \wedge Oxy))$. Where there is no anaphoric connection between subject and predicate, that works fine; we can represent 'Every farmer who owns a donkey is happy' as $\forall x((Fx \wedge \exists y(Dy \wedge Oxy)) \rightarrow Hx)$. Pursuing that strategy for 'Every farmer who owns a donkey beats it,' however, would produce $\forall x((Fx \wedge \exists y(Dy \wedge Oxy)) \rightarrow Bxy)$, which does not serve the purpose at all, since the final 'x' is not bound by any quantifier. Fourteenth-century logicians, as we have seen, worry about such sentences, recognizing that quantifier phrases not only contribute to truth conditions but also, under certain conditions, enable later anaphoric reference. Nothing in Aristotle's theory could explain that function. Nothing in first-order logic explains it either. indexanaphoric

Fourth, Geach draws attention to cases involving such anaphoric connection in which it seems impossible to express the desired connection without an unwanted existential commitment. Consider

> Hob believes that a witch blighted Bob's mare, and Nob believes that she blighted Cob's cow.

The existential expression in the first phrase occurs within a belief context; representing it as something like

> Hob believes that $[\exists x(Wx \wedge Bxm(b))]$

makes no commitment to the existence of witches. Hob believes there is a witch, but, to represent what the sentence says, we do not have to believe it. The anaphoric pronoun 'she' in the second phrase, however, creates a problem similar to that in donkey sentences. Appending another clause containing the variable x leaves it unbound; quantifying it existentially within Nob's belief context conveys that Hob believes some witch has blighted Bob's mare, and that Nob believes that some witch has blighted Cob's cow, but does nothing to convey that Nob believes them to be the same witch. Existentially quantifying outside the belief contexts allows us to capture that, but at the cost of making the sentence commit us to the existence of witches. But surely one can report on the beliefs of Hob and Nob without oneself expressing belief in witches.

Geach's Hob–Nob example is one case in which first-order logic seems to require unwanted existential assumptions. There are many others: 'I want a sloop' [Quine, 1956], which threatens to commit us to the existence of a particular sloop that I want, or 'John is seeking a unicorn' [Montague, 1974], which threatens to commit us to unicorns. Such concerns have led to the development of free logic (logic free of existential commitments with respect to its singular terms) and to Montague grammar.

Finally, Leon Henkin [1961], Jaakko Hintikka [1979], and Jon Barwise [1979] draw attention to branching quantifiers, which seem to defy not only the bounds of first-order logic but any system of linear representation. Consider a sentence such as 'A leader from each tribe and a captain from each regiment met at the peace conference.' Symbolizing this as something beginning with $\forall x \exists y \forall z \exists u$ fails to capture the intended meaning for reasons that become obvious when one thinks about Skolemization. A Skolemized version of a linear representation would replace y with $f(x)$ and u with $g(x, y, z)$ rather than the intended $g(z)$. The choice of leader should depend solely on the tribe; the choice of captain should depend solely on the regiment. A linear representation, however, makes the choice of a value for u depend not only on the value of z but on the values of x, y, and z.

These problems, taken together, suggest that first-order logic does not analyze quantification in a manner adequate to understanding natural language quantification. Though it remains the logic of choice for understanding quantification in mathematics, doubt has begun to arise about its adequacy even there. First-order arithmetic is not categorical; it has non-standard models, with domains consisting not only of the natural numbers but also of one or more Z-chains, that is, chains with the structure of the integers. Second-order arithmetic, in contrast, is categorical. That suggests to many that first-order logic is not adequate even to mathematical reasoning.

5.2 Generalized Quantifiers

First-order logic gains its power, in part, by moving from a monadic conception of predicates to a relational one. It diverges from Aristotelian logic, however, by adopting a monadic theory of quantification. A quantifier $\forall x$ or $\exists x$ operates on a single formula, however complex. In Frege's language, quantifiers are second-level concepts; they act as monadic predicates of monadic predicates — in effect, as terms applying to terms.

This has the effect of making quantifier expressions "disappear" in symbolic representations. In representing 'Every king's donkey is running,' for example, we lose the ability to identify any particular component as corresponding to 'every king's donkey'; the formula $\forall x \forall y((Kx \wedge Dy \wedge Oxy) \rightarrow Ry)$ has no subformula that represents just that noun phrase. More dramatic is the contrast between 'Every farmer who owns a donkey is happy' ($\forall x((Fx \wedge \exists y(Dy \wedge Oxy)) \rightarrow Hx)$) and 'Every farmer who owns a donkey beats it' ($\forall x \forall y((Fx \wedge Dy \wedge Oxy) \rightarrow Bxy)$). This leads Frege and Russell to stress a context principle, that only in the context of a sentence does an expression have meaning. The surface form of a sentence, they hold, is misleading; the true structure of the sentence emerges only at the level of logical form. The distinction between subject and predicate on which logic before the middle of the nineteenth century relies appears, from this point of view, as a midleading artifact of the inadequacies of natural language. What Frege and Russell take as a deep insight, however, might instead be viewed as a defect of their theory.

Suppose we were to combine a relational conception of predicates with a relational theory of quantifiers, seeing determiners as expressing relations between terms or their extensions. The result is the theory of generalized quantifiers, developed in Mostowski [1957], Lindström [1966], Barwise and Cooper [1981], Barwise and Feferman [1985], and many subsequent works. This theory has several advantages over first-order logic as a theory of quantification, including a full range of determiners, handling problem sentences, and respecting compositionality, giving the concepts of subject and predicate a respectable role to play.

We may endorse Frege's thought that quantifiers are second-level concepts and his thought that determiners express relations among concepts without limiting ourselves to quantifiers of Fregean form. *Determiners* are expressions such as 'every,' 'all,' 'some,' 'no,' 'most,' 'at least n,' 'finitely many,' 'uncountably many,' and the like; they denote *quantifiers*. Noun phrases are expressions such as 'every farmer,' 'all donkeys,' 'some European kings of the fourteenth century,' 'no wives of Henry VIII,' 'most donkeys owned by a farmer,' 'at least n students,' 'finitely many numbers,' 'uncountably many functions from numbers to numbers,' and the like. Definite and indefinite articles may be included among the determiners; proper names may be included among the noun phrases.

Let's start, as Aristotle does, with categorical propositions. 'Every S is P,' 'Some S is P,' and so on have the form

Determiner Subject Predicate

We can think of determiners as relations between terms, in effect construing this as

Determiner (Subject, Predicate)

Or, we can see the determiner and subject as forming a noun phrase, and then applying to the predicate:

(Determiner Subject)(Predicate)

The former is more intuitive and closer to most of the logical tradition. Either one, however, can serve as the basis for a theory of generalized quantifiers. Barwise and Cooper [1981], Keenan and Moss [1984], and Keenan and Stavi [1986] adopt the latter strategy; van Benthem [1983; 1984; 1986; 1987; 1989], and Westerståhl [1989] adopt the former.[37]

Letting U represent a universe of discourse, we can think of a categorical proposition, then, as having the structure

$D_U S P$

If we think of quantification as extensional, we can think of terms as standing simply for their extensions, and so think of a quantifier, relative to a universe of discourse, as a relation between sets. The standard examples of quantifiers are binary relations, having the above structure, but there are quantifiers such as 'more Ss than S's' that relate more than two sets. So, generally, we can think of a quantifier as an n-ary relation on sets. Since U is the universe of discourse, we are concerned only with the extensions of S, P, and any other terms involved in this relation within that universe. So, we may see a quantifier Q on U as an n-ary relation on subsets of U: $Q_U \subseteq (\wp(U))^n$.

Not every n-ary relation on subsets of a universe of discourse, however, is a quantifier. Quantifiers satisfy a number of constraints that Barwise and Cooper refer to as natural language universals. For the moment, let's restrict our attention to binary quantifiers.

First, quantifiers satisfy a principle of *Extension*:

$$Q_U S P \wedge U \subset U' \Rightarrow Q_{U'} S P$$

The universe of discourse can expand without affecting the truth of a quantified sentence, provided that the expansion has no effect on the extensions of the related terms.

[37]Good summaries of the theory of generalized quantifiers include [van der Does and van Eijck, 1996; Keenan, 1996; Peters and Westerståhl, 2006; Keenan and Westerståhl, 1997; 2011].

Second, quantifiers *live on* their sets, in the sense that satisfy a principle of *conservativity*:

$$Q_U S P \Leftrightarrow Q_U S (S \cap P)$$

'Every S is P' is equivalent to 'Every S is S-and-P: 'Every man is mortal' is equivalent to 'every man is a mortal man.' As Westerståhl [1989] notes, this distinguishes subjects from predicates, giving subjects "a privileged role" (38), for all that matters to the truth of a quantified sentence are the properties of subsets of the subject's extension. These constraints, together, allow us to drop relativization to the universe of discourse. Conservativity guarantees that we can, without loss of generality, view the predicate's extension as a subset of the subject's extension. Extension guarantees that no universe of discourse larger than the subject's extension affects truth values. So, we can identify the universe with the extension of the subject:

$$S \subseteq U \Rightarrow (Q_U S P \Leftrightarrow Q_S S P)$$

Hereafter, I drop relativization to the universe of discourse and write QSP whenever, in context, the subject term remains constant.

As the term 'quantifier' suggests, the truth value of a quantified proposition depends solely on quantities, not on other aspects of the sets concerned. One way to capture this idea is to say that the truth of a quantified proposition is invariant under permutations. Permute the universe of discourse, so that related subsets change while retaining their cardinalities, and truth values of quantified propositions should remain unaltered. Quantifiers thus satisfy a principle of *permutation* or *isomorphism*: If f is a one-to-one correspondence from U onto U', then

$$Q_U S P \Leftrightarrow Q_{U'} f[S] f[P]$$

This has the effect that the truth values of quantified propositions depend only on the cardinalities of the sets concerned. So, we can view a binary quantifier as a relation between $|S|$ and $|S \cap P|$, or, equivalently, between $|S - P|$ and $|S \cap P|$.

Here, for example, is how we might express some common quantifiers as relations between sets and as relations between cardinalities of sets:

Every S is P $\Leftrightarrow$ $S \subseteq P$ $\Leftrightarrow$ $|S| = |S \cap P|$
Some S is P $\Leftrightarrow$ $S \cap P \neq \emptyset$ $\Leftrightarrow$ $|S \cap P| > 0$
No S is P $\Leftrightarrow$ $S \cap P = \emptyset$ $\Leftrightarrow$ $|S \cap P| = 0$[38]
Most S are P $\Leftrightarrow$ $|S \cap P| > |S - P|$[39]
At least n S are P $\Leftrightarrow$ $|S \cap P| > n$
At most n S are P $\Leftrightarrow$ $|S \cap P| \leq n$
Exactly n S are P $\Leftrightarrow$ $|S \cap P| = n$

[38]This explains why these categorical statement forms could be expressed in Boolean notation as $x = xy$, $xy \neq 0$, and $xy = 0$.

[39]This assumes that 'most' is equivalent to 'more than half.' In fact, it is probably stronger; giving precise truth conditions is difficult. See [Kennan, 1996; Kennan and Westerståhl, 1997; Ariel, 2003; 2004; 2006; Pietroski *et al.*, 2009; Hackl, 2009].

Finitely many S are P $\Leftrightarrow$ $|S \cap P|$ is finite
Uncountably many S are P $\Leftrightarrow$ $|S \cap P|$ is uncountable
All but finitely many S are P $\Leftrightarrow$ $|S - P|$ is finite
All but (exactly) n S are P $\Leftrightarrow$ $|S - P| = \mathrm{n}$

We can do the same for quantifiers that are not binary:

More S than S' are P $\Leftrightarrow$ $|S \cap P| > |S' \cap P|$
At least as many S as S' are P $\Leftrightarrow$ $|S \cap P| \geq |S' \cap P|$
Exactly as many S as S' are P $\Leftrightarrow$ $|S \cap P| = |S' \cap P|$

There are of course many additional quantifiers in natural language, including 'several,' 'many,' 'few,' 'a few,' 'the,' 'this,' 'that,' 'these,' 'those,' 'one,' 'both,' and 'neither.'

Since binary quantifiers are binary relations, we can classify them according to properties as we can other relations. A quantifier Q is reflexive, for example, iff QSS, and symmetric iff $QSP \Leftrightarrow QPS$. 'All' is reflexive; 'some' and 'no' are symmetric. (Aristotle expressed the same thought by saying that the latter convert simply.) Keenan [1987] generalizes symmetry to a property he identifies with weakness, that is, acceptability in 'There is' or 'there are' contexts. Q is *intersective* iff Q is conservative and $QS_1...S_nP \Leftrightarrow Q(S_1 \cap P)...(S_n \cap P)P$. Binary quantifiers are symmetric iff they are intersective [van der Does and van Eijck, 1996, 11].

Barwise and Cooper generalize transitivity to persistence and monotonicity, sometimes called left- and right-monotonicity:

Q is mon$\uparrow$ iff $QSP \wedge P \subseteq P' \Rightarrow QSP'$
Q is mon$\downarrow$ iff $QSP \wedge P' \subseteq P \Rightarrow QSP'$
Q is $\uparrow$mon iff $QSP \wedge S \subseteq S' \Rightarrow QS'P$
Q is $\downarrow$mon iff $QSP \wedge S' \subseteq S \Rightarrow QS'P'$

These properties are responsible for the validity and invalidity of syllogistic inferences. We can summarize the properties of Aristotelian quantifiers straightforwardly:

Every: reflexive, $\downarrow$mon$\uparrow$
Some: intersective, $\uparrow$mon$\uparrow$
No: intersective, $\downarrow$mon$\downarrow$

If we were to associate the left and right arrows with the subject and predicate terms in a categorical proposition, then a downward arrow would correspond to the term's being distributed, and an upward arrow would correspond to its being undistributed. This explains why the medieval concept of distribution could be used to developed rules for syllogistic validity. It also explains why nineteenth century logicians who think that a term must be either distributed or undistributed are making a mistake. Consider the properties of a wider class of quantifiers:

Most: mon$\uparrow$
At least n: intersective, $\uparrow$mon$\uparrow$
At most n: intersective, $\downarrow$mon$\downarrow$
Exactly n: intersective
Finitely many: intersective, $\downarrow$mon$\downarrow$
Uncountably many: intersective, $\uparrow$mon$\uparrow$
All but finitely many: $\downarrow$mon$\uparrow$
All but exactly n:

In 'Most S are P,' S is neither distributed nor undistributed. The same is true of S in 'Exactly one S is P' and 'All but exactly two S are P.' Logicians working in the Aristotelian tradition could easily have expanded their theories to account for determiners such as 'at least n,' 'several,' 'a few,' 'uncountably many' — which would pattern as particular quantifiers — and 'at most n,' 'finitely many,' and 'few' — which would pattern as universal negatives — but did not have the conceptual tools to incorporate nonmonotonic determiners such as 'most' and 'exactly n.'

We have so far been thinking of a quantifier Q on U as an n-ary relation on subsets of U: $Q_U \subseteq (\wp(U))^n$. We can generalize this further to n-ary relations among relations on U. A *local quantifier of type* $\langle k_1, ..., k_n\rangle$ on U is an n-ary relation between subsets of $U^{k_1}, ..., U^{k_n}$: $Q_U \subseteq \wp(U^{k_1}) \times ... \times \wp(U^{k_n})$. A *global quantifier of type* $\langle k_1, ..., k_n\rangle$ is a map from universes U to local quantifier of type $\langle k_1, ..., k_n\rangle$ on U. This generalization permits us to introduce a *Henkin* or *branching quantifier* H of type $\langle 4\rangle$, such that $H = \{R \subseteq U^4 : \exists f, g \subseteq U^2\ \forall x, z \in U\langle x, f(x), z, g(z)\rangle \in R\}$. We can then express a branching quantifier sentence such as 'A leader from every tribe and a captain from every regiment met at the peace conference' as $HxyzuMxyzu$.[40]

The relational conception of quantifiers developed here is equivalent to the conception of Montague [1974] and Barwise and Cooper [1981] according to which a quantifier is a function from subsets of the universe of discourse to sets of subsets, for we may think of a quantifier as mapping a set into the set of all sets to which it relates that set: $Q(S) = \{P : QS\,P\}$. The Montagovian conception, though less intuitive, has the advantage of making it easy to compute semantic values compositionally. Determiners map common nouns into noun phrases. Quantifiers, correspondingly, map sets, the denotations of common nouns, into families of sets, the denotations of noun phrases. We might take a proper noun such as 'Socrates' not as standing for an individual but instead as standing for a family of sets, that is, the family of extensions of predicates that apply to Socrates. Similarly, we might take 'every man' not as standing for the set of all men, as some nineteenth-century logicians were tempted to do, but as standing for a family of sets, the set of extensions of predicates applying to every man.

[40]For a detailed treatment, see [Sher, 1997].

5.3 Quantification and Anaphora

Fourteenth-century logicians such as Burley and Buridan note that quantifiers not only express relations between terms but also introduce items in discourse that can be referred to anaphorically. The problem this raises is not one of truth conditions, as we have seen — it is not difficult to give first-order representations of sentences such as 'an animal is running and it is a man' or 'Every farmer who owns a donkey beats it' — but rather of compositionality. The problem, in other words, is typically not a lack of appropriate truth conditions but a rule-governed way of generating them.

In Aristotelian logic, we can represent 'an animal is running' as 'Some *A* is *R*,' but we then have no way of indicating that the running animal is a man. Appending another categorical proposition will not help. In first-order logic, similarly, we can represent 'an animal is running' as $\exists x(Ax \wedge Rx)$. Having done that, however, we have no way of adding another formula representing that the x in question is a man. What we have to do in both cases is extend the original representation, writing 'Some *A* is *R* and *M*' or $\exists x(Ax \wedge Rx \wedge Mx)$. Since the anaphora occurs within the same sentence, we could perhaps write rules requiring this. But similar anaphora can occur across sentential boundaries. Buridan's example could just as easily have been 'an animal is running. It is a man.' These could moreover be separated by intervening discourse: 'an animal is running. People look on in surprise. It is uncommon to see such a sight in the square, amidst the crowds, in the midday heat. It is a man.' This makes a strategy of waiting to close off the representation of the first sentence until all later anaphors have been collected implausible.

Sometimes, the problem is a lack of appropriate truth conditions. Consider this discourse:

> Mary: "A man fell over the edge!"
> John: "He didn't fall; he jumped."

A quantificational analysis leads to the formula $\exists x$ (x is a man $\wedge$ x fell over the edge $\wedge$ x didn't fall $\wedge$ x jumped) for the discourse, which is contradictory. This seems appropriate enough, for the second assertion contradicts the first. But we have no way of telling whether the second assertion, considered alone, is true or false, for it receives no independent truth conditions. Representing just John's assertion would yield $\exists x$ (x didn't fall $\wedge$ x jumped), which is true if anything jumped and didn't fall. Nothing there ties the assertion to the person spoken of by Mary.

The problem posed by so called donkey sentences is thus quite general. Quantified noun phrases not only relate sets or relations on a universe of discourse but also make subsequent anaphoric connection to certain things or sets possible.

Lauri Karttunen [1976] hypothesizes that quantified noun phrases introduce *discourse referents*, items to which later anaphoric elements can link. Hans Kamp [1981] and Irene Heim [1982] develop that hypothesis into a theory, which has become known as Discourse Representation Theory (DRT). An appearance of a noun phrase (hereafter, NP) that is indefinite, that is, of the form *a(n) F*, establishes a discourse referent if and only if it justifies the occurrence of a coreferential pronoun or definite NP later in the text. For example, in example (1), the indefinite NP *a man* establishes a discourse referent. It justifies the coreferential pronoun *he* in the second sentence, in the sense that the second sentence cannot occur without the first in the absence of something else that would justify the occurrence of such a pronoun — a linguistic context providing other possibilities of anaphoric linkage, for example, or an act of demonstration such as pointing that would make the pronoun deictic rather than anaphoric.

Karttunen's notion, unlike most traditional syntactic and semantic concepts, is multi-sentential. It applies readily to anaphora across sentential boundaries. It is also procedural. Karttunen defines not what a discourse referent is but what it takes to establish one. He thus analyzes indefinite NPs not in terms of what they stand for but in terms of what they do. Dynamic semantics extends this approach to language in general. It sees a sentence as a way of transforming one context into another.

NPs such as 'an animal' do not, however, always establish discourse referents in Karttunen's sense of licensing further anaphora. In particular, when they occur within the scope of quantifiers or negations, they do not permit further anaphoric links:

> Every farmer owns a donkey. *I feed it sometimes.
> It is not true that an animal is running. *It's a man.

An adequate theory based on Karttunen's approach, then, should explain how and when NPs license anaphora. Discourse Representation Theory (DRT) does this in two steps. First, it analyzes indefinites and other NPs with ↑mon↑ determiners — those fitting the pattern of Aristotle's particular quantifiers — as introducing discourse referents in every case. Only sometimes, however, are those referents accessible to later anaphors. Second, it specifies formally a relation of accessibility that determines when anaphoric connections are possible. That relation depends crucially on the Discourse Representation Structure built from the discourse.

Kamp [1981] presents the essentials of Discourse Representation Theory. The general form of the theory is to process a discourse in two steps. First, it parses a sentence syntactically and applies an algorithm to construct a Discourse Representation Structure (DRS). Second, DRSs receive truth conditions by way of a model-theoretic semantics.

The intermediate level of DRSs, or semantic representations, is the key to the theory's novelty. The DRS represents both context and content. Think of prior discourse, for example, as having built up an initial DRS. Then a sentence, processed by the DRS construction algorithm, transforms that initial DRS, acting as content of the previous discourse and context of the current sentence, into a new DRS, representing the content of the entire discourse and the context of further utterances. At each stage of the construction, the DRS determines the truth conditions of the discourse up to that point. But it also constitutes a conceptual level that provides information that cannot be recovered from truth conditions alone.

In short, then, Discourse Representation Theory treats indefinites and other NPs with similar logical properties as referring expressions at the level of the DRS construction algorithm and as quantifiers at the level of the truth definition. Pronouns do not introduce independent elements into a DRS; they refer to items already there. They select their referents from sets of antecedently available entities. Deictic pronouns do so from the real world; anaphoric pronouns select from constituents of the representation — Karttunen's discourse referents. The theory tries to specify sets of referential candidates, by specifying which entities are accessible to a given anaphor. Strategies for selecting referents from among the set of possible referents are complex, relying on semantic, pragmatic, and discourse factors. The theory itself does not spell out these strategies, but is easily supplemented with them.

The DRS construction algorithm provides rules for building or altering DRSs, given the syntactic parse of a sentence — specifically, an analyzed surface structure. To see how it works, let's begin with 'An animal is running.' I will ignore tense, aspect, and other complications; the following DRS would be a component of a richer structure incorporating this additional information.

A DRS $m = \langle U, C \rangle$ consists of a *universe* or *domain* and a *set of conditions*. The universe is a set of entities to be thought of as discourse referents; the conditions provide information about those referents. (For convenience, I shall write a DRS $\langle \{x_1, ..., x_n\}, \{C_1, ..., C_m\} \rangle$ as $\langle x_1, ..., x_n : C_1, ..., C_m \rangle$.) One can thus think of a DRS as an information state or as a partial model, a representation of a situation or part of a world.

Indefinite NPs introduce discourse referents into the universe. They also introduce conditions — entries providing information about discourse referents — expressing the content of the description. 'An animal,' then, introduces a discourse referent and a condition saying that it is an animal. Names and personal pronouns also introduce discourse referents and, often, additional conditions.

Finally, verbs and adjectives introduce conditions. Applying the construction algorithm to *An animal is running*, in the null context, produces the following DRS: $\langle u$: animal(u), u is running$\rangle$. This DRS consists of a domain $\{u\}$ and a set of conditions: {animal(u), u is running}, which express or simply are properties and relations among objects in that domain. It represents the content of 'An animal is running' and acts as a context for 'It's a man.' The anaphoric pronoun*it* must refer to a discourse referent already introduced. We might, as soon as we reach it in the construction algorithm, search for that referent and use it in the conditions to be introduced. To separate the problem of searching among the possible referents for the actual one, however, it is more convenient to introduce another discourse referent, together with a condition identifying it with a previously introduced referent. So, we obtain a DRS of this form: $\langle u, v$: animal(u), u is running, man(v), $v = ?\rangle$. What are the possible referents of v? The theory states that, in the absence of an act of demonstration, all candidates are discourse referents previously introduced. Assuming the earlier DRS to be the entire context for the utterance, nothing but u is available. Identifying v with u then yields $\langle u, v$: animal(u), u is running, man(v), $v = u\rangle$.

We have now set up enough machinery in Discourse Representation Theory to handle simple cases of anaphoric connection. The theory correctly predicts the anaphoric properties of the sentences in Buridan's very simple discourse. After processing the first sentence, we obtain the first DRS above; after processing the second, we obtain the second DRS, which embodies the information conveyed by both sentences.

This situation is quite general. The construction algorithm operates on sentence 1 in context 0, producing DRS 1. It then takes DRS 1 as context in processing sentence 2, yielding DRS 2, and so on. At each stage, the DRS produced embodies the information in all sentences processed up to that stage. Indeed, it must; otherwise it could not serve as context for the next sentence. The meaning of each sentence is not a truth condition that can be stated independently of previous discourse but a function taking discourse contexts into discourse contexts.

Discourse Representation Theory provides a semantics for DRSs and, thereby, for sentences and discourses. A discourse is true if and only if the DRS built from it by the construction algorithm is true. So, primarily, we will speak of DRSs as having truth values.

A DRS is a partial model. It is true in a model $\mathfrak{M}$ if and only if it is a part of $\mathfrak{M}$; there must be a way of embedding the DRS into $\mathfrak{M}$. More formally, a DRS m is true in a model $\mathfrak{M}$ if and only if there is a homomorphic embedding of m into $\mathfrak{M}$. This means that there must be a function from the universe of m into that of $\mathfrak{M}$ preserving the properties and relations m specifies. More formally still, there must be a function f: $U_m \rightarrow U_{\mathfrak{M}}$ such that, if $\langle a_1,, a_n\rangle \in F_m(R)$, then $\langle f(a_1),, f(a_n)\rangle \in F_{\mathfrak{M}}(R)$.

To see how the theory accounts for the quantificational force of indefinites, consider the DRS above, generated by the construction algorithm from the sentence 'An animal is running.' This DRS consists of a domain$\{u\}$ and a set of conditions {animal(u), u is running}. The DRS is true in a model $\mathfrak{M}$, according to our definition, if and only if it can be embedded into $\mathfrak{M}$. This means that there must be a function f from $\{u\}$ into $U_{\mathfrak{M}}$ such that $f(u) \in F_{\mathfrak{M}}$(animal) and $f(u) \in F_{\mathfrak{M}}$(is running). Thus, the DRS, and the sentence that generated it, are true if and only if some animal is running, exactly as we would expect.

The DRS the construction algorithm generates from the entire discourse is true in a model $\mathfrak{M}$ if and only if there is a function f from $\{u, v\}$ into $U_{\mathfrak{M}}$ such that $f(u) \in F_{\mathfrak{M}}$(animal), $f(u) \in F_{\mathfrak{M}}$(is running), $f(v) \in F_{\mathfrak{M}}$(man), and $f(u) = f(v)$. The DRS is true, in other words, if there is an animal that is running and is also a man.

Note that the indefinite *an animal* has quantificational force here, in the sense that the truth conditions for the sentence might appropriately be represented by an existentially quantified formula, even though the indefinite did not introduce a quantifier into the DRS. It introduced nothing but a discourse referent and a condition. In Discourse Representation Theory, indefinite descriptions are referential terms, not existential quantifiers. There is nevertheless no simple answer to the question of what they denote. Their contribution to truth conditions depends on the role played by the clause containing the description, which depends, in turn, on the structure of the DRS. The same is true of any NP with a $\uparrow$mon$\uparrow$ determiner. The theory accounts not only for the quantificational force of such NPs but also for their frequent success in establishing discourse referents, licensing further anaphoric connections.

Discourse Representation Theory thus explains the quantificational and anaphoric characteristics of the indefinite *an animal* at different levels of the theory. The indefinite introduces a discourse referent at the conceptual level of the DRS, which is then accessible to later pronouns. And the truth definition, specifying that the resulting DRS is true if and only if there is a way of embedding it in a model, determines that the discourse is true if and only if there is an animal, specifically a man, who is running.

Other kinds of determiners receive a different treatment. Let's turn to Burley's donkey sentence, 'Every farmer who owns a donkey beats it.' That is equivalent to 'If a farmer owns a donkey, he beats it.' We already know how to understand the antecedent; processing it yields the DRS $\langle u, v$: farmer(u), donkey(v), u owns $v\rangle$. A DRS for a conditional has the form $m \Rightarrow m'$, where m and m' are DRSs for the antecedent and consequent. So, the DRS for this sentence becomes $\langle u, v$: farmer(u), donkey(v), u owns $v\rangle \Rightarrow \langle w, x : w$ beats x, $w = u$, $x = v\rangle$. This is the general strategy for 'every,' which introduces a conditional structure. Notice that the phrase 'farmer who owns a donkey' now corresponds to an identifiable part of the semantic representation; there is no reason to adhere to the Fregean principle that only in the context of a sentence does an expression have meaning, though it is of course true that only in the context of a discourse does a sentence have specific truth conditions.

The semantic condition for conditionals makes it clear why 'every' lives on its subject term. A conditional $m \Rightarrow m'$ is true in a model $\mathfrak{M}$ if and only if every embedding of m into $\mathfrak{M}$ extends to an embedding of m' into $\mathfrak{M}$. This implies that the donkey sentence is true in $\mathfrak{M}$ if and only if every submodel of $\mathfrak{M}$ in which a farmer owns a donkey extends to one in which that farmer beats that donkey. But that is just to say that every farmer-owns-donkey pair in $\mathfrak{M}$ is also a farmer-beats-donkey pair, just as we would expect.

Discourse Representation Theory, by adopting a dynamic strategy in which the meaning of a sentence is a function from discourse contexts to discourse contexts — or, viewed differently, a function from representations to representations, or, from still another point of view, from partial models to partial models — respects surface structure, derives truth conditions compositionally in rule-governed ways, explains anaphoric connections within sentences and across sentence boundaries, and assigns appropriate truth conditions. It does so, however, by treating NPs with different determiners differently. Some, such as 'a(n)' and 'some,' introduce discourse referents; their quantificational force arises from the semantics, but receives no direct representation. Others, such as 'every,' introduce conditionals but also receive no direct representation. Yet others, such as 'no,' 'never,' and so on, introduce negations. And some, such as 'many,' 'at least n,' and 'uncountably many,' introduce plural discourse referents and conditions on them.

The theory of generalized quantifiers and Discourse Representation Theory thus pull in contrary directions. The former seeks a highly abstract theory encompassing all quantifiers and giving them a unified treatment. Discourse Representation Theory treats quantifiers of different kinds as performing very different kinds of tasks. It explains anaphoric behavior that the theory of generalized quantifiers does not address. In the process, however, it views many expressions traditionally viewed as quantificational as doing something else, and having quantificational force only by virtue of the truth conditions governing representations that themselves include nothing explicitly quantificational. Attempts to combine these approaches are underway — see, for example, [Barwise, 1987; Kamp and Reyle, 1993; Muskens, 1996; van Eijck and Kamp, 1997; Kamp *et al.*, 2011] — but remain in their infancy.

The task facing contemporary logicians, then, is to give a fully general and correct of the truth conditions of quantified sentences that explains their anaphoric behavior. It may seem discouraging that we face a situation startlingly like that facing Burley and Buridan in the middle of the fourteenth century — motivated, in fact, by many of the same linguistic examples. But we have, at least, a much richer set of tools with which to attempt the task.

BIBLIOGRAPHY

[Abrreviatio Montana, 1988] In N. Kretzmann and E. Stump (ed.), *The Cambridge Translations of Medieval Philosophical Texts, Volume One, Logic and the Philosophy of Language*. Cambridge: Cambridge University Press, 1988, 40–78.

[Abbreviation Montana, 1967] *Abbreviation Montana*. In De Rijk 1967, 73–107.

[Abelard, 1956] Abelard, Peter, 1956. *Dialectica*. L.M. de Rijk (ed.), Assen: Van Gorcum.

[Alexander of Aphrodisias, 1883] Alexander of Aphrodisias, 1883. *Alexandri in Aristotelis Analyticorum Priorum librum I commentarium*, M. Wallies (ed.). Berolini: G.Reimeri.

[Ariel, 2003] Ariel, M., 2003. "Does *most* mean 'more than half'?" *Berkeley Linguistics Society* 29: 17–30.

[Ariel, 2004] Ariel, M., 2004. "Most," *Language* 80: 658–706.

[Ariel, 2006] Ariel, M., 2006. "A 'just that' lexical meaning for most," in Klaus von Heusinger and Ken Turner (ed.), *Where Semantics Meets Pragmatics*. Amsterdam: Elsevier, 49–91.

[Aristotle, 1984] Aristotle. *De Interpretatione*. Translated by J. L. Ackrill. In J. Barnes (ed.), *The Complete Works of Aristotle*, Volume I, 25–38. Princeton: Princeton University Press, 1984.

[Aristotle, 1984a] Aristotle. *Posterior Analytics*. Translated by J. Barnes. In J. Barnes (ed.), *The Complete Works of Aristotle*, Volume I, 114–166. Princeton: Princeton University Press, 1984.

[Aristotle, 1984b] Aristotle. *Prior Analytics*. Translated by A. J. Jenkinson. In J. Barnes (ed.), *The Complete Works of Aristotle*, Volume I, 39–113. Princeton: Princeton University Press, 1984.

[Aristotle, 1984c] Aristotle. *Sophistical Refutations*. Translated by W. A. Pickard-Cambridge. In J. Barnes (ed.), *The Complete Works of Aristotle*, Volume I, 278–314. Princeton: Princeton University Press, 1984.

[Arnauld and Nicole, 1662] Arnauld, Antoine, and Nicole, Pierre, 1662. *La Logique: Ou, L'art de Penser (Logique de Port-Royal)*. Paris: Charles Savreaux.

[Arnauld and Nicole, 1861] Arnauld, Antoine, and Nicole, Pierre, 1861. *Logic, Or, The Art of Thinking (The Port-Royal Logic)*. Translated by T. S. Baynes. Edinburgh: James Gordon.

[Ars Burana, 1967] *Ars Burana*. In De Rijk 1967, 175–213.

[Ars Emmerana, 1967] *Ars Emmerana*. In De Rijk 1967, 143–174.

[Barnes, 1981] Barnes, Jonathan, 1981. "Proof and the Syllogism," in E. Berti, *Aristotle on Science: the 'Posterior Analytics'*. Padua: Antenore, 17–59.

[Barwise, 1979] Barwise, J. 1979. "On branching quantifiers in English," *Journal of Philosophical Logic* 8: 47–80.

[Barwise and Cooper, 1982] Barwise, J. and Cooper, R., 1981. "Generalized quantifiers and natural language," *Linguistics and Philosophy* 4: 159–219.

[Barwise and Feferman, 1985] Barwise, J. and Feferman, S. (eds.), 1985. *Model-Theoretic Logics* Berlin: Springer-Verlag.

[Barwise, 1987] Barwise, J., 1987."Noun phrases, generalized quantifiers and anaphora," in P. Gärdenfors (ed.), *Generalized Quantifiers: Linguistic and Logical Approaches*. Dordrecht: Reidel, 1–30.

[Bentham, 1827] Bentham, G., 1827. *Outline of a New System of Logic*. London: Hunt and Clarke.

[Bochenski, 1961] Bochenski, I. M., 1961. *A History of Formal Logic*. Notre Dame: Notre Dame Press.

[Boger, 2004] Boger, George, 2004 . "Aristotle's Underlying Logic," In *Handbook of the History of Logic* 1, D. Gabbay and J. Woods (ed.). Amsterdam: Elsevier, 101–246.

[Boole, 1847] Boole, G., 1847. *The Mathematical Analysis of Logic, being an Essay towards a Calculus of Deductive Reasoning*. Cambridge: Macmillan, Barclay, and Macmillan.

[Boole, 1848] Boole, G., 1848. "The Calculus of Logic," *Cambridge and Dublin Mathematical Journal*, 3: 183–198.

[Boole, 1854] Boole, G., 1854. *An Investigation of the Laws of Thought, on Which are Founded the Mathematical Theories of Logic and Probabilities*. London: Walton and Maberly.

[Buriden, 1976] Buridan, John, 1976. *Tractatus de Consequentiis*. H. Hubien (ed.), *Philosophes mèdièvaux* 16, Louvain: Publications Universitaires.

[Buridan, 2001] Buridan, John, 2001. *Summulae de Dialectica*. An annotated translation with a philosophical introduction by Gyula Klima. New Haven: Yale University Press.

[Carlson, 1977] Carlson, G., 1977. *Reference to Kinds in English*. PhD Dissertation, University of Massachusetts at Amherst

[Carroll, 1896] Carroll, Lewis, 1896. *Symbolic Logic*. London: Macmillan.

[Carroll, 1977] Carroll, Lewis, 1977. *Lewis Carroll's Symbolic Logic*. New York: Potter.

[Church, 1936] Church, A., 1936. "A Note on the *Entscheidungsproblem*," *Journal of Symbolic Logic* 1: 40–41.

[Corcoran, 1974] Corcoran, J., 1974, "Aristotle's Natural Deduction System," in J. Corcoran (ed.), *Ancient Logic and its Modern Interpretations*. Dordrecht: Reidel.

[Corcoran, 1972] Corcoran, John, 1972. "Completeness of an Ancient Logic," *Journal of Symbolic Logic* 37: 696–705.

[Corcoran, 1973] Corcoran, John, 1973. "A Mathematical Model of Aristotle's Syllogistic," *Archiv für Geschichte der Philosophie* 55: 191–219.

[Davidson, 1967] Davidson, D., 1967. "Truth and Meaning," *Synthese*, 17: 304–323.

[de Rijk, 1967] de Rijk, L. M., 1967. *Logica Modernorum. A Contribution to the History of Early Terminist Logic. Vol. II*. Assen: Van Gorcum.

[de Rijk, 1968] de Rijk, L. M., 1968. "On the Genuine Text of Peter of Spain's *Summule logicales* I. General problems concerning possible interpolations in the manuscripts," *Vivarium* 6: 1–34.

[de Rijk, 1982] de Rijk, L. M., 1982. "The Origins of the Theory of the Properties of Terms," *The Cambridge History of Later Medieval Philosophy*, edited by Norman Kretzmann, Anthony Kenny and Jan Pinborg. Cambridge: Cambridge University Press, 161–173

[De Morgan, 1847] DeMorgan, A., 1847. *Formal Logic or The Calculus of Inference*. London: Taylor and Walton.

[De Morgan, 1860] DeMorgan, A., 1860. *Syllabus of a Proposed System of Logic*. London: Walton and Malbery.

[Dutilh-Novaes, 2008] Dutilh-Novaes, C., 2008, "Logic in the 14th Century after Ockham," in D. Gabbay and J. Woods (eds.), *Handbook of the History of Logic, vol. 2, Medieval and Renaissance Logic*. Amsterdam: North Holland.

[Frege, 1879] Frege, G., 1879. *Begriffsschrift*. Reprinted in van Heijenoort 1967, 1–82.

[Frege, 1882] Frege, G., 1882. "Über die wissenschaftliche Berechtigung einer Begriffsschrift," *Zeitschrift für Philosophie und Philosophische Kritik*, 81: 48–56.

[Frege, 1892] Frege, G., 1892. "On Concept and Object," originally published as "*Über Begriff und Gegenstand*," in *Vierteljahresschrift für wissenschaftliche Philosophie* 16, 192–205. Translated in Geach, P., and Black, M., *Translations from the Philosophical writings of Gottlob Frege*, Oxford: Oxford University Press, 1952.

[Frege, 1893] Frege, G., 1893. *Grundgesetze der Arithmetik*. Jena: Verlag Hermann Pohle.

[Frege, 1902] Frege, G., 1902. "Letter to Russell," in van Heijenoort 1967, 126–128.

[Gödel, 1930] Gödel, K., 1930. "The Completeness of the Axioms of the Functional Calculus of Logic," in van Heijenoort 1967, 582–591.

[Gödel, 1931] Gödel, K., 1931. "On Formally Undecidable Propositions of *Principia Mathematica* and Related Systems I," in van Heijenoort 1967, 596–616.

[Geach, 1962] Geach, P., 1962. *Reference and Generality*. Ithaca: Cornell University Press.

[Goldfarb, 1979] Goldfarb, W., 1979. "Logic in the Twenties: the Nature of the Quantifier," *Journal of Symbolic Logic* 44: 351–368.

[Hackl, 2009] Hackl, M., 2009. "On the grammar and processing of proportional quantifiers: *most* versus *more than half*," *Natural Language Semantics* 17: 63–98.

[Hailperin, 2004] Hailperin, T., 2004. "Algebraical Logic 1685–1900," in D. Gabbay and J. Woods (ed.), *Handbook of the History of Logic*, volume 3. Amstredam: Elsevier, 323–388.

[Hamilton, 1860] Hamilton, William, 1860. *Lectures on Metaphysics and Logic*. Edinburgh and London: William Blackwood and Sons.

[Heim, 1982] Heim, I., 1982. *The Semantics of Definite and Indefinite Noun Phrases*. Ph.D. thesis, University of Massachusetts, Amherst.

[Henkin, 1961] Henkin, L., 1961. "Some remarks on infinitely long formulas," in *Infinitistic Methods*. Oxford: Pergamon Press, 167–183.

[Heytesbury, 1335] Heytesbury, William, 1335. *Regulae Solvendi Sophismata*.Pavia: Antonius de Carcano, 1481.

[Hilpinen, 2004] Hilpinen, R., 2004. "Peirce's Logic," in D. Gabbay and J. Woods (ed.), *Handbook of the History of Logic*, volume 3. Amstredam: Elsevier, 611–658.

[Hintikka, 1979] Hintikka, J., 1979. "Quantifiers vs. quantification theory," *Linguistics and Philosophy* 5: 49–79.

[Introductiones Montana Minores, 1967] *Introductiones Montana Minores*. In De Rijk 1967, 7–71.

[Irwin, 1988] Irwin, Terence, 1988. *Aristotle's First Principles*. Oxford: Clarendon Press.

[Jungius, 1638] Jungius, Joachim, 1638. *Logica Hamburgensis*. R. Meyer (ed.). Hamburg: J. J. Augustin, 1957.

[Jungius, 1977] Jungius, Joachim, 1977. *Logica Hamburgensis Additamenta*. Edited by Wilhelm Risse. Göttingen: Vandenhoeck and Ruprecht.

[John Philoponus, 1905] John Philoponus, 1905. *In Aristotelis Analytica Priora commentaria*. M. Wallies, ed. Berolini: G.Reimeri.

[Johnson, 1994] Johnson, Fred, 1994. "Apodictic Syllogisms: Deductions and Decision Procedures," *History and Philosophy of Logic* 16: 1–18.

[Kamp, 1981] Kamp, H., 1981. "A theory of truth and semantic representation," in J. A. G. Groenendijk, T. M. V. Janssen, and M. B. J. Stokhof (eds.), *Formal Methods in the Study of Language*. Amsterdam: Mathematical Centre Tracts 135: 277–322.

[Kamp and Reyle, 1993] Kamp, H., and Reyle, U., 1993. *From Discourse to Logic*. Dordrecht: Kluwer.

[Kamp *et al.*, 2011] Kamp, H., Genabith J., and U. Reyle, 2011. "Discourse Representation Theory," in D. Gabbay (ed.), *Handbook of Philosophical Logic* (second edition), Springer.

[Karttunen, 1976] Karttunen, L., 1976. "Discourse Referents," in J. D. McCawley (ed.), *Syntax and Semantics 7: Notes from the Linguistic Underground*. New York: Academic Press, 363–385.

[Keenan and Moss, 1984] Keenan, Edward L., and Moss, L., 1984. "Generalized quantifiers and the expressive power of natural language," in J. van Benthem and A. ter Meulen (eds.), *Generalized quantifiers in Natural Language*. Dordrecht: Foris, 73–124.

[Keenan and Stavi, 1986] Keenan, Edward L., and Stavi, J., 1986. "A Semantic Characterization of Natural Language Determiners," *Linguistics and Philosophy* 9: 253–326.

[Keenan, 1987] Keenan, E., 1987. "A semantic definition of 'Indefinite NP'," in Eric Reuland and Alice ter Meulen (ed.), *The Representation of (In)definiteness*. Cambridge: MIT Press, 286–317.

[Keenan, 1996] Keenan, Edward L., 1996. "The Semantics of Determiners," in Shalom Lappin (ed.), *The Handbook of Contemporary Semantic Theory*. Oxford: Blackwell.

[Keenan and Westerståhl, 1997] Keenan, Edward L., and Westerståhl, D., 1997. "Generalized Quantifiers in Linguistics and Logic," in *Handbook of Logic and Language*. Amsterdam: Elsevier, 837–893.

[Keenan and Westerståhl, 2011] Keenan, E., and Westerståhl, D., 2011. "Generalized quantifiers in Linguistics and Logic (revised version)," in J. van Benthem and A. ter Meulen (eds.), *Handbook of Logic and Language*. Second Edition, Amsterdam: Elsevier, 859–910.

[King, 1985] King, P., 1985. *Jean Buridan's Logic. The Treatise on Supposition. The Treatise on Consequences*. Translation with introduction and notes. Synthese Historical Library 27, Dordrecht: Reidel.

[Kneale and Kneale, 1962] Kneale, William and Kneale, Martha, 1962. *The Development of Logic* Oxford: Oxford University Press

[Kretzmann, 1966] Kretzmann, Norman, 1966. *William of Sherwood's Logic*. Minneapolis: University of Minnesota Press.

[Lear, 1980] Lear, Jonathan, 1980. *Aristotle and Logical Theory*. Cambridge: Cambridge University Press.

[Lewis, 1975] Lewis, D., 1975. "Adverbs of quantification," in E. Keenan (ed.), *Formal Semantics of Natural Language*. Cambridge: Cambridge University Press, 3–15.

[Lindström, 1966] Lindström, P., 1966. "First Order Predicate Logic with Generalized Quantifiers," *Theoria* 32: 186–195.

[Lindström, 1969] Lindström, P., 1969. "On Extensions of Elementary Logic," *Theoria* 35: 1–11.

[Lindström, 1974] Lindström, P., 1974. "On Characterizing Elementary Logic," in S. Stenlund (ed.),*Logical theory and semantic analysis*. Dordrecht: D. Reidel, 129–146.

[Löwenheim, 1915] Löwenheim, L., 1915. "On Possibilities in the Calculus of Relatives," in van Heijenoort 1967, 228–251.

[Lukasiewicz, 1934] Lukasiewicz, J., 1934. "Outlines of the History of the Propositional Logic," *Pr. Fil.* 37.

[Lukasiewicz, 1951] Lukasiewicz, Jan, 1951. *Aristotle's Syllogistic from the Standpoint of Modern Formal Logic*. Oxford: Clarendon Press.

[Martin, 2009] Martin, C., 2009, "The logical text-books and their influence," in J. Marenbon (ed.), *The Cambridge Companion to Boethius*. Cambridge: Cambridge University Press.

[McKirahan, 1992] McKirahan, Richard, 1992. *Principles and Proofs*. Princeton: Princeton University Press.

[Montague, 1973] Montague, R., 1973. "The Proper Treatment of Quantification in Ordinary English," in J. Hintikka, J. Moravcsik, and P. Suppes (ed.), *Approaches to Natural Language*. Dordrecht: D. Reidel, 1973, 221–242. Reprinted in Montague 1974, 247–270.

[Montague, 1974] Montague, R., 1974. *Formal Philosophy* (edited and with an introduction by R. Thomason). Yale University Press, New Haven.

[Mostowski, 1957] Mostowski, A., 1957. "On a generalization of quantifiers," *Fundamenta Mathematicae*, 44: 12–36.

[Muskens, 1996] Muskens, R., 1996. "Combining Montague Semantics and Discourse Representation," *Linguistics and Philosophy* 19: 143–186.

[Normore, 1999] Normore, C., 1999, "Some Aspects of Ockham's Logic," in P. V. Spade (ed.), *The Cambridge Companion to Ockham*. Cambridge: Cambridge University Press

[Ockham, 1974] Ockham, William of, 1974. *Summa Logicae*. Ph. Boehner, G. Gál and S. Brown (eds.), *Opera Philosophica* I, St. Bonaventure, NY: The Franciscan Institute.

[Ockham, 1974a] Ockham, Willliam of, 1974. *Ockham's Theory of Terms: Part I of the Summa Logicae*. Notre Dame: Notre Dame Press.

[Ockham, 1980] Ockham, Willliam of, 1980. *Ockham's Theory of Propositions: Part II of the Summa Logicae*. Notre Dame: Notre Dame Press.

[Patzig, 1969] Patzig, Günther, 1969. *Aristotle's Theory of the Syllogism*. Translation, Jonathan Barnes. Dordrecht: D. Reidel.

[Peano, 1889] Peano, G., 1889. "The Principles of Arithmetic, Presented by a New Method," in van Heijenoort 1967, 83–97.

[Peirce, 1885] Peirce, C. S., 1885, "On the Algebra of Logic: A Contribution to the Philosophy of Notation," *American Journal of Mathematics* 7 (1885), 180–196.

[Peirce, 1893] Peirce, C. S., 1893, *Elements of Logic, Collected Papers of Charles Sanders Peirce*, Volume II, edited by C. Hartshorne and P. Weiss, Cambridge: Belknap Press, 1960.

[Peter of Spain, 1992] Peter of Spain. *Syncategoreumata*. First Critical Edition with an Introduction and Indexes by L.M. de Rijk, with an English Translation by Joke Spruyt. Leiden/Küln/New York, 1992.

[Peter of Spain, 1972] Peter of Spain. *Tractatus called afterwards Summule logicales*. First Critical Edition from the Manuscripts with an Introduction by L.M. de Rijk. Assen, 1972.

[Peters and Westerståhl, 2006] Peters, S. and Westerståhl, D., 2006. *Quantifiers in Language and Logic*. Oxford: Oxford University Press.

[Pietroski *et al.*, 2009] Pietroski, P., Lidz, J., Hunter, T., and Halberda, J., 2009. "The meaning of 'most': Semantics, numerosity and psychology," *Mind and Language* 24: 554–585.

[Quine, 1939] Quine, W. V. O., 1939. "A Logistical Approach to the Ontological Problem," Fifth International Congress for the Unity of Science, Cambridge, Mass., September 9, 1939. Reprinted in *The Ways of Paradox*. Cambridge: Harvard University Press, 1966.

[Quine, 1956] Quine, W. V. O., 1956. "Quantifiers and Propositional Attitudes," *The Journal of Philosophy* 53: 177–187. Reprinted in *The Ways of Paradox*. Cambridge: Harvard University Press.

[Quine, 1960] Quine, W. V. O., 1960. *Word and Object*. Cambridge: MIT Press.

[Rescher, 1964] Rescher, Nicholas, 1964. "Plurality Quantification," *The Journal of Symbolic Logic* 27 (1964): 373–74.

[Ross, 1951] Ross, W. D. (ed.), 1951. *Aristotle's Prior and Posterior Analytics*. Oxford: Clarendon Press.

[Russell, 1902] Russell, B., 1902. "Letter to Frege," in van Heijenoort 1967, 124–125.

[Sher, 1997] Sher, G. 1997. "Partially-ordered (branching) generalized quantifiers: a general definition," *Journal of Philosophical Logic* 26: 1–43.

[Skolem, 1920] Skolem, T., 1920. "Logico-combinatorial Investigations in the Satisfiability or Provability of Mathematical Propositions: A Simplified Proof of a Theorem by L. Löwenheim and Generalizations of the Theorem," in van Heijenoort 1967, 252-263.

[Skolem, 1928] Skolem, T., 1928. "On Mathematical Logic," in van Heijenoort 1967, 508–524.

[Smiley, 1974] Smiley, Timothy, 1974. "What Is a Syllogism?" *Journal of Philosophical Logic* 1: 136–154.

[Smiley, 1994] Smiley, Timothy, 1994. "Aristotle's Completeness Proof," *Ancient Philosophy* 14: 25–38.

[Smith, 1989] Smith, Robin, 1989. *Aristotle's Prior Analytics*. Indianapolis: Hackett.

[Spade, 2002] Spade, P. V., 2002. *Thoughts, Words, and Things: An Introduction to Medieval Logic and Semantic Theory*.

[Striker, 2009] Striker, Gisela (tr.), 2009. *Aristotle, Prior Analytics I*. Translated with commentary by Gisela Striker. Oxford: Clarendon Press.

[Tarski, 1936] Tarski, A., 1936. "Über den Begriff der logischen Folgerung," *Actes du Congrs International de philosophie scientifique*, Sorbonne, Paris 1935, vol. VII, Logique, Paris. Hermann, 1–11. Reprinted as "The Concept of Truth in Formalized Languages" in *Logic, Semantics, Metamathematics: Papers from 1923 to 1938*. Edited and translated by J. H. Woodger. Oxford: Oxford University Press, 1956.

[Tarski and Vaught, 1956] Tarski, A., and Vaught, R., 1956. "Arithmetical Extensions of Relational Systems," *Compositio Mathematica* 13: 81–102.

[Tractatus Anagnini, 1967] *Tractatus Anagnini*. In De Rijk 1967, 215–332.

[Valencia, 2004] Valencia, V. S., 2004. "The Algebra of Logic," in D. Gabbay and J. Woods (ed.), *Handbook of the History of Logic*, volume 3. Amsterdam: Elsevier, 389–544.

[van Benthem, 1983] van Benthem, J., 1983. "Determiners and Logic," *Linguistics and Philosophy* 6: 447–478.

[van Benthem, 1984] van Benthem, J., 1984. " Questions about Quantifiers," *Journal of Symbolic Logic* 49: 443–466.

[van Benthem, 1986] van Benthem, J., 1986. *Essays in Logical Semantics*. Dordrecht: D. Reidel.

[van Benthem, 1987] van Benthem, J., 1987. "Towards a computational semantics," in P. GŁrdenfors (ed.), *Generalized Quantifiers*, Dordrecht: D. Reidel, 31–71.

[van Benthem, 1989] van Benthem, J., 1989. "Polyadic Quantifiers," *Linguistics and Philosophy* 12: 437–464.

[van der Does and van Eijk, 1996] van der Does, J., and van Eijck, J., 1996. "Basic Quantifier Theory," in J. van der Does and J. van Eijck (ed.), *Quantifiers, Logic, and Language*. Stanford: CSLI.

[van Eijck and Kamp, 1997] van Eijck, J., and Kamp, H., 1997. "Representing discourse in context," in J. van Benthem and A. ter Meulen (eds.), *Handbook of Logic and Language*. Amsterdam: Elsevier, 179–237.

[van Heijenoort, 1967] van Heijenoort, J., 1967. *From Frege to Gödel: A Source Book in Mathematical Logic 1879–1931*. Cambridge: Harvard University Press.

[Wallace, 1970] Wallace, J., 1970. "On the Frame of Reference," *Synthese* 22: 117–150.

[Watts, 1725] Watts, Isaac, 1725. *Logic, or The Right Use of Reason in the Enquiry After Truth With a Variety of Rules to Guard Against Error in the Affairs of Religion and Human Life, as well as in the Sciences*. Edinburgh: Printed for Charles Elliot.

[Wedin, 1990] Wedin, Michael V., 1990. "Negation and Quantification in Aristotle," *History and Philosophy of Logic* 11: 131–150.

[Westerståhl, 1987] Westerståhl, D., 1987. "Branching generalized quantifiers and natural language," in P. Gärdenfors (ed.), *Generalized Quantifiers*. Dordrecht: D. Reidel, 269–298.

[Westerståhl, 1989] Westerståhl, D., 1989, "Quantifiers in Formal and Natural Languages," in D. Gabbay and F. Guenthner (eds), *Handbook of Philosophical Logic*, Vol. IV, Dordrecht: D. Reidel, 1–131. 2nd edition, 2007, Berlin: Springer, 223–338.

[Whately, 1826] Whately, Richard, 1826. *The Elements of Logic*. London: Longmans.

[Wilson, 1960] Wilson, C., 1960. *William Heytesbury: Medieval Logic and the Rise of Mathematical Physics*. Madison: University of Wisconsin Press.

[Woods and Irvine, 2004] Woods, J., and Irvine, A., 2004. "Aristotle's Early Logic," in D. Gabbay and J. Woods (ed.), *Handbook of the History of Logic*, volume 1. Amsterdam: Elsevier, 27–100.

HISTORY OF NEGATION

J. L. Speranza and Laurence R. Horn

"I wyl not deny my Greecian ofspring."
Stanyhhurst, *Æneis* II. (Arb.) 1583: 46.

INTRODUCTION: GRICE AS A CATALYST

The American Heritage Dictionary's entry for Grice identifies him as a 'British logician', which for the purposes of this contribution is what he was. (The entry goes on to acknowledge that he is "best known for his studies of the pragmatics of communication and his theory of conversational maxims".) We shall take Grice as a catalyst, since he represents a breakthrough in a rivalry between two groups of philosophers in the history of logic. We hope to demonstrate that he was more of a logician than the history of logic typically recognises.

In choosing Grice as a catalyst and foundation stone, we open with a discussion of Formalism (or Modernism). This we present as giving a "System" for the logic of negation – notably with a syntactic and a semantic component. In the second part, we briefly discuss Neo-Traditionalism (or Informalism) which Grice saw as presenting a challenge to Formalism. We propose, with Grice, that most of the observations made by the Informalists pertain to the pragmatic component of the System, and characterise pragmatic rather than logical inference.

We will try to show that our choice of Grice as a catalyst is general enough to provide a basis for the History of Logic and the treatment of one of its central concepts. We center on the ideas of Grice as an example of a logical treatment of negation, but also as a memorial to a specific chapter in logical historiography. Our focus will be one particular logical feature of negation as it has been conceived in the history of logic as a 'unary' truth-functor. We shall now formulate the Modernist-Neotraditionalist debate.

We take as starting point Grice's opening passage in his epoch-making 'Logic and Conversation' (the second William James lecture), where negation is first cited:

> It is a commonplace of philosophical logic that there are, or appear to be, divergences in meaning between, on the one hand, at least some of what I shall call the formal devices — $\sim, \&, \vee, \supset, (\forall x), (\exists x), (\iota x)$ (when these are given a standard two-valued interpretation) — and, on the other, what are taken to be their analogues or counterparts in natural language – such expressions as 'not', 'and', 'or', 'if', 'all', 'some' (or 'at least one'), 'the'. [Grice, 1989:22]

Handbook of the History of Logic. Volume 11: Logic: A History of its Central Concepts.
Volume editors: Dov M. Gabbay, Francis Jeffry Pelletier and John Woods.
General editors: Dov M. Gabbay and John Woods.

A close reading of the passage will shed light on much of the time-honored debate in the history of logic. Note for example that the contrast Grice makes is between a 'formal device' ("$\sim$") and its vulgar counterpart ("not").

The underlying assumption – which Grice states in the second passage is shared by Modernism, Neo-Traditionalism, and Grice's own Post-Modernism alike — seems to be that there *is* a formal counterpart (in our case, "$\sim$", the squiggle) to a vernacular expression ("not"). This assumption is notably not challenged by Grice's Post-Modernism. He may be a skilful heretic, but his heresy was still of the conservative kind, and he 'can always be relied to upon to rally to the defense of an 'under-dogma"' [Grice, 1989:297]. In this case, the 'under-dogma' is the doctrine concerning the identity between "$\sim$" and "not" ("Some logicians may at some time have wanted to claim that there are in fact no such divergences; but such claims, if made at all, have been somewhat rashly made, and those suspeted of making them have been subjected to some pretty rough handling").

What Grice then takes as a commonplace incorporates the idea that "$\sim$" 'formalizes' "not", or Latin *non* (for the Scholastics), or Greek οὐ. Interestingly, English 'not' shows a complexity that seems absent in the simplicity of "$\sim$". Old English had *na*, and Middle and Modern English 'not' incorporates Old English *na* plus the emphatic *wiht* (cf. [Jespersen, 1917; Horn, 1989]). English was not, however, the language of the schools, and observations by other scholars of the highest intelligence, such as Ælfric, are hardly credited in the history of logic. "Sume [propositions] syndan *abnegativa*, thaet synd, withsacendlice, mid tham we withsacath", as in "Ic *ne* dyde").

As for the squiggle ("$\sim$"), the contention behind its use seems to be the old Pythagorean idea that negation parallels subtraction, whence the minus sign ("$\sim$" being a variant of "-"). The correspondence between "$\sim$" and "not" poses problems of a categorical type. In grammar and scholastic logic, the received opinion seems to have wavered between the idea that 'not' is an adverb (*adverbium negandi* [quod] *denotat negationem*, as Christopher Cooper has it in his *Grammatica Linguae Anglicanae* (1685)) and a *syncategorematon*. A good compromise appears in Robert Bacon's *Sumule dialecticis* with his talk of *"syncategorematon adverbialium"*:

> *Non* est adverbium quod prius et principaliter determinat ipsum... [C]um anima accipit duo incomplexa disconvenientia, ut *hominem* et *asinum*, afficitur quadam dissensione, et huic dissensione, que est intra, respondet hec dictio *non* in sermone extra. Unde illius dissensionis que afficit animam nota est hec dictio *non*.
>
> *Not* is an adverb which first and primarily determines itself... When the mind accepts two simple opposed terms, such as *man* and *donkey*, from this inner disunity, which affects the mind, the word not comes to be: an outer verbal manifestation of the discord within.

In Modernist Logic, on the other hand, "$\sim$" seems to be the only unary truth-functor worth discussing.

A further observation on Grice's use of "$\sim$" as a 'formal device': The idea seems to be not only that "$\sim$" will do duty for "not", which is already a complex assumption. In a narrow interpretation, "$\sim$" is a formal device if (and only if) it can be expressed in the logical form of a natural language counterpart containing "not". On this view, 'formal' would apply, strictly, to a 'formalized' calculus. On a broad view, however, 'formal device' may refer more to logical 'form' understood as an abstraction which may be the result of 'formalization'. ("Not" is still a 'formal device' in, say, Winston Churchill's speeches, even if they never get 'formalized' into first-order predicate logic).

As it happens, while Grice goes on to invoke the debate between the rival groups of the Modernists and the Neo-Traditionalists, he is mainly interested in the latter (as represented by Strawson) and their caveats on 'reading' "$\sim$" as 'not'. For one thing, Strawson had urged a more faithful reading of "$\sim$" as "it is not the case that...", reflecting the sentence-initial external maximal scope position that the sentential operator "$\sim$" necessarily takes in propositional logic ("The identification of "$\sim$" with 'it is not the case' is to be preferred to its identification with 'not' simpliciter" [Strawson, 1952:79]). Grice ignores this and translates "$\sim$" merely as "not". (As a note of interest, Grice does list both 'some' and 'at least one' as the 'translations' of '$\exists x$' in the above-quoted passage. But also note that while "at least one" and "some" do not include each other, "not" is incorporated in the longer "it is not the case".)

In any case, Grice found the commonplace to be wrong, and resulting from a blurring (committed by Modernists and Neo-Traditionalists alike) of logical and pragmatic inference. (Here we rely on Grice's unpublished lectures on Negation; cf. [Chapman, 2005: 87]).

Grice refers to Modernism and Neo-Traditionalism as "two rival groups... which I shall call the formalist and the informalist groups" [Grice, 1989:22]. Grice alludes to "the Modernists, spearheaded by Russell" [1989:372], and it is an interesting project to investigate to what extent works like *Principia Mathematica* commit the mistake that Grice attributes to Modernism. In any case, it seems obvious that Grice was parochially interested in the effect that tenets of Modernism had received in the rather conservative atmosphere of the Oxford of his time, especially as the target of attack by Strawson. Strawson's *Introduction to Logical Theory* had been welcomed by ordinary-language logicians (e.g. Warnock and Urmson) as providing a more faithful characterization of the 'logical behaviour' of certain English expressions.

Indeed, while Grice qualifies a position such as Strawson's as 'Neo-Traditionalist' (and not just 'Traditionalist') the same could hold for 'Modernism'. What Russell spearheaded was indeed a Neo Modernism historically, if we recall that it was authors like Ockham in Oxford who had fought for a *logica moderna* to oppose to the traditionalist *logica vetusta* of his predecessors. But Ockham was perhaps not modern enough. The venerable inceptor whetting his razor (prefiguring Grice's Modified Ockham' Razor here in the form of a dictum "Do not multiply senses of 'not' beyond necessity") had considered whether a proposition featuring 'non' (a *terminus syncategorematicus*) was atomic or molecular, and concluded that the negation of a categorical proposition was still categorical. Thus, in a passage apparently flouting his own razor, he distinguished between propositional negation and term negation. The distinction, he thought, surfaced in cases of 'privation': "Every S is non P" was rendered as "Every S is of a kind K that is naturally P, and no S is P".

A more neutral label for Modernism would be Classicism (as when we speak of first-order predicate logic as 'classical' [Grice, 1986:67]). The issue arises as to what really is "(Neo-) Traditionalist" about Strawson's *Introduction to Logical Theory.* It cannot be Aristotelianism, when Strawson [1950: 344] is eager to grant that neither Modernist ('Russellian') nor 'Aristotelian' logic faithfully represents the logic of ordinary language, which "has no exact logic". Rather, it seems that Aristotle was rather a pre- or proto-Gricean (see [Horn, 1973] on "Greek Grice"). The central source is *De Interpretatione*, which divides indicative-mode declarative sentences into assertion and denial (negation, ἀπόφασις, from ἀποφανεῖν, 'deny, say no'), which respectively affirm or deny something about something [17a25]. As Grice observes, this division of indicative-mode sentences into affirmative and negative "may suggest that the notion of the exhibition of a subject-predicate form enters into the definition of the very concept of a [indicative-mode] declarative sentence or proposition" [Grice, 1988:178; cf. *Cat.* 11b17].

And what are we to do with an author such as Cook Wilson [1926], situated somewhat between the Modernists and the Neo-Traditionalists, who wonders if negation "is a different *species*" from affirmation, "or whether the latter is in some sense the form of all statements, and again whether the negative symbol belongs to the so-called copula". Wilson's target of attack is the 'Modernist' Mill, who listed judgements as being affirmative or negative, "without troubling to find the genus of which they are species — the elementary fallacy of defining by enumeration of species instead of a statement of the genus". Wilson's own view is that 'affirmative' and 'negative' "are not co-ordinate in the strict sense of the term" [Wilson, 1926:264] — the elementary fallacy of defining a thing by what it is not. Symmetricalism is best represented by Ralph Lever when attempting to replace the logician's talk of 'negation' by 'naysay': "every simple shewasay is eyther a yeaysay or a naysay" (*Arte of Reason*, 1573). (Cf. [Horn, 1989: Chapter 1] for a comprehensive chronicle of the "(A)symmetricalist Wars".)

The background: Negation and opposition in Aristotelian logic

The genus of opposition ἀπόφασις, as introduced in Aristotle's *Categories* (11b17), is divided into four distinct species: CONTRARIETY (between two CONTRARIES), e.g. *good* vs. *bad,* CONTRADICTION, or ἀντίφασις, (AFFIRMATIVE to NEGATIVE), e.g. *He sits* vs. *He does not sit,* CORRELATION (between two RELATIVES), e.g. *double* vs. *half*, and PRIVATION (PRIVATIVE to POSITIVE), e.g. *blind* vs. *sighted.* Aristotle proceeds to offer detailed diagnostics for distinguishing "the various senses in which the term 'opposite' is used" (11b16-14a25). Crucially, contradictory opposites *(All pleasure is good, Some pleasure is not good)* are mutually exhaustive as well as mutually exclusive, while contrary opposites *(All pleasure is good, No pleasure is good)* do not mutually exhaust their domain. Contraries cannot be simultaneously true, though they may be simultaneously false. Members of a contradictory pair cannot be true OR false simultaneously; contradictories "divide the true and the false between them." So too, contradictory terms *(black/non-black, odd/even, male/female)* exclude any middle term, an entity satisfying the range of the two opposed terms but falling under neither of them: a shirt which is neither black nor not-black, an integer which is neither odd nor even. Contrary terms, by definition, admit a middle: my shirt may be neither black nor white (but gray), my friend neither happy nor sad (but just blaah). (See [Horn, 2010] for an overview of contradictory opposition and its relatives.)

Privatives and positives always apply to the same subject and are defined in terms of the presence or absence of a default property for that subject:

> We say that that which is capable of some particular faculty or possession has suffered privation when the faculty or possession in question is in no way present in that in which, and at the time in which, it should be naturally present. We do not call that toothless which has not teeth, or that blind which has not sight, but rather that which has not teeth or sight at the time when by nature it should. (*Categories* 12a28-33)

On this understanding, a newborn kitten is no more blind than is a chair, and a baby is not toothless. In later work, privation is taken to be a subcase of contrariety.

One more species of opposition is worth mentioning. Aristotle's early commentators Apuleius and Boethius, in structuring the Aristotelian system in the form of a Square of Opposition (see [Parsons, 2008]), define an additional relation of subcontraries, so called because they are located under the contraries in the geometry of the Square:

As the contradictories of the two contraries, the subcontraries (e.g. *Some pleasure is good, Some pleasure is not good)* can both be true, but cannot both be false. For Aristotle, this was therefore not a true opposition, since subcontraries are "merely verbally opposed" *(Prior Analytics* 63b21-30).

The Aristotelian categories of opposition held sway through generations of logic handbooks. Here, for example, is Edward Bentham [1773: 40-41]:

> An universal affirmative and an universal proposition are termed *Contrary;* They may be both false, but can not be both true. A particular affirmative and particular negative are termed *Subcontrary*; They may be both of them true, but cannot in any instance be both of them false. An universal affirmative, and particular negative, as also an universal negative and particular affirmative are termed *Contradictory.* They can in no instance be both of them true, nor both of them false. The

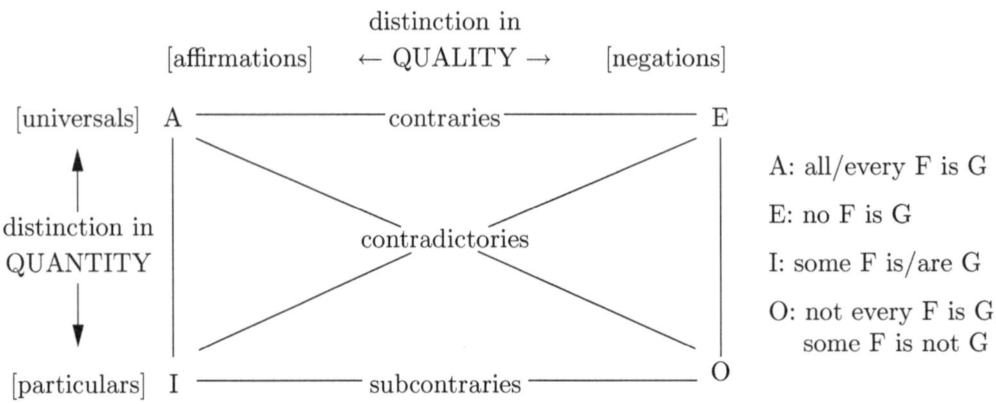

> difference between *contrary* and *contradictory* propositions should be the more carefully observed, as it is common enough to find the two contending parties in a dispute to be both of them mistaken, while they maintain *contrary* positions; which may be both of them false. So likewise as to subcontrary propositions. Men expressing themselves indefinitely sometimes grow angry with each other, supposing that their assertions are inconsistent; when if rightly explained, they may be both of them found to be very true.

In this connection, Bentham provides *Some faith does justify* and *Some faith does not justify* as an example of what ought to be non-fighting words.

Modernists and Neo-Traditionalists in post-Gricean hindsight

It should be easier to catalog logicians as either Modernist or Neo-Traditionalists post-'Logic and Conversation'. A few instances show that this is not so easy. Thus, " '∼' is pronounced 'not'." Hodges [1977:92] puts it: "Given any proposition p, one can form from it another proposition which is its negation. In English this is usually done by insterting *not* in some appropriate place in the sentence expressing it, though ambiguity is better avoided by tediously writing out *It is not the case that* at the front of the sentence" (cf. [Bostock, 1997:17]). He goes on to pose the following exercise: "Discuss that ∼' and 'not' mean the same" [Bostock, 1997:20]. More emphasis comes from Cambridge: "By all means, read "∼" as "not", but remember that it shouldn't be thought as a mere abbreviation of its ordinary-language reading. For a mere abbreviation would inherit the ambiguity of the original, and then what would be the point of introducing "∼"? "∼" is best thought of as a cleaned-up replacement for the vernacular *not*, which unambiguously expresses its original core meaning" (P. Smith, *Logic*, 2003:59).

What Grice saw as the little war between Neo-Traditionalism and Modernism seems to be alive and well. Oxford has now two different chairs for the two kinds of logic: the Wykeham chair of logic (New College, Faculty of Arts) and the chair of "Mathematical Logic" (Merton, and The Mathematical Institute at St. Giles). Other universities have adopted other ways of dissociating the two groups, in a latter-day version of C. P. Snow's nightmare of the 'two cultures'.

Modernism

It was said of Grice that he could always be relied upon to rally to the defense af an underdogma. Modernism was one such underdogma in Oxford [Grice, 1989:297]. Modernism claims that the logic of *not* worth preserving is that expressed by '$\sim$'. Any implicature should be seen as a 'metaphysical excrescence' [Grice, 1989:23]. What we propose is to outline what Modernism offers as the logical behaviour of 'not' (and '$\sim$'): the syntactic and semantic components of a system that will define the class of valid inferences featuring negation.

It makes sense to start the discussion with Modernism not just because it is the first of the rival groups that Grice mentions. It shows that Strawson wasn't really refuting Modernism by bringing attention to the divergence between the formal device and the vulgar counterpart. (As Grice notes, that there is a divergence is a commonplace shared by both Modernism and Neo-Traditionalism.) Second, in adding a pragmatic apparatus to the system in the form of conversational implicature (as well as in proposing a bracketing device to provide a "conventional regimentation" of a pragmatic distinction, a point to which we return below; cf. [Grice, 1989, Chapters 4 and 17]), Grice is ultimately seeking a defense of Modernism.

Curiously, the strawperson here is not Strawson but Quine. Quine had sat in on the seminars on logical form given by Grice and Strawson. Grice indeed presents his System Q as a tribute to Quine. What this System does is incorporate the constraints that the formal device "$\sim$" is supposed to have. Logic may have all started with Aristotle, but many argue that we wouldn't be studying it now if it were not for its 'modern' developments. The term 'modern', as applied to logic, is of course regularly recycled — it was used to refer to the logic of William of Ockham, for example. In the more recent understanding, it applies to classical two-valued systems as found in Whitehead/Russell's *Principia Mathematica;* it is this approach that Grice classifies as 'Modernism' in his 'Retrospective Epilogue' [1989:372]. In the vadcmccum of classical logic, Whitehead and Russell famously introduce the "$\sim$" operator as the 'contradictory function', to be read as having always maximal sentential scope, in the tradition of the Stoics and Frege. "The contradictory function with argument p, where p is any proposition, is the proposition which is the contradictory of p, that is, the proposition asserting that p is not true. This is denoted by '$\sim p$.' Thus '$\sim p$' is the contradictory function with 'p' as argument and means the negation of the proposition p. It will also be referred to as the proposition 'not-p'. Thus '$\sim p$' means 'not-p,' which means the negation of 'p' " [Whitehead/Russell, 1910:6].

If Grice is right, Modernism must find some 'divergence' between 'not' and "$\sim$".

It's not easy to identify one particular feature that may count as an example of this 'divergence' (but cf. [Cohen, 1971]). In any case, the Modernist approach, as viewed by Grice, would consist of dealing with any such divergence as a 'metaphysical excrescence' of "not" — from which "$\sim$" is by definition detached. Russell's particular interest was the role of "$\sim$" in association with a definite description, which he took to yield an essentially scopal ambiguity, with "The king of France is not bald" corresponding either to '$\exists x(Kx \& \forall y(Ky \rightarrow y = x) \& \sim Bx)$' or to '$\sim \exists x(Kx \& \forall y(Ky \rightarrow y = x) \& Bx)$', of which only the latter is a contradictory of "The king of France is bald".

Boole uses "$-$" for contradictory negation, alluding to arithmetical subtraction. "Whatever class of objects is represented by the symbol x, the contrary class will be expressed by '$1 - x$' " ("man, *not*-man"). Boolean negation is classical negation. However, it has been argued that Boole's is Modernism gone wrong, arguably leading to logically uninterpretable expressions in the course of calculating logical equations according to the model of arithmetic.

The Modernist program then is to provide a standard for the class of valid inferences concerning a formal device. This is the idea of a system, such as first-order predicate calculus. The range of valid inferences will be defined by the syntactic and semantic components. We shall proceed accordingly. (With the addition of a pragmatic component, the system can be integrated into a complete 'semiosis'.)

Semiosis for negation

The epitome of a first-order formal predicate calculus (as proposed by the Formalists) is something like System G (System G_{HP}). (We take the idea from George Myro [1987] and offer this 'highly powerful'/'hopefully plausible' version of his system G.) What a system like System G does is to provide a syntactic and semantic component for various formal devices, including "$\sim$".

Syntax for negation

By syntax is understood both the 'formation' rule and the 'inference' rules (introduction and elimination) before they get a semantic interpretation. (Grice's own system relies heavily on Mates's *Elementary Logic*.) A syntax-sensitive formation rule for "$\sim$" indicates the order in which the formal device is introduced with respect to the pre-radical. This will be of use in explaining away the alleged divergence between "$\sim$" and "not" in terms of implicature and 'scopal ambiguity'. Grice [1969:126] mentions similar work on these scope indicators by Charles D. Parsons and George Boolos:

If $\phi_{[n]}$ is a formula, $\sim_{n+m} \phi$ is a formula. The subscripts here indicate scope: the higher the number, the wider the scope. Thus, negation is assigned scope here that will be wider than that of any arguments of the predicate in question.

The apparently simple formation rule already seems to commit the system to a purely sentential account of negation. It suggests that by default negation will be assigned maximal scope. (Non-maximal scope would yield a subscript other than "$n + m$").

What the formation rule does is transform a negation-free 'radical' into a radical containing negation. At this point Grice plays with algebraic and chemico-physical terminology. The "$\sim p$" formula is construed as 'radical' (in fact symbolized as "$\surd \sim p$"; [Grice, 2001:59]). Grice's analogy here is to chemistry, where a radical is "an atom or group of atoms regarded as a (or the) primary constituent of a compound or class of compounds, remaining unaltered during chemical reactions in which other constituents are added or removed...An individual atom or element as a constituent of a compound was formerly termed a simple radical, as distinct from a group or compound radical" (OED). Keeping the chemical analogy, we may refer to "$\sim$" as a *negaton*. As such, "$\sim$", while having maximal scope within the formula, remains internal to any 'illocutionary force' operator that can be appended at a later stage ("$\vdash \surd \sim p$", "$! \surd \sim p$"). Again, having this clearly in mind at the 'formation-rule' stage allows us to explain some tricky interfaces between mode-operators and negation as being ultimately implicatural.

For visual clarification Grice uses here the algebraic radical sign ($\surd$) "$\vdash \surd \sim p$". In Myro's terminology, "p" is the *negatum* of "$\sim p$" and vice versa. The word had been used before to indicate one of the *modi* of the proposition. "Now affirmatum and negatum, verum, ... are ... words of art, for indeed they belong to logic. They call these modals, because the modus is the genus" (Richardson, *The Logician's Schoolmaster*, 1629:261). With relation to "$\vdash \surd \sim p$", Grice follows Moravcsik in referring to $\vdash$ as the mode (*modus*). The modus has been interpreted by some earlier logicians as involving "$\sim$": "The former is termed the dictum; the latter the modus ... And in general, modal propositions are affirmative or *negative*, according as the modus is affirmed, or denied of the dictum" [E. Bentham, 1773:45].

While the root sign obviously arises ultimately from algebra, the direct source is the analogy with the radical in chemistry, since it appears in more complex structures. Note that the Modernist 'formation rule' has "$\sim$" as an external affix to a propositional complex. This is very much in the Stoic and Fregean (rather than the Aristotelian, or later the Montagovian) tradition; cf. [Horn, 1989: Chapter 7].

The negated radical "$\sim$p" is the ἀπόφασις, of the Stoics ("Not: it is day"), which they carefully distinguished from statements like "No one is walking" (ἀρνήσομαι), and "This man is unkind" (στέρεσις, *privatio*). (See [Mates, 1953; Horn 1989: §1.1.2] for more on the Stoic negation.) It is a good exercise to find the negatum of these other types of utterance to see whether they all reduce to a formula containing "$\sim$".

The formation rule also leans towards Asymmetrism. A formula "ϕ", which does not contain negation, is the base for a new formula which does. "ϕ" is far from being a full-fledged 'affirmation'. That is, precisely, the whole point of invoking the notion of 'radical' instead. But the idea is there. In this, System G is traditional; cf. Coke [1654]: "The affirmation is before, and more worthy then

[sic] the negation. Denying or negative is which divideth the consequent subject from the antecedent predicate, as, "Good works do not justifie", "A man is not a stone". That a proposition may be a negative, it is necessary that the particle of denying be either set before the whole proposition: as "No elect are damned", or be immediately added to the coupler [sic], and verb adjective that hath the proce of the coupler or band; as: "Marriage is not a sacrament", "Works justifie not")."

The formation rule presupposes that "$\sim$" applies to a constituent, to form a 'radix', which is still 'mode-less', neither assertoric nor non-assertoric. Furthermore, there is no allusion in the formation rule to anything like the copula. The same formation rule of propositional calculus is used in the formation rule of predicate calculus. Modernism aims at simplifying some of the sempiternal worries of the Traditionalists. Thus, Hobbes famously held that 'not' attaches to the predicate. (For a view of Hobbes as a proto-Gricean, in the analysis of propositions like "That man is *not* a stone" in terms of utterer's intentions, see [Speranza, 1989], following Hacking).

When applying the mode operator to the radical (as in "$\vdash \surd \sim p$" and "$!\surd \sim$p"), Modernism also simplifies an account such as Frege's, for whom "$\vdash$" is a complex sign, with "$|$" and "$-$" representing judgement and content respectively. For that matter, no such complexity is preserved either in systems with mode-operators like "$\dashv$" that some have proposed for the sign of 'illocutionary denegation' — as in "$\dashv p$" (as in "I deny that it's raining") or "¡p" for prohibition (as in "No parking"); cf. Searle [1969] and related work, and Horn [1989] for a critique of the "speech act" of negation.

But 'deny' is a mere 'expositive' used "naturally, but no longer necessarily, to refer to conversational interchange" [Austin, 1962:162]. In Grice's terminology, 'deny' would constitute a 'central', rather than 'peripheral', speech act; indeed, so central that it's part of the radical (Austin's phatic or even rhetic: cf. [Grice, 1989:122]).

In a system like G, the interface between these mode-operators and their negated radicals must be accounted for implicaturally: "$\vdash \surd \sim (p\&q)$", "$\vdash \surd \sim (p \vee q)$", and "$\vdash \surd \sim (p \rightarrow q)$". Their 'neustic' equivalents, "$\surd \dashv (p\&q)$, "$\surd \dashv (p \vee q)$", and "$\sim (\surd \dashv (p \rightarrow q))$", are either ill-formed or misleadingly incorporate what is part of the implicatum into logical form. "I deny that she had a child and got married: she got married and had a child", etc.

The idea of a phrastic ϕ as a *negatum* in a more complex formula "$\sim \phi$" can be traced back at least to Richardson [1629:261]: "Now *negatum* is a word of art, for indeed it belongs to Logic". Cf. "A proposition is negative when the modus is denied of the dictum" [E. Bentham, 1773:45]. Grice's anatomy of negation as a subatomic particle thus has its logical historiography: as Bentham subtly expresses it, a proposition is negative not when the dictum is denied, but rather when "the modus is denied of the dictum" [1773:45]. (As an exercise, the reader may attempt to deny the dictum without denying the modus). Interestingly, the dictum is reintroduced in Oxonian parlance by R. M. Hare (cited by Grice [1986a:50] as a member of the Play-Group) to refer to the phrastic or radical. In other logical treatises, we find, as we do in Richardson's *Logician's School-Master*, negation characterized as a "mode" (*modus*) of a proposition, affecting its quality. The contrast is between the *modus* — what Grice will have as the mode-sign, and the early Hare will have as the 'dictor' — and the dictum, which Grice will have as the radical (the *negatum*). Keeping the chemical analogy, we may refer to "$\sim$" as the *negatron*.

It is not transparent how the radix/modus distinction should apply in contexts like, 'The king of France is not bald'. Hare maintains that "not", as represented by "$\sim$", is part of the radical. Most Formalists would reject as nonsense a formula such as "$\sim\vdash \surd p$". Negation, rather, is 'part of what is asserted'; "that is, that we *assert* either that the cat is on the mat, or that the cat is not on the mat" ([Hare, 2001:25]; cf. [Frege, 1919] for a precursor of this view). Grice's 'radical' connects with Hare's views. Hare had used dictum, dictor and dictive for what he later refers to as the 'phrastic' vs. the 'neustic' ("$\vdash \surd \sim (\iota x)(Kx \& Bx)$").

A phrase containing an illocutionary verb in the first person present indicative active, such as "I *deny* that..." or "I *assert* that..." can be used as an illocutionary-force indicator, and can be negated. However, in a system like G, it cannot be negated *and* used as an illocutionary force indicator.

Formation rules for negation are bound to become more complicated in the predicate calculus by incorporating the copula. For Locke, *is not* disconnects the subject from the predicate, and is "the general mark of the mind, denying" (cf. [Duncan, 1748]). By 'annexing the negative particle *not*' to the copula *is*, the mind "*disjoins* the subject idea from the predicate idea according to the result of its [the mind's] perceptions". The negative particle *not* is inserted after the copula, to signify the disagreement between the subject and the predicate. (Similar ideas date back to Peter of Spain [1972; 1989; 1992]; Kretzmann [1968]; O'Donnell [1941]).

Note that the formation rule of System G gives the squiggle maximal scope, and the issue of whether *not* attaches to the 'copula' is not even raised. But the issue was raised by Duncan [1748], whose solution involves the idea of a negative pregnant, complete with a proto-theory of implicature cancellation:

> Perhaps it may still appear a mystery how the copula can be said to be part of a negative proposition, whose proper business it is to disjoin ideas. This difficulty however will vanish, if we call to mind that a negation *implies* an affirmation. Affirmations are of two kinds, viz. of agreement or disagreement. Where perceptions disagree, there we must call in the negative particle *not*, and this gives us to understand, that the affirmation *implied* in the copula is not of any connection between the subject and predicate, but of their mutual opposition and repugnance.

Logicians working before the advent of Formalism seemed to have difficulties in deciding just what it is that *not* attaches to. Locke terms *not* a particle (*syncategorematon*), rather than an 'adverb'. Again, it's Duncan [1748] who adds the proto-Gricean flavor of psychological intentions. A negation is the disjoining of the subject idea and the predicate idea ("The law is not an ass"): "But as this is the very reverse of *what is intended*, a negative mark is added, to shew that this union does not here take place."

The formation rule for "∼", as it stands, should be qualified for sub-clausal structures (e.g. "The cat which is not black is not on the mat"). Duncan (1748) considers the entailments due to the embedding of *not*. His example: "The man who departs not from an upright behaviour, is beloved by God". The predicate, *beloved of God*, is evidently affirmed of the subject, so that notwithstanding the negative particle in the subordinate, the proposition is still affirmative.

Logicians working with the idea of the copula explain the canonical use of 'not' as an element of 'repugnance' attaching to the copula or a functionally equivalent predicate. Thus Coke [1654:107]:

> That a Proposition may be a Negative, it is necessary that the Particle of denying be either set before the whole Proposition, as, No Elect are damned; or be immediately added to the Coupler [sic], and Verb adjective [sic] that hath the force of the Coupler, or Band, as, Marriage is not a sacrament; Works justifie not. Every true Negation, hangs on a true Affirmation: For it could not rightly be said, *Works justifie not* unless it were true, that Faith onely justifieth.

The second ingredient that System G incorporates in the syntax for *not* is the doublet of introduction and elimination rules. Since Gentzen, it has been the received view of Formalism and Modernism that these are the only two rules or procedures constraining the inferential behaviour of 'not'. These are syntactic procedures in that they precede a full-fledged model interpretation (e.g. via a truth-table for the propositional calculus). Any such pair of introduction and elimination rules should be in principle be harmonious and stable.

Reductio ad absurdum

System G standardly adopts *reductio ad absurdum* as the introduction of negation:

If $\phi^1_{[m]}, \phi^2, \ldots \ \phi^k \vdash \psi_{[n-l]} \ \&_n \quad \sim_{[n-k]} \quad \psi_{[n-k-l]}$, then $\phi^2, \ldots \ \phi^k \vdash \sim_{n+l} \phi$'.

[Grice, 1969:126]; cf. [Myro, 1987:89]

Reductio ad absurdum can be traced back to the Eleatics, particularly Zeno and his *epicheirema*. Aristotle has this as ἡ ἐις τὸ ἀδύνατον ἀπαγωγή —*reductio ad impossibilem*, rather than *ad absurdum*—and it was first used in (Latinized) English by Isaac Watts: "reducing [the respondent] to an absurdity ... is called reductio ad absurdum." (*The Improvement of Mind*, 1741).

Duplex Negatio Affirmat

System G_{HP} also adopts the standard rule for the elimination of "∼":

$\sim_{n+k} \sim_n \phi_{[n-m]} \vdash \phi$.

[Grice, 1969:126], cf. [Myro, 1987:62]

The source here is ὑπεραπόφασις, as used by the Stoics (ὑπεραφότικον [ἔστιν] ἀποφατικόν ἀποφατικôυ (*Stoic* 2.66)), rather than Proclus.

The introduction and the elimination rules should provide for a harmonious system. Griss [1945] has argued that mathematical discourse can do without negation. In other words, negation in mathematical discourse is either introduced in a stronger way than by *reductio ad absurdum* or it cannot simply eliminated via the law of double negation. Faced with what is argued to be a system both unharmonious and unstable, Griss opts for a 'negationless' system.

Dummett [2002: 291] has explored the entailments of a system like G that includes RAA and DNE as introduction and elimination of negation: "Plainly, the classical rule [of elimination, DNE] is not in harmony with the introduction rule". So there is a requirement of harmony that the classical logician thinks she fulfils and the intuitionist logician thinks she does not. Dummett relates the harmony requirement with the stability requirement. "The negation-elimination rule, ... validates negation introduction, which, however, fails to validate negation-elimination. This was a situation we did not envisage when we discussed stability" [2002: 293]. In view of the problems brought about by negation, some intuitionist logicians, such as Griss, speak of 'negationless' systems. We are treating together different accounts of 'anti-realist' negation, taking Griss as the paradigm. In contrast, Grice and system G is realist. "Since different notions of incompatibility are being used [by Griss and Grice], there is no sound objection to the claim that the semantic value of classical negation is determined" [Peacocke, 1987:165].

Semantics for negation

The second component of a formal system for negation is the semantic one. The constraints that system G so far has incorporated are syntactical ones. They characterize *not* by its internal role in the system, regardless of any meaning it may be intended to contribute. The question arises as to whether a purely syntactical account is sufficient. In any case, narrowly construed, the semantic constraint for *not* should create no big metaphysical problem. It must support the classical truth-table for what Whitehead and Russell define as the 'contradictory function'.

It is with the semantic component at play that a system like G provides a defense of truth-functionality. Qua truth-functional operator, and 'when given a standard, classical two-valued interpretation' [Grice, 1989:22], *not* is a toggling operator, reversing the truth-value of its constituent radical:

ϕ is Corr(1) on Z iff if $\phi =\sim_n \psi$, ψ is Corr(0) on Z.
[Grice, 1969:136], cf. [Myro, 1987:43]

('Corr(1)' and 'Corr(0)' are abbreviations, respectively for 'correlated with 1' and 'correlated with 0'). The aim is the standard one, as in the comment "It seems a fair reflection of ordinary usage to identify the negation of p with any statement

"$\sim p$" which is so related to p that if either is true it follows that the other is false." [Ayer, 1952:42].

Note that that System G allows for three other unary truth-functors: "ϕ is Corr(1) on Z iff if $\phi = T_n\psi$, ψ is Corr(1) on Z", "ϕ is Corr(0) on Z iff if $\phi = P_n\psi$, ψ is Corr(1) on Z", and "ϕ is Corr(0) on Z iff if $\phi = Q_n\psi$, ψ is Corr(1) on Z". The choice of "$\sim$" over the other three unary 'formal devices' seems to be a matter of notational convenience or pragmatics.

Neo-Traditionalism has long understood negation in semantic terms, even when dealing with tricks like vacuous descriptors or names. Consider the implicatures (or entanglements) expressed by Oxford logic tutor Simon of Faversham (Simon de Daversiham):

> Ponatur in casu, et est possible, quod Socrates *non* sit, haec ergo est falsa '*Socrates est*', et similiter haec erit falsa '*Socrates non est (iustus).*' Significaretur enim quod Socrates, qui est, non est; sed hoc est falsum; ergo et haec est falsa 'Socrates *non* est'; ergo haec sunt simul falsa 'Socrates est' et 'Socrates non est', et haec sunt contradictoria, ergo contradictoria essent simul falsa; hoc est impossibile Ditto *Nullus homo est animal*, sive 'homo sit' sive 'homo non sit' est falsa.
>
> Let us put it forward that it is possible that Socrates should not be, thus this is false, "Socrates is", and similarly it will be false that Socrates is not (just). For this would signify that Socrates, who is, is *not*, but this is false; thus this is false, "Socrates is *not*'; therefore these are simultaneously false: 'Socrates is' and 'Socrates is *not*', and these are *contradictories*, thus contradictories would be simultaneously false, but this is impossible. Ditto *No man is an animal*, whether 'man is' or 'man is not' is false.

While Grice refers to both Modernism and Neo-Traditionalism as agreeing that there are (or more fastidiously put, 'appear to be') divergences between "$\sim$" (as defined in the calculus so far) and *not*, it is not to easy to come up with an example of a Modernist explicitly alleging so. The straightforward definition of negation as the contradictory function adopted by Whitehead and Russell and later by Quine offers the standard approach.

Grice colourfully ascribes to the Modernist the prejudice that any such divergence will be a 'metaphysical excrescence' [1989:23]. It seems far easier to find illustrations of the alleged divergence from the rival group of Neo-Traditionalism. Historically, too, Grice's reply was explicitly provoked by the Neo-Traditionalist discourse, rather than the Modernist one.

Neo-Traditionalism

If Modernism is concerned with classical logic, Neo-Traditionalism might be viewed as romantic, although in many ways the latter is more faithful to Aristotelian foundations than is the former, at least in spirit. The vademecum here is Strawson's

Introduction to Logical Theory (1952), best seen as a direct attack on Russell's Modernism (but also Quine's logical Puritanism). Interestingly, it was influenced by early views of Grice. In the preface, Strawson explicily cites Grice as the person "from whom I have never ceased to learn about logic since he was my tutor in the subject" [1952: v], and in an important footnote notes that it was "Mr. H. P. Grice" who demonstrated to him "the operation of a pragmatic rule" prefiguring Grice's own maxim of quantity: "One should not make the (logically) less, when one could truthfully (and with greater or equal clarity) make the greater claim" [Strawson, 1952: 179]. Strawson later recollected: "The nature of the logical constants of standard logic was a question that H. P. Grice and I used to discuss in the early 1950s, and I have no doubt that [some of Strawson's essays were] influenced by those discussions" [1991: 10]. And again, later, "I had the pleasure of listening to Grice expounding the essentials of his view [on logical constants] in a paper read to the Philosophical Society in Oxford in the late 1950s" [Strawson, 1991:16].

While it is difficult to pinpoint to an explicit reference by Modernists on the alleged divergence between "not" and "$\sim$", Grice's favourite Neo-Traditionalist is clear on this. Indeed, a sub-section of the *Introduction* (Part II, ch. 3, 2§7) bears the title, "'$\sim$' and 'not"'. Granted, Strawson's main caveats are not with "$\sim$" but with some of the binary truth-functors (notably the conditional). However, his Neo-Traditionalism seems to adopt a focus on 'use' rather than 'meaning'.

By focusing on contradiction as one of the primary and standard uses of "$\sim$", Strawson underlines the role of negation in discourse. The use or function of "$\sim$" (and *not*) is to "exclude". It is "a device used when we wish to *contradict* a previous assertion". But also when we wish "to correct a possible false impression, or to express the contrast between what had been expected, feared, suggested, or hoped, and the reality." [...] "A standard and primary use of *not* is specifically to contradict, or correct; to cancel a suggestion of one's own or another" [1952:7]. And later: "A standard and primary *use* of *not* in a sentence is to assert the contradictory of the statement which would be made by the use, in the same context, of the same sentence without the word *not*" [1952:79].

Strawson does include a caveat aimed apparently at something like the formation rule of Modernism ("if ϕ is a formula, $\sim \phi$ is a formula"): "Of course, we must not suppose that the insertion of *not anywhere* in any sentence always has this effect". His example, adapted from Aristotle: "Some bulls are not dangerous" is not the contradictory, but the subcontrary, of "Some bulls are dangerous". "This is why the identification of '$\sim$' with *it is not the case that* is to be preferred to its identification with *not* simpliciter".

Strawson [1952: 79] continues: "This identification [of *not* and "$\sim$"], then, involves only those minimum departures from the logic of ordinary language which must always result from the formal logician's activity of codifying rules with the help of verbal patterns: viz., the adoption of a rigid rule when ordinary language permits variations and deviations from the standard use ("$\sim\sim p \rightarrow p$", "$p\vee \sim p$") and that stretching of the sense of 'exemplify' which allows us, e.g., to regard 'Tom is not mad' as well as 'Not all bulls are dangerous' as exemplifications of 'not-p"'.

Divergences notwithstanding, Strawson concludes the section, "So we shall call '∼' the negation sign, and read '∼' as 'not"'.

Strawson goes on to attack the Modernist (chiefly Russellian) account of definite descriptions. With Grice, we shall focus our discussion on the negation of vacuous descriptors ("The King of France is not bald"). The attack is premised on the assumption that negation, normally or invariably, leaves the subject 'unimpaired'. For Modernism, as reflected in System G, the negation of a sentence containing a vacuous descriptor comes out as *true*. For Strawson, if what is presupposed is not satisfied, "The king of France is not bald" is not true. It's not false, either, but rather a pointless thing to say, the question of whose truth or falsity fails to arise. For Grice, Modernism is essentially correct: "The king of France is not bald", in the absence of the king of France, is true. (But cf. Grice 1981 for a somewhat more complex picture.) What Strawson takes as *praesuppositum* is merely an *implicatum* (and never an entailment). Strawson does recognize the marginal existence of a non-presupposing negation, as in the exchange "Does he care about it?" "—He neither cares nor does not care. He's dead"' [1952:18]. For the true Modernist, the only correct answer would be: "No, he does not care about it. He is dead". By and large, however, Strawson takes both affirmative and negative statements featuring vacuous descriptors to result in a 'truth-value gap' (in Quine's happy parlance).

It is not obvious that Grice's opposition between the Modernists and the neo-Traditionalists is an exhaustive one. For one, it is difficult to place in either camp an author such as A. J. Ayer, "at [some] time the *enfant terrible* of Oxford philosophy" [Grice, 1986a:48], later Wykeham Professor of Logic, and the author of an influential essay on 'Negation'. In any case, Ayer seems to assume that negation carries something of a metaphysical excrescence, and should not for this be eliminated from a system. "A statement is negative if it states that an object lacks a certain property rather than stating that it possesses the complementary property. A statement is negative if it states that a certain property is not instantiated, rather than stating that the complementary property is universally instantiated" [Ayer, 1952: 61].

Robbing Peter to pay Paul

As noted, Grice's aim is a synthesis between Modernism and Neo-Traditionalism, and one crucial issue is what we may call the identity or isomorphism thesis, the claim that there is no divergence (in terms of entailment and valid inference) between "∼" (as defined by System G) and "not". The challenge is to respond to those Modernists and Neo-Traditionalists who have held such a divergence to exist.

Strawson has been quite explicit as to the origin of his views on negation: his disputes with Grice at the Oxford Philosophical Society, and earlier, in the premises of St. John's, where Grice '— from whom I have never ceased to learn logic' — was his tutor. For Grice, the assumption that there is a divergence between "∼"

and "not" "seemed to me to rest on a blurring of the logical/pragmatic distinction" [Grice, 1989:374]. The "general rule of linguistic conduct" that Strawson [1952: 179] borrows from Grice was developed by the latter in his William James lectures into a fully fledged set of maxims within an overarching Co-operative Principle ("Make your conversational contribution such as is required, at the stage at which it occurs"). These maxims are sorted into four Kantian categories: Quantity ("Make your contribution as informative as is required (for the current purpose of the exchange)", "Do not make your contribution more informative than is required"); Quality ("Try to make your contribution one that is true, Do not say what you believe to be false, Do not say that for which you lack adequate evidence"), Relation ("Be relevant"), and Manner ("Be perspicuous—Avoid obscurity of expression, Avoid ambiguity, Be brief, Be orderly). In his posthumously published retrospective epilogue, Grice [1989:273] adds a tenth submaxim, 'Facilitate your reply', expanding the system into a conversational decalogue; cf. [Speranza, 1991].

This programme is presented within 'Logic and Conversation', the William James lectures Grice presented at Harvard in the spring term of 1967. It is in the 'Prolegomena' that Grice explicitly refers to Strawson [1952] as an "A-philosopher", i.e., as a philosopher who would advance a non-classical semantic or conventional account for various phenomena for which Grice offers an independently required pragmatic, non-conventional approach to supplement the standard Modernist treatment. Among the "A-philosophers", i.e. the Oxford Neo-Traditionalists or Informalists ranged on Strawson's side, we could place other members of the Play-Group, such as H. L. A. Hart, J. O. Urmson, and G. J. Warnock, as well as (on occasion) the group leader J. L. Austin himself.

The dispute concerns the divergence between a given formal device (e.g. "$\supset$, the horseshoe) and its vulgar counterpart ('if') — a topic that was being hotly debated at Oxford and elsewhere (cf. Speranza 1991 on J. F Thomson's 'In defence of the material conditional' and Horn [1989: 378-9] on the role of assertability in Grice's — and Dummett's—characterization of negated conditionals). But for Grice the battle has many fronts of equal importance, examples that "involve an area of special interest to me, namely that of expressions which are candidates for being natural analogues to logical constants and which may, or may not, 'diverge' in meaning from the related constants, considered as elements in a classical logic, standardly interpreted" [Grice, 1989:8].

In the second lecture, as we have seen, Grice explicitly tackles the '$\sim$'/*not* pair. His broadens the target of attack, though. It is not just neo-Traditionalists like Strawson who have claimed that such a divergence exists, but some of the Modernists themselves. In this respect, Grice might be considered post modern. Grice begins his discussion in 'Logic and Conversation' by challenging the acceptance of such a divergence on the part of both Modernism and Neo-Traditionalism as a "commonplace"; rather, he maintains, "the common assumption of the contestants that the divergences do in fact exist is (broadly speaking) a common mistake" [1989:24]. But this divergence is typically seen as relatively innocent in the case of negation, compared to that of the (material) conditional and the other binary connectives.

Over a century ago, Russell provided "∼" (borrowed from Peano) to be read as *not.* The common garden variety of logic manual will typically accept such an identification. Thus note, for example, Benson Mates's treatment of negation in his *Elementary Logic* (1965), and back in Oxford, David Mitchell in *An Introduction to Logic.* In the section dealing with 'The interpretation of the constants' in Chapter 2, Mitchell [1962:59] agrees with Strawson that, in contrast with some of the binary truth functors, ""∼" raises no [major] difficulties as the sign for propositional negation, used in conjunction with either propositions, as '∼(Tom is Australian)', or propositional forms. It may be read as *not*".

Or compare Suppes on "∼(Sugar causes tooth decay)": "The usual method of asserting the negation is to attach *not* to the main verb. We shall use '∼'." He enlists with those who perceive a divergence here, and has a caveat regarding the ∼/*not* identification. "Of course, the rich, variegated character of natural language guarantees that in many contexts ["not" is] used in delicately shaded, non-truth-functional ways. Loss in subtlety of nuance seems a necessary concomitant to developing a precise, symbolic analysis of sentences" [Suppes, 1957:4]. Suppes seems to be one logician who learned from the mistake, if he ever made it, and from Grice's treatment (see his contribution to the Gricean festschrift in Grandy & Warner [1986]).

The strategy of the conversationalist manoeuvre is to explain away the divergence between '∼' and 'not' via conversational implicature. A requirement for this strategy, though, is that we possess a more or less definite account of what "∼" already means in a system like system G. System G is an axiomatic calculus of the type made familiar by Whitehead and Russell which validates the inferential roles that "∼" can take. In this, Grice proves a conservative dissident parting ways with the tradition in Oxford that had been, in matters logical, to follow in the footsteps of the neo-Hegelians or the Informalism of "no calculus" favored by Ryle and Strawson. The Modernism of Whitehead/Russell (and Quine's Logical Puritanism as an offspring) had never taken root in Oxford (cf. Urmson and Warnock), and it has done so only recently with the institution of the chair of "Mathematical Logic" (Merton College and the Mathematical Institute).

What the eye no longer sees the heart no longer grieves for

The defense of Modernism as providing an adequate account of the explicit logical force of "not" proceeds by drawing a distinction between a 'logical inference' and a 'pragmatic inference' [Grice, 1989: 374]. Pragmatic inferences thus fall outside the domain of logic, or if not, they pertain to the terrain of non-monotonic defeasible calculi. In any case, the account they provide concerns the implicit content of 'not' rather than its contribution to truth-conditions.

Overall, Grice's sympathies were with the Modernists. He acknowledges his attraction to 'the tidiness of Modernist logic' [Grice, 1989:374]. More importantly, he notes that his additions are not meant to provide departures from the standard Modernist apparatus: "Even if it should prove necessary to supplement the apparatus of Modernist logic with additional conventional devices, such supplementation [should be regarded as] undramatic and innocuous." These will typically involve "bracketing or scope devices" ("What the eye no longer sees..."). The strategy is to enrich the syntax (rather than the semantics) by adding the formal device. In this way, the alleged divergence between "$\sim$" and "not" will be one of pragmatic import, which can be eliminated once the scope device is put to use.

Grice's observation regarding the credo of both Modernism and Neo-Traditionalism as resting on a mistake may itself be interpreted in a similar way by reducing the alleged semantic property of natural language "not" to a complex of a syntactic property and a pragmatic property.

It should be noted that the terminology of 'pragmatic' is already used by Strawson ("Some will say these points are irrelevant to logic (are 'merely pragmatic')" [Strawson, 1952:179]. On this, and Grice's view, it's only the syntax and the semantics for 'not' as "$\sim$" which account for the 'logical' inferences. Any further divergence will pertain to 'pragmatic inference'. The point is always to simplify the logic, even if the explanation must still be provided. As it has been said [McCawley, 1981: 215], "Grice saves" — but there's no such thing as a free lunch.

At this point, we may provide an inventory of issues which have traditionally been considered part of the logic of negation, but which we can be recalibrated as pragmatic. We need some order for the inventory. We have already touched on those alleged divergences concerning the formal treament of "$\sim$". We continue with other inferences that do not pertain to the formal system of negation but which have nevertheless been included by some as regarding the logic of "not". In all cases, the idea is to provide an *equi*-vocal account of "not".

One issue arises with the negation of sentences containing vacuous descriptors. System G deems "The king of France is not bald" as true in the context of a French republic. This analysis contrasts with the Parmenidean, Platonic and neo-Platonic accounts. Parmenidean negation creates a Meinongian jungle, in which the non-being of the King of France is not directlhy referred to. The problem is not particularly solved by Plato and his proposal to analyse "not" as "other than". "The king of France is not bald, but other than bald" allows for the cancellation, "indeed he is non existent" — but somehow one does not think that is what Plato had in mind.

A related problem which was emphasised by Ryle is the negation of category (or as Grice has them, 'eschatological') mistakes. While 'Virtue is square' is false, 'Virtue is not a square' comes out as trivially true, if only a redundant — if not downright irrelevant – thing to say in most contexts (but cf. 'Virtue is not square, but some windows are' vs. 'Virtue is not square, indeed virtue does not exist'). (See [Horn, 1989: §2.3] for a review of some of the relevant literature.)

"System G" allows for two syntactic devices to signal scopal ambiguity and avoid lexical ambiguity. The central manoevure of the pragmatic component of System G is the introduction of the notion of implicature. Implicature contrasts with entailment but also with explicit content: "Whatever is implied ... is distinct from what is said" [Grice, 1989:24]. When first (re)introduced into the philosophical literature, and thence into the consciousness of linguists, presupposition and implicature each appeared in turn as the meaning relation that dare not speak its name, The Other: an inference licensed in a given context that cannot be identified with logical implication or entailment. Thus Strawson [1950]:

> To say, "The king of France is wise" is, in some sense of "imply" to *imply* that there is a king of France. But this is a very special and odd sense of "imply". "Implies" in this sense is certainly not equivalent to "entails" (or "logically implies").

Two years later, this 'special and odd sense of "imply"' is recast as the neo-Fregean notion of presupposition. Similarly, when Grice first discusses (what would later be called) implicature, he begins by carving out a use of *imply:* "If someone says 'My wife is either in the kitchen or in the bedroom' it would normally be implied that he did not know in *which* of the two rooms she was" [Grice, 1961: 130]. Later, Grice makes a similar move to a more specialized and less ambiguous term: "I wish to introduce as terms of art, the verb 'implicate' and the related nouns 'implicature' (cf. implying) and 'implicatum' (cf. what is implied)" [Grice, 1989:24]. (Latin *implicat* is ambiguous, and Suedonius had already used *implicatura* for 'entanglement' (Short/Lewis, *Latin Dictionary*). Other members of the Play-Group had employed similar notions under various names, e.g. Nowell-Smith [1954] (and later [Hungerland, 1960]) on "contextual implication". (See the discussion in relation to Grice and Strawson in [Horn, 1990; 2012a].)

The distinction among species of implication within the overall genus of inferential relations is of particular relevance for the study of negation. In his influential *Statement and Inference* (posthum., 1926), Cook Wilson considers the distinction between 'asserting' and 'implying' or 'presupposing': "A negative statement normally presupposes the existence of their subjects of attribution; this existence is not *asserted* by the negative statement as such" [1926:259]. Ryle [1929] makes a distinction between 'asserting' and, not implying (or implicating) but presupposing. It is not altogether clear whether these logicians have identified what exactly it is that is implied, presupposed, asserted, or implicated in the issuing of a negation. "Elimination may ensue upon the discovery of the required facts. Still, when what enables me to assert, 'The S is not P' is the knowledge that the S is P', the assertion 'S is not P' is *different* from the assertion "S is P'". If it is known and *presupposed* that the determinable 'coloured' characterizes a hat, the denial of determinate P is not yet the assertion of determinate P"'. When a determinate applies, a negation "*does* assert that the subject is ONE of the determinates of a determinable *other than* the denied determinate." (all quotes from Ryle [1929:87]). This relates to the medieval doctrine of *suppositio* and the 'negative pregnant', discussed below. Part of the manoeuvre is to restate what has been already noted by medieval logicians. Insightful treatments of negative utterances in terms of their presuppositions or implicatures can be found in work by Peter of Spain [1972; 1992; Spruyt, 1987] and William of Sherwood [O'Donnell, 1941; Kretzmann, 1968], among others.

Implicature serves to explain the medieval theory of the 'negative pregnant', "a negative implying or involving an affirmative. As is a man being impleaded, to have done a thing upon such a day, or in such a place, denyeth that he did it *modo & forma declarata*: which implyeth neverthelesse that in some sort he did it" [Cowell, 1607]; (cf. [Bramhall, 1658:155] for a similar characterization). As argued in Horn [1989: Chapter 3], this notion of an affirmative supposition or ground underlying a negative statement must be analyzed as a Gricean implicature, not an entailment (or a Fregean or Strawsonian presupposition).

Negation as otherness

Following a suggestion by Cook Wilson, Grice [1989:75] looks for an analysis [or another term?] that can be specifically assigned to both "$\sim$" and "not". The idea of 'not' as shorthand for 'other than' is taken up by Grice. "As regards *not*: if our language did not contain a unitary device, there would be many things we can now say which we should be then unable to say, unless (1) the language contained some very artificial-looking connective like one or other of the strokes [i.e. the Sheffer 'not both' stroke or the joint denial 'neither-nor'], or (2) we put ourselves to a good deal of trouble to find (more or less case by case) complicated forms of expression involving such expressions as *other than* or *incompatible with*" [Grice, 1989:68].

In the case of 'incompatible with', consider the "reduction" of "$\sim p$" to "$p \mid p$". Rather than "The king of France is not bald" we would have, "The king of France is bald is incompatible with the king of France being bald". In the case of 'other than', philosphers since Plato have tried without notable success to find the appropriate 'complicated' form. In Platonic parlance, it's τὸ διάφορον, or τὸ ἕτερον. Plato saw this as part of a problem for his ideationist theory of meaning. If for any object m, there is a meaning relation such that m means (idea) M, the issue is to find the correlation for a 'judgment' that denies an attribute to an object (*Sophist* 357B). There have been various attempts to formalize what Plato is up to. Wiggins [1971] converts "$\sim$FLY(Theaetetus)" into a Platonic rendering as 'There is a property P such that $P\Delta$(FLY) & P(Theaetetus)', where Δ stands for τὸ διάφορον; cf. also Bostock [1984:115]. The relationship between Parmenidean and Platonic conceptions of negations and non-being is examined in [Pelletier, 1990].

Note that "Theaetetus is not big" cannot simply be reduced to the assertion 'Theaetetus is small'. For one thing, as Plato recognized (Sophist 257Dff.), he may be medium sized, whence the position that 'Theaetetus is not big' *explicitly* means 'Theaetetus is *other than* big' (leaving aside those cases in which Theaetetus does not exist). Negation as otherness is invoked as well by Mill [1843]: "Negative names are employed whenever we have occasion to speak collectively of all things *other than* some thing or class of things." While Grice, and before him, Wilson and Ryle, are more or less in agreement with an account of "not" that allows for its identification with "$\sim$", with any mismatch handled via implicature, the question is trickier for the neo-Idealists or neo-Hegelians. The British neo-Idealist group included Bradley, Bosanquet, Joseph, and Joachim, but a particularly significant and often neglected figure here is the German Sigwart, who anticipated by two decades Frege's notion of presupposition *(Voraussetzung)* in discussing the problem of vacuous subjects.

For Strawson, as for his intellectual predecessor Frege [1892], the notion of presupposition has semantic status as a necessary condition on true or false assertion, but more recent work has taken the commitment to existential import in such cases as constituting a pragmatic presupposition or an implicature (cf. e.g. [Wilson, 1975; Grice, 1989: Essay 17]). In fact, the earliest pragmatic treatments of the failure of existential presupposition predate Frege's analysis by two decades. Here is Christoph Sigwart [1873] on the problem of vacuous subjects:

> **As a rule**, the judgement A is not B presupposes the existence of A in all cases when it would be presupposed in the judgement A is B... 'Socrates is not ill' presupposes in the first place the existence of Socrates, because **only on the presupposition** *[Voraussetuzung]* **of his existence can there be any question of his being ill** . (Sigwart [1873/1895: 122], emphasis added)

Note in particular the contextual nature of the presupposition and the proto-Strawsonian flavor of the conclusion. Further, unlike either Frege or Strawson, Sigwart allows for wide-scope (presupposition-canceling) negation as a real, although marked, possibility, although to be sure a negative singular statement is 'commonly understood' as implying the existence of its subject referent.

> If we answer the question 'Is Socrates ill?' by yes or no, then — **according to our usual way of speaking** — we accept the presupposition *[Voraussetzung]* upon which alone the question is possible; and if we say of a dead man that he is not ill, we are guilty of using our words ambiguously. It may still, however, be claimed that, by calling such an answer ambiguous, we admit that the words do not, in themselves, exclude the other meaning; and that formally, therefore, **the truth of the proposition** [*Socrates is not ill*] **is incontestable** [if Socrates is not alive]. (Sigwart [1873/1895: 152], emphasis added)

The neo-Idealists are generally associated with the idea of 'positive negation', but this cannot be a truth condition (or more broadly a semantic requirement) on negative statements, which can perfectly well be true (if possibly odd) when their positive ground is lacking. Grice observes of a man lighting his cigarette in the ordinary way: "Now it is certainly the case that it would be false to say of the man using a match, 'He is now lighting his cigarette with a 20-dollar bill,' and so it is true that he is *not* lighting his cigarette with a 20-dollar bill." He adds: "So far as I know no philosopher since the demise of the influence of Bradley has been in the least inclined to deny this" [Grice, 1989:15]. (The mocking of Bradley seems to have been a favorite sport with British mid-century philosophers. For example, Ayer [1952:39], quoting from Bradley [1883:115], comments: "Any statement whatsoever which is seriously put forward may be picturesquely described as an attempt to qualify *reality*; and if the statement turns out to be false the attempt may be said to have baffled.")

Armed with this commentary by Grice, let us revisit Bradley's contribution to the history of negation. Formally, Bradley would have endorsed the unpacking of "$\sim Px$" as "$(\exists P')(P'x \& P|P')$". This is consistent with Grice's earlier observation that 'not' has the métier of 'incompatible with' (and not just 'other than'). Bradley's general view on negation focuses on the idea of 'positive ground'. This may again be seen as a development of the aforementioned medieval views on *suppositio* and the notion of a "negative pregnant'. For Bradley [1883:200], any negative proposition presupposes a positive ground:

> Every negation must have a ground, and this ground is positive. Nothing in the world can ever be denied except on the strenght of positive knowledge. We can not deny without also affirming. We should never trust a negative judgement until we have seen its positive ground.

— a.k.a. its *negatum*. For denial to be possible, there must be, first, the "suggestion" of an "affirmative relation":

> If we suppose that, with reference to the tree the utterer has judged, 'This tree is *not* yellow', the judgment would have to be construed as *involving* the affirmative suggestion that the tree *is* yellow. In the negative judgment, the positive relation of 'yellow' to the tree must precede the exclusion of that relation. What gets denied must be something that already has a truth value.

The views of Bradley were influential enough in Oxford to merit a doggerel in *The Oxford Book of Oxford* [Morris, 1978]: "Thou positive negation! negative affirmation! thou great totality of every thing,/that never is, but ever doth become, thee do we sing".

We turn to the second great British neo-Idealist (of whom Grice's assessment was somewhat unenthusiastic: "If we are looking at the work of some relatively minor philosophical figure, such as for example Bosanquet..." — [Grice, 1986a:66]). Bosanquet had his influence on Bradley himself; in his 'Terminal Essay' on negation, Bradley confesses "the chapter [on negation in *Principles of Logic*] contains some serious errors. I have since accepted in the main Dr. Bosanquet's account of negation." Bosanquet starts by citing Jevons, Mill, and Venn and arguing that "In negation, the work of affirmative belief appears to be performed by ignorance" [1911:277].

Of Sigwart's view that every negation presupposes an affirmation, so that 'S is not P' presupposes the affirmation 'S is P', Bosanquet declares: "I think it monstrous. I do not believe that you must find an affirmative standing before you can deny" [1911:277]. Further, "A negation is not a denial of an affirmative judgment, and therefore does not presuppose the affirmation of that which is denied." Yet "a negation does presuppose *some* affirmation" [1911:280]. Bosanquet refines Bradley's position by distinguishing the positive ground of a negative utterance (some *contrary* proposition, whose truth determines the truth of the negative proposition) from the positive consequent (the indeterminate proposition which logically follows from the negative). Thus, the positive ground of 'This shirt is not red' may be 'This shirt is (e.g.) blue', where *blue* is inconsistent with *red.* The positive consequent, however, is simply "There exists a colour P', $P' \neq$ red, such that this surface is P'" [1911:287]. The relation between a negation and its affirmative ground is one of contrariety, while a negation and its affirmative consequent are in contradictory opposition.

While Bradley leaned on Bosanquet, the direction of influence was in fact mutual. Bosanquet, citing Bradley's notion of 'a suggested affirmative relation' endorses the latter's view that 'in the beginning, a negation is a degree more remote from reality than is an affirmation'. But while an affirmation is epistemologically prior to negation, eventually, affirmation and negation alike become double-edged, each involving the other [1911:281]. For Bosanquet, only those negations which presuppose an affirmation can be significant (let alone true). A significant negative utterance "S is not P" can always be analyzed as "S is P' which excludes P". "The surface is not red, but an undetermined colour.' An apparently insignificant or bare negation (e.g. "The lion is not an elephant," "Virtue is not a square") does not posit a correct, true contrary, since it does not limit the sphere of negation [1911:289].

Bosanquet advocates a similar line on negative utterances with vacuous descriptors. Of his example, 'The house on the marsh is not burnt down', Bosanquet allows that the utterance is *true* when there is no house on the marsh, even if "reality excludes the burning down of any such house". Bosanquet confesses a 'strong sympathy' for the objection (straw or real) that the utterance may be said to have 'meaning only if there is a house, and the sentence presupposes [cf. Sigwart above] or asserts that there is one".

From contradiction to contrariety: Bosanquet, Anselm et al.

Bosanquet's insightful comment that "The essence of negation is to invest the contrary with the character of the contradictory" [1911:291] epitomizes his unfortunately overlooked discovery of a class of cases that would now receive a plausible Gricean account of contradictories in contrary clothing. (Cf. [Horn, 1989: Chapter 5; Horn, 2000] for elaborations of a variety of such pragmatic strenthening processes.) One example is "Jones is not good", which appears to represent contradictory negation on "p" ("Jones is good"), and thus being representable as "$\sim p$". However, the ordinary copular negative yields a relatively weak non-informative contradictory that tends to be strengthened, "so that from 'Jones is not good' one may be able to infer something more than that 'it is not true or the case that

Jones is good"'.

As a related illustration of the same tendency to enrich a formal occurrence of a contradictory "∼p" to a contrary, given his premise that "the essence of formal negation is to invest the contrary with the charcter of the contradictory", Bosanquet offers an early account of the phenomenon of negative transportation or neg-raising, i.e. "The habitual use of such a phrases as 'I do not believe it,' which refers grammatically to a fact of my intellectual state, but actually serves as a negation of something ascribed to reality. Compare Gk. οὔ φημί [lit. 'I don't say'], which means 'I deny', or our common phrase, 'I don't think that' — which is really equivalent to 'I think that — not"' [1911:319]. Again this phenomenon can best be accounted by an implicature-based account (see [Horn, 1978, 1989: Chapter 5]). Thus, instead of treating a negative judgement in Bradleyan fashion as aiming at a final affirmative, Bosanquet sees canonical contradictory negation as functioning to express a notional contrary, because when there are only two alternatives, the denial of one is equivalent to, and grounded on, the assertion of the other.

The phenomenon of strengthened contrary readings for apparent contradictory negation has long been recognized, dating back to classical discussions of the figure of litotes, in which an affirmative is indirectly asserted by negating its contrary, has been recognized since the 4th century rhetoricians Servius and Donatus as a figure in which we which we say less and mean more (*minus dicimus et plus significamus,* cited in [Hoffmann, 1987: 28-9]), thus representing one of the first explicitly pragmatic analyses in the Western tradition (cf. [Horn, 1991] for elaboration). Note, however, that litotic interpretations tend to be asymmetrical: it is more likely that calling someone "not happy" or "not optimistic" will convey a contrary (= rather unhappy/pessimistic) than that such virtual contrariety will be signalled by "not sad" or "not pessimistic", which tend to be understood as pure contradictories. Explanations for this asymmetry have been proposed by Ducrot [1972] and Horn [1989: Chapter 5].

For a more formal approach to the strengthening of contradictory negation to virtual contrariety is due to the fourteenth century logician Robert Bacon. Bacon begins the discussion in his *Syncategoreumata* by distinguishing three varieties of interaction between negation and its focus: the ordinary negative name *(nomen negatum)* 'isn't just' *(non est iustus)*, the infinite name *(nomen infinitum)* 'is not-just' *(est non iustus),* and the privative name with incorporated negation *(nomen privatum)* 'is unjust' *(est iniustus).* Technically, he notes, the third of these unilaterally entails the second and the second the first, but ordinary usage is not always consistent with this:

> Ex hiis patet quod bene sequitur argumentum a privato ad infinitum, ut: 'est iniustus; ergo est non iustus.' Similiter: ab infinito ad negatum, ut: 'est non iustus; ergo non est iustus.' Econverso autem non tenet, sed est paralogismus consequentis. [Braakhuis, 1979, Vol. I, 144-45; Spruyt, 1989: 252]

> From these it is apparent that the argument follows validly from the privative to the infinite, thus "s/he is unjust, therefore s/he is not-just." Similarly, from the infinite to the negative, thus "s/he is not-just, therefore s/he isn't just." However, the converse does not hold, but is the fallacy of consequence.

Thus, for Bacon, the move from the contradictory *X isn't just* to the contrary *X is unjust* is an instance of the fallacy of consequence, the deductively invalid but inductively plausible strengthening of a sufficient condition to a necessary-and-sufficient condition (see [Horn, 2000] for an elaboration of this point within a pragmatic account of "conditional perfection").

As for the "neg-raised" reading of "I don't think that *p*" as "I think that not-*p*", while often dismissed as an incidental and deplorable ambiguity or (in Quine's terms) an "idiosyncratic complication" of one language —

> ... the familiar quirk of English whereby 'x does not believe that p' is equated to 'x believes that not p' rather than to 'it is not the case that x believes that p' [Quine, 1960: 145-6]

> [T]he phrase 'a does not believe that p' has a peculiarity... in that it is often used as if it were equivalent to 'a believes that $-p$'. [Hintikka, 1962: 15]

> 'I do not believe that p' can be unfortunately ambiguous between *disbelief* $[B_a - p]$ and *not belief* $[-B_a p]$. [Deutscher 1965: 55]

— its roots and signifance for the study of negation go far deeper. The locus classicus for the phenomenon is St. Anselm's observation in the Lambeth fragments antedating Bosanquet by eight centuries; cf. [Anselm, 1936] and the commentary in [Henry, 1967: 193-94]; [Hopkins, 1972: 231-32], and [Horn, 1978: 200, 1989: 308ff.]. Anselm points out that *'non ... omnis qui facit quod non debet peccat, si proprie consideretur'* — not everyone who does what he *non debet* ('not-should') sins, if the matter is considered strictly (i.e. with the contradictory reading of negation as suggested by the sentence structure); the problem is that we tend to use *'non debere peccare'* to convey *debere non peccare*, rather than its literal contradictory meaning ('it is not a duty to sin'). A man who does what is not his duty does not necessarily sin thereby, but it is hard to stipulate e.g. *non debet ducere uxorem,* the proposition that a man need not marry without seeming to commit oneself to the stronger *debet non ducere uxorem,* an injunction to celibacy [Henry 1967: 193ff.]; cf. [C. J. F. Williams, 1964; Horn 1978b: 200].

For Henry [1967: 193, §6.412], Anselm's observations on modal/negative interaction are "complicated by the quirks of Latin usage. He has become conscious of the fact that, according to that usage, *'non debet',* the logical sense of which is 'It isn't that he ought', is normally used not to mean exactly what it ways, but rather in the sense more correctly expressed by *'debet non'* ('he ought *not*')." In fact, rather than constituting a quirk of English and/or Latin usage, "neg-raising" —

the lower-clause understanding of negation of a *believe-* or *ought*-type predicate—is distributed widely, although systematically, across languages and operators. The raised understanding is always **stronger** than the contradictory (outer) negation; it applies to a proper subset of the situations to which the contradictory applies (is true in a proper subset of possible worlds). Thus neg-raising, as Anselm recognized, always yields a virtual contrariety: the compositional meaning is true but too weak, and the addressee recovers (short-circuited) conversational implicature to 'fill in' the stronger proposition.

In any event, Bosanquet seems to have been the first philosopher to see the general pattern represented by such tendencies in ordinary language, although other instances of his principle (e.g. the generalization that affixal negation in words like *unhappy* or *unjust* tend to develop contrary semantics cross-linguistically) could be cited. One way of putting the point is that contrariety tends to be maximized in natural language, while subcontrariety tends to be minimized.

Asymmetry revisited: the rise and fall of the Neo-Idealists

Returning to the broader question of 'positive negation', we may count H. H. Joachim among its Idealist proponents. For Joachim, another holder of the Wykeham logic professorship, a negative utterance expresses some "knowledge in the making only" [1906:136]. Thus, an utterance like "The diagonal of the square is not commensurable with its side" is not really negative, but has a positive, real import, namely, "to constitute a problem for a certain level of geometrical *knowledge*. We have here a *real* disunion of elements in the *real* whole. The judicial separation expresses a *real* divorce." [1906:128]. The Idealist account was the fare of the Oxford of the period, as in the logic manual by W. H. Joseph, of whom Grice observes that he "was dedicated to the Socratic art of midwifery; he thought to bring forth error and to strangle it at birth" [Grice, 1986a:62]. If "Dead nettles do not sting", they nevertheless should possess some positive quality or other. If no sentient being existed the utterance 'The wall is not blue' could not be true, since it can be uttered only because someone may suppose or believe the wall to be blue. We must accept the negative judgment as expressing the *real* limitation of things, but we must allow that it *presupposes* the affirmative. There is always a positive character as the ground of negation. Snow is not hot because it is cold." [Joseph, 1916:172]. It's not clear why, on this account, the present king of France is not bald. Note also that the square root of 4 would not be 3 even if no human ever existed to express this or any other negative proposition; see Horn 1989: §1.2.2 on arguments for and against the putatively subjective nature of negation.

By the time Grice was writing, neo Idealism had long since been effectively dispelled by Cook Wilson and others. Wilson (1926) sketches a theory of negation that leaves room for presupposition, implicature, and their contexual cancellation. Most of his examples are implicature- or presupposition-carrying examples of negative utterances. "It is not an odd number", someone says, the implication being "It is an even number". Of course this is again cancelable, "In fact, it {is not a number at all/does not exist}". In the realm of non-binary and empirical propositions, the cancellability is more evident and the range of 'implications' fuzzier. With "This man is not a Mohammedan", Wilson [1926: 250], "I cannot thereby determine his religion, nor even that he has any at all". (Due to the use of the demonstrative 'this' an actual cancellation will be awkward here: "In fact this man does not exist".) Wilson goes on to distinguish between the presupposition and the asserted content. When it comes to the latter, he adopts a neo-Platonic position, foreshadowing Grice's later doctrine.

Wilson writes: "Although it is true that ordinary negative statements (e.g. "Nobody in the next room can read Greek") normally *presuppose* the existence of their subjects of attribution, this existence is not *asserted* by the negative statement as such" [1926:259], and it is thus cancelable as an implicature ("... since the next room is empty"). "When is the verbal form of negative statement natural and normal? When do we naturally say '*A* is not *B*'? Clearly when our conception of Aness does not necessarily involve for us the distinction from *B*ness, or the absence of *B*ness." [1926:272]. He distinguishes two scenarios. "The statement may correspond either to the apprehension of something in *A* which excludes Bness, or to the mere observation of the fact that *B*ness is absent from *A*". This first case is of the form "*A* is *C*", where Cness excludes Bness. His example implicitly draws on the epistemological weakness that the felicitous use of negation typically suggests. "This substance does not show blue colour in the flame of the blowpipe" implies "This substance shows a colour *other than blue* in the flame of the blowpipe." We arrive at this by observing that the colour shown in the flame is, say, red. "Why then have the negative statement at all, and not the affirmative which tells us more and is fully adequate to the thought behind the expression? The negative is not adequate, for if I say "The colour is not blue," I do not say what colour it is and I omit besides something which I know, which also is the reason for what I say." (This argument works if I *know* that the flame is red, but not if I merely have sufficient information to rule out blueness.)

The second scenario for a natural negative utterance is "that in which A is merely observed to be without Bness, an attribute compatible with Aness" His example is "Private Atkins is not in the ranks" (implicating that a private *other than* Atkins is in the ranks). "To find out whether Atkins is in the ranks, we have to observe each rank and file and see that he is not Atkins".

Against the then flourishing neo-Hegelian, Idealist dogma, Wilson [1926: 273] is eager to stress the Realist side to his account. In both scenarios, "the apprehension of the negation and of absence is after all the apprehension of two positive realities as different from one another."

The critique of neo-Idealism was redoubled with Ryle's appearance on the Oxford scene. What Ryle brings to the picture is a closer examination of colloquial cases. While Grice emphasizes that both Formalists and Informalists have failed to give adequate attention "to the nature and importance of the conditions governing conversation", he was presumably exempting Ryle, whose efforts he credited for spearheading "the rapid growth of Oxford as a world centre of philosophy" [Grice,

1986a:48].

With respect to negation, the aim of Ryle, as for Grice and Strawson, is to identify some of the "conversational features" of "not". Ryle's example is a familiar one: "Mrs. Smith's hat is not green", allowing the inference that "Mrs. Smith's hat is *other than* green", understood broadly enough to include those scenarios where Mrs. Smith's hat doesn't exist — and therefore it *is* other than green. But is the hat's other-than-greenness part of the content of the negative statement? Ryle writes:

> When I say 'Mrs. Smith's hat is not green', I can equivalently say 'but some other colour'. The 'but some other' is always there, sometimes explicitly, sometimes marked by tone of voice, or simply implied by the context. Without the but clause, negative sentences are elliptical, though still generally interpretable in context. When I say, 'Mrs. Smith's hat is not green but some other colour', I am not stating but presupposing that the hat is coloured. In general, for any 'The S is not P', what gets presupposed is that P belongs to a contextually assumed set, some other member P' of which holds of S. The full explication of what is meant by a negative sentence takes the form of an assertion of otherness as specified or made determinate by mention of the particular disjunctive set to which the other belong to as members. [Ryle, 1929: 89]

Such presuppositions are regarded by Ryle, as they would be for Strawson and as they were for Frege, as pre-conditions for the truth or the falsity of a judgment, rather than merely as conditions on felicitous assertion. This presuppositional approach would deny that "Mrs. Smith's hat is not green" could be (trivially) *true* in the absence of Mrs. Smith's hat, whatever the implicature may be on that occasion. For transcategorial examples, the case is different; "Virtue is not square" is still *true*, even if the continuation, 'but some other shape' confuses rather than illuminates.

Negation, presupposition, and the bracketing device

In the spirit of Modernism, Grice proposes — as we have seen in the formation, inference, and semantic rules of system G — some type of 'formal' indication of the systematic interaction of negation with other elements of the logic. Grice played with two 'scope' devices, and we shall consider them in turn, as they illustrate a pattern in the history of logic. The first is a subscript notation for scopal ambiguity. Quine, for one, found the system "forbiddingly complex" yet, he remarks, "on the whole I am for it" [Quine, 1969:326]. The idea is that any constitutent in a formula gets a subscript to mark its order of arrival. Thus, for the negation of a basic formula "∼p", the device delivers two readings: "$\sim_2 p_1$" and "$\sim_1 p_2$". Consider one of Grice's early examples, "Jones has not left off beating his wife". The default 'logical' reading is the second. Hence the possibility of the cancellation, "He is not married". The default 'pragmatic' reading is the former, hence the paradoxical flavor.

Or consider, "The king of France is not bald". On one reading, the logical one, "$\sim$" has maximal scope: "$\sim_3 ((\exists x)K_1x\&(\forall y)(Ky \rightarrow x = y)\&(B_2x))$". On the pragmatic reading, negation is internal: "$\sim_1 ((\exists x)K_2x\&(\forall y)(Ky \rightarrow x = y)\& (B_3x))$", and gets the cancellation, 'There is no such king" ($\sim_1 ((\exists x)K_2x\&(\forall y) (Ky \rightarrow x = y)\&(B_3x)) \vDash (\exists x)K_2x$). In a notational variant mentioned by Grice, taking up a suggestion by Hans Sluga, the two readings are: $\sim_2 B_1\iota x_3K_1x_2$ and $\sim_4 B_1\iota x_3K_1x_2$. (Grice mentions a notational variant suggested by Charles Parsons: "$[Kx](\sim Bx)$" "$\sim [Kx](Bx)$").

Grice refers to the two readings as the weak reading and the strong reading. The diagnostic is the role played by negation.

> If there were a clear distinction in sense (in English) between, say, 'The king of France is not bald' and 'It is not the case that the king of France is bald' (if the former demanded the strong reading and the latter the weak one), then it would be possible to correlate 'The king of France is bald' with the formal structure that treats the iota-operator like a quantifier. But this does not seem to be the case; I see no such clear semantic distinction. So it seems better to associate 'The king of France is bald' with the formal structure that treats the iota-operator as a term-forming device. [Grice, 1989:272]

The second device is a bracketing device, yielding the representation: "$\sim [(\exists x)Kx\& (\forall y)(Ky \rightarrow x = y)]\&Bx$". Since the square-bracketed material is (normally) scopally immune to negation, this will as a default get rewritten as "$(\exists x)Kx\&(\forall y) (Ky \rightarrow x = y)\& \sim Bx$", with the externalization of the "presupposed" material.

What gets square-bracketed is not so much the 'positive ground' for a given negation, but what Grice calls 'common-ground' status, or noncontroversiality. The square-bracketed material is only implicated, and thus cannot occur freely in monotonic inferences, since implicatures are cancelable (cf. [Wilson, 1975] for a different implementation of a pragmatic account of existential "presuppositions" as conversational implicatures.)

Consider "Jones does not regret Father is ill" [Grice, 1989: 280-1]. A preferred pragmatic inference is again the implicature-carrying one ("Father is ill and Jones thinks Father is ill and Jones is not against Father being ill"), which is cancelable ("Jones does not regret Father is ill. Indeed Father is not ill"). What the square-bracket device does is raise the subordinate clause of 'regret' outside the scope of "$\sim$": "[Father is ill &] Jones thinks Father is ill & $\sim$(Jones is ANTI (Father is ill)"). (See [Horn, 2002: 74-76] for a new look at Grice's bracketing device.)

Another area for implicatural treatment concerns the scope of negation outside the radical ("$\sim \surd p$" rather than "$\surd \sim p$""). In discussing a 'general notion of satisfactoriness' [Grice, 2001:83], while generalized versions for the binary truth-functors are unproblematic — '$\phi\&\psi$' is satisfactory just in case ϕ is satisfactory and ψ is satisfactory, and so on, the "unary truth-functor is not so easily dealt with: "The real crunch comes with negation. '$\sim\vdash p$' might perhaps (contra [Frege, 1919]) be treated as equivalent to '$\vdash\sim p$'. But what about '$\sim!p$'?" — i.e. negation outside the scope of a directive speech act.

In particular, Grice asks,

> "What do we say in cases like, perhaps, 'Let it be that I now put my hand on my head' or 'Let it be that my bicycle faces north', in which (at least on occasion) it seems to be that neither '$!A$' nor '$! \sim A$' is either satisfactory or unsatisfactory? What value do we assign to '$\sim!A$' and to '$\sim!A$'? Do we proscribe the forms altogether (for all cases)? But that would seem to be a pity, since '$\sim! \sim A$' seems to be quite promising as a representation of 'you may (permissive) do A': that is, I signify my refusal to prohibit your doing A. Do we disallow embedding of these forms? But that (again if we use them to represent 'may') seems too restrictive".

The problem "would require careful consideration; but I cannot see that it would prove insoluble, any more than analogous problems connected with presupposition are insoluble; in the latter case the difficulty is not so much to find a solution as to select the best solution from those which present themselves." [2001:89].

Implicature and negation: scales, scopes, and metalinguistic negation

Grice characterizes the complications often introduced by negation in terms of a loss of (logical) innocence. "[I]f rational beings are equipped to assert a certain range of statements, they must also be supposed to be equipped to deny just that range of statements. In that case, the negations of the initial range of logically innocent statements may be supposed to lie within the compass of the speakers of the language; and these statements by virtue of their character as denials, may not wear the same guise of logical innocence." [1989:70]. One illustration involves the interaction of negation with focus or contrast. Grice considers various conversational scenarios here, including one in which an utterer B says (apparently out of the blue), "JONES didn't pay the bill". Grice comments [1989: 52]:

> The remark is not prompted by a previous remark (it is volunteered), and we are inclined to say that the implicature is that someone thinks or might think that Jones did pay the bill. The maxim of Relation requires that B's remark should be relevant to something or other, and B, by speaking as if he would speak in reply to a statement that Jones paid the bill, shows that he has such a statement in mind".

A similar point had been made earlier by Ryle, concerning the example "Jones is not the secretary of the club", where stressing each constituent projects what, in Gricean terms, is a different implicature: "*Jones* is not the secretary of the club" (someone other than Jones is), "Jones is not the *secretary* of the club" (he holds an office other than secretary), "Jones is not the secretary of the *club* (he is the secretary of an institution other than the club), "Jones *is* not the secretary of the

Another much discussed set of examples involves an interchange related to a different exchange Grice presents in this section:

> *I KNEW that* may be contrasted with *I believed that,* and the speaker may implicate not that he would deny *I believed that p,* but that he would not confine himself to such a weaker statement, with the implicit completion *I did not merely believe it.* ([Grice, 1989:52], cf. Proclus, *Parmenides* 913 on *hyperapophasis*)

In such cases, a logically weaker or less informative utterance implicates that the speaker was not in position to have uttered a stronger alternative, salva veritate.

The operative principle is Grice's first submaxim of quantity (or the earlier "rule of strength" attributed by Strawson 1952 to Grice; see discussion above). In fact, this rule and its epistemological constraints on its operation date back at least to Mill [1867:501]:

> If I say to any one, 'I saw some of your children to-day', he might be justified in inferring that I did not see them all, not because the words meant it, but because, if I had seen them all, it is most likely that I should have said so: though even this cannot be presumed unless it is presupposed that I must have known whether the children I saw were all or not."

But as Mill goes on to observe, this cannot be a part of the content or, as we would now call it, the logical form of expressions with "some", contra Sir William Hamilton's arguments for doing just that: "No shadow of justification is shown ... for thus adopting into logic a mere sous-entendu of common conversation in its most unprecise form." Similarly, while disjunctions are naturally taken exclusively — "When we say *A* is either B or *C* we imply that it cannot be both" — this too cannot be a logical inference: "If we assert that a man who has acted in a particular way must be either a knave or a fool, we by no means assert, or intend to assert, that he cannot be both" [Mill, 1867: 512]. (See [Horn, 1972; 1989; 1990; Speranza, 1989] for more on Mill and similar arguments from De Morgan.)

This scalar or *Q*-implicature induced here by the quantitative scale <all, some> arises in a wide range of cases involving both logical operators and ordinary predicates (see [Levinson, 2000] for a comprehensive catalogue). The one alluded to by Grice above involves the scale <know, believe>, with the result that an assertion of belief generally (but non-monotonically) conveys absence of knowledge. In general, for a scale $\langle P_n, P_{n-1}, \ldots, P_2, P_1 \rangle$ and a constituent P_i, the *Q*-implicature is $\sim S(P_i/P_j)$ for all $P_j > P_i (j \neq n)$, where "$\phi(P_i/P_j)$" denotes the result of substituting P_j for P_i within ϕ), $\sim S(Pi/Pn)$, and if $P_k > P_j > P_i, \sim S(P_i/P_j), \sim S(P_i/P_k)$. (See [Gazdar, 1979: 55-62] for a comprehensive formulation.)

But now we are prepared to see that a speaker may choose to convey "I didn't merely believe that p" not by asserting "I KNEW that" (as in Grice's example of contrastive stress) but by apparently denying that she believes that p, especially with the appropriate continuation or rectification: "I didn't believe that p, I KNEW it". This does not require rejecting the inference from knowledge to belief, if we acknowledge a specialized metalinguistic or polemic use of negation in such contexts, and at the same time such an account permits us to retain the "logical innocence" of ordinary negation. Note that while 'He believed it' is *true* if he also knew it, 'He merely believed it' is *false* in the same scenario. Essentially, this specialized use of negation targets not the propositional content (what is said) but the (potential) implicature.

The tacit principle which Mill invokes and which Grice later formulates as the Quantity submaxim, requiring the speaker to use the stronger *all* in place of the weaker *some* when possible and licensing the hearer to draw the corresponding inference when the stronger term is not used, is systematically exploitable to yield upper-bounding generalized conversational implicatures associated with scalar operators. Quantity-based scalar implicature — e.g. my inviting you to infer from my use of *some...* that for all I know *not all...* — is driven by our presumed mutual knowledge that I expressed a weaker proposition in lieu of an equally unmarked utterance that would have expressed a stronger proposition. Thus, what is said in the use of a weaker scalar value like those in *I saw some of your children* or *I believed that p* is the lower bound *(...at least n...)*, with the upper bound *(...at most n...)* implicated as a cancellable implicature. The prima facie alternative view, on which a given scalar predication is lexically ambiguous between weaker and stronger readings, is ruled out by the general metatheoretical consideration that Grice dubs the Modified Occam's Razor principle: "Senses are not to be multiplied beyond necessity" [1989: 47].

Negating such predications normally denies the lower bound: to say that something is not possible is to say that it's impossible, i.e. less than possible. When it is the upper bound that appears to be negated *(It's not possible, it's necessary)*, a range of linguistic and logical evidence indicates that what we are dealing here with is an instance of the metalinguistic (or echoic) use of negation, in which the negative particle is used to object to any aspect of an alternate (actual or envisaged) utterance, including its conventional and conversational implicata, register, morphosyntactic form or pronunciation [Horn, 1989: Chapter 6; Carston, 1996]. If it's hot, it's (a fortiori) warm, but if I know it's hot, the assertion that it's warm can be mentioned and rejected as (not false but) insufficiently informative: "It's not WARM, it's HOT!", "You didn't see SOME of my children, you saw ALL of them", "I didn't BELIEVE that p, I KNEW it".

Such uses of negation effectively scope over other varieties of implicature. Thus, in "It is not the case that she got married and had a child" (after [Grice, 1989: 8]), the utterer may be denying the implicature (generated by the "Be orderly" submaxim) that the events referred to occurred in the order in which they mentioned, conveying something like "Rather, she had a child and [then] got married. If this can be analyzed away as metalinguistic negation, there is no need to depart from the syntax and semantics of System G ("$\sim (p\&q)$").

Or with disjunction: "It is not the case that my wife is in Oxford or in London — she's in Oxford, as you well know." This too can be understood metalinguistically, with the negation scoping over the implicature responsible for the non-truth-functional condition on the felicity (but not the truth!) of *p or q* statements that the speaker should not be in a position to assert either disjunct individually. The logical form is again, however, the classical one: "$\sim (p \vee q)$".

Grice considers a related phenomenon what he calls "substitutive disagreement", as opposed to truth-functional "*contradictory* disagreement" [1989:64]. Once again, we have a negation that does translate into "$\sim p$". In such cases, "I am not contradicting what you say. It is rather that I wish not to assert what you have asserted, but to substitute a different statement which I regard as preferable in the circumstances" [Grice, 1989:64]. This situation arises with both disjunctions and conditionals. Thus,

> If you say "X or Y will be elected", I may reply "That's not so; X or Y *or* Z will be elected." Here ... I am rejecting "X or Y will be elected" not as false but as unassertable. [Grice, 1989: 82]

I do not thereby deny the proposition you have expressed (which would amount to a commitment to the electoral failure of both X and Y), but reject your assertion as epistemologically unwarranted (in ruling out candidate Z). Grice emphasizes that "the possibility of speaking in this way gives no ground for supposing that "or" is not truth-functional"; it does, however, introduce a subtler question on the truth-functionality of "not".

Similarly, to negate a conditional is typically not to assert '$\sim (p \rightarrow q)$': to say "It is not the case that if Jones is given penicillin, he will get better" might be a way of suggesting that the drug might have no effect on X at all" rather than committing the speaker to the truth of "Jones will be given penicillin" and the falsity of "Jones will get better".

> Sometimes the denial of a conditional has the *effect* of a *refusal to assert* the conditional in question, characteristically because the denier does not think that there are adequate *non*-truth-functional grounds for such an expression. In such a case, he denies, in effect, what the thesis represents as an implicature of the utterance of the unnegated conditional. [Grice, 1989:81]

As Grice notes in his 'Retrospective Epilogue', this approach requires an acknowledgement of the possiblity that conversational implicatures need not take wide scope, in particular with respect to negation, a possibility Grice endorses with some diffidence:

> When a sentence which used in isolation standardly carries a certain implicature is embedded in a certain linguistic context, for example appears within the scope of a negation-sign, must the negation sign be interpreted only as working on the conventional import of the embedded sentence, or may it on occasion be interpreted as governing not

> the conventional import but the nonconventional implicatum of the embedded sentence? Only if an embedding operator may on occasion be taken as governing not the conventional import but the nonconventional implicatum standardly carried by the embedded sentence can the first version of my account of such linguistic phenomena as conditionals and definite descriptions be made to work. The denial of a conditional needs to be treated as denying not the conventional import but the standard implicatum attaching to an isolated use of the embedded sentence. [Grice, 1989:375]

Neo-Traditionalists such as Strawson have remained unconvinced by the argument, refusing to concede that the issue of the divergence between "$\sim$" and "not" was now settled, especially with regard to negated conditionals: "The Gricean, though perhaps with a slight air of desperation, could reply that one who denies the condition has no interest in denying what it conventionally and literally means, but only in denying what it standardly and conversationally implies" [Strawson, 1991:15]. Nor is it clear that treating the negatum as a conventional implicatum rather than a conversational one would minimize the air of desperation in the eyes of a Strawson.

Another case that allows for reanalysis involving implicature, as we have noted, concerns the aforementioned case of the disappearing presupposition.

> As far as I can see, in the original version of Strawson's truth-gap theory, he did not recognize any particular asymmetry, as regards the presupposition that there is a king of France, between the two sentences, *The king of France is bald* and *The king of France is not bald;* but it does seem plausible to suppose that there is such an symmetry. I would have thought that the implication that there is a king of France is clearly part of the conventional force of *The king of France is bald;* but that this is not clearly so in the case of *The king of France is not bald*...An implication that there is a king of France is often carried by saying *The king of France is not bald*, but it is tempting to suggest that this is implication is ... a matter of conversational implicature. [Grice 1989: 270]

One key here is the application of the tests for implicature: cancellability (*The king of France isn't bald (because) there is no king of France)* and non-detachability. To demonstrate the latter, Grice notes that whether we use the frame "It is not the case that the King of France is bald", "It is false that the king of France is bald", or "It is not true that the king of France is bald", "Many of what seem to be other ways of saying, approximately, what is asserted by ["The king of France is not bald"] also carry the existential implicature" [Grice, 1989: 271]. Such tests can be taken to support those proposals, including those employing Grice's bracketing device, concur in distinguishing the positive expression (in which existence is entailed) from the negative (in which it is non-monotonically implicated).

Implicature and negation: Subcontrariety and the three-cornered square

Scalar implicature plays another role in the expression of negation in natural language. A Gricean understanding of the relationship between strong and weak scalar values helps motivate a natural account of the lexicalization asymmetry of the Square of Opposition, an asymmetry — a.k.a. the Story of ***O** — displayed in tabular form below (cf. [Horn, 1972: Chap. 4; Horn, 1989: §4.5; Levinson, 2000; Horn, 2012b]).

	DETERMINERS/ QUANTIFIERS	QUANT. ADVERBS	BINARY QUANTIFIERS	CORRELATIVE CONJUNCTIONS	BINARY CONNECTIVES
A:	all α, everyone	always	both (of them)	both...and	and
I:	some α, someone	sometimes	one (of them)	either...or	or
E:	no α, no one (=all¬/¬some)	never (=always¬)	neither (of them) (=both¬/¬either)	neither...nor (=[both...and]¬)	nor (=and¬)
O:	*nall α, *neveryone (= some¬/¬all)	*nalways (= ¬always)	*noth (of them) (= either¬/¬both)	*noth...nand (= [either...or]¬)	*nand (= and¬/¬or)

In fact, this asymmetry was recognized for Latin by St. Thomas, who observed that whereas in the case of the universal negative (**A**) "the word 'no' *[nullus]* has been devised to signify that the predicate is removed from the universal subject according to the whole of what is contained under it", when it comes to the particular negative (**O**), we find that

> there is no designated word, but 'not all' *[non omnis]* can be used. Just as 'no' removes universally, for it signifies the same thing as if we were to say 'not any' [i.e. 'not some'], so also 'not all' removes particularly inasmuch as it excludes universal affirmation. (Aquinas, *in Arist. de Int.,* Lesson X, Oesterle 1962: 82-3)

The Gricean model offers a persuasive motivation for this asymmetry. Although *some* does not contribute the same semantic content as *some not (= not all),* the use of either of the two values typically results in a speaker communicating the same information in a given context, viz. 'some but not all'. The relation of mutual quantity implicature holding between the positive and negative subcontraries results in the superfluity of one of the two for lexical realization and the functional markedness of negation predicts that the unlexicalized subcontrary will always be the one with an **O** rather than **I** meaning. The existence of a lexicalized **O** form implies the existence of a lexicalized **E** counterpart but not vice versa. Additional evidence (see above sources) indicates that even when both forms are attested, as with the negative modalities *can't, mustn't, shouldn't* (**E**) vs. *needn't* (**O**), the lexicalized **E** form tends to be more opaque and semantically and distributionally less constrained than its **O** counterpart. This pragmatic, implicature-based account of the 'three-cornered square' is more general and more explanatory than rival theories that either dismiss the asymmetry as uninteresting or restrict it to the determiners and quantificational operators while bypassing other operator types (e.g. connectives, adverbs, and modalities) along with intermediate values that can be mapped onto the Square of Opposition. (See [Horn, 1989; 2012b] for details and [Jaspers, 2005; Seuren, 2006] for alternative treatments and related discussion.)

Negation and denial

A topic that concerned mathematicians such as Griss [1945] is whether weak negation can be made sense of. He thought not, whence his 'negationless system'. System G is not like that. But the issue of how denial and negation interact is a further problem that must be handled via implicature.

For Grice, one key step occurs when the logical squiggle is internalized into a psychological attitude operator. Grice sets the question in dealing with how negation interacts with the scope of various psychological attitudes in his presidential address to the American Philosophical Association, 'Method in Philosophical Psychology', where Grice explores what an account of negation in terms of the psychological attitude of denial might look like. This has a connection with topics that concerned Austin and specifically with the pattern of inferences in psychological contexts (e.g. If one believes that not-not-p, does one believe that p?). Undertaking the exploration of "not" and '$\sim$' in connection with propositional attitudes is an attempt to examine negation in models of belief- and knowledge-based reasoning. Grice views this as a metaphysical issue. "References to such psychological states will be open to logical operations such as negation" [Grice, 1991:146].

The internalization of the logical operation of negation within the scope of a psychological attitude operator may be seen as involving various stages. Grice distinguishes four:

> At the first stage we have some initial concept, like that expressed by 'not'. We can think of it as, at this stage, an intuitive or unclarified element of our conceptual vocabulary. At the second stage, we reach a specific mental state, in the specification of which it is possible, though maybe not necessary to use the name of the initial concept as an adverbial modifier, 'not-thinking' (or 'rejecting,' or 'denying'). This specific state may be thought of as bound up with, and indeed as generating, some set of responses to the appearance on the scene of an instantiation of the initial concept. At the third stage, a reference to this specific state is replaced by a more general psychological verb, together with an operator corresponding to the particular specific stage which appears within the scope of a general verb ['accept'], but is still allowed only maximal scope within the complement of the verb, and cannot appear in sub-clauses ['thinking not-p', 'accepting not-p']. At the fourth stage, the restriction imposed by the demand that the operator at stage three should be scope-dominant within the complement of the accompanying verb is removed [...]. [1986a:98)].

While Grice uses "He is not lighting his cigarette with a 20-dollar bill" to make an indirect reference to Bradley on negation, he is positively concerned elsewhere with the ontology and logic of non-events (cf. [Gale, 1976; Horn, 1989: 54-55]):

> In many cases, what are to be counted as actions are realized, not in events or happenings, but in non-events or non-happenings. What I do is often a matter of what I do not prevent, what I allow to happen, what I *refrain from* or *abstain from* bringing about — what, when it comes about, I ignore or disregard. I do not interrupt my children's chatter; I ignore the conversational intrusions of my neighbour; I omit the first paragraph of the letter I read aloud; I hold my fire when the rabbit emerges from the burrow, and so on.... Such omissions and forbearances might lead to the admission of negative events, or negative happenstances, with one entity filling the 'event slot' if on a particular occasion I go to Hawaii, or wear a hat, and another entity filling that slot if on that occasion I do not go to Hawaii, or do not wear a hat. [Grice, 1986a:22]

It seems it would be again the conversational context that would advise us as to whether and how to apply the "$\sim$" in a given formalization.

One simple set of implicatures concern *duplex negatio affirmat.* Following Bishop Lowth, Edward Bentham recites the standard principle: "According to the idiom of some languages (Latin and English), two negative particles destroy each others force, and make the proposition affirmative" (cf. [Horn, 1991] for additional references and elaboration). But if, as Jevons avers, "Negatives signify the absence of a quality" [Jevons, 1870:22], a double negative would signify a double absence, which does not obviously yield a presence. The point had been noted by Strawson: "This identification [of *not* and "$\sim$"], then, involves only those minimum departures from the logic of ordinary language which must always result from the formal logician's activity of codifying rules with the help of verbal patterns: viz., the adoption of a rigid rule when ordinary language permits variations and deviations from the standard use ("$\sim\sim p \supset p$", "$p \vee \sim p$")."

The issue was again raised by L. J. Cohen [1971], who argues from the existence of negative concord (in languages like Italian, present-day non-standard dialects of English, or earlier standard English, in which e.g. "I don't want nothing" expresses the same proposition as that expressed in standard English by "I don't want anything") to the incoherence of Grice's identification of "$\sim$" with "not". In such cases *Duplex negatio negat*, although here again, the question arises for the grammarian (if not the logician) as to why a speaker would go out of her way to employ double negation rather than expressing a simple negative (or, in the *Duplex negatio affirmat* cases, a simple affirmative) directly; cf. [Horn, 2012b]. In any case, Cohen's objection seems ill-taken; the natural Gricean rebuttal is to treat negation as being an abstract syntactic constituent or element of semantic representation, rather than to limit the focus to the superficial occurrence of equivalents of "not" in ordinary language. (cf. [Gazdar, 1979: 63-64]).

Implicature also plays a role in motivating the choice of the standard "$\sim$" operator, yielding a truth array of $< 01 >$, over the three other possible unary truth-functors, those yielding arrays of $< 10 >$, $< 11 >$, and $< 00 >$. The choice is based on the fact that "$\sim$" is the only one-place operator that, in collaboration with the Co-Operative Principle and its maxims, yields an intuitively plausible system. The other unary functors can be shown to be either redundant or semantically incoherent [Gazdar, 1979: 76]. System G narrows propositional negation to the contradiction function, while deriving the strengthened reading of contrariety, where appropriate, as an implicature (along the lines of [Horn, 1989, Chapter 5]).

Negation and falsity

One favourite issue for the Oxford Play-Group was the extent to which negation could be identified with falsity. While Grice himself often took "It is false that p" and "It is not the case that p" to count as moves in the same game [1989:271], the situation in reality is somewhat more complex.

For centuries, one popular method for eliminating negation has proceeded by identifying it with and "reducing" it to falsity. One question is whether such a "reduction", if it could be accomplished, would really accomplish anything. But there are in any case strong grounds for rejecting the proposed identification in the first place, without even considering its role within a reductionist program.

That negation and falsity might be conflated, and eventually confused, with each other should not be surprising. Aristotle discusses 'being in the sense of true and non-being in the sense of false' (*Met.* 1027b18), and he seems to explicitly link the negated copula with falsity (as the affirmative copula is linked with truth):

> "To be" and "is" mean that something is true, and "not to be" that it is not true but false... For example, in "Socrates is musical", the "is" means that it is true that Socrates is musical, and in "Socrates is not-white", that this is true; but in "the diagonal is not commensurate with the side" the "is not" means that it is false that the diagonal is commensurate with the side.

Within Aristotle's simple correspondence theory of truth, truth and falsity are interrelated as the two terms of a contradictory opposition. But contradictory negation does not reduce to falsity, since negation and falsity are about different things and operate on different levels: "A falsity is a statement of that which is that it is not, or of that which is not that it is; and a truth is a statement of that which is that it is, or of that which is not that it is not" (Met. 1011b25-27; cf. De Int. 18b2-4).

The equation of negation (often specifically "logical" negation) and falsity is a frequent maneuver among the Idealists of the late 19th and early 20th century, often going hand-in-hand with a view of negation as a second-order comment on a first-order affirmation, and/or as a more subjective act than simple affirmation:

> To say "A is not B" is merely the same as to deny that "A is B", or to assert that "A is B" is false. [Bradley, 1883: 118]

> "A is not B" means "it is false, it must not be believed that A is B"... Immediately and directly, the negation is a judgment concerning

> a positive judgment that has been essayed or passed. [Sigwart, 1895: 122]

> a is not b = that a is b is false. [Baldwin, 1928: 147]

> The pure negative judgment 'A is not B' is equivalent in every case to 'it is false that A is B'... 'Snow is not black' is a shorthand statement for 'snow is black is an erroneous judgment'. [Wood, 1933: 421]

For Russell [1940: 81], too, every negation is a shorthand for some assertion of falsity: "It is unnecessary to have the two words "false" and "not", for, if p is a proposition, "p is false" and "not-p" are strictly synonymous."

Within the modern logical (and linguistic) tradition, the temptation to identify negation and falsity stems directly from the Fregean line that all negation is propositional and reducible to a suitably placed *'it is not true that...'*. In multi-valued logics, there is one form of negation (internal, strong, choice) which does not display the logic of contradictory opposition, being governed by the Law of Non-Contradiction but not the Law of Excluded Middle (see [Horn, 2006]). Within such approaches, at least some negations cannot be "reduced to" assertions of falsity. Similarly, there may be illocutionary distinctions between the negation of a proposition and the statement that that proposition is false, as in Heinemann's differentiation [1944: 143] of not-p ('$\sim p$ is valid') from 'p is not valid'. But even within classical two-valued logic itself, there are sufficient grounds for rejecting the identification of negation and falsity. Philosophers as diverse as Frege [1919], Austin [1950], Quine [1951], and Geach [1972] have observed that the identification of *not* and *false* results from a confusion of language and metalanguage.

Here is Austin's response to the view (represented by Ayer) that 'is true' and 'is false' are logically superfluous:

> An important point about this view is that it confuses falsity with negation: for according to it, it is the same thing to say 'He is not at home' as to say 'It is false that he is at home'... Too many philosophers maintain, when anxious to explain away negation, that a negation is just a second order affirmation (to the effect that a certain first order affirmation is false), yet, when anxious to explain away falsity, maintain that to assert that a statement is false is just to assert its negation (contradictory)... Affirmation and negation are exactly on a level, in this sense, that no language can exist which does not contain conventions for both and that both refer to the world equally directly, not to statements about the world. [Austin, 1950: 128-9]

Quine [1951: 27-8] is also at pains to distinguish the predicates 'is false' and 'is true', which are used to speak about statements, from the connective "$\sim$", which is used to make statements. *'Jones is ill' is false* is a statement about the statement *Jones is ill,* while $\sim$(*Jones is ill),* read 'Jones is not ill', is a statement about Jones. Quine lays the mistaken identification of '$\sim$' with falsehood at Whitehead and Russell's door, but the underlying mistake both antedates and survives the

Principia — as does its rectification. The Stoics were careful to make the same distinction as Quine, that 'between the negation of a proposition and a (meta-linguistic) statement that the proposition is false'; these two operations played different roles in the Stoics' account of syllogistic reasoning [Mates, 1953: 64-5].

In the same vein as Austin and Quine, Geach [1972: 76] inveighs against the 'widespread mistake' of assuming that 'the negation of a statement is a statement that that statement is false, and thus is a statement *about* the original statement and logically secondary to it'. The error of this approach emerges clearly when we look at non-declaratives: '"Do not open the door!" is a command on the same level as "Open the door!" and does not mean (say) "Let the statement that you open the door be false!"'

For symmetricalists like Austin, Geach, and Ayer, conclusions as to the secondary status of negative statements with respect to affirmatives often betoken a confusion of meaning with 'use'. This is how Ayer —and Grice — diagnose Strawson's account of negation. In Ayer's words,

> From the fact that someone asserts that it is not raining one is not entitled to infer that he has ever supposed, or that anyone has ever suggested, that it is, any more than from the fact that someone asserts that it is raining one is entitled to infer that he has ever supposed, or that anyone has ever suggested, that it is not. No doubt negative forms of expression are very frequently used to deny some previous suggestion; it may even be that this is their most common use. But whatever the interest of this fact it cannot be the ground of any viable distinction between different types of statement. [Ayer, 1952: 39]

Ayer goes on to challenge the oft-maintained epistemological worth-less-ness (if not worthlessness) of negative statements: "Why should it not be allowed that the statement that the Atlantic Ocean is not blue is as much a description of the Atlantic as the statement that the Mediterranean Sea is blue is a description of the Mediterranean?" While the negative might well be less informative than its affirmative counterpart, "to say that a description is relatively uninformative is not to say that it is not a description at all" [Ayer, 1952: 47].

"Perhaps", Ayer rhetorically wonders, "there are psychological grounds for sequestering negations as a special class of statements used only for rebuttals or denials? But any statement can be so used" [Ayer,1952:38]. Along the same lines, Grice [1989], while providing "The man at the next table is not lighting his cigarette with a $20 bill" as an instance of a negation that sounds odd if the corresponding affirmative has not been entertained, also brings up cases like "I went to the meeting of my own free will", "I remember my own name", and "Your wife is faithful", where it is the affirmative that is inappropriate in the absence of a specially marked context. When a positive sentence is less informative than the corresponding negation, it is the positive that is odder or presuppositionally richer. (See [Horn, 1989: §3.3.1] for a neo-Gricean derivation of this asymmetry.) Thus, while Strawson may be correct in claiming that "the standard and primary use" of negatives is "to correct and contradict", this cannot be a definitional criterion of the property of negation; use is not meaning.

Envoi: A Gricean program for negation and implicature

While absent from otherwise complex systems of animal communication, negation is a sine qua non of every human language. Indeed, if our species can be dubbed *homo linguisticus,* it is negation that makes us fully human, providing us with the capacity to deny, to contradict, to misrepresent, to lie, and to convey irony. The apparent simplicity of logical negation as a one-place operator that toggles truth and falsity belies the intricate complexity of the expression of negation in natural language. For these reasons, the form and function of negation has engaged the interest and often the passion of scholars for thousands of years. But it is arguably the contributions of Paul Grice that have enabled us to sort out the contributions of general principles of communication and rational interchange to the specific habits of negating and denying that show up in language.

Grice departs from his logician predecessors and from the ordinary language philosophers in the Oxford Play-Group in subsuming his view of linguistic cooperation in the conversational enterprise within a general theory of rationality (see [Kasher, 1982]). But, as he also reminds us, "It is irrational to bite off more than you can chew whether the object of your pursuit is hamburgers or the Truth" [Grice, 1989: 369]. Ever true to the spirit of the Quantity maxim, Grice was always rational enough to bite off neither more nor less than his appetite permitted. But no man lives by meat alone, much less a philosopher of language large enough to bestride the warring camps of Russell's Modernists and Strawson's Neo-Traditionalists; bread is important as well. So it is meet that a healthy portion of the Gricean oeuvre consists not of solutions but of problems, questions, and menus. For, as Grice reminds us elsewhere in offering a defense of absolute value admittedly "bristling with unsolved or incompletely solved problems" [Grice, 1986: 106], "If philosophy generated no new problems it would be dead...Those who still look to philosophy for their bread-and-butter should pray that the supply of new problems never dries up." As the history of work on negation eloquently demonstrates, there is no danger of that dread eventuality coming to pass any time soon.

ACKNOWLEDGMENTS

Our interest in developing the history of negation under a Gricean umbrella results from a convergence of interests. Under the inspiration of his studies with Barbara Partee, David Kaplan, Jim McCawley, George Lakoff, Haj Ross, and Paul Grice in the late 1960s, Horn developed a research program at the union (if not the intersection) of traditional logic, generative semantics, neo-Gricean pragmatic theory, and the analysis of negation. His 1989 book *A Natural History of Negation* (reissued in an expanded version by CSLI in the David Hume series of reprints in

Philosophy and Cognitive Science in 2001) is a comprehensive attempt to extend the Gricean program for non-logical inference to a class of problems arising in the study of negation and its interaction with other logical operators. His interest in Medieval theories on negation, inference, and the semantics of *exponibilia* was sparked by electronic and actual conversations with the late Victor Sánchez Valencia. Meanwhile, Speranza was embarked on a parallel route. His ideas were first formulated in a seminar in the philosophy of logic at the University of Buenos Aires, conducted by Gregorio Klimovsky, where Alberto Moretti, Gladys Palau, and Carlos Alchourron were among the active participants. Speranza later participated fruitfully in a seminar on the history of logic given by Ignacio Angellelli at the University of La Plata. Speranza's rationale for the Gricean approach to the logical operators can be found in chapter iii of his PhD dissertation. Speranza is grateful for commentary and personal correpondence to David Bostock, A. J. P. Kenny, P. H. Nowell-Smith, R. M. Sainsbury, J. O. Urmson, and O. P. Wood. Thanks to Emily Schurr for help with the Latin and Greek. Any faults are the authors'.

Readers interested in current work on negation in philosophy and (especially) linguistics are urged to peruse the papers and extensive bibliography in [Horn, 2012a]. For more background on the history of negation in logic and philosophy, see [Horn, 1989] (and the 2001 reissue with updated introductory and bibliographic material) and [Brann, 2001]. The interaction of negation, contradiction, and the Square of Opposition are surveyed in the Stanford Encyclopedia of Philosophy entries by Horn [2006] and Parsons [2008].

BIBLIOGRAPHY

[Anselm, 1936] Anselm, Saint. The Lambeth fragments. Text collected in F. S. Schmitt (ed.), *Ein neues unvollendetes werk des hl. Anselm von Canterbury.* Münster i. W.: Aschendorff, 1936.

[Ashworth and Spade, 1992] E. J. Ashworth and P. V. Spade. 'Logic', in *The History of the University of Oxford*, ed. J. Catto and Ralph Evans. Oxford: Clarendon Press, pp. 35-64, 1992.

[Austin, 1961] J. L. Austin. *Sense and Sensibilia.* Oxford: Clarendon Press, 1961.

[Austin, 1979] J. L. Austin. *Philosophical Papers.* Oxford: Clarendon Press, 1979.

[Austin, 1980] J. L. Austin. *How to Do Things With Words.* The William James Lectures Delivered at Harvard University in 1955, edited by J. O. Urmson and M. Sbisa. 2nd edn 1975, 3rd edn 1980. Oxford: Clarendon Press, 1980.

[Ayer, 1952] A. J. Ayer. Negation. *Journal of Philosophy* 49: 797-815, 1952. Reprinted in *Philosophical Essays*, 36-65. London: Macmillan, 1963.

[Bentham, 1773] E. Bentham. *Introduction to Logick.* Oxford: W. Jackson and J. Lister, 1773.

[Bostock, 1997] D. Bostock. *Intermediate Logic.* Oxford: Clarendon Press, 1997.

[Braakhuis, 1979] H. A. G. Braakhuis. *De 13de eeuwse tractaten over syncategorematische termen*(2 volumes). University of Leiden dissertation, 1979.

[Bradley, 1883] F. H. Bradley. *Principles of Logic.* London: K. Paul, Trench, 1883.

[Brann, 2001] E. Brann. *The Ways of Naysaying: no, not, nothing, and nonbeing* Lanham, MD: Rowman & Littlefield, 2001.

[Carston, 1996] R. Carston. Metalinguistic negation and echoic use. *Journal of Pragmatics* 25: 309-330, 1996.

[Chapman, 2005] S. Chapman. *Paul Grice: Philosopher and Linguist.* Houndmills: Palgrave Macmillan, 2005.

[Cohen, 1971] L. J. Cohen. Grice's Views on the Logical Particles of Natural Language', 1971. Repr. in *Knowledge and Language.* Dordrecht: Kluwer Academic Press, 2002.

[Coke, 1657] Z. Coke. *Art of Logick.* London: John Streater, 1657.

[Cowell, 1607] J. Cowell. *The Interpreter, or Booke Containing the Signification of Words.* Cambridge: Iohn Legate, 1657.

[Davis, 1986] W. A. Davis. *An introduction to Logic.* Englewood Cliffs, New Jersey: Prentice-Hall, 1986.

[Davis, 1998] W. A. Davis. *Implicature.* Cambridge: Cambridge University Press, 1998.

[Dummett, 2002] M. A. E. Dummett. "Yes', 'No', and 'Can't Say'.' *Mind* 111:289-296, 2002.

[Duncan, 1748] W. Duncan. *Elements of Logick.* London: J. Dodsley, 1748.

[Ebbesen, 1993] S. Ebbesen. Tu non cessas comedere ferrum. *Cahiers de l'Institut du Moyen Age Grec et Latin* 63: 225-230, 1993.

[Flew, 1984] A. G. N. Flew. 'Negation', in *A Dictionary of Philosophy.* London: Pan, 1984.

[Frege, 1919] G. Frege. 'Negation'. *Translations from the Philosophical Writings*, ed. by Peter T. Geach and Max Black, Blacwkell, pp. 117-35, 1919.

[Gadzar, 1979] G. Gazdar. *Pragmatics.* New York: Academic Press, 1979.

[Grandy and Warner, 1967] R. E. Grandy and R. O. Warner, eds. *Philosophical Grounds of Rationality: Intentions, Categories, Ends* [PGRICE]. Oxford: Clarendon Press, 1967.

[Grice, 1957] H. P. Grice. Metaphysics, in D. F. Pears, *The Nature of Metaphysics.* London: Macmillan, 1957.

[Grice, 1961] H. P. Grice. 'The causal theory of perception'. *Proc. Arist. Soc. Supp. Vol.* 35: 121-152, 1961.

[Grice, 1969] H. P. Grice. 'Vacuous Names', in Donald Davidson and J. K. K. Hintikka (eds.), *Words and Objections: Essays on the Work of W. V. O. Quine*, 118-145. Dordrecht: Reidel, 1969.

[Grice, 1986a] H. P. Grice. 'Reply to Richards', in Grandy & Warner (eds.), pp. 43-103m 1986.

[Grice, 1986b] H. P. Grice. Actions and Events. *Pacific Philosophical Quarterly* 67:1-35, 1986.

[Grice, 1988] H. P. Grice. Aristotle on the Multiplicity of Being. *Pacific Philosophical Quarterly* 69:175-200, 1988.

[Grice, 1989] H. P. Grice. *Studies in the Way of Words.* Cambridge, Mass.: Harvard University Press, 1989.

[Grice, 1991] H. P. Grice. *The Conception of Value.* Oxford: Clarendon Press, 1991.

[Grice, 2001] H. P. Grice. *Aspects of Reason.* Oxford: Clarendon Press, 2001.

[Griss, 1945] G. F. Griss. Negationless Intuitionistic Mathematics. *Indagationes Mathematica* 8:675-681, 1945.

[Grize, 1967] J. B. Grize. *Logique moderne.* Paris, Gaulles-Villars, 1967.

[Hare, 1971] R. M. Hare. *Practical Inferences.* London: Macmillan, 1971.

[Hare, 1998] R. M. Hare. *Objective Prescriptions.* Oxford: Clarendon, 1998.

[Harris, 1765] J. Harris. Hermes; or, a philosophical inquiry concerning Language and Universal Grammar. London: H. Woodfall, 1765.

[Hart, 1951] H. L. A. Hart. 'A Logician's Fairy Tale'. *Philosophical Review* 60:98-212, 1951.

[Hobbes, 1962] T. Hobbes. 'Computation', in The English Works, ed. by Sir William Molesworth. Oxford: Clarendon Press, 1962.

[Hodges, 1977] W. Hodges. *Logic.* Hammondsworth, Middlesex: Penguin Books, 1977.

[Henry, 1967] D. P. Henry. *The Logic of St. Anselm.* Oxford: Clarendon, 1967.

[Hoffmann, 1987] M. Hoffmann. *Negatio Contrarii: A Study of Latin Litotes.* Assen: Van Gorcum, 1987.

[Hopkins, 1972] J. Hopkins. *A Companion to the Study of St. Anselm.* Minneapolis: U. of Minnesota Press, 1972.

[Horn, 1969] L. Horn. 'A Presuppositional Analysis of 'Only' and 'Even". In Papers from the Fifth Regional Meeting of the Chicago Linguistics Society, edited by Binnick, Davidson, Green, and Morgan. University of Chicago, pp. 98-107, 1969.

[Horn, 1970] L. Horn. ''Ain't It Hard (Anymore)". Papers from the Sixth Regional Meeting of the Chicago Linguistics Society. University of Chicago, pp. 318-327, 1970.

[Horn, 1971] L. Horn. 'Negative Transportation: Unsafe at Any Speed?' Chicago Linguistics Society 5, 1971.

[Horn, 1972] L. Horn. On the Semantic Properties of Logical Operators in English. PhD UCLA. Distributed 1976 by the Indiana Linguistics Club, 1972.

[Horn, 1973] L. Horn. 'Greek Grice: A Brief Survey of Proto-Conversational Rules in the History of Logic'. Chicago Linguistics Society 9:205-214, 1973.

[Horn, 1975] L. Horn. 'Neg-Raising Predicates: Towards an Explanation'. Chicago Linguistics Society 11:279-94, 1975.

[Horn, 1978] L. Horn. 'Some Aspects of Negation', in Univesals of Human Language, vol. 4. Syntax, edited by J. Greenberg, C. Ferguson, and E. Moravcsik, et al., Stanford: Stanford University Press, pp. 127-210, 1978.

[Horn, 1978b] L. Horn. 'Remarks on Neg-Raising', in Cole, pp. 129-220, 1978.

[Horn, 1984] L. Horn. 'Ambiguity, Negation, and the London School of Parsimony'. Proceedings of NELS 14, pp. 108-131, 1984.

[Horn, 1984b] L. Horn. 'In Defense of Privative Ambiguity'. Berkeley Linguistics Society 10, pp. 141-156, 1984.

[Horn, 1985] L. Horn. 'Metalinguistic Negation and Pragmatic Ambiguity'. Language 61:121-174, 1985.

[Horn, 1989] L. Horn. A Natural History of Negation. Chicago: University of Chicago Press, 1989. Expanded and reissued, Stanford: Center for the Study of Language and Information, 2001.

[Horn, 1990] L. Horn. 'Hamburgers and Truth: Why Gricean Inference is Gricean'. In Parasession on Legacy of Grice. In K. Hall et al, eds. Proceedings of the Berkeley Linguistics Society, 16:454 71, 1990.

[Horn, 1991] L. Horn. 'Duplex Negatio Affirmat...: the Economy of Double Negation', in Dobrin et al, pp. 78-106, 1991.

[Horn, 2000] L. Horn. From *if* to *iff:* Conditional perfection as pragmatic strengthening. *Journal of Pragmatics* 32: 289-326, 2000.

[Horn, 2002] L. Horn. 'Assertoric inertia and NPI licensing.' In *CLS 38, Part 2,* 55-82. Chicago: Chicago Linguistic Society, 2002.

[Horn, 2009] L. Horn. 'WJ-40: Implicature, truth, and meaning.' *International Journal of Pragmatics* 1: 3-34, 2009.

[Horn, 2010] L. Horn. 'Contradiction.' Entry in *Stanford Encyclopedia of Philosophy.* Available at `http://plato.stanford.edu/entries/contradiction/`. Updated entry, 2010.

[Horn, 2010a] L. Horn, ed. *The Expression of Negation.* Berlin: de Gruyter, 2010.

[Horn, 2010b] L. Horn. Double negation in English and other languages. In Horn, ed., 2010a.

[Horn, 2012a] L. Horn. Implicature. In G. Russell and D. Graff Fara, eds., *Routledge Companion to the Philosophy of Language*, pp. 53-66. Routledge, 2012.

[Horn, 2012b] L. Horn. Histoire d *O. In J.-Y. Béziau and G. Payette, eds., *The Square of Opposition: A General Framework for Cognition*, pp. 383-416. Bern: Peter Lang, 2012.

[Horn, to appear, b] L. Horn. 'Lexical pragmatics and the geometry of opposition: The mystery of **nall* and **nand* revisited.' In J.-Y. Béziau (ed.), *Papers from the World Congress on the Square of Opposition.* Available at `http://www.yale.edu/linguist/faculty/horn_pub.html`.

[Horn, to appear, b] L. Horn. 'Double Negation in English and Other Languages', in Horn (ed.).

[Horn, to appear] L. Horn, ed. *The Expression of Negation.* Berlin: Mouton de Gruyter, to appear.

[Hungerland, 1960] I. Hungerland. 'Contextual Implication.' *Inquiry* 3:211-258, 1960.

[Jaspers, 2005] D. Jaspers. *Operators in the Lexicon: On the Negative Logic of Natural Language.* (Universiteit Leiden dissertation.) Utrecht: LOT, 2005.

[Jespersen, 1917] O. Jespersen. *Negation in English and Other Languages.* Copenhagen: A. F. Høst, 1917.

[Jevons, 1870] W. S. Jevons. *Elementary Lessons in Logic.* London: Macmillan, 1870.

[Joachim, 1948] H. H. Joachim. *Logical Studies.* Oxford: Clarendon Press, 1948.

[Joseph, 1916] H. W. B. Joseph. *An introduction to logic.* Oxford: Clarendon Press, 1916.

[Kasher, 1972] A. Kasher. 'Gricean inference revisited', *Philosophica* 29: 25-44, 1972.

[Kirwan, 1995] C. A. Kirwan. 'Negation', in *Oxford Companion to Philosophy*, ed. Ted Honderich. Oxford University Press, 1995.

[Kretzmann, 1968] N. Kretzmann. *William of Sherwood's Treatise on Syncategorematic Words.* Minneapolis: U. of Minnesota Press, 1968.

[Lemmon, 1965] E. J. Lemmon. *Beginning Logic.* London: Nelson, 1965.

[Lever, 1573] R. Lever. *The Arte of Reason*, 1573. Repr. Menston: Scolar Press, 1972.

[Levinson, 2000] S. C. Levinson. *Presumptive Meanings: The Theory of Generalized Conversational Implicature.* Cambridge, MA: MIT Press, 2000.

[Lewry, 1984] P. O. Lewry. 'Logic', in *The History of the University of Oxford*, ed. J. I. Catto. Oxford: Clarendon Press, pp. 401-33, 1984.

[McCawley, 1981] J. D. McCawley. *Everything that linguists have always wanted to know about logic... but were ashamed to ask.* Chicago: U. of Chicago Press, 1981.

[Mates, 1953] B. Mates. *Stoic Logic.* Berkeley: University of California Press, 1953.

[Mates, 1972] B. Mates. *Elementary Logic.* Oxford: Oxford University Press, 1972.

[Mill, 1843] J. S. Mill. *A System of Logic.* London: J. W. Parker, 1843.

[Mill, 1867] J. S. Mill. *An Examination of Sir William Hamilton's Philosophy* (3d edn.). London: Longman, 1867.

[Mitchell, 1962] D. Mitchell. *An Introduction to Logic.* London: Hutchinson, 1962.

[Morris, 1978] J. Morris, ed. *The Oxford Book of Oxford.* Oxford: Oxford University Press, 1978.

[Myro, 1987] G. Myro. *Rudiments of Logic.* With Mark Bedau and Tim Monroe. Englewood Cliffs, New Jersey: Prentice-Hall, 1987.

[Neale, 2001] S. Neale. 'Implicature and colouring', in *Paul Grice's Heritage*, ed. G. Cosenza. Brepols Turnhout, pp. 139-184, 2001.

[Nowell-Smith, 1954] P. H. Nowell-Smith. *Ethics.* Harmondsworth: Pelican Books, 1954.

[O'Donnell, 1941] J. R. O'Donnell. The Syncategoremata of William of Sherwood. *Mediæval Studies* 3: 46-93, 1941.

[Oesterle, 1962] J. Oesterle, ed. *Aristotle: On Interpretation. Commentary by St. Thomas and Cajetan.* Milwaukee: Marquette University Press, 1962.

[Parsons, 2008] T. Parsons. 'The Traditional Square of Opposition', entry in *The Stanford Encyclopedia of Philosophy.* At `http://plato.stanford.edu/archives/fall2008/entries/square/`.

[Peacocke, 1987] C. A. B. Peacocke. 'Understanding Logical Constants: a Realist's Account'. Proceedings of the British Academy 73:153-200, 1987; repr. in T. Smiley, *Logic and Knowledge.* Oxford University Press, 2004.

[Peter of Spain, 1972] Peter of Spain. *Tractatus, called afterward Summule Logicales.* L. M. de Rijk, ed. Assen: Van Gorcum, 1972.

[Peter of Spain, 1992] Peter of Spain. *Syncategoreumata.* L. M. de Rijk, ed.; J. Spruyt, trans. Leiden: E. J. Brill, 1992.

[Pelletier, 1990] F. J. Pelletier. *Parmenides, Plato and the Semantics of Not-Being.* Chicago: U. of Chicago Press, 1990.

[Priest, 1999] G. Priest. 'What not?, in Gabbay and Wansing, pp. 101-120, 1999.

[Quine, 1969] W. V. O. Quine. '[Reply] to Grice', in Davidson & Hintikka, pp. 326-327, 1969.

[Ryle, 1929] G. Ryle. 'Negation' In *Knowledge, Experience, and Realism.* Proceedings of the Aristotelian Society, Supplementary Volume 9, pp. 80-86. London: Harrison and Sons, 1929.

[Sainsbury, 2001] R. M. Sainsbury. *Logical Forms.* Oxford: Blackwell, 2001.

[Schwarz, 1976] D. Schwarz. Notes from the Pragmatic Wastebasket: On a Gricean Explanation of the Preferred interpretation of Negative sentences'. *Pragmatics Microfiches*, 2-E4, 1976.

[Schwarz, 1976b] D. Schwarz. 'Referring, singular terms, and presupposition'. Philosophical Studies 30:63-74, 1976.

[Scott, 1906] R. F. Scott. Edvard Dibgy, the Logician. *The Eagle: St. John's College Magazine*, Oxford, October, pp. 1–24, 1906.

[Searle, 1969] J. Searle. *Speech Acts.* Cambridge: Cambridge University Press, 1969.

[Seuren, 2006] P. Seuren. The natural logic of language and cognition. *Pragmatics* 16: 103-38, 1969.

[Spencer, 1628] T. Spencer. *Logick*, 1628; fasc. repr. Menston, Yorks.: Scolar Press, 1970.

[Speranza, 1984] J. L. Speranza. Plato and the problem of language. Paper presented to seminar by Guillermo Ranea. Department of Philosophy, University of La Plata, 1984.

[Speranza, 1985] J. L. Speranza. The sceptic and language, MS. Paper presented to seminar on Scepticism, organized by Ezequiel de Olaso, Department of Philosophy, University of La Plata, 1985.

[Speranza, 1989] J. L. Speranza. 'J. S. Mill y el problema del mentalismo', – with comments by Horacio Arlo-Costa. Department of Philosophy, University of Buenos Aires, 1989.

[Speranza, 1998] J. L. Speranza. 'Grice Saves — But There's No Free Lunch'. MS. The Grice Circle at the Argentine Society for Philosophical Analysis, Buenos Aires, presented to philosophical logic seminar by Gregorio Klimovsky, Universitity of Buenos Aires, 1998.

[Speranza, 1993] J. L. Speranza. The conversationalist manoeuvre in philosophical logic. MS, 1993.

[Speranza, 1990] J. L. Speranza. 'Strawson and the history of logic', presented to seminar by Ignacio Angelelli, University of La Plata, 1990.

[Speranza, 1993] J. L. Speranza. "'$\sim$' and 'not", MS, 1993.

[Speranza, 1991] J. L. Speranza. Comments on Hare, 'Some subatomic particles of logic'. Boletin Bibliografico, volume 5. Instituto de Filosofia: Buenos Aires, pp. 23-24, 1991.

[Speranza, 1991] J. L. Speranza. Comments on J. F. Thomson, 'In defence of the material conditional'. Boletin Bibliografico, volume 5, Instituto de Filosofia: Buenos Aires, pp. 28, 1991.

[Spruyt, 1987] J. Spruyt, ed. *Peter of Spain on Composition and Negation,* ed. and trans. Nijmegen: Ingenium, 1987.

[Strawson, 1950] P. F. Strawson. On referring. *Mind* 59: 320-44.

[Strawson, 1952] P. F. Strawson. *Introduction to Logical Theory.* London: Methuen, 1952.

[Strawson, 1955] P. F. Strawson. A Logician's Landscape. *Philosophy* 30:229-237, 1955.

[Strawson, 1967] P. F. Strawson. 'Introduction' to *Philosophical Logic.* Oxford: Oxford University Press, pp. 1-16, 1967.

[Strawson, 1971] P. F. Strawson. *Logico-Linguistic Papers.* London: Methuen, 1971.

[Strawson, 1980] P. F. Strawson. 'Replies', to *Philosophical Subjects*, ed. Zak Van Straaten. Oxford University Press, 1980.

[Strawson, 1987] P. F. Strawson. The logic of natural language, in George Englebretsen, *The New Syllogistic.* New York: Peter Lang, 1987.

[Strawson, 1986] P. F. Strawson. If and $\supset$, in Grandy & Warner (eds.), 1986; repr. in (1997).

[Strawson, 1997] P. F. Strawson. *Entity and Identity.* Oxford: Clarendon Press, 1997.

[Strawson, 1998] P. F. Strawson. 'Replies', in *The philosophy of P. F. Strawson*, ed. L. Hahn. Open Court Publishing Company, 1998.

[Suppes, 1957] P. Suppes. *Introduction to Logic.* Princeton, New Jersey: Van Nostrand, 1957.

[Suppes, 1986] P. Suppes. 'The Primacy of Utterer's Meaning', in Grandy & Warner (eds.), pp. 109-119, 1986.

[Swinburne, 1875] A. J. Swinburne. Picture logic. Oxford, 1875.

[Thomas, 1964] I. Thomas. 'Oxford Logic', in *Oxford Studies*, ed. R. W. Southern. Oxford: Clarendon Press, pp. 297-311, 1964.

[Thomason, 1970] R. H. Thomason. *Symbolic Logic.* London: Macmillan, 1970.

[Thomason, 1997] R. H. Thomason. 'Accomodation, Meaning, and Implicature', in Asa Kasher, *Pragmatics.* London: Routledge, 1997.

[Wheeler, 1983] S. C. Wheeler III. Megarian paradoxes as Eleatic arguments. *American Philosophical Quarterly* 20: 287-95, 1983.

[Whitehead and Russell, 1910] A. N. Whitehead and B. Russell. *Principia Mathematica.* Cambridge: Cambridge University Press, 1910.

[Wiggins, 1971] D. Wiggins. 'Sentence Meaning, Negation, and Plato's Problem of Non-Being', in *Plato: A Collection of Critical Essays*, ed. Gregory Vlastos, Notre Dame: Indiana University Press, pp. 268-303, 1971.

[Wiggins, 1996] D. Wiggins. 'Replies', in Sabina Lovibond and S. G. Williams, eds., *Identity, Truth, and Value.* Oxford: Blackwell, pp. 219-284, 1996.

[Wilson, 1926] J. C. Wilson. *Statement and inference.* Oxford: Clarendon Press, 1926.

A HISTORY OF THE CONNECTIVES

Daniel Bonevac and Josh Dever

Contemporary students of logic tend to think of the logic of the connectives as the most basic area of the subject, which can then be extended with a logic of quantifiers. Historically, however, the logic of the quantifiers, in the form of the Aristotelian theory of the syllogism, came first. Truth conditions for negation, conjunction, and disjunction were well understood in ancient times, though not until Leibniz did anyone appreciate the algebraic features of these connectives. Approaches to the conditional, meanwhile, depended on drawing an analogy between conditionals and universal affirmative propositions. That remained true throughout the ancient, medieval, and early modern periods, and extended well into the nineteenth century, when Boole constructed an algebraic theory designed to handle sentential and quantificational phenomena in one go. The strength of the analogy, moreover, undercuts a common and otherwise appealing picture of the history of logic, according to which sentential and quantificational threads developed largely independently and, sometimes, in opposition to each other, until Frege wove them together in what we now consider classical logic. Frege did contribute greatly to our understanding of the connectives as well as the quantifiers. But his contribution consists in something other than unifying them into a single theory.

1 ARISTOTELIAN FOUNDATIONS

Aristotle (384–322 BC), the founder of logic, develops a logic of terms rather than a logic of propositions. He nevertheless argues for and against certain broad metaprinciples with direct relevance to propositional logic. In *Metaphysics* Γ 4, for example, he famously argues in favor of the principle of noncontradiction in the form "it is impossible for anything at the same time to be and not to be," maintaining that anyone who opposes it is "no better than a vegetable." In *De Interpretatione* 9 he argues against the principle of bivalence, contending that, if every proposition is either true or false, everything must be taken to happen by necessity and not by chance. He correspondingly denies that 'There will be a sea battle tomorrow' is either true or false. These are important discussions, raising central logical and philosophical issues. But they fall far short of a logic of propositional connectives.

That has not stopped scholars from finding the core of such a logic in Aristotle. Lear [1980], for example, finds in *Prior Analytics* I, 23's discussion of the reduction of syllogisms to first figure an account of indirect proof, which seems fair enough; Aristotle does rely on the pattern $A, B \vdash C \Leftrightarrow A, \neg C \vdash \neg B$. Slater [1979] notes the parallel between Aristotle's logic of terms and a Boolean propositional logic, constructing an Aristotelian propositional logic by bending the latter to the former rather than the reverse. It is not

Handbook of the History of Logic. Volume 11: Logic: A History of its Central Concepts.
Volume editors: Dov M. Gabbay, Francis Jeffry Pelletier and John Woods.
General editors: Dov M. Gabbay and John Woods.

hard to do; since Aristotle takes universals as having existential import, and Boole takes universals and conditionals as analogues, interpret a conditional $A \rightarrow B$ as holding when $\emptyset \subset [A] \subseteq [B]$, where $[A]$ and $[B]$ are sets of cases in which A and B, respectively, are true. This has the important implication that no true conditional has a necessarily false antecedent. Still, to obtain a logic from this, one needs to take conjunction as corresponding to intersection, disjunction as corresponding to union, and negation as corresponding to complement — none of which is particularly well-motivated within Aristotle's logic as such.

One seemingly propositional argument Aristotle makes is today almost universally considered fallacious. In *Prior Analytics* II, 4 he argues:

> But it is impossible that the same thing should be necessitated by the being and by the not-being of the same thing. I mean, for example, that it is impossible that *B* should necessarily be great if *A* is white and that *B* should necessarily be great if *A* is not white. For whenever if this, *A*, is white it is necessary that, *B*, should be great, and if *B* is great that *C* should not be white, then it is necessary if *A* is white that *C* should not be white. And whenever it is necessary, if one of two things is, that the other should be, it is necessary, if the latter is not, that the former should not be. If then if *B* is not great *A* cannot be white. But if, if *A* is not white, it is necessary that *B* should be great, it necessarily results that if *B* is not great, *B* itself is great. But this is impossible. For if *B* is not great, *A* will necessarily not be white. If then if this is not white *B* must be great, it results that if *B* is not great, it is great, just as if it were proved through three terms. (57a36–57b17)

It is not clear what relation Aristotle has in mind by "necessitated," but let's provisionally represent it symbolically with an arrow. Then Aristotle appears to be arguing for the thesis that $\neg((A \rightarrow B) \wedge (\neg A \rightarrow B))$. In keeping with this propositional schema, let's take A and B as standing for propositions rather than objects, as in Aristotle's text. The argument seems to go as follows:

1. $A \rightarrow B$ (assumption)
2. $\neg A \rightarrow B$ (assumption)
3. $A \rightarrow B, B \rightarrow \neg C \Rightarrow A \rightarrow \neg C$ (transitivity)
4. $A \rightarrow B \Rightarrow \neg B \rightarrow \neg A$ (contraposition)
5. $\neg B \rightarrow \neg A$ (modus ponens, 1, 4)
6. $\neg B \rightarrow B$ (transitivity, 5, 2)
7. $\neg(\neg B \rightarrow B)$ (??)

Aristotle gives no explanation for why a proposition cannot be necessitated by its own negation. Perhaps he has in mind the central idea of connexive implication, that the antecedent of a true conditional must be compatible with the conclusion [McCall, 1966].

But perhaps he simply thinks, given the parallel with universal propositions and their existential import, that antecedents of true conditionals must be possible. Given a necessitation account of the conditional, these are of course equivalent. But they are distinct on other, weaker accounts.

Leave that aspect of the argument aside for a moment. Aristotle does use several interesting principles in this argument, which we might formalize as follows:

- Transitivity of the conditional: $A \to B, B \to C \Rightarrow A \to C$
- Contraposition of the conditional: $A \to B \Rightarrow \neg B \to \neg A$
- Modus ponens on the conditional: $A \to B, A \Rightarrow B$

His use of modus ponens in this passage, however, is metatheoretic. In general, Aristotle's text is ambiguous between a conditional reading and an entailment reading. It would be justifiable to interpret the rules as

- Transitivity of entailment: $A \models B, B \models C \Rightarrow A \models C$
- Contraposition of entailment: $A \models B \Rightarrow \neg B \models \neg A$
- Modus ponens on entailment: $(A \models B, A) \Rightarrow B$

This kind of ambiguity presents a serious problem to anyone seeking to find a propositional logic implicit in Aristotle. It also, as we shall see, presents problems to anyone seeking to interpret medieval discussions of consequences.

2 STOIC LOGIC

The first explicit theory of propositional connectives was developed by a collection of thinkers known as the Stoics. They took logic seriously. Diogenes Laertius reports that when Diodorus Cronos could not solve a logical puzzle the King posed to him at a banquet, he died, heart-broken (*Vitae* II, 111). Philetas of Cos, in the epitaph he asked to be placed on his tombstone, blamed the Liar paradox for his death (*Athen.* IX, 401C).

The Stoic definition of argument is strikingly modern. "An argument is a system consisting of premises and a conclusion. Those propositions which are agreed upon for the establishment of the conclusion are called 'premises,' and the proposition which is established from the premises is called the 'conclusion,' as, for instance, in the following argument:

> If it is day, then it is light.
> It is day.
> Therefore, it is light.

The proposition 'It is light' is the conclusion and the others are premises" (Sextus, *Hyp. Pyrrh.* II, 135ff., in [Mates, 1961, 110]).

The Stoics also offer a general definition of validity: "the criterion for validity is that an argument is valid whenever the conclusion follows logically from the conjunction of the premises" (Sextus, *Adv. Math.*, VIII, 415, in [Mates, 1961, 106]). "Of arguments, some are conclusive and some are inconclusive. Inconclusive arguments are those which are such that the denial of the conclusion is compatible with the conjunction of the premises:

> If it is day, then it is light.
> It is day.
> Therefore, Dion is walking." (Diogenes Laertius, *Vitae* VII, 78, in [Mates, 1961, 114])

This is essentially Aristotle's conception of validity. Like Aristotle, the Stoics sometimes load it down with extraneous considerations. Diogenes Laertius, for example, reports the Stoics as having defined arguments as having two premises (*Vitae* VII, 76, in [Mates, 1961, 114]). Kneale and Kneale take this as accurate (162–163) on grounds that all the basic Stoic inference patterns have two premises.

The basic unit of Stoic logic is not the term, as in Aristotle, but the proposition (*lecton*). A sentence signifies a proposition, a "sayable" that is a bearer of truth value. Sentences are tangible, things spoken or written in a particular language; propositions are intangible and independent of any particular language. They are abstract contents of sentences. 'It's raining,' '*Piove*,' '*Es regnet*,' '*Il pleut*,' '*Det regner*,' '*Esta lloviendo*,' and '*Wa mvua*' are different sentences in different languages, but, if said of the same place and time, have the same content, signifying the same *lecton*. Not all sayables are propositions. Among the *complete* sayables are commands and questions; among the *incomplete* are predicates. Incomplete sayables turn into complete sayables when appended to names. Propositions, according to the Stoics, may change in truth value. 'It is raining' may be false now but true tomorrow.

Propositions, then, are expressible contents that have truth values. The Stoics have a strong commitment to bivalence; every proposition is either true or false. There are no truth value gaps. They are similarly committed to noncontradiction; there are no truth value gluts. They distinguish between these metatheoretic claims and their expression in the law of excluded middle (*A* or not *A*) and noncontradiction (not both *A* and not *A*).

2.1 The Nature of the Conditional

Stoic logic is best known for the controversy over the nature of the conditional. "Even the crows on the rooftops caw about which conditionals are true," Callimachus writes (Sextus, *Adv. Math*, I, 309–310). A conditional, Chrysippus stipulates, is a compound proposition formed with the connective 'if' (*ei, eiper*; Sextus *Adv. Math.* VIII, 109). Under what conditions is such a sentence true? When "the second part follows from the first," as Sextus says (*Adv. Math.* VIII, 111); but under what conditions does the consequent follow from the antecedent?

Sextus outlines four competing answers:

> For Philo says that a true conditional is one which does not have a true antecedent and a false consequent; e.g., when it is day and I am conversing, "If it is day, then I am conversing";

> but Diodorus defines it as one which neither is nor ever was capable of having a true antecedent and a false consequent. According to him, the conditional just mentioned seems to be false, since when it is day and I have become silent, it will have a true antecedent and a false consequent; but the following conditional seems true: "If atomic elements of things do not exist, then atomic elements of things do exist," since it will always have the false antecedent, "Atomic elements of things do not exist," and the true consequent, "Atomic elements of things do exist."
>
> And those who introduce "connection" or "coherence" say that a conditional holds whenever the denial of its consequent is incompatible with its antecedent; so that, according to them, the above-mentioned conditionals do not hold, but the following is true: "If it is day, then it is day."
>
> And those who judge by "suggestion" declare that a conditional is true if its consequent is in effect included in its antecedent. According to these, "If it is day, then it is day," and every repeated conditional will probably be false, for it is impossible for a thing itself to be included in itself. (*HP* II, 110; cf. *Adv. Math.* VIII, 112ff.)

It is hard to understand the final option; what is inclusion? And, if it is impossible for a thing to be included in itself, why are repeated conditionals "probably" rather than necessarily false?[1] The other three, however, seem straightforward:

- Philo: $A \rightarrow B \Leftrightarrow \neg(A \wedge \neg B)$
- Diodorus: $A \rightarrow B \Leftrightarrow$ it is always the case that $\neg(A \wedge \neg B)$
- Chrysippus: $A \rightarrow B \Leftrightarrow$ it is necessary that $\neg(A \wedge \neg B)$

Philo analyzes conditionals as material conditionals in the modern sense: "a conditional holds unless its antecedent is true and its consequent is false" (Sextus, *Adv. Math.* VIII, 332); "... a true conditional is one which does not have a true antecedent and a false consequent" (HP II, 104). Sextus outlines the Philonian position in terms of a truth table:

> Philo said that the conditional is true whenever it is not the case that its antecedent is true and its consequent false; so that, according to him, the conditional is true in three cases and false in one case. For it is true when the antecedent is true and the consequent is true. For example, "If it is day, it is light." Again, it is true when the antecedent is false and the consequent is false. For example, "If the earth flies, then the earth has wings." It is also true whenever the antecedent is false and the consequent is true. For example, "If the earth flies, then the earth exists." It is false only when the antecedent is true and the consequent is false, as, for example, "If it is day, then it is night." (*Adv. Math.* VIII, 112ff.; cf. *Hyp. Pyrrh.* II, 104ff.)

[1] We do not mean to suggest that this option cannot be given interesting interpretations. O'Toole and Jennings [2004], for example, develop a connexivist account. Bonevac, Dever, and Sosa (forthcoming) develop a neighborhood semantics that drops idempotence. But we do not have enough evidence concerning this fourth option to know what its advocates had in mind.

We unfortunately have no record of debates between Philo and Stoics holding contrary positions. So, we do not know what arguments Philo and his followers used to support their position. Nor do we know what arguments Diodorus, Chrysippus, and their followers brought against it.

Chrysippus analyzes conditionals as strict conditionals. The sense of necessity he has in mind is not clear; some examples suggest logical necessity, but some, such as Diogenes's "If it is day, it is light," could be read as physical necessity [Mates, 1961, 48]. Diodorus takes an intermediate position, analyzing conditionals in temporal terms.

The Stoics link argument validity to conditionals in a way that, at first glance, appears familiar from the deduction theorem. In fact, however, its adequacy depends on the interpretation of the conditional. Recall the general definition of validity: "the criterion for validity is that an argument is valid whenever the conclusion follows logically from the conjunction of the premises" (Sextus, *Adv. Math.*, VIII, 415, in [Mates, 1961, 106]). Equivalently, an argument is valid when the truth of the premises is incompatible with the falsehood of the conclusion. Take the conditional with the conjunction of the premises as antecedent and the conclusion as consequent to be an argument's *associated conditional*. Implication, as W. V. Quine stresses, is the validity of the conditional. So, an argument is valid if and only if its associated conditional is valid, that is, a logical truth. The Stoics, however, appear to have held that an argument is valid if and only if its associated conditional is *true*:

> Some arguments are valid and some are not valid: valid, whenever the conditional whose antecedent is the conjunction of the premises and whose consequent is the conclusion, is true (Sextus, *Hyp. Pyrrh.* II, 135ff., in [Mates, 1961, 110]).
>
> So, then, an argument is really valid when, after we have conjoined the premises and formed the conditional having the conjunction of premises as antecedent and the conclusion as consequent, it is found that this conditional is true (Sextus, *Adv. Math.*, VIII, 415, in [Mates, 1961, 107]).
>
> An argument is valid whenever there is a true conditional which has the conjunction of the premises as its antecedent and the conclusion as its consequent (Sextus, *Adv. Math.*, VIII, 426, in [Mates, 1961, 108]; see also *Hyp. Pyrrh.* II, 113, in [Mates, 1961, 110]).
>
> For a proof is held to be valid whenever its conclusion follows from the conjunction of its premises as a consequent follows from its antecedent, such as [for]:
>
> > If it is day, then it is light.
> > It is day.
> > Therefore, it is light.
>
> [we have] "If (if it is day then it is light, and it is day) then it is light." (*Hyp. Pyrrh.* II, 113; see also 135ff.)

These are all from Sextus, so it is possible that he confused logical truth with truth simpliciter. It is also possible that the Stoics did so. If there is no confusion, however, these passages support Chrysippus's understanding of the conditional. The truth of its associated conditional, when that is interpreted materially, hardly suffices for validity. The truth of a strict conditional interpreted as indicating logical necessity, in contrast, does.

2.2 *Stoic Theories of Conjunction, Disjunction, and Negation*

Stoic theories of the relatively uncontroversial connectives — conjunction, disjunction, and negation — are surprisingly contemporary in feel. The Stoics characterize these connectives semantically, in terms of truth conditions, but also in terms of syntactic axioms or rules intended to capture their inferential behavior. They appear to have a concept of scope (*kurieuei*) that matches our contemporary notion as well as a clear distinction between atomic and molecular propositions (Sextus, *Adv. Math.* VIII, 89ff., 93, 108, in [Mates, 1961, 97]; Diogenes Laertius, *Vitae* VII, 68, in [Mates, 1961, 112–113]). The contemporary feel of Stoic doctrines has led many historians of Stoic logic [Łukasiewicz, 1934; Mates, 1961; Kneale and Kneale, 1962; Bochenski, 1963], for example) to treat Stoic logic as a version of contemporary propositional logic, representing Stoic formulas in modern symbolism. As O'Toole and Jennings [2004] remind us, however, this can be dangerous. Stoic concepts often differ from our own. That said, the parallels between Stoic and twentieth-century theories are remarkable.

Negation. Let's begin with what is in some ways the simplest propositional connective, negation. Stoic truth conditions for negation appear to be congruent with contemporary ones: "Among propositions, those are contradictories of one another, with respect to truth and falsehood, of which the one is the negation of the other. For example, 'It is day' and 'It is not day' (Diogenes Laertius, *Vitae* VII, 73, in [Mates, 1961, 113]). The Stoics warn that we must use a negative particle that has scope over the entire proposition in order to obtain a negation. The truth condition they have in mind is evidently

$$[\neg A] = 1 \Leftrightarrow [A] = 0.$$

They have a similarly modern conception of negation's inferential role, advocating a rule of double negation: "A negative proposition is one like 'It is not day.' The double-negative proposition is a kind of negative. For a double negation is the negation of a negation. For example, 'Not: it is not day.' It asserts, 'It is day'" (Diogenes Laertius *Vitae* VII, 70, in [Mates, 1961, 113]). Similarly: "For 'Not: not: it is day' differs from 'It is day' only in manner of speech" (Alexander, *In An. Pr.*, Wallies, 18). We may formulate the rule schematically as

$$\neg\neg A \Leftrightarrow A.$$

No source lists this rule among the basic principles of Stoic logic; it must have been viewed as a derived rule.

Conjunction. Kneale and Kneale are dismissive of Stoic views on conjunction: "Of conjunction the Stoics had not much to say" (160). That is true only in the sense that contemporary logic has "not much to say" about conjunction; what the Stoics say might be viewed as a generalization of typical modern accounts.

The Stoic semantic characterization of conjunction is simple: "a conjunction holds when all the conjuncts are true, but is false when it has at least one false conjunct" (Sextus, *Adv. Math.* VIII, 125, in [Mates, 1961, 98]); "in every conjunction, if one part is false, the whole is said to be false, even if the others are true" (Gellius, *Noctes Atticae* XVI, viii, 1ff; in [Mates, 1961, 122–123]).[2] Stoic conjunction is not binary, as contemporary conjunction is. It is *multigrade*, capable of linking two or more propositions together. In this respect it is closer to the English 'and' than the contemporary binary conjunction connective is. To reflect its truth conditions accurately, we must write

$$[\wedge(A_1, ..., A_n)] = 1 \Leftrightarrow \forall i \in \{1, ..., n\}[A_i] = 1.$$

Surprisingly, one does not find among Stoic rules anything corresponding to modern rules for conjunction:

$$A_1, ..., A_n \Rightarrow \wedge(A_1, ..., A_n)$$

$$\wedge(A_1, ..., A_n) \Rightarrow A_i.$$

Instead, we find conjunction appearing only within the scope of negation.

Disjunction. Stoic logic incorporates at least two conceptions of negation. One corresponds to a multigrade version of our familiar inclusive disjunction:

$$[\vee(A_1, ..., A_n)] = 1 \Leftrightarrow \exists i \in \{1, ..., n\}[A_i] = 1.$$

This, however, is not the primary disjunction connective. Galen, in fact, speaks of it as "pseudo-disjunction":

> "Therefore, in consideration of clarity together with conciseness of teaching, there is no reason not to call propositions containing complete incompatibles 'disjunctions,' and those containing partial incompatibles 'quasi-disjunctions.'... Also, in some propositions, it is possible not only for one part to hold, but several, or even all; but it is necessary for one part to hold. Some call such propositions 'pseudo-disjunctions,' since disjunctions, whether composed of two atomic propositions or of more, have just one true member." (Galen, *Inst. Log.*, 11, 23ff., in [Mates, 1961, 118])

Usually, inclusive disjunction appears not as disjunction at all, but instead as a negated conjunction — indicating that Stoics understood what are often misleading called De Morgan's laws. Philoponus refers to disjunction in this sense as quasi-disjunction, which he defines in terms of a negated conjunction; "It proceeds on the basis of propositions that are not contradictory" (Scholia to Ammonius, *In An Pr.*, Praefatio, xi, in [Mates, 1961, 131]).

The most concise statements of truth conditions for disjunctions in the primary sense appear in Gellius — "Of all the disjuncts, one ought to be true and the others false" (*Noctes Atticae* XVI, viii, 1ff., in [Mates, 1961, 123]) — in Galen: "disjunctions have

[2]Galen restricts the truth of conjunctions to cases in which the conjuncts are neither consequences of one another nor incompatible, but he recognizes that others do not adopt this restriction (*Inst. Log.* 10, 133ff, in [Mates, 1961, 118]; see also 32, 13ff., in [Mates, 1961, 120–121].

one member only true, whether composed of two simple propositions or more than two" (*Inst. Log.* 5.1) — and in Sextus: "a true disjunction announces that one of its disjuncts is true, but the other or others false" (*Hyp. Pyrrh.* 2.191).[3] This suggests a truth condition requiring that exactly one disjunct be true:

$$[\oplus(A_1, ..., A_n)] = 1 \Leftrightarrow \exists! i \in \{1, ..., n\}[A_i] = 1.$$

Kneale and Kneale [1962, 162], Bochenski [1963, 91], Mates [1961, 51], and Łukasiewicz [1967, 74] all take Stoic disjunction in the primary sense as exclusive disjunction. In the binary case, the two are equivalent. In general, however, they are not. As O'Toole and Jennings (2004, 502) observe, linking a sequence of n propositions with exclusive disjunctions yields something weaker than what the Stoics intend, something that is true if and only if an odd number of them are true. The Stoic concept, however, might reasonably be viewed as a multigrade generalization of exclusive disjunction.

Why did Stoic logicians take $\oplus$ rather than $\vee$ as primary? No answer in terms of natural language semantics seems plausible; Greek and Latin, like English, have no connective best understood as expressing $\oplus$ [McCawley, 1980; O'Toole and Jennings, 2004]. One response is that $\vee$ is easily definable in terms of conjunction and negation:

$$\vee(A_1, ..., A_n) \Leftrightarrow \neg \wedge (\neg A_1, ..., \neg A_n)$$

That, indeed, is how the Stoics typically understand it. Defining disjunction in their primary sense, $\oplus$, in contrast, is considerably more complicated. Using inclusive disjunction:

$$\oplus(A_1, ..., A_n) \Leftrightarrow \vee(\wedge(A_1, \neg A_2, ..., \neg A_n), \wedge(\neg A_1, A_2, ..., \neg A_n), ..., \wedge(\neg A_1, \neg A_2, ..., A_n))$$

Using conjunction and negation alone:

$$\oplus(A_1, ..., A_n) \Leftrightarrow \neg\wedge(\neg\wedge(A_1, \neg A_2, ..., \neg A_n), \neg\wedge(\neg A_1, A_2, ..., \neg A_n), ..., \neg\wedge(\neg A_1, \neg A_2, ..., A_n))$$

Adding disjunction in what the Stoics considered its primary sense did not expand the expressive power of the language from a theoretical point of view, but it did make the system capable of expressing certain propositions much more economically.

There may be a third conception of disjunction in the Stoics, relating it to the conditional. Galen describes some Stoics as defining disjunctions in terms of conditionals: 'A or B' is equivalent to 'If not A, then B' (*Inst. Log.* 8, 12ff., in [Mates, 1961, 117–118]). The account of disjunction that results, of course, depends on the analysis of the conditional. If the conditional is Philonian, the result is inclusive disjunction. If the conditional is Diodoran, it is inclusive disjunction prefixed with an "always" operator. If the conditional is Chrysippan, it is inclusive disjunction prefixed with a necessity operator. In no case does such a conception yield $\oplus$.

[3]There is some confusion among sources about the Stoic truth conditions for disjunction. Diogenes Laertius, for example, speaks only of falsehood, and seems to treat disjunction as binary: "This connective announces that one or the other of the propositions is false" (*Vitae* VII, 72, in [Mates, 1961, 113]).

2.3 The Stoic Deduction System

The Stoics develop a deduction system for propositional logic, of which we have substantial fragments. They have a conception of completeness, and, as we have seen, have a formal semantics capable of giving real content to their claim that the system is complete (Diogenes Laertius, *Vitae* VII, 78). Unfortunately, that claim is unfounded. It is not difficult, however, to supplement the Stoic rules to obtain a complete system.

Axioms. The Stoic *indemonstrables*, axioms for the connectives, are attested in multiple sources (Sextus, *Adv. Math.* VIII, 223, in [Mates, 1961, 99–104]; Sextus, *Hyp. Pyrrh.* II, 156ff., in [Mates, 1961, 111–112]; Diogenes Laertius, *Vitae* VII, 79ff., in [Mates, 1961, 114–115]; Galen, *Inst. Log.* 15, 8ff., in [Mates, 1961, 119–120]; *Historia Philosopha* 15, in [Mates, 1961, 121–122]; Cicero, *Topica*, 56-57, in [Mates, 1961, 124–125]; Alexander, *In Top.*, Wallies, 175, 14ff, in [Mates, 1961, 126]; John Philoponus, *In. An. Pr.*, Wallies, 244–245, in [Mates, 1961, 128–129], Ammonius, *In. An. Pr.* 67.32–69.28):

1. If A then B; A; therefore B (modus ponens)
2. If A then B; not B; therefore not-A (modus tollens)
3. It is not the case that both A and B; A; therefore not-B
4. Either A or B; A; therefore not-B
5. Either A or B; not A; therefore B

The fourth axiom indicates that the disjunction intended is not inclusive "pseudo-disjunction" but one implying incompatibility. Also, the use of 'therefore' rather than 'if' indicates that these are what we would now consider rules of inference rather than axioms. So, it seems fair to represent these symbolically as rules

1. $A \to B, A \vdash B$
2. $A \to B, \neg B \vdash \neg A$
3. $\neg(A \wedge B), A \vdash \neg B$
4. $A \oplus B, A \vdash \neg B$
5. $A \oplus B, \neg A \vdash B$

This representation immediately raises a question. Stoic conjunction and disjunction are multigrade, not binary. So, why do the axioms, as stated in all the available sources, treat them as binary? In the case of conjunction, there is no problem, for extended conjunctions are equivalent to conjunctions built up in binary fashion. That is not true for Stoic disjunction; $\oplus(A, B, C)$ is not equivalent to $A \oplus (B \oplus C)$. We may perhaps more accurately capture the intentions of the Stoics by writing multigrade rules, assuming that the binary statement was only a convenience, a shorthand for something harder to state in words. One look at the symbolic representations makes it clear why such a simplification would seem desirable. If we think of conjunction and disjunction as applying to two or more propositions, in fact, we would need to state the third, fourth, and fifth axioms as above in addition to what appears below. We can forego that, however, if we assume that $\wedge(A) = \oplus(A) = A$.

1. $A \rightarrow B, A \vdash B$
2. $A \rightarrow B, \neg B \vdash \neg A$
3. $\neg \wedge (A_1, ..., A_n), A_i \vdash \neg \wedge (A_1, ..., A_{i-1}, ..., A_{i+1}, ..., A_n)$
4. $\oplus(A_1, ..., A_n), A_i \vdash \wedge(\neg A_1, ..., \neg A_{i-1}, \neg A_{i+1}, ..., \neg A_n)$
5. $\oplus(A_1, ..., A_n), \neg A_i \vdash \oplus(A_1, ..., A_{i-1}, A_{i+1}, ..., A_n)$

Rules (Themata). The Stoics are widely reported to have used four rules (Galen, *De Hipp. et Plat. Plac.* ii 3 (92); Alexander, *In Ar. An. Pr. Lib. I Commentarium*, Wallies, 284), of which we have only two [Kneale and Kneale, 169]:

1. $A, B \vdash C \Rightarrow A, \neg C \vdash \neg B$ (and $A, B \vdash C \Rightarrow B, \neg C \vdash \neg A$)
2. $A, B \vdash C$ and $\Gamma \vdash A \Rightarrow \Gamma, B \vdash C$

Just as the Stoic axioms are not what we today consider axioms, but instead simple rules of inference, so these are complex rules of inference. They do not allow us to write formulas of certain shapes if we already have formulas of certain shapes; they instead allow us to infer from a derivation another derivation. In that respect they are more like indirect proof and conditional proof than modus ponens. Indeed, the first rule is a general form of indirect proof; it encodes Aristotle's practice of using reductio proofs for certain syllogistic forms. The second is a version of cut.

What are the missing two? Historians of logic have made a variety of conjectures. Sources describe them as close to the cut rule. Bobzien [1996] reconstructs them as

- $A, B \vdash C$ and $A, C \vdash D \Rightarrow A, B \vdash D$ (and $A, B \vdash C; B, C \vdash D \Rightarrow A, B \vdash D$)
- $A, B \vdash C$ and $\Gamma, A, C \vdash D \Rightarrow \Gamma, A, B \vdash D$ (and $A, B \vdash C$ and $\Gamma, B, C \vdash D \Rightarrow \Gamma, A, B \vdash D$)

If that is correct — and the hypothesis does explain why the sources found it impossible to keep three of the four rules straight — the Stoics could easily have simplified their system by putting cut more generally:

- $\Gamma \vdash A$ and $\Delta, A \vdash B \Rightarrow \Gamma \cup \Delta \vdash B$

Is the Stoic system complete? They claimed it to be (Diogenes Laertius 7.79, Sextus, *Hyp. Pyrrh.* 2.156–157, 166–167, 194). Unlike Aristotle or even Frege, moreover, they had a well-defined semantics, even if the concept of validity itself remained imprecise. We have no record of any argument for completeness, however, much less a proof of it. We do have a record of some of the theorems that the Stoics derived. The most interesting, from the perspective of completeness, is double negation.[4]

[4]Other theorems include:

As it stands, it is obvious that the Stoic system is incomplete; there is nothing corresponding to a rule of conditional introduction. It is easy to show, for example, that $p \rightarrow p$ is not provable (see [Mueller, 1979]). This is not surprising, for it is precisely with respect to such a rule that the Stoic interpretations of the conditional vary. Philo, Diodorus, and Chrysippus can all agree on modus ponens and modus tollens. They can even agree on the general outlines of a method of conditional proof. But they have to disagree about the conditions to be applied. Philo can allow an unrestricted form of conditional proof, adopting a rule of the form $\Gamma, A \vdash B \Rightarrow \Gamma \vdash A \rightarrow B$. Diodorus has to restrict Γ to propositions that are always true; Chrysippus, to propositions that are necessarily true. Without a precise characterization of the semantics for the conditional along Philonian, Diodoran, or Chrysippan lines, and without some form of conditional proof, establishing completeness for the full system would be hopeless.[5]

The Stoic system is obviously incomplete even with respect to the other connectives. There is no way to exploit conjunctions; we cannot get from $A \wedge B$ to A. There is also no way to introduce disjunctions; we cannot go from $A, \neg B$ to $\oplus(A, B)$. As a result, we have no way of pulling apart and then putting together conjunctions and disjunctions, so we cannot prove that either is commutative. We may define conjunction in terms of disjunction as follows: $\wedge(A_1, ..., A_n) \Leftrightarrow \oplus(\neg A_1, ..., \neg A_n, B, \neg B)$, where B, $\neg B$ are not among $A_1, ..., A_n$. The equivalence, however, is not provable. Given $\oplus(\neg A_1, ..., \neg A_n, B, \neg B)$, it is possible to prove $\wedge(A_1, ..., A_n)$, but the other direction is impossible.

If we remedy those obvious defects, we do obtain a complete system. Restrict the language to the connectives $\neg$, $\wedge$, and $\oplus$. Adopt as axioms (where $i \leq n$):

1. $\neg \wedge (A_1, ..., A_n),\ A_i \vdash \neg \wedge (A_1, ..., A_{i-1}, ..., A_{i+1}, ..., A_n)$
2. $\wedge(A_1, ..., A_n) \vdash A_i$
3. $\oplus(A_1, ..., A_{i-1}, A_{i+1}, ..., A_n),\ \neg A_i \vdash \oplus(A_1, ..., A_n)$
4. $\oplus(A_1, ..., A_n),\ A_i \vdash \wedge(\neg A_1, ..., \neg A_{i-1}, \neg A_{i+1}, ..., \neg A_n)$

---

1. $A \rightarrow (A \rightarrow B), A \vdash B$ (Sextus, *Adv. Math.* VIII, 230–233)
2. $(A \wedge B) \rightarrow C, \neg C, A \vdash \neg B$ (234–236)
3. $A \rightarrow A, A \vdash A$ (Alexander, *In Ar. An. Pr. Lib. I Commentarium*, Wallies, 20.
4. $A \vee B \vee C, \neg A, \neg B \vdash C$ (Sextus, *Hyp. Pyrrh.* I 69)
5. $A \rightarrow B, A \rightarrow \neg B \vdash \neg A$
6. $A \rightarrow A, \neg A \rightarrow A, A \vee \neg A \vdash A$
7. $A \vee \neg A, A \vdash \neg\neg A$ (Kneale and Kneale 168)
8. $A \vee \neg A, \neg\neg A \vdash A$
9. $A \rightarrow B \vdash \neg B \rightarrow \neg A$ (Diogenes Laertius 7.194, Philodemus *Sign. PHerc.* 1065, XI.26–XII.14)

[5]This has not stopped people from trying; see [Becker, 1957; Kneale and Kneale, 1962; Mueller, 1979], all of whom interpret Stoic connectives as binary and adopt a Philonian reading of the conditional. In each case, however, some axioms and rules are added to make the system complete. Sometimes, these are artifacts of a Gentzen system that are extraneous to Stoic logic; sometimes, they are found in Stoic sources, but as theorems rather than basic axioms or rules.

5. $\oplus(A_1, \ldots, A_n), \neg A_i \vdash \oplus(A_1, \ldots, A_{i-1}, A_{i+1}, \ldots, A_n)$

Adopt indirect proof and cut as complex rules. Now, construct maximal consistent sets of formulas in standard Henkin fashion. The heart of the proof is to show that, for any maximal consistent set Γ, four proof-theoretic lemmas hold:

1. $\neg A \in \Gamma \Leftrightarrow A \notin \Gamma$. The construction process guarantees that either A or $\neg A$ belongs to Γ; they appear at some stage of the enumeration. At the stage at which A sppears, either A or $\neg A$ is placed in Γ. Say $\neg A \in \Gamma$ and $A \in \Gamma$. Then Γ is inconsistent; contradiction.

2. $\Gamma \vdash A \Rightarrow A \in \Gamma$. Say $\Gamma \vdash A$ but $A \notin \Gamma$. Then, by the preceding lemma, $\neg A \in \Gamma$. But then $\Gamma \vdash A$ and $\Gamma \vdash \neg A$, so Γ is inconsistent; contradiction.

3. $\wedge(A_1, \ldots, A_n) \in \Gamma \Leftrightarrow A_i \in \Gamma$ for each i among $1, \ldots, n$. Say $A_i \in \Gamma$ for each i among $1, \ldots, n$ but $\wedge(A_1, \ldots, A_n) \notin \Gamma$. Then $\neg\wedge(A_1, \ldots, A_n) \in \Gamma$. By Axiom 1, we may deduce $\neg\wedge(A_1, \ldots, A_n), \neg\wedge(A_2, \ldots, A_n)$, and so on, eventually reaching $\neg\wedge(A_n) = \neg A_n$. Since Γ is closed under the axioms and rules, that implies that $A_n, \neg A_n \in \Gamma$, contradicting the previous lemma.

   For the other direction: Say $\wedge(A_1, \ldots, A_n) \in \Gamma$. We can derive each A_i by Axiom 2. So, $A_i \in \Gamma$ for each i among $1, \ldots, n$.

4. $\oplus(A_1, \ldots, A_n) \in \Gamma \Leftrightarrow A_i \in \Gamma$ for exactly one i among $1, \ldots, n$. Say $\oplus(A_1, \ldots, A_n) \in \Gamma$ but none of $A_1, \ldots, A_n \in \Gamma$. Then $\neg A_1, \ldots, \neg A_n \in \Gamma$, by the previous lemma. By Axiom 5, we can derive $\oplus(A_2, \ldots, A_n), \oplus(A_3, \ldots, A_n)$, and so on, until we reach $\oplus(A_n) = A_n$. But then $A_n, \neg A_n \in \Gamma$, contradicting the previous lemma. So, at least one of $A_1, \ldots, A_n \in \Gamma$. Suppose more than one belong to Γ. For convenience, but without loss of generality, assume that $A_i, A_j \in \Gamma$, $i < j$, where no other disjuncts belong to Γ. By two applications of Axiom 4, we obtain $\oplus(A_1, \ldots, A_{i-1}, A_{i+1}, \ldots, A_{j-1}, A_{j+1}, \ldots, A_n)$, but, for all $k \leq n$ such that $k \neq i, j$, $\neg A_k \in \Gamma$. Once again, by repeated applications of Axiom 5, we derive a contradiction.

   For the other direction: Say $A_i \in \Gamma$ for exactly one i among $1, \ldots, n$. Since, for $j \neq i$, $j \leq n$, $A_j \notin \Gamma$, $\neg A_j \in \Gamma$. Begin with $A_i = \oplus(A_i)$. By repeated applications of Axiom 3, we derive $\oplus(A_1, \ldots, A_n)$, so $\oplus(A_1, \ldots, A_n) \in \Gamma$.

Now, given a maximal consistent set Γ and any proposition A, we construct an interpretation v from atomic formulas into truth values such that $A \in \Gamma \Leftrightarrow v \models A$. We stipulate that, for all atomic formulas A, $v(A) = 1$ for all $A \in \Gamma$.

Assume the hypothesis for all propositions of complexity less than that of an arbitrary formula A. We must consider several cases:

1. $A = \neg B$. By hypothesis, $B \in \Gamma \Leftrightarrow v(B) = 1$. Say $\neg B \in \Gamma$. Since Γ is consistent, $B \notin \Gamma$, so $v(B) = 0$. But then $v(\neg B) = 1$. For the other direction, assume that $v(\neg B) = 1$. Then, $v(B) = 0$, so $B \notin \Gamma$. Since Γ is maximal, $\neg B \in \Gamma$.

2. $A = \wedge(B_1, ..., B_n)$. By hypothesis, $B_i \in \Gamma \Leftrightarrow v(B_i) = 1$ for each $i \in \{1, ..., n\}$. Say $\wedge(B_1, ..., B_n) \in \Gamma$. By the above lemma, $B_1, ..., B_n \in \Gamma$, so, by hypothesis, $v(B_1) = ... = v(B_n) = 1$. So, $v(\wedge(B_1, ..., B_n)) = 1$.

   Say $v(\wedge(B_1, ..., B_n)) = 1$. By the truth definition for $\wedge$, $v(B_1) = ... = v(B_n) = 1$. But then $B_1, ..., B_n \in \Gamma$, so, by the above lemma, $\wedge(B_1, ..., B_n) \in \Gamma$.

3. $A = \oplus(B_1, ..., B_n)$. By hypothesis, $B_i \in \Gamma \Leftrightarrow v(B_i) = 1$ for each $i \in \{1, ..., n\}$. Say $\oplus(B_1, ..., B_n) \in \Gamma$. By the above lemma, exactly one of $B_1, ..., B_n$ belongs to Γ. But then v makes exactly one of $B_1, ..., B_n$ true. So, by the truth definition, $v(\oplus(B_1, ..., B_n)) = 1$.

   Say $v(\oplus(B_1, ..., B_n)) = 1$. By the truth definition for $\oplus$, v makes exactly one of $B_1, ..., B_n$ true. By hypothesis, then, exactly one of $B_1, ..., B_n$ belongs to Γ. But then by the above lemma $\oplus(B_1, ..., B_n) \in \Gamma$.

We have constructed an interpretation making everything in Γ true, given Γ's consistency. So, every consistent set of formulas is satisfiable. It follows that, if the set consisting of an argument's premises and the negation of its conclusion is unsatisfiable, it is inconsistent. So, if an argument is valid, its conclusion is provable from its premises.

It is surprising that the Stoics did not include rules of simplification and disjunction introduction. They may have taken the former for granted, but, given the extent of their concern for disjunction, it is unlikely they did the same for the latter. We have no record, however, of an argument that $A, \neg B \vdash \oplus(A, B)$ is admissible in their system. Nor do we have any record of its inclusion among the axioms. The concept of a deduction system as structured around introduction and elimination or exploitation rules was evidently foreign to them.

3 HYPOTHETICAL SYLLOGISMS

Aristotle refers briefly to hypothetical reasoning, but develops no theory of it in any surviving texts. His successor Theophrastus (371–287 BC), however, is reputed to have developed a theory of hypothetical syllogisms — in modern terms, a theory of the conditional. The nature of that theory, and whether it in fact existed, remains a subject of scholarly dispute.[6] Alexander of Aphrodisias (fl. 200 AD) says that Theophrastus discussed the inference pattern commonly known today as hypothetical syllogism: $A \rightarrow B, B \rightarrow C \vdash A \rightarrow C$. Such an interpretation, however, is not that of Alexander, who thinks of the variables as taking the place of terms rather than propositions. His example: "If man is, animal is; if animal is, substance is; if therefore man is, substance is" (*In An. Pr.* 326, 20–327, 18). John Philoponus (490–570), however, gives this example of a simple hypothetical syllogism:

> If the sun is over the earth, then it is day.
> If it is day, then it is light.

[6] See, for example, [Barnes, 1984; Speca, 2001].

> Therefore, if the sun is over the earth, then it is light. (Scholia to Ammonius, *In An. Pr.*, Wallies, *Praefatio*, xi, in [Mates, 1961, 129])

This treats the variables as taking the place of propositions. But Philoponus is not consistent about this; he goes on to substitute terms for the variables.

Alexander gives another form that combines hypothetical syllogism and contraposition: $A \rightarrow B, B \rightarrow C \vdash \neg C \rightarrow \neg A$. He also offers $A \rightarrow C, B \rightarrow \neg C \vdash A \rightarrow \neg B$ and $A \rightarrow B, \neg A \rightarrow C \vdash \neg B \rightarrow C$ (and $\neg C \rightarrow B$).

Alexander reports that Theophrastus introduced other forms of argument that might be considered propositional: by subsumption of a third term, from a disjunctive premise, from denial of a conjunction, by analogy or similarity of relations, and by degrees of a quality (*In Ar. An. Pr.* I [Wallies, 389ff].; see [Kneale and Kneale, 1962, 105]). We do not know what Theophrastus produced concerning these forms, but Boethius (*DSH* 831) says that it was not very substantial.

Alexander, however, gives examples of modus ponens in propositional and generalized forms:

> If the soul always moves, the soul is immortal.
> The soul always moves.
> Therefore the soul is immortal.

> If what appears to be more sufficient for happiness is not in fact sufficient, neither is that which appears to be less sufficient.
> Health appears to be more sufficient for happiness than wealth and yet it is not sufficient.
> Therefore wealth is not sufficient for happiness. (*In Ar. An. Pr. I*, 265; [Kneale and Kneale, 1962, 106])

Here the substituends for the variables are plainly propositions.

The ambiguity about the role of variables in this early theory of hypothetical syllogisms is not merely evidence of confusion. Theophrastus arranges his argument forms into three figures, thinking of them as closely analogous to syllogisms with universal premises and a universal conclusion. He thinks of $A \rightarrow B, B \rightarrow C \vdash A \rightarrow C$, for example, as analogous to *Barbara*: Every S is M, Every M is P, therefore Every S is P. If so, however, his theory of hypothetical syllogisms outstrips Aristotle's theory of categorical syllogisms, for $A \rightarrow B, B \rightarrow C \vdash \neg C \rightarrow \neg A$ would be analogous to Every S is M, Every M is P, therefore Every *nonP* is *nonS*, which goes beyond Aristotle's patterns. Not until Boethius would contraposition be recognized as a legitimate immediate inference and infinite terms such as *nonP* be given their due. The analogy between conditionals and universals is suggestive, though no one before Boole, Peirce, and Frege would fully exploit it.

Suppose, however, we were to think of conditionals $A \rightarrow B$, along Theophrastus's lines, as analogous to universal affirmatives, in effect having truth conditions of the form "Every case in which A is a case in which B." Call this the *universality thesis*. We might also think of particular propositions as corresponding to conjunctions, which are understood to be commutative. And we might think of existential presuppositions as corresponding to distinct assertions. (One might think that 'Every S is P' entails 'Some S

is P,' for example, but it is implausible to think that $S \rightarrow P$ entails $S \wedge P$ without the additional premise S.) Valid syllogistic forms would then generate a set of inference rules for conditionals. Permitting substitutions of negated propositions allows for some simplification:

- *Barbara, Celarent*: $A \rightarrow B, B \rightarrow C \vdash A \rightarrow C$
- *Cesare*: $A \rightarrow B, C \rightarrow \neg B \vdash A \rightarrow \neg C$
- *Camestres*: $A \rightarrow \neg B, C \rightarrow B \vdash A \rightarrow \neg C$
- *Calemes*: $A \rightarrow \neg B, C \rightarrow A \vdash B \rightarrow \neg C$
- *Darii, Ferio, Datisi, Disamis, Ferison, Bocardo, Dimatis*: $A \wedge B, B \rightarrow C \vdash A \wedge C$
- *Festino, Fresison*: $A \wedge B, C \rightarrow \neg B \vdash A \wedge \neg C$
- *Baroco*: $A \wedge \neg B, C \rightarrow B \vdash A \wedge \neg C$
- *Barbari, Celaront, Bamalip*: $A, A \rightarrow B, B \rightarrow C \vdash A \wedge C$
- *Cesaro*: $A, A \rightarrow B, C \rightarrow \neg B \vdash A \wedge \neg C$
- *Camestros*: $A, A \rightarrow \neg B, C \rightarrow B \vdash A \wedge \neg C$
- *Calemos*: $A, B \rightarrow \neg A, C \rightarrow B \vdash A \wedge \neg C$
- *Darapti, Felapton*: $A, A \rightarrow B, A \rightarrow C \vdash B \wedge C$
- *Fesapo*: $A, A \rightarrow B, C \rightarrow \neg A \vdash A \wedge \neg C$

Surprisingly, only one of the three forms mentioned by Alexander is among these. If we were to assume double negation and contraposition, this would of course become far simpler, and all three of Alexander's forms would be readily derivable:

- *Hypothetical Syllogism*: $A \rightarrow B, B \rightarrow C \vdash A \rightarrow C$
- *Conjunctive Modus Ponens*: $A \wedge B, B \rightarrow C \vdash A \wedge C$
- *Chaining*: $A, A \rightarrow B, B \rightarrow C \vdash A \wedge C$
- *Conjunctive Consequents*: $A, A \rightarrow B, A \rightarrow C \vdash B \wedge C$

In any case, Theophrastus's theory of the conditional is plainly incomplete if we construe it as an account of strict implication, even if we credit him with an understanding of modus ponens and the other principles above. There is no way to derive a conditional from non-conditional premises. So, in particular, there is no way to derive $A \rightarrow A$. Nor is there any way to get from $\neg A \rightarrow A$ to A, or from $A \rightarrow \neg A$ to $\neg A$. Of course, Theophrastus may well have rejected all these inferences on Aristotelian grounds, insisting that antecedents must be possible for a conditional to count as true. Even in that case, however, the theory is incomplete, for $\neg(A \rightarrow \neg A)$ is not a theorem.

Boethius (480–524?), a Roman nobleman born just after the fall of Rome, translated Aristotle's logical works into Latin, wrote commentaries on them, and wrote several important logical works of his own, including one on hypothetical syllogisms. His translations of the *Categories* and *De Interpretatione*, as well as his other works, served as the chief sources of information about ancient logic for medieval thinkers until the thirteenth century.

Boethius's *De Syllogismo Hypothetico* uses schemata such as *Si est A, est B*, which suggest a term interpretation but are ambiguous between that and a propositional interpretation (reading '*est*,' in the latter case, as "it is the case that"). He speaks of hypotheticals as complex propositions containing other propositions as components; what is unclear is simply whether *A* or 'est *A*' (or both) should be taken as a proposition.

Boethius discusses the truth conditions of conditionals, distinguishing accidental from natural conditionals. An accidental conditional is one whose consequent is true when its antecedent is true. Thus, he says, 'When fire is hot, the sky is round' is true, because, "at what time fire is hot, the sky is round" (*DSH* 835). A natural conditional expresses a natural consequence, as in "when man is, animal is." His temporal language suggests a Diodoran reading of the conditional, but it is not clear whether 'at what time' is meant as a quantifier.

The term Boethius uses for a conditional connection, accidental or natural, is *consequentia*, which he uses to translate Aristotle's terms *akolouthesis* and *akolouthia*, "following from" [Kneale and Kneale, 1962, 192].

Boethius also has some interesting things to say about negations. "Every negation is indeterminate (*infinita*)," he says (*DSH* 1.3.2–3); negation can separate "contraries, things mediate to contraries, and disparates" — that is, things that are not incompatible but merely different from one another.

His truth conditions for disjunction treat it as inclusive, but perhaps modal:

> The disjunctive proposition that says 'either *A* is not or *B* is not' is true of those things that can in no way co-exist, since it is also not necessary that either one of them should exist; it is equivalent to that compound proposition in which it is said: 'if *A* is, *B* is not.' (*DSH* 875)

The equivalence ($\neg A \vee \neg B \Leftrightarrow A \rightarrow \neg B$) suggests either a Philonian reading for the conditional or a modal reading for disjunction.

Bocthius lists valid inference patterns for conditionals:

1. $A \rightarrow B, A \vdash B$

2. $A \rightarrow B, \neg B \vdash \neg A$

3. $A \rightarrow B, B \rightarrow C \vdash A \rightarrow C$

4. $A \rightarrow B, B \rightarrow C, \neg C \vdash \neg A$

5. $A \rightarrow B, \neg A \rightarrow C \vdash \neg B \rightarrow C$

6. $A \rightarrow \neg B, \neg A \rightarrow \neg C \vdash B \rightarrow \neg C$

7. $B \to A, \neg C \to \neg A \vdash B \to \neg C$

8. $B \to A, \neg C \to \neg A \vdash B \to C$

9. $A \oplus B \vdash A \to \neg B, \neg A \to B, \neg B \to A, B \to \neg A$

10. $\neg A \vee \neg B \Leftrightarrow A \to \neg B$

Boethius uses 'or' (*aut*) for both 9 and 10, but it is important to distinguish them if conditionals are not to become biconditionals.[7] This theory extends that of Theophrastus, by taking account of contraposition and infinite terms.

4 EARLY MEDIEVAL THEORIES

Boethius clearly had access to Aristotle's logical works. So did John Philoponus (490–570), one of our chief sources for Stoic logic, and Simplicius (490–560), both of whom wrote commentaries on Aristotle's logical works. Shortly thereafter, however, the works of Aristotle, except for the *Categories* and *De Interpretatione,* disappeared. After the sixth century, logicians knew logic chiefly through the works of Porphyry and Boethius. The result was the *logica vetus* — the Old Logic.

The Old Logic's chief focus was Aristotelian syllogistic, as presented by Boethius. Propositional logic remained in the background. The Old Logicians nevertheless devote some attention to the logical connectives and have some interesting things to say about them.

The central concept of propositional logic is that of a proposition. The Old Logic has a standard definition, taken from Boethius: "*propositio est oratio verum falsumve significans*" (A proposition is a statement signifying truth or falsehood).[8] Sometimes it appears in the form "*propositio est oratio verum vel falsum significans indicando*" or "*propositio est oratio que cum indicatione significat verum falsumve,*" which might be translated, "a proposition is a statement that purports to signify truth or falsehood."[9] The former definition guarantees bivalence; the latter allows for the possibility that some propositions purport to signify a truth value but do not by virtue of being in a mood other than the indicative. "Sortes to run" and "Plato to dispute" are examples. They are propositions,

[7]Missing from this list is Boethius's most famous thesis, that $\neg((A \to B) \wedge (A \to \neg B)))$. Such a thesis may or may not be defensible, but we fail to find it in Boethius. For discussion of this point, see [Dürr, 1951; Kneale and Kneale, 1962; McCall, 1966; Barnes, 1981; Martin, 1991]. Boethius's thesis might be thought to follow from Aristotle's idea that true conditionals have possible antecedents; the truth of $A \to B$ and $A \to \neg B$ would then require the possible truth of A and thus the possible truth of $B \wedge \neg B$. But we see no evidence that Boethius follows Aristotle on this point. It would in fact be surprising if he holds that view, for the fifth and sixth theses above could then fail; nothing in the premises would guarantee the possibility of B or $\neg B$. In fact, contraposition itself would be problematic, for $A \to B$ would not imply $\neg B \to \neg A$, since $A \to B$ does not guarantee the possibility of $\neg B$. Perhaps this explains why Aristotle did not himself employ contraposition.

[8]Boethius, *De Topicis Differentiis*, 2.22–23. This definition is ubiquitous, found in the *Introductiones Montane Minores*, 18; *Abbrevatio Montana*, 79; *De Arte Dialectica*, 122; *De Arte Disserendi*, 128; *Ars Emmerana*, 152; *Ars Burana*, 183; *Introductiones Parisienses*, 359; *Logica "Cum Sit Nostra,"* 419. It persists into the fourteenth century, appearing in [Buridan, 2001, 21].

[9]This formulation appears, for example, in the *Dialectica Monacensis*, [de Rijk, 1967, 468].

figuring as components in compound propositions such as "Socrates wants Sortes to run" and "Thrasymachus allows Plato to dispute," but they do not in such a context signify a truth value in the way they would in the indicative mood.

Old Logic texts distinguish various kinds of compound propositions (*Ars Emmerana*, 159; cf. *Ars Burana*, 190–191):

- *Conditionals*: If you are running, you are moving.
- *Locals*: You are sitting where I am standing.
- *Causals*: Because Tully is a man, Tully is capable of laughter.
- *Temporals*: While Socrates runs, Plato is moved.
- *Conjunctions*: Socrates is a man and Brunellus is a donkey.
- *Disjunctions*: Someone is running, or nothing is moved.
- *Adjuncts*: The master is reading so that the student might improve.

The Old Logic considers conjunction and disjunction as grammatical conjunctions formed with the words 'and' and 'or,' respectively. They state concise truth conditions for both. A conjunction is true if and only if all (or both) its conjuncts are true; a disjunction is true if and only if some disjunct or other is true.[10] This is disjunction in the modern, inclusive sense. Only in the *Dialectica Monacensis* is it distinguished from exclusive disjunction.[11]

The Old Logicians spend more time on hypothetical propositions, that is, conditionals, which they define as statements formed with the conjunction 'if.'[12] Interestingly, the monks of St. Genevieve use subjunctive conditionals as their paradigm examples: "If to be blind and blindness were the same thing, they would be predicated of the same thing" and "If you had been here, my brother would not have died."[13] They take these to show that not all propositions have subject-predicate form.

Old Logicians generally do not attempt to state truth conditions for hypotheticals. The *Ars Burana*, however, does: "Every conditional is true whose antecedent cannot be true without the consequent, as 'if Socrates is a man, Socrates is an animal.' Also, every conditional is false whose antecedent either can or could or will be able to be true without the consequent, as 'if Socrates is a man, then Socrates is a donkey'" (191). The mixture of modals and tense operators here is curious, suggesting a combination of Diodoran and Chrysippan ideas. But it appears to be equivalent to a strict conditional reading. An

[10]See, for example, *Ars Burana*, 191, *Logica "Cum Sit Nostra,"* 425–426.

[11]That text treats disjunctions as ambiguous between a reading on which one disjunct is true and the other false and another (called subdisjunction) on which one or both is true (485).

[12]Old Logic texts seem to equivocate on hypothetical propositions, sometimes treating them as containing the word 'if' or an equivalent, thus amounting to conditionals in the contemporary sense, and sometimes treating them as including conjunctions, disjunctions, and a variety of other propositions, in which case 'hypothetical' seems equivalent to 'compound' or 'molecular.' See, e.g., *Ars Emmerana* 158–159, which does both in a span of three paragraphs.

[13]*Introductiones Montane Minores*, 39; the former is from Aristotle's *Categories* (10), the latter from John 11:21.

alternative account of truth conditions appears in the *Logica "Cum Sit Nostra,"* which treats conditionals as true when the consequent is understood in the antecedent (*quando consequens intelligitur in antecedente*, 425). This appears to be stronger, requiring that the necessity in question be analytic. Another alternative appears in the *Dialectica Monacensis*, which begins with a strict conditional account — "to posit [the truth of] the antecedent it is necessary to posit [the truth of] the consequent" (484–485) — but then adds *vel saltim probabile*, "or at least probable." The example is 'If this is a mother, she loves." The overall account, then, is that the truth of the antecedent makes the truth of the consequent probable. That intriguing idea, which appears to stem from William of Champeaux (1070–1122), then disappears for several hundred years.

The *Introductiones Norimbergenses* defines a hypothetical as saying something about something *under a condition* (*sub conditione*; 140–141). This is ambiguous between an ordinary view on which a conditional makes an assertion that something holds under a condition and a conditional assertion view that a conditional makes an assertion only conditionally. The latter view would make it difficult to maintain, however, that hypotheticals are propositions, for propositions are, in that work, always true or false.

Old Logicians divide hypotheticals into various kinds. *Simple* hypotheticals have no hypotheticals as components; *composite* hypotheticals do. The Old Logicians take it for granted that there can be embedded conditionals. Indeed, any hypothetical associated with an argument containing as hypothetical premise or conclusion is bound to be composite. Though Old Logicians do not for the most part speak explicitly about associated conditionals, they are aware of them. "An argument can be transformed into a conditional proposition, just as a conditional proposition can be taken as an argument" (*Ars Emmerana*, 164.) The monks of St. Genevieve give an example: "If every man is an animal and every animal is a substance, then every man is a substance" (*Introductiones Montane Minores*, 40). This, they hold, is neither simple nor composite, for it has no hypothetical as a component, but it has a conjoined antecedent. Evidently the definition of simplicity they intend is this: a conditional is simple if and only if its antecedent and consequent are both atomic propositions.

Hypotheticals, strictly speaking, contain the connective 'if,' but there are other connected (*connexa*) propositions that express very similar relationships, such as those formed with the connectives 'when,' 'as often as,' 'as long as,' 'while,' and so on.

Old Logicians are familiar with modus ponens from reading Boethius. The monks follow Boethius in expressing it in a hybrid form:

> If there is A, there is B
> There is A
> Therefore, there is B[14]

They give as an example "As often as Socrates reads, he speaks; but Socrates is reading; therefore, Socrates is speaking." The inference rule they have in mind, then, is a generalized form of modus ponens, one which, in modern form, we might express as $\forall x(Ax \rightarrow Bx), Ac \vdash Bc$. In other places, however, they state a straightforward version: "in

[14] *Introductiones Montane Minores*, 43. As in Boethius, this is ambiguous between, e.g., "There is A" and "It is the case that A."

every hypothetical proposition, if the antecedent is true, the consequent may be inferred" (45).

The monks discuss negations of conditionals and, in particular, the hypothesis, which they attribute to Boethius, that the negation of a conditional is equivalent to a conditional with a negated consequent: $\neg(A \rightarrow B) \Leftrightarrow (A \rightarrow \neg B)$. They carefully distinguish the affirmative conditionals $A \rightarrow B$ and $A \rightarrow \neg B$ from the negated conditionals $\neg(A \rightarrow B)$ and $\neg(A \rightarrow \neg B)$, and observe that only the former license modus ponens (46). They draw an analogy with necessity to make their point about scope: "If a man is living, it is necessary for him to have a heart." Can we apply modus ponens? Suppose a man is living. Can we infer that he necessarily has a heart? If the necessity attaches to the consequent alone, we could, but we would be drawing a false conclusion, indicating that the original conditional is false. If the necessity attaches to the entire hypothetical, however, we can infer only that he has a heart. Similarly, if the negation has scope only over the consequent, we can apply modus ponens. But we cannot if the negation has scope over the entire conditional. So, $\neg(A \rightarrow B)$ and $A \rightarrow \neg B$ are not equivalent; only the latter licenses the move from A to $\neg B$.

5 LATER MEDIEVAL THEORIES

The New Logic arose in the thirteenth century with the rediscovery of Aristotle's logical works and the textbooks of William of Sherwood, Lambert of Auxerre, and Peter of Spain. Peter gives clear truth conditions for conditionals, conjunctions, and disjunctions, with separate clauses for truth and falsehood. He treats conjunction and disjunction as multigrade, and his understanding of disjunction is inclusive:

> For the truth of a conditional it is required that the antecedent cannot be true without the consequent, as in, 'If it's a man, it's an animal.' From this it follows that every true conditional is necessary, and every false conditional is impossible. For falsehood it suffices that the antecedent can be true without the consequent, as 'if Sortes is, he is white.'
>
> For the truth of a conjunction it is required that all parts are true.... For falsehood it suffices that some part or another be false....
>
> For the truth of a disjunction it suffices that some part or other is true.... It is permitted that all parts be true, but it is not so properly [felicitously].... For falsehood it ought to be that all parts are false.... (Peter of Spain, *Tractatus* I, 17, 9–10)

This is the strict conditional analysis of conditionals, which had been championed by Abelard (1079–1142) as well as most twelth-century texts. Unlike most of the Old Logicians, however, Peter follows Abelard (*Dialectica*, 160, 279) to draw the conclusion that all conditionals are necessarily true or necessarily false. That is intriguing evidence that his background conception of modality is S5. He analyzes $A \rightarrow B$, essentially, as $\Box(A \supset B)$, and infers from that $\Box(A \rightarrow B)$, i.e., $\Box\Box(A \supset B)$. Similarly, from $\Diamond(A \wedge \neg B)$ he infers $\Box\Diamond(A \wedge \neg B)$.

Peter's conception of the connectives persists into the fourteenth century, though his multigrade conception begins to yield to a binary conception. William of Ockham (1287–1347), for example, gives similar truth conditions for conjunction, but takes it as binary in the positive half of the definition and multigrade in the negative half:

> Now for the truth of a conjunctive proposition it is required that both parts be true. Therefore, if any part of a conjunctive proposition is false, then the conjunctive proposition itself is false [1980, 187].

John Buridan (1290s–1360?) states the same positive condition, observing that an analogous condition holds for conjunctions of more than two terms [2001, 62–63].

Ockham goes further than Peter of Spain in three ways, however. First, he thinks about the interaction of conjunction and modality. For a conjunction to be necessary, both parts must be necessary. Second, he considers the negation of a conjunction, which is equivalent to a disjunction: $\neg(A \wedge B) \Leftrightarrow (\neg A \vee \neg B)$. Third, he formulates inference rules for conjunction. "Now it is necessary to note that there is always a valid consequence from a conjunctive proposition to either of its parts" [1980, 187]. He warns that one cannot go from just one conjunct to the conjunction, except in the case in which that conjunct entails the other, but fails to point out that one can move from both conjuncts to the conjunction.

Ockham and Buridan follow Peter in treating disjunction as inclusive; for its truth he requires merely that some disjunct be true. The negation of a disjunction, Ockham observes, is equivalent to a conjunction: $\neg(A \vee B) \Leftrightarrow (\neg A \wedge \neg B)$. He formulates inference rules for disjunction as well, stating rules for addition — licensing the move from one disjunct to the disjunction — and disjunctive syllogism: "Socrates is a man or a donkey; Socrates is not a donkey; therefore Socrates is a man" [1980, 189].

Ockham follows Peter, too, in his account of the truth conditions of conditionals. "A conditional proposition is true," he says, "when the antecedent entails the consequent and not otherwise" [1980, 186]. He defers a fuller discussion to his tract on consequences.

Buridan seems to accept a similar condition: "the antecedent cannot be true along with the consequent's not being true, provided both exist simultaneously, or even better, that it is not possible for things to be in the way signified by the antecedent without their being in the way signified by the consequent" [2001, 62]. He draws attention, however, to two exceptions. First, this does not apply to consequences as-of-now (*consequentia ut nunc*), for they depend on additional contingent information. His example: "Gerard is with Buridan; therefore he is in rue du Fouarre" [2001, 62]. This is not a true logical consequence but an enthymeme requiring an additional contingent premise. Second, they do not apply to future contingents, and in particular promises. 'If you visit me, I'll give you a horse' does not require that the visit entail the horse-giving, for of course the promise might be broken. If you visit, and I give you the horse, I have kept my promise; thus, 'If you visit me, I'll give you a horse' turned out to be true. Buridan is noticing that sometimes A and B suffice for the truth of $A \rightarrow B$. But that would not be true on Peter's account, unless B was itself necessary.

The most striking development in fourteenth-century logic relevant to the connectives, however, is the theory of consequences. What is this a theory *of*? Boethius uses *consequentia*, as we have seen, for a conditional connection. But he is translating a term

Aristotle uses for entailment. Abelard distinguishes these, using *consequentia* strictly for the former and *consecutio* for the latter. Abelard's theory is thus a theory of conditionals.[15] His thesis that conditionals are true if and only if the antecedent entails the consequent, however, makes it easy to run the two notions together. Even so great a logician as Buridan defines a consequence as a hypothetical composed of sentences joined by 'if,' 'therefore,' or their equivalent [1985, 1.3.2, 181]. Fourteenth-century accounts, as we shall see, are best seen as theories of entailment.

Walter Burley (1275–1344), for example, outlines a theory of consequences that endorses both the rule that, in a good consequence, "the antecedent *cannot* be true without the consequent" (285, my emphasis) and that "the contradictory of a conditional proposition is equivalent to a proposition that signifies that the opposite of its consequent stands together with its antecedent" (297). For Burley, plainly, a consequence is not a conditional, but a relation of entailment that amounts, roughly, to the necessitation of a conditional. Burley's theory, in other words, is a theory of consecution in Abelard's sense. We shall accordingly represent statements of consequence as consecutions, that is, schemata of the form $A_1, \ldots, A_n \models B$.

The first four of Burley's rules are these, which are those most pertinent to the connectives:

1. The antecedent cannot be true without the consequent: $A \models B \Rightarrow \neg\Diamond(A \wedge \neg B)$ (or, perhaps better, $A \models B \Rightarrow \Box(A \rightarrow B)$. He infers from this rule that what is impossible does not follow from what is contingent ($\Diamond A \wedge \neg\Diamond B \Rightarrow A \not\models B$) and that what is contingent does not follow from what is necessary ($\Box A \wedge \neg\Box B \Rightarrow A \not\models B$).

2. Whatever follows from the consequent follows from the antecedent: $A \models B, B \models C \Rightarrow A \models C$. The relation of entailment, then, is transitive. Burley considers several entertaining potential objections — e.g., "the uglier you are, the more you embellish yourself; the more you embellish yourself, the more attractive you are; therefore, the uglier you are, the more attractive you are" — which, he argues, are based on equivocations. He also draws several corollaries: What follows from the consequent and antecedent follows from the antecedent by itself ($A \models B; A, B \models C \Rightarrow A \models C$) and what follows from the consequent with some addition follows from the antecedent with the same addition ($A \models B, (B \wedge C) \models D \Rightarrow (A \wedge C) \models D$).

3. The contradictory of the antecedent follows from the contradictory of the consequent ($A \models B \Rightarrow \neg B \models \neg A$). Burley points out that Aristotle uses this principle in reducing certain syllogisms to first figure by reductio.

4. The formal element affirmed in one contradictory must be denied in the other. Burley uses this as a generalization covering instances such as $\neg(A \wedge B) \Leftrightarrow (\neg A \vee \neg B)$, $\neg(A \vee B) \Leftrightarrow (\neg A \wedge \neg B)$, and $\neg(A \rightarrow B) \Leftrightarrow (A \wedge \neg B)$.

Buridan and Albert of Saxony (1316–1390) offer a more comprehensive set of rules for consequences, all of which they derives from a definition of consequence like Burley's:

---

[15] Among those who follow him in this are Pseudo-Scot: "A consequence is a hypothetical proposition composed of an antecedent and a consequent by means of a conditional connective...." (*In An. Pr.* I, 10, 7).

1. From the impossible everything follows ($\neg\Diamond A \Rightarrow A \models B$).
2. The necessary follows from anything ($\Box A \Rightarrow B \models A$).
3. Every proposition implies anything whose contradictory is incompatible with it ($\neg\Diamond(A \wedge \neg B) \Rightarrow A \models B$).
4. The contradictory of the consequent implies the contradictory of the antecedent ($A \models B \Rightarrow \neg B \models \neg A$).
5. Transitivity ($A \models B \wedge B \models C \Rightarrow A \models C$).
6. It is impossible for the false to follow from the true, the possible from the impossible, or the contingent from the necessary ($A \wedge \neg B \Rightarrow A \not\models B; \neg\Diamond A \wedge \Diamond B \Rightarrow A \not\models B; \Box A \wedge \neg\Box B \Rightarrow A \not\models B$). Albert deduces that what implies the false is false; the impossible, impossible; and the unnecessary, unnecessary ($A \models B \Rightarrow (\neg B \Rightarrow \neg A); A \models B \Rightarrow (\neg\Diamond B \Rightarrow \neg\Diamond A); A \models B \Rightarrow (\neg\Box B \Rightarrow \neg\Box A)$).
7. If one thing follows from another together with necessary propositions, it follows from that other alone ($\Box C, A \wedge C \models B \Rightarrow A \models B$).
8. Contradictions imply anything ($A, \neg A \models B$). It follows that anything that implies a contradiction implies anything ($A \models B \wedge \neg B \Rightarrow A \models C$).

The fourteenth-century theory of consequences foreshadows abstract logic, which characterizes the relation of entailment from an abstract point of view. Surprisingly, Ockham, Burley, Buridan, and Albert do not remark on the reflexivity of entailment. We cannot find a clear statement of the principle that every proposition implies itself. They do notice transitivity, however, and the behavior of necessary and impossible propositions lets them play the role of top and bottom elements in a lattice.

The last great medieval logician was Paul of Venice (1369–1429), who writes extensively about the propositional connectives. Paul recognizes that conjunction and disjunction apply to terms as well as propositions and discusses each use in detail.

Paul presents a definition of proposition that deviates from the usual medieval definition, characterizing it as a well-formed and complete mental sentence (*congrua et perfecta enuntiatio mentalis*) that signifies truth or falsehood. He distinguishes between a proposition (say, A) and its adequate significate (σA). Paul gives four rules for truth and falsehood. Where a diamond stands for consistency:

1. $T[\sigma(A)] \wedge \Diamond T[A] \Rightarrow T[A]$
2. $T[A] \Rightarrow T[\sigma(A)]$
3. $F[\sigma(A)] \Rightarrow F[A]$
4. $F[A] \wedge \Diamond F[\sigma(A)] \Rightarrow F[\sigma(A)]$

The complexity introduced by the first and fourth rules is designed to handle insolubilia such as the Liar paradox, which end up false, but in a peculiar way — because the consistency clause fails. Suppose $\Diamond T[A]$ and $\Diamond F[\sigma(A)]$. Then these rules imply that $T[A] \Leftrightarrow T[\sigma(A)]$ and $F[A] \Leftrightarrow F[\sigma(A)]$. Assume that the adequate significates observe principles of bivalence and noncontradiction, so that $T[\sigma(A) \Leftrightarrow \neg F[\sigma(A)]$. Consider 'This sentence is false,' which Paul could construe in one of two ways:

- $T[L] \Leftrightarrow F[L]$. Suppose $T[L]$ and $F[L]$; then $T[\sigma(L)]$ and $\Diamond F[\sigma(L)] \Rightarrow F[\sigma(L)]$. But $\neg F[\sigma(L)]$, so $\neg\Diamond F[\sigma(L)]$. Suppose $\neg T[L]$ and $\neg F[L]$. Then $\neg T[\sigma(L)]$ or $\neg\Diamond T[A]$ and $\neg F[\sigma(L)]$. But then $T[\sigma(L)]$; so, $\neg\Diamond T[A]$. So, we can deduce that $\neg\Diamond F[\sigma(L)]$ or $\neg\Diamond T[L]$.

- $T[L] \Leftrightarrow F[\sigma(L)]$. (In effect, this interprets the liar sentence as "The proposition this sentence signifies is false.") Say $T[L]$ and $F[\sigma(L)]$. Then $T[\sigma(L)]$; contradiction. So, say $\neg T[L]$ and $\neg F[\sigma(L)]$. Then $T[\sigma(L)]$. Since $\neg T[L]$, $\neg\Diamond T[L]$.

The upshot is that, on either interpretation, the liar sentence fails to designate consistently an adequate significate.

Paul's account of conjunction as a relation between terms is rich and surprisingly contemporary in feel. He distinguishes collective and distributed (*divisive*) readings. A distributed reading entails a conjunctive proposition: "Socrates and Plato are running. Therefore Socrates is running and Plato is running" [1990, 54]. A collective reading does not: "Socrates and Plato are sufficient to carry stone A. Therefore* Socrates is sufficient to carry stone A and Plato is sufficient to carry stone A" [1990, 55].[16] He advances three theses about contexts that trigger distributed readings:

1. A conjoint term or plural demonstrative pronoun that is subject of a verb having no term in apposition has distributed supposition. (These people are eating, drinking, and loving. Therefore this one is eating, drinking, and loving and that one is eating, drinking, and loving.)[17]

2. A conjoint term or plural demonstrative pronoun that is subject of a substantival verb with a term in apposition has distributed supposition. (These are runners. Therefore this is a runner and that is a runner [1990, 57].)

3. A conjoint term or plural demonstrative pronoun that has supposition in relation to an adjectival verb that determines a verbal composite has distributed supposition. (Socrates and Plato know two propositions to be true. Therefore Socrates knows two propositions to be true and Plato knows two propositions to be true [1990, 61].)[18]

Three other theses describe contexts that trigger collective readings:

[16] We adopt the convention of writing 'therefore*' for inferences that do *not* follow.

[17] 1990, 55. Paul's text has *vel* ('or') in the premise and *et* ('and') in the conclusion, but this is surely a mistake; the same connective should appear in premise and conclusion.

[18] We should read 'two' as 'at least two.'

1. A conjoint term or plural demonstrative pronoun that has supposition in relation to a substantival verb having a singular subject or object has collective supposition. (Thus, "some man is matter and form"; "Shield *A* is white and black" [1990, 62].)

2. A conjoint term or plural demonstrative pronoun that has supposition in relation to a substantival verb with a subject or object and a determinant of it has collective supposition. ("*A* and *B* are heavier than *C*." It does not follow that *A* is heavier than *C* [1990, 64].)

3. A conjoint term or plural demonstrative pronoun that has supposition in relation to an adjectival verb that has a subject or object distinct from the conjoint term or demonstrative pronoun has collective supposition. ("These men know the seven liberal arts" [1990, 65].)

Conjunction is multigrade, according to Paul; conjunctive propositions may consist of two, three, or more parts. His truth conditions are standard: "For the truth of an affirmative conjunctive proposition the truth of each part is necessary and sufficient" [1990, 90]. Paul's presentation goes on to discuss modality as well as conjunctive knowledge and belief. A conjunction is necessary if and only if all its conjuncts are necessary. To know a conjunction, one must know every conjunct. Similarly, to believe a conjunctive proposition is to believe each conjunct. Uncertainty about any conjunct suffices for uncertainty about the whole. Paul presents a rule of simplification [1990, 99] but, like his predecessors, fails to specify a rule for introducing conjunctions.

Disjunction, too, applies to terms as well as to propositions. His rule for collective and distributed readings of disjunctions is simple: disjunctions always receive distributed readings unless "a determinant giving confused supposition, or a demonstrative term, covers the connective" [1990, 121]. Disjunctive propositions are true if and only if at least one disjunct is true. As with conjunction, he states conditions for the modal status of disjunctions as well as for disjunctive knowledge and belief.

Paul devotes the most attention, however, to conditionals. He reviews and rejects many accounts of the truth of conditionals, and despairs of giving a fully adequate and comprehensive account because of the wide variety of conditionals and conditional connections. His basic account, however, is that $A \rightarrow B$ is true if A is incompatible with $\neg B$ [1990b, 12]. He does articulate certain inference rules, including $A \rightarrow B \vdash \neg A \vee B$ (40). He shows that some conditionals are contradictory ("If you are not other than yourself, you are other than yourself") and uses it to show that some conditionals make positive assertions, which he takes as an argument against a conditional assertion account (41–42).

What is most interesting, from our point of view, is that Paul maintains a sharp distinction between conditionals and entailment propositions (*propositiones rationales*), formed with connectives such as 'hence' or 'therefore.' He defines a valid inference as "one in which the contradictory of its conclusion would be incompatible with the premise" (80). Thus, $A \therefore B$ is valid if and only if $\neg\Diamond(A \wedge \neg B)$. This is plainly equivalent to his truth condition for conditionals, however, so it turns out that an inference from A to B is valid (which we will write as $A \models B$) if and only if $A \rightarrow B$ is true.

Paul offers a series of theses that amount to a theory of consequences, together with three principles about propositional attitudes K (knowledge), B (belief), U (understanding), N (denial), and D (doubt):

1. $\neg B \models \neg A \Rightarrow A \models B$
2. $A \models B, A \Rightarrow B$
3. $A \models B, \Box A \Rightarrow \Box B$
4. $A \models B, \Diamond A \Rightarrow \Diamond B$
5. $A \models B, B \models C \Rightarrow A \models C$
6. $A \models B, \Diamond(A \wedge C) \Rightarrow \Diamond(B \wedge C)$
7. $A \models B, K(A \models B), U(A \models B), B(A) \Rightarrow B(B)$
8. $A \models B, K(A \models B), U(A \models B), N(B) \Rightarrow N(A)$
9. $A \models B, K(A \models B), U(A \models B), D(B) \Rightarrow D(A) \vee N(A)$
10. $A \models B, K(A \models B), U(A \models B), K(A) \Rightarrow K(B)$
11. $A \models B, K(A \models B), U(A \models B), D(A) \Rightarrow \neg N(B)$

The first five of these relate closely to rules advanced by Buridan and Albert of Saxony, but the remainder appear to be new. Knowledge and belief are closed under logical consequence, Paul holds, but only provided that the agent understands the inference and knows it to hold.

6 LEIBNIZ'S LOGIC

Gottfried Wilhelm Leibniz (1646–1716), diplomat, philosopher, and inventor of the calculus, developed a number of highly original logical ideas, from his idea of a *characteristica universalis* that would allow the resolution of philosophical and scientific disputes through computation ("*Calculemus*; Leibniz 1849–1863, 7, 200) to his algebra of concepts [Leibniz, 1982]. The algebra of concepts concerns us here, though it falls within a logic of terms, because Leibniz later recognizes that it can be reinterpreted as an algebra of propositions.

Let A, B, etc., be concepts, corresponding to the components of Aristotelian categorical propositions. Leibniz introduces two primitive operations on concepts that yield other concepts: negation ($\bar{A}$) and conjunction (AB). Negating terms to form so-called infinite terms was nothing new; the idea goes back at least to Boethius. But conjoining terms was Leibniz's innovation. He also introduces two primitive relations among concepts: containment ($A \supset B$) and possibility/impossibility ($\Diamond A/\neg\Diamond A$). The former Leibniz intends to capture the relationship between the subject and predicate of a universal affirmative proposition. The latter is new. This collection allows us to define converse containment and identity:

- $A \subset B \Leftrightarrow B \supset A$
- $A = B \Leftrightarrow A \supset B \wedge B \supset A$

Leibniz proceeds to state a number of principles.[19] Some of these principles — transitivity and contraposition, for example — are familiar from ancient and medieval logic. Many, however, are new. Leibniz appears to be the first logician to include principles of idempotence.

- *Idempotence*: $A \supset A$
- *Transitivity*: $(A \supset B) \wedge (B \supset C) \Rightarrow A \supset C$
- *Equivalence*: $A \supset B \Leftrightarrow A = AB$
- *Consequent Conjunction*: $A \supset BC \Leftrightarrow A \supset B \wedge A \supset C$
- *Simplification*: $AB \supset A$
- *Simplification*: $AB \supset B$
- *Idempotence*: $AA = A$
- *Commutativity*: $AB = BA$
- *Double Negation*: $\bar{\bar{A}} = A$
- *Nontriviality*: $A \neq \bar{A}$
- *Contraposition*: $A \supset B \Leftrightarrow \bar{B} \supset \bar{A}$
- *Contraposed Simplification*: $\bar{A} \supset A\bar{}B$
- *Contrary Conditionals*: $[\Diamond(A)\wedge]A \supset B \Rightarrow A \not\supset \bar{B}$
- *Strict Implication*: $\neg\Diamond(A\bar{B}) \Leftrightarrow A \supset B$
- *Possibility*: $A \supset B \wedge \Diamond A \Rightarrow \Diamond B$
- *Noncontradiction*: $\neg\Diamond(A\bar{A})$
- *Explosion*: $A\bar{A} \supset B$

This system is equivalent to Boole's algebra of classes [Lenzen, 1984]. But it actually makes little sense as a theory of categorical propositions, for it gives all universal propositions modal force. Later, however, in the *Notationes Generales*, Leibniz realizes that propositions can be viewed as concepts on the space of possible worlds. That allows him to "*conceive all propositions as terms, and hypotheticals as categoricals....*" ([Leibniz, 1966, 66], quoted in [Lenzen, 2004, 35]). To facilitate a propositional interpretation of the above, we translate negations and conjunctions into modern symbols, and replace identity with the biconditional:

[19] Lenzen [2004, 14–15] contains a useful chart. We use $\supset$ rather than $\in$ for containment to preserve greater continuity with modern usage and to suggest an analogy with propositional logic.

- *Idempotence*: $A \supset A$
- *Transitivity*: $(A \supset B) \wedge (B \supset C) \Rightarrow (A \supset C)$
- *Equivalence*: $A \supset B \Leftrightarrow A \equiv (A \wedge B)$
- *Predicate Conjunction*: $(A \supset B) \wedge C \Leftrightarrow (A \supset B) \wedge (A \supset C)$
- *Simplification*: $(A \wedge B) \supset A$
- *Simplification*: $(A \wedge B) \supset B$
- *Idempotence*: $(A \wedge A) \equiv A$
- *Commutativity*: $(A \wedge B) \equiv (B \wedge A)$
- *Double Negation*: $\neg\neg A \equiv A$
- *Nontriviality*: $\neg(A \equiv \neg A)$
- *Contraposition*: $(A \supset B) \Leftrightarrow (\neg B \supset \neg A)$
- *Contraposed Simplification*: $\neg A \supset \neg(A \wedge B)$
- *Contrary Conditionals*: $([\Diamond(A)\wedge](A \supset B)) \Rightarrow \neg(A \supset \neg B)$
- *Strict Implication*: $\neg\Diamond(A \wedge \neg B) \Leftrightarrow (A \supset B)$
- *Possibility*: $((A \supset B) \wedge \Diamond A) \Rightarrow \Diamond B$
- *Noncontradiction*: $\neg\Diamond(A \wedge \neg A)$
- *Explosion*: $(A \wedge \neg A) \supset B$

It remains to interpret the thick arrows and double arrows. Are these to be understood as object language conditionals and biconditionals, as Lenzen [2004] believes? If so, Leibniz offers a set of axioms, but without any rules of inference. Perhaps we should interpret them as metatheoretic, and construe the schemata in which they appear as rules of inference. That does not leave us in much better position, however, for we still lack modus ponens.

There is another possibility. Leibniz interprets his propositional logic as a logic of entailment: $A \supset B$ is true if and only if B follows from A [Leibniz, 1903, 260, 16]: "*ex A sequi B*"). So, we might replace both the conditional and implication with an arrow that can be read either way. Leibniz does not distinguish object from metalanguage. Indeed, the Stoic-medieval thesis that an argument is valid if and only if its associated conditional is true makes it difficult to see the difference between the conditional and implication. So, we might think of all conditionals or biconditionals as permitting interpretation as rules of inference as well as axioms. We might, in other words, think of this as something like a theory of consequences in the fourteenth-century sense. That effectively builds modus ponens into the system by default, since $A \rightarrow B$ can be read as a conditional or as a rule licensing the move from A to B. Nevertheless, there seems to be no way to get from A and B to $A \wedge B$; a rule of conjunction would have to be added separately.

- *Idempotence*: $A \to A$
- *Transitivity*: $((A \to B) \wedge (B \to C)) \to (A \to C)$
- *Equivalence*: $(A \to B) \leftrightarrow (A \leftrightarrow (A \wedge B))$
- *Consequent Conjunction*: $(A \to (B \wedge C)) \leftrightarrow ((A \to B) \wedge (A \to C))$
- *Simplification*: $(A \wedge B) \to A$
- *Simplification*: $(A \wedge B) \to B$
- *Idempotence*: $(A \wedge A) \leftrightarrow A$
- *Commutativity*: $(A \wedge B) \leftrightarrow (B \wedge A)$
- *Double Negation*: $\neg\neg A \leftrightarrow A$
- *Nontriviality*: $\neg(A \leftrightarrow \neg A)$
- *Contraposition*: $(A \to B) \leftrightarrow (\neg B \to \neg A)$
- *Contraposed Simplification*: $\neg A \to \neg(A \wedge B)$
- *Contrary Conditionals*: $([\Diamond(A)\wedge](A \to B)) \to \neg(A \to \neg B)$
- *Strict Implication*: $\neg\Diamond(A \wedge \neg B) \leftrightarrow (A \to B)$
- *Possibility*: $((A \to B) \wedge \Diamond A) \to \Diamond B$
- *Noncontradiction*: $\neg\Diamond(A \wedge \neg A)$
- *Explosion*: $(A \wedge \neg A) \to B$

Lenzen [1987] shows that this system is approximately Lewis's $S2^o$, although, as he points out, Leibniz surely accepted stronger principles about modality than this calculus represents.

It is striking, given the medieval consensus about inclusive disjunction, that Leibniz's calculus contains no such connective. Nor does it contain some principles familiar from theories of consequences — for example, the principle that the necessary does not imply the contingent (in this notation, that $(\neg\Diamond\neg A \wedge A \to B) \to \neg\Diamond\neg B)$, or that the necessary follows from anything $(\neg\Diamond\neg A \to (B \to A))$, though the latter follows from strict implication, contraposition, contraposed simplification, and possibility. Leibniz appears to have been constructing his algebra of concepts and the resulting propositional logic on a foundation independent of late medieval developments.

7 STANDARD MODERN-ERA LOGIC

Kant famously states in the preface to the second edition of the *Critique of Pure Reason* that "it is remarkable that to the present day this logic has not been able to advance a single step, and is thus to all appearance a closed and completed body of doctrine". This declaration captures much of the mood toward logic during the modern era. From the Renaissance through the beginning of the nineteenth century substantial advances in logic were rare, and typically had little influence when they did occur. Many significant philosophers of this period were dismissive of logic, and discussions of logic tended to be largely displaced by discussions of the psychology of belief formation and revision.

Logic textbooks of this period for the most part follow a common recipe. First, a theory of terms is given. The core of this theory is typically a canonical articulation of sentences into a subject-copula-predicate format:

> Each proposition [contains] two terms; of these terms, that which is spoken of is called the *subject*; that which is said of it, the *predicate*; and these two are called the terms, or extremes, because, logically, the subject is placed *first*, and the predicate *last*; and, in the middle, the copula, which indicates the act of judgment, as by it the predicate is affirmed or denied of the subject. The copula must be either IS or IS NOT; which expressions indicate simply that you affirm or deny the predicate, of the subject. [Whately, 1825, 85]

> In every proposition there are three parts, *viz.* a *Subject*, denoting the thing spoken of; a *Predicate*, denoting that which is asserted of it; and a *Copula*, or the verb, which connects them by affirmation or denial. ... The *Subject* and *predicate*, though they are not always single words, are to be considered as simple terms; for, as in *Grammar* a whole sentence may be the nominative case to a verb, so in *Logick* a whole sentence may be the subject or predicate of a proposition. ... The *copula* is always the verb substantive *am*, *is*, *are*, *am not*, &c. [Benthem, 1773, 36]

This articulation is often accompanied with some discussion of how to massage recalcitrant cases into the proper format:

> It is worth observing, that an infinitive (though it often comes last in the sentence) is *never the predicate*, except when another infinitive is the subject: *e.g.* "I hope to succeed;" *i.e.* "to succeed (subj.) is what I hope (pred.)" [Whately, 1825, 87]

The theory of terms is followed by a theory of propositions. The theory of propositions tends ultimately to center on the Aristotelian division of propositions into the universal affirmative (A), universal negative (E), existential affirmative (I), and existential negative (O). However, along the way to this taxonomy, various forms of molecular propositions are often identified:

> Propositions considered merely as *sentences*, are distinguished into "categorical" and "hypothetical". The categorical asserts simply that the predicate

> does, or does not, apply to the subject. ... The hypothetical makes its assertion under a *condition*, or with an *alternative*; as "if the world is not the work of chance, it must have had an intelligent maker:" "either mankind are capable of rising into civilization unassisted or the first beginning of civilization must have come from above." [Whately, 1825, 89]

The distinction between the categorical and the hypothetical merely taxonomic, in that there is no substantive semantic or logical theory of the hypothetical propositions given at this stage. Negation tends to be subsumed into a predicate feature and placed in the Aristotelian taxonomy, and conjunction is largely ignored.

After the theory of proposition, a theory of syllogistic validity is given, in varying levels of detail. (From this point, textbooks often branch out to less core logical topics, such as inductive reasoning or rhetorical techniques.)

Three noteworthy features of the treatment of connectives then emerge from this standard program:

1. **Core Conditionality**: As noted above, conjunction rarely receives overt attention from the modern-era logicians, and negation tends not to be treated as a sentential connective, but rather as a term modifier. The category of molecular proposition that is most frequently recognized is that of the conditional, or hypothetical, proposition. Disjunctions tend to be subsumed to to the category of hypotheticals, either by identification as a special case (as in the quotation from Whately above), or via an explicit reduction to conditionals:

   > In *Disjunctive* propositions two or more assertions are so connected by a disjunctive particle, that one (and but one of them) must be true, as *Either riches produce happiness, or Avarice is most unreasonable. Either it is day, or it is night.* — This kind of propositions are easily resolvible into *Hypothetical*; as, *If Riches do not produce happiness, then, Avarice is most unreasonable.* [Benthem, 1773, 47][20]

2. **Inferential Categorical Reductionism**: Since among modern-era logicians there is rarely any explicit semantic analysis given to molecular propositions, the analysis of them tends to come via remarks on how their inferential potential can be analyzed into that of categorical propositions whose inferential roles are already specified through the theory of Aristotelian syllogistics.

   The most common technique is to subsume both modus ponens and transitive chaining of conditionals to "Barbara" syllogisms:

   > We must consider every conditional proposition as a universal-affirmative categorical proposition, of which the terms of entire propositions. ... To say, "if Louis is a good king, France is likely to prosper," is equivalent to saying, "The case of Louis being a good king, is a case of France

[20]Because Benthem is reading the disjunction as exclusive, he presumably takes the English "if... then" construction to be biconditional in force, to secure the stated equivalence.

> being likely to prosper:" and if it be granted as a minor premiss to the conditional syllogism, that "Louis is a good king;" that is equivalent to saying, "the present case is the case of Louis being a good king;" from which you will draw a conclusion in *Barbara* (*viz.* "the present case is a case of France being likely to prosper,") exactly equivalent to the original conclusion of the conditional syllogism. [Whately, 1825, 136]

Additionally, modus tollens inferences can in the same manner be subsumed to "Celarent" syllogisms.

3. **Content/Force Ambivalence**: Modern-era logicians tend to describe connectives in a way that is compatible both with the contemporary view that these connectives are content-level operators and also with the view that connectives are indicators of speech-act force. This tendency is most prominent in the case of negation. It is standard to speak of negation as creating *denials*, in a way that leaves unclear whether denials are a particular type of content to be asserted, or are an alternative speech act to assertion:

> Affirmation and Denial belong to the very nature of things; and the distinction, instead of being concealed or disguised to make an imaginary unity, should receive the utmost prominence that the forms of language can bestow. Thus, besides being either universal or particular in quantity, a proposition is either affirmative or negative. [Bain, 1874, 83-84]

Modality also shows this ambivalence — necessity, for example, is often described as a *way of asserting or predicating*, rather than as an aspect of the content asserted:

> The modality of judgments is a quite peculiar function. Its distinguishing characteristic is that it contributes nothing to the content of the judgment (for, besides quantity, quality, and relation, there is nothing that constitutes the content of a judgment), but concerns only the value of the copula in relation to thought in general. [Kant 106]

In general, the modern era contrasts sharply with the ancient and medieval periods in its radically diminished interest in the logic of modality.

At times, even the quantificational forces of universality and existentiality are characterized in speech-act/force terms:

> A *Modal* proposition may be stated as a *pure* one, by *attaching the mode to one of the Terms*: and the Proposition will in all respects fall under the foregoing rules. ... E.g., "man is *necessarily* mortal:" is the same as "*all* men are mortal." ... Indeed every sign (or universality or particularity) may be considered as a *Mode*. [Whately, 1825, 108]

8 BOLZANO

Bernard Bolzano (1781-1840) raises the analysis of logical consequence to a new level of rigor and detail. Bolzano's account of logical consequence begins with an account of propositional structure. Bolzano shares the standard picture of modern-era logic, according to which the paradigm case of a proposition is a subject-predicate proposition. In particular, Bolzano takes the form *A has b*, for terms *A* and *b*, to be the common underlying structure of all propositions:

> The concept of *having*, or still more specifically, the concept signified by the word *has*, is present in all propositions, which are connected with each other by hat *has* in a way indicated by the expression: *A has b*. One of these components, namely the one indicated by *A*, stands as if it was supposed to represent the object with which the proposition is concerned. The other, *b*, stands as if it was supposed to represent the property the proposition ascribes to that object. (§127)

Again in keeping with the standard modern-era picture, Bolzano is committed to this single subject-predicate form subsuming molecular propositions:

> But there are propositions for which it is still less obvious how they are supposed to fall under the form, A has b. Among these are so-called hypothetical propositions of the form, if A is, then B is, and also disjunctive propositions of the form, either A or B or C, etc. I shall consider all of these propositional forms in greater detail in what follows, and it is to be hoped that it will then become clear to the reader that there are no exceptions to my rule in these cases. (§127)

However, Bolzano's conception of the nature of the propositional components *A* and *b* is particularly expansive, which leads him in an interesting direction.

Bolzano anticipates Frege's view that propositions are *timelessly* true:

> But it is undeniable that any given proposition can be only one of the two, and permanently so, either true and then true for ever, or false, and, again, false forever. (§140)

At the same time, he recognizes that ordinary practice involves ascribing changing truth values to propositions:

> So we say that the proposition, wine costs 10 thalers the pitcher, is true at this place and this time, but false at another place or another time. Also that the proposition, this flower has a pleasant smell, is true or false, depending on whether we are using the *this* in reference to a rose or a carrion-flower, and so on. (§140)

Bolzano accounts for the ordinary practice through the idea that "it is not actually one and the same proposition that reveals this diversified relationship to the truth", but rather that "we are considering several propositions, which only have the distinguishing feature of arising out of the same given sentence, in that we regard certain parts of it as variable, and substitute for them sometimes this idea and sometimes that one." (§140)

Once accounting for these broadly contextual features has introduced the idea of treating certain propositional components as variable, that idea then finds profitable application in accounting for logical consequence. Bolzano contrasts the two sentences:

(1) The man Caius is mortal.

(2) The man Caius is omniscient.

When we "look on the idea of Caius... as one that is arbitrarily variable", we discover that (1) is true for many other ideas (the idea of Sempronius, the idea of Titus), but also false for many other ideas (the idea of a rose, the idea of a triangle — because these ideas do not truthfully combine with the idea of a man). On the other hand, (2) is false regardless of what idea is substituted for the idea of Caius.

Sometimes when we pick a proposition P and a collection of constituents $i_1, \ldots, i_n$ of that proposition, we will discover that any substitutions for those constituents will result in a true proposition. In such a case, Bolzano says that P is valid with respect to $i_1, \ldots, i_n$. A proposition that is valid with respect to any collection of constituents is then called analytic. The resulting conception of analyticity is not a perfect match for contemporary notions of analyticity, as Bolzano's own example makes clear:

(3) A morally evil man deserves no respect. (§148) (Valid with respect to the constituent *man*.)

More generally, a proposition is analytic in Bolzano's sense just in case it is a true sentence with a wide-scoped non-vacuous universal second-order quantifier. Many such sentences will be intuitively non-analytic, in that they will depend for their truth on substantive contingencies about the world, as in:

(4) $\forall P$(every senator who owns a P is within $1,000,000$ miles of Washington D.C.)

Bolzano then defines a sequence of relations between propositions using the tools of propositional constituent variability:

1. **Compatibility, Incompatibility**: A collection of propositions $P_1, \ldots, P_n$ is *compatible* with respect to a collection of constituents $i_1, \ldots, i_m$ if and only if there are substitution instances for $i_1, \ldots, i_m$ such that $P_1, \ldots, P_n$ are all true under those substitutions. *Incompatibility* is then lack of compatibility.

2. **Derivability, Equivalence, Subordination**: One set $C_1, \ldots, C_n$ of propositions is *derivable* from another set $P_1, \ldots, P_m$ of propositions with respect to a collection of constituents $i_1, \ldots, i_k$ if and only if $C_1, \ldots, C_n$ are compatible with $P_1, \ldots, P_m$ with respect to $i_1, \ldots, i_k$, and every substution instance for $i_1, \ldots, i_k$ which makes true all of $P_1, \ldots, P_m$ also makes true all of $C_1, \ldots, C_n$. *Equivalence* is then mutual derivability; *subordination* is derivability without equivalence (with premises subordinate to conclusion).

3. **Concatenation**: A collection of propositions $A_1, \ldots, A_n$ is *concatenating with* (or *intersecting with*, or *independent of*) a collection of propositions $B_1, \ldots, B_m$ with respect to a collection of constituents $i_1, \ldots, i_k$ if neither collection is derivable from the other with respect to $i_1, \ldots, i_k$, but the entire collection $A_1, \ldots, A_n$, $B_1, \ldots, B_m$ is compatible with respect to $i_1, \ldots, i_k$.

For each of these relations, its holding is equivalent in an obvious way to the truth of a second-order quantified sentence — the compatibility of $P_1, \ldots, P_n$ with respect to $i_1, \ldots, i_m$ is equivalent to the truth of the m-fold second-order existential quantification of the conjunction of P_1 through P_n with the propositional constituents i_1 through i_m replaced with second-order variables, for example. Bolzano consistently leaves these propositional relations relativized to a choice of collection of propositional constituents — he does not have a notion of "derivability simplicter", but rather a family of derivability notions with no preferred or canonical member. However, if the object language is thought of as a first-order language and derivability is relativized to the collection of standard logical constants, the resulting notion of derivability almost matches the Tarskian model-theoretic definition, except that Bolzano places an additional constraint that premises and conclusion be consistent — thereby ruling out derivations from contradictory premises.

Relationships such as compatibility, derivability, and concatenating with would, in a contemporary setting, be taken as relations in the metalanguage. However, Bolzano makes no object language/metalanguage distinction, and is, for example, explicit that "we make use of [*if* and *then*] to express the relationship of derivability of one proposition from one or more other propositions." (§179) But Bolzano does not simply *equate* derivability and conditionality. He considers the example:

(5) If Caius remains silent on this occasion, he is ungrateful.

and rejects the view that:

> ...there is a certain idea (say the idea of Caius) in those propositions, which can be regarded as variable with the result that every assignment of it which makes the first proposition true also would make the second proposition true. (§179)

Instead, Bolzano's picture is that:

> What [(5)] means to say is that among Caius' circumstances there are some such that they are subject to the general principle that anyone who remains silent under such circumstances is ungrateful. ...For example, if Caius is dead, then Sempronius is a beggar. With these words, too, all we are saying is that there are certain relationships between Caius and Sempronius of which the general principle holds true that of any two men of whom one (in Caius' circumstances) dies the other (in Sempronius' circumstances) is reduced to beggarhood. (§179)

Bolzano thus gives a form of *premise semantics* for conditionals. A conditional of the form *if A, then B* is true just in case there are true supplementary principles which together with A entail B. Bolzano appears to require that these supplementary principles from the combination of which with A, B can be derived.

Because of the close relationship between derivability and conditionality, a sequence of observations Bolzano makes regarding the logic of derivability entail observations regarding the logic of the conditional. Bolzano is thus committed to the following inferences for conditionals:

1. $(A \wedge \top) \rightarrow B \Rightarrow A \rightarrow B$[21]
2. $A \rightarrow B, \neg A \Rightarrow \neg B$
3. $\neg\exists p(\neg p \wedge (A \rightarrow p)) \Rightarrow A$
4. $\Rightarrow \neg(A \rightarrow \neg A)$ [Even for contradictory A, given Bolzano's compatibility requirement on derivability.]
5. $(A \rightarrow B)$, $(\neg A \rightarrow B) \Rightarrow B = \top$
6. $(A \wedge B) \rightarrow C \Rightarrow (\neg C \wedge B) \rightarrow \neg A$
7. $((A \wedge B) \rightarrow C)$, $((A \wedge \neg B) \rightarrow C) \Rightarrow A \rightarrow C$
8. $A \rightarrow B$, $C \rightarrow D \Rightarrow (A \wedge C) \rightarrow (B \wedge D)$
9. $A \rightarrow B$, $(B \wedge C) \rightarrow D \Rightarrow (A \wedge C) \rightarrow D$

Bolzano anticipates Lewis Carroll 1895's regress argument against conflating inference rules with conditionalized premises. In addition to the relation of derivability, Bolzano sets out a hyperintensional relation of *ground and consequence*, which holds between two propositions when the first (partially) grounds or establishes or underlies the truth of the latter. The ground and consequence relation is thus, unlike the derivability relation, asymmetric. Bolzano then considers the ground-consequence relation between the grounds:

(6) Socrates was an Athenian.

(7) Socrates was a philosopher.

and the consequence:

(8) Socrates was an Athenian and a philosopher.

The crucial question is whether (6) and (7) form the *complete* grounds for (8), by virtue of the derivability relation among them, or whether to the complete grounds needs to be added a sentence corresponding to the inference rule:

- A, $B \Rightarrow A \wedge B$

The relevant sentence would then be:

(9) If proposition A is true and proposition B is true, proposition $A \wedge B$ is true.

[21] $\top$ here is an arbitrary tautology.

Bolzano then objects that in adding this sentence as an addition (partial) ground for (8):

> One is really claiming that propositions M, N, O,... are only true because this inference rule is correct and propositions A, B, C, D,... are true. In fact, one is constructing the following inference:
>
> > If propositions A, B, C, D,... are true, then propositions M, N, O,... are also true; now propositions A, B, C, D,... are true; therefore propositions M, N, O,... are also true.
>
> Just as every inference has its inference rule, so does this one. ... Now if one required to start with that the rule of derivation be counted in the complete ground of truths M, N, O,..., along with truths A, B, C, D,..., the same consideration forces one to require that the second inference rule ... also be counter in with that ground. ... One can see for oneself that this sort of reasoning could be continued *ad infinitum* ... But this would seem absurd. (§199)

60 years before Lewis Carroll, Bolzano also sees that a global demand that inferential principles be articulated and deployed as premises will lead to a regress which is incompatible with the presumed well-founded of (in this case) the grounding relation. While a separation between *using* a rule and *stating* a rule need not track with a distinction between derivability and conditionality, Bolzano's sensitivity to the threatened paradox here does suggest some caution on his part regarding a too-tight correlation between the latter two.

Bolzano's propositional relations of compatibility, derivability, and concatenation depend on an underlying notion of truth relative to an *index*, in the form of a quantificationally-bound choice of values for propositional components construed as variable. As such, these relations are intensional relations. Bolzano then recognizes that it is possible to distinguish from these intensional propositional relations a class of *extensional* relations:

> There was no talk of whether the given propositions were true or false so far as the relations among propositions considered [earlier] were concerned. All that was considered was what kind of a relation they maintained, disregarding truth or falsity, when certain ideas considered variable in them were replaced by any other ideas one pleased. But it is as plain as day that it is of the greatest importance in discovering new truths to know whether and how many true — or false propositions there are in a certain set. (§160)

Extensional propositional relations of this sort are a natural home for connectives of conjunction and disjunction. Because Bolzano is happy to talk of a set of propositions being true (and because he does not consider embedded contexts), he has little need for a notion of conjunction. He does, however, set out a number of types of disjunction.

1. **Complementarity**: A collection of propositions is *complementary* or *auxiliary* if at least one member of it is true.

2. **One-Membered Complementarity**: A collection of propositions is *one-membered complementary* if exactly one member of it is true. Bolzano says that this relation "is ordinarily called a *disjunction*" (§160)

3. **Complex Complementarity**: A collection of propositions is *complexly* or *redundantly complementary* if two or more members of it are true.

Bolzano's range of disjunctions are thus explicitly truth-functional, and include both inclusive disjunction (complementarity) and exclusive disjunction (one-membered complementarity), as well as the novel complex complementarity.

Bolzano also recognizes intensionalized versions of disjunction — what he calls "a *formal* auxiliary relation" — in which "this relationship of complementarity can subsist among given propositions M, N, O, ... no matter what ideas we substitute for certain ideas i, j, ... regarded as variable in them." (§160) Using intentionalized disjunction, Bolzano can then define two further varieties of disjunction which cannot be non-trivially specified for extensional disjunction:

4. **Exact Complementarity**: Propositions $P_1, \ldots, P_n$ are *exactly complementary* with respect to propositional constituents $i_1, \ldots, i_m$ if and only if both (a) for any choice of values for $i_1, \ldots, i_m$, at least one member of $P_1, \ldots, P_n$ is true, and (b) there is no proper subset of $P_1, \ldots, P_n$ for which condition (a) also holds.

5. **Conditional Complementarity**: Propositions $P_1, \ldots, P_n$ are *conditionally complementary* with respect to propositions $A_1, \ldots, A_m$ and propositional constituents $i_1, \ldots, i_k$ if and only if for any choice of values for $i_1, \ldots, i_k$ which makes true all of $A_1, \ldots, A_m$, at least one member of $P_1, \ldots, P_n$ is true.

Bolzano characterizes disjunction as a generalization of the subcontrarity relation from the square of opposition. In this context, he notes some preconditions on successful subcontrary relations. "Some *A* are *B*" and "Some *A* are not *B*" are subcontraries only conditional on the assumption that *A* is a denoting concept. Similarly, "If *A* then *B*" and "If Neg. *A* then *B*" are subcontraries only on the assumption that *B* is true. Bolzano thus rejects a principle of *antecedent conditional excluded middle*:

- $\neg(A \to B) \Rightarrow \neg A \to B$

A commitment to the falsity of a conditional with true antecedent and false consequent — or, equivalently, a commitment to the validity of modus ponens — is sufficient to reject antecedent conditional excluded middle.

Bolzano's remarks on negation are relatively cursory. He shares with the standard modern-era logic a tendency to equivocate between a content-level and a speech-act-level analysis of negation. He frequently refers to "Neg. *A*" as the "denial" of *A* in a way that suggests that negation is a speech-act marker. But he also clearly sets out the truth-functional analysis of negation via connections between negation and falsity:

> Where *A* designates a proposition I shall frequently designate its denial, or the proposition that *A* is false, by "Neg. *A*" (§141)

Bolzano regularly appeals to the truth-functional analysis of negation when arguing for metatheoretic results, as when he argues that compatibility of A, B, C, $D, \ldots$ need not be preserved under negation of all elements:

> For it could well be that one of the propositions ... e.g. A, would not only be made true by some of the ideas that also made all the rest B, C, $D, \ldots$ true, but besides that would be made true by all of the ideas by which one of those propositions, e.g. B, is made false and consequently the proposition Neg. B made true. In this case there would be no ideas that made both propositions, Neg. A and Neg. B, true at the same time. (§154)

Bolzano recognizes that equivalence of propositions is preserved under negation (§156), and is explicit that propositions are equivalent to their own double negation:

> I cite the example of 'something', which is equivalent to the double negative concept, 'not not something', and to every similar concept containing an even number of negations. (§96)

9 BOOLE

George Boole (1779-1848) is one of the most radical innovators in the history of logic. He differs sharply from his predecessors in the representational framework he uses for representing logical inferences, in the scope of logical arguments he seeks to analyze, and most markedly in the tools he uses for determining the validity and invalidity of inferences. Boole's approach to logic is heavily *algebraic*, in that he analyzes and solves logical problems by representing those problems as algebraic equations, and then laying a system of permissible algebraic manipulations by which the resulting equations can be reworked into forms that reveal logical consequences. While Boole thus shares with Leibniz a focus on algebraic approaches to logic, Boole's system is considerably more developed and complex than Leibniz's.

Boole's core algebraic representational system consists of four central elements:

1. **Variables**: Boole's logical system admits of two interpretations, the *primary* interpretation and the *secondary* interpretation. On the primary interpretation, variables represent classes. On the secondary interpretation, variables represent propositions. Boole is thus able to model monadic quantificational logic with the primary interpretation and propositional logic with the secondary interpretation.

    - Boole distinguishes between primary propositions, which are propositions which directly represent the state of the world, and secondary propositions, which are "those which concern or relate to Propositions considered as true or false." [Boole, 1854, 160]. As a result, Boole's characterization of molecular propositions is metalinguistic and truth-involving:

        > Secondary propositions also include all judgments by which we express a relation or dependence among propositions. To this class or

division we may refer conditional propositions, as "If the sun shine the day will be fair." Also most disjunctive propositions, as, "Either the sun will shine, or the enterprise will be postponed." ... In the latter we express a relation between the two Propositions, "The sun will shine," "The enterprise will be postponed," implying that the truth of the one excludes the truth of the other.' [Boole, 1854, 160]

Boole also holds that secondary propositions can, in effect, be reduced to primary propositions (and hence the secondary interpretation of the logical algebra to the primary interpretation) by equating each primary proposition with the class of times at which it is true, and then construing secondary propositions as claims about these classes. Thus, for example:

> Let us take, as an instance for examination, the conditional proposition, "If the proposition X is true, the proposition Y is true. An undoubted meaning of this proposition is, that the *time* in which the proposition X is true, is *time* in which the proposition Y is true. [Boole, 1854, 163]

Boole's conception of the conditional is in effect a strict conditional analysis — however, rather than requiring appropriate truth value coordination between antecedent and consequent with respect to every *world*, Boole requires it with respect to every *time*. Boole's willingness to represent molecular claims by classes of truth-supporting indices represents an early form of "possible-worlds" semantics, again with times replacing worlds.

The symbol v is reserved by Boole as the "indefinite" symbol, the role of which is set out below.

2. **Constants**: The constants 0 and 1 play a central role in Boole's system, representing on the primary interpretation the empty class and the universal class, and on the secondary interpretation falsity and truth. Other numerical constants, as we will see below, occasionally play a role in the computational mechanics of his system; these other constants lack a straightforward interpretation.

3. **Operations**: Boole uses a number of arithmetic operations on terms. The two central operations in his system are addition ('+') and multiplication ('×'). Multiplication, under the primary interpretation, represents *simultaneous application* of class terms:

> If x alone stands for "white things", and y for "sheep", let xy stand for "white sheep". [Boole, 1854, 28]

Multiplication thus acts as an intersection operator on classes. Boole's description of the role of multiplication under the secondary interpretation is somewhat baroque:

> Let us further represent by xy the performance in succession of the two operations represented by y and x; i.e. the whole mental operation which

> consists of the following elements, viz., 1st, The mental selection of that portion of time for which the proposition Y is true, 2ndly, The mental selection, our of that portion of time, of such portion as it contains of the time in which the proposition X is true, — the result of these successive processes being the fixing of the mental regard upon the whole of that portion of time for which the propositions of X and Y are true. [Boole, 1854, 165]

Once the excess psychologism is stripped away, however, this amounts to a treatment of multiplication as standard conjunction-as-intersection in an intensional setting. Under the primary interpretation, Boole takes addition to be the operation of "forming the aggregate conception of a group of objects consisting of partial groups, each of which is separately named or described" [Boole, 1854, 32]. This is roughly an operation of set union, but it acts under a presupposition of disjointness of the two sets. Thus when Boole wants to specify a straightforward union, he needs to add some slight epicycles:

> According to the meaning implied, the expression, "Things which are either x's or y's," will have two different symbolic equivalents. If we mean, "Things which are x's, but not y's, or y's, but not x's," the expression will be $x(1 - y) + y(1 - x)$; the symbol x standing for x's, y for y's. If, however, we mean, "Things which are either x's, or, if not x's, then y's,", the expression will be $x + y(1 - x)$. This expression supposes the admissibility of things which are both x's and y's at the same time. [Boole, 1854, 56]

Boole thus distinguishes between disjoint and non-disjoint union, but the expression $x + y$ itself corresponds to neither. On the secondary interpretation, addition similarly corresponds to disjunction with a presupposition of the impossibility of the conjunction.

Multiplication and addition thus serve roughly as meet and join operations on a boolean algebra of classes. But the contemporary algebraic characterization is only an imperfect fit for Boole's actual system — the presupposition of disjointness on addition, the additional arithmetic operations detailed below, and the actual computational mechanisms of Boole's system all distinguish it from contemporary algebraic semantics.

Boole also makes use of inverse operations of subtraction and division to his core operations of addition and multiplication, as well as occasional use of exponentiation. Subtraction is interpreted as "except", with $x - y$ representing the class x, except those member of x which are also y. The interpretation of division is not taken up by Boole. Boole then, based on the arithmetic equivalence between $x - y$ and $-y + x$, allows himself use of a unary operation of negation, which again receives no explicit interpretation.

4. **Equality**: The arithmetic operations recursively combine variables and constants to create complex terms. The mathematical familiarity with recursively embedded

terms creates in Boole a sensitivity to embedded occurrences of sentential connectives that is unusual in the modern-era logicians, as in:

> Suppose that we have $y = xz + vx(1 - z)$ Here it is implied that the time for which the proposition Y is true consists of all the time for which X and Z are together true, together with an indefinite portion of the time for which X is true and Z is false. From this it may been seen, 1st, That if Y is true, either X or Z are true together, or X is true and Z false; secondly, If X and Z are true together, Y is true. [Boole, 154, 173]

Terms, whether simple or complex, are then combined into full expressions with the only predicative element of Boole's system — the identity sign. All full sentences in Boole's algebraic representation are equations, and Boole's symbolic manipulation methods are all techniques for manipulating equations.

The centerpiece of Boole's algebraic manipulation methods is the validity $x^2 = x$. This validity represents for Boole the triviality that the combination of any term with itself picks out the same class as the term itself. $x^2 = x$ marks the primary point of departure of Boole's logical algebra from standard numerical algebra — since from $x^2 = x$ we can derive $x^2 - x = 0$ or $x(x - 1) = 0$, we obtain $x = 0$ and $x = 1$ as the two solutions to the distinctive validity, and establish 0 and 1 as terms of special interest in the logical algebra. For Boole, this special interest manifests in two ways. First, it establishes 0 and 1 as boundary points for the representational system, underwriting their roles as empty class/falsity and universal class/truth. Second, it allows him to appeal to any computational method which is valid for the special cases of 0 and 1, even if it is not globally arithmetically valid.

In deriving logical consequences, Boole uses three central algebraic techniques: *reduction*, for combining multiple premises into a single equation, *elimination*, for removing unwanted variables from equations, and *development*, for putting equations into a canonical form from which conclusions can easily be read off.

1. **Reduction**: Suppose we have two separate equations, such as $x = y$ and $xy = 2x$, and we wish to combine the information represented by these two equations. We first put each equation into a canonical form by setting the right-hand side to 0:

   - $x - y = 0 \Rightarrow x + (-y) = 0$
   - $xy - 2x = 0 \rightarrow xy + (-2x) = 0$

   Since, when we limit our field to 0 and 1, the only way a sum can be 0 is if each term is 0, Boole takes the informational content of each equation to be that each term is equal to 0. The two equations can thus be combined by adding, so long as there is no undue cancellation of terms thereby. When equations are in the properly developed form, as explained below, squaring an equation will have the effect of squaring the constant coefficient of each term, ensuring that all coefficients will be positive and thus that there will be no undue cancellation. Thus Boole's procedure of reduction is to put each equation into canonical form, square each equation, and then add all the squared equations to produce a single equation.

2. **Elimination**: At times, we may be uninterested in information regarding some of the terms in an equation. This is particularly likely in cases involving Boole's *indefinite term* v. Because Boole's sole predicate is equality, he has no direct means for representing subset/containment relations between classes. To indicate that x is a subclass of y Boole thus uses the equation $x = vy$, where the term v represents "a class indefinite in every respect" [Boole, 1854, 61]. Thus the universal generalization "All X are Y" is represented as $x = vy$, as is (on the secondary interpretation) the conditional "If X, then Y". But information about the wholly indefinite class v will be unwanted in the final analysis, so we need a technique for eliminating it from an equation.

   Boole's method of eliminating a variable from an equation proceeds by taking an equation $f(x_1, \ldots, x_n) = 0$ in standard form. A particular variable x_j is targeted for elimination, and a new equation is formed by setting equal to 0 the product of the left-hand term with x_j replaced by 0 with the left-hand term with x_j replaced by 1:

   - $f(x_1, \ldots, x_{j-1}, 0, x_{j+1}, \ldots, x_n) \times f(x_1, \ldots, x_{j-1}, 1, x_{j+1}, \ldots, x_n) = 0$

   The resulting equation is free of x_j.

   Boole justifies the method of elimination via an intricate sequence of algebraic maneuvers, the logical significance of which is far from clear [Boole, 1854, 101-102]. From a contemporary point of view, the crucial fact is that in a boolean algebra, $f(x) = f(1)x + f(0)(-x)$.

3. **Development**: Boole sets out a *canonical form* into which equations are to be placed to allow informational conclusions to be read off of them. When an equation contains only a single variable x, it is in canonical form when it is of the form $ax + b(1 - x) = 0$ — at this point, it gives a characterization both of the class x, and of the complement of the class x, which are the two classes made available by the single class term x.

   To determine the coefficients a and b, we set $a = f(1)$ and $b = f(0)$. Here again, Boole's argument for the value of the coefficient is involved and algebraic (and a second argument is given using Taylor/Maclaurin series). In contemporary terms, the development coefficients are a reflex of the underlying truth-functionality of the logic, which allows propositions to be fully characterized by their behavior when sentence letters are set to true and when sentence letters are set to false.

   In the general case, an equation in n variables $x_1, \ldots, x_n$ is in canonical form when it is in the form:

   - $a_1 x_1 \ldots x_n + a_2(1 - x_1)x_2 \ldots x_n + \ldots + a_{2^n}(1 - x_1) \ldots (1 - x_n)$

   Here each of the 2^n classes that can be produced via intersections of the classes x_1 to x_n and their complements is individually characterized. The development coefficients are then of the form $f(\epsilon_0, \ldots, \epsilon_n)$, where ϵ_j is 0 if the term contains $1 - x_j$, and 1 if the term contains x_j.

Once in canonical form, conclusions can be read off using the nature of the coefficients, via an intricate set of rules that Boole sets out.

Consider now an example worked out in full detail. Let the two premises be "If Socrates is a man, he is an animal", and "If Socrates is an animal, he is mortal". Let x be the class of times at which Socrates is a man, y be the class of times at which Socrates is an animal, and z be the class of times at which Socrates is mortal. Then we begin with the two equations:

- $x = vy$
- $y = vz$

First we put each equation into standard form, and eliminate v from each:[22]

- $x = vy \Rightarrow x - vy = 0 \Rightarrow (x - 1y)(x - 0y) = 0 \Rightarrow (x - y)x = 0 \Rightarrow x^2 - xy = 0 \Rightarrow x - xy = 0$

  [The last step follows from Boole's fundamental principle that $x^2 = x$, which more generally allows elimination of exponents.]
- $y = vz \Rightarrow y - vz = 0 \Rightarrow (y - 1z)(y - 0z) = 0 \Rightarrow (y - z)y = 0 \Rightarrow y^2 - yz = 0 \Rightarrow y - yz = 0$

We then square both equations in preparation for reduction:

- $(x - xy)^2 = 0 \Rightarrow x^2 - 2x^2y + x^2y^2 = 0 \Rightarrow x - 2xy + xy = 0 \Rightarrow x - xy = 0$
- $(y - yz)^2 = 0 \Rightarrow y^2 - 2y^z + y^2z^2 = 0 \Rightarrow y - 2yz + yz = 0 \Rightarrow y - yz = 0$

Adding the two equations together yields:

- $x - xy + y - yz = 0$

Solving for x yields:

- $\frac{y(1-z)}{1-y}$

We then develop the right-hand side in y and z:

- $x = \frac{0}{0}yz + \frac{1}{0}y(1 - z) + 0(1 - y)z + 0(1 - y)(1 - z)$

The coefficient $\frac{0}{0}$ indicates that x is to be set equal to an indefinite portion of that term, and the coefficient $\frac{1}{0}$ indicates that that term is to be set independently equal to zero. The other two terms, with coefficients of 0, disappear and can be disregarded. We thus obtain two results:

1. $y(1 - z) = 0$. This equation shows that there is no time at which Socrates is an animal but not mortal, which simply reconfirms the second conditional.

[22] Note that the elimination of v avoids the need to confront the delicate issue of whether the two occurrences of v represent the same class.

2. $x = vyz$. Since $y = vz$, we have by substitution $x = v^2z^2 = vz$. This equation is the representation of the conditional "If Socrates is a man, he is mortal". The transitive chaining of conditionals is thus derived.

Boole's algebraic system gives him the ability to represent all of the standard truth-functional connectives, and capture their truth-functionally determined inferential properties. His conditional, in particular, is inferentially equivalent to a material conditional. There are, however, some representational peculiarities that derive from his algebraic approach. Conjunctions and disjunctions are repesented as *terms* — xy is the conjunction of x and y, $x(1-y) + y(1-x)$ the exclusive disjunction of x and y, and $x + (1-x)y$ the inclusive disjunction of x and y. Conditionals, on the other hand, are represented by *equations*. The conditional whose antecedent is x and whose consequent is y is represented by the equation $x = vy$. This difference is driven by the fact that Boole's algebraic approach means that *biconditional* claims are really at the foundation of his expressive resources. Boole can, for example, straightforwardly express the biconditional between "x and y" and "y and x" using the equation $xy = yx$. Conditionals come out as equations only because of the peculiar role of the indefinite class v, which allows conditional relations to be expressed with biconditional resources.

Boole's work in logic was tremendously influential, first in the United Kingdom and later more broadly in Europe and America. British logicians such as Augustus DeMorgan (1806-1871), William Jevons (1835-1882), and John Venn (1834-1923) took up Boole's algebraic approach, and made steps toward simplifying and improving it. The followers of Boole clearly see themselves as *enfants terribles* working against a retrograde logical tradition, as can be seen as late as 1881 in the introduction to Venn's *Symbolic Logic*:

> There is so certain to be some prejudice on the part of those logicians who may without offence be designated as anti-mathematical, against any work professing to be a work on Logic, in which free use is made of the symbols $+$ and $-$, $\times$ and $\div$, (I might almost say, in which x and y occur in place of the customary X and Y) that some words of preliminary explanation and justification seem fairly called for. Such persons will without much hesitation pronounce a work which makes large use of symbols of this description to be mathematical and not logical. [Venn, 1881, 1]

Much of the focus of the later Booleans is thus on improving the *interpretability* of Boole's formalism, so that the intermediate computational steps have clearer meaning. As early as the 1860s, DeMorgan and Jevons suggest that the presupposition of disjointness be dropped from the interpretation of $+$, making it into a simple operation of class union. Here the discussion focuses on the status of the law $x + x = x$, which Boole rejects but DeMorgan and Jevons endorse. (This law is difficult for Boole to accept, because unlike his preferred $x^2 = x$, it does not have 0 and 1 as solutions, but only 0.) DeMorgan introduces, and Jevon picks up on, a notational convention in which Boole's use of the subtraction symbol as a monadic class negation operation is replaced by the use of capital and lower-case letters to represent a class and its negation. And Venn famously shows how many of Boole's logical conclusions can be reached diagrammatically with what came to be called *Venn diagrams*. There results a small cottage industry in specifying techniques for producing Venn diagrams for four or more sets. Lewis Carroll, in [Carroll, 1896], shows how overlapping rectangles can produce an elegant four-set diagram:

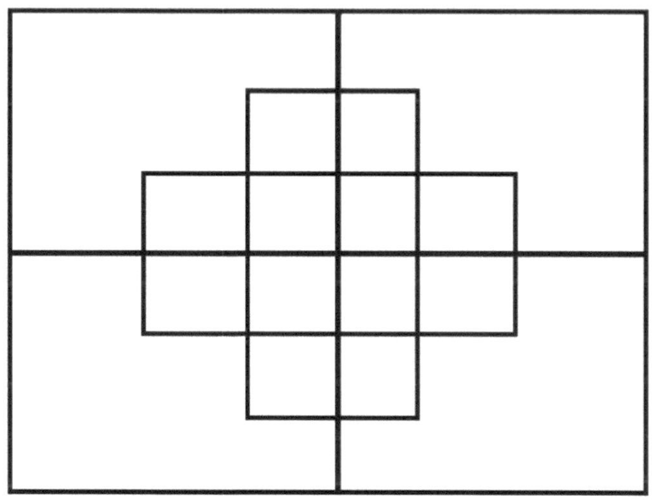

Venn 1881 sets out a clever representational trick in which a specified region represents *non-membership* in a given set, in order to produce a useable five-set diagram:

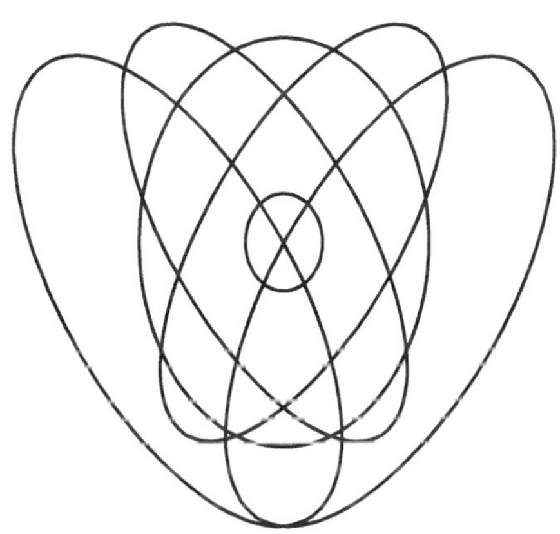

Here the smaller central oval represents objects which are *not* members of the larger central oval.

10 FREGE

Gottlob Frege (1848-1925) brings logic to its contemporary form, albeit in the dress of a rather forbidding notation.[23] In his 1879 *Begriffsschrift*, he sets out the syntax, semantics, and proof theory of first-order quantified logic with a level of detail and precision unmatched in the work of his predecessors and contemporaries.

Frege gives a full recursive syntax for his logical language. A primitive stock of terms for expressing "judgeable contents" are first combined with a *horizontal* or *content* stroke, which "binds the symbols that follow it into a whole" and a *vertical* or *judgment* stroke, which indicates that the stated content is asserted, rather than merely put forth to "arouse in the reader the idea of" that content. (§2) Thus Frege begins with symbolic representations such as:

- 

Four further components are then added to the syntax:

1. **Negation**: The placement of a "small vertical stroke attached to the underside of the content stroke" expresses negation:

    - 

    If we think of the negation stroke as producing a content stroke both before and after it, we have a natural recursive mechanism that allows for embedding of negations:

    - 

2. **Conditional**: Frege expresses conditionals using a two-dimensional branching syntax. Thus:

    - 

    expresses the conditional $A \to B$. Again, the syntax allows recursive embedding, and also combination with negation.

---

[23]The most forbidding aspect of Frege's notation is his extensive use of two-dimensional symbolic representations, as seen below. The two-dimensionality has the immediate benefit of readily representing the hierarchical syntactic structure of logical expressions. In his "On Mr Peano's Conceptual Notation and My Own" (1897), Frege recognizes that a linear notation can also represent the hierarchical relations, but holds that his two-dimensional notation continues to have an advantage in ease of use:

> Because of the two-dimensional expanse of the writing surface, a multitude of dispositions of the written signs with respect to one another is possible, and this can be exploited for the purpose of expressing thoughts. ...For physiological reasons it is more difficult with a long line to take it in at a glance and apprehend its articulation, than it is with shorter lines (disposed one beneath the other) obtained by fragmenting the longer one — provided that this partition corresponds to the articulation of the sense. [Frege, 1897, 236]

Subsequent history has not unambiguously validated this view of Frege's.

- $(A \to B) \to (C \to D)$: 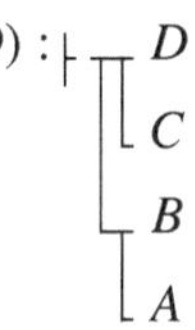
- $\neg A \to \neg B$: 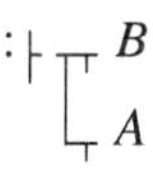

3. **Generality**: Frege expresses universal quantification using a concavity in the content stroke:

   - 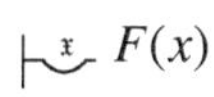

   The recursive specification of the syntax then allows Frege to distinguish scopes for quantifiers, so that he can, for example, distinguish:

   - $\forall x(F(x) \to G(x))$: 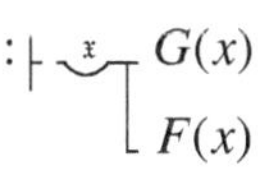
   - $\forall x F(x) \to \forall x G(x)$: 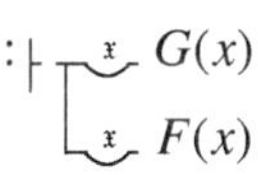

   or place one quantifier within the scope of another:

   - $\forall x(\forall y R(x, y) \to F(x))$: 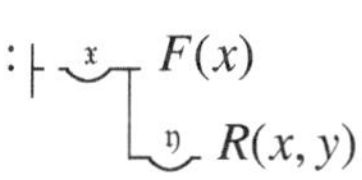

4. **Identity**: Frege makes broad use of the identity sign $\equiv$ to express sameness of semantic content between expressions. When placed between sentential expressions, $\equiv$ acts as a biconditional, allowing the expression of claims such as:

   - $(A \to B) \leftrightarrow (\neg B \to \neg A)$: 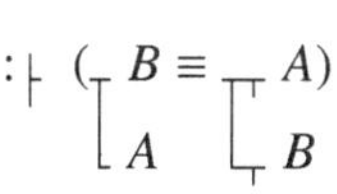

Frege 1879 gives explicitly truth-functional semantics for the conditional and for negation:

> If A and B denote judgeable contents, then there are the following four possibilities:
>
> 1. A is affirmed and B is affirmed;
> 2. A is affirmed and B is denied;
> 3. A is denied and B is affirmed;
> 4. A is denied and B is denied.

 now denotes the judgement that *the third of these possibilities does not obtain, but one of the other three does.* (§5)

In [Frege, 1893], Frege goes further, and rather than just specifying truth conditions for negations and conditionals in terms of the truth conditions of their parts, holds that the connectives themselves express functions from truth values to truth values:

> The value of the function $\top\xi$ shall be the False for every argument for which the value of the function $-\xi$ is the True, and shall be the True for all other arguments. (§6, see also §12 for similar remarks about the conditional)

Frege also shows that disjunction and conjunction can be expressed using appropriate combinations of the conditional and negation. He distinguishes between inclusive and exclusive disjunction, with inclusive disjunction represented by:

- 

and exclusive disjunction represented by:

- 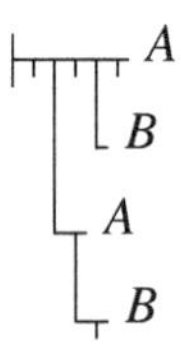

Conjunction is represented by:

- 

Frege's proof theory is axiomatic — a collection of nine axioms combine with rules of modus ponens and universal substitution to give a logic which is sound and complete on its propositional and its first-order fragments. The propositional axioms are:

1. $A \to (B \to A)$: 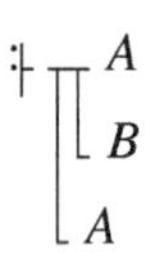

2. $(C \to (B \to A)) \to ((C \to B) \to (C \to A))$: 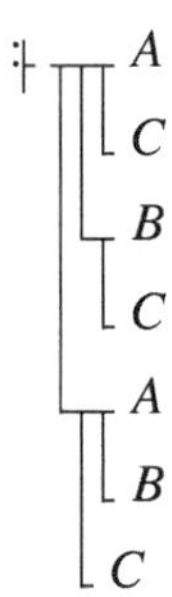

3. $(D \to (B \to A)) \to (B \to (D \to A))$: 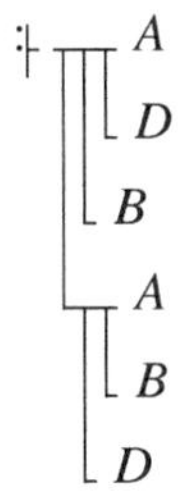

4. $(B \to A) \to (\neg A \to \neg B)$: 

5. $\neg\neg A \to A$: 

6. $A \to \neg\neg A$: 

7. $C \equiv D \Rightarrow \Sigma(C) \to \Sigma(D)$: 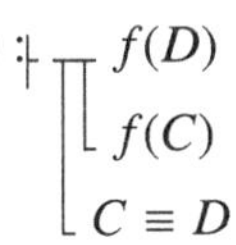

8. $C \equiv C$: $\vdash C \equiv C$

Frege shares with the standard modern-era logician a dismissive attitude toward modality — essentially his only remark on modality occurs in [Frege, 1879], in which he simply says:

> The apodictic judgment differs from the assertory in that it suggests the existence of universal judgments from which the proposition can be inferred, while int he case of the assertory such a suggestion is lacking. By saying that a proposition is necessary I give a hint about the grounds for my judgment. *But, since this does not affect the conceptual content of the judgment, the form of the apodictic judgment has no significance for us.* (§4)

11 PEIRCE AND PEANO

Although Frege's treatment of connectives, as with his logical system in general, is a very close match for contemporary logic, it had only a minimal impact on other logicians at the time. In a brief review of the *Begriffsschrift* in 1880, Venn describes it as a "somewhat novel kind of Symbolic Logic, dealing much more in diagrammatic or geometric forms than Boole's", and concludes that its central logical observations are "all points which must have forced themselves upon the attention of those who have studied this development of Logic" and that Frege's logic is "cumbrous and inconvenient" and "cannot for a moment compare with that of Boole". Frege's influence on contemporary logic begins to emerge only at the turn of the twentieth century, as Bertrand Russell discovers and is influenced by his work.

An earlier and powerful line of influence on Russell and twentieth century mathematical logicians more generally begins with Charles Sanders Peirce (1839-1914) and from there is passed through Ernst Schröder (1841-1902) to Giuseppe Peano (1858-1932). Peirce, beginning in Peirce 1870 and culminating in Peirce 1885, develops a quantified first-order logic that is formally equivalent to that set forth in [Frege, 1879]. Peirce begins in a Boolean framework, which he then modifies extensively. He shares with Boole a fascination with tracing out elaborate relations between mathematics and a logical "arithmetic" — in Peirce 1870, he explores the logical significance of the binomial theorem, of infinitesimals, of Taylor and Maclaurin series, of quaternions, and of Lobachevskian spherical non-Euclidean geometry.[24]

Peirce 1870 represents Peirce's first attempts to extend Boole's treatment of quantification to cover relational and hence polyadic quantification. This first attempt is complex and awkward, but by the time of Peirce 1885, a much simpler and more elegant system involving the use of subscripts to mark variable binding relations has been developed. The details of the treatment of quantification lie outside the scope of this piece, but several points of interest in the treatment of connectives emerge along the way. Peirce follows Jevons in recognizing the unsatisfactory nature of Boole's addition operation, with its presupposition of class disjointness, and replaces it with what he calls a "non-invertible" addition operation symbolized by '+,'. When interpreted as an operation on propositions, rather than classes, '+,' is then an inclusive disjunction. Peirce innovates more significantly by augmenting Boole's uniform treatment of logic as a system of *equations* with the use of *inequalities*. Peirce uses the symbol '$-\!\!<$' as his symbol for inequality. When terms are interpreted as representing classes, '$-<$' then represents the subset relation. Peirce 1885 emphasizes the role of '$-\!\!<$' as a conditional connective. The conditional is taken primarily as a strict conditional:

> The peculiarity of the hypothetical proposition is that it goes out beyond the actual state of things and declares what *would* happen were things other than they are or may be. ... There can be no doubt that the Possible, in its primary meaning, is that which may be true for aught we know, that whose falsity we do not know. The purpose is subserved, then, if, throughout the whole range of possibility, in every state of things in which *A* is true, *B* is true too. [Peirce, 1885, 186-187]

[24]He argues, for example, that when the parallel postulate is given its quaternion interpretation as the claim that the square of a vector is a scalar, and when terms are expressed using what Peirce calls "elementary relatives" in a way that models the arithmetic of quaternions, we find that

> Just as in the non-Euclidean or imaginary geometry of Lobatchewsky the axiom concerning parallels holds good only with the ultimate elements of space, so its logical equivalent holds good only for elementary relatives. [Peirce, 1870, 366]

But Peirce then immediately converts this strict conditional into a material conditional by noting that "the range of possibility is in one case taken wider, in another narrower; in the present case it is limited tot he actual state of things" [Peirce, 1885, 187]. Peirce then gives the truth conditions for the material conditional in terms of the truth values of antecedent and consequent, and uses these truth conditions to justify axioms for the conditional.

Peirce gives five axioms for the conditional:

1. $A -\!\!< B$
2. $(A -\!\!< (B -\!\!< C)) -\!\!< (B -\!\!< (A -\!\!< C))$
3. $(A -\!\!< B) -\!\!< ((B -\!\!< C) -\!\!< (A -\!\!< C))$
4. $\neg A -\!\!< (A -\!\!< B)$
5. $((A -\!\!< B) -\!\!< A) -\!\!< A$

The fifth axiom has come to be known as Peirce's Law; it is equivalent to the principle of the excluded middle, and like that principle, it fails in intuitionistic logic.

Peirce's use of inequalities allows him to avoid the use of Boole's indefinite class symbol v in representing conditionals, and his treatment of quantification, including the use of explicit existential ('Σ') and universal ('Π') quantifiers allows him also to avoid the use of v in representing indefinite claims, thus fully removing v from the logic. Peirce notes the relation between existential and universal quantification, on the one hand, and (potentially infinitary) disjunctions and conjunctions, on the other hand.

In later work, Peirce developed what he called "existential graphs" — a system of logic based on diagrammatic, rather than linguistic, representations. The simplest system of existential graphs is the so-called *alpha system*. In the alpha system, the placement of two graphs next to each other represents the conjunction of their contents, and the "cutting", or encircling, of a graph represents the negation of its content. The alpha system is thus primarily a minor representational reworking of standard propositional logic, although the choice of conjunction and negation as the sole logical operations is interesting in light of the later proof by Post 1921 that these two connectives are jointly functionally complete. Peirce's beta system extends the alpha system to include quantification. The beta system adds a novel sort of connective — variable coindexing is treated as a connective, by using a line to connect monadic predicates to represent the fact that the two predicates are predicated of the same object. Peirce's gamma system then extends the logic to modality, by using a "broken cut" diagrammatic representation which represents the possibility of falsity. Although Peirce's existential graphs had relatively little impact on the subsequent development of logic, they have in recent years been the focus of increased attention.[25]

Ernst Schröder (1841-1902)'s [1895] work '*Vorlesungen über die Algebra der Logik* serves as a point of entry for the logic of Peirce and Boole into the continental European mathematical logic community. A standardized logical system, which serves as the framework in which much of the mathematical logic of the early twentieth century is done, is

[25]See, e.g., [Shin, 2002].

then produced by Peano. Peano 1893 inherits from Schröder an approach on which logic is treated as a system of classes. However, he adds to Boole's system a symbol ϵ for the relation between individuals and classes, and allows a limited form of quantification over individuals — he allows variables to be subscripted to his conditional symbol, by which the universal quantification of the conditional with respect to those variables is represented. Like Frege, he uses a syntax that explicitly allows recursive embedding of connectives, and observes the definability of disjunction out of negation and conjunction. Unlike Frege, he does not give explicit truth conditions for his connectives. He takes the conditional to express deducibility of the consequent from the antecedent, rather than as a material conditional.

12 ON TO THE TWENTIETH CENTURY

As the nineteenth century draws to a close, the pieces are in place for the standard contemporary treatment of the connectives — the recursive syntax, the truth-functional semantics, and a well-formulated proof theory — but approaches vary considerably from logician to logician. Hugh MacColl (1837-1909) says:

> Various logicians have adopted various symbols, each giving some reason founded on some mathematical analogy for his own special choice. Boole adopts $P = \frac{0}{0}Q$ or $P = vQ$; Pierce [sic.] takes $P —\ < Q$; Schröder uses $P = (= Q$; and there are many others; each writer, as I have said, justifying his choice on the ground of some real or fancied mathematical analogy. My own choice has been the symbol $P : Q$ *not* (as has been erroneously supposed) on the ground of any analogy to a ratio or division, but simply because a *colon* symbol is easily formed, occupies but little space — two important considerations — and — though this is less important — because it is not unpleasing to the eye. [MacColl, 1895, 494]

MacColl also anticipates two aspects of twentieth century work that had receded to the background in most modern-era work in logic: an intense interest in the logic of modality, and a willingness to experiment with non-classical logics. MacColl suggests that a two-valued logic be replaced with a three-valued one:

> In dealing with implications of the higher degrees (*i.e.* implications of implications) a calculus of two dimensions (unity and zero) is too limited, and that for such cases we must adopt a *three*-divisional classification of our statements. We have often to consider not merely whether a statement is *true* or *false*, but whether it is a *certainty*, like $2 + 3 = 5$; an *impossibility*, like $2 + 3 = 8$; or a *variable* (neither always true nor always false), like $x = 4$. [MacColl, 1895, 496]

MacColl thus introduces the symbols ϵ for certainty, η for impossibility, and θ for variability. These symbols play three separate roles for MacColl (with the roles often run together). They are names for semantic values, sentential operators indicating that a sentence has the named semantic value, and sentential constants picking out sentences which always have that semantic value. Thus MacColl will say that sentence α is of class ϵ, will say that $(\alpha\beta)^\theta$ implies $\alpha^\epsilon\beta^\theta + \alpha^\theta\beta^\epsilon$, and will also say that the sentence $\alpha > \eta$, asserting that α implies η, is an impossibility.

MacColl's three-valued treatment of modal notions anticipates Łukasiewicz's [1920] three-valued system. However, a crucial difference is that MacColl does not take up in any systematic way the manner in which his three semantic values combine under various connectives. MacColl uses a superscript notation to indicate that a sentence has a particular semantic value — α^η means, for example, the impossibility of α. Elsewhere in MacColl [1895], he also uses the superscript notion to indicate logical consequence, so that α^β means that β implies α. When using the superscript notation in this way, he investigates the analogies between exponentiation and his representation of implication, and endorses the principle that $(\alpha\beta)^\gamma = \alpha^\gamma\beta^\gamma$. (This principle amounts to a rule of conjunction elimination — if γ implies the conjunction of α and β, then it also implies each of the individual conjuncts.) When superscripts change to indications of semantic value, however, the question of the correctness of that principle is not re-opened. He endorses some other points of the analogy — he holds that $(\alpha^\eta)^\epsilon = \alpha^{\eta\epsilon}$ (and that both represent the certainty of the impossibility of α, indicating a willingness to consider iterated modals) — and he gives the specific instance that $(\alpha\beta)^\tau = \alpha^\tau\beta^\tau$, where τ is truth, but he does not raise the more general question.

The interactions between modal semantic values and conjunction and disjunction are thus not taken up systematically by MacColl, but he does make a systematic pronouncement on the modal-conditional interaction. Here he says:

> No implication $\alpha : \beta$ can be a *variable* when ... its antecedent α and consequent β are both *singulars*; that is to say, when each letter denotes only *one statement, and always the same statement*, be it of the class ϵ or η or θ. Hence, $\alpha : \beta$ being synonymous with $\alpha\beta')^\eta$ [β' is the negation of β] must be either $= \epsilon$ or else $= \eta$; it cannot be $= \theta$. [MacColl, 1895, 503]

The view that conditionals, as themselves modal claims (the impossibility of the conjunction of the antecedent with the negation of the consequent), are either necessary or impossible, but never contingent, is an endorsement of the **4** axiom $\Box A \to \Box\Box A$ and the **5** axiom $\Diamond A \to \Box\Diamond A$ as basic modal principles.

MacColl closes his paper by suggesting that we might also want a nine-valued logic, in which the three modal values ϵ, η, and θ are combined with three epistemic values κ ("known to be true"), λ ("known to be false"), and μ ("neither known to be true nor known to be false"), to allow claims such as $\alpha^{\epsilon\kappa}$ ("it is known that α is always true") and $\alpha^{\kappa c}$ ("it is always true that α is known"). The resulting proliferation of options anticipates the proliferation of logical options the twentieth century would witness as the mathematical methods introduced in the nineteenth century came to fruition. The twentieth century development of logic lies outside the scope of this historical overview, but some major lines of investigation springing from the nineteenth century work include:

1. The emergence of metalogic in the twentieth century resulted in Post's [1921] proof of the functional completeness of (various subsets of) the standard connectives, and Sheffer 1913's observation that the Sheffer stroke, or NOR, connective is by itself

functionally complete. The development of the method of truth tables by Post and in [Wittgenstein, 1922] also shows the influence of an increased metalogical interest.

2. Modal logic, and the modal operators of necessity and possibility, becomes a topic of greatly increased interest. The work in Lewis [1918] and Lewis [1932] in characterizing a range of systems of modal logic did much to re-initiate interest in the area, and the subsequent development in Kripke [1959] and Kripke [1963] of possible worlds and frame-based semantics showed the considerable logical power of these systems.

3. Multivalued logics and treatments of the logical connectives became a topic of increased interest with the work of Post [1921] and Łukasiewcz [1920]. Important subsequent work on multivalued logic occurs in [Gödel, 1932; Bochvar, 1938; Kleene, 1938].

4. Nonclassical logics become a topic of intense interest in twentieth century work on logic. The intuitionistic logic of Brouwer [1907] and Heyting [1956] is the first major nonclassical system, and is then followed by the relevance and entailment logics surveyed in Anderson and Belnap [1975], and by other substructural logics. A central concern in much discussion of nonclassical logics is finding a treatment of conditionals which allows them to be properly integrated with other connectives in a given nonclassical system.

BIBLIOGRAPHY

[Abbreviatio Montana, 1998] *Abbreviatio Montana*. In N. Kretzmann and E. Stump (ed.), *The Cambridge Translations of Medieval Philosophical Texts, Volume One, Logic and the Philosophy of Language*. Cambridge: Cambridge University Press, 1988, 40–78.

[Abbreviatio Montana, 1967] *Abbreviatio Montana*. In De Rijk 1967, 73–107.

[Abelard, 1956] Abelard, Peter, 1956. *Dialectica*. L.M. de Rijk (ed.), Assen: Van Gorcum.

[Alexander of Aphrodisias, 1883] Alexander of Aphrodisias, 1883. *Alexandri in Aristotelis Analyticorum Priorum librum I commentarium*, M. Wallies (ed.). Berolini: G.Reimeri.

[Aristotle, 1984] Aristotle. *De Interpretatione*. Translated by J. L. Ackrill. In J. Barnes (ed.), *The Complete Works of Aristotle*, Volume I, 25–38. Princeton: Princeton University Press, 1984.

[Aristotle, 1984b] Aristotle. *Posterior Analytics*. Translated by J. Barnes. In J. Barnes (ed.), *The Complete Works of Aristotle*, Volume I, 114–166. Princeton: Princeton University Press, 1984.

[Aristotle, 1984c] Aristotle. *Prior Analytics*. Translated by A. J. Jenkinson. In J. Barnes (ed.), *The Complete Works of Aristotle*, Volume I, 39–113. Princeton: Princeton University Press, 1984.

[Aristotle, 1984d] Aristotle. *Sophistical Refutations*. Translated by W. A. Pickard-Cambridge. In J. Barnes (ed.), *The Complete Works of Aristotle*, Volume I, 278–314. Princeton: Princeton University Press, 1984.

[Ars Burana, 1967] *Ars Burana*. In De Rijk 1967, 175–213.

[Ars Emmerana, 1967] *Ars Emmerana*. In De Rijk 1967, 143–174.

[Bain, 1874] A. Bain. *Logic: Deductive and Inductive*. D. Appleton and Company, 1874.

[Barnes, 1981] J. Barnes. "Boethius and the Study of Logic," in M. Gibson (ed.), *Boethius: His Life, Thought, and Influence*. Oxford: Blackwell, 73–89, 1981.

[Barnes, 1984] J. Barnes. *Terms and Sentences: Theophrastus on Hypothetical Syllogisms*. London: British Academy, 1984.

[Benthem, 1773] E. Bentham. *An Introduction to Logick*. W. Jackson & J. Lister, 1773.

[Bochenski, 1961] I. M. Bochenski. *A History of Formal Logic*. Notre Dame: Notre Dame Press, 1961.

[Bochvar, 1938] D. A. Bochvar. On a Three-Valued Calculus and its Application in the Analysis of the Paradoxes of the Extended Functional Calculus. *Matamaticheskii Sbornik* 4, 287-308, 1938.

[Bolzano, 1973] B. Bolzano. *Theory of Science*. Translated by Burnham Terrell. Dordrecht, 1973.

[Boole, 1854] G. Boole. *An Investigation of the Laws of Thought, On Which Are Founded The Mathematical Theories of Logic and Probabilities*. Dover Publications, Inc, 1854.

[Buridan, 1976] J. Buridan. *Tractatus de consequentiis*. H. Hubien (ed.), *Philosophes mèdièvaux* 16, Louvain: Publications Universitaires, 1976.

[Buridan, 2001] J. Buridan. *Summulae de Dialectica*. An annotated translation with a philosophical introduction by Gyula Klima. New Haven: Yale University Press. 2001.

[Burley, 1988] W. Burley. "Consequences," in N. Kretzmann and E. Stump (ed.), *The Cambridge Translations of Medieval Philosophical Texts, Volume One: Logic and the Philosophy of Language*. Cambridge: Cambridge University Press, 284–311, 1988.

[Carroll, 1895] L. Carroll. What the Tortoise Said to Achilles. *Mind* 4, 278-280, 1895.

[Carroll, 1896] L. Carroll. *Symbolic Logic*. Dover Publications, Inc, 1896.

[De Arte Disserendi, 1967] *De Arte Disserendi*. in de Rijk 1967, 127–133.

[de Rijk, 1967] L. M. de Rijk. *Logica Modernorum. A Contribution to the History of Early Terminist Logic. Vol. II*. Assen: Van Gorcum, 1967.

[de Rijk, 1968] L. M. de Rijk. "On the Genuine Text of Peter of Spain's *Summule logicales* I. General problems concerning possible interpolations in the manuscripts," *Vivarium* 6: 1–34, 1968.

[de Rijk, 1982] L. M. de Rijk. "The Origins of the Theory of the Properties of Terms," *The Cambridge History of Later Medieval Philosophy*, edited by Norman Kretzmann, Anthony Kenny and Jan Pinborg. Cambridge: Cambridge University Press, 161–173, 1982.

[Dialectica Monacensis, 1967] *Dialectica Monacensis*. In de Rijk 1967, 453–638.

[Dürr, 1951] K. Dürr. *The Propositional Logic of Boethius*. Amsterdam: North Holland, 1951.

[Dutilh-Novaes, 2008] C. Dutilh-Novaes, "Logic in the 14th Century after Ockham," in D. Gabbay and J. Woods (eds.), *Handbook of the History of Logic, vol. 2, Medieval and Renaissance Logic*. Amsterdam: North Holland, 2008.

[Frege, 1879] G. Frege. Begrisschrift. In Jan van Heijenoort, ed., *From Frege to Gödel: A Sourcebook in Mathematical Logic, 1879-1931*. Harvard University Press, 1879.

[Frege, 1893] G. Frege. *Grundgesetze der Arithmetik*. Translated by Montgomery Furth. University of California Press, 1893.

[Frege, 1897] G. Frege. On Mr. Peano's Conceptual Notation and My Own. In *Collected Papers on Mathematics, Logic, and Philosophy*, Wiley-Blackwell, 1897.

[Gödel, 1932] K. Gödel. On the Intuitionistic Propositional Calculus. In S. Feferman *et al.*, eds. *Collected Works Volume I*. Oxford University Press, 1932.

[Introductiones Montana Minores, 1967] *Introductiones Montana Minores*. In De Rijk 1967, 7–71.

[Introductiones Norimbergenses, 1967] *Introductiones Norimbergenses*. In De Rijk 1967, 134–141.

[Introductiones Parisienses, 1967] *Introductiones Parisienses*. In De Rijk 1967, 353–373.

[Irwin, 1988] T. Irwin. *Aristotle's First Principles*. Oxford: Clarendon Press, 1988.

[John Philoponus, 1905] John Philoponus, 1905. *In Aristotelis Analytica Priora commentaria*. M. Wallies, ed. Berolini: G.Reimeri.

[Kant, 1787] I. Kant. *Critique of Pure Reason (Second Edition)*. Translated by Norman Kemp Smith. MacMillan and Co., Ltd, 1787.

[King, 1985] P. King. *Jean Buridan's Logic. The Treatise on Supposition. The Treatise on Consequences*. Translation with introduction and notes. Synthese Historical Library 27, Dordrecht: Reidel, 1985.

[Kneale and Kneale, 1962] W. Kneale and M. Kneale. *The Development of Logic* Oxford: Oxford University Press, 1962.

[Kretzmann, 1966] N. Kretzmann. *William of Sherwood's Logic*. Minneapolis: University of Minnesota Press, 1966.

[Kripke, 1959] S. Kripke. A Completeness Theorem in Modal Logic. *Journal of Symbolic Logic* 24(1), 1-14, 1959.

[Kripke, 1963] S. Kripke. Semantic Considerations on Modal Logic. *Acta Philosophica Fennica* 16, 83-94, 1963.

[Lear, 1980] J. Lear. *Aristotle and Logical Theory*. Cambridge: Cambridge University Press, 1980.

[Leibniz, 1849–1863] G. W. Leibniz. *Die philosophischen Schriften von G. W. Leibniz*. C. I. Gerhardt, ed. Berlin/Halle, 1849–1863.

[Leibniz, 1903] G. W. Leibniz. *Opuscules et fragments inédits de Leibniz*. L. Couturat, ed. Paris, 1903; reprint Hildesheim, 1960.

[Leibniz, 1966] G. W. Leibniz. *Logical Papers*. G. H. R. Parkinson, ed. Oxford: Clarendon Press, 1966.

[Leibniz, 1982] G. W. Leibniz. *Generales Inquisitiones de Analysi Notionum et Veritatum – Allgemeine Untersuchungen über die Analyse der Begriffe und Wahrheiten*. F. Schupp, ed. Hamburg: Meiner, 1982.

[Lenzen, 1984] W. Lenzen. "*Leibniz und die Boolesche Algebra*," *Studia Leibnitiana* 16, 187–203, 1984.

[Lenzen, 1987] W. Lenzen. "Leibniz's Calculus of Strict Implication," in J. Srzednicki (ed.), *Initiatives in Logic*. Dordrecht: D. Reidel, 1987.

[Lenzen, 2004] W. Lenzen. "Leibniz's Logic," in D. Gabbay and J. Woods (ed.), *Handbook of the History of Logic, Vol. 3: The Rise of Modern Logic*. Amsterdam: Elsevier, 2004.

[Lewis and Langford, 1832] C. I. Lewis and C. H. Langford. *Symbolic Logic*. Dover Publications, 1832.

[Logica, 1967] Logica "Cum Sit Nostra." In de Rijk 1967, 413–451.

[Lukasiewicz, 1920] J. Łukasiewicz. On Three-Valued Logics. In L. Borkowski, ed., *Selected Works by Jan Łukasiewicz*, North-Holland, 1920.

[Łukasiewicz, 1934] J. Łukasiewicz. "Outlines of the History of the Propositional Logic," *Pr. Fil.* 37, 1934.

[Łukasiewicz, 1951] J. Łukasiewicz. *Aristotle's Syllogistic from the Standpoint of Modern Formal Logic*. Oxford: Clarendon Press, 1951.

[MacColl, 1897] H. MacColl. Symbolic Reasoning III. *Mind* 6, 493-510, 1897.

[Martin, 1991] C. Martin. "The Logic of Negation in Boethius," *Phronesis* 34: 277–30, 1991.

[Martin, 2009] C. Martin. "The logical text-books and their influence," in J. Marenbon (ed.), *The Cambridge Companion to Boethius*. Cambridge: Cambridge University Press, 2009.

[Mates, 1961] B. Mates. *Stoic Logic*. Berkeley: University of California Press, 1961.

[McCall, 1966] S. McCall. "Connexive Implication," *Journal of Symbolic Logic* 31: 415–433, 1966.

[McKirahan, 1992] R. McKirahan. *Principles and Proofs*. Princeton: Princeton University Press, 1992.

[Normore, 1999] C. Normore. "Some Aspects of Ockham's Logic," in P. V. Spade (ed.), *The Cambridge Companion to Ockham*. Cambridge: Cambridge University Press, 1999.

[O'Toole and Jennings, 2004] R. R. O'Toole and R. E. Jennings. "The Megarians and the Stoics," in D. Gabbay and J. Woods (ed.), *Handbook of the History of Logic, Volume 1: Greek, Indian, and Arabic Logic*. Amsterdam: Elsevier, 2004.

[Ockham, 1974] Ockham, William of, 1974. *Summa Logicae*. Ph. Boehner, G. Gál and S. Brown (eds.), *Opera Philosophica* I, St. Bonaventure, NY: The Franciscan Institute.

[Ockham, 1974b] Ockham, Willliam of, 1974. *Ockham's Theory of Terms: Part I of the Summa Logicae*. Notre Dame: Notre Dame Press.

[Ockham, 1980] Ockham, Willliam of, 1980. *Ockham's Theory of Propositions: Part II of the Summa Logicae*. Notre Dame: Notre Dame Press.

[Paul of Venice, 1990] Paul of Venice, 1990. *Logica Magna, Secunda Pars: Tractatus de Hypotheticis*. Ed. with an English translation by Alexander Broadie. Oxford: Oxford University Press.

[Paul of Venice, 1990b] Paul of Venice, 1990b. *Logica Magna, Secunda Pars: Capitula de Conditionali et de Rationali*. Ed. with an English translation by G. E. Hughes. Oxford: Oxford University Press.

[Peano, 1889] G. Peano. The Principles of Arithmetic, Presented By a New Method. In Jan van Heijenoort, ed., *From Frege to Gödel: A Sourcebook in Mathematical Logic, 1879-1931*. Harvard University Press, 1889.

[Peirce, 1870] C. S. Peirce. Description of a Notation for the Logic of Relatives, Resulting from an Amplication of the Conceptions of Boole's Calculus. *Memoirs of the American Academy* 9, 317-378, 1870.

[Peirce, 1885] C. S. Peirce. On the Algebra of Logic: A Contribution to the Philosophy of Notation. *American Journal of Mathematics* 7(2), 180-202, 1885.

[Peter of Spain, 1992] Peter of Spain. *Syncategoreumata*. First Critical Edition with an Introduction and Indexes by L.M. de Rijk, with an English Translation by Joke Spruyt. Leiden/Kln/New York, 1992.

[Peter of Spain, 1972] Peter of Spain. *Tractatus called afterwards Summule logicales*. First Critical Edition from the Manuscripts with an Introduction by L.M. de Rijk. Assen, 1972.

[Post, 1921] E. Post. Introduction to a General Theory of Elementary Propositions. *American Journal of Mathematics* 43(1), 163-185, 1921.

[Schröder, 1895] E. Schröder. *Vorlesungen uber die Algebra der Logik*. B.G. Teubner, 1895.

[Sheer, 1913] H. Sheer. A Set of Five Independent Postulates for Boolean Algebras, With Application to Logical Constants. *Transactions of the American Mathematical Society* 14, 481-488, 1913.

[Shin, 2002] S.-J. Shin. *The Iconic Logic of Peirce's Graphs*. Bradford Books, 2002.

[Slater, 1979] B. H. Slater. "Aristotle's Propositional Logic," *Philosophical Studies* 36:35–49, 1979.

[Smith, 1989] R. Smith. *Aristotle's Prior Analytics*. Indianapolis: Hackett, 1989.

[Space, 2002] P. V. Spade. *Thoughts, Words, and Things: An Introduction to Medieval Logic and Semantic Theory*, 2002.

[Speca, 2001] A. Speca. *Hypothetical Syllogistic and Stoic Logic*. Leiden: Brill, 2001.

[Striker, 2009] G. Striker, tr. *Aristotle, Prior Analytics I*. Translated with commentary by Gisela Striker. Oxford: Clarendon Press, 2009.

[Tractatus Anagini, 1967] *Tractatus Anagnini*. In De Rijk 1967, 215–332.

[Venn, 1880] J. Venn. Review of Frege's Conceptual Notation. *Mind* 5, 297, 1880.

[Venn, 1881] J. Venn. *Symbolic Logic*. MacMillan and Co., Ltd, 1881.
[Whately, 1825] R. Whately. *Elements of Logic*. Scholars' Facsimiles and Reprints, 1825.
[Wittgenstein, 1922] L. Wittgenstein. *Tractatus Logico-Philosophicus*. Routledge, 1922.
[Woods and Irvine, 2004] J. Woods and A. Irvine. "Aristotle's Early Logic," in D. Gabbay and J. Woods (ed.), *Handbook of the History of Logic*, volume 1. Amsterdam: Elsevier, 27–100, 2004.

A HISTORY OF TRUTH-VALUES

Jean-Yves Béziau

1 AN EMBLEMATIC CONCEPT OF MODERN LOGIC

1.1 The whole story

The compound word "truth-value", sometimes written "truth value", is a bit monstrous and ambiguous. It is the name of a central concept of modern logic, but has not yet invaded everyday language. An ordinary man will say: it is true that Paris is the capital of France, rather than: the truth-value of "Paris is the capital of France" is true. And a mathematician also will say: it is true that 2+3=5, rather than the truth-value of "2+3=5" is true. We don't even find "truth-values" in postmodern or new age discussions side by side with "quantum leap", "imaginary number", "betacognition". It seems that "truth-value" is exclusively used by logicians, philosophers of logic and analytic philosophers. In this paper we will examine the origin of this strange way of speaking and the concept related to it.

Logical philosophers generally don't define the expression "truth-value", they take its definition for granted, known to everybody who has attended a first course on propositional logic. However if we have a close look at a textbook of logic, it is difficult to find a precise definition of such a notion. Truth-values will appear in the so-called truth tables, as T and F, or 1 and 0. The expression "truth tables" also is generally not explained, but this makes sense as a picture and to put truth in a table may look amusing.

One may think that by drawing the following table representing the 16 connectives of classical bivalent propositional logic one has said everything that can be said about truth-values:

0	0	0	0	0	0	1	1	1	1	0	0	1	1	1	1
0	1	0	0	0	1	0	0	1	0	1	1	0	1	1	1
1	0	0	0	1	0	0	0	0	1	1	1	1	0	1	1
1	1	0	1	0	0	0	1	0	0	0	1	1	1	0	1

16 BINARY CONNECTIVES

But when doing that, we are just writing the first 16 natural numbers using the binary numeral system. Does the concept of truth-value reduce to that? One can argue that there are more than two truth-values, not everything is black and white.

Handbook of the History of Logic. Volume 11: Logic: A History of its Central Concepts.
Volume editors: Dov M. Gabbay, Francis Jeffry Pelletier and John Woods.
General editors: Dov M. Gabbay and John Woods.

In fact more truth-values have been introduced, more that we may imagine. On the other hand people also have argued in favor of gaps and gluts: some propositions may have no truth-values and some other luckier ones may have more than one truth-value. The mechanism of bivalent truth-values can be generalized, this was done right at the beginning by people like Peirce, Post, Łukasiewicz and Bernays. Moreover the notion of truth-value does not reduce to propositional logic. The expression "truth-value semantics" is used as a name for an alternative semantics to first-order logic, but this is not just a question of name, truth-values are also present in the standard Tarskian semantics for first-order logic and also in Kripkean possible worlds semantics, etc. They are in fact everywhere ...

For this reason to write the history of the concept of truth-value is a thrilling but arduous task, exciting but almost impossible. The notion of truth-value is pervading the whole subject of modern logic, it is a central concept at the heart of modern logic. It is transversal not only from the viewpoint of technical constructions (beyond dichotomies such as propositional/first-order, bivalent/many-valued, proof/model), but also from the viewpoint of the interdisciplinarity of logic with all the richness and flavor of its philosophical, linguistical, mathematical and computational aspects. To write the history of the concept of truth-value one has to traverse in depth and breadth the whole history of modern logic. It is a very interesting task in the sense that by so-doing one is able to have a global vision of modern logic and to really reflect about its development.

But to write a complete history of the concept of truth-value would take several decades and this would lead to a book of one thousand and one pages. So one needs to simplify, in an intelligent way if possible. One dubious simplification would be to make the history of truth tables. This is not the option we have chosen, neither are we presenting here a history of propositional logic or many-valued logic. These are complementary stories.

1.2 What are we talking about?

The simplest answer would be to say that we are talking first of all of the expression "truth-value". But we are making the history of a concept, the history of a word is another story. However in the case of truth-value there is a strong connection between the word and the concept, a connection much stronger than, for example, between the word "energy" and the concept of energy. "Energy" is a word as old as Aristotle. The concept of energy of modern physics is quite different from the original Greek one, but the word is literally the same. The situation with truth-value is different, there is no corresponding word in the Aristotelian corpus. The word "truth-value" does not come from a remote Greek era but is the translation of the German word "Wahrheitswert", a word promoted by Frege in the XIXth century.

Can our history of the concept of truth-value reduce to the history of the word "Wahrheitswert"? No, for at least two opposite reasons. Firstly, Frege did not create the word "Wahrheitswert", it was used before him, in particular by Windelband in a meaning which is indirectly connected with the Fregean sense and logic (see [Gabriel, 1984]). Secondly, even if there is an intimate link between Frege's "Wahrheitswert" and the logical concept of truth-value, the mature concept of truth-value is quite different from Frege's original concept and also is connected with previous works of other modern logicians like Boole and Peirce not using the terminology "truth-value", whether in English or German. The German word "Wahrheitswert" in its Fregean sense was translated in English as "truth-value" by Bertrand Russell in the Appendix A of *The Principles of Mathematics* (1903) presenting the work of Frege. This is one of the reasons we will use the hyphenated version and not the two-word expression "truth value", also frequently used in logic textbooks. The other reason is to avoid any ambiguity with the moral sense of the word "value" that can be found at the same period for example in a paper by A.W.Moore entitled "Truth value" [Moore, 1908] or in a paper by another Russell (John) entitled "Truth as value and the value of truth" [Russell, 1911].

We are nevertheless lucky that there is a strong connection between the word "truth-value" and the design(at)ed concept. A new word has been used for a new concept: we will look neither for the concept, nor for the word, in the work of Aristotle or some remote Indian thinker, like Medhatithi Gautama. Let us note however that "truth-value" is a compound name. Its head, "truth", is a central concept of philosophy. Its tail, "value", pushing aside the moral ambiguity, is a mathematical concept. A central feature that differentiates modern logic from traditional logic is that it is mathematical. This radical difference, this conceptual jump, makes our task easier. To do the history of the concept of truth-value is much easier than to do the history of a philosophical concept such as truth, or of a mathematical concept such as the concept of function or the number 1. The history of 1 and truth can both be traced back to the beginning of history, but from the perspective of the history of modern logic, the story of truth(-value) before the mathematization of logic is prehistory, and we will not be concerned here with prehistory. Truth-value is the birth of the mathematical conception of truth, bearing mathematical value in its proper name.

On the other hand our history of the concept of truth-value can be considered as part of the history of the concept of function, since people like Boole and Frege, by using this concept in logic, have changed its meaning and were influential in its evolution. The concept of "propositional function", not so much in use nowadays, central for the development of the concept of truth-value (another central concept, still in use, is "truth-functionality"), was a radical shift for the mathematical conceptualization of the notion of function. The history of 1 has also been transfigured by modern logicians, either considering it as a Boolean truth value such that $1 + 1 = 1$, changing a fundamental law of arithmetics, or as a logical object from Frege's logicist perspective. We can also say that our present contribution is part of the history of the philosophical concept of truth: putting truth in tables, structures and possible worlds, multiplying the truth-values up to the infinite, considering that a proposition can both be true and false, certainly have changed the philosophical face of truth.

1.3 The main character of the story: MTV

For our story the central character will be the mathematical conception of truth-value. We will give to him the nickname MTV. But who is MTV? MTV is a folkloric concept that can hardly be found in a logic textbook. Like for many mathematical concepts, this is due in part because it has various faces. At the level of a general definition it is pure abstract nonsense, trivial for a mathematician, beyond understanding for a philosopher whether analytic or synthetic. Let us have a look at MTV. It is a double structure: on the one hand we have an absolutely free algebra, on the other hand, facing it, an algebra of a similar type, finite or not, and between them, the central notion establishing relations between mathematical structures, the notion of morphism. The elements of the free algebra are called *propositions* and its functions *connectives*, the elements of the facing structure are called *truth-values* and its functions *truth-functions*, and finally the morphisms between the two structures *valuations*.

This terminology is quite standard but to avoid any ambiguity we will give more formal names for the three kinds of functions we have here: connectives will be called *P-functions*, truth-functions *T-functions* and valuations *V-functions*. This will allow us to say without ambiguity for example that, John and James being two characters of our story, John uses the expression *truth-functions* to speak about V-functions contrary to James using the expression *truth-functions* to speak about T-functions. To harmonize the whole setting we will call *P-structure* the free algebra of propositions and *T-structure* the algebra of truth-values. We have therefore the following picture of MTV and terminology to describe it:

P-structure	$\longrightarrow$	T-structure
propositions	V-functions	truth-values
P-functions	$\longrightarrow$	T-functions

MTV in one shot

According to this picture truth-values are just a part of MTV. Why not then saying that MTV, the mathematical concept of truth-value, is just the T-structure? This would be indeed more trivial that triviality because the T-structure is any abstract algebra. It is not because we give the name "truth-values" to its elements and the name "truth-functions" to its functions that we have properly defined the concept of truth-value. A truth-value is something that turns true or not a proposition, so it makes sense only interacting with the P-structure, via the V-functions. To generate truth, the elements of the T-structure are divided in two classes: *designated* and *non-designated* values. Given a proposition p, if $v(p)$ is designated, p is said to be true (according to the morphism/valuation v) and if for all v, $v(p)$ is designated, p is said to be *logically true*. This general presentation of truth-values that we call MTV can also be called truth-functional semantics (for a detailed account of truth-functionality see [Marcos, 2009]). The bivalent semantics of classical propositional logic is the particular case where the T-structure is the Boolean algebra on $\{0, 1\}$, 1 being the unique designated value and 0 the unique non-designated value.

A pivotal figure for MTV is the Polish logician Roman Suszko (1919-1979). Pivotal because Suszko presented MTV, discussed it, criticized it and suggested alternatives to it. Roman Suszko is one of the most prominent figures of post-war logic in Poland. Jointly with Jerzy Łoś, he proved some basic theorems of model theory in the 1950s, he is the creator of abstract logic, that he developed with Bloom and Brown, and also the creator of non-Fregean logic (see [Omyła and Zygmunt, 1984], [Jansana, 2012]). But his main work is related to truth-values, running from one of his first papers entitled "Formal theory of logical values" published in volume 6 of *Studia Logica* (1957) to his last paper entitled "The Fregean axiom and Polish mathematical logic in the 1920s" published twenty years later in volume 36 of *Studia Logica* (1977). His work is still not very well-known outside of Poland, in part due to the isolation of Poland during the communist period. His 1957 paper has been published in Polish ("Formalna teoria wartości logicznych. I") and has not yet been translated in English, the same with Łoś monograph on logical matrices [Łoś, 1949].

2 FROM TARSKI TO SUSZKO

2.1 *Tarski and the origin of logical matrices*

MTV is much connected with the development of logical matrices theory. We can see three important works in the development of this theory:

- "Untersuchungen über des Aussagenkalküls", J.Łukasiewicz and A.Tarski, 1930
- *O matrycach logicznych*, J.Łoś, 1949
- "Remarks on sentential logics", J.Łoś and R.Suszko, 1958.

Matrices theory has been developed mainly by Tarski, Lindenbaum, Łoś and Suszko. Lindenbaum had many ideas and proved many important theorems but he didn't spend much time writing and publishing and was killed during the second world war. His contribution to matrices theory is very important and his main results were published by Łoś in 1949. Matrices theory is a central part of Polish logic, not very well-known abroad. Many people don't even know the expression "logical matrices" or they will assimilate it to truth tables or/and to many-valued logic. There is a connection, but logical matrices are at a superior level of abstraction and were developed as such by Alfred Tarski.

Here is how Tarski describes the situation in a footnote added to [Łukasiewicz and Tarski, 1930] when translated and published in *Logic, semantics, metamathematics*:

> The origin of this method is to be sought in the well-known verification procedure for the usual two-valued sentential calculus (see below Def 5), which was used by Peirce ('On the algebra of logic' *Am. Journ. of Math.* vol. 7 (1885). p.191) and Schröder. This was thoroughly treated in Łukasiewicz J. (50 a). Łukasiewicz was also the first to define by means of a matrix a system of the sentential calculus different from the usual one, namely his three-valued system (see below, p. 47 note 2). This he did in the year 1920. Many-valued systems defined by matrices, were also known to Post (see Post, E.L. (60j, pp. 180 ff.). The method used by P.Bernays ('Axiomatische Untersuchung des Aussagenkalküs der *Principia Mathematica*', *Math. Z.*, vol. 25, 1926, pp.305-20) for the proof of his theorems on independence also rests on matrix formation. The view of matrix formation as a general method of constructing systems is due to Tarski. [Tarski, 1983, p. 40]

This is a good description stressing the fact that people like Peirce, Schröder, Post, Bernays and Łukasiewicz were developing special cases of logical matrices and ending with the fact that the general method is due to Tarski. This is indeed supported by the definition which follows ([Tarski, 1983] p.41), the presentation of an abstract definition of matrix (due to Tarski as stated), where V-functions are called *value functions* and are defined as homomorphisms between the P-structure and the T-structure. This is expressed by the equation

$$h(c(x,y)) = f(h(x), h(y))$$

where h is a a V-function, c a P-function and f a T-function. But if the T-structure is clearly presented as such, this is not the case of the P-structure, it is not explicitly presented as an algebraic structure and V-functions also are not explicitly presented as homomorphisms (but curiously the letter h is used to denote a V-function).

All this is probably due to some philosophical problems related to the nature of propositions and sentences. Under the influence of Leśniewski, the Polish logicians developed a materialistic approach: according to the Lesniewskian definition, a sentence is a set of equiform inscriptions (equiformity was formulated in particular by Presburger — see [Zygmunt, 1991]). Łukasiewicz and Tarski are making the distinction between a sentence and the name of the sentence, the latter being considered as a mathematical object in the perspective of metamathematics:

> The symbol "*Cpg*", which in the sentential calculus *expresses* the implication between "*p*" and "*q*", is to be clearly distinguished from the metamathematical symbol: $c(z, y)$, which *denotes* the implication with the antecedent x and the consequent y. The expression "*Cpq*" is a

sentence (in the sentential calculus) whilst the expression "$c(x, y)$" is the *name of a sentence* (in the metasentential calculus). [Tarski, 1983, p. 39]

Łukasiewicz and Tarski don't use in this paper the standard semantic network: truth-value, true, truth, logical truth. When the value of a sentence is always designated, they say that the sentence is *satisfied*, meaning therefore logical truth. In model theory Tarski is known as the satisfaction guy, but in model theory satisfaction is used to speak of the truth of a sentence in one model, not in all models. These variations are tricky and if one is not aware of them, one may easily be led to misinterpretation.

2.2 *Logical matrix vs truth table*

The two Poles also don't use the expression "truth table", although the way they speak clearly emphasized a decision procedure: "The origin of this method is to be sought in the well-known verification procedure for the usual two-valued sentential calculus (see below Def 5)" [Tarski, 1983, p. 40] This verification procedure is presented as follows in Definition 5:

$f(0,0) = f(0,1) = f(1,1) = 1, f(1,0) = 0$

$g(0) = 1, g(1) = 0.$

where f is the T-function for implication and g the T-function for negation [Tarski, 1983, p. 43]

In a discussion about the origin of truth tables, Shosky makes the distinction between *truth-table technique* and *truth-table device*:

> A truth-table technique is a logically exhaustive analysis of the truth-functions of a given proposition, examining each possible set of truth values, and deriving definitions or arguments based on the results of this logically exhaustive process. A truth-table device is the mechanical creation of vertical columns of possibilities, measured against horizontal rows of logically exhaustive options, with actual lines of valid argumentation tested against the horizontal and vertical matrix. In short, the truth-table technique is a reference to the logical process of examining all truth values for a proposition; the truth-table device is the creation of an actual truth table. [Shosky, 1997, p. 13]

This distinction may be considered as a good starting point, but it bears some ambiguities which are present in discussions and uses of the notions of truth table and truth-value. Both expressions "truth-table technique" and "truth table device" are based on "truth-table". To go deeper we can consider the distinction between *logical matrix*, i.e. the abstract definition of logical matrix, and *truth table*, a concrete tabular diagram connected with a decision procedure. On the one hand a logical matrix does not necessarily lead to a table or/and a decision procedure, on the other hand a truth table is not necessarily based on an abstract conceptual framework. But from the historical point of view the two notions are intertwined.

Tarski is the one who has introduced the word "matrix" in logic to speak about the T-structure, as presented in his joint paper with Łukasiewicz. The paper is in German, in Definition 3 the words used are "Die (logische) Matrix" [Łukasiewicz and Tarski, 1930, p. 33]. The German word is the same as the English word "matrix", which was introduced in mathematics by James Joseph Sylvester in ([Sylvester, 1850]) to denote a rectangular array of symbols[1]. Not surprisingly this word was also used by Peirce, since Sylvester was a main figure for the development of mathematics in the USA in the second half of the XIXth century. But Peirce had no definition of logical matrix approaching the one given by Tarski, he rather had a presentation of a truth-table device or/and technique which was influential in the development of the Tarskian notion of matrix, as Tarski himself claims. Peirce, Schröder, Post, Bernays and Łukasiewicz certainly had some ideas connected to abstract logical matrices, but they were not reaching Tarski's high level of abstraction (that we can also see in his development of the consequence operator [Zygmunt, 2012b]). Their ideas are more connected to truth-table device, or, at best, to truth-table technique, but always related to "table", they are not dealing with an abstract structure in the sky.

In Poland, before the appearance of the concept of matrix, Łukasiewicz [1929] and Sleszynski [1925] used in Polish the term "tabelka", which means "table". The concept of matrix was translated in Polish by two different words: "matryca" or "macierz", the first being a word with a more technical meaning, used by printers, a mold for shaping letters. This is the word used by J.Łoś in his 1949 monography *O matrycach logicznych*. But the word which prevailed was finally the word "macierz" preferred by Suszko and the one used by Polish mathematicians in the Sylvesterian sense. The Polish word "macierz", like the English word "matrix", has a polysemous sense related to generation.

Tarski sometimes was going down to the earth identifying truth tables with matrices, like for example in the American edition of his famous book *Introduction to logic and the methodology of deductive science*. This book was originally published in Polish in 1936 [Tarski, 1936] and immediately translated in German [Tarski, 1937]. There are four American editions, the later, prepared by the son of Tarski, to which we refer for pagination [Tarski, 1994]. The first American edition was published in 1941 with a section on truth tables not to be found in the original Polish and German editions. This is section 13, entitled "The symbolism of sentential calculus; sentential functions and truth tables". It starts as follows:

> "There is a simple and general method, called the METHOD OF TRUTH TABLES or of MATRICES, which enables us to determine whether any given sentence from the domain of sentential calculus is true, and therefore, whether it can be counted as a law of this calculus (footnote: this method originates with PEIRCE)." [Tarski, 1994, p.

[1] Sylvester also introduced the expression "universal algebra".

34]

Tarski assimilates here the notion of matrix to the notion of truth table via the mathematical notion of matrix, something tabular. The book of Tarski is for beginners, including philosophers having nearly no background in mathematics, this is therefore a good way to picture the scene. Probably for Tarski a truth table is an elementary and concrete presentation of an abstract logical matrix, so that the difference is a question of degree of abstraction, but while presenting truth tables Tarski also certainly had in mind the abstract version of a logical matrix.

Nevertheless Tarski makes here two confusions which were also common among his predecessors and were cleared only later on at the level of explicit MTV. These two confusions go together and are related to the ambiguous notion of propositional function promoted by Russell. The first confusion is about the notion of sentential variable, the second one about the mixture of P-function and T-function. Tarski writes:

> We can easily convince ourselves that every sentential function which is constructed in such a way is a so-called TRUTH FUNCTION. This means the following: if a sentence is obtained from that function by substituting whole sentences for the variables, its truth or falsity depends exclusively on the truth or falsity of the sentences which have been substituted. (footnote: A truth function is therefore analogous to what may be called a NUMERICAL FUNCTION, i.e., to a designatory function whose designation (a number) depends exclusively on the numerical constants which have been substituted for the variables. However, the term "*truth function*" is not widely used, and in practice one would refer to a formula like "$p \rightarrow (p \wedge q)$" as a *compound sentential function*, or as an *expression of sentential calculus* .) [Tarski, 1994, p. 35]

In this description Tarski is not making a clear distinction between a P-function and a T-function, calling indeed a sentential function a truth function. However he is making a move by rejecting the terminology *truth function* in favor of *compound sentential function* and even in favor of *expression of sentential calculus.* We see here that Tarski is not able to define a P-function independently of a T-function, not able to properly conceive the P-structure. In fact the concept of truth function, which was the subject of Tarski's thesis [1923a] (published as [Tarski, 1923b], [Tarski, 1924]), will be perfectly cleared up only by Łoś and Suszko in 1958 introducing the concept of structural consequence operator.

Tarski goes on writing:

> Let us elaborate on the analogy with numbers and numerical functions (cf. the preceding footnote). In the latter case, we start by substituting numerical constants for variables, and instead of talking about these constants, we can also refer to numbers and use the term "numerical values". From these values we obtain the numerical values of

> the successive combinations, and finally, the numerical value of the original function. We see that the the truth values of the expression "$(p \vee q) \to (p \wedge r)$" are determined in much the same way as the numerical values of, say, "$x^2 + 2x + y$". ([Tarski, 1994] p.36)

This is a classical mixture of P-structure and T-structure, P-functions and T-functions, which is still present in many textbooks due in particular to the fact that the same symbols are used for connectives (P-functions) and T-functions. The difference with numerical values is that variables like x and y are ranging over numbers, but "sentential variables" are ranging over sentences which moreover can be seen as ranging over truth-values such as 0 and 1. We have here two different meanings of "variable", the first is connected with the notion of undetermined constant. The shortcut consisting of saying that sentential variables are ranging over truth-values is rather confused, a confusion cleared up with the full MTV picture.

This imprecision was pointed out by the meticulous Quine in his review of the 1937 German edition of Tarski's book:

> It is only in criticizing this book by its own high standards that a few imperfections of formulation are to be found. Tarski distinguishes sharply between terms (Bezeichnungen) and statements (Sätze), and explains that whereas the variables "x", "y", ... , supplant terms, the variables "p", "q", ... , supplant statements. This is excellent, and calculated to eliminate questions which arise from the confused use of "p", "q", ... , as nouns: the question "What sort of objects do the variables "p", "q", ... , denote, or take as values?" ... The reader should be warned against identifying Tarski's Satzfunktion with the propositional function of *Principia Mathematica*. The former is not a function at all, in the mathematical sense of a one-many relation, but is rather a statement form or matrix, an expression derived from a statement by putting variables for constants." ([Quine, 1938] p. 317)

2.3 Suszko and abstract nonsense

Let us recall that MTV is for us an abbreviation for *Mathematical conception of Truth-Value*. The "M" is a bit ambiguous: one may interpret MTV as *Matricial concept of Truth-Value*. We are pleased with the ambiguity because the full-fledged concept of truth-value is intimately connected with logical matrices theory, but at the same time the mathematical theory of truth-values does not reduce to logical matrices. Suszko, after consolidating the matrix concept together with Łoś, went on destroying it presenting non matricial truth-values — as we will see in the final round — also as a mathematical concept.

It is important to point out two important differences between Suszko's concept of truth-value in the 1950s and Tarskian matrices theory: one mathematical, one philosophical. From the mathematical viewpoint, Suszko's approach is totally in the Tarskian spirit, going on with the development of "a general method of constructing systems". The difference is just that Suszko with Jerzy Łoś in their paper [Łoś and Suszko, 1958], following Tarski and Lindenbaum, are going higher in general abstract nonsense, in particular considering the P-structure as an absolutely free algebra and talking explicitly of morphisms: considering V-functions as homomorphisms between the P-structure and the T-structure, and defining substitutions as endomorphisms of the P-structure. It is not clear when the expression "absolutely free algebra" and the corresponding concept appeared for the first time. But we certainly have here a close link between the history of universal algebra and logic, since such algebras are also known as "algebras of terms". Łoś and Suszko don't use the adjective "absolutely", but, right at the beginning, use a property which characterizes this notion, according to which an absolutely free algebra is the initial object of the category of algebras of similar type (see [Béziau, 1999c]). At this point it seems that they are just writing down a concept part of the folklore (of Polish logic or/and universal algebra). They write:

> Let S be the set of all the formulas of a sentential logic formed by means of variables p_t and connectives $F_1, ..., F_n$ where F_i is a k_i-ary connective. As known S is a free algebra of the type $\langle k_1, ..., k_n\rangle$ and the variables p_t are free generators of it. Therefore it follows that every function which maps the variables p_t into an algebra of a same type as S, may be extended to a homomorphism. [Łoś and Suszko, 1958, p. 177]

However this approach is revolutionary in the sense that the fact that a function attributing truth-values to atomic propositions uniquely extends to a V-function from the full set of propositions to the set of truth-values can be seen as an axiom defining the P-structure rather than a theorem. Łoś and Suszko don't explicitly state this situation in their paper. Their main contribution is the definition of structural consequence operator based on the definition of substitutions as endomorphisms ("Especially if we assign to every variable p_t a formula $\varepsilon p_t = \alpha_t$ then we can extend this mapping to an endomorphism ε of S into itself [Łoś and Suszko, 1958, p. 177], according to which the substitution theorem becomes an axiom:

$$\varepsilon Cn(X) \subset Cn(\varepsilon X)$$

This is the final clarification of the concept of substitution which was formulated through the ambiguous concept of truth-function by Whitehead and Russell and which Tarski started to clarify in his doctoral dissertation, but without reaching the level of abstraction of a P-structure. Here Łoś and Suszko are extending Tarski's theory of consequence operator, defining the consequence operator over an absolutely free algebra (later on Suszko will define the consequence operator over an algebra, this will be his definition of "abstract logic" — see [Suszko *et al.*, 1973]). This notion of substitution as endomorphism allows them to nicely reformulate Lindenbaum's theorem (already stated in Łukasiewicz and Tarski's 1930 paper, and a proof of which was presented by Łoś in his 1949 monograph —

see [Łoś, 1949] and also [Łoś, 1948a]) and to generalize it to consequence operators (full generalization was given later on by Wójcicki). These results are explained by Jan Zygmunt in his paper "Structural consequence operations and logical matrices adequate for them" [Zygmunt, 2012a], an introduction of Łoś and Suszko's 1958 paper for the anthology of universal logic where it is reprinted [Béziau, 2012].

Lindenbaum's theorem (an equivocal expression since Lindenbaum proved many theorems) says that any denumerable system of logic can be characterized by a matrix. The philosophy of such a theorem is important because matrix here does not mean truth table and the matrix is endogamic, since an endomorphism is a morphism within a structure: values for propositions here are not truth-values but propositions. According to Lindenbaum's theorem any logic as a matricial semantics. This is connected with a philosophical perspective Suszko was interesting in right at the beginning, which led him to develop his attack again the main philosophical figure of Polish logic, the godfather Jan Łukasiewicz, as we will see in the final part of our study.

For the present time let us have a look at Suszko's analysis of truth-value expressed in his 1957 paper. Suszko, following the Polish tradition, speaks rather of "logical values" than "truth-values", and following this tradition a logical value is a member of the domain of a logical matrix. Here is the definition given by Suszko in his 1957 paper and quoted as a definition of "logical value" in W.A.Pogorzelski's dictionary of logic (*Notions and theorems of elementary formal logic*, 1994 — in this dictionary there is no entry for "truth-value"):

> Generally speaking, the term logical value in a metalanguage of a certain propositional logic refers to every element of a characteristic matrix of that logic or (more generally) an arbitrary element of a universum of an arbitrary logical matrix of the language.

Suszko goes on describing the evolution of the concept:

> In the early stages of the development of modern logic, truth and falseness (which were referred only to propositions or propositional expressions) were called logical values. When it was assumed that a sentence α was true, then it was said that it had a logical value of truth and it was possible to replace it by the symbol 1 or V; when it was false, it was possible to be replaced by the symbol 0 or the sign F (0,1 — symbols taken from Boolean algebra). However, this point of view was not uniform: as logical values were treated both truth and falseness (intuitively understood), and the symbols 0, 1, or as symbols of a fixed but arbitrary — false or true sentence. In the last case, it was a linguistic understanding of logical values.
>
> Good foundation of such a formulation of the notion of logical value was Post's proof showing that classical propositional logic in axiomatic form has two-valued characteristic matrix. However, almost at the same time logics with characteristic matrices of numbers bigger than

> two occurred — e.g. Łukasiewicz's three-valued logic (the term logic is understood here as set of tautologies of a matrix). This third element of universum of this matrix was traditionally interpreted as possibility. Logical values have quickly lost their philosophically-intuitive interpretation when logics, whose characteristic matrices were n-ary, occurred (with an arbitrary natural n) or could even be of infinite cardinality. Obviously, logical values (many-valued logics, in contradistinction to twovalued logic) were — and are — discussed but these logical values have lost their intuitive sense, although an opinion was stated that they could be degrees of truthfulness of propositions.
>
> As the propositional logics and their metatheories were developing, the variety of characteristic models (matrices) and elements of their carriers (from rational numbers to topological spaces) increased and, according to that, it is stated that by logical values are understood elements of a universum of logical matrix, which because of intuitive or philosophical reasons, are identified with the notion of truth, falseness or their variants (necessity, possibility, randomness). However, when such identifications are not made and intentions of founders of a certain logic do not lead to a simple split of the notion of truth or falseness and so making a logic with many logical values instead of two, then a certain logic is not called many-valued and elements of universum of characteristic matrix are not always called logical values. Thus, this notion is not described properly and the fact whether a certain element of characteristic matrix of a certain logic is called logical value or not depends on non-formal means, intuitions or a predilection of a founder of a logic or its main users. ([Suszko, 1957] as translated in [Pogorzelski, 1994], pp.276-277)

Suszko rightly points out here the continuous ambiguity between a mathematical tool (logical matrices) and possible philosophical interpretations (truth-values). We will examine the development of MTV and then come back to Suszko's attack against matrices and talk about his generalization to non truth-functional semantics.

3 THE INITIAL BOUILLON: THREE WISE MEN

3.1 Boole

George Boole didn't use the expression "truth-value", but his work is strongly connected with the development of modern logic, in particular with MTV. He introduced in logic the use of 0 and 1 and is the forerunner of the so-called Boolean algebras, a central feature of MTV. MTV is mathematical and Boole is the chief operator of the mathematization of logic. By using mathematics to develop a theory of reasoning the work of Boole was also seminal for the evolution of mathematics, considering operations applying to things other than quantities (numbers), a decisive step towards abstract and universal algebra. Boole writes:

> We might justly assign it as the definitive character of a true Calculus, that it is a method resting upon the employment of Symbols, whose laws of combination are known and general, and whose results admit of a consistent interpretation. That to the existing forms of Analysis a quantitative interpretation is assigned, is the result of the circumstances by which those forms were determined, and is not to be construed into a universal condition of Analysis. It is upon the foundation of this general principle, that I purpose to establish the Calculus of Logic, and that I claim for it a place among the acknowledged forms of Mathematical Analysis, regardless that in its object and in its instruments it must at present stand alone. [Boole, 1847, p. 4]

The work of Boole in logic is presented in two books and a paper which is a summary of his first book:

- *A mathematical analysis of logic*, 1847
- "The calculus of logic", 1848
- *The laws of thought*, 1854

The subtitle of the 1847 book is *Being an essay towards a calculus of deductive reasoning* but Boole will rather use instead of *calculus of deductive reasoning* the shorter expression *calculus of logic* as in the title of his 1848 paper and along his 1854 book. What is Boole's calculus of logic? It is a calculus in the sense of basic arithmetic, rules are applied to compute. By applying rules of arithmetic to describe reasoning Boole was the first to really develop a *calculus raciocinator* as Leibniz wanted to do.[2] And Jevons, a follower of Boole, presented a logical machine in his paper entitled "The mechanical performance of logical inference" [Jevons, 1870], a paper he starts by saying that *calculus* as well as the Greek work *arithmos* are connected with operations on pebbles. So with Boole we are on the way to compute truth, a central characteristic of MTV. Boole didn't present truth tables, but his ideas are close to this concept.

The main idea of Boole is to consider a calculus with operations over classes — let us called it BCC — which can be seen as the backbone of his calculus of logic. These classes are supposed to represent concepts and the operations between them combinations of concepts. At this level he introduces the signs "0" and "1" to represent the empty class and the class of everything (the universe). Boole can be considered as the first to have introduced these two classes and a forerunner of set theory. Much later the Bourbachic mathematician André Weil introduced

[2]Boole was not inspired by Leibniz whose work on the subject was quite unknown at this time in England. The work of Boole is rather connected with the philosophy of Locke, see [Durand-Richard, 2007] .

the sign "$\emptyset$" to denote the empty set, which is a very good choice, distinguishing the number 0 from the empty set, but at the same time keeping the connection between the two objects using notations which are similar.

Can we say that Boole is considering a field of sets? Or a powerset Boolean algebra? Not exactly, because he is too much preoccupied to apply BCC to Aristotle's logic. With his algebra on sets he wants to describe categorical propositions of syllogistic. In BCC "0" and "1" do not correspond to truth-values of MTV. Truth of propositions is described by operations over classes. and it is in this perspective that operations over classes are considered. The truth of a proposition is expressed by an equation whose terms are operations over classes: the equation

$$xy = 0$$

where x is man, y is white, xy the multiplication of x and y, means "no men are white".

The exact structure corresponding to BCC has been described by Hailpern, it is a commutative ring with unit having no additive or multiplicative nilpotents and Hailpern says: "by restricting himself to the idempotents of his algebra, Boole would have had Boolean algebra, and we rightly honor him by attaching his name to this algebra" [Hailpern, 1981, p. 178]

Boole is also close to consider the Boolean algebra over {0,1}, the semantics of truth-values of classical propositional logic, precisely the T-structure of classical MTV. For Boole a central feature of reasoning is the idempotent law according to which the combination of a concept with itself is the concept itself: "good, good" is equivalent to "good". So he puts the equation:

$$x^2 = x$$

This law is true in arithmetic only for 0 and 1, Boole had therefore the idea to consider a calculus reduced to these two values:

> We have seen that the symbols of Logic are subject to the special law $x^2 = x$ Now of the symbols of Number there are but two, viz. 0 and 1, which are subject to the same formal law. We know that $0^2 = 0$, and that $1^2 = 1$; and the equation $x^2 = x$, considered as algebraic, has no other roots than 0 and 1. Hence, instead of determining the measure of formal agreement of the symbols of Logic with those of Number generally, it is more immediately suggested to us to compare them with symbols of quantity admitting only of the values 0 and 1. Let us conceive, then, of an Algebra in which the symbols x, y, z, etc. admit indifferently of the values 0 and 1, and of these values alone. The laws, the axioms, and the processes, of such an Algebra will be identical in their whole extent with the laws, the axioms, and the processes of an Algebra of Logic. Difference of interpretation will alone divide them. Upon this principle the method of the following work is established." ([Boole, 1854], pp.26-27)

But Boole didn't consider the Boolean algebra over {0,1} probably because, too much connected with the nature of 1, he is not able to conceive an algebra where 1+1=1, and he is also unable to jump into abstraction considering an algebra with two objects similar but different to 0 and 1. Boole developed his algebra of logic in the context of the British algebraic school, much influenced by George Peacock and Duncan Gregory. In this school people were making a difference between *arithmetical algebra* and *symbolical algebra*. The latter being a generalization of the former for undetermined objects. We are on the way to consider algebraic systems with different interpretations, but the full level of abstraction has not yet been reached.

Nevertheless Boole was able to conceive different interpretations of his calculus, the one for example where the variables x and y stand for propositions. He was led to do that by dealing with hypothetical propositions, those are for Boole secondary propositions, which are formed combining primary propositions (the ones constructed by algebra on classes). And Boole is considering the same operations for combination of propositions and combination of classes.

Kneale and Kneale wrote: "we have here all that is needed for an interpretation of Boole's system in terms of the truth-values of propositions with the symbols 1 and 0 standing respectively for truth and falsity" [Kneale and Kneale, 1962, p. 413]. But when Boole writes "x is true" where x is a proposition, it expresses all the "cases" in which x true. Later on Wittgenstein will use the expression "truth-possibilities", in MTV it is the set of valuations in which x is true. The following table drawn by Boole [1947, p.50] has to be understood in this sense:

X true, Y true	xy
X true, Y false	$x(1-y)$
X false, Y true	$(1-x)y$
X false, Y false	$(1-x)(1-y)$

We are on the way here to the algebra over the sets of valuations, which leads to the so-called Lindenbaum-Tarski algebra of classical propositional logic by identifying propositions true in the same cases, to use Boole terminology. And as proved by Tarski [1935-36] later on, such algebra is not surprisingly a Boolean algebra.

The connection between the two interpretations of Boole's calculus of logic was clearly expressed by Couturat:

> The algebra in question is susceptible of two distinct interpretations, the parallelism between them being almost perfect, according as the letters represent concepts or propositions. Doubtless we can, with Boole and Schröder, reduce the two interpretations to one, by considering the concepts on the one hand and the propositions on the other as corresponding to assemblages or classes; since a concept determines the class of objects to which it is applied (and which in logic is called its extension), and a proposition determines the class of the instances or

> moments of time in which it is true (and which by analogy can also be called its extension). Accordingly the calculus of concepts and the calculus of propositions become reduced to but one, the calculus of classes, or, as Leibniz called it, the theory of the whole and part, of that which contains and that which is contained." [Couturat, 1905, p. 2]

We have seen that Boole had some basic elements of the MTV machinery, especially in connection with the T-structure of classical logic but also indirectly considering operations over propositions, which is a first step towards a P-structure. He can therefore be regarded as one of the fathers of the concept of truth-values even if he was not using the name.

3.2 Peirce

During many years the work of Charles Sanders Peirce was almost unknown inclusive his work in logic. Almost no credits to the history of logic was given to him, in particular following the line of Jean van Heijenoort (JvH hereafter). But then things started to change up to a diametrically opposed viewpoint. Peirce would have been the creator of most of the concepts of modern logic. The new perspective was developed in particular by JvH's former student Irving Anellis.

Let us see in which sense truth-values can be credited to Peirce, independently of the fact that, like Boole, he didn't use the expression "truth-value". What is generally credited to Peirce are truth-tables and the discovery of three-valued logic. But having in mind Shosky's distinction between truth-table technique and truth-table device and our previous distinction between logical matrix and truth table (section 2.2), this does not necessarily mean that the full-fledged concept of truth-value can be credited to Peirce. Fisch and Turquette write: "Charles Peirce is generally recognized as an originator of what is often called the matrix method for constructing the usual two-valued sentential calculus" [Fisch and Turquette, 1966, p. 71]. The most reasonable reading of this claim is to interpret "matrix" here as matrix in the sense of Sylvester and/or of truth table, a different concept than the one of abstract matrix *à la* Tarski.

But first of all let us emphasized the work of Peirce on "Boolian algebras" as he nicely calls them. Peirce was much influenced by Boole and he is considered as having improved Boole's work, in particular being the first to conceive what we now call Boolean algebra. This is what is claimed by E.Michael: "by 1870 Peirce presented a complete set of operations for what is commonly called Boolean algebra ... It is not Boole's algebra but fundamentally the Boolean system as revised by Peirce that is what is currently referred to as Boolean algebra" [Michael, 1979, pp. 637–638].

This can be confirmed by the following writing by Peirce, in §13 of the "The logic of relatives" [1897], entitled "Introduction to the logic of quantity":

> Jevons wrote a book entitled *Pure logic*, the science of quality, which expounded, with a certain modification, the logical algebra of Boole.

> But it is a mistake to regard that algebra as one in which there is no system of quantity. As Boole rightly holds, there is a quadratic equation which is fundamental in it. The meaning of that equation may be expressed as follows: Every proposition has one or other of two values, being either true (which gives it one value) or false (which gives it the other). So stated, we see that the algebra of Boole is nothing but the algebra of that system of quantities which has but two values – the simplest conceivable system of quantity. (CP 3.561)

What Peirce had clearly in mind was the Boolean algebra over $\{0, 1\}$. In the "The logic of quantity" [1893], §9 entitled "The logical and the quantitative algebra", he writes:

> Thus, quantitative algebra is only a special development of logical algebra. On the other hand, it is equally true that the Boolian algebra is nothing but the mathematics of numerical congruences having 2 for their modulus. (CP 4.136)

Peirce prefers to use v and f rather than 0 and 1, because for him the algebra of logic is not just an application of arithmetic to logic. In "On the algebra of logic: a contribution to the philosophy of notation" [1885], he writes:

> For my part, I prefer for the present not to assign determinate values to f and v, nor to identify the logical operations with any special arithmetical ones, leaving myself free to do so hereafter in the manner which may be found most convenient. Besides, the whole system of importing arithmetic into the subject is artificial, and modern Boolians do not use it. The algebra of logic should be self-developed, and arithmetic should spring out of logic instead of reverting to it. (CP 3.372)

For Peirce however numerical values are worth using in the perspective of probability. In "The doctrine of chances" [1878], he writes:

> The theory of probabilities is simply the science of logic quantitatively treated. There are two conceivable certainties with reference to any hypothesis, the certainty of its truth and the certainty of its falsity. The numbers one and zero are appropriated, in this calculus, to marking these extremes of knowledge; while fractions having values intermediate between them indicate, as we may vaguely say, the degrees in which the evidence leans toward one or the other. (CP 2.647)

From these different quotations we can see that Peirce is still far from a mathematical construction such as MTV, although he is getting closer to it than Boole, by envisioning a Boolean algebra as a T-structure and clearly distinguishing it from other possible interpretations.

Let us come back to the seminal 1885 paper by Peirce, about which Alonzo Church claims: "The explicit use of two truth-values appears for the first time in a paper by C.S.Peirce in the *American Journal of Mathematics*, vol. 7 (1885)" [Church, 1956, p. 25] Here is an important extract:

> A proposition of the form $x \prec y$ is true if $x =$ f or $y =$ v. It is only false if $y =$ f and $x =$ v. A proposition written in the form $x \sim\prec y$ is true if x = v and $y =$ f, and is false if either $x =$ f or $y =$ v. Accordingly, to find whether a formula is necessarily true substitute f and v for the letters and see whether it can be supposed false by any such assignment of values. Take, for example, the formula $(x \prec y) \prec \{(y \prec z) \prec (x \prec z)\}$. To make this false we must take $(x \prec y) =$ v, $\{(y \prec z) \prec (x \prec z)\} =$ f. The last gives $(y \prec z) =$ v, $(x \prec z) =$ f, $x =$ v, $z =$ f. Substituting these values in $(x \prec y) =$ v, $(y \prec z) =$ v we have $(v \prec y) =$ v, $(y \prec f)$ = v, which cannot be satisfied together. (CP 3.387)

Using Shosky's distinction, have we here a truth-table technique or/and a truth-table device? Certainly we have here a truth-functional analysis even if it is not yet carried out from an abstract mathematical perspective, but rather from a computational one. And Peirce was mastering the truth-functional analysis in many ways, having proved before Sheffer that all binary connectives can be defined on the basis of only one of them, either the NAND or the NOR operators as we now call them — Peirce used the word "amphek" (CP 4.264). A student of Peirce, Christine Ladd, developed a systematic study of bivalent truth-functions in her essay "On the algebra of logic" [Ladd, 1883], as pointed out by Anellis.

About the above crucial quotation of Peirce, Anellis makes the following comments:

> The citation from the 1885 article, "On the Algebra of Logic: a Contribution to the Philosophy of Notation", in providing a definition of necessary truth and a technique for determining whether a complex proposition is necessarily true in terms of arranging its combinations of truth values, if read anachronistically, in effect gives what is unmistakenly an explanation for the use of an indirect truth table. In fact, it is a truth-functional analysis, and satisfies the condition for a truth-table technique, but yet not a truth-table device. The question is whether Peirce ever, expressed in Shoskys terminology, arrived at a truth-table device or merely the truth-table technique. Berry [1952] held that Peirce developed the technique, but not the device. [Anellis, 2004, p. 61]

In fact we don't have here a tabular truth table, but we can say that we have a computable device, close to what was systematically developed many years later as the tableaux technique as rightly pointed out by Hilpinen as well as the fact that Peirce is clearly making the distinction between truth and logical truth:

> This method is based on the model-theoretic conception of logical truth as truth under all interpretations (in the present case, assignments of

> truth-values to propositional letters), and it resembles the method of one-sided (signed) semantic tableaux. Like the tableau methods, it consists in a systematic search for a counterexample to a given formula or inference. [Hilpinen, 2004, p. 618].

Let us now turn to "real" truth tables, that Peirce indeed did produce. Peirce presents in fact the two kinds of truth tables: the ones which look like arithmetical table (*arithmetical truth table* hereafter) and the ones most commonly presented in textbooks nowadays, similar to the one presented by Wittgenstein in the *Tractatus*. Peirce also presents truth tables both for two values and three values. But all these diagrams do not directly embed in MTV. As with Frege's stroke, that we will examine next, we must be careful not to be driven by the superficiality of drawings.

About Peirce's truth tables, Robert Lane describes the situation as follows:

> For many years, commentators have recognized that Peirce anticipated the truth-table method for deciding whether a wff is a tautology (3.387, 1885). And it has long been known that he gave an example of a two-valued truth table (4.262, 1902). But his truth-table definitions of three-valued operators were discovered only about thirty years ago, by Max Fisch and Atwell Turquette [1966]. [Lane, 1999, p. 284]

Here is the two-valued 1902 Peirce's truth table that can be found in *The simplest mathematics* [1902] (CP 4.262):

x	y	z
v	v	v
v	f	f
f	v	f
f	f	v

Though it apparently looks like a modern truth table, there are several important differences: this is not a definition of a connective, but a relation between three propositions x, y, z, this is not clearly a definition of a T-function on the set {v, f}, otherwise it could have been better presented as an arithmetical truth table:

	v	f
v	v	f
f	f	v

The three-valued truth tables of Peirce discovered by Fisch and Turquette are from 1909 and are arithmetical truth tables. We reproduce some specimens:

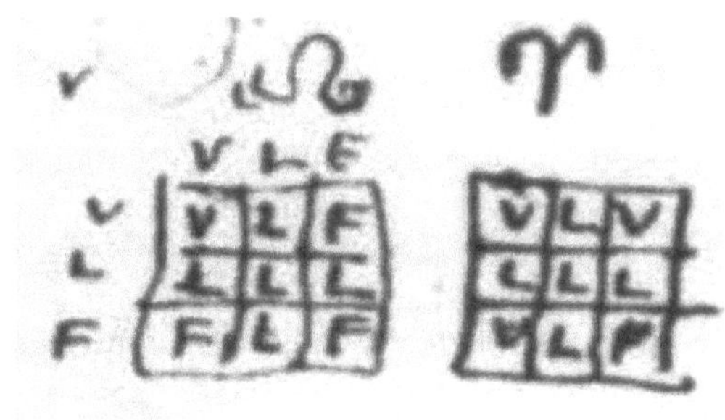

Here we can interpret such a table as a function on the set {v, l, f }, but it not clear what is the corresponding connective, if there is any P-structure. In fact all the discussion about the interpretations of the value l developed by Lane ([Lane, 1999]) shows that only specific propositions may have the value l.

Anellis says: "The source of Peirce's development of the truth-table device were (1) his work on truth-table technique and (2) his work on linear algebra and matrix theory" [Anellis, 2004, p. 62]. That sounds good. It seems in fact that Peirce's technique is a mixture of Boolean algebra over $\{0, 1\}$ and matrix in the sense of Sylvester. This how we can also interpret Peirce's table representing the 16 binary connectives dated from 1902, discovered in 1999 (cf. [Anellis, 2004]):

1		2	3	4	5		6	7	8	9	10	11		12	13	14	15		16
F		F	F	F	T		T	T	T	F	F	F		F	T	T	T		T
F		F	F	T	F		T	F	F	T	T	F		T	F	T	T		T
F		F	T	F	F		F	T	F	T	F	T		T	T	F	T		T
F		T	F	F	F		F	F	T	F	T	T		T	T	T	F		T

To conclude we can say that Peirce has developed both the truth-table technique and the truth-table device, but he didn't reach the abstract level of MTV. His perspective was more computational, as illustrated by the following quotation where Peirce saw reasoning as an algorithm like a multiplication table:

> It would seem, then, that in the simplest cases, the natural judgments of the mind in regard to what is logically necessary are as nearly infallible as are our ready judgments upon the simplest parts of the multiplication-table. They seem to be formed, too, in much the same way. Namely, in regard to the multiplication-table, we form our little diagram, perform our little experiment, and generalize the result under a conviction that what is true of one diagram of skeletal simplicity will be equally true, however complicated the nature of the units counted may be. (Peirce [1893], *General and historical survey of logic — elements of logic*, Chap. 3 Why study logic? §6 The fallibility of reasoning and the feeling of rationality, CP 2.160)

3.3 Frege

The *Begriffsschrift* [Frege, 1879], the first published writing of Frege, is the first item in *From Frege to Gödel* [Heijenoort, 1967b], an anthology of papers skillfully selected and translated in English by Jean van Heijenoort. Each paper is preceded by a presentation written by JvH himself. Since JvH was the first historian of modern logic his influence has been significant, in particular he promoted the idea that modern logic started with Frege from scratch.

In the preface of *From Frege to Gödel* JvH says that the *Begriffsschrift* "presented to the world, in full-fledged form, the propositional calculus and quantification theory" [Heijenoort, 1967b, vi]. Few pages later in his presentation of the *Begriffsschrift* JvH says more precisely that this booklet presents the *truth-functional propositional calculus*. Comparing the propositional calculus as presented by Frege in the *Begriffsschrift* with MTV is quite interesting to assess JvH's claim.

In the *Begriffsschrift* Frege didn't speak about truth-value, nor about truth *tout court* (as an object), only about true considered as a predicate. In introducing the famous judgment stroke he writes:

> A judgment will always be expressed by means of the sign $\vdash$, which stands to the left of the sign, or the combination of signs, indicating the content of the judgment. If we omit the small vertical stroke at the left end of the horizontal one, the judgment will be transformed into a mere combination of ideas of which the writer does not state whether he acknowledges it to be true or not. ([Frege, 1879, p. 11]

From this we can deduce that for Frege "$\vdash A$" means "the proposition A is true." Can we interpret this as a function attributing to the proposition A the truth-value true, a V-function in MTV? This would be a quite artificial and anachronistic interpretation since Frege does not introduce here truth as an object, does not use the word "truth-value" and does not use the notion of function at this level.

In the *Begriffsschrift* Frege's emphasize on function in breaking the Aristotelian distinction subject/predicate is rather at the level of P-functions, when he writes:

> Suppose that a simple or complex symbol occurs in one or more places in an expression (whose content need not be a possible content of judgment). If we imagine this symbol as replaceable by another (the same one each time) at one or more of its occurrences, then the part of the expression that shows itself invariant under such replacement is called the function; and the replaceable part, the argument of the function. [Frege, 1879, p. 22]

With such an idea Frege is on the way to a P-structure, but he is not concerned only with a P-structure of propositional logic as we can clearly see, since what is replaceable is not necessarily a proposition. This concept of "propositional function" (as it will later be called by Russell) does not make the difference between propositional logic and higher-order logic.

In the *Begriffsschrift* Frege considers $\vdash$ as a predicate: "the symbol $\vdash$ is the common predicate of all judgments" [Frege, 1879, p. 13], later on he rejected this view. Anyway what is important to stress is that in the *Begriffsschrift* and later on for Frege "$\vdash A$" does not mean: "the proposition A is true for all valuations", or "A is logically true", as we would now say, according to the present use of "$\vdash$", modulo the completeness theorem. Frege is not making the distinction between truth and logical truth, a fundamental distinction of modern logic, which was made clear by Wittgenstein in particular with his concept of tautology.

And Frege is also not making the distinction between proof theory and model theory (semantics). Nowadays "$\vdash$" is rather used in proof theory by opposition to "$\models$", used in model theory (the name of this symbol in Latex is "models"). For Frege "$\vdash$" means neither the present proof-theoretical "$\vdash$", nor the model-theoretical "$\models$", unless we consider a specific use of "$\models$" in model theory. In model theory the symbol "$\models$" is used in two different senses: 1) "$\models A$" means "A is true in all models", this is rather used in first-order logic, but we can use it in propositional logic as meaning "A is true in all valuations", 2) "$\models A$" means "A is true in a model $\mathcal{M}$", in this case generally the model is specified, and one writes something like "$\mathcal{M} \models A$" or "$\models_{\mathcal{M}} A$" so that there is no ambiguity between the two senses of the symbol.

For Frege "$\vdash A$" is connected with this second sense of "$\models$", which is generally expressed in propositional logic without such a symbol, just saying "the proposition A is true in a valuation v", or using a functional notation: "$v(A) = 1$". Let us have a look at how Frege introduces the conditional :

> If A and B stand for contents that can become judgments, there are the following four possibilities:
>
> (1) A is affirmed and B is affirmed;
>
> (2) A is affirmed and B is denied;
>
> (3) A is denied and B is affirmed;
>
> (4) A is denied and B is denied.
>
> Now $\begin{array}{l}\vdash\!\!\!\top A\\ \;\;\llcorner B\end{array}$ stands for the judgment that the third of these possibilities does not take place, but one of the three others does. [Frege, 1879, p. 13]

Interpreting $\begin{array}{l}\vdash\!\!\!\top A\\ \;\;\llcorner B\end{array}$ as "$B \to A$ is true" and "A (or B) is affirmed" as "A (or B) is true" and "A (or B)" is denied as "A (or B) is false", this means $B \to A$ is true iff (1) A is true and B is true or (2) A is true and B is false or (4) A is false and B is false.

This is close to the definition of MTV, which can be stated as follows: for any valuation v, $v(B \to A) = 1$ iff (1) $v(A) = 1$ and $v(B) = 1$, (2) $v(A) = 1$ and $v(B) = 0$, (4) $v(A) = 0$ and $v(B) = 0$. We can be generous with Frege, supposing that "for any valuation v" is implicit, but to claim that for Frege $\begin{array}{l}\vdash\!\!\!\top A\\ \;\;\;\llcorner B\end{array}$ means for any valuation v, $v(B \to A) = 1$ is one more step.

JvH says: "Frege's definition of the conditional is purely truth-functional and it leads him to the rule of detachment" [Heijenoort, 1967b, p. 2]. Having in mind that the notion of function is central in MTV at three levels — P-function, T-function and V-function — it seems a bit anachronistic to speak about truth-functionality here. Anachronistic also to speak of "rule of detachment" as if we had a proof-theoretical system. Interpreting "$\vdash X$" as it is interpreted nowadays in proof theory, this can be seen indeed as the rule of detachment. But Frege is not doing proof theory even when speaking of inference and presenting his proof using a horizontal stroke between conclusions and premises (cf. [Frege, 1879, p. 16]). We must not confuse similarity of scriptures with similarity of the concepts which are behind. Frege writes:

> The definition given in 5 makes it apparent that from the two judgments the new judgment follows. Of the four cases enumerated above, the third is excluded by $\begin{array}{l}\vdash\!\!\!\top A\\ \;\;\;\llcorner B\end{array}$ and the second and fourth by $\vdash B$, so that only the first remains. [Frege, 1879, p. 16]

The most natural interpretation of this proof is the following one in MTV: given one valuation v, if $v(B \to A) = 1$ and $v(B) = 1$ it follows that $v(A) = 1$, which is different from: if for all v, $v(B \to A) = 1$ and for all v, $v(B) = 1$, then for all v, $v(A) = 1$. This second formulation is the semantic version of the rule of detachment. Frege is at least two steps away from the rule of detachment because he is not making the distinction between truth and logical truth and between logical truth and proof.

In *Über Sinn in Bedeutung* [Frege, 1892], published 13 years later than the *Begriffsschrift*, Frege introduced the notion of *truth-value*, in German *Wahrheitswert*. As pointed out by Gabriel [1984], this expression was already used by Windelband and probably Frege borrowed it from him, but Windelband was not using this expression in any sense connected with the way it is used in MTV. Frege introduced truth-values as references (*Bedeutung*) of sentences (*Sätzen*) and there are only two truth-values the True and the False, considered as objects:

> We are therefore driven into accepting the truth-value of a sentence as constituting its reference. By the truth-value of a sentence I understand the circumstance that it is true or false. There are no further truth-values. For brevity I call the one the True, the other the False. Every declarative sentence concerned with the reference of its words is therefore to be regarded as a proper name, and its reference, if it has one, is either the True or the False. These two objects are recognized, if only implicitly, by everybody who judges something to be true and so even by a sceptic. The designation of the truth-values as objects may

> appear to be an arbitrary fancy or perhaps a mere play upon words, from which no profound consequences could be drawn. What I mean by an object can be more exactly discussed only in connection with concept and relation." ([Frege, 1892], pp.63–64)

In his talk *Funktion und Begriff* Frege gives the following examples:

> I now say: 'the value of our function is a truth-value' and distinguish between the truth-values of what is true and what is false. I call the first, for short, the True; and the second, the False. Consequently, e.g., '$2^2 = 4$' stands for the True as, say, '2^2' stands for 4. And '$2^2 = 1$' stands for the False. ([Frege, 1891], pp.28–29)

We will not be concerned here with the philosophical discussion about whether it makes sense or not to consider a truth-value as a reference of a sentence, in the same way that the famous statute in New-York is the reference of the expression "La liberté éclairant le monde", the original name of Frédéric Bartholdy's sculpture. Thousands of papers have been written about this question (a good starting point for Bedeutungology is [Angelelli, 1982]). Here we will examine truth-values as references of sentences from the viewpoint of MTV, something which may bring some new light to the understanding of Frege's work.

When reading *Über Sinn in Bedeutung* we are facing a confusion similar to the one showing up when interpreting the *Begriffsschrift*, related to the fact that Frege didn't make the distinction between truth and logical truth. Let us examine a fundamental argument of Frege in *Über Sinn in Bedeutung*:

> If our supposition that the reference of a sentence is its truth-value is correct, the latter must remain unchanged when a part of the sentence is replaced by an expression having the same reference. And this is in fact the case. Leibniz gives the definition: *Eadem sunt, quae sibi mutuo substitui, salva veritate*. What else but the truth-value could be found, that belongs quite generally to every sentence if the reference of its components is relevant, and remains unchanged by substitutions of the kind in question? [Frege, 1892, p. 64]

Let us extract from this the following:

- *Frege's substitution principle (SF)*: It two sentences A and A' have the same truth-value , thus the sentence B containing A as a subsentence has the same truth-value as the sentence B' that we get from B substituting A' for A.

This is only one aspect of substitution, the case where the *Satzteil* is itself a *Satz*. Frege had a more general view about substitution because his emphasize is not on propositional logic. Nevertheless he considers this specific case: "the truth-value of a sentence containing another as part must remain unchanged when the part is replaced by another sentence having the same truth value" [Frege, 1892, p. 65]. So our above extraction makes sense.

Frege's substitution principle can be interpreted in two different ways connected with the distinction between truth and logical truth:

- *Substitution's principle 1 (S1)*: It two sentences A and A' have the same truth-value in a given valuation (or model), thus the sentence B containing A as a subsentence has the same truth-value in this valuation (or model) as the sentence B' that we get from B substituting A' for A.

- *Substitution's principle 2 (S2)*: It two sentences A and A' are logically equivalent, i.e. are true exactly in the same valuations (or models), thus the sentence B containing A as a subsentence is logically equivalent, i.e. true exactly in the same valuations (or models), to the sentence B' that we get from B substituting A' for A.

$S2$ is now generally called the *Replacement theorem* (see e.g. [Kleene, 1952]), but some people like Church [1956], a follower of Frege, called it *substitution theorem.* Anyway it is important to make the distinction between $S1$ and $S2$, which are also both different from another principle also called the *substitution theorem*: the fact that if, in a tautology, we substitute a given sentence for all the occurrences of an atomic sentence, it is still a tautology — let us call this principle $S0$. There are some relations between these three principles: $S0$ is a consequence of $S1$, but $S2$ is not in general a consequence of $S1$, only in the case of bivalent truth-functional semantics. Using the technical machinery of MTV and a Polish way of speaking this means: truth-functional logics are structural but not necessarily extensional.

In [Béziau, 1999a] it has been shown how an interpretation of SF as an incoherent mixture of $S1$ and $S1$ has been presented by Sengupta [1983] to refute Frege's view of truth-values as references of sentences. It seems reasonable to interpret Frege's substitution principle SF as $S1$, since he had in mind truth rather than logical truth. But, according to $S1$, the set of valuations (or models) in which a sentence is true can also do the job, and can be considered as the reference of the sentence.

The idea that the meaning of a complex expression is determined by the meanings of its constituents is sometimes called *Frege's principle of compositionality.* To consider that the principle of compositionality is $S1$ is a fairly reductive interpretation of "determine". There are many other possible constructions, for example Kripke's semantics based on double recurrence can be viewed as a more sophisticated concept of determination.

Frege asks: "What else but the truth-value could be found, that belongs quite generally to every sentence if the reference of its components is relevant, and remains unchanged by substitutions of the kind in question?" [Frege, 1892, p. 64]. Nowadays one can say that any truth-functional semantics can do the job, any kind of objects can be considered as values of V-functions, provided the T-structure is of the same type as the P-structure. Frege is considering something like V-functions, but without considering explicitly neither P-functions, nor T-functions. This confusion prevails up to now, especially among philosophers, but a definition of V-function makes properly sense only when P-functions and T-functions are defined, unless we consider non truth-functional semantics.

That Frege is considering V-functions can be sustained by the fact that he is using the word "value" in the expression "truth-value" as value of a function. But these functions are far to be completely defined and maybe Frege is not just considering the True and the False as mathematical values of a function. Maybe, as suggested by Gabriel [1984], he is considering also the True and the False as values in the original moral sense of the world that he borrowed from Windelband.

In *Über Sinn in Bedeutung* Frege does not make directly the connection with the *Begriffsschrift*, in particular he does not use the judgment stroke but he says: "Judgments can be regarded as advances from a thought to a truth value" [Frege, 1892, p. 65]. The connection is established in 1893 in the *Grundgesetze*:

> I have already said above that a mere equation is to make no assertion as yet; '2 + 3 = 5' just designates a truth-value, and it is not stated which of the two it is. Again, if I write '(2 + 3 = 5) = (2 = 2)' and took for granted that we know that 2 = 2 is the True, yet I should not thereby have asserted that the sum of 2 and 3 is 5, but only designated the truth-value of: '2 + 3 = 5' stands for the same as '2 = 2.' Thus we need a special sign to assert that something or other is true. For this purpose I write the sign '$\vdash$' before the name of the truth-value, so that in '$\vdash 2^2 = 4$' it is asserted that the square of 2 is 4. ([Frege, 1893], pp.63–64)

This fits with the natural interpretation of "$\vdash$" in the *Begriffsschrift*, even if "$\vdash$" was then considered as a predicate and there were no truth-values. This combination of truth-value with "$\vdash$" shows how far Frege's views are from the way a bivalent system of logic is nowadays presented.

From a MTV viewpoint, the main contribution of Frege is on the one hand the functional analysis of the proposition (anticipation of the P-structure), on the other hand the introduction of truth-values as objects (anticipation of the T-structure), but the whole framework is still quite rough, the correspondence between the two ideas being described via substitutionality/compositionality (anticipation of V-functions).

4 DEVELOPING STAGE

4.1 Russell and Wittgenstein

The influence of Frege on Russell concerning truth values is obvious. In the Appendix A of *The Principles of Mathematics* [1903], entitled "The logical and arithmetical doctrines of Frege", Russell presents the ideas of Frege. The paragraph 477 is called "Truth-values and judgment".

The word "truth-value" does not appear in the main part of *The Principles of Mathematics*. It is introduced in the Appendix, probably the first appearance of the expression "truth-value" in English, as the translation of the German word *Wahrheitswert*, properly hyphenated.

Russell does not literally translate Frege in English, he translates it using his own semantic network. He translates for example "Gedanke" not by "Thought" but by "propositional concept". He then summarizes as follows Frege's theory: "we have first a propositional concept, next its truth or falsity as the case may be, and finally the assertion of its truth or falsity" ([Russell, 1903], §477).

Russell had developed his thought before reading Frege and had produced his own concepts. He had in particular introduced the concept of "propositional function". This expression is not really used today and it is quite ambiguous: does this mean that propositions are inputs (arguments) or outputs (values) of a function? In MTV a P-function is a function which has propositions as inputs and outputs, a V-function which has propositions as inputs only and a T-function which as propositions neither as inputs, nor as outputs. But for Russell in 1903, like for Frege (dealing with this notion but no using the expression "propositional function"), a propositional function is neither a P-function nor a V-function, It has to do in particular with variables for individuals, related with what is nowadays called an open formula in first-order logic, something which is neither true nor false independently of an assignation to free variables.

Here is how Russell defines the notion of proposition and propositional function in *The Principles of Mathematics*:

> A proposition, we may say, is anything that is true or that is false. An expression such as x is a man is therefore not a proposition, for it is neither true nor false. If we give to x any constant value whatever, the expression becomes a proposition: it is thus as it were a schematic form standing for any one of a whole class of propositions. And when we say x is a man implies x is a mortal for all values of x, we are not asserting a single implication, but a class of implications; we have now a genuine proposition, in which, though the letter x appears, there is no real variable ... I shall speak of propositions exclusively where there is no real variable: where there are one or more real variables, and for all values of the variables the expression involved is a proposition, I shall call the expression a *propositional function.* [Russell, 1903, p. 477]

The notion of truth-function does not appear in *The Principle of Mathematics*, it will appear only in *Principia Mathematica.* Truth-values appear page 8 of the first edition of *Principia Mathematica* [1910], on the introduction, in a section called *Truth-values*, following the sections *Variables*, *The uses of various letters*, *The fundamental functions o propositions*, *Equivalence.* The section is one page and has two paragraphs. The first starts with: "The "truth-value" of a proposition is truth if it is true, and falsehood if it is false*". The asterisk refers to the following footnote: "This phrase is due to Frege.". No bibliographical reference is indicated, but this is quite faithful to what Frege wrote in "Über Sinn in Bedeutung". Whitehead and Russell don't make any specific comments about this word, they introduce it through quotation marks, a definitional way to introduce a word. We have to see in which sense exactly Whitehead and Russell are departing from Frege and are getting closer to MTV. They go on and write:

> It will be observed that the truth-values of $p \vee q$, $p.q$, $p \supset q$, $\sim p$, $p \equiv q$ depend only upon those of p and q, namely the truth-value of $p \vee q$ is truth if the truth-value of either p or q is truth, and is falsehood otherwise; that of $p.q$ is truth if that of both p and q is truth, and is falsehood otherwise; that of $p \supset q$ is truth if either that of p is falsehood or that of q is truth; that of $\sim p$ is the opposite of that of p; and that of $p \equiv q$ is truth if p and q have the same truth-value, and is falsehood otherwise.

This is not very different from Frege, if we consider as minor the difference of notation (borrowed from Peano) and the choice of connectives. As we have seen Frege in the *Begriffsschrift* talks about implication and say something similar.

We may say that Whitehead and Russell are considering here a truth-table technique. What about a truth-table device? We can see drawings of truth-table on the back on one leaf of a manuscript of Russell, one by Russell and one by Wittgenstein in 1912 (cf. [Shosky, 1997]) and on notes by T.S.Eliot on Russell's logic course at Harvard, April 1914, this time as arithmetical truth tables:

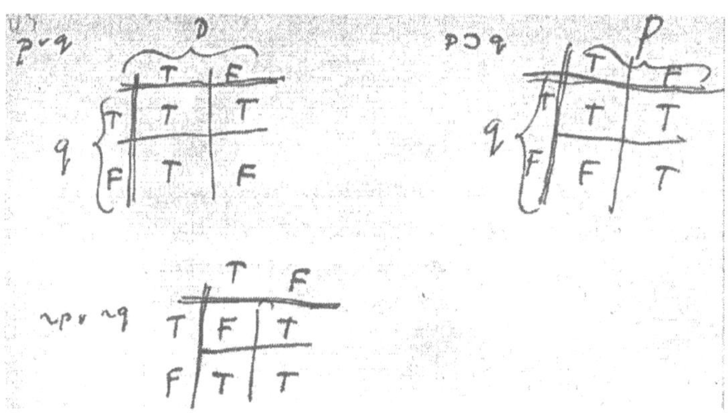

But facing tables or not one may ask: are Whitehead and Russell considering P-functions, T-functions, V-functions? About P-functions, let us see what they say in the previous section, *The fundamental functions of propositions*: "An aggregation of propositions (...) into a single proposition more complex than its constituents, is a function with propositions as arguments" [Whitehead and Russell, 1910, p. 6]. We are here much closer to the P-structure than with Frege, because Whitehead and Russell do not only consider the couple function/argument but consider explicitly propositions as arguments. They say "the general idea of such an aggregation of propositions, or of variables representing propositions, will not be employed in this work" [Whitehead and Russell, 1910, p. 6]. To do so would be in fact to consider a P-structure, this still will take years before someone does it.

Whitehead and Russell use the expression "propositional function" to speak of the logical sum. They write:

> The Logical Sum is a propositional function with two arguments p and q, and is the proposition asserting p or q disjunctively, that is, asserting that at least one of the two p and q is true. This is denoted by $p \vee q$. Thus $p \vee q$ is the logical sum with p and q as arguments. It is also called the logical sum of p and q. Accordingly $p \vee q$ means that at least p or q is true, not excluding the case in which both are true. [Whitehead and Russell, 1910, p. 6]

Let us note that they say at the same time that the logical sum is a propositional function and a proposition. This ambiguous way of speaking makes sense in the perspective of a P-structure: the disjunction is a P-function and the disjunction of two propositions p and q is the proposition $p \vee q$ which is the value of this function.

At this stage Whitehead and Russell are still talking informally about truth and falsity, but two pages later they introduce the terminology "truth-value". In this context one has the impression that these truth-values are values of the propositional functions, which have as argument propositions. This kind of confusion has prevailed during many years and the situation is still not clear in many introductory books to propositional logic. Nowadays people speak of *connectives* rather than propositional functions, which is at the same time a quality and a defect. A quality because people are able to think of the logic syntax independently of the semantics of truth-values, a defect because rarely logic syntax is considered in its full functional sense. A connective is generally not considered as a function. Whitehead and Russell don't use the word "connective".

After having introduced the expressions "propositional function" and "truth-value", Whitehead and Russell introduce the expression "truth-function" which reinforces the mixture and confusion between P-function, V-function and T-function: "We may call a function $f(p)$ a "truth-function" when its argument p is a proposition, and the truth-value of $f(p)$ depends only upon the truth-value of p" [Whitehead and Russell, 1910, p. 8]. For them truth-functions are a particular case of propositional functions:

> Such functions are by no means the only common functions of propositions. For example, "A believes p" is a function of p which will vary its truth-value for different arguments having the same truth-value: A may believe one true proposition without believing another, and may believe one false proposition without believing another. Such functions are not excluded from our consideration, and are included in the scope of any general propositions we may make about functions; but the particular functions of propositions which we shall have occasion to construct or to consider explicitly are all truth-functions. [Whitehead and Russell, 1910, p. 8]

Taking this into consideration, it is clear that neither truth-functions, nor propositional functions, are for Whitehead and Russell P-functions. But is it not usual nowadays to say that for example conjunction is a truth-functional connective? A rigorous interpretation of this runs as follows: the connective of conjunction which is a P-function, having as arguments and values propositions can be interpreted as a T-function having as arguments and values truth-values, the bridge between conjunction as P-function and as T-function being established via V-functions.

At this point it is important to recall that *Principia Mathematica* has been written by both Whitehead and Russell. For Russell it was the sequel of *The Principles of Mathematics* [1903] and for Whitehead the sequel of *A Treatise on Universal Algebra* [1898]. Whitehead was much influenced by Boole and the algebra of logic tradition. It would be misleading to consider that *Principia Mathematica* has been written only under the influence of Frege (and Peano). Whitehead and Russell speak of "logic sum" and use many concepts from the Boolean tradition. And their development of truth-values can also be seen from this perspective. This is what we can check in the following extract, which is the second passage referring to truth-values in *Principia Mathematica* quoted by Post in his dissertation ([1921], footnote 6 — the first reference is p.8 that we have examined above). This quotation clearly shows the perspective of the Boolean algebra over $\{0,1\}$:

> In this number, we shall be concerned, with rules analogous, more or less, to those of ordinary algebra. It is from these rules that the usual "calculus of formal logic" starts ... One very simple interpretation of the "calculus" is as follows: The entities considered are to be numbers which are all either 0 or 1; "$p \supset q$" is to have the value 0 if p is 1 and q is 0; otherwise it is to have the value 1; $\sim p$ is to be 1 if p is 0, and 0 if p is 1 ; $p.q$ is to be 1 if p and q are both 1, and is to be 0 in any other case; $p \vee q$ is to be 0 if p and q are both 0, and is to be 1 in any other case. [Whitehead and Russell, 1910, p. 120]

The work of Wittgenstein on truth-values in the *Tractatus* is clearly a continuation of the work of Frege and Russell. At the same time he had some fundamental new ideas, his conceptual apparatus is much closer to MTV than the ones of Frege and Russell.

In the *Tractatus* Wittgenstein introduces three fundamental interrelated notions: atomic proposition, truth-possibilities, tautology. The expression "propositional function" is never used in the *Tractatus* and there are only two occurrences of the word "truth-values", but Wittgenstein uses many times the expressions "truth-function", which is used as meaning ambiguously a P-function and its correlated T-function. Wittgenstein also does not use the expression "atomic proposition" which was promoted by Bertrand Russell, the mentor but also first follower of Wittgenstein, in a series of lectures later on published together under the title "The philosophy of logical atomism":

> The following [is the text] of a course of eight lectures delivered in [Gordon Square] London, in the first months of 1918, [which] are very

> largely concerned with explaining certain ideas which I learnt from my friend and former pupil Ludwig Wittgenstein. [Russell, 1918-19, p. 177]

Russell understood immediately the importance of the ideas of Wittgenstein and presented a very good sketch of them in his preface to the *Tractatus*

> Given this totality of atomic facts, every true proposition, however complex, can theoretically be inferred. A proposition (true or false) asserting an atomic fact is called an atomic proposition. All atomic propositions are logically independent of each other. No atomic proposition implies any other or is inconsistent with any other. Thus the whole business of logical inference is concerned with propositions which are not atomic. Such propositions may be called molecular. Wittgenstein's theory of molecular propositions turns upon his theory of the construction of truth-functions. A truth-function of a proposition p is a proposition containing p and such that its truth or falsehood depends only upon the truth or falsehood of p and similarly a truth-function of several propositions p, q, r, ... is one containing p, q, r, ... and such that its truth or falsehood depends only upon the truth or falsehood of p, q, r, ... ([Wittgenstein, 1921], xiii — xiv)

The fact that Russell presents this picture as a summary of Wittgenstein's ideas proves that such framework, which can be seen as a skeleton of MTV, is neither due to Frege nor himself.

About atomic propositions, we can emphasize the following items of the *Tractatus*:

> 4.21 The simplest kind of proposition, an elementary proposition, asserts the existence of a state of affairs.

> 4.26 If all true elementary propositions are given, the result is a complete description of the world. The world is completely described by giving all elementary propositions, and adding which of them are true and which false.

Wittgenstein introduces the concept of *truth-possibility* (in German: *Wahrheits möglichkeiten*), which are V-functions. Like his predecessors, Peirce and Russell, he does not make a clear distinction between a truth-table device and a truth-table technique, and he explains what is truth-possibility by drawing a table:

> 4.31 We can represent truth-possibilities by schemata of the following kind ('*T*' means 'true', '*F*' means 'false'; the rows of '*T*'s' and '*F*'s' under the row of elementary propositions symbolize their truth-possibilities in a way that can easily be understood):

p	q	r
T	T	T
F	T	T
T	F	T
T	T	F
F	F	T
F	T	F
T	F	F
F	F	F

,

p	q
T	T
F	T
T	F
F	F

,

p
T
F

.

After having clearly distinguished elementary propositions and having given a hint of what is a truth-possibility, Wittgenstein claims

> 4.41 Truth-possibilities of elementary propositions are the conditions of the truth and falsity of propositions.

This is one of the main contributions of Wittgenstein. And on the basis of such framework, he defines the now famous notion of tautology, rightly contrasting it to contradiction:

> 4.46 Among the possible groups of truth-conditions there are two extreme cases. In one of these cases the proposition is true for all the truth-possibilities of the elementary propositions. We say that the truth-conditions are *tautological*. In the second case the proposition is false for all the truth-possibilities: the truth-conditions are contradictory. In the first case we call the proposition a tautology; In the second, a contradiction.

But the confusion between P-function and T-functions appears in the following items:

> 5. A proposition is a truth-function of elementary propositions. (An elementary proposition is a truth-function of itself.)

> 5.01 Elementary propositions are the truth-arguments of propositions.

A fundamental feature of the *Tractatus* is the introduction of the notion of semantical consequence. Wittgenstein doesn't use such an expression, but his definition corresponds exactly to the present definition. He starts by introducing what he calls *truth-grounds* (*Wahrheitsgründe* in German). The truth-grounds of a given propositions are the the truth-possibilities (V-function) according to which it is true:

> 5.101 I will give the name truth-grounds of a proposition to those truth-possibilities of its truth-arguments that make it true.

Then Wittenstein introduces the idea of semantical consequence using the terminology *follow* (*folge* in German):

> 5.12 the truth of a proposition 'p' follows from the truth of another proposition 'q' if all the truth-grounds of the latter are truth-grounds of the former.

Just after, in 5.121, Wittgenstein abbreviated "the truth of p follows from the truth of q" just by "p follows from q", and in 5.132 Wittgenstein wrote: "If p follows from q, I can make an inference from q to p deduce p from q". If we interpret here inference/deduction by inference/deduction in a formal proof-system, this statement looks like the completeness theorem. However Wittgenstein neither proves, nor states the completeness theorem, since there is no definition of inference/deduction in the sense of proof theory. What Wittgenstein claims is that inference/deduction is nothing else than semantical consequence: "Laws of inference, which are supposed to justify inferences, as in the works of Frege and Russell, have no sense, and would be superfluous" (5.132). The completeness theorem justifies Wittgenstein's claim of superfluousness, but he didn't prove it.

In the case of Frege and Russell the situation is confused and it is not clear if they are using rules of deduction or/and semantical consequence. For Wittgenstein inference clearly does not mean "if q is true then p is true" but "the truth of p follows from the truth of q", meaning all truth-grounds of q are also truth-grounds of p. On the basis of this definition he states that:

> 5.134 One elementary proposition cannot be deduced from another.

Further on there is an item of the *Tractatus* to which few people have paid attention. Wittgenstein writes:

> 5.141 If p follows from q and q from p, then they are one and the same proposition.

This means that for him $p \vee q$ is the same proposition as $q \vee p$. From a philosophical viewpoint this means that for Wittgenstein a proposition can be seen as the class of its models (truth-grounds), since two propositions are the same if they have the same models. From a technical viewpoint this means that Wittgenstein is considering the Tarski-Lindenbaum algebra, that is to say a Boolean algebra, since he is dealing with classical connectives. Of course Wittgenstein didn't consider explicitly the Tarski-Lindenbaum algebra, this was done later by Tarski [1935-36]. To do so one has to check that logical equivalence is a congruence relation.

4.2 Bernays and Post

The first important results about truth-values have been proved by Paul Bernays and Emil Post, both extremely talented logicians. At the beginning of the history of truth-values there are many geniuses. We can split them into two groups: the initiators, Boole, Peirce, Frege, Russell, Wittgenstein, Łukasiewicz; and those who started to prove theorems: Bernays, Post, Tarski, Gödel, Lindenbaum. But the second group of people are not mere theorem provers, they were also great conceptualizators.

Bernays developed the truth-value technique, proved the completeness theorem and introduced many-valued truth-functions, not on a philosophical basis but to prove the independence of axioms - a properly original contribution. These results were obtained by Bernays independently of Post and before. Here is how Richard Zach describes the situation:

> These results include: explicit semantics for propositional logic using truth values, decidability of the set of valid propositional formulas, completeness of the axiom systems considered relative to that semantics, as well as what is now called Post completeness, consistency and independence results, general three- and four-valued matrices, and rule-based derivation systems. All these results were obtained independently of logicians to whom they are usually credited (notably Pierce, Wittgenstein, Post, and Łukasiewicz). [Zach, 1999, p. 332]

As rightly described by Zach, Bernays was working in the spirit of Hilbert's program, with emphasis on consistency and independence. The work of Bernays has long been unknown, only part of his dissertation [Bernays, 1918] was published, in German eight years later [Bernays, 1926]. This has been recently translated by Zach and presented by Walter Carnielli [2012a] in the *Anthology of universal logic* [Béziau, 2012]. Carnielli strongly stresses the Hilbertian model-theoretical approach of Bernays in his development of many-valued truth-tables.

Bernays proves the independence of the following five axioms:

- *(Taut)* $p \vee p \supset p$
- *(Add)* $q \supset p \vee q$
- *(Perm)* $p \vee q \supset q \vee p$
- *(Assoc)* $p \vee (q \vee r) \supset q \vee (p \vee r)$
- *(Sum)* $(q \supset r) \supset (p \vee q \supset p \vee r)$

To do so he uses three or four truth-values, not drawing truth tables but using a truth-table device that can be easily translated in truth tables. For example to prove that *Taut* cannot be derived from *Add*, *Perm*, *Assoc*, *Sum*, Bernays uses the following device where 1 and 2 are designated and 0 non-designated:

$\vee$	0	1	2
0	0	0	0
1	0	1	2
2	0	2	0

	0	1	2
$\sim$	1	0	2

Concluding his analysis of Bernays's work, Carnielli reaches the following sharp conclusion, developing a relevant parallel with geometry:

> In this sense, the *soi-disant* precursors of many-valued logics, including MacColl, Peirce, Vasil'ev, Post and Łukasiewicz, are just looking for "non-straight lines", what the geometers did not do, while Bernays, in the pure geometrical spirit, is making room for broader mathematical concepts. [Carnielli, 2012a]

Post proved three important results about classical propositional logic in his doctoral dissertation

- completeness theorem
- truth-functional completeness
- maximality

and also introduced many-valued truth tables, all this published later on in a 22 page papers, the best ratio in all the history of truth-values. An interesting thing about the work of Post is that it stands at the junction between truth table as computational device (Post was a main forerunner of the theory of recursion, see [De Mol, 2006]) and abstract algebra.

Post was probably the first to isolate propositional logic, developing a systematic theory about it. To do so he was mainly influenced by *Principia Mathematica* which can be considered as a first step towards propositional logic:

> In the general theory of logic built up by Whitehead and Russell (1910, 1912, 1913) to furnish a basis for all mathematics there is a certain subtheory which is unique in its simplicity and precision, and, though all other portions of the work have their roots in this subtheory, it itself is completely independent of them. Whereas the complete theory requires for the enunciation of its propositions real and apparent variables, which represent both individuals and propositional functions of different kinds, and as a result necessitates the introduction of the cumbersome theory of types, this subtheory uses only real variables, and these real variables represent but one kind of entity, which the authors have chosen to call elementary propositions. The most general statements are formed by merely combining these variables by means of the two primitive propositional functions of propositions, negation and disjunction; and the entire theory is concerned with the process of asserting those combinations which it regards as true propositions, employing for this purpose a few general rules, which tell how to assert new combinations from old, and a certain number of primitive assertions from which to begin. ([Post, 1921] p.265)

Moreover Post was the first to make a clear distinction between logic and metalogic:

> We here wish to emphasize that the theorems of this paper are about the logic of pro- positions but are not included therein. More particularly, whereas the propositions of Principia are particular assertions introduced for their interest and usefulness in later portions of the work, those of the present paper are about the set of all such possible assertions. Our most important theorem gives a uniform method for testing the truth of any proposition of the system. [Post, 1921, p. 265]

The first important point of Post is to give a recursive definition of the set of propositional formulas:

> If p is an elementary proposition, $\sim p$ is an elementary proposition, If p and q are elementary propositions, $p \vee q$ is an elementary proposition. [Post, 1921, p. 266]

This a first approximation of the P-structure. What is missing here is that noting else is a formula. Without this close, we may have non-standard formulas similar to non-standard numbers in arithmetic.

The semantics of propositional logic, the T-structure is still mainly conceived starting with truth tables, the truth-values being denoted neither as 0 and 1, nor as V and F, but as + and −, another way to arithmetize logic!

> Let us denote the truth value of any proposition p by + if it is true and by − if it is false. This meaning of + and − is convenient to bear in mind as a guide to thought, but in the actual development that follows they are to be considered merely as symbols which we manipulate in a certain way. Then if we attach these two primitive truth tables to $\sim$ and $\vee$

p	$\sim p$
+	−
−	+

p	q	$p \vee q$
+	+	+
+	−	+
−	+	+
−	−	−

> we have a means of calculating the truth values of $\sim p$ and $p \vee q$ from those of their arguments. [Post, 1921, p. 267]

Post then conceives T-functions directly from P-functions with the following setting:

> Now consider any function $f(p_1, p_2, ..., p_n)$ in our system of functions, which we will designate by F. Then since f is built up of combinations of $\sim$'s and $\vee$'s, if we assign any particular set of truth values to the p's, successive application of the above two primitive tables will enable us to calculate the corresponding truth value of f. So corresponding to each of the 2^n possible truth configurations of the p's a definite truth value of f is determined. The relation thus effected we shall call the truth table of f. [Post, 1921, p. 268]

In his first theorem, Post proves the correspondence between truth table and T-functions, a fundamental step to connect truth-table device to a mathematical conceptualization of it, that it would be misleading to only call a truth-table technique:

> THEOREM. To every truth table of whatever order there corresponds at least one function of F which has it for its truth table. [Post, 1921, p. 268]

Post, after proving the completeness theorem, introduces m-valued truth tables and T-functions, being the first to present a full generalization of bivalent logic. He doesn't develop a systematic theory of it: "The problem of determining all possible systems of m-dimensional truth tables, however, is one we have not considered, though its solution would throw considerable light on the ordinary problem" [Post, 1921, p. 282] but shows that this method generates new logics: "We have thus verified our statement that we obtain essentially new logical systems" [Post, 1921, p. 283].

4.3 Quine and Gödel

Quine, a former PhD student of Whitehead, became one of the most famous American philosopher of the XXth century, but he was first of all a logician with an acute ability for conceptual analysis. His 1934 paper "Ontological Remarks on the Propositional Calculus" was much influential, for the good or for the bad, to promote the expression *sentential calculus* against the expression *propositional calculus*. Quine's analysis is a step back in the development of MTV, since he argues against considering connectives as functions, therefore against a P-structure homomorphic to the T-structure:

> The expression "$\sim p$" is construed in the propositional calculus as denoting the contradictory of the proposition p; here again we may short-cut the passage through propositions, by construing the sign "$\sim p$" as shorthand for the word "Not" or for the words "It is false that". Thus, where "p" is an abbreviation for a sentence " — " the sign "$\sim p$" becomes an abbreviation for the sentence "It is false that" ... Thus reconstrued the theory of deduction remains unchanged in structure, but ceases to be a system in the usual sense. The usual sort of system

> treats of some manner of elements, say cardinal numbers or geometrical points, which are denoted ambiguously by variables ; operative upon these elements are certain operations or relations, appropriately expressed within the language of the system. The theory of deduction, when construed as a calculus of propositions, is a system of this kind; its elements are propositions denoted by the variables "p", "q" etc., and its operations are the propositional operations of denial, alternation, material im-plication etc., denoted by prefixture or interfixture of the signs $\sim$, $\vee$, $\supset$, etc. When, on the other hand, the theory of deduction is reconstrued in the foregoing manner as a mere organon of sentences, it ceases to be concerned with elements sub-ject to operations; the former proposition-variables "p", "q", etc., become ambiguous sentences, symbols of nothing, and the signs $\sim$, $\vee$, $\supset$, etc., become connectives of sentences, innocent of operational correlates in the realm of denotations. [Quine, 1934, p. 474]

And Quine describes as follows the gain of the loss of proposition: "The theory of deduction becomes a paradigm depicting the use of the connectives "or", "if-then" etc., with a view to the truthfulness of the sentences which they generate. There are no inferred entities, no flights of abstraction beyond the realm of everyday uses of words" [Quine, 1934, p. 475]. According to him we have then a kind of demathematization of logic: "a theory of deduction proceeding merely on the basis of abbreviations of sentences and making no pretense to being a system in the ordinary sense" [Quine, 1934, p. 475]

By opposition to Quine's perspective which has generated no progress in the development of the theory of truth-values, we can consider Gödel's generalized proof of the completeness theorem. Gödel's proof is based on a generalization of Lindenbaum's theorem. Quine wrote the comments on Gödel's paper in the *Collected Works* of the famous logician. Here is how he explains the situation using his own terminology, calling atomic propositions "letters":

> We easily impose a true-false dichotomy on the truth-functional formulas of propositional logic by assigning arbitrary truth values to all the propositional letters and then letting the truth tables determine the truth values of all compound formulas. But now let us imagine, instead of the familiar formulas, a well-ordered but perhaps indenumerable supply. It is still to be closed under the operations '$\sim$' and '$\supset$', but there need no longer be minimal units such as letters. Can we specify a true-false dichotomy for these formulas, conformable still to the truth-functional constraints on '$\sim$' and '$\supset$'? Gödel does so by a transfinite recursion based on Łukasiewicz's three axioms and modus ponens. Lindenbaum had shown (see [Tarski, 1930]) how any consistent set of closed formulas of first-order predicate logic can be extended to a true-false dichotomy. Gödel's result, though limited to propositional logic, is stronger in applying to the case of indenumerably many

> formulas. His method of proof is a transfinite extension of that used by Lindenbaum. The paper is a pioneer venture in the metalogic of indenumerable notation. ([Gödel, 1986], pp.238-239)

This result was transposed later at a higher abstract level in the Polish school, showing that Lindenbaum lemma at the abstract level is equivalent to the axiom of choice (see [Łoś, 1954; Dzik, 1981]). But Gödel is already working at a very abstract level considering valuations as sets of formulas — maybe he was the first to do so. He writes:

> That is, if an arbitrary realization of the axioms of the propositional calculus is given, the elements ("propositions") can be divided into two disjoint classes that behave exactly like the classes of the true and the false propositions of the ordinary propositional calculus. [Gödel, 1932b, p. 241]

What Gödel proves is that a set of propositions closed under the axioms and rules of classical propositional logic can be extended to a set of propositions whose characteristic function is a classical bivaluation, proof carried out without assuming much about the structure of the set of propositions.

5 MANY TRUTH-VALUES

5.1 Indeterminacy

Łukasiewicz is a main figure of modern logic, a man concerned with deep philosophical problems and creating mathematical tools to deal with them. He developed three-valued logic in this perspective, not just as a generalization of two-valued logic or a tool to prove results about it.

He perfectly describes his approach as follows:

> I can assume without contradiction that my presence in Warsaw at a certain moment of next year, e.g. at noon on 21 December, is at the present time determined neither positively nor negatively. Hence it is *possible*, but not *necessary* ... On this assumption the proposition "I shall be in Warsaw at noon on 21 December of next year", can at the present time be neither true nor false ... and must possess a third value, different from '0' or falsity and '1' or truth. This value we can designate by '$\frac{1}{2}$'. It represents the "possible" and joins "the true" and "the false" as a third value. The three-valued system of propositional logic owes its origin to this line of thought. ([Łukasiewicz, 1930, p. 53]

So for Łukasiewicz possibility is a truth-value. A position opposed to the one of Wittgenstein and the possible worlds tradition, where possibility is a metalogical notion, appearing through the concepts of truth-possibility and possible world. However Łukasiewicz was also considering possibility as a connective but Tarski showed that it was possible to defined it as $\neg p \rightarrow p$, so at the end Łukasiewicz used a system with only negation and implication. Anyway Łukasiewicz can be considered as the first logician to have created a semantics for modalities. C.I.Lewis the father of formal modal logic did consider also modal notions as truth-values and more than two truth-values, but he didn't build a matrix semantics on this basis and there is a confusion between connectives and truth-values, syntax and semantics:

> Systems previously developed, except MacColl's, have only two truth-values, "true" and "false". The addition of the idea of impossibility gives us five truth-values, all of which are familiar logical ideas:
> (1) p, "p is true".
> (2) $-p$, "p is false".
> (3) $\sim p$, "p is impossible".
> (4) $- \sim p$, "It is false that p is impossible", i.e., "p is possible".
> (5) $\sim -p$, "It is impossible that p be false", i.e., "p is necessarily true".
> [Lewis, 1918, p. 292]

For Lukasiewicz the third truth-value is non-designated, so that the law of excluded middle $p \vee \neg p$ is not valid in his system, since the value of p is $\frac{1}{2}$ when the value of $\neg p$ is $\frac{1}{2}$, according to the following Łukasiewiczian truth-table for negation:

p	$\neg p$
0	1
$\frac{1}{2}$	$\frac{1}{2}$
1	0

If the notion of semantic consequence is defined on the basis of the dichotomy designated/non-designated, there is in Łukasiewicz's logic L3 a discrepancy between the two forms of the principle of contradiction: $\neg(p \wedge \neg p)$ is not a tautology in L3 but we have $p, \neg p \models q$. If we consider the third value as designated we have the reverse phenomenon: $\neg(p \wedge \neg p)$ is a tautology but we have $p, \neg p \not\models q$.

This latter approach has been developed by people working in paraconsistent logic: Asenjo [1954], D'Ottaviano and da Costa [1970] and Priest [1979]. Curiously this strategy was not developed by Łukasiewicz who is generally considered as a forerunner of paraconsistent logic due to his lengthy and detailed criticism of Aristotle's justification of the principle of non-contradiction presented in his 1910 book [Łukasiewicz, 1910].

In three-valued paraconsistent logic the natural way to interpret the third value is "both true and false". If one wants to follow the modal approach of Łukasiewicz, one may want to call the third value "impossible", but none of these paraconsistent authors have developed such an interpretation probably because modalities were more and more exclusively considered as connectives, not as truth-values, in particular due to the failure of Łukasiewicz's programme and Dugundji's negative result about the impossibility to characterize a standard modal logic by a finite matrix [Dugundji, 1948].

Paraconsistent logicians are facing anyway the paradox of interpreting the third value in relation with the dichotomy designated/non-designated. If they consider the third value as designated, they can claim that in paraconsistent logic a proposition and its negation can both be true, but in this case it is not possible to claim that negation is a contradictory forming functor, contrarily to a trick presented by Priest and unveiled by Slater (see [Slater, 1995], [Priest and Routley, 1989], [Béziau, 2006]).

At the end of his life, in his Irish period, Łukasiewicz came back to modal logic and many-valued logic, developing a logic with two modalities as connectives and with four truth-values. However he didn't present a coherent interpretation of his system which has several defects. In particular the fact that $\diamond a \wedge \diamond b \rightarrow \diamond(a \wedge b)$ is valid in this logic is counterintuitive, because everybody agrees that it is possible for Łukasiewicz to be present in Warsaw for Christmas and that it is possible for him not to be present in Warsaw for Christmas, but on the other hand, even for Santa Claus, it may sound absurd that it is possible for Łukasiewicz to be present and not to be present in Warsaw for Christmas. A way to solve Łukasiewicz's paradox is to consider a partial order between the four truth-values (see [Béziau, 2011]).

Paulette Février (Paulette Destouches-Février, after her marriage with Jean-Louis Destouches, the main pupil of Louis de Broglie),[3] extended the framework of many-valued logic in another direction. She developed an interesting three-valued logic in 1936-37 (cf. [Février, 1937a], [Février, 1937b], [Février, 1937c]) motivated by Heisenberg uncertainty relations, a logic that she considered as a logic of complementarity, as explained in particular in her paper "Manifestations et sens de la notion de compémentarité" published in a special issue of the journal *Dialectica* on complementarity in 1948 including papers by L.de Broglie, A.Einstein, N.Bohr, W.Heisenberg (*Dialectica* was founded in 1947 by F.Gonseth jointly with P.Bernays and G.Bachelard).

A striking feature of this logic is that it challenges the structural feature of MTV and leads to what Destouches calls a quasi-formal logic:

> Let us consider the proposition p, "the corpuscule c has the position a" and the proposition q, "the corpuscule c has the speed v". In quantum mechanics we cannot assert simultaneously these two propositions without contradictions; as a consequence the logical product of p and q cannot follow the common rules; either we will have to not apply the logical operation to certain pairs of propositions, either we will have to modify the definition so that applied in such pairs the product is

[3]Tarski was a good friend of Paulette and the couple Destouches-Février was very active for the organization of the congress on axiomatization organized by Suppes at Berkeley in 1957, a preliminary event of the LMPS series — see details in Fefermans' biography of Tarski [Feferman and Feferman, 2004].

not true. It is very inconvenient to consider operations which do not apply to all the pairs of elements of a set (here, the set of propositions), we thus have advantage to adopt the second possibility. But in both cases we see that the operation & cannot apply to all the pairs of propositions by giving the same result and by obeying the classic rules. Therefore the logic which we shall use in the quantum theories cannot be a logic with universal rules because rules on the product are not universal. Such a logic will not thus be purely formal because the contents of the propositions will intervene in the statements of rules. However we can build a logic adapted to this case as shown by Miss Février, such a logic as a symbolic logic can be called a *quasi-formal logic.* The pairs (p, q) of propositions must be divided into two classes: the class of the *composables pairs* which follow the usual rules *p&q* is true if p and q are both true) and the class of *incomposables pairs* for these *p&q* is never true. For this reason the operation & is defined by two matrices: one for each class of pairs of propositions. The proposition *p&q* for a pair (p, q) of incomposables propositions is never true, but we cannot consider it as false in the usual sense (allowing to assert the negation of p or of q); we have to give it a new logical value A meaning: "absolute falsity"; this leads to consider for propositions (at least) three logical values which are: V (true), F (falsity), A (absolute falsity). [Destouches, 1948, pp. 80–81]

What is interesting is that to develop this logic Paulette Février constructed a general framework for logical systems in a paper symptomatically entitled "Sur la forme générale de la définition d'une logique" (On a general form to define a logic) [Février, 1937b] presented by Louis de Broglie at the *Académie*:

A set of abstract elements $\mathcal{E}$ is said to be a set of propositions:
1. If there is an operator Vl, called logical value, which attributes to each element of $\mathcal{E}$, an element o $\mathcal{F}$, the set of logical values, which has at least two elements, V, F
2. There exists a least a binary symmetric relation R which transforms each pair of elements of $\mathcal{E}$, into an element of $\mathcal{E}$, the logical value of such element being a function of logical value of the elements of the pair, this function can then be written as a binary operation Ω_R on the set of logical values, associated to R. We thus have

$$Vl(aRb) = Vl(a)\Omega_R Vl(b).$$

A function being defined by the sets of its values, when the set of logical values is finite, the operation will be defined by a finite matrix. A logic will be:
1. A set of propositions
2. A set of operators
3. A set of binary operators. [Février, 1937b, p. 398]

We clearly see here a general description of MTV: on the one hand P-functions, on the other hand T-functions, and V-functions described as homomorphisms. A curiosity however is to consider the relation R, which is a P-function (a connective), as syntactically symmetric. But Février does not stop here, she goes on making distinctions bewteen standard and non-standard many-valued systems in which different tables may be used to define the same connective.

We can make here a comparison to a technology introduced by Jaśkowski [1936] and presented also in Paris at the same time: the use of products of matrices to provide a semantics for intuitionistic logic, after the negative result by Gödel [1932a] showing that intuitionistic logic cannot be characterized by a single matrix.

5.2 Gaps and Gluts

While it is possible to consider that there are propositions which are neither true nor false, it is also possible to consider that there are propositions which are both true and false. This makes sense from a philosophical point of view, and from a mathematical point of view this is nicely symmetrical. There are three different ways to develop such framework:
(1) Considering a standard T-structure with two elements, truth and false, but dropping V-functions for relations: to an element of the P-structure, a proposition, it is possible to associate nothing, one value or two values.
(2) Considering a T-structure whose elements are elements of the power set of {T, F}: ∅, {T}, {F}, {T, F}.
(3) Considering a T-structure with four elements: truth, false, neither-true-nor-false, both-true-and-false.

The first option has not been developed because it is quite mathematically intricate and the second option expresses the same philosophical idea via sets. Now what is interesting is that mathematically speaking it is equivalent to deal with (2) or (3). Option (2) has been developed by Michael Dunn since 1966 ([Dunn, 1966], [Dunn, 1967], [Dunn, 1969], [Dunn, 1971], [Dunn, 1976]) but it became famous mainly through the paper of his PhD advisor Nuel Belnap based on the third option: "A useful four-valued logic", published much later in the book *Modern uses of multiple-valued logic* edited by Dunn himself [Belnap, 1977b]. Kit Fine also contributed to popularize this framework with a comics terminology: "Gaps and Gluts" [Fine, 1974]. (The terminology "gaps" was already there before, used in particular by van Fraassen, author also of "supervaluation" [van Fraassen, 1966], but gaps alone are less than half fun.)

Belnap's popularization of gaps and gluts through the computer scene [Belnap, 1977a] is in fact the continuity of Kleene's motivation for the development of three-valued logic. The posture of Kleene is well-described by Fitting as follows:

> Multiple-valued logics have been introduced for many reasons: purely mathematical, as with Post, or philosophical, as with Łukasiewicz. Few have been as useful or as natural as the three-valued logics of Kleene introduced for computer science purposes or at least they would have

> been if computer science had existed at the time. Kleene thought of the third truth value as undefined or undetermined, rather than as contingent or of probability 1/2. This reading suggests a natural condition: the behavior of the third truth value should be compatible with any increase in information. That is, if the value of some propositional letter, P say, is changed from undefined to either true or false, the value of any formula with P as a component should never change from true to false or from false to true, though a change from undefined to one of false or true is allowed. Kleene referred to this as regularity; today we phrase it in terms of monotonicity in an ordering that places undefined below both false and true. [Fitting, 1994, p. 113]

However the arguments of Belnap about the usefulness of four-valued logic for computers has been challenged by Fox [1990] and more recently by the fuzzy logician Didier Dubois:

> We claim that we cannot consistently reason under incomplete or conflicting information about propositions by augmenting the set of Boolean truth-values true and false with epistemic notions like "unknown" or "contradictory", modeling them as additional genuine truth-values of their own. [Dubois, 2008b, p. 113]

Details of Dubois's criticisms can be found in his paper 'On ignorance and contradiction considered as truth-values" [Dubois, 2008a]; Wansing and Belnap [2008] have replied to him. Wansing has developed with Shramko a generalization of Dunn-Belnap's logic, a 16-valued logic, describing how a computer network should think (see [Shramko and Wansing, 2005] and [Shramko and Wansing, 2011]).

What about fuzzy logic? Many people have the idea that it is based on a continuous interval of truth-values between $[0,1]$. However the position of Lotfi Zadeh is quite original, for him fuzzy logic is not numerical, but based on linguistic variable which are ranging on discrete concepts. This is what he explains in his fundamental paper "The concept of a linguistic variable and its application to approximate reasoning I" [Zadeh, 1975a] (see also his paper in *Synthese* [Zadeh, 1975b] the same year):

> By a linguistic variable we mean a variable whose values are words or sentences in a natural or artificial language. For example, *Age* is a linguistic variable if its values are linguistic rather than numerical, i.e., young, not young, very young, quite young, old, not very old and not very young, etc., rather than 20, 21, 22, 23. ... Treating truth as a linguistic variable with values such as true, very true, completely true, not very true, untrue, etc., leads to what is called fuzzy logic. [Zadeh, 1975a, p. 199]

People working in technological applications prefer to use the terminology *Multiple-valued logic* often used abbreviated as MVL — since 1971 there is a yearly congress on the subject. One of the main figures of MTV, Michael Miller from Canada, describes the situation as follows:

> Multiple-Valued Logic (MVL) as discussed in this book is the study of discrete p-valued systems where $p > 2$, or in other words, non binary-valued systems. In the most general sense, both binary valued (or digital) and discrete variables with an infinite number of values can be considered as MVL systems. In keeping with current convention, anytime we refer to an MVL system, we consider the system to utilize variables that can take on a discrete set of values with cardinality of three or more. Depending on the application, these systems can be considered to be mathematical structures, non binary discrete logic circuits, or symbolic logic constructs. In the context of this latter application, many logicians have referred to this topic as "many-valued logic" and this may well be a more precise name for the field since we are interested in systems that can evaluate many (meaning > 2) different but distinct logic values in contrast to systems that can take on multiple logic values simultaneously. [Miller and Thornthone, 2008, p. 1]

It is in fact quite easy to understand the difference between MVL and many-valued logic: many-valued logic makes sense from the viewpoint of full MTV, by contrast MVL is the study of T-structures from the viewpoint of some technological applications, the members of the T-structures do not necessarily represent truth-values but may represent some electronic devices (and, as we have already pointed out, at the more general level T-structures are just abstract algebras in the sense of Birkhoff).

5.3 Order

An underlying idea behind the framework for truth-values developed by Dunn and Belnap is the idea of *De Morgan algebras* which were not created by Augustus De Morgan but by Grigore Moisil (1906-1973), from Romenia, a main character in the history of many-valued logic, author of the big book *Essai sur les logiques non-chrysipiennes* [Moisil, 1972].

The idea of De Morgan algebra can be traced back to Moisil's paper "Recherche sur l'algèbre, de la logique" [Moisil, 1935]. They were later on studied in Poland and called *quasi-boolean algebras* by Rasiowa [Bialynicki-Birula and Rasiowa, 1957]. They also have been called *distributive i-lattices* by Kalman [1958]. They have been especially studied by the algebraic logic school of Antonio Monteiro in Bahía Blanca, Argentina. In his paper "Matrices de Morgan caractéristiques pour le calcul propositionnel classique" [1960], Monteiro showed that they can be used as a semantics for classical logic, a result which looked at first sight paradoxical to Alonzo Church.

A De Morgan algebra obeys the following De Morgan laws:

$$\overline{a \vee b} = \overline{a} \wedge \overline{b}$$
$$\overline{a \wedge b} = \overline{a} \vee \overline{b}$$
$$\overline{\overline{a}} = a$$

and nothing more.

Dunn has been working on De Morgan algebras [Dunn, 1967] and the idea of Dunn's four-valued semantics is to consider a De Morgan algebra among these values. This done by considering the following partial order:

false $\prec$ neither-true-nor-false $\prec$ true

false $\prec$ both-true-and-false $\prec$ true

neither-true-nor-false and both-true-and-false are not comparable.

The (truth-function of) conjunction is then defined as *inf* and disjunction as *sup*, negation of false leads to truth, negation of truth leads to false, negation of neither-true-nor-false leads to neither-true-nor-false, and negation of true-and-false leads to true-and-false. Truth and true-and-false being the only designated values, this leads to a logic in which neither $p \vee \neg p$ nor $\neg(p \wedge \neg p)$ are tautologies, but in which all the above De Morgan laws hold.

Before such construction, order between truth-values was used implicitly in three-valued logic, the third value being below, between or above truth and false, according to the desired interpretation. For more than three values, the idea to consider a partial order between the truth-values is due to Birkhoff [1940], considering the most famous partial order, the one of lattice theory, but was properly developed by Alan Rose later on. In his 1951 paper entitled "Systems of logic whose truth-values form lattices", he describes the situation as follows:

> In 1920–22 LUKASIEWICZ and POST independently generalised the classical 2-valued logic to systems with finite simply ordered sets of truth-values. It has been suggested by BIRKHOFF (Lattice Theory, AMS) that it would be of value to develop systems in which the truth-values form lattices. He points out that J.M.KEYNES has suggested that the modes of probability form a partly ordered system and that this view is shared by B.O.KOOPMAN (Annals of Math. 41, 271–292 1940)). We shall consider these systems, which include those of POST and LUKASIEWICZ as special cases. Illustrations will be given mainly with reference to one of the two non-distributive lattices with five elements. The diagram for this lattice is given in Fig. 3. Since our methods apply to non-distributive lattices our systems include some in which a distributive formula fails. This may be of interest in connection with quantum mechanics since BIRKHOFF and von NEUMANN have found non-distributive modes in this subject. The functions "or", "and", "implies" and "not" of two-valued logic have been generalised to simply-ordered m-valued systems and we generalise these definitions further here so that we have truth-tables for these functions for any system in which the truth-values form a lattice. [Rose, 1951, p. 152]

The relation between the truth-values of the four-valued system of Łukasiewicz [1953] is a partial order, but he did not use it explicitly and moreover he confusingly denoted the truth-values by 0, 1, 2 , 3, so that one can mistakenly think that there is the usual linear order between them. Rescher [1965] presented a four-valued semantics based on a De Morgan lattice. This does not appear very clearly because he uses the values 1, 2, 3, 4. On the other hand Rescher defines a negation which transforms designated values in non-designated values so that that this semantics defines classical negation. His idea is not to define a non-classical negation but to develop modal logic. Following Prior [1955] (and this is also connected with the idea of C.I.Lewis [1918]), he describes the four values as follows:

Truth-value	Interpretation
1	necessarily true
2	contingently true
3	contingently false
4	necessarily false

He explains then that to avoid paradoxes, one has to consider that the value of the conjunction of a contingently true and a contingently false proposition has to be necessarily false. He further describes various possibilities to define the connective of possibility and says that the best solution is the following:

p	$\Diamond p$
1	1
2	1
3	1
4	4

This idea has systematically developed in [Béziau, 2011], considering first a better notation to express the partial order and the differences between designated and non-designated values:

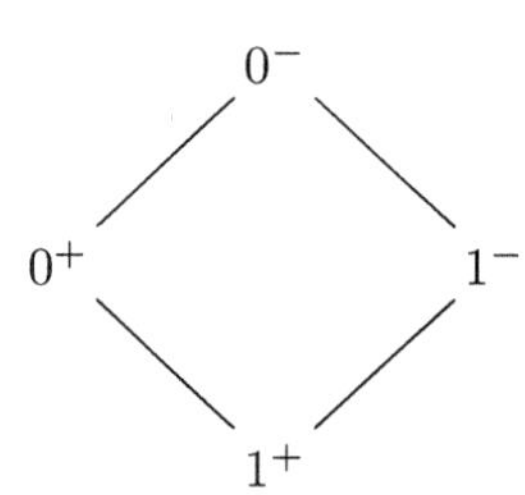

The truth table for conjunction is then described as follows

$\wedge$	0^-	0^+	1^-	1^+
0^-	0^-	0^-	0^-	0^-
0^+	0^-	0^+	0^-	0^+
1^-	0^-	0^-	1^-	1^-
1^+	0^-	0^+	1^-	1^+

and the one for possibility as follows:

p	$\diamond p$
0^-	0^-
0^+	1^+
1^-	1^+
1^+	1^+

This solves Łukasiewicz's paradox: $\diamond(a \wedge b)$ is not a logical consequence of $\diamond p \wedge \diamond q$. The following table for necessity, allowing nice properties, is also presented:

p	$\Box p$
0^-	0^-
0^+	0^-
1^-	0^-
1^+	1^+

De Morgan algebra seems therefore a good tool to develop a quite natural four-valued semantics for modal logic in the spirit of C.I.Lewis, A.N.Prior and N.Rescher.

6 STRUCTURES, MODELS, WORLDS

6.1 *Truth, Models, and Truth-values*

The place and role of truth-value in first-order logic is not clear if we stick too close to the word "truth-value". In standard model theory this word is not necessarily in the outpost, but this does not mean that the concept is not there. People tend to think that truth-value is rather a notion of propositional logic, but in the same way that the concept of proposition is present in first-order logic even if people don't use much the word in this context — they prefer to speak about formulas, the concept of truth-value is also there. However the situation is quite confused due to the fact that the expression "truth-value semantics" has been used by Leblanc as a name for an alternative semantics for first-order logic (cf. [Leblanc, 1976]). One could think that there are truth-values in Leblanc's semantics but not in Tarski's semantics. This will be the occasion to clarify the relations between these two semantics and also between propositional logic and first-order logic.

Even Tarski is not clear on the subject, for example in his famous 1944 paper "The semantic conception of truth: and the foundations of semantics" which is an informal presentation and discussion of his 1935 piece, he writes the following after having presented his definition of truth:

> We can deduce from our definition various laws of a general nature. In particular, we can prove with its help the *laws of contradiction and of excluded middle*, which are so characteristic of the Aristotelian conception of truth; i.e., we can show that one and only one of any two contradictory sentences is true. These semantic laws should not be identified with the related logical laws of contradiction and excluded middle; the latter belong to the sentential calculus, i.e., to the most elementary part of logic, and do not involve the term "*true*" at all. [Tarski, 1944, p. 354]

People generally speak of truth-values at the level of propositional logic and truth at the level of first-order logic, using the notation $\models$. Instead of saying that the *proposition* $\exists x \forall y Rxy$ has the *truth-value* true in a structure $\mathcal{M}$ where R is interpreted as an order relation with a first element, they will say that the *formula* $\exists x \forall y Rxy$ is true in $\mathcal{M}$. In the case of propositional logic, truth-values are distributed arbitrary in all the possible ways to the atomic propositions and then these distributions extend through the MTV procedure to all propositions. In first-order logic, atomic propositions are further on atomized and the definition of truth is related to this subatomic realm. This procedure developed by Tarski in 1935 is based on a neo-Aristotelian correspondence theory of truth, with, on the one hand formulas, on the other hand structures. In the case of the "truth-value semantics" promoted by Leblanc, we are also at the subatomic level, but truth (or falsity) are attributed to formulas without going through models, considering substitution as the main device. In the case of Tarski's approach we merge in the subatomic level through structures but at the end we also give truth-values to atomic formulas.

Although model theory is one of the greatest achievements of modern logic, the definition of truth developed by Tarski (or Leblanc) has no relation with how we generally attribute truth and falsity to atomic propositions. Patrick Suppes claims:

> One of the scandals of both philosophy and linguistics, as well as psychology, is the absence of any detailed theory of how the most elementary empirical sentences are judged true or false, or to put it more directly, how their truth value is computed. Consider atomic sentences. Tarski's semantic theory of truth offers no help in determining their truth. He was not concerned to give a theory of how we compute or know that individual atomic sentences such as Paris is north of Rome are true or false. [Suppes, 2008, p. 21]

In fact, developing his theory, Tarski was concerned mainly with mathematical structures, and his definition would be better called "truth in mathematical structures". It took a long time for Tarski to reach the model-theoretical definition of truth. As pointed out by Hodges [1985-86], one cannot find it in his 1935 paper, which is only the first step with the notion of *satisfaction*. The excellent paper of Hodges points the many difficulties related to the conceptualization of logic. If we look at the famous paper of Henkin, "The completeness of the first-order functional calculus", we find a way of speaking which is nearly beyond understanding for someone who knows nothing about the history of logic. "Functional" is used here following the Fregeo-Russellian tradition, according to which propositions are conceived as functions. Probably due to this tradition Henkin calls the symbols of relations that appear inside formulas, symbols of functions:

> Elementary well-formed formulas are the propositional symbols and all formulas of the form $G(x_i, ..., x_n)$ where G is a functional symbol of degree n ...
>
> The functional constants (variables) of degree n are to denote (range over) subsets of the set of all ordered n-tuples of I. $G(x_i, ..., x_i)$ is to have the value T or F according as the n-tuple $G(x_i, ..., x_i)$ of individuals is or is not in the set G. [Henkin, 1949, pp. 159–160]

In this definition we see that Henkin describes the situation of atomic formulas of first-order logic using the truth-values T and F. His presentation is thus uniformized since curiously he also defines the set of first-order formulas considering unanalysed propositions to which truth-values are attributed.

Chang and Keisler are also later on using the notion of truth-values at the level of first-logic. It is very interesting to examine the similarities/differences between truth-values in propositional logic and first order logic in the book *Model theory* by Chang and Keisler first published in 1973. The book of Chang and Keisler is emblematic of Tarski's idea and school. This was the first book on model theory, as Chang and Kreisel themselves state: "up to now no book of this sort has been written" [Chang and Keisler, 1973, p. v].[4] This book is symbolic in many senses. It is the crystallization of central ideas, not only of model theory, but of modern logic, considering that model theory is the final stage of the development of modern logic (post-modern logic is another story). In particular in model theory the bridge between truth and structure is rigorously established.

Chang and Keisler were both students of Alfred Tarski, the founder of model theory. Some people are tracing back model theory to Löwenheim, Skolem and even Peirce, this is not necessarily conflicting with the fact that the expression "theory of models" was created by Tarski only in the 1950s, but Tarski didn't just invent the name, he consolidated the whole theory, properly defining the central notion of model theory: the notion of truth in a structure; and developing many other ideas and concepts.

[4]Note however that Kreisel and Krivine did publish a book entitled *Elements of Mathematical Logic (Model Theory)*, original version in French [Kreisel and Krivine, 1966].

Contrary to what many people think the notion of truth in first-order logic was not defined in [Tarski, 1935]. Wilfrid Hodges in his seminal paper "Truth in a structure" writes:

> A few years ago I had a disconcerting experience. I read Tarski's famous monograph 'The concept of truth in formalized languages' [1935] to see what he says himself about the notion of truth in a structure. The notion was simply not there. This seemed curious, so I looked in other papers of Tarski ... I believe that the first time Tarski explicitly presented his mathematical definition of truth in a structure was his joint paper [1957] with Robert Vaught. This seems remarkably late. Putting Tarski's 'Concept of truth' paper side by side with mathematical work of the time, both Tarski's and other people's, I think there is no doubt that Tarski had in his hand all the ingredients for the definition of truth in a structure by 1931, twenty-six years before he published it. What held him back? In a moment I shall analyse exactly what parts of the notion are missing from Tarski's earlier papers. I believe there were some genuine difficulties, not all of them completely resolved today, and they fully justify Tarski's caution. [Hodges, 1985-86, pp. 137–138]

The difficulty pointed out by Hodges is the following; "A sentence of a first-order language isn't true or false outright. It only becomes true or false when we have interpreted the non-logical constants and the quantifiers" [Hodges, 1985-86, pp. 144]. And according to Hodges this issue is a problem from a Fregean perspective:

> He (Frege) packed his objections into one short dictum: em Ein Satz, der nur unter Umstanden gilt, ist kein eigentlicher Satz. The issue is simply this. If a sentence contains symbols without a fixed interpretation, then the sentence is meaningless and doesn't express a determinate thought. But then we can't properly call it true or false. [Hodges, 1985-86, pp. 147–148]

In zero-order logic we can say that propositions are vacuously interpreted: propositional variables are ranging over propositions that are supposed to be already interpreted. In first-order logic a proposition needs to be interpreted and we can say that the main achievement of Tarski in model theory was to properly define the notion of interpretation, a notion complementary but different from the notion of satisfaction previously defined by Tarski in 1935.

Let us have now a look at the mechanism of interpretation as described by Chang and Keisler. First, they speak about truth definition:

> To arrive at model theory, we set up our formal language, the first-order logic with identity. We specify a list of symbols and then give precise rules by which sentences can be built up from the symbols. The reason for setting up a formal language is that we wish to use

> the sentences to say things about the models. This is accomplished by giving a basic *truth definition*, which specifies for each pair consisting of a sentence and a model one of the truth values *true* or *false*. The truth definition is the bridge connecting the formal language with its interpretation by means of models. If the truth value 'true' goes with the sentence φ and model $\mathfrak{A}$, we say that φ is *true in* $\mathfrak{A}$ and also that $\mathfrak{A}$ is a *model of* φ. Otherwise we say that φ is *false in* $\mathfrak{A}$ and that $\mathfrak{A}$ is not a model of φ. [Chang and Keisler, 1973, pp. 1–2]

At this stage there is not much difference with zero-order logic, even if the terminology is slightly different, but we can also say: "φ has the truth-value true in $\mathfrak{A}$" instead of "φ is true in $\mathfrak{A}$", this latter formulation is simply a natural abbreviation of the former. Instead also of the usual notation of model theory "$\mathfrak{A} \models \varphi$", we can write "$v(\mathfrak{A}, \varphi) = 1$". The difference is that the truth-value of a proposition depends on a model. But what is a model? Are there no models in zero-order logic? Chang and Keisler are not afraid to speak about models at the ground level of zero-order logic:

> We shall start from scratch, in order to show what sentential logic looks like when it is developed in the spirit of model theory. Classical sentential logic is designed to study a set $\mathcal{S}$ of simple statements, and the compound statements built up from them. At the most intuitive level, an intended interpretation of these statements is a 'possible world', in which each statement is either true or false. We wish to replace these intuitive interpretations by a collection of precise mathematical objects which we may use as our models. The first thing which comes to mind is a function F which associates with each simple statement S one of the truth values 'true' or 'false'. Stripping away the inessential, we shall instead take a model to be a subset A of $\mathcal{S}$; the idea is that $S \in A$ indicates that the simple statement S is true, and $S \notin A$ indicates that the simple statement S is false. [Chang and Keisler, 1973, p. 4]

And then Chang and Keisler go on to first-order logic:

> We begin here the development of first-order languages in a way parallel to the treatment of sentential logic ... Finally, we give the key definition of a sentence being true in a model for the language $\mathcal{L}$. The precise formulation of this definition is much more of a challenge in first-order logic than it was for sentential logic. [Chang and Keisler, 1973, p. 18]

Where is the challenge? In zero-order logic models are sets of formulas, and in fact it is the same to consider bivaluations as to consider sets of formulas. In first-order logic the situation is more complex, we have models which are not simply sets of formulas, we have to interpret the formulas. Chang and Keisler define therefore a model as a set A with an interpretation function, which is mapping the symbols of a given language to corresponding objects of A, individuals, relations, functions accordingly to the types of the symbols:

> Turning now to the models for a given language $\mathcal{L}$, we first point out that the situation here is more complicated than for the sentential logic. There each $S \in \mathcal{S}$ could take on at most two values, true or false. Thus the set of intended interpretations for $\mathcal{S}$ has rather simple properties, as the reader discovered. This time, each n-placed relation symbol has as its intended interpretations all n-placed relations among the objects, each m-placed function symbol has as its intended interpretations all m-placed functions from objects to objects, and, finally, each constant symbol has as intended interpretations fixed or constant objects. Therefore, a 'possible world', or model for $\mathcal{L}$ consists, first of all, of a *universe* A, a nonempty set. In this universe, each n-placed P corresponds to an n-placed *relation* $R \subset A^n$ on A, each m-placed F corresponds to an m-placed *function* $G : A^m \rightarrow A$ on A, and each constant symbol c corresponds to a constant $x \in A$. This correspondence is given by an *interpretation* function $\mathcal{I}$ mapping the symbols of $\mathcal{L}$ to appropriate relations, functions and *constants* in A. A *model* for $\mathcal{L}$ is a pair $\langle A, \mathcal{I} \rangle$. We use Gothic letters to range over models. Thus we write $\mathfrak{A} = \langle A, \mathcal{I} \rangle$. [Chang and Keisler, 1973, pp. 19–20]

Let us note that the formulation of Chang and Keisler is ambiguous here when they say "at most two values", because the difference between zero-oder logic and first-order logic is not a question of *plus-value* in the sense of additional values, the plus-value is about interpretation.

After the notion of interpretation is defined it is possible to properly define the notion of truth in a structure through the notion of satisfaction:

> We now come to the key definition of this section. In fact, the following definition of satisfaction is the cornerstone of model theory. We first give the motivation for the definition in a few remarks. If we compare the models of Section 1.2 and the models discussed here, we see that with the former we were only concerned with whether a statement is true or false in it, while here the situation is more complicated because the sentences of $\mathcal{L}$ say something about the individual elements of the model. The whole question of the (first-order) truths or falsities of a possible world (i.e., model) is just not a simple problem. For instance, there is no way to decide whether a given sentence of $\mathcal{L} = \{+, ., S, O\}$ is true or false in the standard model $\{+, ., S, O\}$ of arithmetic (where S is the successor function). Whereas we have already seen in Section 1.2. that there is such a decision procedure for every model for $\mathcal{S}$ and for every sentence of $\mathcal{S}$. [Chang and Keisler, 1973, p. 26]

Here again we can see an ambiguity in the sense that when Chang and Keisler are talking about "not a simple problem" they are mixing two aspects of the difference between zero-order and first-order logic: decidability and the subatomic level. As we know the subatomic level does not necessarily yield to undecidability and vice-versa some propositional logics may be undecidable.

Kleene, twenty years earlier, was not afraid to present the semantics of first-order logic using truth tables that he presents on pages 170-171 of *Introduction to metamathematics* after having explained the following:

> We now interpret $\forall$ and $\exists$ as fixed functions over the logical functions of one variable taking values from the domain $\{t, f\}$, where the variable x in $\forall x$ or $\exists x$ indicates of what individual variable the operand shall be considered as logical function. These two fixed logical functions are defined as follows. For a given logical function $A(x)$, the value of $\forall x A(x)$ is t, if $A(x)$ has the value t for every value of x in the domain $\{1, ..., k\}$; otherwise the value is f. The value of $\exists x A(x)$ is t, if $A(x)$ has the value t for some value of x in the domain $\{1, ..., k\}$; and otherwise the value is f.
>
> Given a predicate letter formula, we are now in a position to compute a table expressing the values of the function, of the distinct free individual variables and predicate letters occurring in it (or of these and other variables and letters), which the formula represents. [Kleene, 1952, p. 169]

6.2 Truth, Worlds and Truth-values

Possible worlds semantics was mainly developed by Kripke. Here we find truth-values, although there are in the shadow of the stars: possible worlds. In this cosmology "models", "frames", "structures" are also part of the lexicon.

Let us recall that, in anticipation to possible worlds semantics, Wittgenstein was calling "truth-possibilities" valuations attributing truth and falsity to propositions and considering them as representing "state of affairs". Carnap made the connections between Leibniz and Wittgenstein in *Meaning and Necessity* introducing the notion of "state-description":

> A class of sentences in S_I which contains for every atomic sentence either this sentence or its negation, but not both, and no other sentences, is called a *state-description* in S_I, because it obviously gives a complete description of a possible state of the universe of individuals with respect to all properties and relations expressed by predicates of the system. Thus the state descriptions represent Leibniz' possible worlds or Wittgenstein's possible states of affairs. [Carnap, 1947, p. 9]

Later on Quine introduced the terminology "literals" as a common name for atomic formulas and negations of atomic formulas: "Let us speak of single letters and negations of single letters collectively as literals" [Quine, 1950, p. 50]; and pointed out the connection with computation (see [Quine, 1952]). Literals are a way not to speak about truth-values and are quite popular in proof-theory and in logic programming where a finite set of literals is call a "clause" (disjunction of its members) and where the notion of "Horn clause" is a central notion (a clause with at most one positive literal, notion introduced by Albert Horn [Horn, 1951]). In first order logic literals lead to the diagram method of Abraham Robinson which is also a way to avoid speaking of truth-values (the method was presented by Robinson during the same period (see [Robinson, 1951, p. 74]).

In possible worlds semantics the idea of Kripke was not to throw away truth-values. His idea was to introduce possible worlds as additional objects and to define truth-values for propositions in possible worlds through valuations and relations between possible worlds, to characterize necessity as a unary modal operator. Kripke writes:

> The basis of the informal analysis which motivated these definitions is that a proposition is necessary if and only if it is true in all "possible worlds." [Kripke, 1959, p. 2]

Contrary to what is done nowadays, Kripke started to define possible worlds semantics for first-order logic, this is probably why he is using the expression "model":

> Given a non-empty domain **D** and a formula A, we define a complete assignment for A in **D** as a function which to every free individual variable of A assigns an element of **D**, to every propositional variable of A assigns either T or F, and to every n-adic predicate variable of A assigns a set of ordered n-tuples of members of **D**. We define a *model* of A in **D** as an ordered pair (**G**, **K**), where **G** is a complete assignment for A in **D** and **K** is a set of complete assignments for A in **D** such that $\mathbf{G} \in \mathbf{K}$ and such that every member of **K** agrees with **G** in its assignments for free individual variables of A (but not necessarily in its assignments for propositional and predicate variables of A). [Kripke, 1959, p. 2]

We can note here a similarity with what Henkin [1949] was doing in first-order classical logic, Kripke is considering propositional variables.

> An ordinary classical truth table is a set of possible valuations of the propositional variables; each set of possible valuations for each propositional variable is determined by a row of the table. We then evaluate a formula using the usual rules. For S5 truth tables we adopt a similar definition, except that in any table some (but not all) rows may be omitted. Thus a formula has many truth tables, depending on how many rows are omitted. We evaluate "$\wedge$" and "$\sim$" according to the usual methods. In any truth table $\Box A$ is assigned **T** in every row if A is assigned **T** in every row; otherwise $\Box A$ is assigned **F** in every row. A formula B is a tautology of S5 if and only if it is assigned **T** in every row of each of its tables. [Kripke, 1959, p. 11]

Such original method described by Kripke is not much used but has been presented for example by Gerald J. Massey in his interesting paper entitled "The Theory of

Truth Tabular Connectives, both Truth Functional and Modal" [1966]. He draws the following picture for $\Box(p \supset q)$:

p	q	$\Box(p \supset q)$
t	t	t

p	q	$\Box(p \supset q)$
t	f	f

p	q	$\Box(p \supset q)$
f	t	t

p	q	$\Box(p \supset q)$
f	f	t

p	q	$\Box(p \supset q)$
t	t	f
t	f	f

p	q	$\Box(p \supset q)$
t	t	t
f	t	t

p	q	$\Box(p \supset q)$
t	t	t
f	f	t

p	q	$\Box(p \supset q)$
t	f	f
f	t	f

p	q	$\Box(p \supset q)$
t	f	f
f	f	f

p	q	$\Box(p \supset q)$
f	t	t
f	f	t

p	q	$\Box(p \supset q)$
t	t	f
t	f	f
f	t	f

p	q	$\Box(p \supset q)$
t	t	f
t	f	f
f	f	f

p	q	$\Box(p \supset q)$
t	t	t
f	t	t
f	f	t

p	q	$\Box(p \supset q)$
t	f	f
f	t	f
f	f	f

p	q	$\Box(p \supset q)$
t	t	f
t	f	f
f	t	f
f	f	f

As he says, this is "a complete set of truth tables which completely codifies the semantic relations of the compound proposition $\Box(p \supset)q$ to its component propositions p and q" [Massey, 1966, p. 595] In such a set of truth-tables, we cannot see any *possible worlds*, just *truth-possibilities*, i.e. valuations. This way of doing shows that in fact what matter in possible worlds semantics are not possible worlds but relations between valuations, the reason why the terminology "relational semantics" may be more appropriate, or "Kripke semantics", since this idea is mainly due to Kripke as emphasized by Goldblatt (see [Goldblatt, 2006; Copeland, 2002]). Kripke extended his idea of *model* to the idea of model structure, i.e. considering a triple $(\mathbf{G}, \mathbf{K}, \mathbf{R})$ where $\mathbf{R}$ is a binary relation on $\mathbf{K}$. Later on Krister Segerberg [1971] introduced the notion of *frame* to speak about the pair $(\mathbf{K}, \mathbf{R})$ (suggestion for the name was given by Dana Scott whom together with Lemmon threw out the real world $\mathbf{G}$ of Kripke [Lemmon and Scott, 1966]). Segerberg also introduced new notations, in particular using the symbol "$\models$" of model theory.

What is interesting concerning the evolution of truth-value in modal logic is a change of perspective. Instead of considering a function assigning a truth-value to a proposition in a given world, $\Phi(p, W) = T$, to use Kripke's notation, people have started to define a function attributing a set of worlds to a proposition: the worlds in which it is true, or, one might say, its models. This approach has been in particular canonized in the textbook of Brain F.Chellas [1980].

It is possible in fact to do the same at the level of truth-values for propositions in classical propositional logic: instead of considering valuations attributing truth-values, we can consider sets of formulas obeying certain conditions such as:

- $(F \wedge G) \in \Gamma$ iff $F \in \Gamma$ and $G \in \Gamma$

- $\neg F \in \Gamma$ iff $F \notin \Gamma$.

The key to the equivalence of these two approaches is to identify valuations with sets of formulas, a fundamental methodology as we will see in the next session.

7 NON TRUTH-FUNCTIONAL TRUTH-VALUES

7.1 Back to Suszko

In his last published paper entitled "The Fregean Axiom and Polish mathematics in the 1920s" Suszko criticizes in the following way Łukasiewicz, one of the godfathers of many-valued logic and MTV:

> Lukasiewicz is the chief perpetrator of a magnificent conceptual deceit lasting out in mathematical logic to the present day ... Any multiplication of logical values is a mad idea and in fact, Łukasiewicz did not actualize it. [Suszko, 1977, p. 377]

Quine in his classical book *Philosophy of logic* already criticized many-valued logic:

> One setting where classical negation and alternation fall away is many-valued logic. This kind of logic was developed somewhat by C.S.Peirce in the last century, and independently later by Łukasiewicz. It is like the logic of truth functions except that it recognizes three or more so-called truth values instead of truth and falsity. Primarily the motivation of these studies has been abstractly mathematical: the pursuit of analogy and generalization. Studied in this spirit, many-valued logic is logic only analogically speaking; it is uninterpreted theory, abstract algebra." [Quine, 1970, p. 84]

Quine is certainly right in criticizing the proliferation of truth-values, as a kind of mathematical game, without serious philosophical motivations or/and interpretations. However Quine's criticism is superficial in the sense that he does not make the difference here between a many-valued T-structure and the whole framework of MTV. A T-structure considered separately is in fact just an abstract algebra, but in the framework of MTV it is much more, it permits to define connectives, inference relations, logics.

The idea of Suszko was to emphasize the ambiguity of considering elements of many-valued matrices as truth-values and to show that a logic defined by a many-valued matrix can in general be defined by a bivalent non truth-functional semantics. Suszko makes a distinction between *logical value* and *algebraic values.*

When there are more than two values in the T-structure, Suszko calls a V-function, an *algebraic valuation* and a semantics based on V-functions, a *referential semantics*. The set of referential values is cut into two parts to define an inference relation. This is the dichotomy between designated and non-designated values. We are going from the referential to the inferential and we are back to truth and falsity, the logical values.

Suszko presents the situation as follows:

> In case of any logic considered as an inference relation ... one can find sets V of zero-one valued functions defined for all formulas and, called here logical valuations, with the following adequacy property: where $a_1, ..., a_n \vdash b$ are arbitrary formulas with $n = 0, 1, 2, \ldots$, then $a_1, ..., a_n \vdash b$ if and only if for all t in V, $t(b) = 1$ whenever $t(a_1) = ... = t(a_n) = 1$. In short, every logic is (logically) two-valued. This general statement can be easily exemplified in case of Łukasiewicz three-valued sentential logic. [Suszko, 1977, p. 378]

Suszko has in fact provided a bivalent semantics for Łukasiewicz's logic L3 [Suszko, 1975a]. This may seem absurd since Ł3 is called a trivalent logic, not only because it can be characterized by a three-valued matrix (this is also the case of classical bivalent logic), but also because it cannot be characterized by a two-valued logical matrix. So what Suszko is talking about? A bivalent semantics is for him just a set of functions from the set of propositions to $\{0, 1\}$. It is not necessarily truth-functional.

The terminology "Suszko's thesis" has been put forward (see e.g. [Malinowski, 1993]). This is a quite ambiguous terminology. First all there is a result according to which every logic is bivalent. Such a result is not a thesis. What can be considered as a thesis is rather the interpretation of this result [5]. Suszko's result was never really clearly stated, sometimes (cf. [Malinowski, 1993]) it is formulated in the context of structural consequence relation, a notion promoted by Suszko and as Woleński puts it: "Suszko tacitly assumes this property (structurality)" [Woleński, 1999, p. 99]. But for reducing a matricial many-valued semantics to a non truth-functional bivalent semantics (NTB hereafter), it is not necessary to presuppose structurality as pointed out in [da Costa *et al.*, 1996], where more general reduction theorems are described.

The only known NTB presented by Suszko is a symbolical one concerning the mother of many-valued logic, Łukasiewicz's logic L3. Suszko didn't work much the details. Suszko's conditions are presented as follows and are nearly beyond understanding (see [Malinowski, 1993]):

(a) $\beta(a) = 0$ or $\beta(\neg a) = 0$
(b) if $\beta(b) = 1$ then $\beta(a \supset b) = 1$

[5] People arguing against Suszko's thesis are presenting logic systems where the consequence relation is defined differently not allowing the bivalent reduction (see [Malinowski, 1990], [Wansing and Shramko, 2008]).

(c) if $\beta(a) = 1$ and $\beta(b) = 0$, then $\beta(a \supset b) = 0$
(d) if $\beta(a) = \beta(b)$ and $\beta(\neg a) = \beta(\neg b)$, then $\beta(a \supset b) = 1$
(e) if $\beta(a) = \beta(b) = 0$ and $\beta(\neg a) = 6\beta(\neg b)$, then $\beta(a \supset b) = \beta(\neg a)$
(f) if $\beta(\neg a) = 0$, then $\beta(\neg\neg a) = \beta(a)$
(g) if $\beta(a) = 1$ and $\beta(b) = 0$, then $\beta(\neg(a \supset b)) = \beta(\neg b)$
(h) if $\beta(a) = \beta(\neg a) = \beta(b)$ and $\beta(\neg b) = 1$, then $\beta(\neg(a \supset b)) = 0$.

What is not clear is how to interpret these bivalent conditions and how to use them. These conditions can be presented in a different but equivalent way:
(1) if $\beta(a) = 0$ and $\beta(\neg a) = 0$ and $\beta(b) = 0$ and $\beta(\neg b) = 0$, then $\beta(a \supset b) = 1$
(2) if $\beta(a) = 0$ and $\beta(\neg a) = 0$ and $\beta(b) = 0$ and $\beta(\neg b) = 1$, then $\beta(a \supset b) = 0$
(3) if $\beta(a) = 0$ and $\beta(\neg a) = 0$ and $\beta(b) = 1$ and $\beta(\neg b) = 0$, then $\beta(a \supset b) = 1$
(4) if $\beta(a) = 0$ and $\beta(\neg a) = 1$ and $\beta(b) = 0$ and $\beta(\neg b) = 0$, then $\beta(a \supset b) = 1$
(5) if $\beta(a) = 0$ and $\beta(\neg a) = 1$ and $\beta(b) = 0$ and $\beta(\neg b) = 1$, then $\beta(a \supset b) = 1$
(6) if $\beta(a) = 0$ and $\beta(\neg a) = 1$ and $\beta(b) = 1$ and $\beta(\neg b) = 0$, then $\beta(a \supset b) = 1$
(7) if $\beta(a) = 1$ and $\beta(\neg a) = 0$ and $\beta(b) = 0$ and $\beta(\neg b) = 0$, then $\beta(a \supset b) = 0$
(8) if $\beta(a) = 1$ and $\beta(\neg a) = 0$ and $\beta(b) = 0$ and $\beta(\neg b) = 1$, then $\beta(a \supset b) = 0$
(9) if $\beta(a) = 1$ and $\beta(\neg a) = 0$ and $\beta(b) = 1$ and $\beta(\neg b) = 0$, then $\beta(a \supset b) = 1$
(10) if $\beta(a) = 1$, then $\beta(\neg a) = 0$
(11) if $\beta(a) = 0$, then $\beta(\neg\neg a) = 0$
(12) if $\beta(a) = 1$, then $\beta(\neg\neg a) = 1$
(13) if $\beta(a \supset b) = 0$ and $[\beta(a) = 1$ or $\beta(\neg a) = 1]$ and $[\beta(b) = 1$ or $\beta(\neg b) = 1]$, then $\beta(\neg(a \supset b)) = 1$
(14) if $\beta(a \supset b) = 0$, and $\beta(a) = 0$ and $\beta(\neg a) = 0$ then $\beta(\neg(a \supset b)) = 0$
(15) if $\beta(a \supset b) = 0$, and $\beta(b) = 0$ and $\beta(\neg b) = 0$ then $\beta(\neg(a \supset b)) = 0$.

This has been present in [Béziau, 1999b] where such conditions are called *tabular conditions* because they can be used to develop some kinds of truth tables (which can be used as a device for decidability). In [Béziau, 2009] is also presented a bivalent semantics for the four-valued semantics of Dunn and Belnap. This is based on some ideas developed by da Costa that we will examine in the next section. Systematic way to effectively transformed many-valued matricial semantics into NTB has been developed by João Marcos (see e.g. [Caleiro *et al.*, 2007; Caleiro and Marcos, 2010]).

7.2 Newton da Costa's theory of valuation

Although Suszko was the first to present a non truth-functional bivalent semantics and to stress the possible reduction of many-valued matricial semantics to NTB, he did not systematically develop this framework, neither was it developed in the Polish school. A systematic development of NTB was performed by Newton da Costa and his school. It seems that the development of NTB in Brazil was quite independent of the Polish school, although interaction between the Polish and Brazilian schools was active: da Costa visited Poland in the 1970s and met Suszko. Main figures of Polish logic were invited to Brazil in the 1970s and 1980s:

Tarski, Wójcicki, Dubikajtis, Kotas, Malinowski, Perzanowski. NTB was first used by da Costa to provide a semantics for his paraconsistent logic C1. He then used it for many other logical systems and developed a general semantics theory based on NTB, the so-called *theory of valuation*, including a generalization of the notion of truth table (see [da Costa and Béziau 1994a-b], [Grana, 1990]).

The first presentation of a NTB by da Costa is [da Costa and Alves, 1976] and [da Costa and Alves, 1977]. This is essentially the same work: [da Costa and Alves, 1976] is a note in French subsequently developed and published in English as [da Costa and Alves, 1977]. The NTB presented here is a semantics for the paraconsistent logic C1, (such logic was presented in a note in French [da Costa, 1963], subsequently developed and published in English as [da Costa, 1974]).

Here is how this NTB is presented by da Costa and Alves [1977, pp. 622–623]:

> Definition 5 A valuation of C1 is a function $v : \mathcal{F} \longrightarrow \{0,1\}$ such that:
> 1. $v(A) = 0 \Rightarrow v(\neg A) = 1$;
> 2. $v(\neg\neg A) = 1 \Rightarrow v(A) = 1$;
> 3. $v(B^o) = v(A \supset B) = v(A \supset \neg B) = 1 \Rightarrow v(A) = 0$;
> 4. $v(A \supset B) = 1 \Leftrightarrow v(A) = 0$ or $v(A) = 1$;
> 5. $v(A \& B) = 1 \Leftrightarrow v(A) = 1$ and $v(A) = 1$;
> 6. $v(A \vee B) = 1 \Leftrightarrow v(A) = 1$ or $v(A) = 1$;
> 7. $v(A^o) = v(B^o) = 1 \Rightarrow v((A \vee B)^o) = v((A\&B)^o) = v((A \supset B)^o) = 1$
>
> where A^o is an abbreviation for $\neg(A\&\neg A)$.

According to such NTB, when the truth-value of an atomic proposition p is 1, the truth-value of its negation $\neg p$ can be truth or falsity. But despite this indeterminacy, it is possible to construct some tables which are a tool for decidability, such as the following presented by da Costa and Alves [1977, p. 627]:

1. $\urcorner(A \vee B) \supset \urcorner A \ \& \urcorner B$ is not valid in $\mathbf{C}_1$:

A	B	$\urcorner A$	$\urcorner B$	$A \vee B$	$\urcorner(A \vee B)$	$\urcorner A \ \& \urcorner B$	$\urcorner(A \vee B) \supset \urcorner A \ \& \urcorner B$
0	0	1	1	0	1	1	1
1	0	0	1	1	0	0	1
		1	1	1	0	1	1
					1	1	1
0	1	1	0	1	0	0	1
			1	1	0	1	1
					1	1	1
1	1	0	0	1	0	0	1
			1	1	0	0	1
					1	0	0
		1	0	1	0	0	1
					1	0	0
			1	1	0	1	1
					1	1	1

In this paper, da Costa and Alves call such tables "quasi-matrices" later on da Costa used the terminology "valuation tableaux".

From the point of view of this NTB semantics, the negation of da Costa's system C1 is not truth-functional, it is not a T-function, just a P-function. The important point is that a valuation here is a function from the *whole* set of formulas to {0,1}. The set of C1-valuations cannot be defined from a set of distributions of truth-values over atomic propositions which naturally extends to a set of V-functions. Strictly speaking there are here no V-functions. The valuations are not homomorphisms between a P-structure and a T-structure. We have no T-structure and no T-functions. This is a characteristic feature of NTB. This clearly shows how much NTB is deviating from MTV.

But the NTB semantics of C1 is not completely different from the semantics of classical logic, they have something in common: in both cases a bivaluation is a characteristic function of a maximal non trivial set of formula. This feature is used to prove the completeness theorem, as in the classical case. This theorem can be proved using Lindenbaum's extension lemma showing that any given non trivial theory can be extended in a maximal non trivial one.

Da Costa has developed a general theory of NTB playing with this other side of V-functions. He has called this theory, *theory of valuation.* Da Costa has developed this theory at a general abstract level, in particular with Andrea Loparić, and applying it to develop many systems of non-classical logics. A seminal paper is their paper entitled "Parconsistency, paracompletness and valuation" [Loparić and da Costa, 1984]. In this paper Loparić and da Costa present some general results supporting the bivalent claim: "every logical system whatever has a two-valued semantics of valuations" [Loparić and da Costa, 1984, p. 119], and present a NTB for a logic with a negation obeying neither the principle of contradiction nor the principle of excluded middle.

Using NTB Loparić and da Costa are able to give a clear-cut definition of paraconsisteny and paracompleteness:

> We can define precisely the notion of paraconsistent logic; a logic is paraconsistent if it can be the underlying logic of theories containing contradictory theorems which are both true. Such theories we call paraconsistent. Similarly, we define the concept of paracomplete logic: a logical system is paracomplete if it can function as the underlying logic of theories in which there are formulas such that their negations are simultaneously false. [Loparić and da Costa, 1984, p. 119]

The theory of valuation permits to avoid some classical confusions concerning the relation between the principle of bivalence (PB), the principle of non contradiction (PC) and the principle of excluded middle (EM). Loparić and da Costa [1984] present a system π which falsifies the equation PB = PC + EM. A way to save this equation would be to formulate PC as "a formula cannot both be true and false" and EM as "a formula cannot be neither true nor false". But then PC and EM are principles about truth and falsity, involving no negation, only the difference between two values. Moreover these formulations just mean that a V-function is properly a function (see [Béziau, 2003]).

From the point of view of the theory of valuation, it is possible not to speak explicitly of truth and falsity, but just of theories (since a bivaluation can be considered as the characteristic function of a theory). This mathematical perspective is interesting because it shows that the usual distinction between syntax and semantics is quite artificial.

Wójcicki in his 1984 book presents the situation as follows:

> Still, one must agree that the difference between syntax and semantics is much more of philosophical than technical nature. Note, for instance, that the analyzes carried out in terms of truth can be carried out in terms of theories; the admissible valuations need not be interpreted as functions that assign truth-values to sentences but as characteristic functions of certain theories (call them for instance "admissible"). The quest for truth, that seems to be the cornerstone of any scientific activity, can be replaced by the quest for a "good" theories. This seems to be the gist of program put forward by R.K.Meyer, 1978, who launched attack on Tarskian concept of truth, with an attempt to show that Tarskian style semantics may have workable alternatives based just on the notion of a theory. [Wójcicki, 1984, p. 13]

This book entitled *Lectures on propositional calculus* is the first Bible of Polish logic and was written by R. Wójcicki, a former student of Suszko, when in Brazil. The reference to the work of the late Bob Meyer is rather connected with first-order logic. Da Costa has also developed the theory of valuation for first-order logic working on the semantics of paraconsistent logic, discussed and developed in [Béziau, 1996].

The vanishing of the distinction between syntax and semantics has been strengthened in [Béziau, 1995] and [Béziau, 2001], establishing a close relation between bivaluations (theories) and sequent rules. It is shown how it is possible to establish a strict correspondence between sequent rules and valuations, allowing to faithfully translate a sequent rule in a bivaluation condition and vice-versa. This method was applied in different ways, for example to provide a sequent system for Łukasiewicz's logic L3 based on Suszko's bivalent semantic for it [Béziau, 1999b].

These results are based on a general completeness theorem according to which (relatively maximal) theories form a sound and complete semantics for any finite consequence relation, which is a strengthening of Gentzen's first theorem [1932] showing that the class of closed theories form a sound and complete semantics for any consequence relation.

It is worth emphasizing that non truth-functionality does not in general mean modality or intensionality. Moreover many logic systems which can be characterized by a NTB cannot be characterized by a Kripkean semantics, this in particular the case of C1. Most of the time the confusion is based on the identification between extensionality and truth-functionality, on this last question see "The philosophical import of Polish logic" [Béziau, 2002].

Non truth-functional does not also necessarily mean bivalency. Non truth-functional many-valued semantics have also been developed (see [Béziau, 2004]). In these semantics there is, as in standard matricial many-valued semantics, the dichotomy designated/non-designated, but the set of values which can be of any cardinality is not a T-structure, and therefore the semantics is not defined with morphisms between the structure of the propositions and a structure of truth-values. This framework was applied to da Costa's system providing a methodology where the value of a complex formulas depends only on subformulas as in the standard case.

Arnon Avron with his collaborators (in particular Beata Konikowska and Anna Zamanski) has developed what he calls *non-deterministic matrices*, semantics which are a generalized version of MTV, not strictly speaking truth-functional (see [Avron, 2007a], [Avron, 2007b] — about what is truth-functional and what is not truth-functional see [Marcos, 2009]). This is connected with another generalization proposed by Walter Carnielli [Carnielli and Lima-Marques, 1999] and successfully developed by Marcos ([Marcos, 1999], [Marcos, 2005]), the so-called *possible translation semantics* [Carnielli, 2012b]. These new theories provide new perspectives for the development of the theory of truth-values, full of applications.

ACKNOWLEDGMENTS

Thanks you to Irving Anellis, Frank Brown, Walter Carnielli, Newton da Costa, Alexandre Costa Leite, Michael Dunn, Dov Gabbay, Dirk Greimann, Ivo Ibri, Kevin Klement, Bob Lane, Andrea Loparić, João Marcos, Amirouche Moktefi, Shahid Rahman, Juan Redmomd, Jacques Riche, Matthias Schirn, Yaroslav Shramko, Jane Spurr, Patrick Suppes, Arthur de Vallauris Buchsbaum, Heinrich Wansing, and last but not least John Woods (for his support to write this chapter) and Jan Zygmunt (supplying crucial information about the Polish school).

Work partially supported by a grant CAPES/DAAD, 1892/11-3.

BIBLIOGRAPHY

[Anellis, 2004] I.Anellis, "The genesis of the truth-table device", *Russell: The Journal of the Bertrand Russell Archives*, **24**, 55-70, 2004.

[Anellis, 2011] I.Anellis, "Peirce's truth-functional analysis and the origin of the truth table", *History and Philosophy of Logic*, 2011.

[Angelelli, 1982] I.Angelelli, "Frege's notion of *Bedeutung*", in *Logic, methodology and philosophy of science, VI*, L.J.Cohen *et al.*(eds), North-Holland, Amsterdam, pp.735–753, 1982.

[Arielli and Avron, 1998] O.Arielli and A.Avron, "The value of four values", *Artificial Intelligence*, **102**, 97–141, 1998.

[Asenjo, 1954] F.Asenjo, "La idea de un cálculo de antinomies", Seminario Matemático, Universidad de La Plata, 1954.

[Asenjo, 1966] F.Asenjo, "A calculus of antinomies", *Notre Dame Journal of Formal Logic*, **7**, 103-105, 1966.

[Asenjo, 1975] F.Asenjo and J.Tamburino, "A calculus of antinomies", *Notre Dame Journal of Formal Logic*, **16**, 17–44, 1975.

[Avron, 2007a] A.Avron, "Non-deterministic matrices and modular semantics of rules", in [Béziau, 2007], pp.155–174, 2007.

[Avron, 2007b] A.Avron, "Non-deterministic semantics for families of paraconsistent logics", in J.-Y,Béziau, W.A.Carnielli, D.M.Gabbay *Handbook of paraconsistency*, College Publication, London, pp.285–320, 2007.

[Belnap, 1977a] N.D.Belnap, "How a computer should think", in G. Ryle (ed.), *Contemporary Aspects of Philosophy*, Oriel Press, Stocksfield, pp.30-56, 1977.

[Belnap, 1977b] N.D.Belnap, "A useful four-valued logic", in M.Dunn (ed), *Modern uses of multiple-valued logic*, Reidel, Boston, 1977, pp.8–37, 1977.

[Bernays, 1918] P.Bernays, *Beiträge zur axiomatischen Behandlung des Logik-Kalküls*, Habilitationsschrift, University of Göttingen, 1918.

[Bernays, 1926] P.Bernays, "Axiomatische Untersuchung des Aussagenkalküls der Principia Mathematica", *Mathematische Zeitschriftt*, **25**, 305-320, 1926. Reprinted and translated by R.Zach in [Béziau, 2012], pp-43–58.

[Berry, 1952] D.W.Berry, "Peirces contributions to the logic of statements and quantifiers", in P.P.Wiener and F.H.Young (eds), *Studies in the philosophy of Charles Sanders Peirce*, Harvard University Press, Cambridge, Mass., pp. 152–165.

[Béziau, 1995] J.-Y.Béziau, *Recherches sur la logique universelle*, PhD thesis, Department of Mathematics, University of Paris 7 — Denis Diderot, 1995.

[Béziau, 1996] J.-Y.Béziau, *Sobre a verdade lógica*, PhD thesis, Department of Philosophy, University of São Paulo, 1996.

[Béziau, 1999a] J.-Y.Béziau, "Was Frege wrong when identifying reference with truth-value?", *Sorites*, **11**, 15–23, 1999.

[Béziau, 1999b] J.-Y.Béziau, "A sequent calculus for Łukasiewicz's three-valued logic based on Suszko's bivalent semantics", *Bulletin of the Section of Logic*, **28**, 89–97, 1999.

[Béziau, 1999c] J.-Y.Béziau, "The mathematical structure of logical syntax", in W.A.Carnielli and I.M.L.D'Ottaviano (eds), *Advances in contemporary logic and computer science*, American Mathematical Society, Providence, pp.1–17, 1999.

[Béziau, 2001] J.-Y.Béziau, "Sequents and bivaluations", *Logique et Analyse*, **176** 373–394, 2001.

[Béziau, 2002] J.-Y.Béziau, "The philosophical import of Polish logic", in M.Talasiewicz (ed), *Methodology and philosophy of science at Warsaw University*, Semper, Warsaw, pp.109–124, 2002.

[Béziau, 2003] J.-Y.Béziau, "Bivalence, exluded middle and non contradiction", *The Logica Yearbook 2003*, in L.Behounek (ed), Academy of Sciences, Prague, pp.73–84, 2003.

[Béziau, 2004] J.-Y.Béziau, "Non truth-functional many-valued semantics", in *Aspects of Universal Logic*, J.-Y.Béziau, A.Costa-Leite and A.Facchini (eds), University of Neuchâtel, Neuchâtel, pp.199–218, 2004.

[Béziau, 2006] J.-Y.Béziau, "Paraconsistent logic! (A reply to Slater)", *Sorites*, **17**, pp.17–25, 2006.

[Béziau, 2007] J.-Y.Béziau (ed), *Logica Universalis — Towards a general theory of logic*, Second Edition, Birkhäuser, Basel, pp.175–194, 2007.

[Béziau, 2009] J.-Y.Béziau, "Bivalent semantics for De Morgan logic (the uselessness of four-valuedness)", in W.A.Carnielli, M.E.Coniglio, I.M.L.D'Ottaviano (eds), *The many sides of logic*, College Publication, London, pp.391–402, 2009.

[Béziau, 2011] J.-Y.Béziau, "A new four-valued approach to modal logic", *Logique et Analyse*, **54**, 18–33, 2011.

[Béziau, 2012] J.-Y.Béziau (ed), 2012, *Universal logic: an anthology — From Paul Hertz to Dov Gabbay*, Birkhäuser, Basel, 2012.

[Bialynicki-Birula and Rasiowa, 1957] A.Bialynicki-Birula and H.Rasiowa, "On the representation of quasi-boolean algebra", *Bulletin de l'Académie Polonaise des Sciences, Classe III*, **5**, 259–261, 1957.

[Birkhoff, 1940] G.Birkhoff, *Lattice theory*, American Mathematical Society, New York, 1940.

[Birkhoff, 1946] G. Birkhoff, "Universal algebra", in *Comptes Rendus du Premier Congrès Canadien de Mathématiques*, University of Toronto Press, Toronto, 310–326, 1946.

[Bochvar, 1938] D.A.Bochvar, "On a three-valued calculus and its application to analysis of paradoxes of classical extended functional calculus" (Russian), *Matématičeskij Sbornik*, **4**, 387-308, 1938.

[Boole, 1847] G.Boole, *The Mathematical analysis of logic, being an essay towards a calculus of deductive reasoning*, Macmilan, Barclay, & Macmillan, Cambridge and George Bell, London, 1847. Page references, reprinted edition, The philosophical library, New-York 1948.

[Boole, 1848] G.Boole, "The Calculus of Logic", *The Cambridge and Dublin Mathematical Journal*, **3**, 183-198, 1848.

[Boole, 1854] G.Boole, *An Investigation of the laws of thought on which are founded the mathematical theories of logic and probabilities*, Macmilan, London, 1854.

[Bourbaki, 1968] N.Bourbaki, *Theory of Sets*, Addison-Wesley, Boston, 1968.

[Brady, 2000] G.Brady, *From Peirce to Skolem A neglected chapter in the history of logic*, Elsevier, Amsterdam, 2000.

[Caleiro and Marcos, 2010] C.Caleiro and J.Marcos, "Two many values: An algorithmic outlook on Suszkos Thesis", in *Proceedings of the XL International Symposium on Multiple-Valued Logic*, IEEE Computer Society, Los Alamos, pp.93–97, 2010.

[Caleiro *et al.*, 2007] C.Caleiro, W.Carnielli, M.Coniglio and J.Marcos, "The Humbug of many logical values", in [Béziau, 2007], pp.175–194.

[Carnap, 1947] R.Carnap, *Meaning and necessity: a study in semantics and modal logic*, The University of Chicago Press, Chicago, 1947.

[Carnielli, 2012a] W.A.Carnielli, "Paul Bernays and the eve of non-standard models in logic", in [Béziau, 2012], pp.33–42.

[Carnielli, 2012b] W.A.Carnielli, *Possible-translations semantics: their scope, limits and capabilities*, Birkhäuser, Basel, 2012.

[Carnielli and Lima-Marques, 1999] W.A. Carnielli and M. Lima-Marques, "Society semantics and multiple-valued logics", in *Advances in contemporary logic and computer science*, AMS, Providence, pp.33–52, 1999.

[Chang and Keisler, 1973] C.C.Chang and H.J.Keisler, *Model theory*, North-Holland, Amsterdam, 1973.

[Chellas, 1980] B.F.Chellas, *Modal logic — an introduction*, Cambridge University Press, Cambridge, 1980.

[Church, 1956] A.Church, *Introduction to mathematical logic*, Princeton University Press, Princeton, 1956.

[Copeland, 2002] B.J.Copeland "The genenis of possible world semantics", *Journal of Philosophical Logic*, **31**, 99–137, 2002.

[da Costa, 1963] N.C.A. da Costa, 1963, "Calculs propositionels pour les systèmes formels inconsistants", *Comptes Rendus de l'Académie des Sciences de Paris*, **257**, 3790–3793, 1963.

[da Costa, 1974] N.C.A. da Costa, "On the theory of inconsistent formal systems", *Notre Dame Journal of Formal Logic*, **15**, 497–510, 1974.

[da Costa and Alves, 1976] N.C.A. da Costa and E.H.Alves, "Une sémantique pour le calcul C1", *Comptes Rendus d l'Académie des Sciences de Paris*, **283**, 729–731, 1976.

[da Costa and Alves, 1977] N.C.A. da Costa and E.H.Alves, "A semantical analysis of the Calculi C_n", *Notre Dame Journal of Formal Logic*, **18**, 621–630, 1977.

[da Costa and Béziau, 1994a] N.C.A da Costa and J.-Y.Béziau, 1994, "La théorie de la valuation en question", in *Proceedings of the IX Latin American Symposium on Mathematical Logic Vol.2*, Universidad del Sur, Bahía Blanca, pp.95–104, 1994.

[da Costa and Béziau, 1994b] N.C.A da Costa and J.-Y.Béziau, 1994, "Théorie de la valuation", *Logique et Analyse*, **146**, 95–117.

[da Costa *et al.*, 1996] N.C.A da Costa, J.-Y.Béziau, O.A.S.Bueno, "Malinowski and Suszko on many-valuedness : on the reduction of many-valuedness to two-valuedness", *Modern Logic*, **6**, 272–299, 1996.

[Couturat, 1905] L.Couturat, *L'algèbre de la logique*, Paris, Gauthier-Villars, 1905. English translation: *The algebra of logic*, Open Court, London and Chicago, 1914.

[De Mol, 2006] L.De Mol, "An analysis of Emil Post's early work", *The Bulletin of Symbolic Logic*, **12**, 257–289, 2006.

[Destouches, 1948] J.-L.Destouches, *Cours de logique et philosophie gnrale*, Centre de document universitaire, Fournier & Constane, Paris, 1948.

[Destouches-Février, 1948] P.Destouches-Février, "Manifestations et sens de la notion de compémentarité", *Dialectica*, **2**, 383–412, 1948.

[D'Ottaviano and da Costa, 1970] I.M.L.D'Ottaviano and N.C.A. da Costa, "Sur un problème de Jaśkowski", *Comptes Rendus de l'Académie des Sciences de Paris*, **270**, 1349–1353, 1970.

[Dubois, 2008a] D.Dubois, "On ignorance and contradiction considered as truth-values", *Logic Journal of IGPL*, **16**, 195–216, 2008.

[Dubois, 2008b] D.Dubois, "On degrees of truth, partial ignorance and contradiction", in L.Magdalena, M.Ojeda-Aciego, J.L.Verdegay (eds) *Proceedings of IPMU'08*, Torremolinos, Málaga, pp.31–38, 2008.

[Dugundji, 1948] J.Dugundji, "Note on a property of matrices for Lewis and Langford's calculi of propositions", *Journal of Symbolic Logic*, **5**, 150–151, 1940.

[Dunn, 1966] J.M.Dunn, *The algebra of intensional logics*, Ph.D. thesis, University of Pittsburgh, Ann Arbor, 1966.

[Dunn, 1967] J.M.Dunn, 'The effective equivalence of certain propositions about De Morgan lattices', *The Journal of Symbolic Logic*, **32**, 433–434, 1967.

[Dunn, 1969] J.M.Dunn, *Natural language versus formal language*,, unpublished manuscript. Presented at the joint APA-ASL symposium, New York, Dec. 27, 1969.

[Dunn, 1971] J.M.Dunn, "An intuitive semantics for first degree relevant implications"' (abstract), *The Journal of Symbolic Logic*, **36**, 362–363, 1971.

[Dunn, 1976] J.M.Dunn, "Intuitive semantics for first-degree entailments and coupled trees", *Philosophical Studies* , **29**, 149–168, 1976.

[Dunn, 2000] J.M.Dunn, "Partiality and its dual", *Studia Logica*, **65**, 5–40, 2000.

[Durand-Richard, 2007] M.-J.Durand-Richard, "Opération, fonction et signification de Boole Frege", *Cahiers Critiques de Philosophie*, **3**, 99–128, 2007.

[Dzik, 1981] W.Dzik, "The Existence of Lindenbaums Extensions is Equivalent to the Axiom of Choice", *Reports on Mathematical Logic*, **12**, 29–31, 1981.

[Epstein, 1990] R.L.Epstein, *The semantic foundations of logic*, Kluwer, Dordrecht, 1990.

[Feferman and Feferman, 2004] S.Feferman and A.B.Feferman, 2004, *Tarski: Life and Logic*, Cambridge University Press, Cambridge.

[Février, 1937a] P.Février, "Les relations d'incertitude de Heisenberg et la logique", *Comptes Rendus de l'Académie des Sciences de Paris*, **204**, 481–483, 1937.

[Février, 1937b] P.Février, 1937, "Sur la forme générale de la définition d'une logique", *Comptes Rendus de l'Académie des Sciences de Paris*, **204**, 958–959, 1937.

[Février, 1937c] P.Février, "Les relations d'incertitude d'Heisenberg et la logique", in *Travaux du IXe Congres International de Philosophie, Vol. VI*, Hermann, Paris, 88–94, 1973.

[Fisch and Turquette, 1966] M.Fisch and A.Turquette, "Peirce's triadic logic", *Transactions of the Charles S. Peirce Society*, **2**, 71–85, 1966.

[Fine, 1974] K.Fine, "Models For Entailment", *Journal of Philosophical Logic*, **3**, 347–372, 1974.

[Fitting, 1994] M.Fitting, "Kleene's Three Valued Logics and Their Children", *Fundamenta Informaticaey*, **20**, 113–131, 1994.

[Font, 1997] J.-M.Font, "Belnap's four-valued logic and De Morgan lattices", *Logic Journal of the IGPL*, **5**, 413–440, 1997.

[Font and Hájek, 2002] J.M.Font and P.Hájek, "On Łukasiewicz's four-valued logic", *Studia Logica*,70, 157–182, 2002.

[Font and Verdú, 1997] J.M.Font and V.Verdú, "Completeness theorems for a four-valued logic related to De Morgan lattices", *Mathematics Preprint Series*, **57**, Barcelonna, 1989.

[Fox, 1990] J.Fox, "Motivation and demotivation of a four-valued Logic", *Notre Dame Journal of Formal Logic*, **31**, 76–80, 1990.

[van Fraassen, 1966] B.C. van Fraassen, "Singular terms, truth-value gaps, and free logic", *Journal of Philosophy*, **63**, 481–495, 1966.

[Frege, 1879] G.Frege, *Begriffsschrift, eine der arithmetischen nachgebildete Formelsprache des reinen Denkens*, L.Nebert, Halle, 1879. English Translation (reference for pagination): *Begriffsschrift, a formula language, modeled upon that of arithmetic, for pure thought* in [Heijenoort, 1967b], pp.1–82.

[Frege, 1891] G.Frege, "Function und Begriff", *Vortrag gehalten in der Sitzung vom 9. Januar 1891 der Jenaischen Gesellschaft fur Medicin und Naturwissenschaft.* English Translation: *Function and concept* in [Geach and Black, 1952], pp.21–41.

[Frege, 1892] G.Frege, "Über Sinn und Bedeutung", *Zeitschrift fur Philosophie und philosophische Kritik*, new series, **100**, 25–50, 1893. English Translation: *On sense and reference* in [Geach and Black, 1952], pp.56–78.

[Frege, 1893] G.Frege, *Grundgesetze der Arithmetik, begriffsschriftlich abgeleitet , vol.1*, Hermann Pohle, Jena, 1893. Partial English Translation: *The basic laws of arithemtic* in [Geach and Black, 1952], pp.137-158.

[Gabriel, 1984] G.Gabriel, "Fregean connection: *Bedeutung*, value and truth-value", *The Philosophical Quarterly*, **34**,372–376, 1984..

[Geach and Black, 1952] P.Geach and M.Black, *Translations from the philosophical writings of Gottlob Frege*, Basil Blackwell, Oxford, 1952, second edition 1960.

[Gentzen, 1932] K.Gentzen, "Über die Existenz unabhängiger Axiomensysteme zu unendlichen Satzsystemen", *Matematische Annalen*, **107**, 329–350, 1932.

[Gödel, 1932a] K.Gödel, "Zum intuitionisticischen Aussagenkalkül", *Akademie der Wissenschaften in Wien, Mathematisch-natuirwissenschaft Klasse*, **64**, 65–66, 1932.

[Gödel, 1932b] K.Gödel, "Eine Eigenschaft der Realisierungen des Aussagenkalküls", *Ergebnisse eines mathematischen Kolloquiums*, reprinted and translated in [Gödel, 1986], pp.238–241, 1932.

[Gödel, 1986] K.Gödel, *Collected Works — Volume 1*, S.Feferman et al. (eds), Oxford University Press, Oxford, 1986.

[Goldblatt, 2006] R.Goldblatt, "Mathematical modal logic: a view of its evolution", in D.M.Gabbay and J.Woods (ed), *Handbook of the history of logic — Volume 7: Logic and the modalities in the twentieth century*, North-Holland, Amsterdam, pp.1–98, 2006.

[Grana, 1990] N.Grana, *Sulla teoria delle valuationi di N.C.A da Costa*, Ligori, Napoli, 1990.

[Grattan-Guiness, 2000] I.Grattan-Guinness, *The search for mathematical roots 1870-1940 (Logics, set theories and the foundations of mathematics from Carnot through Russell to Gödel)* , Princeton University Press, Princeton, 2000.

[Grattan-Guiness, 2004-5] I.Grattan-Guinness, "Comments on Stevens's Review of the Cambrige Companion and Anellis on truth tables", *Russell: The Journal of the Bertrand Russell Archives*, **24**, 185–88, 2004-5.

[Green, 1994] J.Green, "The algebra of logic: what Boole really started", *Modern Logic*, **4**, 48–62, 1994.

[Hailpern, 1981] T.Hailperin, "Boole's algebra isn't boolean algebra", *Mathematics Magazine*, **54**, 172–184, 1981.

[Halmos, 1956] P.Halmos, "The basic concepts of algebraic logic", *American Mathematic Monthly*, **63**, 363–387, 1956.

[Heijenoort, 1967a] J.van Heijenoort, "Logic as calculus and logic as language", *Synthese*, **17**, 324–330, 1967.

[Heijenoort, 1967b] J.van Heijenoort, *From Frege to Gödel, A Source Book in Mathematical Logic, 1879-1931*, Harvard University Press, Cambridge, Mass. 1967.

[Henkin, 1949] L.Henkin, "The Completeness of the First-Order Functional Calculus", *The Journal of Symbolic Logic*, **14**, 159–166, 1949.

[Henkin, 1996] L.Henkin, "The Discovery of My Completeness Proofs", *The Bulletin of Symbolic Logic*, **2**, 127–158, 1996.

[Hodges, 1985-86] W.Hodges, "Truth in a structure", *Proceedings of the Aristotelian Society* , **86**, 135–151, 1985-86.

[Hilpinen, 2004] R.Hilpinen, "Peirce's logic", in D.M.Gabbay and J.Woods (ed), *Handbook of the history of logic . Volume 3 The rise of modern logic I: Leibniz to Frege* , North-Holland, Amsterdam, pp.611–658, 2004.

[Horn, 1951] A.Horn, "On Sentences Which are True of Direct Unions of Algebras", *The Journal of Symbolic Logic*, **16**, 14–21, 1951.

[Jaśkowski, 1936] S.Jaśkowski, "Recherches sur le système de la logique intuitionniste", in *Actes du Congrés International de Philosophie Scientifique, Vol.6* , Hermann, Paris, pp.58–61, 1936.

[Jansana, 2012] R.Jansana, "Bloom, Brown and Suszko's work on asbtract logics", in [Béziau, 2012].

[Jevons, 1864] S.Jevons, *Pure logic, or, the logic of quality apart from quantity — with remarks on Boole system and on the relation of logic and mathematics*, Edward Stanford, London, 1864.

[Jevons, 1870] S.Jevons, "The mechanical performance of logical inference", *Philosophical Transactions*, **160**, 497–518, 1870.

[Kalman, 1958] J.A.Kalman, "Lattices with involution", *Transactions of the American Mathematical Society*, **87**, 485–491, 1958.

[Kleene, 1938] S.C.Kleene, "On a notation for ordinal numbers", *The Journal of Symbolic Logic*, **3**, 150–155, 1948.

[Kleene, 1952] S.C.Kleene, *Introduction to metamathematics*, North-Holland, Amsterdam, 1952.

[Kneale and Kneale, 1962] W.Kneale and M.Kneale, *The development of logic*, Clarendon, Oxford, 1962.

[Kreisel and Krivine, 1966] G.Kreisel and J.-L.Krivine, *Eléments de logique mathématique (Théorie des modèles)*, Dunod, Paris, 1966. English translation: *Elements of Mathematical Logic (Model Theory)*, North-Holland, Amsterdam, 1967.
[Kripke, 1959] S.Kripke, "A completeness theorem in modal logic", *The Journal of Symbolic Logic*, **24**, 1–14, 1959.
[Kripke, 1963a] S.Kripke, "Semantical analysis of modal logic I, Normal modal propositional calculi", *Zeitschrift fr mathematische Logik und Grundlagen der Mathematik*, **9**, 67–96, 1963.
[Kripke, 1963b] S.Kripke, "Semantical Considerations on Modal Logic", *Acta Philosophica Fennica*, **16**, 83–94, 1963.
[Kripke, 1965a] S.Kripke, "Semantical analysis of intuitionistic logic I", in J.N.Crossley and M.A.E.Dummett, eds, *Formal Systems and Recursive Functions*, North-Holland, Amsterdam, pp.92-130, 1965.
[Kripke, 1965b] S.Kripke, "Semantical analysis of modal logic II. Non-normal modal propositional calculi", In J.W.Addison, L.Henkin, and A. Tarski, eds, *The Theory of Models*, North-Holland, Amsterdam, pp.206–220, 1965.
[Ladd, 1883] C.Ladd, "On the algebra of logic", in [Peirce, 1883], pp.17–71.
[Lane, 1999] R.Lane, "Peirce's triadic logic revisited", *Transactions of the Charles S. Peirce Society*, **35**, 284–311, 1999.
[Leblanc, 1973] H.Leblanc (ed), *Truth, syntax and modality*, North-Holland, Amsterdam, 1973.
[Leblanc, 1976] H.Leblanc, *Truth-value semantics*, North-Holland, Amsterdam, 1976.
[Lemmon and Scott, 1966] E.J.Lemmon and D.S.Scott, *Intensional logics*, Unpublished manuscript known as "The Lemmon notes", 1966.
[Lewis, 1918] C.I.Lewis (ed), *A survey of symbolic logic*, University of California Press, Berkeley, 1918.
[Loparić, 1977] A.Loparić, "Une étude sémantique de quelques calculs propositionnels", *Comptes Rendus de l'Académie des Sciences de Paris*, **284a**, 835–838, 1977.
[Loparić, 1986] A.Loparić, "A semantical study of some propositional calculi", *The Journal of Non-Classical Logic*, **3**, 74–95, 1986.
[Loparić, 2010] A.Loparić, "Valuation semantics for intuitionistic propositional calculus and some of its subcalculi", *Principia*, **14**, 125–133, 2010.
[Loparić and Alves, 1980] A.Loparić and E.H.Alves, "The semantics of systems Cn of da Costa", In A. I. Arruda, N. C. A. da Costa, and A. M. Sette, (eds), *Proceedings of the Third Brazilian Conference on Mathematical Logic*,, Sociedade Brasileira de Lógica, São Paulo, pp.161–172, 1980.
[Loparić and da Costa, 1984] A.Loparić and N.C.A da Costa, 1984, "Parconsistency, paracompletness and valuation", *Logique et Analyse*, **106**, 119–131, 1984. Reprinted in [Béziau, 2012], pp.373–388.
[Łoś, 1948a] J.Łoś, "Logiki wielowartosciowe a formalizacja funkcji intensjonalnych", *Kwartalnik Filozoficzny* , **17**, 59–78, 1948.
[Łoś, 1948b] J.Łoś, "Sur les matrices logiques", *Colloquium Mathematicum* , **1**, 337–339, 1948.
[Łoś, 1949] J.Łoś, *O matrycach logicznych*, Travaux de la Société des Sciences et des Lettres de Wrocław, Série B, 19, 1949.
[Łoś, 1951] J.Łoś, "On algebraic proof of completeness for the two-valued propositional calculus", *Colloquium Mathematicum* , **2**, 271–271, 1951.
[Łoś, 1954] J.Łoś, "Sur le theorème de Gödel pour les théories indénombrables", *Bulletin of the Polish Academy of Sciences* , **2**, 319–320, 1954.
[Łoś and Suszko, 1958] J.Łoś and R.Suszko, "Remarks on sentential logics", *Indagationes Mathematicae* , **20**, 177–183, 1958. Reprinted in [Béziau, 2012], pp.177–186.
[Łukasiewicz, 1910] J.Łukasiewicz, *O zasadzie sprzeczności u Arystotelesa*, Studium Krytyczne, Kraków, 1910.
[Łukasiewicz, 1920] J.Łukasiewicz, "O logice trójwartościowej", *Ruch Filozoficny*, **5**, 170–171, 1920.
[Łukasiewicz, 1929] J.Łukasiewicz, *Elementy logiki matematycznej*, University of Warsaw, Warsaw, 1929.
[Łukasiewicz, 1930] J.Łukasiewicz, "Philosophische Bemerkungen zu mehrwertigen Systemen des Ausagenkalküls", *Comptes rendus des séances de la Société des lettres et des sciences de Varsovie. Classe III*, **23**, 51–77, 1930. English translation in [McCall, 1967].
[Łukasiewicz, 1953] J.Łukasiewicz, "A system of modal logic", *Journal of Computing Systems*, **1**, 111–149, 1953.

[Łukasiewicz and Tarski, 1930] J.Łukasiewicz and A.Tarski, "Untersuchungen über des Aussagenkalküls", *Comptes rendu des séances de la Société des lettres et des sciences de Varsovie. Classe III*, **23**, 23–50, 1930. English translation in [Tarski, 1983], pp.38–59.

[Malinowski, 1990] G.Malinowski, "Towards the concept of logical many-valuedness", *Folia Philosophica*, **7**, 97–103, 1990.

[Malinowski, 1993] G.Malinowski, *Many-valued logics*, Clarendon, Oxford, 1993.

[Malinowski, 1994] G.Malinowski, "Inferential many-valuedness", in J.Wolenski (ed), *Philosophical logic in Poland*, Kluwer, Dordrecht, pp.75–84, 1994.

[MacColl, 1877-1879] H.MacColl, "The calculus of equivalent statements and integration limits I-III", *Proceedings of the London Mathematical Society*, **9** (1877-78), 9-20; 177–186; **10** (1878-79), 16–28.

[MacColl, 1880-1906] H.MacColl, "Symbolic(al) reasoning I — VIII" *Mind*, **17** (1880), 45–60; **24** (1897), 493–510; **33** (1900), 75–84; **43** (1902), 352–368; **47** (1903), 355–364; **53** (1905), 74–81; **55** (1905), 390–397; **60** (1906), 504–518.

[MacColl, 1906] H.MacColl, *Symbolic logic and its application*, Longmans, Green and Co, New York, 1906.

[Marcos, 1999] J.Marcos, *Possible-translations semantics*, MD, State University of Campinas, Campinas, 1999.

[Marcos, 2005] J.Marcos, *Logics of Formal Inconsistency* , PhD, State University of Campinas, Campinas, 2005.

[Marcos, 2009] J.Marcos, "What is a Non-truth-functional Logic?", *Studia Logica* , **92**, 215–240, 2009.

[Massey, 1966] G.J.Massey, "The theory of truth tabular connectives, both truth functional and modal" *The Journal of Symbolic Logic*, **31**, 593–608, 1966.

[McCall, 1967] S.McCall, *Polish logic 1920-1939*, Clarendon, Oxford, 1967.

[Michael, 1979] E.Michael, "A note on Peirce on Boole's algebra of logic", *Notre Dame Journal of Formal Logic*, **20**, 1979, 636–639, 1979.

[Miller and Thornthone, 2008] D.M.Miller and M.A.Thornthone, *Multiple valued logic: concepts and representations*, Morgan & Claypool, 2008.

[Moisil, 1935] G.C.Moisil, "Recherche sur l'algèbre, de la logique", *Annales Scientifiques de l'Université de Jassy*, **22**, 1–117, 1935.

[Moisil, 1972] G.Moisil, *Essais sur les logiques non-chrysipiennes*, Académie de la République Socialiste de Roumanie, Bucarest, 1972.

[Monteiro, 1960] A.Monteiro, "Matrices de Morgan caractéristiques pour le calcul propositionnel classique", *Anais da Academia Brasileira de Cienciais*, **32**, 1–7, 1960.

[Moore, 1908] A.W.Moore, "Truth value", *The Journal of Philosophy, Psychology and Scientific Methods*, **5**, 429–436, 1908.

[Omyła and Zygmunt, 1984] M.Omyła and J.Zygmunt, "Roman Suszko (1919–1979): Bibliography of the Published Work with an Outline of His Logical Investigations", *Studia Logica*, **43**, 421–441, 1984.

[Peckhaus, 1999] V.Peckhaus, "Hugh MacColl and the German algebra of logic", *Nordic Journal of Philosophical Logic*, **3**, 17–34, 1999.

[Peirce, 1880a] C.S.Peirce, "The doctrine of chances", *Popular Science Monthly*, **12**, 604–615, 1880.

[Peirce, 1880b] C.S.Peirce, "On the algebra of logic", *American Journal of Mathematics*, **3**, 15–57, 1880.

[Peirce, 1883] C.S.Peirce (ed), *Studies in logic*, Little, Brown and Company, Boston, 1883.

[Peirce, 1885] C.S.Peirce, "On the algebra of logic: a contribution to the philosophy of notation", *American Journal of Mathematics*, **7**, 180–202, 1885.

[Peirce, 1897] C.S.Peirce, "The logic of relatives", *The Monist*, **7**, 161–217, 1897.

[Peirce, 1893a] C.S.Peirce, *General and historical survey of logic — elements of logic*, posthumously published in Volume 2 of [Peirce, 1931-35].

[Peirce, 1893b] C.S.Peirce, "The logic of quantity", posthumously published in Volume 3 of [Peirce, 1931-35].

[Peirce, 1902] C.S.Peirce, *The simplest mathematics*, posthumously published in Volume 4 of [Peirce, 1931-35].

[Peirce, 1931-35] C.S.Peirce, *Collected papers*, edited by C.Hartshorne and P.Weiss, Harvard University Press, Cambridge, Mass. 1931-35.

[Pogorzelski, 1994] W.A.Pogorzelski, 1994 *Notions and theorems of elementary formal logica*, Warsaw University — Białystok Branch, Białystok, 1994.

[Pogorzelski and Wojtylak, 2008] W.A.Pogorzelski and P.Wojtylak, *Completeness theory for propositional logics*, Birkhäuser, Basel, 2008.

[Post, 1921] E.Post, "Introduction to a general theory of elementary propositions", in *American Journal of Mathematics*, **13**, 163–185, 1921. Reproduced in [Heijenoort, 1967b], pp.264–283 (reference for pagination).

[Priest, 1979] G.Priest, "Logic of paradox", *Journal of Philosophical Logic*, **8**, 219–241, 1979.

[Priest and Routley, 1989] G.Priest and R.Routley, "Systems of paraconsistent logic", in *Paraconsistent logic: Essays on the inconsistent*, Philosophia, Munich, pp.151–186, 1989.

[Prior, 1955] A.N.Prior, "Many-Valued and Modal Systems: An Intuitive Approach", *The Philosophical Review*, **64**, 626–630, 1955.

[Quine, 1934] W.V.O.Quine, "Ontological remarks on the propositional calculus", *Mind*, **43**, 472–476, 1934.

[Quine, 1938] W.V.O.Quine, "Review of [Tarski, 1937], *Bulletin of the American Mathematical Society*, **44**, 317–318, 1938.

[Quine, 1950] W.V.O.Quine, *Methods of logic*, Holt, Rinehart and Winston, New York, 1950.

[Quine, 1952] W. V. Quine, "The Problem of Simplifying Truth Functions", *The American Mathematical Monthlyy*, **59**, 521–531, 1952.

[Quine, 1970] W.V.O.Quine, *Philosophy of logic*, Englewood Cliffs, Prentice-Hall, 1970.

[Rahman and Redmond, 2007] S.Rahman and J.Redmond, *Hugh MacColl: an overview of his logical work with anthology*, College Publications, London, 2007.

[Rasiowa and Sikorski, 1963] H.Rasiowa and R.Sikorski, 1963, *The mathematics of metamathematics*, Polish Academy of Science, Warsaw, 1963.

[Rescher, 1962] N.Rescher, "Quasi-truth-functional systems of propositional logic", *The Journal of Symbolic Logic*, **27**, 1–10, 1962.

[Rescher, 1965] N.Rescher, "An intuitive interpretation of systems of four-valued logic", *Notre Dame Journal of Formal Logic*, **6**, 154–156, 1965.

[Rescher, 1969] N.Rescher, *Many-valued logic*, McGraw-Hill, New York, 1969.

[Robinson, 1951] A.Robinson, *On the metamathematics of algebra*, North-Holland, Amsterdam, 1951.

[Rose, 1951] A.Rose, "Systems of logic whose truth-values form lattices", *Mathematische Annalen*, **123**, 152–165, 1951.

[Russell, 1903] B.Russell, *The principles of mathematics*, Cambridge University Press, Cambridge, 1903.

[Russell, 1918-19] B.Russell, "The philosophy of logical atomism", *The Monist*, **28**, 495–527, **29**, 32–63, 190–222, 345–380, 1918-19. Reprinted in B.Russell, *Logic and knowledge*, Allen and Unwirn. London 1956, pp.177–281.

[Russell, 1911] J.E. Russell, "Truth as Value and the Value of Truth", *Mind*, **20**, 538–539, 1911.

[Schirn, 1996] M.Schirn (ed), *Frege: importance and legacy*, Walter de Gruyter, Berlin, 1996.

[Segerberg, 1971] K.Segerberg, *An essay in classical modal logic*, PhD, Stanford University, Stanford, 1971.

[Sengupta, 1983] G.Sengupta, "On identifying reference with truth-value", *Analysis*, **43**, 72–74, 1983.

[Sher, 2001] G.Sher, "Truth, logical structure and compositionality", *Synthese*, **126**, 195–219, 2001.

[Shosky, 1997] J.Shosky, "Russell's use of truth tables", *Russell: The Journal of the Bertrand Russell Archives*, **17**, 11–26, 1997.

[Shramko and Wansing, 2005] Y.Shramko and H.Wansing, "Some useful 16-valued logics: how a computer network should think", *Journal of Philosophical Logic*, **34**, 121–153, 2005.

[Shramko and Wansing, 2009] Y.Shramko and H.Wansing, "Truth values", special issues of *Studia Logica*, **91**, **92**, 2009.

[Shramko and Wansing, 2010] Y.Shramko and H.Wansing, "Truth values", *The Stanford Encyclopedia of Philosophy*, 2010.

[Shramko and Wansing, 2011] Y.Shramko and H.Wansing, *Truth and falsehood*, Springer, Berlin, 2011.

[Schröder, 1890-1895] E.Schröder, *Vorlesungen über die Algebra der Logik (exakte Logik)*, I (1890), II (1891) , III (1895), B.G.Teubner, Leipzig.

[Slater, 1995] B.H.Slater, "Paraconsistent logics?", *Journal of Philosophical logic*, **24**, 451–454, 1995.

[Sleszyńskiego, 1925] J.Sleszyńskiego, *Teorja dowodu*, Jagiellonian University, Kraków, 1925.

[Suppes, 2008] P.Suppes, "A revised agenda for philsophy of mind (and brain)", in *Themes from Suppes*, Ontos, Frankfurt, pp.19–51, 2008.

[Suppes and Béziau, 2004] P. Suppes and J.-Y. Béziau, "Semantic computation of truth based on associations already learned", *Journal of Applied Logic*, **2**, 457–467, 2004.

[Suszko, 1957] R.Suszko, "Formalna teoria wartości logicznych. I", *Studia Logica*, **6**, 145–237, 1957.

[Suszko, 1975a] R.Suszko, "Remarks on Łukasiewicz's three-valued logic", *Bulletin of the Section of logic*, **4**, 87–90, 1975.

[Suszko, 1975b] R.Suszko, "Abolition of the Fregean axiom", in R.Pahrik, *Logic Colloquium — Symposium on Logic Held at Boston, 1972-73*, Springer, Berlin, pp.169–239, 1975.

[Suszko, 1977] R.Suszko, "The Fregean axiom and Polish mathematical logic in the 1920s", *Studia Logica*, **36**, 377–380, 1977.

[Suszko *et al.*, 1973] R.Suszko, S.L.Bloom and D.J.Brown, "Abstract logics", "Classical abstract logics", *Dissertationes mathematicae*, **102**, 1973.

[Sylvester, 1850] J.J.Sylvester, "Additions to the Articles On a new class of theorems, and On Pascals theorem", *Philosophical Magazine*, 363–370, 1850.

[Tarski, 1923a] A.Tarski, "0 wyrazie pierwotnym logistyki" (Doctoral dissertation), *Przeglqd Filozoficzny*, **26**, 68–89, 1923.

[Tarski, 1923b] A.Tarski, "Sur le terme primitif de la logistique", *Fundamenta Mathematicaea*, **4**, 196–200, 1923.

[Tarski, 1924] A.Tarski, "Sur les truth-functions au sens de MM. Russell et Whitehead", *Fundamenta Mathematicae*, **5**, 59–74, 1924.

[Tarski, 1935] A.Tarski, "Der Wahrheitsbegriff in den formalisierten Sprachen", *Studia Philosophica*, **1**, 261–405, 1935.

[Tarski, 1936] A.Tarski, *Wprowadzenie do logiki i do metodologii nauk dedukcyjnych*, Bibljoteczka Matematyczna, vol. 3–5, Ksiaznica-Atlas, Lwow and Warsaw, 1936.

[Tarski, 1935-36] A.Tarski, "Grundzüge des Systemenkalküls. *Fundamenta Mathematicaea*, Erster Teil **25** (1935), 503–526, Zweiter Teil **26** (1936), 283–301.

[Tarski, 1937] A.Tarski, *Einführung in die mathematische Logik und in die Methodologie der Mathematik*, Springer, Vienna, 1937. Exact translation of [Tarski, 1936].

[Tarski, 1944] A.Tarski, "The semantic conception of truth and the foundations of semantics', *Philosophy and Phenomenological Research*, **4**, 341–376, 1944.

[Tarski, 1954-55] A.Tarski, "Contributions to the theory of models. I, II, III", *Indigationes Mathematicae*, **16**, pp.572-581, pp.582-588, 1954, **17**, pp.56-64, 1955.

[Tarski, 1983] A.Tarski, *Logic, semantics, metamathematics*, Hackett, Indianapolis, second edition, 1983.

[Tarski, 1994] A.Tarski, *Introduction to logic and the methodology of deductive science*, Oxford University Press, New-York, 1994. Fourth american edition of [Tarski, 1936] prepared by Jan Tarski.

[Tarski and Vaught, 1957] A.Tarski and R.L.Vaught, "Arithmetical extensions of relational systems", *Compositio Mathematica*, **13**, 81–102, 1957.

[Tsuji, 1998] M.Tsuji, "Many-valued logics and Suszko's Thesis revisited", *Studia Logica*, **60**, 299–309, 1998.

[Wansing and Belnap, 2008] H.Wansing and N.D.Belnap, "Generalized truth values. A reply to Dubois", *Logic Journal of IGPL*, **16**, 921–935, 2008.

[Wansing and Shramko, 2008] H.Wansing and Y.Shramko, "Suszkos thesis, inferential many-valuedness, and the notion of a logical system", *Studia Logica*, **88**, 405–429, 2008.

[Whitehead, 1898] A.N.Whitehead, *A treatise on universal algebra*, Cambridge University Press, Cambridge, 1898.

[Whitehead and Russell, 1910] A.N.Whitehead and B.Russell, 1910, *Principia Mathematica*, vol.1, Cambridge University Press, Cambridge, 1910.

[Wittgenstein, 1921] L.Wittgenstein, "Logisch-philosophische Abhandlung", *Annalen der Naturphilosophie*, **14**, 185–262, 1921. Translated as *Tractatus Logico-Philosophicus*, Kegan Paul, London, 1922.

[Wójcicki, 1984] R.Wójcicki, *Lectures on propositional logic*, Ossolineum, Wrocław, 1984.

[Wójcicki, 1988] R.Wójcicki, *Theory of logical calculi*, Reidel, Dordrecht, 1988.

[Woleński, 1999] J.Woleński, "The principle of bivalence and Suszko's thesis", *Bulletin of the Section of Logic*, **28**, 99–110, 1999.
[Zach, 1999] R.Zach, "Completeness before Post: Bernays, Hilbert, and the development of propositional logic", *The Bulletin of Symbolic Logic*, **5**, 331–366, 1999.
[Zadeh, 1975a] L.Zadeh, "The concept of a linguistic variable and its Application to Approximate Reasoning -I", *Information sciences*, **8**, 199–249, 1975.
[Zadeh, 1975b] L.Zadeh, "Fuzzy logic and approximate reasoning", *Synthese*, **30**, 407–428, 1975.
[Zygmunt, 1991] J.Zygmunt, "Mojżesz Presburger, life and work", *History and Philosophy of Logic*, **12**, 211–223, 1991.
[Zygmunt, 2012a] J.Zygmunt, "Structural consequence operations and logical matrices adequate for them", in [Béziau, 2012], pp.163–176.
[Zygmunt, 2012b] J.Zygmunt, "Tarski's first published contribution to general metamathematics", in [Béziau, 2012], pp.59–67.

A HISTORY OF MODAL TRADITIONS

Simo Knuuttila

My aim is to shed light on long-lived assumptions in Western modal conceptions as well as some clashes between them. In the first section, I shall deal with those trends in ancient and medieval modal thought which were inclined to codify the meaning of modal notions in frequency terms. The second section concentrates on the emergence of a different paradigm in which the analysis of necessity and possibility is separated from a one world model, and which considerably modified late medieval modal logic. The third section is about the interplay of these traditions in the early modern period and their influence up to the nineteenth century.

1 EXTENSIONAL MODAL CONCEPTIONS IN ANCIENT AND MEDIEVAL PHILOSOPHY

Ancient philosophers assumed that possibilities refer to actualization in history, and they found it natural to think that no genuine generic possibility remains eternally unrealized — whether there might be unrealized singular possibilities was more controversial. This habit of thinking was famously called the principle of plenitude by Arthur O. Lovejoy [1936]. Some commentators, e.g., [Hintikka, 1981, p. 6], have wondered whether the principle of the paucity of possibilities would be a more proper term for this balance between possibility and actuality because the types of things and events which are never exemplified are labelled as impossible and the invariant features of reality as necessary. There are well-known examples of these assumptions, such as Plato's doctrine of the ideal models of being which are exhaustively imitated in the world by the Demiurge, Aristotle's metaphysical theory of the priority of actuality over potentiality, the Stoic doctrine of the divine world-order and the cosmic cycles in which everything which can be actual is eternally repeated, Plotinus' metaphysics of emanation, which actualizes all levels of being until the limit of the possible, or the atomist idea of the atoms which bring about all possible configurations during the infinity of time.[1] David Sedley has recently discussed the theories of Plato and the Stoics as examples of ancient philosophical creationism and the atomist arguments for innumerable worlds in which no possibility goes unrealized as an answer to Plato's design argument. He sees here an ancient analogy to the contemporary multiverse response to the fine-tuning argument [Sedley, 2007]. Some scholars have asked whether the principle

[1]See Plato, *Timaeus*, 30c-32d; 55c; Aristotle, *Metaphysics* IX.8; Origen, *Contra Celsum*, 4.20; 4.68 (on the Stoic theory of the eternal return), Diogenes Laertius IX.31, 44, Lucretius, *De rerum natura* II, 1067-1089 (infinite worlds); Plotinus, *Enneads* IV, 8.5, 33-35; 8.6, 12-13.

Handbook of the History of Logic. Volume 11: Logic: A History of its Central Concepts.
Volume editors: Dov M. Gabbay, Francis Jeffry Pelletier and John Woods.
General editors: Dov M. Gabbay and John Woods.

of plenitude was derived from the special metaphysical assumptions of ancient philosophy, which influenced the use of modal terms, or whether it was associated with the meaning of modal terms as such. (See [Waterlow, 1982b, pp. 25-6, 31-5]; [van Rijen, 1989, pp. 30-55]). There is no obvious evidence bearing on this question in the sources or for the distinction between logical and natural modalities which was introduced in medieval times.

One Aristotelian modal paradigm could be called the 'statistical' or 'frequency' model of modality according to which what is necessary is always or in all instances actual, what is impossible is never actual or in no instance, and what is possible is actual at least sometimes or in some instances. Even though Aristotle did not define modal terms with the help of temporal or extensional notions, this model can be found in his discussion of eternal beings, the natures of things, and the types of events or statements about such things. (See, for example, *Metaphysics* IX.10, 1051b10-17.) This way of dealing with modalities was supported by his taking it for granted that modal terms refer to the one and only world of ours and classify the types of things and events on the basis of their occurrence.[2] Because of the one world frame, the idea of modality as alternativeness was largely absent in Aristotle and in ancient modal thought in general. This has been particularly argued by Jaakko Hintikka in studies inspired by the difference between ancient modalities and the assumptions of possible worlds semantics in which the alternativeness relation is constitutive of the meaning of modal terms [Hintikka, 1957b].[3]

There are lots of examples of the application of the frequency paradigm in ancient and medieval modal thought. According to the temporal version found in Boethius's commentaries on Aristotle's *De interpretatione*, what always is, is by necessity, and what never is, is impossible. Possibility is interpreted as expressing what is at least sometimes actual. Correspondingly, a property which belongs to all members of a group is necessary with respect to that group. An impossible property does not belong to any members of a group and a possible property belongs to at least one member.[4] Those medieval thinkers who followed Boethius often restricted these definitions to the general features of nature because they believed that divine omnipotence required a different analysis. One influential application of the frequency model was the late ancient doctrine of the threefold matter of statements, according to which temporally unrestricted universal affirmative and negative statements are false and particular affirmative and negative statements are true in contingent matter; i.e., when the statements pertain to contingent states of affairs. Aquinas made use of this classification, which was known

[2]In *Posterior Analytics* I.6 Aristotle writes that certain predicates belong to their subjects at all times without belonging to them necessarily. Since he denies this elsewhere, he may have in mind a distinction between essential *per se* necessities (*Posterior Analytics* I.4) and weaker necessities of non-essential invariances, such as inseparable accidents. The degrees of necessity were widely discussed in later ancient philosophy.

[3]In the same volume he published an article which included an early formulation of the principles of possible worlds semantics; see [Hintikka, 1957a]. See also [Hintikka, 1973].

[4]Boethius, *Commentarii in librum Aristotelis Perihermeneias* I-II, ed. C. Meiser (Leipzig: Teubner, 1877-80), I, 120.24-121.16; 124.30-125.14; 200.20-201.3; II, 237.1-5.

in medieval times through Boethius.[5]

The same frequency view that was applied in the discussion of the square of opposition for universal and particular assertoric statements was employed in the description of modal propositions in the ancient commentaries on Aristotle's *Prior Analytics* by Alexander Aphrodisias and John Philoponus. Alexander describes the necessity and impossibility of syllogistic premises as follows:

> Of those terms which hold of something, some hold always while others sometimes hold and sometimes do not hold. If what is said to hold holds always and is taken to hold always, the proposition saying this is a necessary true affirmative; but a necessary negative true proposition is one which takes what by nature never holds of something as never holding of it.

He goes on that if X holds of Y now but not always, 'X holds of Y' is an actual true affirmative, and similarly, if X does not hold of Y now, 'X does not hold of Y' is an actual true negative; if X does not hold of Y but can hold of Y, the proposition indicating this is a true contingent affirmative, and if X holds or does not hold of Y now but can both hold and not hold, the proposition which says that it is contingent that it does not hold is a true contingent negative.[6] The contingency is also said to cover what holds for the most part, in equal parts, or for the lesser part.[7]

The temporal frequency paradigm was not quite unproblematic for Aristotle because he criticized various kinds of determinism, and the idea of actualization as a general criterion of possibility was not innocent in this respect (*De interpretatione 9, Metaphysics* IX, 3). Another Aristotelian modal conception was that of potency which seemingly was more suitable for speaking about unrealized possibilities without deterministic implications. In *Metaphysics* V.12 and IX.1, Aristotle characterizes potency as the principle of motion or change either as the activator or the receptor of a relevant influence, these two aspects forming one full potency (1046a19-26).[8] The frequency model is associated with this paradigm through

[5]Thomas Aquinas, *In Aristotelis libros Peri Hermeneias et Posteriorum analyticorum expositio*, ed. R. Spiazzi (Turin: Marietti, 1964), I, lect. 13, n. 168; see also *In duodecim libros Metaphysicorum Aristotelis expositio*, ed. M.-R. Cathala and R. Spiazzi (Turin: Marietti, 1977), IX, lect. 11, n. 1900. For the doctrine of the modal matter of propositions in Ammonius and Boethius, see [Knuuttila, 2008, pp. 508–509].

[6]Alexander of Aphrodisias, *In Aristotelis Analyticorum priorum librum I commentarium*, ed. M. Wallies, Commentaria in Aristotelem Graeca 2.1 (Berlin, Reimer, 1883), 26, 3-14, translated in [Barnes *et al.*, 1991, p.79]. For a discussion of this passage, see [Sorabji, 2004, pp. 275–276]. For temporal interpretation of modality in Alexander, see also [Mueller, 1999].

[7]Alexander, *In An. pr.* 39.19-40.5; translated in [Barnes *et al.*, 1991, pp.97–98]. For temporal views of modality in other late ancient commentators, see Ammonius, *In Aristotelis De interpretatone commentaries*, ed. A. Busse, Commentaria in Aristotelem Graeca, 4.5 (Berlin, Reimer, 1897), 88.7 19, 153, 13-25, 215, 11-14; Philoponus, *In Aristotelis Analytica priora commentaria*, ed. M. Wallies, Commentaria in Aristotelem Graeca 13.2 (Berlin, Reimer, 1905), 42.4-9. Philoponus says here that contrary particular statements are true in contingent manner; cf. Ammonius, op. cit., 91.30–32, and note 5 above.

[8]For agent and patient in Aristotle's natural philosophy in general, see [Waterlow, 1982a].

Aristotle's principle that the types of potency-based possibilities belonging to a species are recognized as possibilities because of their actualization — no natural potency type remains eternally frustrated because there is nothing in vain in nature.[9] Aristotle also says that when the agent and the patient come together as being capable, the one must act and the other must be acted on (*Metaphysics* IX.5, 1048a5-7). These specifications do not sound very promising with respect to conceptual indeterminism. The model of possibility as potency allowed Aristotle to speak about all kinds of unrealized singular possibilities by referring separately to passive or active potencies, as they were later called, but these are partial possibilities which do not guarantee that their actualization can take place. When the further requirements are added, such as the contact between the active and passive factor and the absence of an external hindrance, the potency is immediately actualized. It seems that it *can* be actualized in the full sense only when it is in fact actualized (*Met.* IX.5, *Phys.* VIII.1).[10]

The potency model was widely used in the natural philosophy of medieval Aristotelianism. In early medieval writers, the common course of nature was said to result from natural potencies with respect to which things were called possible according to inferior causes. Miraculous possibilities deviating from these were possible through the superior supranatural power. The types of natural potency based possibilities were later systematized using Averroes's frequential classification of the causes into necessary ones which always produce their effect when they act as causes, and contingent ones, the efficiency of which may be prevented to various degrees. A particular effect is necessary with respect to its cause but it can be regarded as contingent if its cause is contingent and does not always produce the effect in similar cases. Many natural causes bring about their effect in most cases, being contingent *ut in pluribus*, or they are not bound to a defined end, being causes *ad utrumlibet*, as for example the power of free will. Following Boethius, it was usually believed that natural causal chains are sometimes initiated by chance events which have no proper cause [Knuuttila, 2008, pp. 514–15].

Aristotle's difficulties with unrealized singular possibilities are seen in *De Caelo* I.12, where he asks whether a thing can have contrary potencies one of which is continuously actualized. He argues that the continuously non-actualized potency cannot be real: since its opposite is always actual, one cannot assume it to be realized at any time without contradiction. Aristotle applies the model of possibility

[9]Aristotle often refers to the principle that nature does nothing in vain. In *De caelo* I.4, 271a28-33, he argues that a heavenly vault the potential movement of which is never actualised is in vain and therefore non-existent. This was later applied to the types of natural potencies in general; see, for example, Boethius, *In Periherm.* II, 236.11-18; see also 243.13-15.

[10]Aristotle sometimes refers to second-order potencies as preceding first-order potencies; for example, that which is cold may be potentially burning because when it has become fire, it necessarily burns, unless something prevents this (*Phys.* VIII.4, 255b4-7). The necessity of the movements may be compulsory or in accordance with a thing's natural tendency; see *Posterior Analytics* II.11 (94b36-95a2): "by necessity a stone is borne upwards and downwards, but not by the same necessity". In *Parts of Animals* he refers to these modes of necessity as "those which are set forth in our philosophical treatises", adding that there is also hypothetical necessity of what is required for an end (I.1, 642a3-13).

as non-contradictoriness here, which he defines in *Prior Analytics* I.13 as follows: when a possibility is assumed to be realized, it results in nothing impossible. In speaking about the assumed non-contradictory actualization of a possibility, Aristotle thinks that it is realized in our one and only history. Therefore the argument in *De caelo* excludes from genuine possibilities those which remain eternally unrealized. (See also *Met.* IX.4.) John of Jandun, an early fourteenth-century Aristotelian with Averroist sympathies, comments on the argument in *De caelo* as follows:

> Such a thing cannot be destroyed from the assumed extinction of which something impossible follows since nothing impossible follows from assuming that a possibility is actual, although something false may follow from it. But if an omnitemporal thing were destroyed, an impossibility follows, namely that the same thing simultaneously is and is not It follows from these that whatever can be destroyed will be destroyed by necessity, and if this is a demonstrative argument, we can deduce by it that what can be generated will be generated by necessity.

He continues that the principle that what can be generated will be generated and what can be destroyed will be destroyed applies to all sorts of entities — simple and composite substances and real or intentional accidents.[11]

In discussing future contingent statements in Chapter 9 of *De interpretatione*, Aristotle states that while things necessarily are when they are, it does not follow that what is actual is necessary without qualification (19a23-27). The necessity of the present follows from the absence of the idea of simultaneous alternatives. One example of how this influential thesis was understood in medieval Aristotelianism is found in thirteenth-century treatises on obligations logic for dialectical disputations in which various statements are put forward by an opponent and evaluated by a respondent.[12] In the most usual case, a false and contingent proposition was first put forward and accepted. The rules of consistency defined how other proposed propositions were to be evaluated by answering 'I grant it', 'I deny it' or 'I don't know'. While irrelevant propositions were evaluated as such, relevant propositions were treated in terms of their logical relations with respect to the acceptance of the original position and other previous answers. Since the initial proposition was false but not impossible, thirteenth-century versions of this logic included a rule about the time of its possible truth:

1. When a contingently false statement referring to a present instant is posited, one must deny that it is [now].[13]

[11]John of Jandun, *In libros Aristotelis De caelo et mundo quae extant quaestiones*, I, q. 34 (Venice, 1552), 21vb.

[12]For medieval obligations logic, see [Yrjönsuuri, 1994; Yrjönsuuri, 2001; Keffer, 2001; Dutilh Novaes, 2007].

[13]Anonymous, *De obligationibus*, in Romuald Green in *The Logical Treatise 'De obligationibus': An Introduction with Critical Texts of William of Sherwood (?) and Walter Burley*, Ph. D. diss., University of Louvain, 1963, 8.32-33.

Like Aristotle and other ancient logicians, the anonymous author treats 'You are in Rome' as a statement which is true or false depending on how things are at various moments at which it is uttered.[14] He assumes that when such a statement is true/false, it is necessarily true/false at that time because of the impossibility of change at an instant of time. This is meant to be a proof for (1).[15]

Medieval discussions of *De interpretatione* 9 were concentrated on the question of what Aristotle might have meant by his remark that the necessity of the present does not represent simple necessity. One interpretation derived from Boethius's explanation that the temporal necessity of a present event does not imply that similar events necessarily take place in similar circumstances. This is a quasi-statistical attempt to avoid the problem that changeability as a criterion of contingency makes temporally definite singular events necessary. Another traditional interpretation which is also found in Boethius is that the necessity of an event at a certain time does not imply that it would have been antecedently necessary. Aristotle discusses such singular diachronic modalities in some places (*Met.* VI.3; *EN* III.5, 1114a17-21; *De int.* 19a13-17) in which he seems to assume that the conditions which at t_1 are sufficient for the possibility that p obtains at a later time t_2 may be changed before t_2 so that the possibility no longer exists. Aristotle did not elaborate on these ideas, which were developed further in later ancient discussions.

Diodorus Cronus defined modal notions as follows: what is possible is that which is or will be true; what is impossible is that which is false and will not be true; necessity is that which is true and will not be false; non-necessary is or will be false.[16] He treated these modalities as predicable of temporally indefinite token reflexive propositions. A proposition 'p (now)' is possible if it is true now or later and a non-necessary proposition is false now or in the future. Since the disjunctions in the above definition are not exclusive, possibility does not exclude necessity (always true from now on) and non-necessity does not exclude impossibility (always false from now on). Possible propositions may become impossible and non-necessary may become necessary. Those which are neither necessary nor impossible have a changing truth-value. It seems to follow that all true past tense propositions are necessary; in fact this is included in Diodorus' argument for logical determinism, the so-called Master argument, which is described by Epictetus (*Dissertationes* II.19.1). Its conclusion is: 'Nothing is possible that neither is nor will be true'. According to Epictetus, the first premise runs: 'Everything past and true is necessary'. The second premise is: 'The impossible does not follow from the possible'. One way of interpreting this argument is as follows. If the second premise is taken to mean that an impossible proposition does not follow from a possible proposition, let us consider a possible proposition which is in agreement with the thesis that everything does not take place by necessity, i.e., a singular present tense proposition p which neither is nor will be true and which is not

[14]See, for example, Aristotle, *Categories* 5, 4a23-26; and [Bobzien, 1998, pp. 66, 109].

[15]Anonymous, *De obligationibus*, 8.33-9.8.

[16]Boethius, *In Periherm.* II, 234.22-26; see [Bobzien, 1998, pp. 102–105].

impossible. Since p is false and it has been true that p is going to be false, the statement 'It has been true that p is going to be false' is necessarily true and 'It has been true that p is going to be true' is necessarily false. This follows from the premise that p neither is nor will be true, which consequently is not possible [Bobzien, 1999a, pp. 90-92].

Very little is known of Philo, who deviated from Diodorus by redefining a possible proposition as 'that which is capable of truth according to the proposition's own nature', necessary as that which in itself is never capable of falsity, non-necessary as that which in itself is capable of falsity, and impossible as that which is never capable of truth according to its own nature.[17] Philo seems to regard a temporally indefinite proposition as possible if it is capable of truth at some time. According to him, 'This piece of wood burns' is possible even if the wood will never burn, which suggests that he had in mind some sort of generic conceptual possibilities.[18] Singular propositions which are always false may be possible in the sense in which possibilities are abstracted from their external conditions. This is how the Stoics understood Philo's notions of possibility, distancing themselves from Diodorus' position by using Philo's definition of possibility with the addition 'and not hindered from being true by external circumstances'.[19]

In this discussion, temporally indefinite propositions about the present are said to be possible, necessary, or impossible now (n), depending on whether from now on they are always or sometimes true or false (T, F) or are capable (CAP) of truth or falsity with or without an external hindrance (H). These alternative conceptions of possibility can be characterized by as follows:[20]

Diodorus:	Philo:	Chrysippus:
$M_n p = \exists t(Tp_t)$	$M_n p = \exists t(\text{CAP}Tp_t)$	$M_n p = \exists t(\text{CAP}Tp_t \ \& \ \neg HTp_t)$
$N_n p = t(Tp_t)$	$N_n p = \neg\exists(\text{CAP}Fp_t)$	$N_n p = \neg\exists(\text{CAP}Fp_t) \vee t(HFp_t)$

The ancient discussion of these formulations was mainly associated with disagreements about determinism and freedom. While Diodorus argued that the meaning of modal notions implied that nothing which is not actualized was possible, the Stoics tried to defend their determinist philosophy against the charges of fatalism by redefining modal concepts, particularly by accepting that there are possible propositions which are never true. It remains somewhat unclear what they meant by 'capable of truth' and 'external hindrance' which apparently are associated with their temporalized conception of truth. The critics argued that the Stoics inconsistently maintained that there are individual possibilities which are never realized and that fate as cosmic reason necessitates everything which takes place.[21]

[17] Boethius, *In Periherm.* II, 234.10-21.

[18] See Alexander of Aphrodisias, In *An. pr.* 184.6-12, Philoponus, In *An. pr.* 169.19-20.

[19] Diogenes Laertius, VII.75; [Bobzien, 1998, p. 112].

[20] This is a somewhat modified version of the formulations in [Bobzien, 1998, pp. 103, 110, 115].

[21] [Bobzien, 1998, pp. 122–131]. In criticizing the second premise of the Master Argument,

Influenced by the Stoic conception of diachronic modalities, Alexander of Aphrodisias argued for what he considered as Aristotle's view that there are undetermined prospective alternatives which remain open options until the moment of time to which they refer. Following this tradition, Boethius sketches diachronic prospective possibilities. A temporally determinate prospective possibility may not be realized at the time to which it refers, in which case it ceases to be a possibility. Boethius did not develop the idea of simultaneous possibilities which would remain intact even when diachronic possibilities had vanished, insisting that only what is actual at a particular time is possible at that time [Knuuttila, 2008, pp.516–17].

Apart from diachronic modalities, Aristotle's indirect proof with a modal premise incited discussion among ancient thinkers. According to Aristotle, 'the impossible only follows from the impossible' (*De caelo* I.12, 281b15-16), which was applied in the Master Argument of Diodorus as well. In his *Physics*, Aristotle often uses this principle in demonstrating that various competitive positions are not acceptable. In these contexts, he added auxiliary premises which he himself considered impossible elsewhere. This was found strange and led to considerations of the nature of the auxiliary impossible premises which should not be responsible for the impossible conclusion. In addition to indirect proofs, Aristotle also employed impossible premises in various constructive arguments; for example, in considering things which cannot be separated but are separated 'in thought' (*Metaphysics* VII.3). (See [Knuuttila and Kukkonen, 2011]). These ideas belong to the background of what was called the Eudemian procedure, in which one assumes something impossible 'in order to see what follows'. Counterfactual hypotheses of this kind were not regarded as formulations of possibilities in the sense of what could be actual and were called impossible hypotheses by Philoponus and Boethius [Martin, 1999]. These ideas were applied in the discussion of Aristotle's indirect proof with impossible premises by Averroes and Thomas Aquinas. Averroes writes that when an accidentally impossible proposition is used in an argument, it is supposed to be true 'in so far as it is possible, not in so far as it is impossible'. For example, when Aristotle employs the impossible premise that there is a body larger than the heavens, this is possible as such, with reference to a body qua body, but accidentally impossible with reference to the universe [Knuuttila and Kukkonen, 2011, pp. 87–94]. A combination is accidentally impossible if incompatible elements thought abstractly as such, without their particular conditions, would not exclude each others. Aquinas explains the abstract possibilities as follows:

> For example, in speaking of animals I can state that it is contingent that every animal is winged, but if I descend to the consideration of

---

Chrysippus argued that 'Dio is dead' is possible, but it follows from this that this man is dead, which is impossible. While 'Dio is dead' is possible because it will be truly said in the future, 'This man is dead' is impossible because what is referred to is not a man. See [Bobzien, 1999b, pp. 116–117]. Alexander of Aphrodisias remarked that the Stoics cannot argue in this way because they believe in the cyclical return of history and 'This man has died' is true with respect to previous cycles (*In An. pr.* 180.28–181.19); [Sorabji, 2004, pp. 280–282].

> human beings, it would be impossible for this animal to be winged. Now Aristotle is speaking here about movers and mobile objects in general ... Therefore, he states that it is contingent that all moving things are continuous with each other. This, however, is impossible if moving things are considered according to their determinate natures.[22]

The abstract possibilities of Averroes and Aquinas were impossible in the sense that they could not be thought of as actualized. Both thought that if this was considered problematic, one could also formulate Aristotle's reduction arguments by using conditional propositions. Averroes formulates Aristotle's indirect argument against self-movers as follows: 'If it is true that the body of the heavens stands, if one part of it stands, then it is moved by something else'. These conditionals are true, but the antecedent conditional is denied by the assumption that the heavenly body is in motion by itself.[23] Aquinas follows Averroes and uses conditionalizing as an answer to those who might find abstract possibilities problematic (*Phys.* VII, lect. 1, n. 889).

2 MODALITY AS ALTERNATIVENESS

While early medieval discussions of modal questions were strongly influenced by ancient paradigms which were known through Boethius's works, it was also thought that the Christian doctrine demanded some further considerations. The main source here was Augustine's doctrine of the creation by God's choice and omnipotence, which involved an intuitive idea of modality as alternativeness. According to Augustine, God freely chose the content of creation and the providential history of the world, which could have been otherwise. (See [Knuuttila, 2001]). Some medieval writers regarded God's possibilities as a special theological matter which did not affect the use of traditional ideas in other contexts; natural possibilities could be discussed in accordance with ancient extensional paradigms with the proviso that things were different with respect to the superior power of divine omnipotence. However, there were some twelfth-century thinkers who realized the philosophical significance of the new modal conception which was employed in dealing with the question of whether the possibility of events having happened otherwise than God had foreseen them implied the possibility of error in God. In his influential *Sententiae*, Peter Lombard presented his solution was based on the distinction between modal statements in compound and divided sense (*de dicto* vs. *de re*), with minor changes. Since Lombard's work was used as the basic theological textbook through the Middle Ages, it contributed to the popularity of this modal distinction analysed in a more detail in Abelard's logical treatises.[24]

[22]Thomas Aquinas, *In octo libros Physicorum Aristotelis expositio*, ed. P. Maggiòlo (Turin: Marietti, 1965), VII, lect. 2, n. 896.

[23]Averroes, *In Aristotelis de Physico audito libri octo*, Aristotelis opera cum Averrois commentariis (Venice, 1562-1574), vol. IV, f. 308ra.

[24]Peter Abelard, *Philosophische Schriften I. Die Logica 'Ingredientibus'*, ed. B. Geyer, Beiträge zur Geschichte der Philosophie und Theologie des Mittelalters, 21,1-3 (Münster: As-

Abelard thought that what was actual was temporally necessary as being no longer avoidable, but he argued that mutually exclusive alternatives are possible at the same time in the sense that one or another of them could have happened at that time.[25] As distinct from Boethius's diachronic modalities, Abelard also operates with simultaneous alternatives of the present. He seems to think that one might analyse the example 'A standing man can sit' as

2. $\exists x(\neg\varphi_t x \& \Diamond\varphi_t x)$.[26]

The possibility is not temporalized here as in Boethius's answers to the question of what Aristotle might have meant when he said that what is necessarily is when it is but is not therefore necessary *simpliciter*:

3. $\exists x(\neg\varphi_t x \& \Diamond_{\text{earlier than}t}\varphi_t x)$.[27]

Boethius assumed that the definite truth-values of future contingent propositions imply that all future things are necessary because the unchanging antecedent truth is no less necessary than the present truth. In order to avoid this, he held that future contingent propositions are merely true-or-false, not determinately true or false. Abelard argued that future contingent statements are either true or false although they are not determinately true or determinately false, by which he means that their truth or falsity does not imply that the things which make them true or false are determinate.[28] Twelfth-century interest in modal questions is shown by Gilbert of Poitiers's explanation of Plato's 'Platonitas', which is said to include all that Plato was, is and will be as well as what he could be.[29] The modal element of the individual concept was probably needed in order to speak about persons in alternative providential histories, for example, about the Son of God as having been begotten by another mother or the apostle Peter as saved or not saved.[30] A more elaborated analysis of the philosophical aspects of the Augustinian theological modalities was put forward by Robert Grosseteste in his *De libero arbitrio*

---

chendorff, 1919-27) 429.26-430.36; *Dialectica*, ed. L. M. de Rijk, Wijsgerige teksten en studies, 1 (Assen: van Gorcum, 1956), 217.27-219.24; Peter Lombard, *Sententiae in IV libris distinctae, I-II* (Grottaferrata: Collegium S. Bonaventurae ad Claras Aquas, 1971), I, d. 38, a. 2.

[25] *Logica 'Ingredientibus'*, 273.39–274.19; see [Martin, 2003, pp. 238–9].

[26] For simultaneous alternatives of the present in Abelard, see [Martin, 2001; Martin, 2003], [Pinzani, 2003, pp.189–192], [Knuuttila, 2008, pp.537–538]. See also Peter Abelard, *Super Periermenias XII-XIV*, ed. L. Minio-Paluello, in *Twelfth Century Logic: Texts and Studies II: Abaelardiana inedita* (Rome: Edizioni di Storia e Letteratura, 1958), 41.23-42.6.

[27] See Boethius, *In Periherm.* II, 245.4-246.19.

[28] See [Lewis, 1987]. Later medieval authors usually regarded future contingent statements as either true or false. While Abelard ascribed this view to Aristotle as well, the majority of later writers believed that Aristotle gave up bivalence with respect to these statements; see [Knuuttila, 2010a].

[29] Gilbert of Poitiers, *The Commentaries on Boethius*, ed. N.M. Häring (Toronto: Pontifical Institute of Mediaeval Studies, 1966), 177.77–88; 274.75–76; [Marenbon, 2007, pp. 158–159].

[30] See Anselm of Canterbury, *De conceptu virginali et de originali peccato*, ed. F.S. Schmitt, Opera omnia, 2 (Edinburgh: Nelson, 1946), 159.13-16; Peter of Poitiers, *Sententiae*, ed. P.S. Moore and M. Dulong, Publications in Mediaeval Studies, 7 (Notre Dame, Ind.: The University of Notre Dame Press, 1961), 14.350-353.

(c. 1230), distinguishing between necessities and possibilities 'from eternity and without beginning' and those pertaining to the actual history.[31] Twelfth-century authors who were later called *nominales* preferred to treat singular propositions as temporally definite and as having an unchanging truth-value, one of their slogans being 'what is once true is always true'. They argued that while tensed statements about definite singular events have a changing truth-value, the corresponding non-tensed statements are unchangingly true or false, without being necessarily true or false for this reason. This was in agreement with Abelard's view that future contingent statements are true or false without being necessarily true or necessarily false [Knuuttila, 2008, 522–523].

Late medieval discussions of modality were largely influenced by the works of Duns Scotus, who criticized the central assumptions of the ancient modal views. As for the time rule of obligations logic (1), he suggested that it may be omitted without changing other rules,[32] which was in agreement with his refutation of the necessity of the present as a logical or metaphysical principle. Scotus defines the notion of contingency as follows in explaining the contingency of an individual state of affairs:

> I do not call something contingent because it is not always or necessarily the case, but because the opposite of it could be actual at the very moment when it occurs.[33]

In this passage, Scotus first distances himself from the frequency division between necessity and contingency and then defines contingency by referring to simultaneous alternatives. As for the time rule and its proof in the anonymous obligations treatise quoted above, Scotus states that both are false since even if 'You are in Rome' is false now, it can be true now.[34] In defending the consistency of

4. $\neg p_t \ \& \ \Diamond p_t$,

Scotus remarks that when a present unactualized possibility is thought to be actual, nothing impossible should follow, except an incompossibility with what is actual.[35] Mutually exclusive possibilities refer to actuality in alternative scenarios.

The idea of simultaneous alternatives plays a central role in Scotus's theistic metaphysics. God's omniscience involves knowledge of all intelligible things, whether finite or infinite. One set of compossible possibilities forms God's providential plan of creation and will receive actual being. Although possibilities

---

[31] Robert Grosseteste, *De libero arbitrio*, ed. in N. Lewis, 'The First Recension of Robert Grosseteste's *De libero arbitrio*', *Mediaeval Studies* 53 (1991), 168.26–170.33, 178.28–29; [Lewis, 1997].

[32] *Lectura* I, d. 39, q. 1-5, n. 59, in *Opera omnia*, ed. Commissio Scotistica, vol. 17 (Vatican City: Typis polyglottis, 1960), 499.

[33] *Ordinatio* I, d. 2, p. 1, q. 1-2, n. 86 in *Opera omnia*, ed. Commissio Scotistica, vol. 2 (Vatican City: Typis polyglottis, 1950), 178.

[34] *Lectura* I, d. 39, q. 1-5, n. 56 (498).

[35] *Lectura* I, d. 39, q. 1-5, n. 72 (504).

necessarily are what they are, the actualizations of non-necessary possibilities are contingent. Since all finite things are contingently actual when they are actual and most of their properties are contingent, the actual state of affairs in the world is associated with a huge number of alternative possibilities with respect to the same time. Impossibilities are incompossibilities between possible components, such as Socrates's sitting at a certain time and Socrates's not sitting at that same time. Extensive discussion was provoked by Scotus's idea that modal propositions would be true or false even when nothing else existed apart from an intellect which forms these propositions. God is not needed for modality but for actualization in Scotus's metaphysics.[36]

Thirteenth-century authors began to describe divine omnipotence as being determined by absolute possibilities which are expressed by statements which are not contradictory.[37] Scotus applied the notion of 'logical possibility' (*possibile logicum*) to these absolute possibilities; he was first to use the term *possibile logicum*. The notion of logical possibility is conceptually prior to that of metaphysical possibility (*potentia metaphysica*) which is associated with the power of actualizing logical possibilities. Anything which can be posed as actual without incoherence is logically possible; logical impossibilities are inconsistent combinations of logically possible constituents [Honnefelder, 1991, pp. 45–74].

Scotus's modal theory brought together many of the semantic insights of modality as alternativeness in his predecessors and had a strong impact on interpreting modality in late medieval philosophy and logic. After Scotus's revision of the time rule of obligations logic (1), it was seldom mentioned [Yrjönsuuri, 1994, p. 74]. In his approach, the answers of standard obligations games could be interpreted as descriptions of possible states of affairs which differed from the actual ones because of the counterfactual original position and its implications. The semantic application of the Scotist obligations logic was employed in various fourteenth-century theological discussions [Gelber, 2004]. A different approach was developed by Richard Kilvington, who argued that one should answer in a way which would be reasonable if the counterfactual position were true and things would otherwise differ from the actual world as little as possible. For example, if the false original position is 'You are in Rome', and the opponent proposes: "You are in Rome' and 'You are a bishop' are alike', one should deny this if one is not a bishop [Kretzmann and Kretzmann, 1990a], [Kretzmann and Kretzmann, 1990b, 47q]. The propositions are alike, both being false, but conceding this would commit one to concede 'You are a bishop' as alike to what one has accepted. Any false contingent proposition could be proved in the same way. Kilvington's approach shows some similarities to the theories of subjunctive conditionals which are based on possible worlds semantics.[38] Another innovation was Richard Swyneshed's revision according to

[36]For Scotus's modal theory, see [Honnefelder, 1991; Knuuttila, 1996; Normore, 2003; Hoffmann, 2009].

[37]See Aquinas, *Summa theologiae*, ed. P. Caramello (Turin: Marietti, 1948-50), I.25.3.

[38]See [Spade, 1982]. Historians have been more skeptical on Spade's attempt to interpret obligations logic in general as a theory of counterfactual conditionals; see, for example, [Yrjönsuuri, 1994].

which relevant and irrelevant statements should be evaluated independently. This theory of two column book-keeping could be understood as an attempt to deal with an actual and possible domain simultaneously [Yrjönsuuri, 1994, pp.89–101]. In criticizing the abstract possibilities of Aquinas and Averroes, Buridan also uses obligations terminology, assuming that investigating possibilities is to think about them as actualized in a coherent context of compossibilities. Since the abstract possibilities in Averroes and Aquinas cannot be treated in this way, calling them possibilities is based on a conceptual confusion. (See [Knuuttila and Kukkonen, 2011, 96–99].)

Ancient and Medieval Modal Logic and Modal Syllogistic

In *Prior Analytics* I.15, 34a5-7, 22-4 Aristotle puts forward the rules of modal inference

5. $(p \rightarrow q) \rightarrow (\Box p \rightarrow \Box q)$

and

6. $(p \rightarrow q) \rightarrow (\Diamond p \rightarrow \Diamond q)$

He did not separately comment on these principles of propositional modal logic, which he employs in many places. These were commonly known and accepted in ancient and medieval logic, one exception being Chrysippus's argument against the Master Argument of the Diodorus Chronus. Other principles of propositional modal logic which were found in Aristotle and were assumed in ancient and medieval logic include equivalent combinations of various modal terms and negations as follows:

7. $\Box p \leftrightarrow \neg\Diamond\neg p$
8. $\neg\Diamond p \leftrightarrow \Box\neg p$
9. $\Diamond p \leftrightarrow \neg\Box\neg p$
10. $Qp \leftrightarrow \Diamond p \;\&\; \Diamond\neg p$
11. $Qp \leftrightarrow Q\neg p$.[39]

In *De interpretatione* 13, Aristotle mentions the principles

12. $\Box p \rightarrow p$
13. $p \rightarrow \Diamond p$

and

[39] For Aristotle discussions of the interrelations between modalities, see *De int.* 12–13, *An. pr.* 13, *Met.* V, 5, 1014a34–35; 12, 1019b20–29.

14. $\Box p \rightarrow \Diamond p$

($\Box$ stands for necessity, $\Diamond$ for possibility and Q for contingency).

The main part of Aristotle's modal logic is his theory of modal syllogistics. In Chapters 8–22 of the *Prior Analytics*, he discusses the syllogistic moods in which the premises are modalized by necessity or contingency (neither necessary nor impossible) and in which one is assertoric ($-$). The combinations are as follows: $\Box\Box, \Box-, -\Box, QQ, Q-, -Q\Box$ and $\Box Q$. In some cases Aristotle mentions that the conclusion is possible ($\Diamond$), i.e., not impossible. Modal syllogistics is organized after the model of assertoric syllogistic. The following first-figure combinations yield immediately obvious perfect moods: $\Box\Box\Box, \Box-\Box, -\Box-, Q-Q, Q\Box Q$. First-figure QQQ moods are also perfect when the first premise is understood as ampliated: A contingently belongs to all things to which B contingently belongs (I.13, 32b25-32). The remaining first-figure mixed moods with contingent minor premises Aristotle treated as imperfect. These as well as second- and third-figure moods are proved by reducing them to the perfect first-figure moods by conversion, by a *reductio ad impossibile* or by *ecthesis*. The first procedure takes place by converting the terms, so that the discussion of these conversions plays a central role in Aristotelian modal syllogistic. According to Aristotle, necessity premises are converted in the same way as the corresponding assertoric premises: 'Every A is B' implies 'Some B is A', 'Some A is B' is equivalent to 'Some B is A' as well as 'No A is B' to 'No B is A'. Negative contingency premises are converted into affirmative contingency premises of the same quantity (universal, particular) and these by the conversion of terms into particular contingency propositions (*An. pr.* I.3, 13). One interpretational problem in Aristotle's theory is that while these conversion rules are not problematic for modal premises *de dicto*, in which the modal term qualifies the whole sentence, reading the premises of modal moods in this way leads to problems. Many commentators think that modal terms are treated as modifiers of predication (*de re*) as is shown, for example, by Aristotle's acceptance of the first-figure $\Box - \Box$ moods as valid and refuting the corresponding $-\Box\Box$ moods. This interpretation is problematic with respect to the conversion because actuality is then upgraded to necessity and contingency to actuality. Aristotle did not systematically analyse the fine structure of modal premises, apparently having various intuitions about how modal premises should be understood in various contexts. He says the subject of 'An A is contingently B' may be 'an A' or 'what is contingently A' but does not tell when one should take the subject as modalized. As for the assertoric premises in the mixed moods, he says that they must be taken without temporal limitations — how this should be understood remains unclear.[40]

There are several recent works on Aristotle's modal syllogistics, but no generally accepted historical construction which would make it coherent. Some commentators think that there are various background assumptions in Aristotle's theory

[40]See *An. pr.* I.13, 32b24-37; I.15, 34b7-18; Aristotle, *Prior Analytics*, trans. with introduction and notes by R. Smith (Indianapolis: Hackett, 1989), 128, 132-3.

which are partially incompatible.[41] Others have been interested in finding coherent layers of the theory by explicating them in terms of Aristotle's other philosophical works. These discussions have avoided the use of contemporary predicate logic or modal logic as potentially anachronistic interpretative tools in historical reconstruction [van Rijen, 1989; Patterson, 1995]. In addition, some logicians have constructed formal semantics for Aristotle's work, which is taken to contain a formal deductive system. These technical approaches continue McCall's non-interpreted axiomatization of Aristotle's syllogistic [McCall, 1963]. Some authors have employed the interpretative frame of modern modal predicate logic and possible worlds semantics,[42] while others have preferred set theoretical constructions which are taken to be closer to Aristotle's one world metaphysics.[43] Historians who are interested in what Aristotle probably meant have not found these approaches helpful for this purpose, whatever their eventual philosophical interest might be. (See, for example, [Striker, 2009, p. xvi]).

The commentaries on *Prior Analytics* by Alexander of Aphrodisias and John Philoponus involve references to ancient discussions, such as Theophrastus's criticism of $\Box - \Box$ moods and the attempts to define modal notions in various philosophical schools.[44] Both authors distinguish between simple and qualified necessity statements and maintain that syllogistic necessity premises are of the first type, holding invariantly between the terms which are never empty. The main reason for this restriction were various counter-examples to the conversion rules for necessity premises, such as 'All literate beings are humans' or 'All walking beings are moving', which are necessary but not convertible, and related reasons for explaining that the conclusions of mixed necessity and assertoric syllogisms are not necessary *simpliciter*. Syllogistic necessity premises were regarded necessary *per se* because of the relationship between the terms, as explained in Aristotle's *Posterior Analytics* I.4. While simple necessity premises themselves have no temporal limitations, their essential nature could be taken to make the conclusion of the mixed necessity and assertoric syllogisms necessary 'as long as the subjects are actual'. This is less than simple necessity but more than the weaker temporal necessity which is an

[41] The thesis of the incompatible assumptions behind the modal moods and conversion rules was put forward in [Becker, 1933]. For this and other problems, see also [Hintikka, 1973; Smith, 1989; Striker, 2009] . While modern authors often refer to Becker when dealing with Aristotle's shortcomings with respect to the analysis of the structure of modal propositions, the same observations were also made by fourteenth-century interpreters.

[42] [Nortmann, 1996; Schmidt, 1989; Ebert and Nortmann, 2007]. Aristotle's modal syllogistic is interpreted using standard predicate logic and a distinction between necessary and contingent terms in [Rini, 2010].

[43] [Johnson, 1989; Johnson, 1995; Thomason, 1993]. A set theoretical formal interpretation on the basis of the works by Johnson and Thomason is [Thom, 1996]. While operating with formal set theoretical tools, Thom tries to develop a formal semantics for the deductive system which he thinks is contained in Aristotle's texts and to evaluate how this semantics can be given a metaphysical turn. [Malink, 2006] has developed a related formal mereological semantics.

[44] Theophrastus argued that the conclusion of a modal syllogism should not be stronger than the 'weaker' premise, the actual being weaker than the necessary and the contingent weaker than the actual. For Theophrastus's modal logic, see the texts in [Fortenbaugh, 1992, pp.98–109], [Huby, 2007], and [Flannery, 1995, 53–108].

extension of the necessity of the present and applied to any proposition as long as it is true.[45]

There are other examples of the temporal frequency view of modal terms in these works which had some influence on medieval Arabic logic. Avicenna wrote an Arabic summary of Aristotle's modal syllogistic, but he also developed an independent theory of his own in which necessity is treated as invariant actuality and possibility as sometime actuality [Lagerlund, 2009]. While Averroes often equates modal and temporal terms as well, he thinks that 'necessity' in Aristotelian syllogistic necessity premises cannot simply refer to invariant combinations or divisions of things. Such propositions are necessary, to be sure, but they may be accidentally necessary and not necessary *per se* as the syllogistic necessity premises should be, both terms being essential. Since Averroes takes syllogistic modal premises to be of the divided type, he states that assertoric premises in Aristotelian mixed necessity-assertoric syllogisms must have a predicate term which is essential. The same applies to the subject term of the first premise in mixed assertoric-necessity syllogisms. In the former case, the minor premise and the conclusion are accidentally necessary. This is meant to explain Aristotle's asymmetrical treatment of mixed necessity-assertoric syllogisms and mixed assertoric-necessity syllogisms.[46]

The first known Latin commentary on *Prior Analytics* is an anonymous late twelfth-century treatise which involves detailed discussions of modal conversion and modal syllogisms as well as many problems dealt with in ancient commentaries. Some questions of Aristotle's modal syllogistics were studied in early thirteenth-century logical treatises before Robert Kilwardby's commentary (c. 1240) which became an authoritative thirteenth-century work on the *Prior Analytics*.[47] Like Averroes, Kilwardby understands modal syllogistic as pertaining to modal premises in a divided sense, arguing that syllogistic necessary premises are convertible because the terms are essential and they are essentially or *per se* connected or divided, as distinct from merely invariant *per accidens* necessities. In dealing with contingency premises (neither necessary nor impossible), Kilwardby states that a negative one implies a corresponding affirmative one. While indefinite (*utrumlibet*) contingency premises are converted into premises of the same type of contingency, the conversion of natural contingency premises (possible in most cases) results in statements of possibility proper (not impossible). Referring to Aristotle's remark in *An. pr.* I.13, 32b23-32, Kilwardby states that the subject term in a contingency premise may be 'ampliated' as follows: 'What-

[45] Alexander of Aphrodisias, *In An. pr.* 26.3–14; 36.25-32; 130.1–15; 155.20–25; 201.21-24; Philoponus, *In An. pr.* 43.8-18; 119.13–18; 126.7-29; 146.22; Flannery 1995, pp. 62–5, 99–106. For a related distinction between simple necessity and temporal or conditional necessity in Boethius, see *In Periherm.* I, 121.20-122.5; II, 241.1-242.15.

[46][Thom, 2003, pp. 81–85]. Averroes's considerations were influenced by the discussion of various necessity propositions in ancient commentators; see note 45 above.

[47][Knuuttila, 2008, pp.544–545]. Kilwardby's *In libros Priorum Analyticorum expositio* was printed in Venice in 1516 under the name of Giles of Rome (reprint Frankfurt am Main: Minerva, 1968).

ever is contingently A is contingently B'. There were extensive discussions of the kinds of contingency based on various philosophical ideas in the *Prior Analytics* commentaries by Kilwardby and Albert the Great and in other treatises by their contemporaries. Albert's commentary is largely dependent on Kilwardby [Thom, 2007, pp. 18-40]; [Lagerlund, 2000, pp.23–52].

According to Kilwardby, the modal status of the conclusion of the perfect first-figure syllogism follows that of the first premise, which involves the whole syllogism in accordance with the *dici de omni et nullo* principle. In mixed first-figure syllogisms with a major necessity premise and a minor assertoric premise, the non-modalized premise is *simpliciter* assertoric, by which Kilwardby means that it is necessarily true *per se*. In the corresponding syllogism with a contingent major and assertoric minor premise, the assertoric premise is also *simpliciter* assertoric, but this time the predicate is said to belong to the subject *per se*, invariably or by natural contingency. In the first case, a first-figure major premise 'appropriates' to itself a minor which is necessary *per se* and in a latter case a minor premise which is true in most cases, if not necessary. He develops a great number of similar appropriation rules for all modal figures [Thom, 2007, pp. 160–1, 165–6, 172–4, 219–20].

Kilwardby and his followers believed that understanding Aristotle's modal syllogistic demanded considerable metaphysical considerations, such as restricting the necessity premises to those involving essential terms, distinguishing between various kinds of assertoric and contingency premises, and identifying the appropriation relations between syllogistic premises. The theory of appropriation is a particularly clear example of the close link between logical and metaphysical ideas in this approach. Some of the metaphysical assumptions were eliminated in Richard Campsall's commentary on the *Prior Analytics* from the early fourteenth-century, in which the syllogisms with modalities *de dicto* and *de re* modals were discussed separately. This became usual in late medieval logic. Campsall argued that *de re* necessity with respect to actual things equates to unchanging predication, and contingency to changing predication.[48]

William Ockham, John Buridan and many other fourteenth-century logicians largely dropped the thirteenth-century essentialist assumptions from their modal logic, the basic concept of which was what Scotus called logical possibility, i.e.,

[48]Richard Campsall, *Questiones super librum Priorum Analeticorum*, ed. Edward A. Synan in *The Works of Richard of Campsall*, vol. 1 (Toronto: Pontifical Institute of Mediaeval Studies, 1968) 5.38; 5.43-45; 6.25; 9.19; 12.31. Campsall takes the possibility proper (not impossible) as a basic notion. An affirmative *de re* possibility statement with terms standing for actual things implies the corresponding assertoric statement (5.40) and a negative *de re* possibility statement about the present implies the corresponding *de re* necessity statement (5.50). These formulations are meant to be in agreement with the definition of a *de re* contingency statement as the conjunction of an affirmative and a corresponding negative possibility statement (7.34-36). Things cannot be otherwise in the present because all true present tense negative statements are necessarily true. Campsall seems to think that while contingent actual combinations may be changed in the future, being non-necessary in this sense, there is no similar qualification of the necessity of non-existent combinations. For various interpretations, see [Lagerlund, 2000, pp. 87–90], [Thom, 2007, pp. 117–123], and [Knuuttila, 2008, pp. 542–544].

something which could be actual without a contradiction. Questions of modal logic were discussed separately with respect to modal propositions in a compound sense (*de dicto*), a modal term qualifying the content (*dictum*) of an assertion, and a divided sense (*de re*), a modal term qualifying the copula. The syllogistic of *de dicto* premises was simple because necessary propositions were compossible with all possible premises. There were no valid syllogisms with two possibility or contingency premises or their mixtures with actual premises without an extra assumption of compossibility. Modal propositions *de re* were further divided into two groups depending on whether the subject terms referred to actual or possible beings. Ockham's approach differed from Buridan in restricting necessity statements to those having an actual subject, which made his theory less systematic. It was thought logicians should analyse the relationships between these readings and, furthermore, the consequences with various types of modal propositions as their parts. Aristotle's modal syllogistics was regarded as a fragmentary theory in which the distinctions between different types of fine structure were not explicated. These authors did not try to reconstruct it as a uniform system without using the new tools which were not derived from Aristotle's logic ([Lagerlund, 2000, pp. 91-201], [Thom, 2003, pp. 141–191], [Knuuttila, 2008, pp. 551–559]).

According to Ockham and Buridan, the truth of 'A white thing can be black' demands the truth of 'This can be black' and, furthermore, 'This can be black' and "This is black' is possible' mean the same.[49] The latter statement exemplifies a compound reading and the former a divided reading. These forms are equated in the same way as in the so-called Barcan formula at the basic level with involves demonstrative pronouns, but they are separated in the discussion of quantified universal and particular statements. The truth of quantified divided modals demands the truth of all or some relevant singular statements of the type just mentioned, demonstrative pronouns then being imagined to refer to possible beings.[50] Buridan thought that there were two assertoric copulas, affirmative and negative, and modalized copulas in de re modal propositions. In describing his identity view of predication, Buridan suggests that 'An A was B' means that a past thing had the (past) predicates A and B and that 'An A is possibly B' means that a possible thing has the (possible) predicates A and B:

> Furthermore, it is also clear that if we say 'A is B', then provided that the terms are not ampliated to the past or future, it follows that 'A is B' is equivalent to 'A is the same as B' ... in the sense that some A should be posited to be the same as some B. And the same goes for the past and the future. For there is no difference in saying 'Aristotle

[49]William Ockham, *Summa logicae*, ed. Ph. Boehner, G. Gàl and S. Brown, Guillelmi de Ockham Opera philosophica, 1 (St Bonaventure, NY: St Bonaventure University, 1974), II.10 (276-9); III-1, 32 (448); III-3, 10 (632-4); John Buridan, *Tractatus de consequentiis*, ed. H. Hubien, Philosophes médiévaux, 16 (Louvain: Publications Universitaires; Paris: Vander-Oyez, 1976) II.7, 16 (75-6).

[50]William Ockham, *Summa logicae* I, c. 72 (215-6); III-3, (634); John Buridan, *Tractatus de consequentiis* II.6, 5 (66-7).

> was someone disputing' and 'Aristotle was the same as someone disputing' ... not because Aristotle and someone disputing are the same, but because they were the same, and the case is similar with the future and the possible. [51]

The truth of 'An A is necessarily B' would mean that a possible being which is or can be A is B in all possible states of affairs in which the thing occurs.[52] If a counterfactual state of affairs is possible, it can be coherently imagined as actual. Buridan remarks that the matter of water can take the form of air in the sense that this matter could be informed by the form of air instead of the form of water.[53] It is presupposed here, as in the Scotist revision of the obligational time rule, that the same being can be considered in alternative situations. Many scholars have followed George Hughes in regarding Buridan's modal logic as congenial with the philosophical assumptions of possible worlds semantics, but there are, of course, historical differences as well ([Hughes, 1989, p. 97]; [Knuuttila, forthcoming]).

One part of Buridan's modal logic was his octagon of opposition for divided modal statements. The equivalent formulations of modals with the different order of quantifying terms (every, some), negation, and modal terms, are combined into eight groups (universal and particular affirmative and negative forms of necessity propositions and possibility propositions), each of these presented in nine equivalent formulae and arranged into a diagram which shows the relations of contradiction, contrariety, subcontrariety, and subalternation between them ([Hughes, 1989, pp. 109-110], [Karger, 2003], [Reed, forthcoming]).

The new modal semantics also influenced late medieval theories of epistemic logic and deontic logic. Some twelfth-century thinkers asked whether the basic rules for modal sentences *de dicto*, viz. (5) and (6), held for other concepts showing similarities the notions of necessity and possibility, for example knowledge or obligation. This approach was developed for epistemic and normative notions by many fourteenth-century authors. The following equivalences analogous to those between modal concepts were used in fourteenth-century discussions of the norms:

15. $\neg O \neg p \leftrightarrow Pp$

16. $\neg P \neg p \leftrightarrow Op$

17. $\neg O \neg p \leftrightarrow P \neg p$

18. $\neg Pp \leftrightarrow O \neg p$

19. $Op \leftrightarrow F \neg p$

20. $Fp \leftrightarrow O \neg p$.

---

[51] John Buridan, *Sophismata* 2, concl. 10, translated in [Klima, 2001, pp. 855–856].

[52] For the relative or conditional notion of necessity, see John Buridan, *Tractatus de consequenciis* IV (112).

[53] *Tractatus de consequentiis*, II.7, 16 (76).

O stands here for obligation (*obligatum*), P for permission (*licitum*), and F for prohibition (*illicitum*). An extensive discussion of the logical properties of deontic concepts is to be found in the second article of the first question of Roger Roseth's *Lecture on the Sentences* which he finished ca. 1335.[54] Roseth defines the rationality of a system of norms and discusses various possible objections to his rules of rationality. He analyses conditional norms which are called contrary-to-duty imperatives and questions which are labelled as the paradox of the Good Samaritan in twentieth-century deontic logic. The most sophisticated part of Roseth's treatise deals with the question of how conditional obligation should be formulated. His final proposal is [Knuuttila, 2008, pp. 563–567], [Knuuttila and Hallamaa, 1995]:

21. $(p \rightarrow Oq) \mathbin{\&} \neg\Box(p \rightarrow q) \mathbin{\&} \Diamond(p \mathbin{\&} q)$.

Knowledge and belief were considered as partially analogous to necessity and possibility, but the inference rules of modal logic *de dicto* were usually not accepted as rules for knowledge and belief. Most medieval authors did not operate with the conception of logical omniscience which is included in some modern theories, treating the logic of epistemic notions from the point of view of factual attitudes. In the fourteenth century, there were various views of the distinction between the notions of apprehension and assent and the relationship between knowledge and belief, which Ockham formulated as follows

22. $K_a p = B_a p \mathbin{\&} p \mathbin{\&} J_a p$.

($J_a p$ stands for 'the person a is justified in believing that p'. [Boh, 2000; Boh, 1993, pp. 89–125].)

As for the relationship between epistemic propositions *de dicto* and *de re*, it was commonly thought that knowledge statements *de re* did not simply follow from knowledge statements *de dicto* or *vice versa*. Buridan adds that when Socrates knows that some A is B, then of something which is A he knows that it is B. Buridan agrees that the *de re* reading does not follow from the *de dicto* reading in the sense which implies that Socrates knows what or who A is, but there is an intermediate reading between pure *de dicto* and *de re* readings which does follow. According to Buridan, statements of the type

23. $K_a \exists x(Fx)$

imply that there are individuals who have property F, although a does not necessarily know which they are. In principle they are identifiable, however, and if we suppose that one of them is z, we can write [Knuuttila, 2008, pp. 561–563]; *Summulae de Dialectica*, trans. in [Klima, 2001, pp. 900–902]:

24. $K_a(\exists x)(Fx) \rightarrow (\exists x)((x = z) \mathbin{\&} K_a(Fx))$.

---

[54]The article is printed in the first question of *Determinationes magistri Roberti Holcot* at Lyon in 1518. The printed version is abridged and not very reliable.

Modalities in Early Modern Philosophy

The first part of the fourteenth century was a period of exceptional creativity in the history of modal logic and Buridan's theory, which was its most sophisticated achievement, strongly influenced the discussions of modal syllogistic until the beginning of the sixteenth century. It was employed in the fourteenth-century treatises of Albert of Saxony, *Perutilis logica*, and Marsilius of Inghen, *Qaestiones super librum Priorum Analyticorum*, both of which were printed in early sixteenth century. At that time, Buridan's approach was also known through Jodocus Trutfetter's *Summulae totius logicae* and George of Brussels's commentary on Aristotle's logic. In addition, there was an abridgement of Buridan's *Summulae de dialectica* with commentaries by John Dorp, first printed in 1499. William Ockham's *Summa logicae* was first printed in 1488. In addition, there were late fifteenth-century and early sixteenth-century treatises on truth and falsity of modal propositions by several nominalist logicians, such as Jerónimo Pardo, Robert Caubraith, and Juan de Celaya. (See [Lagerlund, 2000, pp. 202–227]; [Coombs, 1990].) Contrary to what one might assume judging by these publications, late medieval theories of modal syllogistic and modal consequences did not play any great role in later sixteenth-century logic. The interest in the analysis of various reading of modal propositions and modal syllogistic was to some extent revived in the so-called second scholasticism of the seventeenth-century [Roncaglia, 2003].

In spite of the decline of medieval modal logic, medieval discussions of modal metaphysics continued in sixteenth- and seventeenth-century thought. Many scholars commented on Scotus's idea that asserting something as logically possible would be true even if there were nothing but one intellect which formulates such a statement. Some writers regarded logical possibilities as independent preconditions of anything which is or is understood. This view was held by writers from different traditions, such as the Dominican Cardinal Cajetan and the Jesuit Francisco Suárez, who believed, like Scotus, that the intelligibility of things is neither derived from actual beings nor created by any mind. ([Coombs, 2003, p. 203]; for Suárez's view of the foundations of modality, see [Doyle, 2010]; [Honnefelder, 1991, pp.247–294].) Others, such as Francisco Albertinus, Zaccaria Pasqualigo and the influential Franciscan and Scotist Bartholomew Mastri, argued that logical modalities were embedded in the eternal divine intellect in such a way that if God did not exist, nothing would be possible or impossible. This was meant to be an interpretation of Scotus's view and continued the interpretation of some earlier Scotists, who found this view easier to understand than the more abstract idea of logical modalities as universal preconditions of being and understanding without any kind of actuality of their own. ([Coombs, 2003, pp. 229–234]; [Doyle, 2006]; see also [Hoffmann, 2002].) A third position was defended by John Punch, who took Scotus to mean that there is a special domain of possible beings between non-existence and existence. Cajetan argued that Scotus also claimed that possibilities have some kind of independent being, but this was denied by others such as Suárez, Vasquez, and Mastri [Coombs, 2003, p. 202]. According to Mastri, the modal nature of all possible states of affairs is dependent on God's necessary knowledge. This view of the ontological foundation of modalities was held by many others, including Leibniz:

> The late Jacob Thomasius ... made the apt observation ... that one ought not to say, with some Scotists, that the eternal truths would subsist, even if there were no intellect, not even that of God. For it is, in my opinion, the divine intellect which makes the eternal truths real, although his will has no part in it. Whatever is real must be founded in something existent. It is true that an atheist may be a geometer, but if there were no God, there would be no object of geometry, and without God there would neither be anything existent nor anything possible. (*Essais de Théodicée*, 184; in [Gerhardt, 1885, p. 226])

In his *Disputationes metaphysicae*, Francisco Suárez criticized the assumption that primary truths are dependent on God's will.[55] Now, Anselm of Canterbury wrote about God that all necessity and impossibility are subject to divine will, but it is not subject to any necessity or impossibility, 'for nothing is necessary or impossible without his willing it to be so.'[56] Referring to Anselm's view in his *De causa Dei* (1344), Thomas Bradwardine explains that there are absolute necessities which are not dependent on divine will, but all natural necessities and potencies as constituents of the created order are ultimately voluntary. This seems to be what Anselm had in mind as well.[57] Criticizing Scotus's view of autonomous modalities, Bradwardine also argues, as Mastrius and others later did, that if there were no God, there would not be any kind of modality. It seems clear that this was not the approach which Suárez criticized in referring to the view that there are no limitations to divine will and omnipotence. He could mean the modal voluntarism of Martin Luther, who argued that the same things are not possible in philosophy and theology and those of the latter, i.e., divine possibilities, are not limited by anything at all. What is necessary or possible in philosophy is so because of God's sovereign decision. This view found some Lutheran adherents but was refuted in seventeenth century philosophy of Lutheran orthodoxy [Knuuttila, 2003; Knuuttila, 2010b]. Similar ideas were formulated in the early seventeenth-century by the Jesuit Hurtado de Mendoza and more famously by Descartes, who taught that propositions are necessarily or contingently true or false because of God's eternal decision.[58] Descartes's position was taken to mean, not mistakenly, that there are no necessary truths which would be so with respect to God's voluntary thinking. This view was heavily criticized by Leibniz:

[55] *Disputationes metaphysicae*, 31.12.40 in *Opera omnia*, vol. 26 (Paris, 1861), 295.

[56] *Cur Deus homo*, ed. F.S. Schmitt, Opera omnia, 2 (Edinburgh: Nelson, 1946), 2.17; *Meditatio Redemptionis Humanae*, ed. F.S. Schmitt, Opera omnia, 3 (Edinburgh: Nelson, 1946), 86.60-2.

[57] Thomas Bradwardine, *De causa Dei contra Pelagium et de virtute causarum* (London, 1618, reprint Frankfurt am Main: Minnerva, 1964), 209, 214, 231; see also [Frost, forthcoming].

[58] For Hurtado's somewhat ambivalent formulations, see [Coombs, 2003, pp. 213–217]; for Descartes, see [Alanen, 1991].

> an unheard of paradox by which Descartes showed how great can be the errors of great men, as if the reason that a triangle has three sides or that two contrary propositions are incompatible, or that God himself exists, is that God has willed so. [Riley, 1972, pp.71–72]

The same authors who were interested in the metaphysical foundations of modalities also discussed the various types of modality. The division between metaphysical, physical, and moral necessities and possibilities was largely employed by the Spanish representatives of the second scholasticism, who taught that moral possibility entailed physical possibility and this entailed metaphysical possibility, but not vice versa. The distinction between metaphysical and physical modalities was understood in accordance with the late medieval doctrine in which the invariant connections between things defined physical necessities which could be regarded as metaphysically contingent. Moral modalities had a traditional background in theological considerations about the extent to which people can live without sin. According to Thomas Aquinas, they can avoid a particular sin but not all of them. This was systematized in detailed discussions of moral modalities, which were increasingly interpreted using statistical probabilities.[59]

Medieval and Renaissance theologians usually thought that divine foreknowledge, however it was understood, presupposed bivalence. It was assumed from Boethius to Aquinas that God's timelessness involved the simultaneous presence of the whole history to God. Human predictions of future contingents are true or false from the point of view of God's eternal knowledge, which is not temporally located. Scotus and other critics found the idea of the simultaneous presence of all historical events to God's eternal vision problematic. Ockham thought that divine foreknowledge could be treated as temporally past. His answer to the question of how God's foreknowledge as something past and fixed is compatible with the contingency of future things was that even though foreknowledge is past, its content is future, and as far as it is about future contingents, it is itself contingent. (See [Knuuttila, 2010a]). The next question of how God can foreknow the free decisions of people was dealt with in the theory of middle knowledge by Luis de Molina (1535-1600). In addition to the general knowledge of metaphysical possibilities and historical actualizations in the chosen world, God has a third kind of knowledge (*scientia media*) which comprises the hypothetical truths about possible beings. In creating the world, God knows of possible free creatures what they would do in various possible situations. Molina's 'middle knowledge' theory about the counterfactuals of freedom was actively debated in the sixteenth and seventeenth centuries [Dekker, 2000].—indexOckham, W.

The Scotist interpretation of modality as alternativeness was well known to early modern times as such and also through the version developed in Francisco Suárez's very influential *Disputationes metaphysicae*. The writers of the second scholasticism took it for granted that only a small portion of metaphysical possibilities will be actualized, some of them beginning to use the notion of possible worlds

[59] [Knebel, 2003]; the article is based on [Knebel, 2000].

in this context.[60] Hobbes and Spinoza distanced themselves from this tradition. They argued for determinism and the Diodorean view of modalities. Descartes's remarks on modal metaphysics are more controversial. (See [Leijenhorst, 2002; Koistinen, 2003; Normore, 1991; Normore, 2006; Alanen, 2008].)

In an early treatise, Leibniz also defined modal terms with frequential terms:

> Possible is whatever can happen or what is true in some cases or what is understood clearly and distinctly;
> Impossible is whatever cannot happen or what is true in no case or not true in any case or what is not understood clearly and distinctly;
> Necessary is whatever cannot happen or what is true in every case or not in any case not true or the opposition of which is not understood clearly and distinctly;
> Contingent is whatever cannot happen or what is not true in some case or the opposition of which is understood clearly and distinctly.[61]

These formulations involve reminiscences from the traditional square of modal opposition and a combination of statistical classification of modal terms with Cartesian epistemic notions. 'Contingency' is somewhat unusually understood as the denial of necessity. While Leibniz treats necessity and possibility as analogous to universality and particularity even later, he associates them with possible domains and hence not extensional quantifiers. In fact he regarded the modal views of Hobbes and Spinoza as seriously mistaken.[62]

The famous part of Leibniz's view is his conception of possible worlds as described by maximal compossible sets of propositions. There is an infinite number of such combinations about infinite possible beings in God's intellect, the best of these being chosen to be actualized. The interest in the properties of the possible worlds and the theory of the best possible world as a vectorial compromise between several values, such as richness, economic laws, justice, and happiness, are typically Leibnizian themes with a background in his interest in the theories of combinatorics and rational choice. The conception of possible worlds was associated with the notion of possible individuals as having infinitely long conceptions which were compossible with all other concepts of things in the same world, all things being different because of the famous principle of the identity of indiscernability. Leibniz describes the difference between necessary and contingent features of things by arguing that while the proofs of necessary propositions are finite, the proofs of contingent propositions are infinite. This prevents the knowledge of the truth of future contingent propositions, although there might be knowledge of their probability. Even though individual finite beings are contingent, all their properties are necessary in the sense that if they were

[60] For the Scotist background of the discussions of modality in seventeenth-century theology, see also [Goudriaan, 2006; Bac, 2010]. For the history of possible world terminology see [Schmutz, 2006].

[61] For the text and its variants, see [Poser, 1969, pp. 16–18].

[62] See Letter to Philipp, January 1680 in [Gerhardt, 1880, pp. 283–284], translated in [Loemker, 1969, p. 273].

different, the subject would not be the same. Possible beings are consequently world-bound; when it is said that Julius Caesar might have not crossed Rubicon, this proposition refers to another possible world in which Caesar's counterpart would act in this way. The idea of the world-bound individuals is one of the differences between Leibniz and his late medieval predecessors [Adams, 1994; Nachtomy, 2007].

Leibniz's best-known contribution to logic consists of his attempts to develop the calculus of concepts and elements of the calculus of propositions of strict implication. Leibniz's remarks on alethic modal logic are scattered and not very original in comparison with late medieval achievements. The conception of possible worlds was not relevant in this context; it served for the metaphysical comparison between the worlds, and Leibniz did not have ideas corresponding to the accessibility relations of possible worlds semantics. He also sketched elements of deontic logic in the same spirit as Roger Roseth and some other fourteenth-century authors. (See [Lenzen, 2004; Lenzen, 2005]; for Roseth, see note 11 above.)

Leibniz's modal ideas were used by Christian Wolff, who applied Diodorean modalities to the actual history as one of the divine alternatives.[63] Kant followed this metaphysics in his early writings, but he later preferred to associate modal terms with epistemic notions. Many nineteenth-century thinkers followed this approach, often with a psychological emphasis.[64] While these developments did not produce new ideas of modal logic, there was an increasing interest in the theory of probability, which some logicians took to replace traditional modal discussions. John Venn reduced statistically understood physical modalities to frequency probabilities. Bertrand Russell's view was similar to this. Charles Peirce also dealt with probability and modality, but he associated probability with objective possibilities for which the principle of plenitude was false. (See [Niiniluoto, 1988].) In comparison to the relative low profile of nineteenth-century modal logic, the increasing interest in formal and theoretical aspects of modal concepts in twentieth-century philosophy was a remarkable reorientation which, as shown above, was not without predecessors in pre-Kantian philosophy.

PRIMARY LITERATURE

[Alexander of Aphrodisias, 1883] Alexander of Aphrodisias. In *Aristotelis Analyticorum priorum librum I commentarium*, ed. M. Wallies. Commentaria in Aristotelem Graeca 2.1. Reimer, Berlin, 1883.

[Ammonius, 1897] Ammonius. In *Aristotelis De interpretatone commentaries*, ed. A. Busse. Commentaria in Aristotelem Graeca, 4.5. Reimer, Berlin, 1897.

[Anonymous,] Anonymous. *De obligationibus*, ed. by Romuald Green in *The Logical Treatise 'De obligationibus': An Introduction with Critical Texts of William of Sherwood (?) and Walter Burley*. Ph. D. diss., University of Louvain.

[63] See *Vernnüfftige Gedancken über Gott, der Welt und der Seele des Menschen*, 7th ed. (Frankfurt and Leipzig, 1738), § 572. Leibniz says that this follows when one looks at the compossibilities of the actual world [Gerhardt, 1887, p. 572]; translated in [Loemker, 1969, p. 661].

[64] For some studies, see [Haaparanta, 1988; Korte *et al.*, 2009].

[Anselm of Canterbury, 1946] Anselm of Canterbury. *De concordia praescientiae et praedestinationis et gratiae Dei cum libero arbitrio*, ed. F.S. Schmitt. Opera omnia, 2. Nelson, Edinburgh, 1946.

[Anselm of Canterbury, 1946a] Anselm of Canterbury. *De conceptu virginali et de originali peccato*, ed. F.S. Schmitt, Opera omnia, 2. Nelson, Edinburgh, 1946.

[Anselm of Canterbury, 1946b] Anselm of Canterbury. *Cur Deus homo*, ed. F.S. Schmitt. Opera omnia, 2. Nelson, Edinburgh, 1946.

[Anselm of Canterbury, 1946c] Anselm of Canterbury. *Meditatio Redemptionis Humanae*, ed. F.S. Schmitt. Opera omnia, 3. Nelson, Edinburgh, 1946.

[Aristotle, 1949] Aristotle.*Categoriae et Liber de interpretatione*, ed. L. Minio-Paluello. Clarendon Press, Oxford, 1949.

[Aristotle, 1936] Aristotle. *De caelo*, ed. D.J. Allan. Clarendon Press, Oxford, 1936.

[Aristotle, 1924] Aristotle. *Metaphysics*. A revised text with introduction and commentary by W.D. Ross. Clarendon Press, Oxford, 1924.

[Aristotle, 1937] Aristotle. Parts of Animals, ed. A.L. Beck. Loeb Classical Library. Harvard University Press, Cambridge, MA, 1937.

[Aristotle, 1936a] Aristotle. *Physics*. A revised text with introduction and commentary by W.D. Ross. Clarendon Press, Oxford, 1936.

[Aristotle, 1949] Aristotle. *Prior and Posterior Analytics*. A revised text with introduction and commentary by W.D. Ross. Clarendon Press, Oxford, 1949.

[Aristotle, 1989] Aristotle. *Prior Analytics*, trans. with introduction and notes by R. Smith. Hackett, Indianapolis, 1989.

[Averroes, 1962] Averroes.*Aristotelis Opera cum Averrois Commentariis*. Venice, 1562-1574; reprint Minerva, Frankfurt am Main, 1962.

[Boethius, 1887] Boethius. *Commentarii in librum Aristotelis Perihermeneias I-II*, ed. C. Meiser. Teubner, Leipzig, 1877-1880.

[Diogenes Laertius, 1925] Diogenes Laertius. *Lives of Eminent Philosophers*, ed. R.D. Hicks. Loeb Classical Library. Harvard University Press, Cambridge, MA, 1925.

[Gilbert of Poitiers, 1966] Gilbert of Poitiers. *The Commentaries on Boethius*, ed. Nicholas M. Häring. Pontifical Institute of Mediaeval Studies, Toronto, 1966.

[Buridan, 1964] John Buridan. *Quaestiones super octo Physicorum libros Aristotelis*. Paris, 1509, reprint Minerva, Franfurt am Main: 1964.

[Buridan, 1976] John Buridan. *Tractatus de consequentiis*, ed. Hubert Hubien, Philosophes mdivaux, 16. Publications universitaires, Paris; Vander-Oyez, Louvain, 1976.

[Buridan, 2001] John Buridan. *Summulae de Dialectica*, an annotated translation with a philosophical introduction by G. Klima. Yale University Press, New Haven–London, 2001.

[Scotus, 1950] John Duns Scotus. *Ordinatio I*, d. 1-2. Opera omnia, 2. Typis Polyglottis Vaticanis, Civitas Vaticana, 1950.

[Scotus, 1960] John Duns Scotus. *Lectura I*, d. 8-45. Opera omnia, 17. Typis Polyglottis Vaticanis, Civitas Vaticana, 1960.

[John of Jandun, 1552] John of Jandun. *In libros Aristotelis De caelo et mundo quae extant quaestiones*. Apud Iuntas, Venice, 1552.

[Leibniz, 1885] G. W. Leibniz. *Essais de Théodicée. Die Philosophischen Schriften von G.W. Leibniz*, ed. C.I. Gerhardt, vol. 6. Berlin, 1885.

[Leibniz, 1880] G. W. Leibniz. *Die Philosophischen Schriften von G.W. Leibniz*, ed. C.I. Gerhardt, vol. 4. Berlin, 1880.

[Leibniz, 1887] G. W. Leibniz. *Die Philosophischen Schriften von G.W. Leibniz*, ed. C.I. Gerhardt, vol. 3. Berlin, 1887.

[Leibniz, 1972] G. W. Leibniz. *The Political Writings of Leibniz*, translated by P. Riley. Cambridge University Press, Cambridge, 1972.

[Lucretius Carus, 1955] Lucretius Carus. *De rerum natura*, ed. C. Bailey. Oxford University Press, Oxford, 1955.

[Origen, 2001] Origen. *Contra Celsum*, ed. M. Marcovich. Brill, Leiden, 2001.

[Peter Abelard, 1919] Peter Abelard. *Philosophische Schriften I. Die Logica 'Ingredientibus'*, ed. Bernhard Geyer. Beiträge zur Geschichte der Philosophie und Theologie des Mittelalters, 21,1-3. Aschendorff, Mnster, 1919-1927.

[Peter Abelard, 1956] Peter Abelard. *Dialectica*, ed. Lambert M. de Rijk, Wijsgerige teksten en studies, 1. van Gorcum, Assen, 1956.

[Peter Abelard, 1958] Peter Abelard. *Super Periermenias XII-XIV*, ed. L. Minio-Paluello, in Twelfth Century Logic: Texts and Studies II: Abaelardiana inedita. Edizioni di Storia e Letteratura, Rome,1958.

[Peter Lombard, 1971] Peter Lombard. *Sententiae in IV libris distinctae, I-II.* Collegium S. Bonaventurae ad Claras Aquas, Grottaferrata, 1971.

[Peter of Poitiers, 1961] Peter of Poitiers.*Sententiae*, ed. P.S. Moore and M. Dulong. Publications in Mediaeval Studies, 7. The University of Notre Dame Press, Notre Dame, Ind., 1961.

[Philoponus, 1905] Philoponus. *In Aristotelis Analytica priora commentaria*, ed. M. Wallies. Commentaria in Aristotelem Graeca 13.2. Reimer, Berlin, 1905.

[Plato, 1900] Plato. , *Opera*, ed. J. Burnet. Oxford University Press, Oxford, 1900-1907.

[P;otinus, 1964] Plotinus. *Opera I-III*, ed. P. Henry and H.R. Schwyzer. Clarendon Press, Oxford,1964-1982.

[Richard Campsall, 1968] Richard Campsall. *Questiones super librum Priorum Analeticorum*, ed. Edward A. Synan in The Works of Richard of Campsall, vol. 1. Pontifical Institute of Mediaeval Studies, Toronto, 1968.

[Richard Kilvington, 1990] Richard Kilvington. *The Sophismata of Richard Kilvington*, ed. Norman Kretzmann and Barbara E. Kretzmann. Oxford University Press, Oxford, 1990; The Sophismata of Richard Kilvington, Introduction, Translation, and Commentary by Norman Kretzmann and Barbara E. Kretzmann. Cambridge University Press, Cambridge, 1990.

[Robert Grosseteste, 1991] Robert Grosseteste. *De libero arbitrio*, ed. in Neil Lewis, 'The First Recension of Robert Grosseteste's De libero arbitrio', Mediaeval Studies 53 (1991), 1-88.

[Robert Kilwardy, 1968] Robert Kilwardby. *In libros Priorum Analyticorum expositio*, Venice, 1516 (under the name of Giles of Rome), reprint Minerva, Frankfurt am Main, 1968.

[Roger Rosoth, 1518] Roger Roseth. *Lecture on the Sentences*, q. 1, in Determinationes magistri Roberti Holcot, Lyon, 1518.

[Suárez, 1861] F. Suárez. *Disputationes metaphysicae.* Opera omnia 25-26. Vivs, Paris, 1861.

[Thomas Aquinas, 1948] Thomas Aquinas.*Summa theologiae*, ed. P. Caramello. Marietti, Turin, 1948-1950.

[Thomas Aquinas, 1965] Thomas Aquinas. , *In octo libros Physicorum Aristotelis expositio*, ed. M. Maggilo. Marietti, Turin, 1965.

[Thomas Aquinas, 1964] Thomas Aquinas. *In Aristotelis libros Peri Hermeneias et Posteriorum analyticorum expositio*, ed. R. Spiazzi. Marietti, Turin, 1964.

[Thomas Aquinas, 1977] Thomas Aquinas. *In duodecim libros Metaphysicorum Aristotelis expositio*, ed. M.-R. Cathala and R. Spiazzi. Marietti, Turin, 1977.

[Thomas Bradwardine, 1964] Thomas Bradwardine. *De causa Dei contra Pelagium et de virtute causarum.* London, 1618, reprint Minnerva, Frankfurt am Main, 1964.

[William Ockham, 1974] William Ockham. *Summa logicae*, ed. Ph. Boehner, G. Gàl, S. Brown. Guillelmi de Ockham Opera philosophica, 1. St. Bonaventure University, St. Bonaventure, N.Y., 1974.

[Wolff, 1738] C. Wolff. *Vernünfftige Gedancken über Gott, der Welt und der Seele des Menschen*, 7th ed. Frankfurt and Leipzig, 1738.

SECONDARY LITERATURE

[Adams, 1994] R. Adams. *Leibniz: Determinist, Theist, Idealist.* Oxford University Press, Oxford, 1994.

[Alanen, 1991] L. Alanen. Descartes, conceivability, and logical modality. In T. Horowitz and G.J. Massey, editors, *Thought Experiments in Science and Philosophy*, pages 65–84. Rowman and Littlefield, Savage, MD, 1991.

[Alanen, 2008] L. Alanen. Omnipotence, modality, and conceivability. In J. Broughton and J. Carriero, editors, *A Companion to Descartes*, pages 351–371. Blackwell, Oxford, 2008.

[Bac, 2010] M. Bac. *Perfect Will Theology: Divine Agency in Reformed Scholasticism as Against Suarez, Episcopius, Descartes, and Spinoza.* Brill, Leiden, 2010.

[Barnes *et al.*, 1991] J. Barnes, S. Bobzien, K. Flannery, and K. Ierodiakonou. *Alexander of Aphrodisias on Aristotle's Prior Analytics.* Duckworth, London, 1991.

[Becker, 1933] A. Becker. *Die aristotelische Theorie der Möglichkeitschlsse.* Junker, Berlin, 1933.

[Bobzien, 1998] S. Bobzien. *Determinism and Freedom in Stoic Philosophy.* Clarendon Press, Oxford, 1998.

[Bobzien, 1999a] S. Bobzien. Logic II: The 'Megarics'. In K. Algra, J. Barnes, J. Mansfeld, and M. Schofield, editors, *The Cambridge History of Hellenistic Philosophy*, pages 83–92. Cambridge University Press, Cambridge, 1999.

[Bobzien, 1999b] S. Bobzien. Logic III: The Stoics. In K. Algra, J. Barnes, J. Mansfeld, and M. Schofield, editors, *The Cambridge History of Hellenistic Philosophy*, pages 93–157. Cambridge University Press, Cambridge, 1999.

[Boh, 1993] I. Boh. *Epistemic Logic in the Later Middle Ages.* Routledge, London, 1993.

[Boh, 2000] I. Boh. Four phases of medieval epistemic logic. *Theoria*, 66:129–149, 2000.

[Coombs, 1990] J. Coombs. *Truth and Falsity of Modal Propositions in Renaissance Nominalism.* PhD thesis, University of Texas at Austin, 1990.

[Coombs, 2003] J. Coombs. The ontological source of logical possibility in Catholic second scholasticism. In R.L. Friedman and L.O. Nielsen, editors, *The Medieval Heritage in Early Modern Metaphysics and Modal Theory, 1400–1700*, pages 191–229. Kluwer, Dordrecht, 2003.

[Dekker, 2000] E. Dekker. *Middle Knowledge.* Peeters, Leuven, 2000.

[Doyle, 2006] J.P. Doyle. Mastri and some Jesuits on possible and impossible objects of God's knowledge and power. In M. Forlivesi, editor, *Rem in seipsa cernere. Saggi sul pensiero filosofico di Bartolomeo Mastri (1602-1673)*, pages 439–468. Il Poligrafo, Padua, 2006.

[Doyle, 2010] J.P. Doyle. *Collected Studies on Suárez S.J. (1548-1617).* Leuven University Press, Leuven, 2010.

[Dutilh Novaes, 2007] C. Dutilh Novaes. *Formalizing Medieval Logical Theories: Suppositio, Obligationes and Consequentia.* Springer, Dordrecht, 2007.

[Ebert and Nortmann, 2007] T. Ebert and U. Nortmann. *Aristoteles,* Analytica Priora*, Buch I.* Akademie Verlag, Berlin, 2007. Translation with commentary.

[Flannery, 1995] K.L. Flannery. *Ways into the Logic of Alexander of Aphrodisias.* Brill, Leiden, 1995.

[Fortenbaugh, 1992] W. Fortenbaugh, editor. *Theophrastus of Eresus: Sources for His Life, Writings, Thought and Influence, vol. 1.* Brill, Leiden, 1992.

[Frost, forthcoming] G. Frost. Thomas Bradwardine on God and the foundations of modality. *British Journal for the History of Philosophy*, forthcoming.

[Gelber, 2004] H. Gelber. *It Could Have Been Otherwise: Contingency and Necessity in Dominican Theology at Oxford, 1300–1350.* Brill, Leiden, 2004.

[Gerhardt, 1880] C.I. Gerhardt, editor. *Die philosophischen Schriften von Gottfried Wilhelm Leibniz; Band 4 (1663–1671).* Weidmannsche Buchhandlung, Berlin, 1880.

[Gerhardt, 1885] C.I. Gerhardt, editor. *Die philosophischen Schriften von Gottfried Wilhelm Leibniz; Band 6 (1702–1716).* Weidmannsche Buchhandlung, Berlin, 1885.

[Gerhardt, 1887] C.I. Gerhardt, editor. *Die philosophischen Schriften von Gottfried Wilhelm Leibniz; Band 3 (Correspondence).* Weidmannsche Buchhandlung, Berlin, 1887.

[Goudriaan, 2006] A. Goudriaan. *Reformed Orthodoxy and Philosophy, 1625-1750: Gisbertus Voetius, Petrus van Mastricht, and Anthonius Driessen.* Brill, Leiden, 2006.

[Haaparanta, 1988] L. Haaparanta. Frege and his German contemporaries on alethic modalities. In S. Knuuttila, editor, *Modern Modalities: Studies of the History of Modal Theories from Medieval Nominalism to Logical Positivism*, pages 239–274. Kluwer, Dordrecht, 1988.

[Hintikka, 1957a] J. Hintikka. Modality as referential multiplicity. *Ajatus*, 20:49–64, 1957.

[Hintikka, 1957b] J. Hintikka. Necessity, universality, and time in Aristotle. *Ajatus*, 20:65–90, 1957.

[Hintikka, 1973] J. Hintikka. *Time and Necessity: Studies in Aristotle's Theory of Modality.* Clarendon Press, Oxford, 1973.

[Hintikka, 1981] J. Hintikka. Gaps in the great chain of being: An exercise in the methodology of the history of ideas. In S. Knuuttila, editor, *Reforging the Great Chain of Being*, pages 1–17. Reidel, Dordrecht, 1981.

[Hoffmann, 2002] T. Hoffmann. *Creatura intellecta. Die Ideen und Possibilien bei Duns Scotus mit Ausblick auf Franz von Mayronis, Poncius und Mastrius.* Aschendorff, Münster, 2002.

[Hoffmann, 2009] T. Hoffmann. Duns Scotus on the origin of the possibles in the divine intellect. In S.F. Brown, T. Dewender, and T. Kobusch, editors, *Philosophical Debates at Paris in the Early Fourteenth Century*, pages 359–379. Brill, Leiden, 2009.

[Honnefelder, 1991] L. Honnefelder. *Scientia transcendens. Die formale Bestimmung der Seiendheit und Realität in der Metaphysik des Mittelalters und der Neuzeit.* Felix Meiner, Hamburg, 1991.

[Huby, 2007] P. Huby. *Theophrastus of Eresus: Sources for His Life, Writings, Thought and Influence. Commentary; vol. 2.* Leiden, Leiden, 2007.

[Hughes, 1989] G. Hughes. The modal logic of John Buridan. In G. Corsi, C. Mangione, and M. Mugnai, editors, *Atti del convegno Internazionale di Storia della Logica. Le Teorie della Modalità*, pages 93–111. CLUEB, Bologna, 1989.

[Johnson, 1989] F. Johnson. Models for modal syllogisms. *Notre Dame Journal of Formal Logic*, 30:271–284, 1989.

[Johnson, 1995] F. Johnson. Apodeictic syllogisms: deductions and decision procedures. *History and Philosophy of Logic*, 16:1–18, 1995.

[Karger, 2003] E. Karger. John Buridan's theory of the logical relations between general modal formulae. In H. Braakhuis and C. H. Kneepkens, editors, *Aristotle's* Peri Hermeneias *in the Latin Middle Ages: Essays on the Commentary Tradition*, pages 429–49. Ingenium Publishers, Groningen/Haren, 2003.

[Keffer, 2001] H. Keffer. *De obligationibus: Rekonstruktion einer spätmittelalterlichen Disputationstheorie.* Brill, Leiden, 2001.

[Klima, 2001] G. Klima, editor. John Buridan, *Summulae de Dialectica*. Yale University Press, New Haven, 2001. Translated with a philosophical introduction by G. Klima.

[Knebel, 2000] S. Knebel. *Wille, Würfel und Wahrscheinlichkeit. Das System der moralischen Notwendigkeit in der Jesuitenscholastik 1550-1700.* Meiner, Hamburg, 2000.

[Knebel, 2003] S. Knebel. The renaissance of statistical modalities in early modern Scholasticism. In R.L. Friedman and L.O . Nielsen, editors, *The Medieval Heritage in Early Modern Metaphysics and Modal Logic 1400-1800*, pages 231–251. Kluwer, Dordrecht, 2003.

[Knuuttila and Hallamaa, 1995] S. Knuuttila and O. Hallamaa. Roger Roseth and medieval deontic logic. *Logique & Analyse*, 149:75–87, 1995.

[Knuuttila and Kukkonen, 2011] S. Knuuttila and T. Kukkonen. Thought experiments and indirect proofs in Averroes, Aquinas, and Buridan. In K. Ierodiakonou and S. Roux, editors, *Thought Experiments in Methodological and Historical Contexts*, pages 83–99. Brill, Leiden, 2011.

[Knuuttila, 1996] S. Knuuttila. Duns Scotus and the foundations of logical modalities. In L. Honnefelder, R. Wood, and M. Dreyer, editors, *John Duns Scotus: Metaphysics and Ethics*, pages 127–146. Brill, Leiden, 1996.

[Knuuttila, 2001] S. Knuuttila. Augustine on time and creation. In N. Kretzmann and E. Stump, editors, *The Cambridge Companion to Augustine*, pages 103–115. Cambridge University Press, Cambridge, 2001.

[Knuuttila, 2003] S. Knuuttila. The question of the validity of logic in late medieval thought. In R. Friedman and L. Nielsen, editors, *The Medieval Heritage in Early Modern Metaphysics and Modal Logic 1400-1800*, pages 121–142. Kluwer, Dordrecht, 2003.

[Knuuttila, 2008] S. Knuuttila. Medieval modal theories and modal logic. In D. M. Gabbay and J. Woods, editors, *Handbook of the History of Logic. Vol. 2: Mediaeval and Renaissance Logic*, pages 505–578. Elsevier, Amsterdam, 2008.

[Knuuttila, 2010a] S. Knuuttila. Medieval commentators on future contingents in *de interpretatione* 9. *Vivarium*, 48:75–95, 2010.

[Knuuttila, 2010b] S. Knuuttila. Philosophy and theology in seventeenth-century Lutheranism. In S. Knuuttila and R. Saarinen, editors, *Theology and Early Modern Philosophy 1550-1750*, pages 41–54. Finnish Academy of Science and Letters, Helsinki, 2010.

[Knuuttila, forthcoming] S. Knuuttila. Modality. In J. Marenbon, editor, *Oxford Handbook of Medieval Philosophy.* Oxford University Press, Oxford, forthcoming.

[Koistinen, 2003] O. Koistinen. Spinoza's proof of necessitarianism. *Philosophy and Phenomenological Research*, 67:283–310, 2003.

[Korte *et al.*, 2009] T. Korte, A. Maunu, and T. Aho. Modal logic from Kant to possible worlds semantics. In L. Haaparanta, editor, *The Development of Modern Logic*, pages 516–562. Oxford University Press, Oxford, 2009.

[Kretzmann and Kretzmann, 1990a] N. Kretzmann and B. Kretzmann, editors. *The* Sophismata *of Richard Kilvington.* Oxford University Press, Oxford, 1990.

[Kretzmann and Kretzmann, 1990b] N. Kretzmann and B. Kretzmann, editors. *The* Sophismata *of Richard Kilvington.* Cambridge University Press, Cambridge, 1990. Introduction, translation, and commentary.

[Lagerlund, 2000] H. Lagerlund. *Modal Syllogistics in the Middle Ages.* Brill, Leiden, 2000.

[Lagerlund, 2009] H. Lagerlund. Avicenna and and Tusi on modal logic. *History and Philosophy of Logic*, 30:227–239, 2009.

[Leijenhorst, 2002] C. Leijenhorst. *The Mechanisation of Aristotelianism: The Late Aristotelian Setting of Thomas Hobbes' Natural Philosophy.* Brill, Leiden, 2002.

[Lenzen, 2004] W. Lenzen. Leibniz's logic. In D.M. Gabbay and J. Woods, editors, *Handbook of the History of Logic, Volume 3. The Rise of Modern Logic: From Leibniz to Frege*, pages 40–56. Elsevier, Amsterdam, 2004.

[Lenzen, 2005] W. Lenzen. Leibniz on alethic and deontic modal logic. In D. Berlioz and F. Nef, editors, *Leibniz et les puissances du language*, pages 341–62. Vrin, Pris, 2005.

[Lewis, 1987] N. Lewis. Determinate truth in Abelard. *Vivarium*, 25:81–109, 1987.

[Lewis, 1997] N. Lewis. Power and contingency in Robert Grosseteste and Duns Scotus. In L. Honnefelder, R. Wood, and M. Dreyer, editors, *John Duns Scotus: Metaphysics and Ethics*, pages 205–225. Brill, Leiden, 1997.

[Loemker, 1969] L.E. Loemker, editor. *G.W. Leibniz: Philosophical Papers and Letters.* Reidel, Dordrecht, 1969. L.E. Loemker, translator.

[Lovejoy, 1936] A.O. Lovejoy. *The Great Chain of Being: A Study of the History of an Idea.* Harvard University Press,, Cambridge, Mass., 1936.

[Malink, 2006] M. Malink. A reconstruction of Aristotle's modal syllogistic. *History and Philosophy of Logic*, 27:95–141, 2006.

[Marenbon, 2007] J. Marenbon. *Medieval Philosophy: An Historical and Philosophical Introduction.* Routledge, London, 2007.

[Martin, 1999] C. Martin. Non-reductive arguments from impossible hypotheses in Boethius and Philoponus. *Oxford Studies in Ancient Philosophy*, 17:279–302, 1999.

[Martin, 2001] C. Martin. Abaelard on modality: Some possibilities and some puzzles. In Thomas Buchheim, C.H. Kneepkens, and K. Lorenz, editors, *Potentialität und Possibilität: Modalaussagen in der Geschichte der Metaphysik*, pages 97–134. Frommann-Holzbook, Stuttgart, 2001.

[Martin, 2003] C. Martin. An amputee is bipedal: The role of the categories in the development of Abelard's theory of possibility. In J. Biard and I. Rosier-Catach, editors, *La Tradition médiévale des Catégories (XIIe-XIVe siècles)*, pages 225–242. Peeters, Louvain and Paris, 2003.

[McCall, 1963] S. McCall. *Aristotle's Modal Syllogisms.* North-Holland, Amsterdam, 1963.

[Mueller, 1999] I. Mueller. Introduction. In *Alexander of Aphrodisias on Aristotle's Prior Analytics*, pages v. I, 8–13. Duckworth, London, 1999.

[Nachtomy, 2007] O. Nachtomy. *Possibility, Agency, and Individuality in Leibniz's Metaphysics.* Springer, Dordrecht, 2007.

[Niiniluoto, 1988] I. Niiniluoto. From possibility to probability: British discussions on modality in nineteenth century'. In S. Knuuttila, editor, *Modern Modalities: Studies of the History of Modal Theories from Medieval Nominalism to Logical Positivism*, pages 275–309. Kluwer, Dordrecht, 1988.

[Normore, 1991] C. Normore. Descartes's possibilities. In G.J.D. Moyal, editor, *René Descartes: Critical Assessments, vol. 3*, pages 68–83. Routledge, London, 1991.

[Normore, 2003] C. Normore. Duns Scotus's modal theory. In T. Williams, editor, *The Cambridge Companion to Duns Scotus*, pages 129–160. Cambridge University Press, Cambridge, 2003.

[Normore, 2006] C. Normore. Necessity, immutability, and Descartes. In V. Hirvonen, T. Holopainen, and M. Tuominen, editors, *Mind and Modality*, pages 257–283. Brill, Leiden, 2006.

[Nortmann, 1996] U. Nortmann. *Modale Syllogismen, mögliche Welten, Essentialismus. Eine Analyse der aristotelischen Modallogik.* de Gruyter, Berlin, 1996.

[Patterson, 1995] R. Patterson. *Aristotle's Modal Logic: Essence and Entailment in the Organon.* Cambridge University Press, Cambridge, 1995.

[Pinzani, 2003] R. Pinzani. *The Logical Grammar of Abelard.* Kluwer, Dordrecht, 2003.

[Poser, 1969] H. Poser. *Zur Theorie der Modalbegriffe bei G.W. Leibniz.* Steiner, Wiesbaden, 1969.

[Reed, forthcoming] S. Reed. John Buridan's theory of consequence and his octagons of opposition. manuscript, forthcoming.

[Riley, 1972] P. Riley, editor. *The Political Writings of Leibniz.* Cambridge University Press, Cambridge, 1972. P. Riley, translator.

[Rini, 2010] R. Rini. *Aristotle's Modal Proofs: Prior Analytics A8-22 in Predicate Logic.* Springer, Dordrecht, 2010.

[Roncaglia, 2003] G. Roncaglia. Modal logic in Germany at the beginning of seventeenth century: Christoph Scheibler's Opus logicum. In R.L. Friedman and L.O. Nielsen, editors, *The Medieval Heritage in Early Modern Metaphysics and Modal Theory, 1400–1700*, pages 253–307. Kluwer, Dordrecht, 2003.

[Schmidt, 1989] K. J. Schmidt. *Die modale Syllogistik des Aristoteles. Eine modal-prädikatenlogische Interpretation.* Mentis, Padenborn, 1989.

[Schmutz, 2006] J. Schmutz. Qui a inventé les mondes possibles. In J.-C. Bardout and V. Juillien, editors, *Les mondes possibles*, pages 9–45. Presses universitaires de Caen, Caen, 2006.

[Sedley, 2007] D. Sedley. *Creationism and Its Critics in Antiquity.* University of California Press, Berkeley and Los Angeles, 2007.

[Smith, 1989] R. Smith. *Aristotle's* Prior Analytics. Hackett, Indianapolis, 1989. Translated with introduction and notes by Robin Smith.

[Sorabji, 2004] R. Sorabji. *The Philosophy of the Commentators, 200-600 AD: A Sourcebook, vol. 3: Logic and Metaphysics.* Duckworth, London, 2004.

[Spade, 1982] P. Spade. Three theories of obligationes: Burley, Kilvington and Swyneshed on counterfactual reasoning. *History and Philosophy of Logic*, 3:19–28, 1982.

[Striker, 2009] G. Striker. *Aristotle:* Prior Analytics*, Book I.* Clarendon Press, Oxford, 2009.

[Thom, 1996] P. Thom. *The Logic of Essentialism: An Interpretation of Aristotle's Modal Syllogistic.* Kluwer, Dordrecht, 1996.

[Thom, 2003] P. Thom. *Medieval Modal Systems: Problems and Concepts.* Ashgate, Aldershot, 2003.

[Thom, 2007] P. Thom. *Logic and Ontology in the Syllogistic of Robert Kilwardby.* Brill, Leiden, 2007.

[Thomason, 1993] S. Thomason. Semantic analysis of the modal syllogistic. *Journal of Philosophical Logic*, 22:111–128, 1993.

[van Rijen, 1989] J. van Rijen. *Aspects of Aristotle's Logic of Modalities.* Kluwer, Dordrecht, 1989.

[Waterlow, 1982a] S. Waterlow. *Nature, Change, and Agency in Aristotle's Physics.* Clarendon Press, Oxford, 1982.

[Waterlow, 1982b] S. Waterlow. *Passage and Possibility: A Study of Aristotle's Modal Concepts.* Clarendon Press, Oxford, 1982.

[Yrjönsuuri, 1994] M. Yrjönsuuri. *Obligationes: 14th Century Logic of Disputational Duties.* Acta Philosophica Fennica, Societas Philosophica Fennica, Helsinki, 1994.

[Yrjönsuuri, 2001] M. Yrjönsuuri, editor. *Medieval Formal Logic: Obligations, Insolubles and Consequences.* Kluwer, Dordrecht, 2001.

A HISTORY OF NATURAL DEDUCTION

Francis Jeffry Pelletier and Allen P. Hazen

1 INTRODUCTION

Work that is called 'natural deduction' is carried out in two ways: first, as an object-language method to prove theorems and to demonstrate the validity of arguments; and secondly, as a metatheoretic investigation into the properties of these types of proofs and the use of properties of this sort to demonstrate results about other systems (such as the consistency of arithmetic or analysis). In the former realm, we turn to elementary textbooks that introduce logic to philosophy students; in the latter realm, we turn to the topic of proof theory.

2 OBJECT LANGUAGE NATURAL DEDUCTION

Natural deduction for classical logic is the type of logical system that almost all philosophy departments in North America teach as their first and (often) second course in logic.[1] Since this one- or two-course sequence is all that is required by most North American philosophy departments, most non-logician philosophers educated in North America know only this about logic. Of course, there are those who take further courses, or take logic in mathematics or computer science departments. But the large majority of North American philosophers have experienced only natural deduction as taught using one or another of the myriad elementary logic textbooks (or professor's classroom notes, which are usually just "better ways to explain" what one or another of the textbooks put forth as logic).

But when a student finishes these one- or two-semester courses, he or she is often unable to understand a different elementary logic textbook, even though it and the textbook from the course are both "natural deduction". Part of the reason for this — besides the students' not yet having an understanding of what logic is — concerns the fact that many different ideas have gone into these different books, and from the point of view of an elementary student, there can seem to be very little in common in these books. There is, of course, the basic issue of the specific language under consideration: will it have upper- or lower-case letters? will it use arrows or horseshoes? will it have free variables in premises? etc. But these issues of the choice of language are not what we are pointing towards here; instead, we

[1]This is still true, despite the encroachment of semantic tableaux methods that are now often taught alongside natural deduction.

Handbook of the History of Logic. Volume 11: Logic: A History of its Central Concepts.
Volume editors: Dov M. Gabbay, Francis Jeffry Pelletier and John Woods.
General editors: Dov M. Gabbay and John Woods.

think of the differing ways that different textbooks represent proofs, differing rules of inference, and so on. How can all these differences be accommodated under the single term 'natural deduction'? What are the essential properties that make a proof system be 'natural deduction'? What is 'natural' about them all?

Our view is that there are quite a number of characteristics that contribute to a proof system's being called 'natural deduction', but a system need not have them all in order to be a natural deduction system. We think that the current connotation of the term functions rather like a prototype: there is/are some exemplar(s) that the term most clearly applies to and which manifest(s) a number of characteristics. But there are other proof systems that differ from this prototypical natural deduction system and are nevertheless correctly characterized as being natural deduction. It is not clear just how many of the properties that the prototype exemplifies can be omitted and still have a system that is correctly characterized as a natural deduction system, and we will not try to give an answer. Instead we focus on a number of features that are manifested to different degrees by the various natural deduction systems. The picture is that if a system ranks "low" on one of these features, it can "make up for it" by ranking high on different features. And it is somehow an overall rating of the total amount of conformity to the entire range of these different features that determines whether any specific logical system will be called a natural deduction system. Some of these features stem from the initial introduction of natural deduction in 1934 [Jaśkowski, 1934; Gentzen, 1934]; but even more strongly, in our opinion, is the effect that elementary textbooks from the 1950s had.

This mixture of features lends itself to identifying both a wide and a narrow notion of 'natural deduction'. The narrow one comes from the formal characterization of proof systems by (especially) Gentzen, and to some extent from the elementary textbook authors. The wide notion comes by following up some informal remarks that Gentzen and Jaśkowski made and which have been repeated in the elementary textbooks. Also thrown into this are remarks that have been made by researchers in related disciplines... such as in mathematics and computer science... when they want to distinguish natural deduction from their own, different, logical proof systems.

2.1 The Wider Notion of Natural Deduction

Before moving onto distinctions among types of proof systems, we mention features that are, by some, associated with natural deduction. As we have said, we do not think of these features as *defining* natural deduction, but rather as contributing to the general mixture of properties that are commonly invoked when one thinks of natural deduction.

One meaning of 'natural deduction' — especially in the writings from computer science and mathematics, where there is often a restriction to a small set of connectives or to a normal form — focuses on the notion that systems employing it will retain the 'natural form' of first-order logic and will not restrict itself to any subset of the connectives nor any normal form representation. Although this is clearly a feature of the modern textbooks, we can easily see that such a definition is neither necessary nor sufficient for a logical system's being a natural deduction system. For, surely we can give natural deduction accounts for logics that have restricted sets of connectives, so it is not necessary. And we can have non-natural-deduction systems (e.g., axiomatic systems) that contain all the usual connectives, so it is not sufficient.

Another feature in the minds of many is that the inference rules are "natural" or "pretheoretically accepted." To show how widely accepted this feature is, here is what five elementary natural deduction textbooks across a fifty year span have to say. [Suppes, 1957, p. viii] says: "The system of inference... has been designed to correspond as closely as possible to the author's conception of the most natural techniques of informal proof." [Kalish and Montague, 1964, p. 38] says that these systems "are said to employ natural deduction and, as this designation indicates, are intended to reflect intuitive forms of reasoning." [Bonevac, 1987, p. 89] says: "we'll develop a system designed to simulate people's construction of arguments.... it is natural in the sense that it approaches... the way people argue." [Chellas, 1997, p. 134] says "Because the rules of inference closely resemble patterns of reasoning found in natural language discourse, the deductive system is of a kind called natural deduction." And [Goldfarb, 2003, p. 181] says "What we shall present is a system for *deductions*, sometimes called a system of *natural deduction*, because to a certain extent it mimics certain natural ways we reason informally." These authors are echoing [Gentzen, 1934, p. 74] "We wish to set up a formalism that reflects as accurately as possible the actual logical reasoning involved in mathematical proofs."

But this also is neither necessary nor sufficient. An axiom system with only modus ponens as a rule of inference obeys the restriction that all the rules of inference are "natural", yet no one wants to call such a system 'natural deduction', so it is not a sufficient condition. And we can invent rules of inference that we would happily call natural deduction even when they do not correspond to particularly normal modes of thought (such as could be done if the basis had unusual connectives like the Sheffer stroke[2], and is often done in modal logics, many-valued logics, relevant logics, and other non-standard logics).

As we have said, the notion of a rule of inference "being natural" or "pretheoretically accepted" is often connected with formal systems of natural deduction; but as we also said, the two notions are not synonymous or even co-extensive. This means that there is an interesting area of research open to those who wish to investigate what "natural reasoning" is in ordinary, non-trained people, and its relationship with what logicians call 'natural deduction'. This sort of investigation is being carried out by a group of cognitive scientists, but their results are far from universally accepted [Rips, 1994; Johnson-Laird and Byrne, 1991]. (See also the papers in [Adler and Rips, 2008, esp. §3].)

[2]See [Price, 1961] for a nice, but unusual, trio of rules for the Sheffer stroke.

2.2 Different Proof Systems

Another way to distinguish natural deduction from other methods is to compare what these competing proof systems offer or require, compared with natural deduction. Proof systems can be characterized by the way proofs of theorems and arguments proceed. The syllogistic, for example, is characterized by the basic proofs consisting of two premises and a conclusion, and which obeyed certain constraints on the items that made up these sentences. And all extended proofs consist of concatenations of basic proofs, so long as they obey certain other constraints. An axiomatic system contains a set of "foundational" statements (axioms) and some rules that characterize how these axioms can be transformed into other statements. A tableaux system consists of a number of rules that describe how to "decompose" a formula to be proved, and does so with the intent of discovering whether or not there can be a proof of the formula.

The foregoing descriptions are merely rough-and-ready. For one thing, they do not always distinguish one type of proof system from another, as we will see. For another thing, all of them are rather too informal to clearly demarcate a unique set of proof systems. Part of the explanation of these shortcomings is that the different types of systems in fact *do* overlap, and that (with a bit of latitude) one can see some systems as manifesting more than one type of proof theory.

Nonetheless, we think that each type has a central core — a prototypical manifestation — and that the resulting cores are in fact different from one another. We think that there are at least five[3] different types of modern systems of logic: axiomatic, resolution, tableaux, sequent calculus, and natural deduction. We do not intend to spend much time describing the first two types, merely enough to identify them to the reader who is already familiar with some different types of systems. Later on (§3.2) we will discuss the relation between sequent calculus and natural deduction, and also describe the way tableaux methods are related to both sequent calculus and natural deduction. Our main goal is to describe natural deduction, and it is to that end that our accounts of the other types of systems are directed: as signposts that describe what is *not* considered a natural deduction system.

Axiomatic Systems of logic[4] are characterized by having axioms — formulas that are not proved other than by a citation of the formula itself — although this

[3]There are others, no doubt, such as "the inverse method". But these lesser-known systems won't be mentioned further. (The inverse method is generally credited to Sergey Maslov [1964; 1969]. Nice introductions to the inverse method are [Lifschitz, 1989; Degtyarev and Voronkov, 2001].) There are also algebraic systems, where formulas are taken as terms and which use substitution of equals for equals as rules in an equation calculus. And there is also the interesting case of Peirce's "existential graphs."

[4]Often called Frege or Hilbert systems, although the idea of an axiomatic system seems to go back to Euclid. The name 'Hilbert system' is perhaps due to [Gentzen, 1934], who mentioned "einem dem Hilbertschen Formalismus angeglichenen Kalkül"; and its use was solidified by Kleene in his [1952, §15] when he called them "Hilbert-type systems". If one wanted to use a name more modern than 'Euclid', it would be historically more accurate to call them "Frege systems" (as some logicians and historians in fact do).

is not a sufficient characterization because natural deduction systems also often have axioms, as we will see below. Prototypically, axiom systems are characterized by having only a small number of rules, often merely modus ponens (detachment; conditional elimination) plus a rule of substitution (or schema formulation of the axioms). And then a proof of formula φ is a sequence of lines, each one of which is either an axiom or follows from preceding lines by one of the rules of inference, and whose last line is φ. Premises can also be accommodated, usually by allowing them to be entered at any point in the proof, but not allowing the rule of substitution to apply to them.

Resolution Systems of logic[5] are refutation-based in that they start by assuming that the to-be-proved is false, that is, starting with its negation (in classical logic). It also employs a normal form: this negation, plus all the premises (if any), are converted to clausal normal form. In the propositional logic clausal normal form is just conjunctive normal form (a conjunction of disjunctions of literals [which are atomic sentences or their negations]). In the predicate logic one first converts to a prenex normal form where all quantifiers have scope over the remainder of a formula, and then existential quantifiers are eliminated by use of Skolem functions. The remaining universal quantifiers are dropped, and variables are assumed to be universally quantified. These formulas are then converted to conjunctive normal form. Each conjunct is called a clause, and is considered as a separate formula. The resulting set of clauses then has two rules applied to them (and to the clauses that result from the application of these rules): resolution and unification. Resolution is the propositional rule:

DEFINITION 1. **Resolution:**

From $(p_1 \vee p_2 \vee \cdots \vee r \vee \cdots \vee p_n)$ and $(q_1 \vee q_2 \vee \cdots \vee \neg r \vee \cdots \vee q_m)$
infer $(p_1 \vee p_2 \vee \cdots \vee p_n \vee q_1 \vee q_2 \vee \cdots \vee q_m)$. [With no occurrence of r or $\neg r$].

In the predicate logic case, if one of r and $\neg r$ contains a free variable (hence universally quantified) x while the other contains some term τ in that position, then infer the clause in accordance with the rule of resolution but where all occurrences of x in the resolvent clause are replaced by the most general unifier of x and τ. The goal of a resolution proof is to generate the null clause, which happens when (the most general unifiers of) the two parent clauses are singletons (disjunctions with only one disjunct) that are negations of each other. If the empty clause is generated, then the initial set of formulas (with its unnegated conclusion) is a valid argument.

Tableau Systems of logic[6] are characterized by being decompositional in nature, and often are constructed so as to mimic the semantic evaluation one would employ in determining whether some given formula is true. (Because of this one

[5]Initially described by [Robinson, 1965], these are the most commonly-taught methods of logic in computer science departments, and are embedded into the declarative programming language Prolog.

[6]Usually these are credited to [Beth, 1955] and [Hintikka, 1953; Hintikka, 1955a; Hintikka, 1955b], although the ideas can be teased out of [Herbrand, 1930] and [Gentzen, 1934], as done in [Craig, 1957]. See also [Anellis, 1990] for a historical overview.

often sees the term 'semantic tableaux method', indicating to some that these methods are not really proof procedures but rather are ways to semantically evaluate formulas. We take the view here that the system can also be viewed as purely formal methods to manipulate formulas, just as much as axiomatic systems seem to do. And therefore they count as proof systems.) Generally speaking, these systems start with a formula asserting that the formula to be proved is not a designated formula. This can be done in a number of different ways, such as claiming that it is false or that its negation is true (depending on whether the system is bivalent and whether negation works "normally"). The decomposition rules are constructed so as to "preserve truth": "if the formula to be decomposed were true, then at least one of the following would be true." But if the initial formula were provable, then there can be no string of decompositions thathas all true sentences in it. (Various adjustments to this characterization are necessary if the system is not bivalent or not 'extensional'.) Most developments of tableaux systems express proofs in terms of decomposition trees, and when a branch of the tree turns out to be impossible, then it is marked 'closed'; if all branches are closed, then the initial argument is valid. We return to tableaux systems in §3.7.

Sequent Calculus was invented by Gerhard Gentzen [1934], who used it as a stepping-stone in his characterization of natural deduction, as we will outline in some detail in §3.2. It is a very general characterization of a proof; the basic notation being $\varphi_1, \ldots, \varphi_n \vdash \psi_1, \ldots, \psi_m$, which means that it is a consequence of the premises $\varphi_1, \ldots, \varphi_n$ that at least one of $\psi_1, \ldots, \psi_m$ holds.[7] If Γ and Σ are sets of formulas, then $\Gamma \vdash \Sigma$ means that it is a consequence of all the formulas of Γ that at least one of the formulas in Σ holds. Sequent systems take basic sequents such as $\varphi \vdash \varphi$ as axiomatic, and then a set of rules that allow one to modify proofs, or combine proofs. The modification rules are normally stated in pairs, 'x on the left' and 'x on the right': how to do something to the premise set Γ and how to do it to the conclusion set Σ. So we can understand the rules as saying "if there is a consequence of such-and-so form, then there is also a consequence of thus-and-so form". These rules can be seen as being of two types: structural rules that characterize the notion of a proof, and logical rules that characterize the behavior of connectives. For example, the rule that from $\Gamma \vdash \Sigma$ one can infer $\Gamma, \varphi \vdash \Sigma$ ("thinning on left") characterizes the notion of a proof (in classical logic), while the rule that from $\Gamma, \varphi \vdash \Sigma$ one can infer $\Gamma, (\varphi \wedge \psi) \vdash \Sigma$ ("$\wedge$-Introduction on left") characterizes (part of) the behavior of the logical connective $\wedge$ when it is a premise. We expand considerably on this preliminary characterization of sequent calculi in §3.2.

[7]One can also read this as asserting the existence of proofs: the just-mentioned sequent would then be understood as saying that there is a proof from the premises $\varphi_1, \ldots, \varphi_n$ to the disjunction of $\psi_1, \ldots, \psi_m$. Basic sequents, described just below, would then be understood as asserting the existence of "basic proofs". In such an interpretation of sequents, the inference rules would then be seen as asserting that "if there is a proof of such-and-such type, then there is also a proof of so-and-so type."

2.3 The Beginnings of Natural Deduction: Jaśkowski and Gentzen (and Suppes) on Representing Natural Deduction Proofs

[Gentzen, 1934] coined the term 'natural deduction', or rather the German *das natürliche Schließen*.[8] But [Jaśkowski, 1934] might have a better claim to have been the first to invent a system that embodies what we now recognize as natural deduction, calling it a "method of suppositions".

According to [Jaśkowski, 1934], Jan Łukasiewicz had raised the issue in his 1926 seminars that mathematicians do not construct their proofs by means of an axiomatic theory (the systems of logic that had been developed at the time) but rather made use of other reasoning methods; especially they allow themselves to make "arbitrary assumptions" and see where they lead. Łukasiewicz wondered whether there could be a logical theory that embodied this insight but which yielded the same set of theorems as the axiomatic systems then in existence. Again according to [Jaśkowski, 1934], he (Jaśkowski) developed such a system and presented it to the First Polish Mathematical Congress in 1927 at Lvov, and it was mentioned in their published proceedings of 1929. There seems to be no copies of Jaśkowski's original paper in circulation, and our knowledge of the system derives from a lengthy footnote in [Jaśkowski, 1934]. (This is also where he said that it was presented and an abstract published in the Proceedings. Jan Woleński, in personal communication, tells us that in his copy of the Proceedings, Jaśkowski's work [Jaśkowski, 1929] was reported by title in the Proceedings.) Although the footnote describes the earlier use of a graphical method to represent these proofs, the main method described in [Jaśkowski, 1934] is rather different — what we below call a bookkeeping method. [Cellucci, 1995] recounts Quine's visit to Warsaw in 1933, and his meeting with Jaśkowski, and he suggests that the change in representational method might be due to a suggestion of Quine (who also used a version of this bookkeeping method in his own later system, [Quine, 1950]).

This earlier graphical method consists in drawing boxes or rectangles around portions of a proof; the other method amounts to tracking the assumptions and their consequences by means of a bookkeeping annotation alongside the sequence of formulas that constitutes a proof. In both methods the restrictions on completion of subproofs (as we now call them) are enforced by restrictions on how the boxes or bookkeeping annotations can be drawn. We would now say that Jaśkowski's system had two subproof methods: conditional-proof (conditional-introduction)[9]

[8]For more on Gentzen, and how natural deduction fit into his broader concerns about the consistency of arithmetic, etc., see [von Plato, 2008b; von Plato, 2008a].

[9]Obviously, this rule of Conditional-Introduction is closely related to the deduction theorem, that from the fact that $\Gamma, \varphi \vdash \psi$ it follows that $\Gamma \vdash (\varphi \rightarrow \psi)$. The difference is primarily that Conditional Introduction is a rule of inference in the object language, whereas the deduction theorem is a metalinguistic theorem that guarantees that proofs of one sort could be converted into proofs of the other sort. According to [Kleene, 1967, p.39fn33], "The deduction theorem as an informal theorem proved about particular systems like the propositional calculus and the predicate calculus...first appears explicitly in [Herbrand, 1930] (and without proof in [Herbrand, 1928]); and as a general methodological principle for axiomatic-deductive systems in [Tarski, 1930b]. According to [Tarski, 1956, p.32fn], it was known and applied by Tarski since 1921."

and reductio ad absurdum (indirect proof). It also had rules for the direct manipulation of formulas (e.g., Modus Ponens). After formulating his set of rules, Jaśkowski remarks (p.238) that the system "has the peculiarity of requiring no axioms" but that he can prove it equivalent to the established axiomatic systems of the time. (He shows this for various axiom systems of Łukasiewicz, Frege, and Hilbert). He also remarks (p.258) that his system is "more suited to the purposes of formalizing practical [mathematical] proofs" than were the then-accepted systems, which are "so burdensome that [they are] avoided even by the authors of logical [axiomatic] systems." Furthermore, "in even more complicated theories the use of [the axiomatic method] would be completely unproductive." Given all this, one could say that Jaśkowski was the inventor of natural deduction as a complete logical theory.

Working independently of Łukasiewicz and Jaśkowski, Gerhard Gentzen published an amazingly general and amazingly modern-sounding two-part paper in (1934/35). The opening remarks of [Gentzen, 1934] are

> My starting point was this: The formalization of logical deduction, especially as it has been developed by Frege, Russell, and Hilbert, is rather far removed from the forms of deduction used in practice in mathematical proofs. Considerable formal advantages are achieved in return.
>
> In contrast, I intended first to set up a formal system which comes as close as possible to actual reasoning. The result was a 'calculus of natural deduction' ('NJ' for intuitionist, 'NK' for classical predicate logic). ...

Like Jaśkowski, Gentzen sees the notion of making an assumption to be the leading idea of his natural deduction systems:

> ...the essential difference between NJ-derivations and derivations in the systems of Russell, Hilbert, and Heyting is the following: In the latter systems true formulae are derived from a sequence of 'basic logical formulae' by means of a few forms of inference. Natural deduction, however, does not, in general, start from basic logical propositions, but rather from assumptions to which logical deductions are applied. By means of a later inference the result is then again made independent of the assumption.

These two founding fathers of natural deduction were faced with the question of how this method of "making an arbitrary assumption and seeing where it leads" could be represented. As remarked above, Jaśkowski gave two methods; Gentzen also contributed a method. There is one further method that was introduced some 20 years later in [Suppes, 1957]. All of the representational methods used in today's natural deduction systems are variants on one of these four.

The differences and similarities among these methods of representing natural deduction proofs are easiest to see in an example of a proof done in the different methods, with the proof's ebb and flow of assumptions and cancellation of assumptions. Of course, our writers had different rules of inference in mind as they described their systems, but we need not pause over that at the moment. We will use names in common use for these rules. (And also, since Jaśkowski did not have $\wedge$ in his language, we are employing a certain latitude. But it is clear how his systems would use it.) We consider the way a proof of the theorem $(((p \supset q) \wedge (\neg r \supset \neg q)) \supset (p \supset r))$ is represented in the four different formats.

Since the main connective is a conditional, the most likely strategy will be to prove it by a rule of conditional introduction. But to apply this rule one must have a subproof that assumes the conditional's antecedent and ends with the conditional's consequent. All the methods will follow this strategy; the differences among them concern only how to represent the strategy.

In Jaśkowski's graphical method, each time an assumption is made it starts a new portion of the proof which is to be enclosed with a rectangle (a "subproof"). The first line of this subproof is the assumption... in the case of trying to apply conditional introduction, the assumption will be the antecedent of the conditional to be proved and the remainder of this subproof will be an attempt to generate the consequent of that conditional. If this can be done, then Jaśkowski's rule conditionalization says that the conditional can be asserted as proved *in the subproof level of the box that surrounds the one just completed.* So the present proof will assume the antecedent, $((p \supset q) \wedge (\neg r \supset \neg q))$, thereby starting a subproof trying to generate the consequent, $(p \supset r)$. But this consequent itself has a conditional as main connective, and so it too should be proved by conditionalization with a yet-further-embedded subproof that assumes its antecedent, p, and tries to generate its consequent, r. As it turns out, this subproof calls for a yet further embedded subproof using Jaśkowski's reductio ad absurdum. Here is how this proof would be represented in his graphical method.

1.	$((P \supset Q)\&(\sim R \supset \sim Q))$	Supposition
2.	P	Supposition
3.	$((P \supset Q)\&(\sim R \supset \sim Q))$	1, Repeat
4.	$(P \supset Q)$	3, Simplification
5.	Q	2,4 Modus Ponens
6.	$(\sim R \supset \sim Q)$	3, Simplification
7.	$\sim R$	Supposition
8.	$(\sim R \supset \sim Q)$	6, Repeat
9.	$\sim Q$	7,8 Modus Ponens
10.	Q	5, Repeat
11.	R	7-10 Reductio ad Absurdum
12.	$P \supset R$	2-11 Conditionalization
13.	$(((P \supset Q)\&(\sim R \supset \sim Q)) \supset (P \supset R))$	1-12 Conditionalization

Jaśkowski's second method (which he had hit upon later than the graphical method, and was the main method of [Jaśkowski, 1934]) was to make a numerical annotation on the left-side of the formulas in a proof. Again, this is best seen by example; and so we re-present the previous proof. In this new method, Jaśkowski changed the statements of various of the rules and he gave them new names: Rule I is now the name for making a supposition, Rule II is the name for conditionalization, Rule III is the name for modus ponens, and Rule IV is the name for reductio ad absurdum. (Rules V, VI, and VII have to do with quantifier elimination and introduction). Some of the details of these changes to the rules are such that it is no longer required that all the preconditions for the applicability of a rule of inference must be in the same "scope level" (in the new method this means being in the same depth of numerical annotation), and hence there is no longer any requirement for a rule of Repetition. To indicate that a formula is a supposition, Jaśkowski now prefixes it with 'S'.

1.1.	S $((p \supset q) \wedge (\neg r \supset \neg q))$	I
2.1.	$(p \supset q)$	&E 1
3.1.	$(\neg r \supset \neg q)$	&E 1
4.1.1.	S p	I
5.1.1.	q	III 4,2
6.1.1.1.	S $\neg r$	I
7.1.1.1.	$\neg q$	III 6,3
8.1.1.	r	IV 5,7,6
9.1.	$(p \supset r)$	II 4,8
10.	$(((p \supset q) \wedge (\neg r \supset \neg q)) \supset (p \supset r))$	II 1,9

It can be seen that this second method is very closely related to the method of rectangles. (And much easier to typeset!) Its only real drawback concerns whether we can distinguish different subproofs which are at the same level of embedding. A confusion can arise when one subproof is completed and then another started, both at the same level of embedding. In the graphical method there will be a closing of one rectangle and the beginning of another; but here it could get confused. Jaśkowski's solution is to mark the second such subproof as having '2' as its rightmost numerical prefix. This makes numerals be superior to using other symbols in this role, such as an asterisk. As we will see in §2.4, this representational method was adopted by [Quine, 1950], who used asterisks rather than numerals thus leading to the shortcoming just noted.

A third method was given by Gentzen. Proofs in the N calculi (the natural deduction calculi) are given in a tree format with formulas appearing as nodes of the tree. The root of the tree is the formula to be proved, and the "suppositions" are at the leaves of the tree. The following is a tree corresponding to the example we have been looking at, although it should be mentioned that Gentzen's main rule for indirect proofs first generated $\perp$ ("the absurd proposition") from the two parts of a contradiction, and then generated the negation of the relevant assumption.[10]

$$
\dfrac{\dfrac{\dfrac{\dfrac{\overset{\cancel{3}}{\neg r} \quad \dfrac{\overset{\cancel{1}}{((p \supset q) \wedge (\neg r \supset \neg q))}}{(\neg r \supset \neg q)}\,\wedge\text{-E}}{\neg q}\,\supset\text{-E} \qquad \dfrac{\dfrac{\overset{\cancel{1}}{((p \supset q) \wedge (\neg r \supset \neg q))}}{(p \supset q)}\,\wedge\text{-E} \quad \overset{\cancel{2}}{p}}{q}\,\supset\text{-E}}{\dfrac{\bot}{r}\,\bot\text{-E (3)}}\,\bot\text{-I}}{(p \supset r)}\,\supset\text{-I (2)}}{(((p \supset q) \wedge (\neg r \supset \neg q)) \supset (p \supset r))}\,\supset\text{-I (1)}
$$

The lines indicate a transition from the upper formula(s) to the one just beneath the line, using the rule of inference indicated on the right edge of the line. (We might replace these horizontal lines with vertical or splitting lines to more clearly indicate tree-branches, and label these branches with the rule of inference responsible, and the result would look even more tree-like). Gentzen uses the numerals on the leaves as a way to keep track of subproofs. Here the main antecedent of the conditional to be proved is entered (twice, since there are two separate things to do with it) with the numeral '1', the antecedent of the consequent of the main theorem is entered with numeral '2', and the formula $\neg r$ (to be used in the reductio part of the proof) is entered with numeral '3'. When the relevant "scope changing" rule is applied (indicated by citing the numeral of that branch as part of the citation of the rule of inference, in parentheses) this numeral gets "crossed out", indicating that this subproof is finished.

One reason that Jaśkowski's (and Quine's) bookkeeping method did not become more common is that [Suppes, 1957] introduced a method using the line numbers of the assumptions which any given line in the proof depended upon, rather than asterisks or arbitrary numerals. The method retained the ease of typesetting that the bookkeeping method enjoyed over the graphical and tree representations, but was much clearer in its view of how new subproofs were started. In this fourth method, when an assumption is made its line number is put in set braces to the left of the line (its "dependency set"). The application of "ordinary rules" such as &E and Modus Ponens make the resulting formula inherit the union of the dependencies of the lines to which they are applied, whereas the "scope changing" rules like $\supset$I and Reductio delete the relevant assumption's line number from the dependencies. In this way, the "scope" of an assumption is not the continuous sequence of lines that occurs until the assumption is discharged by a $\supset$I or $\neg$I rule, but rather consists of just those (possibly non-contiguous) lines that "depend upon" the assumption. But the fact that the lines in a given subproof are no longer necessarily contiguous marks a break from the idea that we are "making an assumption and seeing where it leads" — or at least, one might argue this idea has been lost. In fact, if one views the numbers in a dependency set as just an

[10] He also considered a double negation rule.

abbreviation for the actual formula that occurs on the line with that number, and the line number itself is just to be replaced with '$\vdash$', it can appear that the system is actually representing a sequent calculus proof (except for the fact that there is always just one formula on the right side). This is explored in detail in §3.3.

Without using Suppes's specific rules, we can get the flavor of this style of representation by presenting the above theorem as proved in a Suppes-like manner.

{1}	1.	$((p \supset q) \wedge (\neg r \supset \neg q))$	
{1}	2.	$(p \supset q)$	&E 1
{1}	3.	$(\neg r \supset \neg q)$	&E 1
{4}	4.	p	
{1,4}	5.	q	$\supset$E 4,2
{6}	6.	$\neg r$	
{1,6}	7.	$\neg q$	$\supset$E 6,3
{1,4}	8.	r	Reductio 5,7,6
{1}	9.	$(p \supset r)$	$\supset$I 4,8
$\emptyset$	10.	$(((p \supset q) \wedge (\neg r \supset \neg q)) \supset (p \supset r))$	$\supset$I 1,9

These four methods of representing natural deduction proofs continue to characterize natural deduction, although it should be remarked that neither the bookkeeping method nor the Gentzen method became very popular. ([Quine, 1950] used the bookkeeping method, and because of Quine's stature, some few other textbooks retained it, but this became very rare. Gentzen's method finds its place in more technical discussions of natural deduction, and in a few intermediate-level textbooks. It is quite rare in beginning textbooks, although it is used in [van Dalen, 1980; Tennant, 1978] if these are seen as elementary.)

We turn now to the way natural deduction became the dominant logical method that was taught to generations of (mostly philosophy) undergraduates (in North America).

2.4 Natural Deduction in Elementary Textbooks

In §2.3 we looked at the four ways that have been used in proofs to visually display the notion of "making an assumption and seeing where it leads". One reason elementary students find the generic notion of natural deduction confusing is because they find it difficult to understand how systems that have such diverse representational methods can really be "the same type of logical system".

These four ways have, each to a greater or lesser extent, been adopted by the elementary logic textbooks that started to appear in the early 1950s. [Quine, 1950] was the first of these textbooks, and as we have said, it used the Jaśkowski bookkeeping method. As we also said, this way of representing natural deduction proofs was not very popular in the textbook genre, being followed by only four further textbooks in the following 60 years.[11] Gentzen's style of proof was also not

[11]This number, as well as the ones to come, are derived from [Pelletier, 2000, p. 135], and augmented with another 17 textbooks. As is remarked there, the figures are not "scientific" in

very popular in the textbooks, finding its way into only five of the 50 textbooks, and as mentioned above, these are mostly claimed to be "intermediate logic books" or "intended for graduate students in philosophy".

Despite the fact that Jaśkowski apparently preferred his bookkeeping to his earlier graphical method, it was the graphical method that caught on. The second elementary logic textbook to use natural deduction was [Fitch, 1952], and the popularity of his version of the graphical method has made this method be called "the Fitch method", despite the fact that it is really Jaśkowski's method. (Fitch [p. vii] says merely "The method of subordinate proofs was suggested by techniques due to Gentzen and Jaśkowski.") The main difference between Jaśkowski's original method and Fitch's is that Fitch does not completely draw the whole rectangle around the embedded subproof (but only the left side of the rectangle), and he underlines the assumption. The same proof displayed above using Jaśkowski's graphical method is done like the following in Fitch's representation (with a little laxness on identifying the exact rules Fitch employs).

1	$\mid\ \underline{((p \supset q) \wedge (\neg r \supset \neg q))}$	
2	$\mid\ \mid\ \underline{p}$	
3	$\mid\ \mid\ ((p \supset q) \wedge (\neg r \supset \neg q))$	1, R
4	$\mid\ \mid\ (p \supset q)$	3,$\wedge$E
5	$\mid\ \mid\ q$	2,4 $\supset$E
6	$\mid\ \mid\ (\neg r \supset \neg q))$	3,$\wedge$E
7	$\mid\ \mid\ \mid\ \underline{\neg r}$	
8	$\mid\ \mid\ \mid\ (\neg r \supset \neg q))$	6,R
9	$\mid\ \mid\ \mid\ \neg q$	7,8 $\supset$E
10	$\mid\ \mid\ \mid\ q$	5,R
11	$\mid\ \mid\ r$	7–10, $\neg$E
12	$\mid\ (p \supset r)$	2–11, $\supset$I
13	$(((p \supset q) \wedge (\neg r \supset \neg q)))$	1–12,$\supset$I

The graphical method, in one form or another, is in use in very many elementary textbooks. Some authors do not make an explicit indication of the assumption, other than that it is the first formula in the scope level; and some authors do not even use the vertical lines, but rather merely employ indentation (e.g., [Hurley,

any clear sense, and represent only the 33 (at that time, and 50 now) elementary logic textbooks using natural deduction that Pelletier "had on his bookshelf". Nonetheless, they certainly seem to give the correct flavor of the way that elementary logic textbooks dealt with natural deduction in the 1950–2010 timeframe under consideration. More details of the contents of these books is given in Table I below.

1982]). Other authors, e.g., [Kalish and Montague, 1964; Kalish *et al.*, 1980; Bonevac, 1987] keep the rectangles, but put the conclusion at the beginning (both as a way to guide what a legitimate assumption is and also to help in the statement of restrictions on existential elimination and universal introduction rules). And there are authors who retain still different parts of the rectangle, such as [Copi, 1954], who keeps the bottom and left side parts, and ends the top part with an arrowhead pointing at the assumption. (Similar variations are in [Gamut, 1991; Harrison, 1992; Bessie and Glennan, 2000; Arthur, 2010]). But clearly, all these deviations from Jaśkowski's original boxes are minor, and all these should be seen as embracing the graphical method.

Table I below reports that 60% of the elementary logic textbooks that were surveyed used some form of this graphical method, making it by far the most common representation of natural deduction proofs. The bookkeeping and Gentzen-tree methods accounted for 10% each, and the remaining 20% employed the Suppes method. (Some might think this latter percentage must be higher, but this perception is probably due to the immense popularity of [Lemmon, 1965] and [Mates, 1965], which both used the Suppes-method.)

2.5 More Features of the Prototype of Natural Deduction

In §2.1 we looked at two aspects of systems of natural deduction that are in the minds of many researchers... especially those from computer science and mathematics... when they think about natural deduction. Although as we said there, these do not define natural deduction, it is still the case that they form a part of the prototypical natural deduction systems. In this section we look at some more features of this sort, but where these are more closely tied to the remarks that Gentzen and Jaśkowski made in their initial discussions.

A third feature of natural deduction systems, at least in the minds of some, is that they will have two rules for each connective: an introduction rule and an elimination rule. But again this can't be necessary, because there are many systems we happily call natural deduction which do not have rules organized in this manner. The data in Table I below report that only 23 of the 50 texts surveyed even pretended to organize their natural deduction rules as matching int-elim rules (and some of these 23 acknowledged that they were "not quite pure" int-elim). Furthermore, since they are textbooks for classical (as opposed to intuitionistic) logic, these systems all have to have extra features that lead to classical logic, such as some "axioms" (see below). And anyway, even if we concocted an axiomatic system that did have rules of this nature, this would not make such a system become a natural deduction system. So it is not sufficient either.

A fourth idea in the minds of many, especially when they consider the difference between natural deduction and axiomatic systems, is that natural deduction does not have axioms. (Recall Jaśkowski's remark that his system has no requirement of axioms.) But despite the fact that Jaśkowski found no need for axioms, Gentzen *did* have them in his NK, the natural deduction system for classical logic; it was only his NJ, the intuitionistic logic, that did not have them. And many of the authors of more modern textbooks endorse methods that are difficult to distinguish from having axioms. For example, as a primitive rule many authors have a set of 'tautologies' that can be entered anywhere in a proof. This is surely the same as having axioms. Other authors have such a set of tautological implications together with a rule that allows a line in a proof to be replaced by a formula which it implies according to a member of this set of implications. So, if φ is a line in a proof, and one of the tautological implications in the antecedently-given list is $\varphi \rightarrow \psi$, then ψ can be entered as a line (with whatever dependencies the φ line had). And in the world of "having axioms" it is but a short step from here to add to the primitive formulation of the system a set of 'equivalences' and the algebraically-inspired rule that one side of the equivalence that can be substituted for a *subpart* of an existing line that matches the other side of the equivalence, as many authors do. A highly generalized form of this method is adopted by [Quine, 1950], [Suppes, 1957], and others, which have a rule TF ("truth functional inference") that allows one to infer "any schema which is truth-functionally implied by the given line(s)".[12] Although one can detect certain differences amongst all these variants just mentioned here, they seem all to be ways of adopting axioms.[13] Table I lists 22 of the 50 textbooks (44%) as allowing either a TF inference or else tautologies or replacement from a list of equivalences in the primitive portion of their system... in effect, containing axioms.

A fifth idea held by some is that "real" natural deduction will have introduction and elimination rules for every connective, and there will be no further rules. But even Gentzen, to whom this int-elim ideal is usually traced, thought that he had a natural deduction system for classical logic. And as we remarked above, a difference between the natural deduction system for intuitionistic logic and that for classical logic was that the latter had every instance of $\varphi \vee \neg\varphi$ as axioms. So, not even Gentzen followed this dictum.[14] We can get around the requirement of axioms for classical logic if we have a bit more laxity in what counts as a "pure" introduction/elimination rule — for example, allowing double negation rules to be int-elim. But even with this laxity, the number of elementary textbooks that even try to be int-elim is nowhere near universal: Table I shows less than half to be of this nature, and many of them have *very* lax interpretations of what an int-elim pair is. Other textbooks happily pair modus ponens with modus tollens, pair biconditional modus ponens with two-conditionals-gives-a-biconditional, pair

[12] Their TF rules allow one to infer anything that follows from the conjunction of lines already in the proof.

[13] One might separate systems with the TF rule from these other "axiomatic" systems in that the former do not have any list of tautological implications to employ, and instead this is formulated as a rule. Note that, in the propositional logic, there would be no real need for any other rules, including the rule of conditionalization. For, everything can be proved by the rule TF. (Any propositional theorem follows from the null set of formulas by the rule TF).

[14] However, many theorists believe that this is a reason to think that intuitionistic logic is the "real" logic and that classical logic is somehow defective. Perhaps Gentzen thought so too, many people claim. See §§4.3–4.4 below for some thoughts on this topic.

modus tollendo ponens (= unit resolution, disjunctive syllogism) and separation of cases with *or*-introduction, and so forth. In fact, one elegant version of natural deduction has two *pairs* of rules per connective. [Fitch, 1952] supplements standard rules of, e.g., Conjunction Introduction and Elimination with rules of Negative Conjunction Introduction and Elimination embodying the principle that a negated conjunction is equivalent to a disjunction. In Fitch's book, these are stated as rules making the negated conjunction inter-inferable with the equivalent disjunction: Negative Conjunction Introduction licenses the inference from $\neg\phi \vee \neg\psi$ to $\neg(\phi \wedge \psi)$, Negative Conjunction Elimination licenses the converse inference. It may be that this formulation is easier for beginning students to memorize, but, at least for theoretical purposes, the explicit disjunctive formula is redundant: a more streamlined version of the system would make these rules formally parallel to the disjunction rules, so Negative Conjunction Introduction would license the inference of $\neg(\phi \wedge \psi)$ from either $\neg\phi$ or $\neg\psi$, and Negative Conjunction Elimination would allow a conclusion θ to be inferred from three items, a negated conjunction $\neg(\phi \wedge \psi)$, a subproof deriving θ from the hypothesis $\neg\phi$, and a subproof deriving θ from the hypothesis $\neg\psi$. Similarly, rules of Negative Disjunction Introduction and Negative Disjunction Elimination would parallel the standard ("positive") rules for conjunction; one could (though Fitch, due to the peculiarities of the type-free theory of classes his book presents, does not) add conjunction-like rules of Negative Implication Introduction and Negative Implication Elimination. Since there are dualities in quantificational as well as propositional logic, we can have rules of Negative Universal Quantifier Introduction and Elimination (parallel to the positive rules for the Existential Quantifier) and Negative Existential Quantifier Introduction and Elimination (parallel to the positive rules for the Universal Quantifier). All these negative rules are, of course, classically valid, and would be derivable rules in any complete natural deduction system for classical logic. In application, though, they are useful enough in shortening classical derivations to be worth making explicit. In a natural deduction formulation of any of a number of three- and four-valued logics (the Strong 3-valued logic of [Kleene, 1952], the Logic of Paradox of [Priest, 1979], the logic of First Degree Entailment of [Anderson and Belnap, 1975]), or of Intuitionistic Logic with the constructible negation (also called strong negation) introduced in [Nelson, 1949], it would be appropriate to take both positive and negative rules as primitive. We will refer to them below in describing a number of systems related to natural deduction.

A sixth idea, perhaps the one that most will gravitate to when the other ideas are shown not to distinguish natural deduction from other frameworks, is that the concept of a subproof is unique to natural deduction. Of course, one needs to take some care here: an axiomatic or resolution (etc.) proof could have a contiguous portion that can be seen as a subproof... in fact, it is pretty clear that one can find pairs of axiomatic and natural deduction proofs such that, if one were to lay the proofs side-by-side, the subproof portions of the natural deduction proof would correspond to contiguous sections of the axiomatic proof. Still, there is a feeling that the concept of "make an assumption and see where it leads, and discharge the assumption at a latter point in the proof" is a distinguishing characteristic of natural deduction, for, although *some* axiomatic/resolution/etc. proofs might have this property, it might be said that *all* natural deduction proofs do.

Actually, though, *not* all natural deduction proofs have this property. First, not all particular natural deduction proofs even make assumptions[15], and second, the Suppes representational method does *not* follow the dictum of "make an assumption and see where it leads", because the contiguous lines of a proof need not have the same dependency set of assumptions. So it is not quite clear that this can be used as a defining characteristic, as opposed to being merely a typical characteristic of natural deduction. (There is also the issue that some textbooks that self-describe as natural deduction do not have *any* mechanism for making assumptions, such as [Barker, 1965; Kilgore, 1968; Robison, 1969].[16] [Gustason and Ulrich, 1973] have a system where their only subproof-requiring rule, $\supset I$, 'is dispensable' (they have $(\varphi \vee \neg\varphi)$ as an axiom scheme).[17]

Jaśkowski and Gentzen both had (what we now call) $\supset I$, and all of the textbooks in Table I have such a rule (even if it is said to be "dispensable"). Most natural deduction textbooks have further subproof-requiring rules (Table I lists 82% of the books as having some subproof-requiring rule besides $\supset I$). Gentzen for example used a subproof for his $\vee E$ rule, as do some 44% of our textbook authors. This latter was not necessary, since in the presence of $\supset I$, the $\vee E$ rule could simply have been Separation of Cases:

$$\frac{\phi \vee \psi \qquad \phi \supset \theta \qquad \psi \supset \theta}{\theta} \text{ (SC)}$$

which it is in about 42% of the elementary textbooks surveyed in Table I; these books usually also have Disjunctive Syllogism (= unit resolution or *modus tollendo ponens*: the rule from $\phi \vee \psi$, $\neg\phi$ to infer ψ) together with SC. The remaining either had Disjunctive Syllogism alone or else used equivalences/tautologies for their reasoning concerning how to eliminate disjunctions.

We have seen above Gentzen's rules for negation, which used $\bot$. This subproof-using method is still in use in some six textbooks, but most elementary textbooks (64%) have this instead as their subproof-using $\neg I$ rule:

[15] At least, if we don't think of the premises of an argument as assumptions. Some books treat them as assumptions — e.g., those using the Fitch style of proof, where they are all listed at the beginning of the outermost scope line; but others don't, for example, those books that simply allow a premise to be entered anywhere in a proof, as in [Kalish and Montague, 1964].

[16] We chose not to count [Barker, 1965; Kilgore, 1968; Robison, 1969] as natural deduction despite those authors' claims.

[17] We chose to count [Gustason and Ulrich, 1973] as natural deduction.

$$\dfrac{\begin{matrix}[\phi] \\ \vdots \\ \psi \\ \vdots \\ \neg\psi\end{matrix}}{\neg\phi}\ (\neg I)$$

(or some version where $(\psi \wedge \neg\psi)$ is on one line). And 24% of the texts have a $\neg E$ rule of the same sort, except that the assumed formula is a negation to be eliminated. 20% have an $\equiv I$-rule that requires subproofs, the rest just directly infer $(\varphi \equiv \psi)$ from $(\varphi \supset \psi)$ and $(\psi \supset \varphi)$. Two books have other subproof-requiring rules: [Bostock, 1999; Goodstein, 1957] have, respectively,

$$\dfrac{\begin{matrix}[\phi] \\ \vdots \\ \psi\end{matrix} \qquad \begin{matrix}[\neg\phi] \\ \vdots \\ \psi\end{matrix}}{\psi} \qquad\qquad \dfrac{\begin{matrix}[\phi] \\ \vdots \\ \psi\end{matrix}}{\neg\psi \supset \neg\phi}$$

There are further important differences among our textbook authors, having to do with the ways they handle quantification. But even at the level of propositional logic, we can see that there are no clear and unequivocal criteria according to which we can always label a logical system as natural deduction. There is instead a set of characteristics that a system can have to a greater or lesser degree, and the more of them that a system has, the more happily it wears the title of 'natural deduction'.

2.6 Natural Deduction Quantificational Rules

The story of elementary natural deduction and textbooks is not complete without a discussion of the difficulties involved with the existential quantifier (and related issues with the universal quantifier) in natural deduction systems. If a smart student who has not yet looked at the quantification chapter of his or her textbook were asked what s/he thinks ought to be the elim-rule for existential sentences like $\exists xFx$, s/he will doubtless answer that it should be eliminated in favor of Fa — "so long as a is arbitrary". That is, the rule should be

$$\frac{\exists xFx}{Fa}\ \text{(EI); } a \text{ is arbitrary}$$

This is the rule Existential Instantiation (by analogy with Universal Instantiation, but here with the condition of arbitrariness). Since it was used by [Quine, 1950], sometimes the resulting systems are called Quine-systems. [Quine, 1950b, fn.3] says that Gentzen "had a more devious rule in place of EI". This "more devious" rule, which we call $\exists E$, is

$$\frac{\exists x Fx \qquad \begin{matrix}[Fa] \\ \vdots \\ \phi\end{matrix}}{\phi} \; (\exists E)\text{; with restrictions on } a, \phi, \text{ and all "active assumptions"}$$

Although it may be devious, the Gentzen $\exists E$ has the property that if all the assumptions on all branches that dominate a given formula are true, then that formula must be true also; and this certainly makes a soundness proof be straightforward. (In the linear forms — the bookkeeping or graphical methods — this becomes: if all the active assumptions above a given formula are true then so must that formula be. In the Suppes-style form this becomes: if all the formulas mentioned in the dependency set are true, then so is the given formula.) The reason for this is that the "arbitrary instance" of the existentially quantified formula becomes yet another assumption, and the restrictions on this instance ensure that it cannot occur outside the subproof. The same cannot be said about the Quine *EI* rule where the arbitrary instance is in the same subproof as the existentially quantified formula, and Quine goes to prodigious efforts to have the result come out right: in his earliest version [first edition], he introduces total ordering on the variables and decrees that multiple *EI*s must use this ordering[18], and he introduces the notion of an "unfinished proof" — where all the rules are applied correctly and yet the proof still might not be valid — in which some postprocessing needs to take place to determine validity.

Quine's method, complex though it was, did not torture students and professors as much as the third published natural deduction textbook, [Copi, 1954], which also had a rule of *EI*. Quine's restrictions actually separated valid arguments from invalid ones, but Copi's allowed invalid arguments to be generated by the proof theory. There were many articles written about this, and the proper way to fix it, but it remained an enormously popular textbook that was the most commonly-used textbook of the 1950s and 1960s, despite its unsoundness. The conditions on the rules changed even within the different first edition printings and again in the second edition, but the difficulties did not get resolved until the 1965 third edition, where he adopted the Gentzen $\exists E$ rule.[19] One might think that with all the difficulties surrounding *EI* — particularly the specification of the restrictions on the instantiated variable and the difficulties that arise from having an inferred line that is not guaranteed to be true even if its premise is true — almost all textbooks would employ $\exists E$. But Table I shows that, of the 49 books that had

[18]The second edition changed this to "there is a total ordering on the variables that are used in the proof". See the discussion surrounding [Pelletier, 1999, p.16–17fn18,20]. Other interesting aspects of this can be gathered from [Cellucci, 1995] and [Anellis, 1991].

[19]However, the "proof" later in the book of the soundness of the system was not changed to accommodate this new rule. It continued to have the incorrect proof that wrongly dealt with *EI*. It was not until the 1972 fourth edition that this was corrected — some 18 years of irritated professors and confused students, who nonetheless continued using the book! The circumstances surrounding all this are amusingly related in [Anellis, 1991]. See also [Pelletier, 1999, §5] for a documentation of the various changes in the different printings and editions.

a rule for eliminating existential quantifiers[20], the percentage was only 61% using $\exists E$: 30 used $\exists E$ while 19 used EI.

Related to the choice of having a subproof-introducing $\exists E$ rule vs. non-subproof EI rule, is the choice of requiring a subproof for universal quantifier introduction. Table I lists only eleven of 49 texts[21] as having a subproof-requiring $\forall I$; the other 38 had generalization on the same proof-level (UG: universal generalization), with further restrictions on the variable or name being generalized, so as to ensure "arbitrariness". Of the eleven that required a $\forall I$ subproof, six had $\exists E$ subproofs and four had EI non-subproofs. (The remaining one had no existential elimination at all.) So, there seems to be no particular tie between $\exists E$ subproofs and $\forall I$ subproofs... nor between EI and $\forall I$ subproofs, nor between any other combination, in elementary logic textbooks.

One further note should be added to the discussion of the quantifier rules in natural deduction, and that concerns the description of EI, $\exists E$, and $\forall I$ as requiring "arbitrary objects" for their values. In most textbooks, at least when the proof techniques are being taught, this is merely some sort of *façon de parler*. It plays no actual role in proofs; it is just a way of emphasizing to students that there are conditions on the choice of singular term that is being used in these rules. But sometimes in these textbooks, in the chapters that have some metatheoretical proofs, the notion of an arbitrary object is also appealed to. Here one might find difficulties, since some student may ask embarrassing questions concerning the nature of these objects ("Is this arbitrary person a male? A female? Or neither, since it wouldn't be arbitrary otherwise? But aren't all people either male or female?" Or perhaps: "Our domain doesn't contain this object... it only has this-and-that sort of thing as members... There are no abstract objects in it." Or perhaps from a more sophisticated student: "Objects aren't arbitrary! It is the way that a specific object is *chosen* that is arbitrary!"[22]

Some of our textbook authors were advocates of "the substitution interpretation" of quantification, and so they naturally thought that the singular term that was used in finding "the arbitrary instance" was actually a proper name of some object in the domain. For them, the arbitrariness came into play in the way this name was chosen — it had to be "foreign" to the proof as thus far constructed.

[20]One of the 50 textbooks did not have any rule for eliminating existential quantifiers, other than by appeal to "axioms".

[21]One of the 50 books did not have any generalization rules at all, and thus no universal introduction.

[22]Echoing [Frege, 1904]:

> But are there indefinite numbers? Are numbers to be divided into definite and indefinite numbers? Are there indefinite men? Must not every object be definite? But is not the number n indefinite? I do not know the number n. 'n' is not the proper name of some number, definite or indefinite. And yet one sometimes says 'the number n'. How is that possible? Such an expression must be considered in context. ... One writes the letter 'n', in order to achieve generality....
>
> Naturally one may speak of indefinite here; but here the word 'indefinite' is not an adjective of 'number', but an adverb of, for example, to 'signify'.

And they then had a proof in their metatheory chapter to showed that this restriction was sufficient: if the conclusion could be drawn from such a substitution instance then it could be drawn from the existentially quantified formula, because these formulas were true (in their substitutional interpretation) just in case some instance with a name being substituted in the formula made it true. The "objectual interpretation" version of quantification couldn't make use of this ploy, and they naturally gravitated toward using free variables in their statements of $\forall I$. A formula with a free variable, occurring in a proof, is semantically interpreted as a universally-quantified formula — unless the free variable is due to an EI rule or is introduced in an assumption or is "dependent on" some EI-introduced variable. There are (to judge from the textbooks) a myriad of ways of saying this that have the effect that only the variables that are in fact able to be interpreted as universal are allowed to show up in the precondition to the rule. However, the informal talk in their metatheoretical discussions continued, in many of these authors, to make use of the notion of 'arbitrary object', ignoring the philosophical objections.[23]

But there is one more group: those who use a special vocabulary category of *parameter*. These are not names, but they are not variables either. In fact, the metatheory in the textbooks that use this tends to ignore parameters altogether. In some books the rules are that existential quantifiers can never introduce them, but universal instantiation can, since a universal can instantiate to *anything*; and a legitimate $\forall I$ will require that the item being generalized be one of these parameters. Other books take it that the existential instantiation or elimination rule will make use of these parameters and that the universal generalization rule cannot generalize on such a term. It can be seen that there are two notions here of "arbitrary object" being employed by these two different versions — the former notion of an arbitrary object is "any object at all, it doesn't matter which", while the latter is "a particular object, but you can't know anything about which one". It is not obvious that either conception of 'arbitrary object' makes much sense, except as a *façon de parler*.

2.7 A Summary of Elementary Natural Deduction Textbooks

In this section we will compare how some 50 elementary logic textbooks deal with natural deduction proofs. Every textbook author surveyed here (and reported in Table I) describes their book as being an instance of natural deduction, and so to some extent contributes to the image that philosophers have of what "real" natural deduction is. Many textbooks have gone through many editions over the years, and sometimes this has resulted in differences to the sort of features we are considering here. Whenever we could find the first edition of the book, we used the statements of the rules in that edition.

Here is some background that is relevant to understanding Table I. First, column headings are the names of the first author of textbook, and the textbooks

[23]But [Fine, 1985] set about to make this notion have a more philosophically satisfactory interpretation. See [Hazen, 1987] for further discussion.

can be found in the special bibliographic listing of textbooks after the regular bibliography, below. Second, the features described belong to the *primitive basis* of the textbook. Most textbooks, no matter what their primitive basis, provide a list of "equivalences" or "derived rules." However, these features are not considered in the Table (unless indicated otherwise). We are instead interested in the basis of the natural deduction system.

The Table identifies eight different general areas where the textbooks might differ. And within some of these areas, there are sub-points along which the differences might manifest. The explanation of the eight general areas is the following.

I. Proof Style This area describes which of the four different styles of representation — which we discussed above in §2.3 — is employed by the textbook.
II. "Axiomatic" Describes whether, in the primitive basis of the system, there are unproved formulas that are taken for granted and allowed to be entered or used in the proof without further justification. This is subdivided into (a) having a list of tautologies that are allowed to be entered anywhere in a proof, for instance allowing the formulas $\neg\phi \supset (\phi \supset \psi)$ or $\neg\neg\phi \supset \phi$ to be entered anywhere in a proof; (b) having a list of equivalences that justify the replacement of a subformula of some formula in the proof with what it is equivalent to according to the list, for example, having $\neg(\phi \wedge \psi) \equiv (\neg\phi \vee \neg\psi)$ on the list and allowing formulas in a proof that contain as a subformula (a substitution instance of) the left side to be inferred by the same formula but where the subformula is replaced by (the same substitution instance of) the right side, as in algebraic systems of logic; (c) having a general rule of "truth functionally follows", TF, that allows the inference of a new line in a proof if it follows truth functionally from some or all the antecedent lines of the proof.
III. All Int-Elim This category is supposed to pick out those systems which self-consciously attempt to make their rules of inference come in pairs of 'introduce a connective' and 'eliminate that connective'. We have allowed quite a bit of laxity in asserting that the rules are int-elim in nature, mostly following whether the authors *think* they are giving int-elim rules. As discussed above in §2.5, some authors call rule-sets 'int-elim' which would not be thus considered by other logicians.
IV. Sub-Proofs This tracks the propositional rules of inference which are required to have a subproof as part of their preconditions. Again, keep in mind that this is for the *primitive* portion of the systems: many authors introduce derived rules that might (or not) require subproofs, but this information is not recorded in Table I.
V. One Quantifier This answers the question of whether the system has only one quantifier in its primitive basis (and introduces the other by means of some definition).
VI. Arbitrary Names The question being answered is whether the system makes use of an orthographically distinct set of terms ("parameters" or whatever) being employed in one or both of Universal Quantifier Introduction and Existential Quantifier Elimination. As we said above, most authors use for this purpose either the "names" or the "free variables" that are employed elsewhere in the proof system; but some authors have a distinct set of symbols for this purpose. Sur-

prisingly, it can be difficult to discern whether a system does or doesn't employ orthographically different items for this — witness [Teller, 1989, p.75], who uses letters with circumflex accents for this purpose: "A name with a hat on it is not a new kind of name. A name is a name is a name, and two occurrences of the same name, one with and one without a hat, are two occurrences of the same name. A hat on a name is a kind of flag to remind us that at that point the name is occurring arbitrarily." We have chosen to classify this as using an orthographically different set of terms for use in $\forall I$ and $\exists E$, despite the informal interpretation that Teller puts on it.

VII. Existential Rule The issue here is whether the system employs the Gentzen-like Existential Elimination $\exists E$ rule that requires a subproof, or uses the Quine-like Existential Instantiation (EI) rule which stays at the same proof level.

VIII. Universal Rule Here we track whether the Universal Introduction rule $\forall I$ requires a subproof or not.

2.8 *Exploiting Natural Deduction Techniques: Modal Logic*

The modern study of modal logic was initiated by C. I. Lewis [1918] in an effort to distinguish two concepts of "implication" which he thought had been confused in *Principia Mathematica*: the relation between propositions expressed by the material implication connective and the relation of logical consequence. It can be argued that natural deduction formulations of logic, in which logical *axioms* formulated as material implications are replaced by logical *rules*, go a long way toward clarifying the relation between these concepts; but natural deduction formulations were first published in 1934, almost two decades after Lewis began his study of his systems of "strict implication." In any event, he presented modal logic in a series of basically *axiomatic* systems (differing, however, from the usual axiomatic formulations of non-modal logic by including an algebraically inspired, primitive rule of substitution of proved equivalents). The modal logics he formulated include some that are still central to the study of modality (S_4 and S_5), but also others (which he thought better motivated by his philosophical project) now regarded as having at best artificial interpretations, and he failed to discover yet others (like the basic alethic logic T) that now seem obvious and natural.[24] It seems plausible that the logical technology he was familiar with is what led him to prefer logics like S_2 and S_1, and arguable that this preference was a philosophical mistake: these logics do not count even logical theorems as necessarily necessary, but his rule of substitution allows proven equivalents to be substituted for each other even in positions in the scope of multiple modal operators. This combination of features seems unlikely for a well-motivated necessity concept.

[24]The term 'normal modal logic' was originally introduced by [Kripke, 1963] for a class of logics extending T, and only later extended to include such weaker systems as K when it came to be appreciated how naturally Kripke's model-theoretic approach extended to them.

	Anderson	Arthur	Barwise	Bergmann	Bessie	Bonevac	Bostock	Byerly	Carter	Cauman	Chellas	Copi	Curry	DeHaven	Fitch	Forbes	Gamut	Georgacarakos	Goldfarb	Goodstein	Gustason	Guttenplan	Harrison	Hurley
I. Proof Style																								
bookkeeping										x								x						
graphical	x	x	x	x	x	x		x	x			x		x	x		x				x		x	x(2)
Gentzen							x						x							x				
Suppes											x					x			x			x		
II. "Axiomatic"?																								
tautologies									x			x									x		x	
Equivalences		x	(d)		x			x	x			x						x			x		x	x
"TF inference"																			x					
III. All Int-Elim?(e)	x		x	x		x	x(g)			x			x		x(h)	x	x(g)			x		x		
IV. Sub-Proof for:																								
$\supset$I	x	x	x	x	x	x	x	x	x	x	x	x	x	x	x	x	x	x	x	x	x(k)	x	x	x
$\vee$E	x		x	x			x			x			x		x	x	x			x		x		
~I	x	x	x	x		x			x	x		x	x	x		x	x	x		x		x	x	x
~E				x	x				x		x			x										
$\equiv$I			x	x											x									
other							x(g)													x(l)				
V. One Quantifier?													x							x				
VI. Arbitrary Names?		x								x											x			
VII. Existential Rule																								
$\exists$E	x		x	x	x		x				x		x	x	x	x	x(n)	x	x			x		
$\exists$I		x				x		x	x	x		x								x	x		x	x
VIII. Universal Rule																								
VIII. Universal Rule																								
$\forall$I subproof?			x			x									x									

Table 1: Characterization of 50 elementary natural deduction textbooks

Iseminger	Jacquette	Jennings	Kalish	Klenk	Kozy	Leblanc	Lemmon	Machina	Martin	Massey	Mates	Myro	Pollock	Purtill	Quine	Resnick	Simco[a]	Simpson	Suppes	Tapscott	Teller	Tennant	Thomason	van Dalen	Wilson	Total (of 50)	Percent
										x				x	x											**5**	10%
x	x		x	x	x	x		x	x							x	x	x		x	x		x		x	**30**	60%
																						x		x		**5**	10%
		x					x				x	x	x						x							**10**	20%
																											22[c] of 50=
													x				x			x		x				**8**	44%
				x			x						x	x			x		x	x					x	**17**	
										x	x				x				x							**5**	
x	x	x				x		x	x									x			x	x[f]	x	x[i]		**23**	46%
																											41[j]=82%
x	x	x	x	x	x	x	x	x	x	x	x	x	x	x	x	x	x[k]	x	x	x	x	x	x	x	x	**50**	100%
x	x	x			x	x	x		x			x						x				x	x			**22**	44%
x	x	x	x	x	x	x	x	x	x			x				x	x	x		x	x	x	x		x	**36**	72%
	x		x					x	x								x							x	x	**12**	24%
x	x					x			x			x						x					x			**10**	20%
																										2	4%
											x													x		**4**	8%
		x					x				x			x				x	x		x		x			**11**	22%
													x[m]														of 49
	x	x			x	x	x	x	x		x[o]	x				x		x		x	x	x	x	x[o]		**30**	61%
x			x	x		p)				x				x	x		x		x						x	**19**	39%
																											of 49
																									(q)		of 49
x			x	x		(p)							x			x		x			x		x			**11**	22%

Table 1: Characterization of 50 elementary natural deduction textbooks, continued

Footnotes to Table I

(a) Reporting "the derived system". The main system is axiomatic.

(b) Hurley uses indentation only, without any other "scope marking".

(c) Meaning "allows at least one type of 'axiom'."

(d) Although equivalences are used throughout the book, apparently the intent is that they are derived, even though there is no place where they were derived. Authors claim that the system is int-elim.

(e) We allow as int-elim for a rule of Repetition, the "impure" DN rules, the "impure" negation-of-a-binary connective rules, and others — including various ways to convert intuitionistic systems to classical ones.

(f) Tennant considers a number of alternatives for converting the intuitionistic system to classical, including axioms, rules such as $[\phi]\cdots\psi, [\neg\phi]\cdots\psi \vdash \psi$, etc.

(g) Has $\phi, \neg\phi \vdash \psi$ as well as $[\phi]\cdots\psi, [\neg\phi]\cdots\psi \vdash \psi$, so the rules are not quite int-elim, although we are counting them so.

(h) Also has "impure" $\neg\wedge$ and $\neg\forall$ rules (etc.).

(i) Not strictly int-elim, because the two $\perp$ rules are both elim-rules. ($\neg$ is defined in terms of $\supset$ and $\perp$.)

(j) Meaning "has some subproof-requiring propositional rule besides $\supset I$

(k) Says the $\supset I$ rule is theoretically dispensable.

(l) Has $[\phi]\cdots\psi \vdash (\neg\psi \supset \neg\phi)$.

(m) A basic tautology is $\forall x(Fx \supset P) \supset (\exists xFx \supset P)$, if P does not contain x. Rather than an $\exists$-rule, if $\exists xFx$ is in the proof, Pollock recommends proving the antecedent of this conditional by $\forall I$, then appeal to this tautology to infer the consequent by $\supset E$, and then use $\supset E$ again with the $\exists xFx$ and thereby infer P. We count this as neither $\exists E$ nor as EI; hence totals are out of 49.

(n) The actual rule is $\exists xFx, (F\alpha/x \supset \psi) \vdash \psi$. Since there will be no way to get this conditional other than by $\supset I$, and thereby assuming the antecedent, this rule is essentially $\exists E$.

(o) After introducing the defined symbol, $\exists$

(p) The basic system is $\exists E$ and no subproof for $\forall$, but gives alternative system with EI and a subproof for $\forall I$. We are counting the basic system here.

(q) Wilson's system has only UI, EI and QN — no generalization rules. (So, totals are out of 49.)

Around 1950, several logicians, among them the most notable perhaps being H.B. Curry and F.B. Fitch[25], had the idea of formulating modal logic[26] in terms of Introduction and Elimination rules for modal operators. Possibility Introduction and Necessity Elimination are simple and obvious: $\Diamond\phi$ may be inferred from ϕ, and ϕ may be inferred from $\Box\phi$. For the other two rules, natural deduction's idea of subproofs — that in some rules a conclusion is inferred, not from one or more statements, but from the presentation of a derivation of some specified kind — seems to have provided useful inspiration. Now, the subproofs used in standard rules for non-modal connectives differ from the main proof in that within them we can appeal to an additional assumption, the hypothesis of the subproof, but for modal logic a more important feature is that reasoning in a subproof can be subject to special restrictions, as is the case in some quantificational rules. $\forall$-Introduction allows us to assert $\forall x Fx$ if we have presented a derivation of $F\alpha$, with the restriction that no (undischarged) assumptions containing the parameter α have been used. Fitch's rule of $\Box$-Introduction allows us, similarly, to assert $\Box\phi$ if we have been able to give a proof of ϕ in which only (what we have already accepted to be) necessary truths have been assumed. More formally: $\Box\phi$ can be asserted after a subproof, with no hypothesis, in which ϕ occurs as a line, under the restriction that a statement, ψ, from outside the subproof, can only be appealed to within it if (not just ψ itself, but) $\Box\psi$ occurs in the main proof before the beginning of the subproof. In the following proof, the modal subproof comprises lines 3–5, in which q is deduced from the premises $(p \supset q)$ and p, premises which are asserted to be necessary truths by the hypotheses of the non-modal, $\supset I$, subproofs.

1	$\vert\ \Box(p \supset q)$	
2	$\vert\ \vert\ \Box p$	
3	$\vert\ \vert\ {}^{\Box}\vert\ (p \supset q)$	1, $\Box R$
4	$\vert\ \vert\ \ \vert\ p$	1,$\Box R$
5	$\vert\ \vert\ \ \vert\ q$	3,4 $\supset$E
6	$\vert\ \vert\ \Box q$	3–5, $\Box I$
7	$\vert\ (\Box p \supset \Box q)$	2–6, $\supset I$
8	$\Box(p \supset q) \supset (\Box p \supset \Box q)$	1–7, $\supset I$

[25][Curry, 1950] gave a formulation of S_4; [Curry, 1963] gives a more satisfactory discussion, although he does not seem to have given much consideration to any other system. Nonetheless, the changes to his rules needed for a formulation of T are quite simple. Fitch's earlier contributions are in a series of articles, summarized in his textbook [Fitch, 1952, Ch. 3].

[26]Alethic modal logic, that is. The idea that deontic, etc., concepts could be represented by "modal" operators seems not to have received significant attention until several years later, despite the publication of [von Wright, 1951a].

This rule, together with $\Box$-Elimination, gives a formulation of the basic alethic logic T (with necessity as sole modal primitive).[27] Varying the condition under which a formula is allowed to occur as a "reiterated" premise in a modal subproof yields formulations of several other important logics. S_4, for example, with its characteristic principle that any necessary proposition is necessarily necessary, is obtained by allowing $\Box\psi$ to occur in a modal subproof if $\Box\psi$ occurs above it, and S_5 by allowing $\neg\Box\psi$ to occur in a modal subproof if it occurs above the subproof (or alternatively, if $\Diamond$ is in the language, to allow $\Diamond p$ to appear in a modal subproof if it appears in an encompassing one).

Fitch also gave a corresponding $\Diamond$-Elimination rule, using subproofs with both restrictions and hypotheses: $\Diamond\phi$ may be inferred from $\Diamond\psi$ together with a subproof in which $\Diamond\phi$ is derived from the hypothesis ψ, the subproof having the same restrictions as to reiterated premises as the subproofs used in $\Box I$. This rule is, however, somewhat anomalous: obviously, in that the operator, $\Diamond$, supposedly being eliminated occurs in the conclusion of the rule, less obviously in being deductively weak. The possibility rules are derivable from the necessity rules if we define $\Diamond\phi$ as $\neg\Box\neg\phi$, but not conversely: the necessity rules cannot be derived from the possibility rules and the definition of $\Box\psi$ as $\neg\Diamond\neg\psi$. Again, if both modal operators are taken as primitive, the definitions of $\Diamond$ in terms of $\Box$ and of $\Box$ in terms of $\Diamond$ are not derivable from the four modal rules and the standard classical negation rules. Fitch compensates for these weaknesses by imposing Negative introduction and elimination rules for the modal operators, by which $\neg\Box\phi$ is interderivable with $\Diamond\neg\phi$ and $\neg\Diamond\psi$ with $\Box\neg\psi$.[28]

Fitch's rules for the necessity and possibility operators have a more than passing formal resemblance to the rules for the Universal and Existential quantifiers: depending on one's point of view, the parallels between the logical behaviors of the modal operators and the quantifiers can be seen as explained by or as justifying the Leibniz-Kripke paraphrases of modal statements in terms of quantification over possible worlds. Since S_5 has the simplest possible worlds semantics (no accessibility relation is needed), it is to be expected that it can be given a natural deduction formulation in which its modal rules mirror the quantifier rules particularly well. Think of ordinary, non-modal, formulas as containing an invisible term referring to the actual world. If we think of modalized formulas as disguised quantifications of possible worlds, this term is changed, in $\Box I$ and $\Diamond E$ subproofs, into the proper parameter of a quantificational subproof: hence the restriction on reiteration. The condition for legitimate reiteration is that the invisible world-term should have no free occurrences in the reiterated formula, so: a formula ϕ occurring above a modal subproof may be reiterated in it if every occurrence sentence letter in ϕ is inside

[27] T was first presented in [Feys, 1937], but is often attributed to Feys and von Wright. Given that von Wright didn't publish the system until [1951b], it seems a bit churlish not to credit Fitch as well, since Fitch presents the natural deduction formulation in his [1952] treatise/textbook. Unlike Feys and von Wright, Fitch didn't give the system a name, saying only that it was "similar to" S_2.

[28] Fitch also notes that, if the rules for the modal operators are added to Intuitionistic rather than Classical logic, the inferability of $(\Diamond\phi \vee \Diamond\psi)$ from $\Diamond(\phi \vee \psi)$ has to be postulated specially.

the scope of some modal operator. Call such an ϕ modally closed. We can now give the $\Diamond E$ rule a formulation closer to the standard $\exists E$ rule: any modally closed formula ϕ may be inferred from $\Diamond\psi$ together with a modally restricted subproof in which ϕ is derived from the hypothesis ψ. In S_5 with these rules, the Negative modal rules are derivable from the positive ones and standard negation rules, and the necessity rules are derivable from the possibility rules when $\Box\phi$ is defined as $\neg\Diamond\neg\phi$.[29]

Not all modal logics have nice natural deduction formulations, but those that do include many of the most important for applications. [Fitch, 1966] gives natural deduction formulations for a number of alethic and deontic logics and systems with both alethic and deontic operators. [Fitting, 1983] describes natural deduction formulations of a large number of logics, in addition to tableau versions. Several introductory logic texts (for example, [Anderson and Johnstone, 1962; Iseminger, 1968; Purtill, 1971; Bonevac, 1987; Gamut, 1991; Forbes, 1994; Bessie and Glennan, 2000]) include natural deduction formulations of one or more systems of modal logic, and [Garson, 2006], intended for students who have already done a bit of logic, introduces many modal logics, both propositional and quantified, with both natural deduction and tableau formulations.

3 THE METATHEORY OF NATURAL DEDUCTION

3.1 *Normalizing Natural Deduction*

Gentzen's rules of his system of natural deduction come in pairs of an introduction and an elimination rule. The rules in such a pair tend to be — in an intuitively obvious way whose precise definition isn't easy to formulate — *inverses* of each other. This is clearest with the rules for conjunction: conjunction introduction infers a conjunction from its conjuncts, conjunction elimination infers the conjuncts from the conjunction. In other cases the inverse relationship is not as direct: disjunction introduction infers a disjunction from one of its disjuncts, but disjunction elimination can't infer a disjunct from the disjunction (that would be unsound, as neither disjunct is in general entailed by the disjunction). The effect of the disjunction elimination rule is rather to reduce the problem of inferring a conclusion from a disjunction to that of deriving it from the disjuncts. With other operators the inversion can take yet other forms, and negation tends to be a special case.

What this inverse relationship between rules suggests (and in particular, what it suggested to Gentzen when he was working on his thesis, which formed the basis of [Gentzen, 1934]) is that if, in the course of a proof, you infer a conclusion by an introduction rule and then use that conclusion as the major[30] premise for

[29][Fitch, 1952] treats modal operators before quantifiers. Despite the reverse relative importance of the two topics, this may make pedagogical sense, as it accustoms students to the use of subproofs with restrictions on reiteration before they have to cope with the technicalities of free and bound occurrences of variables.

[30]In an elimination rule needing more than one premise, the major is the one containing the

an inference by the corresponding elimination rule, you have made an avoidable detour: you have inserted a bit of conceptual complexity (represented by the operator introduced by the introduction rule) into your proof, but pointlessly, since after the elimination inference your proof uses as premises only things you had reached before the detour. Leading to the conjecture — since proven as a theorem for many systems — that if you can prove something in a system of natural deduction, you can prove it in a way that avoids all such detours. Or, as Gentzen put it in the published version of his thesis [Gentzen, 1934, p.289]

> ...every purely logical proof can be reduced to a determinate, though not unique, normal form. Perhaps we may express the essential properties of such a normal proof by saying 'it is not roundabout.' No concepts enter into the proof other than those contained in its final result, and their use was therefore essential to the achievement of that result.

Let us be a bit more formal. Define a proof in a natural deduction system as *normal* if no formula in it is both the conclusion of an inference by an introduction rule and the (major) premise of an inference by an elimination rule.[31] Normal proofs avoid the useless conceptual complexity involved in introduction-elimination detours: they proceed by first applying elimination rules to premises or hypotheses (reducing their complexity by stripping off the main logical operator at each step) and then monotonically building up the complexity of the final conclusion by introduction rules. They are thus in a sense conceptually minimal: no piece of logical construction occurs in them unless it is already contained in the original premises or the final conclusion. In technical terms, a normal proof has the *subformula property*: every formula occurring in it is a subformula[32] of the conclusion or of some premise. This feature of normal proofs has proven valuable for consistency proofs, proof-search algorithms, and decision procedures, among other properties.

We can now formulate stronger and weaker claims that might be made about a system of natural deduction. A *normal form theorem* would say that for every proof in the system there is a corresponding normal proof with the same conclusion and (a subset of) the same premises. A stronger *normalization theorem* would say that there is a (sensible[33]) procedure which, applied to a given proof, will convert it into a corresponding normal proof. Both theorems hold for nice systems.

operator mentioned in the name of the rule: e.g., when B is inferred from A and $(A \to B)$ by $\to$ elimination, the conditional is the major premise.

[31]This is a rough preliminary definition. A technical refinement is added below in connection with "permutative reductions," and for some systems a variety of subtly different notions of normality have been studied.

[32]This is literally true for standard systems of intuitionistic logic. For classical systems it is often necessary to allow the negations of subformulas of a given formula to count as honorary subformulas of it. And in, e.g., a system of modal logic one might want to allow formulas obtained by adding some limited number of modal operators to a real subformula.

[33]Not every algorithm for converting proofs qualifies. The normal form theorem would imply the existence of an idiot's algorithm: grind out all the possible normal proofs in some systematic

Gentzen saw this, at least for intuitionistic logic, and in a preliminary draft of his thesis that remained generally unknown until recently [von Plato, 2008b], he proved the normalization theorem for his system of intuitionistic logic. Gentzen's proof is essentially the same as the one published thirty years later in [Prawitz, 1965], and the normalization procedure is simple and intuitive: look for a "detour" and excise it (this is called a reduction); repeat until done[34]. Thus, for example, suppose that in the given proof $A \vee B$ is inferred from A by disjunction introduction and C then inferred from $A \vee B$ by disjunction elimination. One of the subproofs (to use our Fitch-inspired terminology) of the disjunction elimination will contain a derivation of C from the hypothesis A. So delete the occurrence of $A \vee B$ and the second subproof, and use the steps of the A-subproof to derive C from the occurrence of A in the main proof. That's basically it, although there's a technical glitch. Suppose, in the proof to be normalized, you used an introduction rule inside the subproof of an application of existential quantifier elimination (or inside one of the subproofs of an application of disjunction elimination) to obtain the final formula of the subproof, wrote another copy of the same formula as the conclusion of the existential quantifier elimination, and then applied the corresponding elimination rule in the main proof. Under a literal reading of the definition of a normal proof given above, this would not qualify as a detour (no one occurrence of the formula is both the conclusion of the introduction rule and premise of the elimination rule), and it won't be excised by the reduction steps described. Clearly, however, it would be cheating to allow things like this in a "normal" proof (and allowing it would mean that "normal" proofs might not have the subformula property). An honest normalization procedure will eliminate this sort of "two stage" detours. To do this Gentzen and Prawitz included a second sort of step in the procedure that Prawitz called a permutative reduction: when something like what is described above occurs, rearrange the steps so the elimination rule is applied within the subproof, reducing the two-stage detour to an ordinary detour in the subproof. So the full normalization procedure is: look for a detour or a two-stage detour, excise it by an ordinary or permutative reduction, repeat until done.

Gentzen despaired, however, of proving a normalization theorem for classical logic: in the published version of his thesis, [Gentzen, 1934, p.289], he says:

> For, though [his natural deduction system] already contains the properties essential to the validity of the *Hauptsatz*, it does so only with respect to its intuitionist form, in view of the fact that the law of excluded middle ... occupies a special position in relation to these properties.

He therefore defined a new sort of system, his L-systems or sequent calculi (de-

---

order and stop when you get one with the right premises and conclusion. Logicians are interested in normalization procedures in which each step in modifying the given proof into a normal one intuitively leaves the proof "idea" the same or is a simplification.

[34]Gentzen's proof that the procedure works, like that in [Prawitz, 1965], depends on attacking the different detours in a particular order. But the order doesn't actually matter, as was proved by the strong normalization theorem of [Prawitz, 1971].

scribed below), for which he could prove an analogue of normalization for both intuitionistic and classical systems, and left the details of normal forms and normalization of natural deduction out of the published version of the thesis.

Since classical logic is in some ways simpler than intuitionistic logic (think of truth tables!), it may seem surprising that proof-theoretic properties like the normalization theorem are harder to establish for it, and details will differ with different formulations of classical logic. Prawitz (and Gentzen in the early draft of his thesis) worked with a natural deduction system obtained from that for intuitionistic logic by adding the classicizing rule of *indirect proof*: ϕ may be asserted if a contradiction is derivable from the hypothesis $\neg\phi$. Note that this rule can be broken down into two steps: first, given the derivation of a contradiction from $\neg\phi$, assert $\neg\neg\phi$ by the intuitionistically valid form of *reductio* (that is, by the standard negation introduction rule of natural deduction systems for intuitionistic logic), and then infer ϕ by (another possible choice for a classicizing rule) *double negation elimination*. Thought of this way, there is an immediate problem: the characteristic inference pattern that allows for classical proofs that are not intuitionistically valid seems to be essentially a "detour", namely, an introduction rule is used to infer a double negation which is then used as a premise for an elimination rule, with the result that on this naïve construal of 'normal', no intuitionistically invalid classical proof can possibly be normalized!

But there is another problem, much deeper than this rather legalistic objection. Recall that the interest of normal proofs lies largely in their conceptual simplicity: in the fact that they possess the subformula property. If we ignored the legalistic objection and simply allowed indirect proof (as, perhaps, a special rule outside the pattern of introduction and elimination rules) we could produce proofs that failed this simplicity criterion very badly: working from simple premises we could build up (perhaps without any obvious detours) to an indirect proof of some immensely long and complex formula, and then (again by normal-looking means) derive a simple conclusion from it. In other words, whether we think of the classicizing rule *de jure* normal or not, an interesting notion of normal proof for classical logic must put some restriction on its use.

Prawitz (and Gentzen before him) chose an obvious restriction which would give normal proofs the subformula property[35]: in a normal proof, indirect proof is used only to establish atomic formulas.[36] With such a definition of a normal proof, a normal form theorem will hold only if the system with its classicizing rule restricted to the atomic case is complete for classical logic; that is, if, by using the classicizing rule in the restricted form we can prove everything we could by using it without restriction. Now, in some cases it is possible to "reduce" non-atomic uses of indirect proof (e.g.) to simpler cases. Suppose we used indirect proof to

[35]Well, "weak subformula" property: negations of genuine subformulas allowed.

[36]If a different classicizing rule is chosen, it will also have to be restricted somehow to get an "honest" normalization theorem. Results about a system with a restricted form of one rule don't carry over automatically to systems with comparably restricted forms of other rules. Some, otherwise attractive, natural deduction systems for classical logic don't seem to have good normalization theorems.

establish a conjunction, $\phi \wedge \psi$. $\neg(\phi \wedge \psi)$ is implied (intuitionistically) by each of $\neg\phi$ and $\neg\psi$, so the contradiction derived from the hypothesis $\neg(\phi \wedge \psi)$ could have been derived in turn from each of the two hypotheses $\neg\phi$ and $\neg\psi$. So we could have used indirect proof to get the simpler formulas ϕ and ψ instead, and then inferred $\phi \wedge \psi$ by conjunction introduction.

In other cases, however, this is not possible. $\forall x(\exists y \neg Fy \vee Fx)$, where the variable x does not occurring free in Fy, classically (but not intuitionistically) implies $(\exists y \neg Fy \vee \forall x Fx)$: this is an inference that a natural deduction system for classical logic must allow. Assuming classical logic (so allowing indirect proof or excluded middle) for formulas only of the complexity of Fx, will not help. (For, excluded middle is provable for quantifier-free formulas of intuitionistic (Heyting) arithmetic, so if we could go from having excluded middle for atomic instances to excluded middle for quantifications, then Heyting Arithmetic would collapse into classical Peano Arithmetic.) So to get full classical first-order logic *some* classicizing rule must be allowed to work on complex formulas.

On closer examination, however, it turned out that this difficulty only arises in connection with formulas containing disjunction operators or existential quantifiers. Prawitz was able to prove a normalization theorem (the definition of "normal" including the proviso that indirect proof be used only for atomic formulas) for the fragment of classical logic having only the universal quantifier and only the conjunction, conditional and negation connectives. For some purposes this is sufficient: disjunction and existential quantification are classically definable in terms of the operators included in the fragment, so to every proof in full classical logic there corresponds a proof, with trivially equivalent premises and a trivially equivalent conclusion, in the fragment Prawitz showed to be normalizable. Thus, for example, if one were to want to use the normalization theorem and the subformula property of normal proofs as part of a consistency proof for some classical mathematical theory, Prawitz's result would suffice even though it doesn't extend to classical logic with a full complement of operators.

This leaves open the question of whether, with a more carefully worded definition of normal proofs, a reasonable normalization theorem would be provable for full classical logic. Various results have been obtained; the current state of the art is perhaps represented by [Stålmarck, 1991]'s normalization theorem: here normal proofs, in addition to avoiding explicit detours, are not allowed to use indirect proof to obtain complex formulas which are then used as major premises for elimination rules.

The normalization theorems described so far have been proven by methods which are finitistic in the sense of Hilbert's program: Gentzen's hope in studying natural deduction systems was that normalization theorems would contribute to that program's goal of finding a finitistic consistency proof for arithmetic. In contrast, stronger set-theoretic methods have also been used in studying normalization. These have yielded strong normalization theorems (that any reduction process, applying normalizing transformations in any order to a proof, will terminate in a finite number of steps with a normal proof) and confluence theorems (that the normal proofs produced by different reduction processes to a given proof will, up to trivialities like relettering variables, be identical). [Prawitz, 1971]'s strong normalization theorem for intuitionistic logic is an early example; [Stålmarck, 1991] gives a set-theoretic proof of a strong normalization theorem as well as an elementary proof of normalization for classical logic.

3.2 Natural Deduction and Sequent Calculus

Gentzen's own response to the problem of normalization for classical logic was to define a new kind of formal system, closely related to natural deduction, for which he was able to prove something similar to normalization for full classical logic. This involved ideas (and useful notations!) which have proved to be of independent interest. As a first step (in fact applying only to intuitionistic logic), consider a formal system in which the lines of a proof are thought of not as formulas but as *sequents*. For the moment (it will get a bit more complex later when we actually get to classical logic), a sequent may be thought of as an ordered pair of a finite sequence of formulas (the antecedent formulas of the sequent) and a formula (its succedent formula). This is usually written

$$\phi_1, \phi_2, \ldots, \phi_n \vdash \psi$$

with commas separating the antecedent formulas, and different authors choosing for the symbol separating them from the succedent a turnstile (as here) or a colon or an arrow (if this is not used for the conditional connective within formulas). In schematic presentations of rules for manipulating sequents it is standard to use lower-case Greek letters as schematic for individual formulas and capital Greek letters for arbitrary (possibly empty) finite sequences of formulas. A sequent is interpreted as signifying that its antecedent formulas collectively imply its succedent: if the antecedent formulas are all true, so is the succedent. (So, if one wants, one can think of a sequent as an alternative notation for the theoremhood of a conditional with a conjunctive antecedent.) There are special cases: a sequent with a null sequence of antecedents is interpreted as asserting the succedent, a sequent with no succedent formula is interpreted as meaning that the antecedents imply a contradiction, and an empty sequent, with neither antecedent nor succedent formulas, is taken to be a contradiction: consistency, for formal systems based on sequents, is often defined as the unprovability of the empty sequent.

3.3 Sequent Natural Deduction

In an axiomatic system of logic each formula occurring as a line of a proof is asserted as a logical truth: it is either an axiom or follows from the axioms. The significance of a line in a natural deduction proof is less obvious: the formula may not be valid, for it may be asserted only as following from some hypothesis or hypotheses which, depending on the geometry of the proof presentation, may be hard to find! One way of avoiding this difficulty would be to reformulate a natural deduction system so that lines of a proof are sequents instead of formulas: sequents with the formula occurring at the corresponding line of a conventional proof as succedent and a list of the hypotheses on which it depends as antecedent. (Such a reformulation might seem to risk writer's cramp, since the hypotheses have to be copied down on each new line until they are discharged, but abbreviations are possible: [Suppes, 1957], by numbering formulas and writing only the numerals instead of the formulas in the antecedent, made a usable text-book system of sequent-based natural deduction that was followed by many others, as we have seen above.) It is straightforward to define such a system of *sequent natural deduction* corresponding to a given conventional natural deduction system:

- Corresponding to the use of hypotheses in natural deduction, and to the premises of proofs from premises, sequent natural deduction will start proofs with identity sequents, sequents of the form $\phi \vdash \phi$, which can be thought of as logical axioms.

- Corresponding to each rule of the conventional system there will be a rule for inferring sequents from sequents. For example, the reformulated rule of conjunction introduction will allow the inference of $\Gamma \vdash (\phi_1 \wedge \phi_2)$ from the two sequents $\Gamma \vdash \phi_1$ and $\Gamma \vdash \phi_2$; conjunction elimination will allow the inference of either of the latter sequents from the former. Corresponding to a rule of the conventional system that discharges a hypothesis there will be a rule deleting a formula from the antecedent of a sequent: conditional introduction will allow the inference of $\Gamma \vdash (\phi_1 \rightarrow \phi_2)$ from $\Gamma, \phi_1 \vdash \phi_2$.

- There will be a few book-keeping, or structural rules. One, which Gentzen called thinning[37], allows extra formulas to be inserted in a sequent. To see the use of this, note that as stated above the conjunction introduction rule requires the same antecedent in its two premises. To derive $(\phi_1 \wedge \phi_2)$ from the two hypotheses ϕ_1 and ϕ_2, then, one would start with the identity sequents $\phi_1 \vdash \phi_1$ and $\phi_2 \vdash \phi_2$, use thinning to get $\phi_1, \phi_2 \vdash \phi_1$ and $\phi_1, \phi_2 \vdash \phi_2$, and then apply conjunction introduction to these sequents. Two other structural rules would be demoted to notational conventions if one was willing (as Gentzen was not[38]) to regard the antecedents of sequents as unstructured *sets* of formulas. *Permutation* allows the order of the antecedent formulas to be changed. *Contraction*[39] allows multiple occurrences of a single formula in the antecedent to be reduced to a single occurrence.

[37]German *Verdünnung*. This is thinning in the sense in which one thins paint: some logicians translate it as *dilution*.

[38]The logics he considered — classical and intuitionistic — can both be described by sequents with sets of antecedent formulas. Sequents with more structured antecedents are useful in formulating things like *relevance logics* or Girard's *linear logic*: systems which, precisely because their formulation omits or restricts some of Gentzen's structural rules, are often called substructural logics. See [Restall, 2000].

[39]Following Curry, many logicians refer to contraction as rule W. Unfortunately other logicians call Thinning *Weakening*, which also abbreviates to W.

As an example of a proof in sequent natural deduction, consider the commutativity of conjunction:

$\phi_1 \wedge \phi_2$	$\vdash \phi_1 \wedge \phi_2$	Axiom
$\phi_1 \wedge \phi_2$	$\vdash \phi_1$	by conjunction elimination
$\phi_1 \wedge \phi_2$	$\vdash \phi_2$	also by conjunction elimination
$\phi_1 \wedge \phi_2$	$\vdash \phi_2 \wedge \phi_1$	by conjunction introduction

3.4 From Sequent Natural Deduction to Sequent Calculus

The action, so to speak, in a sequent natural deduction proof is all in the succedents of the sequents: whole formulas can be added or subtracted from the antecedents, or rearranged within them, by the structural rules, but the rules for the connectives only add or subtract operators from succedent formulas. (Not surprisingly, since it is the succedent formulas that correspond to the formulas occurring in the conventional variant of the natural deduction proof!) What Gentzen saw was that, though this is natural enough for the introduction rules, the elimination rules could also be seen as corresponding to manipulations of antecedent formulas: in particular, just as the conclusion of the sequent form of an introduction rule is a sequent whose succedent formula has one more operator than the succedent formulas of the premise sequents, analogues of the elimination rules can be given in which the conclusion is a sequent containing an antecedent formula with an extra operator. Since, for example, the effect of the conjunction elimination rule is to allow the derivation from a conjunction of any formula derivable from either of its conjuncts, a sequent natural deduction analogue (henceforth called conjunction left; the analogue of conjunction introduction stated above will be called conjunction right) will tell us that $\Gamma, \phi_1 \wedge \phi_2 \vdash \psi$ can be inferred from either $\Gamma, \phi_1 \vdash \psi$ or $\Gamma, \phi_2 \vdash \psi$. As an illustration of how this works, let us see the commutativity of conjunction in this framework:

ϕ_1	$\vdash \phi_1$	Axiom
$\phi_1 \wedge \phi_2$	$\vdash \phi_1$	By conjunction left
ϕ_2	$\vdash \phi_2$	Axiom
$\phi_1 \wedge \phi_2$	$\vdash \phi_2$	By conjunction left (second axiom)
$\phi_1 \wedge \phi_2$	$\vdash \phi_2 \wedge \phi_1$	By conjunction right (2nd and 4th lines)[40]

Our trivial example about the commutativity of conjunction has involved reformulations of a *normal* natural deduction proof: conjuncts are inferred from an assumed conjunction by an elimination rule and a new conjunction is then inferred from them by an introduction rule. What if we *wanted* to make a detour, to infer something by an introduction rule and then use it as the major premise for an application of the corresponding elimination rule? Sequent natural deduction

[40]Obviously the identity sequent $\phi_1 \wedge \phi_2 \vdash \phi_1 \wedge \phi_2$ can be derived in a similar fashion. A useful exercise for newcomers to Gentzen's rules is to show that all identity sequents are derivable when only the identities for atomic formulas are assumed as axioms.

amounts to ordinary natural deduction with hypothesis-recording antecedents decorating the formulas, so there is no difficulty here. In Gentzen's *sequent calculi* (or *L-systems*: he named his sequent formulations of intuitionistic and classical logic LJ and LK respectively) things aren't as straightforward. Suppose, for example, that we can derive both ϕ and ψ from some list of hypotheses (so: start with the sequents $\Gamma \vdash \phi$ and $\Gamma \vdash \psi$) and we decide to infer $\phi \wedge \psi$ from them by conjunction introduction, and then (having somehow forgotten how we got $\phi \wedge \psi$) decide to infer ϕ from it by conjunction elimination. The introduction step is easy: the sequent

(i) $\Gamma \vdash \phi \wedge \psi$

follows immediately from the two given sequents by conjunction right. And we can prove something corresponding to the elimination step,

(ii) $\Gamma, \phi \wedge \psi \vdash \phi$

almost as immediately:

ϕ	$\vdash \phi$	Axiom
$\phi \wedge \psi$	$\vdash \phi$	By conjunction left
$\Gamma, \phi \wedge \psi$	$\vdash \phi$	By as many thinnings as there are formulas in Γ

But how do we put them together and infer (forgetting we already have it!)

(iii) $\Gamma \vdash \phi$

from (i) and (ii)? The short answer is that we can't: sequent (iii) is shorter than (i) or (ii) and (except for the irrelevant Contraction) none of the rules stated so far allow us to infer a shorter sequent from a longer. To allow the inference, Gentzen appeals to a fourth structural rule, *Cut*. Cut says that, given a sequent with a formula θ in its antecedent and another sequent with θ as succedent formula, we may infer a sequent with the same succedent as the first and an antecedent containing the formulas other than θ from the antecedent of the first sequent along with the formulas from the antecedent of the second sequent.[41] Schematically, from $\Gamma, \theta \vdash \delta$ and $\Delta \vdash \theta$ we may infer $\Gamma, \Delta \vdash \delta$. The inference from (ii) and (i) above to (iii) is an instance of Cut.[42]

Interpreting sequents as statements of what is implied by various hypotheses, it is clear that Cut is a valid rule: it amounts to a generalization of the principle of the transitivity of implication. On the other hand it plays a somewhat anomalous role in the system. Normal natural deduction proofs can be translated into L-system proofs in which the Cut rule is never used, and — our example is typical

[41]Technically, if the first sequent has multiple copies of θ in its antecedent, Gentzen's cut only removes one. This complicates the details of his proof of the Hauptsatz, but can be overlooked in an informal survey: the version stated here is closer to his rule *Mix*, which he shows equivalent to Cut in the presence of the other structural rules.

[42]For convenience we have shown proofs in sequent calculi as simple columns of sequents. Gentzen displayed such proofs as trees of sequents (this is probably the most perspicuous representation for theoretical purposes.). If we were to use LJ or LK in practice for writing out proofs, it would be inconvenient because the same sequent will often have to have multiple occurrences in an arboriform proof, and converting a columnar proof into a tree can lead to a superpolynomial increase in proof size.

— any "detour" in a non-normal proof corresponds to an application of Cut.[43] It is also analogous to detours in its effect on conceptual complexity in proofs: in a proof in which Cut is not used, every formula occurring in any sequent of the proof is a subformula of some formula in the final sequent. Cut is the only rule which allows material to be removed from a sequent: Thinning adds whole formulas, the left and right rules for the different logical operators all add an operator to what is present in their premise sequents, Permutation just rearranges things, and even Contraction only deletes extra copies. Possible L-system proofs of a sequent not using Cut, in other words, are constrained in the same way that normal natural deduction derivations of the sequent's succedent from its antecedent formulas are, and Cut relaxes the constraints in the same way that allowing detours does.

In the published version of his thesis, then, Gentzen, after presenting his natural deduction systems NJ and NK for intutitionistic and classical logic, sets them aside in favor of the sequent calculi LJ and LK. In place of the normalization theorem for the natural deduction systems he proves his *Hauptsatz* (literally: 'principal theorem', but now used as a proper name for this particular result) or *Cut-elimination theorem*: any sequent provable in the L-systems is provable in them without use of Cut, and indeed there is a procedure, similar to the reduction procedure for normalizing natural deduction proofs, which allows any L-system proof to be converted into a Cut-free proof of the same final sequent.[44]

3.5 The Sequent Calculus LK

So far we have mentioned the classical sequent calculus LK, but not described it. What Gentzen saw was that a very simple modification of the intuitionistic sequent calculus LJ — one which, unlike the classical system of natural deduction, did not involve adding a new rule for some logical operator — produced a formulation of classical logic. Recall that the difficulties in defining a normal form and proving a normalization theorem for classical logic centered on the treatment of disjunction (and the existential quantifier). The classically but not intuitionistically valid inference from $\forall x(P \vee Fx)$ (x not occurring free in P) to $(P \vee \forall xFx)$ is typical. For any particular object in the domain we can (intuitionistically) infer that either it is F or else P is true: Universal Quantifier Elimination applied to the premise gives us $(P \vee Fa)$. Applying Disjunction Elimination, we adopt the two disjuncts in succession as hypotheses and see what they give us. P is helpful: Disjunction Introduction gives us the conclusion, $(P \vee \forall xFx)$. Fa, however, isn't: we want to be able to infer $\forall xFx$, but being told that one object in the domain is F doesn't tell us they all are! What we would like to do, classically, is to consider the two

[43]Though not every application of Cut corresponds to a normality-destroying manoeuvre in a natural deduction proof.

[44]Gentzen defined his L-systems with the Cut rule, so this is his formulation of the result. Some subsequent writers have defined their sequent calculi as *not* containing Cut. They thus rephrase the Hauptsatz as: adding Cut as an additional rule would not enable us to prove any sequents we couldn't prove in their official systems. This yields another name for Gentzen's result: *Admissibility of Cut.*

disjuncts, but somehow to still be generalizing when we think about Fa.[45] This is, of course, objectionable from the point of view of intuitionistic philosophy (it assumes that the entire, possibly infinite, domain exists as a given whole, whereas the premise only tells us that, whatever individuals there are in the domain, a certain disjunction holds of each one). There is also no obvious way of fitting this kind of thinking into the framework of natural deduction: introduction rules install new main operators in formulas, so we can only "generalize" — formally, use Universal Quantifier Introduction to insert a new universal quantifier — if we are in a position to assert an (arbitrary) instance of the quantification as an independent formula, but in this situation we are only able present it as an alternative: as one disjunct of a disjunction.

Gentzen's solution was to generalize the notion of a sequent to allow one sequent to have multiple succedent formulas, interpreted disjunctively, and to allow the analogues of natural deduction's Introduction rules — the rules that add an operator in the succedent — to act on one succedent formula while ignoring the rest. Formally, a sequent is redefined as an ordered pair of sequences of formulas (again allowing the special cases of one or the other or both being the null sequence),

$$\Gamma \vdash \Delta$$

As before, the interpretation is a sort of implication: this time, that *if* all the antecedent formulas are true, *then* at least one of the succedent formulas is. (Thus it can be thought of as a notational variant of a conditional with a conjunctive antecedent and a disjunctive consequent.) But, now that we are looking in more detail about how quantificational inferences are handled in the L systems, it is worth being more precise about the interpretation. The formulas in a sequent may, after all, contain free variables (= parameters), and they are to be given a *generality* interpretation: the sequent says that, for any values assigned to the free variables (assigning values uniformly, so if the same variable occurs free in two or more formulas of the sequent they will all be taken to be about the same object), if all the antecedent formulas are true for those values, then at least one of the sucedent formulas will be true for the same values. (So, if we are to think of a sequent as a notational variant of some sentence, it will have to be, not just an implication, but a generalized implication: variables free in the sequent are to be thought of as implicitly bound by universal quantifiers having the entire implication in their scope.)

The rules for the classical sequent calculus LK are virtually identical to those of the intuitionistic LJ, but now applied to sequents with multiple succedent formulas. In particular, the structural rules of Permutation and Contraction now apply to succedents ("Permutation right," etc) as well as to antecedents, and Thinning can add additional formulas to non-null succedents. (Thinning was allowed on the right in LJ, but only if the premise had a null succedent: the inference from $\Gamma \vdash$

[45] One possibility would be to use something like Quine's rule of U.G., letting a be an "arbitrary." Unless serious restrictions are imposed, use of rules like E.I. and U.G. in combination with intuitionistic propositional rules will yield a superintuitionistic quantified logic.

to $\Gamma \vdash \phi$ is the formalization in LJ of the *ex falso quodlibet* principle.) Cut takes a more general form: from $\Gamma, \theta \vdash \Delta$ and $\Phi \vdash \Psi, \theta$ to infer $\Gamma, \Phi \vdash \Psi, \Delta$[46] The rules for the logical operators are unchanged, but now the right rules as well as the left ones tolerate neighbors to the formulas they act on. We illustrate the workings of the system by showing how to derive some standard examples of intuitionistically invalid classical validities.

The negation rules in the L-systems switch a formula from one side to the other, adding a negation to it in the process. (If the validity of any of these rules, on the interpretation described above for sequents, is not obvious, it is a useful exercise to think it through until it becomes obvious: none are hard.) In the intuitionistic case, LJ allows us to prove

ϕ	$\vdash \phi$	Axiom
$\phi, \neg\phi$	$\vdash$	Negation left

(contradictions are repugnant to intuitionists as well as to classical logicians). Since LJ allows at most a single succedent formula, the move in the opposite direction is impossible, but in LK we have

ϕ	$\vdash \phi$	Axiom
	$\vdash \neg\phi, \phi$	Negation right

which, on the disjunctive interpretation given to multiple succedents, amounts to the Law of Excluded Middle. Since the rule for Disjunction on the right corresponds to the Disjunction Introduction rule of natural deduction, we can continue this derivation to get a more explicit statement of the Law:

$\vdash \neg\phi, (\phi \vee \neg\phi)$	Disjunction right
$\vdash (\phi \vee \neg\phi), (\phi \vee \neg\phi)$	Disjunction right
$\vdash (\phi \vee \neg\phi)$	Contraction right[47]

As an even simpler example,

ϕ	$\vdash \phi$	Axiom
	$\vdash \neg\phi, \phi$	Negation right
$\neg\neg\phi$	$\vdash \phi$	Negation left.

And, finally, our example of restricting a universal quantification to one disjunct:

ϕ	$\vdash \phi$	Axiom
ϕ	$\vdash \phi, Fa$	Thinning right
Fa	$\vdash Fa$	Axiom
Fa	$\vdash \phi, Fa$	Thinning right
$(\phi \vee Fa)$	$\vdash \phi, Fa$	Disjunction left, from 2nd and 4th sequents

[46]The classical validity of this rule is easily seen. Suppose there is an interpretation (and assignment to free variables) falsifying the conclusion: making all the formulas in Γ and Φ true and all those in Ψ and Δ false. If formula θ is true on this interpretation, the first premise sequent is falsified, and if it is false the second is.

[47]Gentzen actually formulated his Disjunction right rule as applying only to the last formula of the succedent, so technically there should have been a Permutation right inference between the two Disjunction rights: a complication we will ignore in examples.

Note here an interesting general feature of LK: we can prove a sequent saying that an explicit disjunction implies the pair of its disjuncts (interpreted disjunctively!). The dual principle, that a pair of conjuncts implies their conjunction, is expressed by a sequent with a single succedent formula, provable in both LJ and LK. Continuing our example,

$\forall x(P \vee Fx) \vdash P, Fa$ Universal quantifier left.

The condition on the Universal quantifier right rule — corresponding to the condition on Universal Quantifier Introduction, that the free variable to be replaced by a universally quantified one may not occur free in assumptions on which the instance to be generalized depends — is that the free variable cannot occur in any other formula, left or right, of the premise sequent: but we now have a sequent satisfying this, so we may continue with

$\forall x(P \vee Fx) \vdash P, \forall x Fx$ Universal quantifier right,

after which a couple of Disjunction rights and a contraction will give us the desired

$\forall x(P \vee Fx) \vdash (P \vee \forall x Fx)$.

What gets pulled out of the hat probably got put into it earlier. The rules which allowed us to prove a sequent expressing the inference from $\forall x(P \vee Fx)$ to $(P \vee \forall x Fx)$ can be seen as a generalization of that very inference. As mentioned above, a sequent can be thought of as a notational variant of a universally quantified implication, with a disjunction as its consequent (= the succedent of the sequent). Viewed this way, the rule of Universal Quantifier Right amounts to taking a universal quantifier from the front of a sentence, binding a variable that occurs in only one disjunct of the matrix, and moving it in to that disjunct! So what, exactly, has been gained? The problem with classical logic, when it came to normalizing natural deduction proofs, was that classicizing rules had to be applied to logically complex formulas. (It's easy to derive $(P \vee \forall x Fx)$ from $\forall x(P \vee Fx)$ by indirect proof, deriving a contradiction from the premise and the hypothesis $\neg(P \vee \forall x Fx)$!) In order to get around this problem, the sequent calculus has rules that in effect amount to rules for manipulating formulas (viz., quantified conditionals with disjunctions as consequents) more complex than the formulas explicitly displayed. What keeps this from being a trivial cheat is the fact that the additional complexity is strictly limited. The formulas occurring in the antecedents and succedents of the sequents in an LK or LJ proof are the ones we are really interested in. The sequents themselves are equivalent to more complex formulas built up from them, but not arbitrarily more complex: only a single layer each, so to speak, of conjunctive (in the antecedent), disjunctive (in the succedent), implicational and universal-quantificational structure is added beyond the formulas we are interested in. This is the justification for using different notation rather than writing sequents as quantified implications: in building up formulas with normal connectives and quantifiers we can iterate ad infinitem, but in writing commas and turnstiles instead, we are noting the strictly limited amount of additional structure required by Gentzen's rules. And it is because of the limitation on this additional structure that Gentzen's Hauptsatz and the subformula property of Cut-free proofs are "honest": yes, there is a bit more structure than is contained in the formulas whose implicational relations are demonstrated in the proof, but it is strictly constrained. We have what we really need from a subformula property: a limit on the complexity of structure that can appear in the proof of a given logical validity.

3.6 Variants of Sequent Calculi

Following Gentzen's initial introduction of sequent calculi, many variants have been been defined. Gentzen himself noted that alternative formulations of the rules for the logical operators are possible. For example, Gentzen's Disjunction left rule requires that the premise sequents be identical except for containing the left and right disjuncts in their antecedents:

From $\phi, \Gamma \vdash \Theta$ and $\psi, \Gamma \vdash \Theta$ to infer $(\phi \vee \psi), \Gamma \vdash \Theta$

and his Disjunction right rule adds a disjunct to a single formula in the succedent:

From $\Gamma \vdash \Theta, \phi$ to infer $\Gamma \vdash \Theta, (\phi \vee \psi)$ (or $\Gamma \vdash \Theta, (\psi \vee \phi)$

One could equally well have a Disjunction left rule that combined the extra material in the two premises:

From $\delta, \Gamma \vdash \Delta$ and $\theta, \Phi \vdash \Psi$ to infer $(\delta \vee \theta), \Gamma, \Phi \vdash \Delta, \Psi$

and a Disjunction right rule that joined two distinct formulas in the succedent into a disjunction:

From $\Gamma \vdash \Delta, \phi, \psi$ to infer $\Gamma \vdash \Delta, (\phi \vee \psi)$.

Which to choose? In the presence of the structural rules, the two pairs of rules are equivalent. Without them, or with only some of them, the two pairs of rules can be seen as defining different connectives: what the Relevance Logic tradition calls *extensional* (Gentzen's rules) and *intensional* (the alternative pair) disjunction and the tradition stemming from Girard's work on Linear Logic calls *additive* and *multiplicative* disjunction. The conjunction rules are exact duals of these: Gentzen's version defining, when structural rules are restricted or omitted, what the two traditions call extensional or additive conjunction, the alternative pair intensional or multiplicative conjunction.[48] Gentzen's rules for the conditional are more nearly parallel to the alternative (intensional/multiplicative) versions of the conjunction and disjunction rules than to Gentzen's own:

From $\Gamma \vdash \Theta, \phi$ and $\psi, \Delta \vdash \Phi$ to infer $(\phi \to \psi), \Gamma, \Delta \vdash \Theta, \Phi$

on the left, and

From $\phi, \Gamma \vdash \Theta, \psi$ to infer $\Gamma \vdash \Theta, (\phi \to \psi)$

[48]The Relevance Logic tradition also calls intensional disjunction and conjunction *fission* and *fusion*, respectively.

on the right. Here again, an alternative, "extensional", pair of rules is possible:

From $\phi, \Gamma \vdash \Delta$ to infer $\Gamma \vdash \Delta, (\phi \rightarrow \psi)$

and

From $\Gamma \vdash \Delta, \psi$ to infer $\Gamma \vdash \Delta, (\phi \rightarrow \psi)$

on the right, and

From $\Gamma \vdash \Delta, \phi$ and $\psi, \Gamma \vdash \Delta$ to infer $(\phi \rightarrow \psi), \Gamma \vdash \Delta$

on the left. Once again, in the presence of all the structural rules, the two rule-pairs are equivalent.

Other choices and combinations of rules can "absorb" some of the functions of the structural rules, making possible variant sequent calculi in which the use of some structural rules is largely or completely avoided. Thus, for example, if the axioms are allowed to contain extraneous formulas (that is, to take the form $\phi, \Gamma \vdash \Delta, \phi$ instead of just $\phi \vdash \phi$), applications of Thinning can be avoided. In what is called the "Ketonen" version of LK [Ketonen, 1944], these generalized axioms are combined with what we have called above the "extensional" forms of rules with two premises and the "intensional" forms of the rules with a single premise. In the propositional case, this permits proofs of classically valid sequents containing no applications of Thinning or Contraction: the system, therefore, makes searching for proofs comparatively easy. [Curry, 1963, §5C–5E] gives detailed coverage of this and some other variant forms of sequent calculus, including systems for classical logic with single formulas in the succedent and for intuitionistic logic with multiple succedent formulas: an additional rule is needed in one case, restrictions on the rules in the other.

Other modifications lead to a simplification of the structure of sequents. Among Gentzen's rules, only those for negation and the conditional move material from one side of the sequent to the other.[49] By changing the rules for these connectives we can have a variant of LK in which everything stays on its own side of the fence. In detail, this involves four changes to Gentzen's system. First, keeping his rules for operators other than negation and the conditional, we add negative rules, analogous to the negative introduction and elimination rules of Fitch's version of natural deduction. Thus, for example, Negated Disjunction left,

[49]Which makes it easy to see that Cut-free proofs have a certain strengthening of the subformula property. Define *positive* and *negative* subformulas by simultaneous induction. Every formula is a positive subformula of itself, if it is a quantification, conjunction or disjunction its positive (negative) subformulas also include the positive (negative) subformulas of its instances, conjuncts and disjuncts respectively, if it is a negation its positive (negative) subformulas include the negative (positive) subformulas of the negated formula, and if it is a conditional its positive (negative) subformulas include the positive (negative) subformulas of its consequent and the negative (positive) subformulas of its antecedent. Then, in a Cut-free proof, every formula occurring in the antecedent of some sequent of the proof is either a positive subformula of a formula in the antecedent of the final sequent or a negative subformula of a formula in the succedent of the final sequent, and every formula in the succedent of a sequent of the proof is either a positive subformula of a succedent formula of the final sequent or a negative subformula of an antecedent formula of the final sequent.

From either $\neg\phi, \Gamma \vdash \Delta$ or $\neg\psi, \Gamma \vdash \Delta$ to infer $\neg(\phi \vee \psi), \Gamma \vdash \Delta$

and Negated Disjunction right,

From the two premises $\Gamma \vdash \Delta, \neg\phi$ and $\Gamma \vdash \Delta, \neg\psi$ to infer $\Gamma \vdash \Delta, \neg(\phi \vee \psi)$,

treat negated disjunctions like the conjunctions to which they are classically equivalent, and similarly for Negated Conjunction, Negated Existential Quantifier and Negated Universal Quantifier rules. Second, add Negated Negation (double negation) rules:

From $\phi, \Gamma \vdash \Delta$ to infer $\neg\neg\phi, \Gamma \vdash \Delta$

and

From $\Gamma \vdash \Delta, \phi$ to infer $\Gamma \vdash \Delta, \neg\neg\phi$.

Third, add "extensional" rules, both positive and negative, for the conditional, treating $(\phi \rightarrow \psi)$ like the classically equivalent $(\neg\phi \vee \psi)$. Finally, take as axioms (all with α atomic: corresponding forms with complex formulas are derivable) not just Gentzen's original $\alpha \vdash \alpha$ but also (n) $\neg\alpha \vdash \neg\alpha$, (em) $\vdash \alpha, \neg\alpha$ and (efq) $\alpha, \neg\alpha \vdash$.[50] The new rules are of forms parallel to inferences already available in LK, so essentially Gentzen's original reasoning proves the eliminability of Cut: not just standard Cut, but also "Right Handed Cut" (from $\Gamma \vdash \Delta, \phi$ and $\Phi \vdash \Psi, \neg\phi$ to infer $\Gamma, \Phi \vdash \Delta, \Psi$ and "Left Handed Cut" (from $\phi, \Gamma \vdash \Delta$ and $\neg\phi, \Phi \vdash \Psi$ to infer $\Gamma, \Phi \vdash \Delta, \Psi$) [51] Cut free proofs have an appropriately weakened subformula property.

Once the rules have been modified so that nothing ever changes side, it is possible to have proofs in which every sequent has one side empty: a logical theorem can be given a proof in which no sequent has any antecedent formulas, and a refutation — showing that some collection of formulas is contradictory — can be formulated as a proof with no succedent formulas. Both of these possibilities have been exploited in defining interesting and widely-studied systems. The deductive apparatus considered in [Schütte, 1960] , and in the large volume of work influenced by Schütte, is a right-handed sequent calculus (with its succedent-only sequents written as disjunctions).

3.7 Sequent Calculi and Tableaux

Left-handed (and so refutational) sequent calculi — in somewhat disguised notation and under another name — have proven very popular: they may now be the

[50] Leave out the axioms of form (em) and you get an L-system for Kleene's (strong) three-valued logic [Kleene, 1952]. Leave out the axioms of form (efq) and you get a formulation of Graham Priest's three-valued "Logic of Paradox" [Priest, 1979]. Leave out axioms of both these forms and you get a formulation of Anderson and Belnap's First Degree Entailment [Anderson and Belnap, 1975].

[51] Right Handed Cut is a derivable rule of both LJ and LK: use standard Cut and provable sequents of the form $\phi, \neg\phi \vdash$. Left Handed Cut is similarly derivable in LK. Addition of Left Handed Cut as a new rule to LJ would yield a formulation of Classical logic.

deductive systems most often taught to undergraduates! [Beth, 1955] developed his systems of semantic tableaux as a simplification of sequent calculi.[52] Actually writing out proofs in Gentzen's L systems is laborious, in part because the "inactive" formulas in sequents (the ones represented by the Greek capitals in the rule schemas) have to be copied from one line of the proof to the next, with the result that a proof is likely to contain many copies of such formulas. One way of looking at Beth's systems is that they optimize Gentzen's LK (which, after all, was originally introduced only for theoretical purposes) for use by incorporating notational conventions that minimize this repetition. This is perhaps easiest to explain using a modern version of tableaux.[53] A tableau, in this formulation, is a tree of formulas, constructed in accordance with certain rules, and constitutes a refutation of a finite set of formulas. To construct a propositional tableau, start with the set of formulas to be refuted, arranged in an initial, linear, part of the tree. One then extends the tree by applying, in some order, branch extension rules to formulas occurring on the tree, "checking off" (notationally, writing a check beside) a formula when a rule is applied to it. A branch of the tree (i.e., a maximal part of the tree linearly ordered by the tree partial ordering) is closed if it contains some formula and its negation; otherwise it is open. A tableau is closed, and constitutes a refutation, just in case all its branches are closed. The branch extension rules are analogous to the (positive and negative) elimination rules of natural deduction:

- if a conjunction-like formula (i.e. a conjunction, negated disjunction, or negated conditional) occurs in the tree and is not yet checked off, extend each open branch of the tree passing through it by adding its conjuncts (or the negations of the disjuncts, or the antecedent and the negation of the consequent, as the case may be) and check the formula off

- if a disjunction-like formula (i.e. a disjunction, conditional, or negated conjunction) occurs unchecked, split each open branch passing through it in two, adding one disjunct to each of the new branches so formed (or: adding the consequent to one and the negation of the antecedent to the other, or: adding the negation of one conjunct to each) and check the formula off

- if a double negation occurs unchecked, add the un-doubly-negated formula to each open branch passing through it and check the double negation off.

[52]Beth describes a tableau as a "natural deduction" proof, but we find the analogy with sequent calculi clearer, although given the close relationship between the two kinds of system, the question may be moot. (Recall from our discussion above in §2 that many logicians use the term 'natural deduction' in a broad sense, thereby including sequent calculi as well as the systems we are calling natural deduction.)

[53]Beth's original notation for tableaux is a bit clumsy. A variety of improvements were defined by several logicians in the 1960s: the best and most influential seems to be due to [Smullyan, 1968] and [Jeffery, 1967]. For more on the history of tableaux, mentioning especially the role Jaakko Hintikka [1953; 1955a; 1955b] might have played in developing it, see [Anellis, 1990]

The finished product, a closed tableau, is a much "skinnier" array of symbols than the corresponding LK proof would be, but the logic is essentially the same. To see this, consider a record of the stages of the construction of a closed tableau:

- for the starting point, write out the formulas the tableau starts with horizontally, separated by commas,
- for any stage of construction involving a non-splitting rule, write, above the top of the representation of any branch of the tableau affected, the list of the formulas remaining unchecked on that branch, horizontally and separated by commas,
- for any stage of construction involving a splitting rule, write side-by-side, above the top of the representation of any branch affected, the lists of unchecked formulas on each of the branches introduced by the splitting rule.

The result (ignoring Permutation) will be a proof in a left-handed sequent calculus, using Ketonen forms of the rules and axioms with extra formulas! The tree structure of the two proofs is nearly the same (the tableau's tree has extra nodes if it starts with more than one formula and in places where a conjunction-like rule has added two formulas to a branch), but where the tableau has a single formula at a node of the tree, the sequent proof has a (left-handed) sequent made up of the formulas remaining unchecked on the branch when that node was reached in the process of tableau construction.

Tableaux with quantifiers can be construed as similarly abbreviated presentations of sequent calculus proofs, but with an additional modification to Gentzen's rules. The branch extension rule for existential and negated universal quantifications

- if $\exists xFx$ (or $\neg\forall xFx$) occurs unchecked, extend each branch through it by adding Fa (or $\neg Fa$), where a is a parameter (=free variable, instantial term, dummy name) that does not yet occur on the branch and check the quantified formula off,

presents no problems: it corresponds exactly to the usual sequent calculus rule for introducing an existential quantifier (negated universal quantifier) on the left, with the proviso that the parameter be new corresponding to the condition on variables in this rule. The branch extension rule for universal and negated existential quantifications, on the other hand, introduces something new:

- if $\forall xFx$ (or $\neg\exists xFx$) occurs, then, for any term t occurring free in a formula on the branch through it (and for an arbitrarily chosen parameter if there are none) extend any branch through it by adding Ft (or $\neg Ft$), and do not check the quantified formula off.

The quantified formula, in other words, "stays active": at a later stage in the construction of the tableau (after, perhaps, new terms have appeared through the use of the existential/negated universal rule) we can re-apply the rule to add a new instance of the same quantified formula. Thinking of the finished tableau as an abbreviated presentation of a sequent proof with the sequents containing the formulas active at a given stage of tableau construction, this corresponds to a new rule:

From $\Gamma, Ft, \forall x Fx \vdash$ to infer $\Gamma, \forall x Fx \vdash$
(and from $\Gamma, \neg Ft, \neg\exists x Fx \vdash$ to infer $\Gamma, \neg\exists x Fx \vdash$)

and the "extra formulas" in the axiom sequents at the top of the sequent proof will include all the universal (and negated existential) quantifications occurring in sequents below them. This variant form of the quantifier rules, however, is like the Ketonen variant of the propositional rules in "absorbing" some of the functions of structural rules into rules for the logical operators, and so simplifying the search for proofs. Often, in proving a sequent containing quantified formulas in versions of sequent calculus with rules more like Gentzen's it is necessary to apply a quantifier rule (universal quantifier on the left, existential quantifier on the right) more than once, generalizing on different free variables, to produce duplicate copies of the same quantified formula, with the duplication then eliminated by use of Contraction.[54] The variant quantifier rules given in this paragraph get the effect of this sort of use of Contraction — contracting, as it were, the newly inferred quantified formulas into the spare copies already present — and allow proofs in which Contraction is not otherwise used.

That comparison with (single-sided) sequent calculi seems to us the most informative way of comparing tableaux to other systems, but — illustrating again the close relationship between natural deduction and sequent calculi — a tableau can also be seen as a special sort of natural deduction derivation. Textbook versions of tableaux often include the instruction that some special symbol be written at the bottom of each closed branch to mark its closed status. We can pretend that this symbol is a propositional *Falsum* constant $\perp$, and then interpret a closed tableau (which is, after all, a *refutation* of the tableau's initial formula(s)) as a derivation of $\perp$ from the set of formulas the tableau started with. Some of the branch extension rules for tableau construction are virtually identical to standard natural deduction rules: the rules for conjunctions and Universal Quantifications simply extend branches by adding formulas derivable by Conjunction Elimination and Universal Quantifier Elimination. The, formally parallel, branch extension rules for negated disjunctions, negated conditionals and negated Existential Quantifications correspond in the same way to natural deduction inferences by the Negative Disjunction (Conditional, Existential Quantifier) Elimination rules. The branch extension rule for disjunction splits the branch: the two new sub-branches formed amount to the two subproofs of an application of Disjunction Elimination, the two formulas heading the new sub-branches being the hypotheses of the two subproofs: requiring that

[54]This reflects a deep, essential, property of First-Order Logic that shows up in different ways in different deductive systems: it is responsible for the undecidability of First-Order Logic, in the sense that provability without use of Contraction in standard LK is a decidable property. To see a simple example, write out a proof of $\forall x(Fx \wedge \neg\forall y Fy) \vdash$.

all branches be closed (contain $\perp$) corresponds to the requirement, for Disjunction Elimination, that the conclusion be derived in each subproof. The splitting branch extension rule for negated conjunctions corresponds to the parallel rule of Negative Conjunction Elimination, and that for conditionals to a non-standard Conditional Elimination rule treating material implications like the disjunctions to which they are semantically equivalent. Finally, the branch extension rules for Existential (and Negated Universal) Quantifications can be seen as initiating the subproofs for inferences by Existential (and Negative Universal) Quantifier Elimination: the requirement, in the branch extension rule, that the instantial term (free variable, parameter, dummy name...) in the "hypothesis" of the subproof be new to the branch is precisely what is needed to guarantee that any formula occurring earlier on the branch can be appealed to within the subproof. Writing in the "branch closed" symbol after a pair of contradictory formulas has appeared on a branch can be seen as a use of Negation Elimination (i.e., *ex falso quodlibet*) to infer $\perp$. In other words, allowing for the graphical presentation (which is not very explicit in showing which lines depend on hypotheses), a closed tableau is a deduction of $\perp$ from its initial formulas in a variant of natural deduction with Negative rules and a non-standard (but classically valid) rule for the conditional. A deduction, moreover, which satisfies strict normality conditions: only elimination rules are used, and Negation Elimination is used *only* to infer $\perp$.

Variant forms of tableaux are, as one should expect, known, and tableaux formulations can be given for a wide range of logics, including intuitionistic and modal logics. [Fitting, 1983] has given an encyclopedic survey of such systems. [Toledo, 1975] has used tableaux, rather than [Schütte, 1960]'s own sort of sequent calculus, in expounding much of the content of [Schütte, 1960].[55]

The popularity of tableaux in recent undergraduate textbooks is, in the opinion of the present writers, unfortunate. By casting all proofs into the form of refutations and then putting refutations into the specialized form of tree diagrams they obscure the relation between formal derivations and naturally occurring, informal, intuitive deductive argument. This surely detracts from the philosophical value of an introductory logic course! It might be forgivable if tableau systems had outstanding practical virtues, but in fact they (like any system permitting only normal or Cut-free derivations) are, in application to complex examples, very inefficient in terms of proof size. Their academic advantage seems to be in connection with assessment: proof-search with tableaux is routine enough, and tableaux for very simple examples small enough, that a reasonable proportion of undergraduates can be taught to produce them on an examination.

[55]Toledo's book is based on the first edition of Schütte's, and so includes material — for instance, on Ramified Type Theory and on the Type-free logic of Ackermann — that was dropped from the second edition and the English translation based on the second edition.

3.8 Natural Deduction with Sequences

It is also possible to have a sort of hybrid system, using disjunctively-interpreted sequences of formulas as the lines of a proof, with structural rules to manipulate them, as in a right-handed sequent calculus, but using both introduction and elimination rules, as in standard natural deduction. In such a system, Disjunction Elimination literally breaks a disjunction down into its component disjuncts: the rule licenses the inference from something like $\Gamma, (\phi \vee \psi), \Delta$ to $\Gamma, \phi, \psi, \Delta$. Such systems have been proposed by [Boričić, 1985] and [Cellucci, 1987; Cellucci, 1992]. These systems are typically called systems of *sequent natural deduction*, but they are radically unlike the systems described in §3.3. The systems of that section are simply reformulations of standard natural deduction systems, using the antecedent formulas of a sequent to record what hypotheses the single succedent formula depends on. In contrast, the systems described here use sequents with no antecedent formulas (dependence on hypotheses is represented in Gentzen's way, by writing proofs in tree form), but they exploit the disjunctive interpretation of multiple succedent formulas in the way LK does. Thus, to use a familiar illustrative example, start with the hypothesis $\forall x(P \vee Fx)$, thought of now as a single-formula sequent. Infer $(P \vee Fa)$, another single-formula sequent, by Universal Quantifier Elimination. Now the novel step: Disjunction Elimination allows us to infer the two-formula sequent P, Fa. The parameter a occurs only in the second formula, and does not occur in the hypothesis from which this sequent is derived, so Universal Quantifier Introduction allows us to infer the $P, \forall xFx$, and Disjunction Introduction (twice, with a Contraction) yields $(P \vee \forall xFx)$ as a final, single-formula, sequent.

The system shares with sequent calculi the laborious-seeming feature that "inactive" formulas have to be copied repeatedly from one line to the next, but Cellucci's system is very efficient: on many test problems (though there are exceptions) a proof in Cellucci's system is smaller (contains fewer symbols) than one in, for example, Fitch's version of natural deduction. This is partly due to careful choice of rules: Cellucci's Conjunction Introduction, for example, unlike Gentzen's Conjunction Right rule for LK (but like the alternative rule for "intensional" conjunction), allows different sets of inactive formulas in the two premises; the total rule-set is perhaps comparable to Ketonen's [1944] version of sequent calculus. (Boričić's system, with a different choice of rule formulations, seems to be much less efficient.) For further efficiency, Cellucci considers quantifier rules like the EI described in §2.6. The whole package is perhaps the best available version of natural deduction for anyone intending actually to write out large numbers of serious complex derivations (particularly since even minimal editing computer software can automate the task of recopying inactive formulas)!

For theoretical purposes, systems of this sort can be as well-behaved as any natural deduction or sequent calculi: Cellucci has proved normalization theorems for them.

3.9 Size of Normal Proofs

Normalizing a natural deduction proof (or eliminating the applications of Cut from a sequent calculus proof) *simplifies* it in an obvious intuitive sense: it gets rid of detours. Why, one might wonder, would one ever want to make use of a *non-normal* proof? Why prove something in a more complicated manner than necessary? The surprising answer is that detours can sometimes be shortcuts! Getting rid of a detour sometimes makes a proof smaller (in terms of number of lines, or number of symbols), but not always. If a detour involves proving a Universal Quantification, for example, which is then used several times to give a series of its instances by Universal Quantifier Elimination, normalizing will yield a proof in which the steps used in deriving the universal formula will be repeated several times, once for each instance. The resulting proof will still be a simplification — the repeated bits will be near duplicates, differing only in the term occurring in them, and all will be copies of the Universal Quantifier Introduction subproof with new terms substituted for its proper parameter — but it may be larger, and indeed much larger.[56] A simple example illustrates the problem. Consider a sequence of arguments, Argument_0, Argument_1, $\text{Argument}_2 \cdots$. For each n, Argument_n is formulated in a First-Order language with one monadic predicate, F, and $n+1$ dyadic predicates, $R_0, \cdots, R_n$, and $n+1$ premises. All share the common premise

$$\forall x \forall y((R_0(xy) \wedge Fx) \rightarrow Fy).$$

For each k, Argument_{k+1} adds the premise

$$\forall x \forall y(R_{k+1}(xy) \rightarrow \exists s \exists t(R_k(xs) \wedge R_k(st) \wedge R_k(ty)))$$

to the premises of the preceding argument. The conclusion of Argument_n is[57]

$$\forall x \forall y((R_n(xy) \wedge Fx) \rightarrow Fy).$$

(To put it in English, the first premise says that the property of F-ness is hereditary along the R_0 relation and the rest say that any two objects related by one of the "higher" relations are connected by a chain of objects related by "lower" relations: note that the length of this chain roughly triples for each succeeding relation. The conclusion is that F-ness is hereditary along R_n.)

A normal proof of the conclusion of Argument_n from its premises will have an Existential Quantifier Elimination subproof for each object in the chain of R_0-related objects connecting a pair of R_n-related objects, applying the common premise in each one to establish that each object in the chain is an F-object. Since the length of the chain increases exponentially as one moves to higher relations,[58]

[56] [Toledo, 1975] discusses this and shows that Cut elimination can be thought of as "repetition introduction."

[57] So Argument_0 is a trivial argument, with the conclusion identical to the premise.

[58] Looking at the $(n+1)^{st}$ case, $R_n(xy)$, we can see that the chain of R_0-related objects starting with x and ending with y is $(3^n)+1$ objects long.

the number of lines in the normal proof increases exponentially as one goes to later arguments in the series. In contrast, if we allow "detours" the length of proofs is much more civilized. To prove the conclusion of Argument$_n$ for any large n, one should first use the first two premises (the premises, that is, of Argument$_1$) and prove (using two Existential Quantifier Elimination subproofs) the conclusion of Argument$_1$:

$$\forall x \forall y((R_1(xy) \wedge Fx) \rightarrow Fy)$$

This "lemma" has the same syntactic form as the common premise, but it is a distinct formula, and deriving it is a detour and a conceptual complication in the proof: it is not a subformula of any premises or of the conclusion of Argument$_n$ (for $n > 1$). Since it is of the same form as the common premise, however, an entirely similar series of steps leads from it and the premise

$$\forall x \forall y(R_2(xy) \rightarrow \exists s \exists t(R_1(xs) \wedge R_1(st) \wedge R_1(ty)))$$

(i.e., the new premise first added in Argument$_2$) to obtain the conclusion of Argument$_2$,

$$\forall x \forall y((R_2(xy) \wedge Fx) \rightarrow Fy)$$

And we continue in this way until we reach the conclusion of Argument$_n$. The number of lines of the proofs obtained in this way for the different Arguments$_k$ grows only linearly in k. Very quickly[59] we reach arguments that can be shown valid by formal deduction if we use lemmas but would be humanly impossible if normal proofs were required.[60]

Since the 1970s there has been much theoretical investigation of the comparative efficiency of different kinds of logical proof-procedure,[61] and the topic is linked to one of the most famous open questions of contemporary mathematics and theoretical computer science: if there is a proof procedure in which any valid formula of classical propositional logic containing n symbols has a proof containing less than $f(n)$ symbols, where f is a polynomial function, then NP=co-NP. In this rarified theoretical context, two systems are counted as tied in efficiency if each can p-simulate the other: if, that is, there is a polynomial function g such that, for any proof containing n symbols of a valid formula in one there is a proof of the same formula in the other containing less than $g(n)$ symbols. By this standard, all typical natural deduction systems are equivalent to each other and to axiomatic systems.[62] On the other hand, as the example above suggests, systems restricted

[59]The normal proof is longer than the lemma-using proof from around Argument$_3$ on. For Argument$_3$, each method — with reasonably-sized handwriting — requires around a half-dozen pages. After that, each succeeding argument requires another two pages with "detours" and about three times as many with normal proofs.

[60]The example is adapted from [Hazen, 1999]; it was inspired by the example in [Boolos, 1984], a very readable paper strongly recommended to anyone wanting further discussion of the issue.

[61]For an introduction, see the review article [Urquhart, 1995].

[62]This result — that the proof procedures for propositional logic of Frege and Gentzen and the myriad variants of each are, up to p-simulation, of equivalent efficiency — seems to be due to Robert Reckhow (see [Cook and Reckhow, 1979]). The equivalence is not obvious: although

to *normal* proofs — tableaux, sequent calculi without cut — have been proven *not* to *p*-simulate natural deduction.

The differences in efficiency between systems that can *p*-simulate each other, though irrelevant to abstract complexity theory, can be practically important. Quine's proof of the soundness of his formulation of First Order Logic [Quine, 1950], using an Existential Instantiation rule, suggests an algorithm for translating proofs in his system into proofs in a system with something like Gentzen's Existential Quantifier Elimination with a theoretically trivial increase in the number of lines: multiplication by a constant factor. Since the factor is > 2, however, the difference in efficiency can be of considerable practical relevance.

4 PROBLEMS AND PROJECTS

Our investigations have in passing brought up a number of topics that deserve further study, and in this section we discuss them with an eye to presenting enough information about each so that the motivated reader would be equipped to pursue it.

4.1 The Concept of Natural Deduction: Further Informal Thoughts

In §§2.1 and 2.5 we discussed "the wider notion of natural deduction." We discussed the sorts of features that are associated — or thought to be associated — with natural deduction, and we argued that none of these informal features were either necessary or sufficient for a natural deduction system — even though they are generally found in the elementary logic textbooks.

One aspect that is often thought to be central to natural deduction is that it doesn't have axioms from which one reasons towards a conclusion. Rather, the thought goes, natural deduction systems are comprised of sets of *rules*. Now, we have pointed out that very many systems that we are happy to call natural deduction in fact *do* have axioms. But even setting that thought aside, the mere fact that a logical system contains only rules does not on its own confer the title of "natural deduction". And it is here, we think, that some scholars of the logic of antiquity have been misled. [Łukasiewicz, 1951] reconstructed the Aristotelian syllogistic by means of an axiomatic system, but many researchers have thought that the particular axioms Łukasiewicz employed were alien to Aristotle [Thom, 1981; 1993]. This has led other scholars to try a reconstruction along some other lines, and many of them claim to discern "natural deduction" in the syllogism... on the grounds that it is merely a set of rules. (See [Corcoran, 1972; Corcoran, 1973; Corcoran, 1974; Martin, 1997; Andrade and Becerra, 2008].) However, these reconstructions do not contain any way to "make an assumption and see where it leads"; instead, they just apply the set of rules to premises. Those of us who see

the algorithm for converting natural deduction proofs into axiomatic ones suggested by textbook proofs of the deduction system seems to lead to exponential increases in proof size, more efficient algorithms are available.

the making of assumptions and then their discharge as being crucial to natural deduction will not wish to have Aristotelian syllogistic be categorized as natural deduction. On the other hand, there is the metatheoretic aspect of Aristotelian logic, where he shows that all the other valid syllogisms can be "reduced" to those in the first figure. In doing this, Aristotle makes use of both the method of *ecthesis* — which seems to be a kind of $\exists E$, with its special use of arbitrary variables (see [Smith, 1982; Thom, 1993]) — and Reductio ad Absurdum. There is a reading of Aristotle where his use of these rules seems to involve assumptions from which conclusions are drawn, leading to new conclusions based on what seems to be an embedded subproof. So, the background metatheory maybe could be thought to be, or presuppose, some sort of natural deduction framework.

We now turn our attention to some other general features of various systems of logic (or rather, features often claimed to belong to one or another type of system). For example as we noted in §2.1, natural deduction systems of logic are said to be "natural" *because* the rules "are intended to reflect intuitive forms of reasoning" and *because* they "mimic certain natural ways we reason informally." In fact, though, research into the psychology of reasoning (e.g., [Evans *et al.*, 1993; Manktelow, 1999; Manktelow *et al.*, 2010]) has uniformly shown that people do not in general reason in accordance with the rules of natural deduction. The only ones of these rules that are clearly accepted are MP (from $\phi, \phi \supset \psi$ infer ψ), Biconditional MP (from $\phi, \phi \equiv \psi$ infer ψ, and symmetrically the reverse), $\wedge I$, and $\wedge E$.

The majority of "ordinary people" will deny $\vee I$. Now, it is probably correct that their denial is due to "pragmatic" or "conversational" reasons rather than to "logical" ones. But this just confirms that $\vee I$ is *not* "pretheoretically accepted." Study after study has shown that "ordinary people" do not find MT (from $\neg\psi, \phi \supset \psi$ infer $\neg\phi$) any more logically convincing than denying the antecedent or affirming the consequent ($\neg\phi, \phi \supset \psi$ infer $\neg\psi$; $\psi, \phi \supset \psi$ infer ϕ). Similar remarks hold for $\vee E$, Disjunctive Syllogism (from $\neg\phi, \phi \vee \psi$ infer ψ), and Reductio ad Absurdum (in any of its forms).

So, maybe the more accurate assessment would be what Jaśkowski and Gentzen originally claimed: that natural deduction (as developed in elementary textbooks) is a formalization of certain informal methods of reasoning employed by *mathematicians*... and not of "ordinary people."

Another claim that used to be more common than now is that natural deduction is the easiest logical method to teach to unsophisticated students. This usually was interpreted as saying that natural deduction was easier than "the logistic method" (axiomatic logic). Nowadays it is somewhat more common to hear that tableaux methods are the easiest logical method to teach to unsophisticated students. Now, Tableaux are *intrinsically* simpler, as shown in §3.7 where they are presented as a special case of natural deduction derivations, so it would not be surprising if they were more easily learned. But so far as we are aware, there has been no empirical study of this, although anecdotal evidence does seem to point in that direction. (It is also not so clear how to test this claim; but it would be quite a pedagogically useful thing to know, even though ease of learning is not the only consideration relevant to the choice of what to cover in a logic course. Should we stop teaching natural deduction in favor of tableaux methods, as seems to be a trend in the most recent textbooks?)

And again, axiomatic systems are usually alleged to be more suited to metatheoretic analysis than are natural deduction systems (and also tableaux systems?). Gentzen agreed with this in his opening remarks of [Gentzen, 1934]:

> ...The formalization of logical deduction, especially as it has been developed by Frege, Russell, and Hilbert, is rather far removed from the forms of deduction used in practice in mathematical proofs. Considerable formal advantages are achieved in return.

Even in some textbooks that teach natural deduction we will find a shift to an axiomatic system in the proof of soundness and completeness (for instance, in [Kalish and Montague, 1964; Thomason, 1970; Kalish *et al.*, 1980], where reference is made to "equivalent axiomatic systems"). But we might in fact wonder just how much easier it really is to prove that (say) propositional *Principia Mathematica* with its five axioms and two rules of inference is complete than is a propositional natural deduction system with nine rules of inference. (And if we wanted really to have a "fair" comparison, the two systems ought to have the same connectives.) Or for that matter, is it really easier to show that five axioms are universally true and that $\supset E$ plus Substitution are truth-preserving than it is to show that some nine rules "preserve truth, if all active assumptions are true"? It would seem that before such claims are asserted by teachers of elementary logic on the basis of their intuitions, there should be some controlled psychological/educational studies to gather serious empirical data relevant to the claim. Again, so far as we are aware, this has not been done.

4.2 Natural Deduction and Computers

As mentioned in §2.2, resolution is the standard proof method employed in computer-oriented automated proof systems. There have been some automated theorem proving systems that employ natural deduction techniques; however, some of the systems, especially the early ones, that called themselves natural deduction (e.g., [Bledsoe, 1971; Bledsoe, 1977; Nevins, 1974; Kerber and Präcklein, 1996]) were really various different sorts of "front end programs" that had the effect of classifying problems as being of one or another type, breaking them down into simpler problems based on this classification, and then feeding the various simpler problems to standard resolution provers. Still, there have been some more clearly natural deduction automated theorem provers, such as [Li, 1992; Pelletier, 1998; Pollock, 1992; Sieg and Byrnes, 1998; Pastre, 2002].

In the mid-1990s, a contest pitting automated theorem proving systems against one another was inaugurated.[63] Although various of the natural deduction theorem

[63]CASC: the CADe ATP System Competition — CADe stands for "Conference on Automated

proving systems have been entered into the contests over the last 15 years, none of them has come close to the performance of the highest-rated resolution-based provers. In the most recent contest in which a natural deduction system was entered (2008)[64], it came in second-to-last in a field of 13 for the competition in which it would be expected to show its strength best: the first-order format (FOF) category which has no normal forms that might give resolution systems an edge. Even though all the resolution-based systems had to first convert the problems to clause form before starting their proofs, they managed to win handily over MUSCADET. For example, of the 200 problems to be solved within the 300 second time window, the winning system solved 169 whereas MUSCADET solved only 38. Despite the fact that MUSCADET solved some problems that none of the other systems could solve, it seems that the lesson learned from computerized chess also applies to theorem proving: systems that follow human patterns of heuristic reasoning about solving problems cannot successfully compete against brute force algorithms that employ massive amounts of memory and extremely fast deep and broad search mechanisms. (See [Slate and Atkin, 1977], for a typical account of this view in the realm of chess). It just may be that the goal set by the founders of natural deduction, and its many followers, of presenting logical proofs "in the way that ordinary mathematicians construct their proofs" is not really the most effective way to prove logic problems, even if it is the way that mathematicians proceed when they are proving mathematical problems. (A different rationale for the relatively poor showing of natural deduction systems over the last 15 years that direct comparisons have been made has pointed to the amount of effort that has been poured into resolution theorem proving since the mid-1960s as compared to natural deduction. See for instance [Pelletier, 1998, p.33].)

Of course, there have all along been computer programs that were designed to help elementary logic students learn natural deduction. Probably the earliest of these were the suite developed under the guidance of Patrick Suppes and designed to help school students learn "the new math" of the 1960s, partially summarized in [Goldberg and Suppes, 1972] for learning natural deduction. In more recent times, it has become more or less of a requirement that a beginning natural deduction logic textbook have some computer assistance, and this includes a "proof assistant" that will help the student in the construction of proofs in the chosen natural deduction system. Some of the more long-standing systems that are associated with textbooks are BERTIE3, DAEMON, PLATO, SYMLOG, PANDORA, FITCH, INFERENCE ENGINE, JAPE (see [Bergmann *et al.*, 2008; Allen and Hand, 2001; Bonevac, 1987; Portoraro, 1994; Broda *et al.*, 1994; Barwise and Etchemendy, 2002; Bessie and Glennan, 2000; Bornat, 2005] respectively)[65]. The student constructs

---

Deduction" and ATP stands for "automated theorem proving". CASC in general is described in [Sutcliffe and Suttner, 2006; Pelletier *et al.*, 2002]; the most recent contest (2009) is reported in [Sutcliffe, 2009]. See also the website `http://www.cs.miami.edu/~tptp/CASC`.

[64]The system is called MUSCADET; see [Pastre, 2002].

[65]The Association for Symbolic Logic maintains a website for educational logic software, listing some 46 programs (not all of which are natural deduction). See `http://www.ucalgary.ca/aslcle/logic-courseware`.

a proof as far as s/he can, and the assistant will help by giving a plausible next step (or explain some mistake the student has made). Most of these systems also have an "automatic mode" that will generate a complete proof, rather than merely suggesting a next step or correcting previous steps, and in this mode they can be considered automated theorem proving systems along the lines of the ones just considered in the preceding paragraphs. (However, since they were not carefully and especially tuned for this role, it is easy to understand why they can be "stumped" by problems that the earlier-mentioned systems can handle). There are a number of further systems on the web which are not associated with any particular textbook: [Saetti, 2010; McGuire, 2010; Kaliszyk, 2010; Andrews, 2010; Christensen, 2010; Gottschall, 2010; Frické, 2010; von Sydow, 2010].

One of the initial motivations for programming a natural deduction automated theorem proving system was to assist mathematicians in the quest for a proof of some new theorems. (Such a view is stated clearly in [Benzmüller, 2006; Pastre, 1978; Pastre, 1993; Siekmann *et al.*, 2006; Autexier *et al.*, 2008]; and the history of this motivation from the point of view of one project is in [Matuszewski and Rudnicki, 2005]). The rationale for using the natural deduction framework for this purpose was (channeling Jaśkowski and Gentzen) that natural deduction was the way that "ordinary mathematicians" reasoned and gave their proofs. Thus, the sort of help desired was of a natural deduction type, and the relevant kinds of hints and strategic recommendations that a mathematician might give to the automated assistant would be of a nature to construct a natural deduction proof. A related motivation has been the search for "real proofs" of accepted mathematical theorems, as opposed to the "informal proofs" that mathematicians give. (In this latter regard, see the QED project, outlined in [Anonymous, 1994], and the earlier MIZAR project at both `http://mizar.org/project` and `http://webdocs.cs.ualberta.ca/~piotr/Mizar/` which is now seen as one of the leading forces within the QED initiative. Many further aspects are considered by Freek Wiedijk on his webpage `http://www.cs.ru.nl/~freek`, from which there are many links to initiatives that are involved in the task of automating and "formalizing" mathematics — for instance, `http://www.cs.ru.nl/~freek/digimath/index.html`). A very nice summary of the goals, obstacles and the issues still outstanding is in [Wiedijk, 2007], in which Wiedijk ranks the systems in terms of how many well-known theorems each has proved. The three systems judged to be "state of the art" are HOL-systems[66], the COQ system[67], and MIZAR. MIZAR history is recounted in [Matuszewski and Rudnicki, 2005]. The crucial Mizar Mathematical Library (MML) contains the axioms of the theory (set-theoretic axioms) added to the basic underlying natural deduction proving engine, and also a number of works written using the system. These works undergo a verification of their re-

[66]This Higher-Order Logic family consists of three different research efforts: HOL-LIGHT, PROOFPOWER, ISABELLE-HOL. As Wiedijk points out, of these three only ISABELLE-HOL employs a clearly natural deduction system. HOL-LIGHT is described in [Harrison, 2007]; PROOFPOWER at `http://www.lemma-one.com/ProofPower`; and ISABELLE-HOL is described in [Nipkow *et al.*, 2002] and with the natural deduction language ISAR in [Wenzel, 2002].

[67]See `http://pauillac.inria.fr/coq/doc/main.html`.

sults, extracting facts and definitions for the Library that can then be used by new submissions to the Library.

A study of the successes of giving a complete formalization of mathematics by means of derivations from first principles of logic and axioms of set theory will show quite slow progress. A number of writers have thought that the entire QED goal to be unreachable for a variety of reasons, such as disagreement over the underlying logical language, the unreadability of machine-generated proofs, the implausibility of creating a suitably large background of previously-proved theorems with the ability to know which should be used and when, and the general shortcomings of automated theorem proving systems on difficult problems. [Wiedijk, 2007] evaluates many of these types of problems but nonetheless ends on a positive note that the QED system "will happen earlier than we now expect... in a reasonable time".

4.3 Natural Deduction and Semantics

Natural deduction, at least for standard logics with standard operators, is, well, natural and in use in most informal mathematical reasoning. The rules are intuitive: most people, or at least, most mathematicians, after a bit of reflective thought, can come to see them as obviously valid. It is therefore plausible to think that the rules are, somehow and in some sense, closely tied to our understanding of the logical operators. Almost from the start the study of natural deduction has been accompanied by the search for some philosophical pay-off from this connection. In one way or another, many philosophers have thought that the natural deduction rules for the logical operators are semantically informative: they reveal or determine the meanings of the operators. Since the rules, thought of naïvely, govern a practice — reasoning, or arguing, or proof presentation — it has seemed that a success in grounding the semantics of the operators on the rules would be a success in a more general philosophical project: that of seeking (an explanation of) meanings (of at least these bits of language) in *use*. The first published suggestion along these lines is in [Gentzen, 1934], where, discussing the natural deduction calculus after presenting it, Gentzen writes:

> The introductions represent, as it were, the 'definitions' of the symbols concerned, and the eliminations are no more, in the final analysis, than the consequences of these definitions. This fact may be expressed as follows: In eliminating a symbol, the formula... may be used 'only in the sense afforded it by the introduction of that symbol'.[68]

Other writers have thought of the rules as *collectively* giving the meanings of the operators. Perhaps the first to bring such a proposal to the attention of a wide philosophical audience was Karl Popper, in [Popper, 1947a; Popper, 1947b].

There are problems with these suggestions, but they are attractive and the problems have, at least in part, been overcome. One problem is that not every

[68]This resembles Heyting's account of the meaning of intuitionistic logical operators closely enough that some writers speak of the *Gentzen-Heyting* account of the logical operators.

imaginable collection of rules can be taken to define a connective. The classic statement of this problem (in an era when philosophers were less verbose than now) is [Prior, 1950]. Prior considers an alleged connective, **tonk**, with an introduction rule allowing (A **tonk** B) to be inferred from A and an elimination rule licensing the inference of B from (A **tonk** B). Together the two rules allow any proposition to be inferred from any other: since, whatever propositions may be, some are true and others false, there can be no operation on them satisfying these rules. Prior suggests that rules cannot define a connective, but rather must be responsive to its prior meaning. The equally classic reply is [Belnap, 1962]. Definitions, no less than assertions, are subject to logical restriction, as shown by the familiar fallacious "proofs" that smuggle in an illicit assumption by concealing it in the definition of an unusual arithmetic operator. The fundamental restriction, violated by Prior's **tonk**, is that a definition should be non-creative: should not allow us to deduce conclusions which could not have been derived without using the defined expression. In the particular case of definitions of logical operators by introduction and elimination rules, Belnap gives a precise formulation of this restriction: assuming a language with a notion of deduction (involving logical operators already in the language and/or non-logical inference rules) satisfying the structural rules of Gentzen's L-calculus, the result of adding the new operator with its rules must be a conservative extension: it should yield no inference, whose premises and conclusion are stated in the original language, which could not be derived in the original system. A sufficient condition for this is that Gentzen's rule Cut should be admissible in the extended calculus.[69]

If Prior's objection is that rules can do too much, other challenges claim they cannot do enough: even the best pairs of introduction and elimination rules can fail to determine the meaning of the operator they supposedly define. This can happen in different ways, illustrated by two examples.

> **EXAMPLE A.** Two different notions of necessity — say, logical necessity and physical necessity — might have the same formal logic — perhaps S5. The introduction and elimination rules of that logic's necessity operator will then be neutral between the two interpretations, and so cannot by themselves determine a unique meaning for the $\Box$.

> **EXAMPLE B.** We ordinarily think of disjunction as truth-functional: a disjunction is true if and only if at least one of its disjuncts is. As Rudolf Carnap [1944] observed, however, the rules of classical logic do not require this interpretation: the entire deductive apparatus of classical logic will also be validated by an interpretation on which sentences can take values in a Boolean Algebra with more than two elements,

[69] Prior's objection applies to proposals on which the introduction and elimination rules jointly define the new operator. Gentzen's proposal, on which it is the introduction rules which are definitional and the elimination rules are required to respect the meaning so defined, seems to avoid it: Belnap's discussion can be seen as a precise working out of the details of Gentzen's sketchily-presented suggestion.

with only those taking the top value counted as *true*. On such an interpretation a disjunction can be true even if neither of its disjuncts is: in particular, assuming negation is interpreted by Boolean complement, all instances of the Law of Excluded Middle will be true, but most will not have true disjuncts.[70]

Carnap's own response to the problem in the second example was to consider an enriched logical framework. The problem arises when a logic is understood as defining a consequence relation in the sense of [Tarski, 1930a; Tarski, 1930b; Tarski, 1935]: a relation holding between a finite set of premises and a given conclusion if and only if the inference from the premises to the conclusion is valid. It disappears if a logic is thought of as providing a multiple conclusion consequence relation: a relation holding between two finite sets of sentences just in case the truth of all the members of the first set implies that at least one of the second is true. These abstract relations relate in an obvious way to syntactic notions we have seen: an ordinary, Tarskian, consequence relation is (if we identify relations with sets of ordered pairs) a set of the sort of single-succedent sequents used in Gentzen's LJ, and the generalized consequence relation Carnap moved to is a set of generalized sequents of the sort introduced for LK.[71] If classical logic is defined as a generalized consequence relation, then it excludes non-truth-functional interpretations of the sort Carnap noted: the validity of the sequent $\phi \vee \psi \vdash \phi, \psi$ means that a disjunction can only be true if at least one of its disjuncts is. Critics, notably [Church, 1944], have thought this was cheating: the move to multiple-succedent sequents amounts to introducing a second notation for disjunction, *stipulating* that it is to be given a truth-functional interpretation, and then defining the interpretation of the ordinary disjunction connective in terms of it.

So far we have discussed whether the consequence relation induced by the logical rules can determine the interpretation of the logical operators. Natural deduction, with rules allowing inferences from subproofs and not merely from (n-tuples of) formulas, has a richer structure, and we can ask whether attention to this structure can close the gap.[72] In fact it seems to come tantalizingly close. The disjunction elimination rule allows a conclusion to be inferred from a disjunction together with two subproofs, one deriving the conclusion from one disjunct and the other from the other, but we have to be careful in specifying the conditions the subproofs must satisfy. Define a subproof to be *de facto truth preserving* iff it satisfies the condition that either the conclusion derived in it is true or the hypothesis (or one of the reiterated premises appealed to in the subproof) is not true. If

[70]Readers familiar with van Fraassen's notion of *supervaluations* will recognize here another consequence of the same fact about Boolean algebras.

[71]The correspondence between multiple conclusion consequence relations and semantic interpretations of logics has been widely studied since the 1970s. Such relations are often called *Scott consequence relations* in reference to [Scott, 1974].

[72]Readers familiar with supervaluations will have seen other examples where these rules introduce novelties. Supervaluational interpretations (of, e.g., a language with self-reference and a truth predicate) validate the same consequence relation as truth-functional, but do not validate all the subproof-using rules of classical logic.

disjunction elimination is postulated for all de facto truth preserving subproofs, then its validity forces the truth-functional interpretation of disjunction! (The rule, with this kind of subproof allowed, would not be sound on a non-truth-functional interpretation. To see this, let $\phi \vee \psi$ be a true disjunction with two non-true disjuncts, and θ some untruth. θ is implied by $\phi \vee \psi$ on this version of the rule, since the two degenerate subproofs in which θ is directly inferred from ϕ and ψ are, trivially, de facto truth preserving: since their hypotheses are not true, they have, so to speak, no truth *to* preserve.) If, however, we require that the subproofs embody *formally* valid reasoning, disjunction elimination doesn't do any more toward ruling out a non-truth-functional interpretation of disjunction than imposing the consequence relation of classical logic does. (By the Deduction Theorem, if there are formally valid subproofs from ϕ to θ and from ψ to θ, we would have proofs of $(\phi \supset \theta)$ and $(\psi \supset \theta)$, and the rule of disjunction elimination tells us no more than that the consequence relation includes $(\phi \vee \psi), (\phi \supset \theta), (\psi \supset \theta) \vdash \theta$.) The philosophical significance of this contrast depends, obviously, on whether or not there is a principled reason for preferring one or the other class of admissible subproofs. If we see the project of defining logical operators by their rules as a contribution to a more general explication of meaning in terms of "use", however, it would seem that only rules which could be adopted or learned by reasoners are relevant. From this point of view it would seem that the restriction to formally valid subproofs is appropriate: recognizing de facto truth preserving subproofs is not something a (non-omniscient) being could learn in the way it can learn to reason in accordance with formal rules.

Philosophical discussion of the significance of Example B, and in particular of the legitimacy of Carnap's appeal to a "multiple conclusion" consequence relation, continues: cf., e.g., [Restall, 2005; Rumfitt, 2008]. At least for those who share Church's intuitions about "multiple conclusion" consequence, however, Carnap's observation would appear to set limits to what we can hope for. If you insist that semantics must include something like a model-theoretic interpretation, assigning truth values (or other values) to formulas, then their logical rules cannot fully determine the semantics of the logical operators.[73] It remains possible, however, that the logical rules determine some more general kind of meaning, or some aspect of the meaning, of the logical operators. Here, for a change, there is an impressive positive result! Think of the meanings of logical operators abstractly, without requiring any particular connection between their meanings and the assignment of truth values (or similar) to complex sentences. Say that two operators are semantically *equivalent* if any sentence formed using one is logically equivalent to the

[73]It follows from the completeness theorem that the possible interpretations of a classically consistent theory will include standard model theoretic ones, in which the connectives are truth-functional. Whether, in all cases, any of the models mathematically possible deserve to be thought of as possible semantic interpretations, in a strong sense of semantic, is a debated philosophical question: certainly the Löwenheim-Skolem theorem and Non-Standard models of arithmetic suggest that not all models are semantically relevant.

sentence formed in the same way from the same constituents using the other.[74] Then the standard introduction and elimination rules determine the meanings of the operators at least up to equivalence. Thus, for example, suppose we had two operators, $\wedge_1$ and $\wedge_2$, each governed by "copies" of the $\wedge I$ and $\wedge E$ rules. Then we could derive $\phi \wedge_1 \psi$ from $\phi \wedge_2 \psi$ and conversely. (Left to right: use $\wedge_1 E$ to get ϕ and ψ, then combine them by $\wedge_2 I$. Right to left similarly.) Parallel results hold for the other standard connectives $\vee, \supset, \neg$, in both Classical and Intuitionistic logic. Parallel results hold for the quantifiers in both logics, though here there is an added subtlety: after all, in Higher Order logics (as in other many-sorted logics), the quantifier rules are (or at least can be formulated so as to be) the same for different orders of quantification, but nobody wants to say that First and Second Order universal quantifiers are equivalent! The key here is that the rules for different quantifiers are differentiated, not by formal differences in the rule schemas, but by the classes of terms substitutable for their bound variables: in order to show two universal quantifiers to be equivalent, we must be able to instantiate each, by its $\forall E$ rule, to the parameters (free variables, arbitrary names,...) used in the other's $\forall I$ rule. Leading us, at last, back to **Example A**. If we think of modal operators as quantifiers over possible worlds, it would seem that the case of distinct modalities with formally identical worlds *ought* to be explained in a way analogous to our treatment of many-sorted quantificational logic... but the "terms" for possible worlds are "invisible." Looking back to our discussion of natural deduction formulations of modal logics (§2.8), the idea of "invisible" terms for possible worlds was used to motivate the restrictions on reiteration into modal subproofs. So we can say that for two necessity operators to have "the same" introduction and elimination rules, something more than identity of the schematic presentation of the rules is needed: the two versions of $\Box I$ must have the same restrictions on what can be reiterated into a modal subproof. And, in fact, two necessity operators governed by the $\Box I$ and $\Box E$ rules of, say, the logic T will be equivalent if formulas governed by each can be reiterated into the modal subproofs of the other.

These results — showing that Introduction and Elimination rules characterize the meanings of the logical operators up to "equivalence" — are impressive, and it is hard not to see them as relevant to semantics, in some sense of 'semantics'. They depend on what is often referred to as *harmony* between the Introduction and Elimination rules. Avoiding Prior's problem with **tonk** required that the elimination rules for an operator not be too strong: in Gentzen's words, they must use the operator only in the sense afforded it by its introduction rules. Conversely, the uniqueness up to equivalence results require elimination rules that are strong enough: they must make full use of the sense afforded the operator by the introduction rules. Working in the context of sequent calculus, Belnap proposed a

[74] We will not discuss the question of whether equivalent operators are synonymous here. Certainly many philosophers — notably Bertrand Russell in his logical writings from the first decade of the 20th Century, e.g., [Russell, 1906, p.201] — have wanted to allow equivalence to be properly weaker than synonymy.

formal test of this harmony: cut must be admissible if the elimination rules are not too strong, and the identity sequents for compound formulas must be derivable when only identity sequents for atoms are taken as axiomatic if they are not too weak.

It is worth noting possible limitation on what can be achieved along these lines. Even if we restrict our attention to operators broadly analogous to standard logical ones, it is not clear that all operators can be characterized by their logical rules. The vocabulary of an intuitionistic theory of real numbers, for example, can be enriched with two different *strong (constructible) negation* operators with the same logical rules. This is possible because the purely schematic rules for Nelson's negation [Nelson, 1949], though determining how complex formulas containing negations relate to each other, do not afford a sense to the negation of an atomic formula: there is no negation introduction rule allowing the proof of a negated atom. In adding two strong negations to an intuitionistic theory, then, we would have to postulate rules giving the content of negated atoms in terms of non-logical, specifically mathematical, ideas: allowing us to infer the negation (in the sense of one strong negation) of an identity if its terms are non-identical (in the sense of ordinary Heyting negation) but to infer its negation (in the sense of the other strong negation) only if they are separated.[75] Natural deduction was first introduced for Intuitionistic and Classical logics with their standard operators, and works best on its "native turf": natural deduction techniques can be extended beyond this, but it cannot be assumed that the extension will have all thenice properties of the original. One nice property is that the natural deduction rules can, to a degree, be seen as defining the logical operators, and what we have seen is that this does not always extend, even to non-standard negation operators. The success, such as it is, of the idea that logical operators are defined by their rules has helped to inspire larger programs of *inferential semantics*[76] in the general philosophy of language. The case of strong negation can perhaps be taken as an illustration of how difficult it may be to extend this sort of treatment beyond the strictly logical domain.

The possibility of characterizing a logical operator in terms of its Introduction and Elimination rules has made possible a precise formulation of an interesting question. One of the properties of classical logic that elementary students are often told about is *functional completeness*: every possible truth-functional connective (of any arity) is explicitly definable in terms of the standard ones. The question *should* present itself of whether there is any comparable result for intuitionistic logic, but this can't be addressed until we have some definite idea of what counts as a possible intuitionistic connective. We now have a proposal: a possible intuitionistic connective is one that can be added (conservatively) to a formula-

[75]Formally, the negation operator of classical logic has rules extending those of both Heyting's negation and Nelson's. It is usually taken to be a logical operator generalizing Heyting's negation, but Bertrand Russell's discussion of negative facts, in his [Russell, 1918], suggests treating it as less than purely logical, on the analogy of Nelson's negation.

[76]Cf. [Brandom, 1994] and the literature it has stimulated, for example [Peregrin, 2008].

tion of intuitionistic logic by giving an introduction rule (and an appropriately matched, not too strong and not too weak) elimination rule for it. Appealing to this concept of a possible connective, [Zucker and Tragesser, 1978] prove a kind of functional completeness theorem. They give a general format for stating introduction rules, and show that any operator that can be added to intuitionistic logic by a rule fitting this format can be defined in terms of the usual intuitionistic operators. Unexpectedly, the converse seems not to hold: there are operators, explicitly definable from standard intuitionistic ones, which do *not* have natural deduction rules of the usual sort. For a simple example, consider the connective $\ddot{\vee}$ defined in intuitionistic logic by the equivalence:[77]

$$(\phi \mathbin{\ddot{\vee}} \psi) =_{df} ((\phi \supset \psi) \supset \psi).$$

(In classical logic, this equivalence is a well-known possible definition for disjunction, but intuitionistically $(\phi \mathbin{\ddot{\vee}} \psi)$ is much weaker than $(\phi \vee \psi)$.) The introduction and elimination rules for the standard operators of intuitionistic logic are pure, in the sense that no operator other than the one the rules are for appears in the schematic presentation of the rule, and $\ddot{\vee}$ has no pure introduction and elimination rules. (Trivially, it has impure rules: an introduction rule allowing $(\phi \mathbin{\ddot{\vee}} \psi)$ to be inferred from its definiens and a converse elimination rule.) To get around this problem, [Schroeder-Heister, 1984b] introduces a generalization of natural deduction: subproofs may have inferences instead of (or in addition to) formulas as hypotheses.[78] In this framework we can have rules of $\ddot{\vee}I$ allowing the inference of $(\phi \mathbin{\ddot{\vee}} \psi)$ from a subproof in which ψ is derived on the hypothesis that $\phi \vdash \psi$ is valid, and $\ddot{\vee}E$ allowing ψ to be inferred from $(\phi \mathbin{\ddot{\vee}} \psi)$ and a subproof in which ψ is derived on the hypothesis ϕ. Schroeder-Heister proves that any connective characterized by introduction and elimination rules of this generalized sort is definable in terms of the standard intuitionistic connectives and vice versa. [Schroeder-Heister, 1984a] proves a similar result for intuitionistic logic with quantifiers.

4.4 The One True Logic? Some Philosophical Reflections

Here we very briefly survey some points at which natural deduction's format for proofs has been employed to argue for some more philosophical-logic conclusions, or at least to lend support or clarity to certain doctrines. Although we do not follow up on the literature surrounding these topics, we commend the issues to further study by those interested in natural deduction.

As we have mentioned at various places, the straightforward int-elim rules for natural deduction as chosen by Gentzen generate intuitionistic logic. Further additions[79] are needed to extend the rules to describe classical logic. Some might

[77]This connective was suggested to Allen Hazen by Lloyd Humberstone.

[78][Fitch, 1966] defines a similarly generalized sort of subproof, but uses it only to abbreviate proofs.

[79]As we said above, Gentzen added (all instances of) $(\phi \vee \neg\phi)$ as "axioms", yielding proof structures that look like the *Law of the Excluded Middle*, LEM, left-most of the following five rules. But other writers would prefer to use $\neg\neg E$ as in the second-left rule, or perhaps $\neg E$ as shown in the middle of the five rules. Or perhaps a rule that embodies "Pierce's Law", or a

— indeed, some have — taken the simplicity and aesthetic qualities of the intuitionistic rules over the kinds of rules identified in the preceding footnote to be an argument in favor of intuitionistic logic over classical logic. (One argument among many, perhaps.[80] But some think it is quite strong.) In this sort of view, the other logics are generated for special purposes by adding some axiom or new rule of inference to show how some connective will act in the new, "unnatural" environment where it is being employed. Intuitionism becomes the Pure Logic of Inference to which special additions might be appended. Views like this are often associated with Michael Dummett (for example, [Dummett, 1978]), although Dummett also has further views concerning verification that make intuitionism a natural choice for a logic that correctly describes his view.

Looking more closely at the possible ways of strengthening a set of intuitionistic rules, we see that classical logic can be obtained by adding rules for negation or the conditional, and a superintuitionistic First Order logic (the logic of "constant domains") can be obtained by adding a rule for distributing the universal quantifier over disjunction. On the other hand, there is a sense in which the standard intuitionistic rules for conjunction, disjunction and the existential quantifier already imply all the classically valid rules for these operators.[81] To see this, note that Gentzen's intuitionistic sequent calculus LJ can, without strengthening intuitionistic logic, be modified to allow multiple succedent formulas as long as the rules for negation, conditional and universal quantifier on the right are applied only to premise-sequents with at most a single succedent formula: the rules for conjunction, disjunction and existential quantification in the modified system are identical to those for the classical LK. This suggests that the two sets of operators are somehow different in status. Perhaps we could say that there is a *basic*

contraposition law in the form given rightmost:

$$\dfrac{\begin{matrix}[\phi]\\ \vdots\\ \psi\end{matrix}\qquad\begin{matrix}[\neg\phi]\\ \vdots\\ \psi\end{matrix}}{\psi}\,(\text{LEM})\qquad \dfrac{\neg\neg\phi}{\phi}\,(\neg\neg E)\qquad \dfrac{\begin{matrix}[\neg\phi]\\ \vdots\\ \psi\\ \vdots\\ \neg\psi\end{matrix}}{\phi}\,(\neg E)\qquad \dfrac{\begin{matrix}[\phi\supset\psi]\\ \vdots\\ \phi\end{matrix}}{\phi}\,(\text{Pierce})\qquad \dfrac{\begin{matrix}[\neg\phi]\\ \vdots\\ \neg\psi\end{matrix}}{\psi\supset\phi}\,(\text{C-pose})$$

and various other rules could be used, to effect the extension to classical logic.

[80]In some theorists' minds this aesthetic argument is just a "follow-on argument" that started with the considerations we brought out in §4.3. That argument starts with the a view on how to define the "meaning" of logical connectives and culminates in the conclusion that the meanings of the natural language 'and', 'or', 'not', 'if–then–', 'if and only if' are precisely given by the int-elim rules. It is yet another step to infer from that the conclusion that intuitionism is The One True Logic. But one could. And in the course of doing so the present aesthetic consideration might be raised.

[81]Cf. [Belnap and Thomason, 1963; Belnap *et al.*, 1963]. The result seems at odds with the common assumption that intuitionism incorporates a distinctive, strong, "sense" of disjunction: see the discussion in [Hazen, 1990].

logic[82] with only conjunction, disjunction and existential quantification as logical operators, and that the distinction between classical and intuitionistic logics only applies to systems extending the basic logic with other operators.

Gentzen separated the rules he described into two sorts: the logical rules and the structural rules. The former give the ways to employ the logical connectives, while the latter characterize the form that proofs may take. From the point of view of natural deduction, the way to characterize the differences between some pairs of logics — e.g., between classical logic and some particular modal logic — is to point to the existence of some new logical operators and the (int-elim) rules that govern them. The way to characterize the differences between other pairs of logics — e.g., between classical logic and intuitionistic logic — is to talk about the differences between their respective int-elim rules for the same logical operators. This brings into the foreground that there could be other pairs of logics — e.g., intuitionistic logic and some relevant logic — that differ instead in their structural rules governing proofs. And in fact, that is a fruitful way to look at a whole group of logics that were initially developed axiomatically: the various relevant logics. (See [Restall, 2000]). This is arguably a clearer viewpoint from which to evaluate their properties than the axiomatic versions (in, e.g., [Anderson and Belnap, 1975], or the semantic characterizations (in, e.g., [Routley and Meyer, 1972; Routley and Meyer, 1973]), or the algebraic perspectives (as lattices with residuated families of operators in, e.g., [Dunn, 1993]), and the existence of such a fruitful viewpoint gives some reason to appreciate natural deduction (and sequent calculi) as an approach to logic generally.

Logical pluralists (see [Beall and Restall, 2006]) take the view that there are many differing notions of validity of arguments. In turn, this leads them to interpret the view just outlined about the structural versus logical rules of a proof theory in such a way that the *language of logic*, including the connectives and the rules governing them, stays constant in the different views of logic, but the *fine structure of proofs* as described by the structural rules is allowed to vary from one application to another. It is this variability of the structural rules that allows for the distinctively different features of the consequence relations in the differing logics. They are all "legal" views about consequence, just employed for different purposes. Again, it is the viewpoint offered by natural deduction (and sequent calculi) that make this approach viable.

Finally, the history of thought is replete with confrontations between, on the one hand, those thinkers who wish to explain complex phenomena in terms of the parts of the complex and the ways these parts interact with one another, and on the other hand, those thinkers who seek to explain parts of a complex system in terms of the roles that the parts play within the system. [Pelletier, 2012], reflecting what

[82] Terminology due to Fitch: conjunction, disjunction and existential quantification are the logical operators of the systems considered in [Fitch, 1942] and subsequent papers. First Order logic restricted to these three basic operators has an interesting recursion-theoretic property that might make it useful in expositions of recursion theory: if the primitive predicates of such a language all express recursively enumerable sets or relations, so do all its formulas.

is common language in philosophy, calls the former mode of explanation "atomism" and the latter "holism."

When applied to the topic of proof theory and model theory, some thinkers have called the atomistic method of explanation "the analytic mode" and the holistic method "the synthetic mode" (see [Belnap, 1962]). Applying the distinction to the logical connectives, the idea is that the analytic mode would wish to define or explain a connective — as well as correct inferences/deductions that involve that connective — in terms of antecedently given features, such as truth tables, or preservation of some property (such as truth, or truth-in-a-model, or truth-in-a-possible-world, and so on) that is thought to be antecedently known or understood. The synthetic mode, which [Belnap, 1962] favored over the analytic mode of [Prior, 1960], takes the notion of a good type of derivation as the given and defines the properties of the connectives in terms of how they contribute to these types of derivation. (See our discussion in §4.3 for Belnap's considerations.)

There has not been much interaction between the philosophical writings on the atomism/holism debate and the debate within the philosophy of logic about the analytic/synthetic modes of explanation. We think that a clearer understanding of the issue, prompted by consideration of natural deduction rules, could open the door to some fruitful interchange. It might also open some interesting interaction with the position known as structuralism in logic (e.g., [Koslow, 1992]), which currently seems quite out of the mainstream discussions in the philosophy of logic.

The precision possible in logical metatheory has made it an attractive laboratory for the philosophy of language: success in sustaining the claim that logical operators are defined by their rules doesn't necessarily imply that inferential semantics will succeed elsewhere, but it is the work in this area, stemming from Gentzen's suggestions, that inspires the hope that the inferentialist project can be carried out rigorously and in detail.

ACKNOWLEDGMENTS

We are grateful for discussions, assistance, and advice from Bernard Linsky, Jack MacIntosh, Greg Restall, Piotr Rudnicki, Jane Spurr, Geoff Sutcliffe, and Alasdair Urquhart. Pelletier also acknowledges the help of (Canadian) NSERC grant A5525.

BIBLIOGRAPHY

[Adler and Rips, 2008] Jonathan Adler and Lance Rips. *Reasoning: Studies of Human Inference and Its Foundations.* Cambridge University Press, New York, 2008.

[Allen and Hand, 2001] Colin Allen and Michael Hand. *Logic Primer.* MIT Press, Cambridge, 2001.

[Anderson and Belnap, 1975] Alan Anderson and Nuel Belnap. *Entailment: Vol. 1.* Princeton University Press, Princeton, NJ, 1975.

[Anderson and Johnstone, 1962] John Anderson and Henry Johnstone. *Natural Deduction: The Logical Basis of Axiom Systems.* Wadsworth Pub. Co., Belmont, CA, 1962.

[Andrade and Becerra, 2008] Edgar Andrade and Edward Becerra. Establishing connections between Aristotle's natural deduction and first-order logic. *History and Philosophy of Logic*, 29:309–325, 2008.

[Andrews, 2010] Peter Andrews. (educational) theorem proving system, 2010. `http://gtps.math.cmu.edu/tps.html`.

[Anellis, 1990] Irving Anellis. From semantic tableaux to Smullyan trees: A history of the development of the falsifiability tree method. *Modern Logic*, 1:36–69, 1990. See also the errata in *Modern Logic* **1:** 263 and **2:** 219.

[Anellis, 1991] Irving Anellis. Forty years of "unnatural" natural deduction and quantification: A history of first-order systems of natural deduction from Gentzen to Copi. *Modern Logic*, 2:113–152, 1991. See also the "Correction" in *Modern Logic* (1992) **3:** 98.

[Anonymous, 1994] Anonymous. The QED manifesto. In Alan Bundy, editor, *Automated Deduction — CADE 12*, pages 238–251. Springer-Verlag, Berlin, 1994. Volume **814** of *Lecture Notes in Artificial Intelligence*.

[Arthur, 2010] Richard Arthur. *Natural Deduction: An Introduction to Logic*. Wadsworth Pub. Co., Belmont, CA, 2010.

[Autexier *et al.*, 2008] Serge Autexier, Christoph Benzmüller, Dominik Dietrich, and Marc Wagner. Organisation, transformation, and propagation of mathematical knowledge in ΩMEGA. *Mathematics in Computer Science*, 2:253–277, 2008.

[Barker, 1965] Stephen Barker. *Elements of Logic*. McGraw-Hill, NY, 1965.

[Barwise and Etchemendy, 2002] Jon Barwise and John Etchemendy. *Language, Proof and Logic*. CSLI Press, Stanford, 2002.

[Beall and Restall, 2006] J.C Beall and Greg Restall. *Logical Pluralism*. Oxford University Press, Oxford, 2006.

[Belnap and Thomason, 1963] Nuel Belnap and Richmond Thomason. A rule-completeness theorem. *Notre Dame Journal of Formal Logic*, 4:39–43, 1963.

[Belnap *et al.*, 1963] Nuel Belnap, Hugues Leblanc, and Richmond Thomason. On not strengthening intuitionistic logic. *Notre Dame Journal of Formal Logic*, 4:313–320, 1963.

[Belnap, 1962] Nuel Belnap. Tonk, plonk and plink. *Analysis*, 22:130–134, 1962.

[Benzmüller, 2006] Christoph Benzmüller. Towards computer aided mathematics. *Journal of Applied Mathematics*, 4:359–365, 2006.

[Bergmann *et al.*, 2008] Merrie Bergmann, James Moor, and Jack Nelson. *The Logic Book, Fifth Edition*. Random House, New York, 2008.

[Bessie and Glennan, 2000] Joseph Bessie and Stuart Glennan. *Elements of Deductive Inference*. Wadsworth Pub. Co., Belmont, CA, 2000.

[Beth, 1955] Evert Willem Beth. Semantic entailment and formal derivability. *Mededlingen van den Koninklijke Nederlandse Akademie van Wetenschappen*, 18:309–342, 1955. Reprinted in J. Hintikka (ed.) (1969) *The Philosophy of Mathematics*, Oxford UP, pp. 9–41.

[Bledsoe, 1971] Woody Bledsoe. Splitting and reduction heuristics in automatic theorem proving. *Artificial Intelligence*, 2:55–78, 1971.

[Bledsoe, 1977] Woody Bledsoe. Non-resolution theorem proving. *Artificial Intelligence*, 9:1–35, 1977.

[Bonevac, 1987] Daniel Bonevac. *Deduction*. Mayfield Press, Mountain View, CA, 1987.

[Boolos, 1984] George Boolos. Don't eliminate cut. *Journal of Philosophical Logic*, 13:373–378, 1984.

[Boričić, 1985] Branislav Boričić. On sequence-conclusion natural deduction systems. *Journal of Philosophical Logic*, 14:359–377, 1985.

[Bornat, 2005] Richard Bornat. *Proof and Disproof in Formal Logic*. Oxford Univ. Press, Oxford, 2005.

[Bostock, 1999] David Bostock. *Intermediate Logic*. Oxford UP, Oxford, 1999.

[Brandom, 1994] Robert Brandom. *Making it Explicit*. Harvard University Press, Cambridge, MA, 1994.

[Broda *et al.*, 1994] Krysia Broda, Susan Eisenbach, Hessam Khoshnevisan, and Steve Vickers. *Reasoned Programming*. Prentice-Hall, Englewood Cliffs, NJ, 1994.

[Carnap, 1944] Rudolf Carnap. *The Formalization of Logic*. Harvard University Press, Cambridge, 1944.

[Cellucci, 1987] Carlo Cellucci. Efficient natural deduction. In C. Cellucci and G. Sambin, editors, *Atti del Congresso Temi e Prospettive della Logica e della Filosofia della Scienza Contemporanee, Volume I — Logica*, pages 29–57, Bologna, 1987. Cooperativa Libraria Universitaria Editrice Bologna.

[Cellucci, 1992] Carlo Cellucci. Existential instantiation and normalization in sequent natural deduction. *Annals of Pure and Applied Logic*, 58:111–148, 1992.

[Cellucci, 1995] Carlo Cellucci. On Quine's approach to natural deduction. In Paolo Leonardi and Marco Santambrogio, editors, *On Quine: New Essays*, pages 314–335. Cambridge UP, Cambridge, 1995.

[Chellas, 1997] Brian Chellas. *Elementary Formal Logic*. Penny Lane Press, Calgary, 1997.

[Christensen, 2010] Dan Christensen. DC Proof, 2010. `http://www.dcproof.com`.

[Church, 1944] Alonzo Church. Review of [Carnap, 1944]. *Philosophical Review*, 53:493–498, 1944.

[Cook and Reckhow, 1979] Stephen Cook and Robert Reckhow. The relative efficiency of propositional proof systems. *Journal of Symbolic Logic*, 44:36–50, 1979.

[Copi, 1954] Irving Copi. *Symbolic Logic*. Macmillan Co., NY, 1954.

[Corcoran, 1972] John Corcoran. Completeness of an ancient logic. *Journal of Symbolic Logic*, 37:696–702, 1972.

[Corcoran, 1973] John Corcoran. A mathematical model of Aristotle's syllogistic. *Archiv für Geschichte der Philosophie*, 55:191–219, 1973.

[Corcoran, 1974] John Corcoran. Aristotle's natural deduction system. In J. Corcoran, editor, *Ancient Logic and Its Modern Interpretations*, pages 85–131. Reidel, Dordrecht, 1974.

[Craig, 1957] William Craig. Review of E.W. Beth, 'Remarks on natural deduction', 'Semantic entailment and formal derivability'; K.J.J. Hintikka, 'A new approach to sentential logic'. *Journal of Symbolic Logic*, 22:360–363, 1957.

[Curry, 1950] Haskell Curry. *A Theory of Formal Deducibility*. Univ. Notre Dame Press, Notre Dame, IN, 1950.

[Curry, 1963] Haskell Curry. *Foundations of Mathematical Logic*. McGraw-Hill, New York, 1963.

[Degtyarev and Voronkov, 2001] Anatoli Degtyarev and Andrei Voronkov. The inverse method. In Alan Robinson and Andrei Voronkov, editors, *Handbook of Automated Reasoning*, pages 178–272. Elsevier, Amsterdam, 2001.

[Dummett, 1978] Michael Dummett. The philosophical basis of intuitionistic logic. In *Truth and Other Enigmas*, pages 215–247. Duckworth, London, 1978.

[Dunn, 1993] Michael Dunn. Partial-gaggles applied to logics with restricted structural rules. In P. Schroeder-Heister and Kosta Dosen, editors, *Substructural Logics*, pages 63–108. Oxford UP, Oxford, 1993.

[Evans *et al.*, 1993] Jonathan St.B. T. Evans, Stephen Newstead, and Ruth Byrne. *Human Reasoning: The Psychology of Deduction*. Lawrence Erlbaum Associates, Hove, East Sussex, 1993.

[Feys, 1937] Robert Feys. Les logiques nouvelles des modalités. *Revue Néoscholastique de Philosophie*, 40:517–553, 1937.

[Fine, 1985] Kit Fine. *Reasoning with Arbitrary Objects*. Blackwell, Oxford, 1985.

[Fitch, 1942] Fredric Fitch. A basic logic. *Journal of Symbolic Logic*, 7:105–114, 1942.

[Fitch, 1952] Fredric Fitch. *Symbolic Logic*. Roland Press, NY, 1952.

[Fitch, 1966] Fredric Fitch. Natural deduction rules for obligation. *American Philosophical Quarterly*, 3:27–28, 1966.

[Fitting, 1983] Melvin Fitting. *Proof Methods for Modal and Intutionistic Logics*. D. Reidel, Dordrecht, 1983.

[Forbes, 1994] Graeme Forbes. *Modern Logic*. Oxford UP, Oxford, 1994.

[Frege, 1904] Gottlob Frege. What is a function? In S. Meyer, editor, *Festschrift Ludwig Boltzmann gewidmet zum sechzigsten Geburtstage*, pages 656–666. Barth, Leipzig, 1904. Translation in P. Geach & M. Black (1952) *Translations from the Philosophical Writings of Gottlob Frege* [Blackwell: Oxford], pp. 107–116.

[Frické, 2010] Martin Frické. SoftOption, 2010. `http://www.softoption.us`.

[Gamut, 1991] L.T.F. Gamut. *Logic, Language and Meaning, Vol. I*. University of Chicago Press, Chicago, 1991.

[Garson, 2006] James Garson. *Modal Logic for Philosophers*. Cambridge Univ. Press, Cambridge, 2006.

[Gentzen, 1934] Gerhard Gentzen. Untersuchungen über das logische Schließen, I and II. *Mathematische Zeitschrift*, 39:176–210, 405–431, 1934. English translation "Investigations into Logical Deduction" published in *American Philosophical Quarterly*, **1:** 288–306 (1964), and **2:** 204–218 (1965). Reprinted in M. E. Szabo (ed.) (1969) *The Collected Papers of Gerhard Gentzen*, North-Holland, Amsterdam, pp. 68–131. Page references to the original for German and the *APQ* version for English.

[Goldberg and Suppes, 1972] Adele Goldberg and Patrick Suppes. A computer-assisted instruction program for exercises on finding axioms. *Educational Studies in Mathematics*, 4:429–449, 1972.

[Goldfarb, 2003] Warren Goldfarb. *Deductive Logic*. Hackett Pub. Co., Indianapolis, 2003.

[Goodstein, 1957] Rueben Goodstein. *Mathematical Logic*. Leicester UP, Leicester, UK, 1957.

[Gottschall, 2010] Christian Gottschall. Gateway to logic, 2010. `http://logik.phl.univie.ac.at/~chris/gateway/formular-uk.html`.

[Gustason and Ulrich, 1973] William Gustason and Dolph Ulrich. *Elementary Symbolic Logic*. Holt, Rinehart & Winston, NY, 1973.

[Harrison, 1992] Frank Harrison. *Logic and Rational Thought*. West Pub. Co., St. Paul, 1992.

[Harrison, 2007] John Harrison. *Introduction to Logic and Automated Theorem Proving*. Cambridge University Press, Cambridge, 2007.

[Hazen, 1987] Allen Hazen. Natural deduction and Hilbert's ϵ-operator. *Journal of Philosophical Logic*, 16:411–421, 1987.

[Hazen, 1990] Allen Hazen. The myth of the intuitionistic 'or'. In J.M. Dunn and A.K. Gupta, editors, *Truth or Consequences: Essays in Honor of Nuel Belnap*, pages 177–195. Kluwer, Dordrecht, 1990.

[Hazen, 1999] Allen Hazen. Logic and analyticity. In A. Varzi, editor, *European Review of Philosophy, Vol. 4: The Nature of Logic*, pages 79–110. CSLI Press, Stanford, 1999.

[Herbrand, 1928] Jacques Herbrand. Sur la théorie de la démonstration. *Comptes rendus hebdomadaires des séances de l'Académie des Sciences (Paris)*, 186:1274–1276, 1928.

[Herbrand, 1930] Jacques Herbrand. *Recherches sur la théorie de la démonstration*. PhD thesis, University of Paris, 1930. Reprinted in Warren Goldfarb (ed. & trans.) (1971) *Logical Writings*, D. Reidel, Dordrecht.

[Hintikka, 1953] K. Jaakko Hintikka. A new approach to sentential logic. *Societas Scientiarum Fennica. Commentationes Physico-Mathematicae*, 17:1–14, 1953.

[Hintikka, 1955a] K. Jaakko Hintikka. Form and content in quantification theory. *Acta Philosophica Fennica*, 8:7–55, 1955.

[Hintikka, 1955b] K. Jaakko Hintikka. Notes on quantification theory. *Societas Scientiarum Fennica: Commentationes Physico-Mathematicae*, 17:1–13, 1955.

[Hurley, 1982] Patrick Hurley. *A Concise Introduction to Logic*. Wadsworth, Belmont, CA, 1982.

[Iseminger, 1968] Gary Iseminger. *Introduction to Deductive Logic*. Appleton-Century-Crofts, New York, 1968.

[Jaśkowski, 1929] Stanisław Jaśkowski. Teoria dedukcji oparta na regułach założeniowych (Theory of deduction based on suppositional rules). In *Księga pamiątkowa pierwszego polskiego zjazdu matematycznego (Proceedings of the First Polish Mathematical Congress), 1927*, page 36, Krakow, 1929. Polish Mathematical Society.

[Jaśkowski, 1934] Stanisław Jaśkowski. On the rules of suppositions in formal logic. *Studia Logica*, 1:5–32, 1934. Reprinted in S. McCall (1967) *Polish Logic 1920–1939* Oxford UP, pp. 232–258.

[Jeffery, 1967] Richard Jeffery. *Formal Logic: Its Scope and Limits*. McGraw-Hill, New York, 1967.

[Johnson-Laird and Byrne, 1991] Philip Johnson-Laird and Ruth Byrne. *Deduction*. Lawrence Erlbaum Assoc., East Sussex, 1991.

[Kalish and Montague, 1964] Donald Kalish and Richard Montague. *Logic: Techniques of Formal Reasoning*. Harcourt, Brace, World, NY, 1964.

[Kalish *et al.*, 1980] Donald Kalish, Richard Montague, and Gary Mar. *Logic: Techniques of Formal Reasoning, Second Edition*. Harcourt, Brace, World, NY, 1980.

[Kaliszyk, 2010] Cezary Kaliszyk. Proofweb, 2010. `http://proofweb.cs.ru.nl/login.php`.

[Kerber and Präcklein, 1996] Manfred Kerber and Axel Präcklein. Using tactics to reformulate formulae for resolution theorem proving. *Annals of Mathematics and Artificial Intelligence*, 18:221–241, 1996.

[Ketonen, 1944] Oiva Ketonen. Untersuchungen zum Prädikatenkalkül. *Annales Academiae Scientiarum Fennicae, Series A*, 1(23):1–77, 1944.

[Kilgore, 1968] William Kilgore. *An Introductory Logic*. Holt, Rinehart & Winston, NY, 1968.

[Kleene, 1952] Stephen Kleene. *Introduction to Metamathematics*. van Nostrand, NY, 1952.

[Kleene, 1967] Stephen Kleene. *Elementary Logic*. Wiley, NY, 1967.

[Koslow, 1992] Arnold Koslow. *A Structuralist Theory of Logic*. Cambridge Univ. Press, Cambridge, 1992.

[Kripke, 1963] Saul Kripke. Semantical analysis of modal logic I: Normal propositional calculi. *Zeitschrift für Mathematische Logik und Grundlagen der Mathematik*, 9:67–96, 1963.

[Lemmon, 1965] E. John Lemmon. *Beginning Logic*. Nelson, London, 1965.

[Lewis, 1918] C. I. Lewis. *A Survey of Symbolic Logic*. University of California Press, Berkeley, 1918.

[Li, 1992] Dafa Li. A natural deduction automated theorem proving system. In *Automated Deduction — CADe–11*, pages 668–672. Springer, Berlin, 1992. Volume **607/1992** of *Lecture Notes in Computer Science*.

[Lifschitz, 1989] Vladimir Lifschitz. What is the inverse method? *Journal of Automated Reasoning*, 5:1–23, 1989.

[Łukasiewicz, 1951] Jan Łukasiewicz. *Aristotle's Syllogistic from the Standpoint of Modern Formal Logic*. Clarendon Press, Oxford, 1951.

[Manktelow *et al.*, 2010] Ken Manktelow, David Over, and Shira Elqayam. *The Science of Reason*. Psychology Press, Hove, East Sussex, 2010.

[Manktelow, 1999] Ken Manktelow. *Thinking and Reasoning*. Psychology Press, Hove, East Sussex, 1999.

[Martin, 1997] John Martin. Aristotle's natural deduction reconsidered. *History and Philosophy of Logic*, 18:1–15, 1997.

[Maslov, 1964] Sergey Yu. Maslov. An inverse method for establishing deducibility in classical predicate calculus. *Doklady AN SSSR*, 159:1420–1424, 1964. In Russian. English translation in J. Siekmann & G. Wrightson (1983) *The Automation of Reasoning, Vol. 2* (Springer: Berlin), pp. 48–54.

[Maslov, 1969] Sergey Yu. Maslov. Relationship between tactics of the inverse method and the resolution method. In A. Slisenko, editor, *Zapiski Nauchnykh Seminarov LOMI v.16*, pages 139–146. 1969. In Russian. English translation in J. Siekmann & G. Wrightson (1983) *The Automation of Reasoning, Vol. 2* (Springer: Berlin), pp. 264–272.

[Mates, 1965] Benson Mates. *Elementary Logic*. Oxford UP, NY, 1965.

[Matuszewski and Rudnicki, 2005] Roman Matuszewski and Piotr Rudnicki. MIZAR: The first 30 years. *Mechanized Mathematics and Its Applications*, 4:3–24, 2005.

[McGuire, 2010] Hugh McGuire. Proofbuilder, 2010. `http://www.cis.gvsu.edu/~mcguire/ProofBuilder/`.

[Nelson, 1949] David Nelson. Constructible falsity. *Journal of Symbolic Logic*, 14:16–26, 1949.

[Nevins, 1974] Arthur Nevins. A human oriented logic for automatic theorem-proving. *Journal of the ACM*, 21:606–621, 1974.

[Nipkow *et al.*, 2002] Tobias Nipkow, Larry Paulson, and Markus Wenzel. ISABELLE/HOL — *A Proof Assistant for Higher-Order Logic*. Springer, Berlin, 2002. Volume **2283** of *Lecture Notes in Computer Science*.

[Pastre, 1978] Dominique Pastre. Automated theorem proving in set theory. *Artificial Intelligence*, 10:1–27, 1978.

[Pastre, 1993] Dominique Pastre. Automated theorem proving in mathematics. *Annals of Mathematics and Artificial Intelligence*, 8:425–447, 1993.

[Pastre, 2002] Dominique Pastre. Strong and weak points of the MUSCADET theorem prover — examples from CASC-JC. *AI Communications*, 15:147–160, 2002.

[Pelletier *et al.*, 2002] Francis Jeffry Pelletier, Geoff Sutcliffe, and Christian Suttner. The Development of CASC. *AI Communications*, 15(2-3):79–90, 2002.

[Pelletier, 1998] Francis Jeffry Pelletier. Natural deduction theorem proving in THINKER. *Studia Logica*, 60:3–43, 1998.

[Pelletier, 1999] Francis Jeffry Pelletier. A brief history of natural deduction. *History and Philosophy of Logic*, 20:1–31, 1999.

[Pelletier, 2000] Francis Jeffry Pelletier. A history of natural deduction and elementary logic textbooks. In J. Woods and B. Brown, editors, *Logical Consequence: Rival Approaches, Vol. 1*, pages 105–138. Hermes Science Pubs., Oxford, 2000.

[Pelletier, 2012] Francis Jeffry Pelletier. Holism and compositionality. In M. Werning, W. Hinzen, and E. Machery, editors, *Handbook of Compositionality*, pp. 149–174. Oxford Univ. Press, Oxford, 2012.

[Peregrin, 2008] Jaroslav Peregrin. What is *the* logic of inference? *Studia Logica*, 88:263–294, 2008.

[Pollock, 1992] John Pollock. Interest-driven suppositional reasoning. *Journal of Automated Reasoning*, 6:419–462, 1992.

[Popper, 1947a] Karl Popper. Logic without assumptions. *Proceedings of the Aristotelian Society*, 47:251–292, 1947.

[Popper, 1947b] Karl Popper. New foundations for logic. *Mind*, 56:193–235, 1947. Corrigenda in Vol. 57 (1948, pp. 69-70).

[Portoraro, 1994] Frederic Portoraro. *Logic with Symlog*. Prentice-Hall, Englewood Cliffs, NJ, 1994.

[Prawitz, 1965] Dag Prawitz. *Natural Deduction: A Proof-theoretical Study*. Almqvist & Wicksell, Stockholm, 1965.

[Prawitz, 1971] Dag Prawitz. Ideas and results in proof theory. In J.E. Fenstad, editor, *Proceedings of the Second Scandinavian Logic Symposium*, pages 235–307, Amsterdam, 1971. North-Holland.

[Price, 1961] Richard Price. The stroke function and natural deduction. *Zeitschrift für mathematische Logik und Grundlagen der Mathematik*, 7:117–123, 1961.

[Priest, 1979] Graham Priest. The logic of paradox. *Journal of Philosophical Logic*, 8:219–241, 1979.

[Prior, 1960] Arthur Prior. The runabout inference ticket. *Analysis*, 21:38–39, 1960.

[Purtill, 1971] Richard Purtill. *Logic for Philosophers*. Harper & Row, New York, 1971.

[Quine, 1950a] Willard V. Quine. *Methods of Logic*. Henry Holt & Co., New York, 1950.

[Quine, 1950b] Willard V. Quine. On natural deduction. *Journal of Symbolic Logic*, 15:93–102, 1950.

[Restall, 2000] Greg Restall. *An Introduction to Substructural Logics*. Routledge, London and New York, 2000.

[Restall, 2005] Greg Restall. Multiple conclusions. In P. Hájek, L. Valdes-Villanueva, and Dag Westerståhl, editors, *Logic, Methodology and Philosophy of Science: Proceedings of the Twelfth International Congress*, pages 189–205, London, 2005. Kings' College Publications.

[Rips, 1994] Lance Rips. *The Psychology of Proof: Deduction in Human Thinking*. MIT Press, Cambridge, 1994.

[Robinson, 1965] J. Alan Robinson. A machine-oriented logic based on the resolution principle. *Journal of the ACM*, 12:23–41, 1965.

[Robison, 1969] Gerson Robison. *An Introduction to Mathematical Logic*. Prentice-Hall, Englewood Cliffs, NJ, 1969.

[Routley and Meyer, 1972] Richard Routley and Robert Meyer. The semantics of entailment – II. *Journal of Philosophical Logic*, 1:53–73, 1972.

[Routley and Meyer, 1973] Richard Routley and Robert Meyer. The semantics of entailment. In Hugues Leblanc, editor, *Truth, Syntax and Modality*, pages 194–243. North Holland, Amsterdam, 1973.

[Rumfitt, 2008] I. Rumfitt. Knowledge by deduction. *Grazer philosophische Studien*, 77:61–84, 2008.

[Russell, 1906] Bertrand Russell. The theory of implication. *American Journal of Mathematics*, 28:159–202, 1906.

[Russell, 1918] Bertrand Russell. Lectures on logical atomism. *The Monist*, 28:495–527, 1918. Continues in Volume 29: pp. 32–63, 190–222, 345–380.

[Saetti, 2010] John Saetti. Logic toolbox, 2010. `http://philosophy.lander.edu/~jsaetti/Welcome.html`.

[Schroeder-Heister, 1984a] Peter Schroeder-Heister. Generalized rules for quantifiers and the completeness of the intuitionistic operators $\wedge, \vee, \supset, \bot, \forall, \exists$. In M. Richter, E. Börger, W. Oberschelp, B. Schinzel, and W. Thomas, editors, *Computation and Proof Theory: Proceedings of the Logic Colloquium Held in Aachen, July 18–23, 1983, Part II*, pages 399–426, Berlin, 1984. Springer-Verlag. Volume **1104** of *Lecture Notes in Mathematics*.

[Schroeder-Heister, 1984b] Peter Schroeder-Heister. A natural extension of natural deduction. *Journal of Symbolic Logic*, 49:1284–1300, 1984.

[Schütte, 1960] Kurt Schütte. *Beweistheorie.* Springer-Verlag, Berlin, 1960. English translation *Proof Theory*, Springer-Verlag: New York, 1977.

[Scott, 1974] Dana Scott. Completeness and axiomatizability in many-valued logic. In L. Henken, J. Addison, W. Craig, C.C. Chang, D. Scott, and R. Vaught, editors, *Proceedings of the Tarski Symposium: Proceedings of Symposia in Pure Mathematics* **25**, pages 411–435, Providence, 1974. American Mathematical Society.

[Sieg and Byrnes, 1998] Wilfried Sieg and John Byrnes. Normal natural deduction proofs (in classical logic). *Studia Logica*, 60:67–106, 1998.

[Siekmann *et al.*, 2006] Jörg Siekmann, Christoph Benzmüller, and Serge Autexier. Computer supported mathematics with ΩMEGA. *Journal of Applied Logic*, 4:533–559, 2006.

[Slate and Atkin, 1977] David Slate and Larry Atkin. The northwestern university chess program. In P. Frey, editor, *Chess Skill in Man and Machine*, pages 82–118. Springer, 1977.

[Smith, 1982] Robin Smith. What is Aristotelian echthesis? *History and Philosophy of Logic*, 3:113–127, 1982.

[Smullyan, 1968] Raymond Smullyan. *First Order Logic.* Springer-Verlag, New York, 1968.

[Stålmarck, 1991] Gunnar Stålmarck. Normalization theorems for full first-order classical natural deduction. *Journal of Symbolic Logic*, 56:129–149, 1991.

[Suppes, 1957] Patrick Suppes. *Introduction to Logic.* Van Nostrand/Reinhold Press, Princeton, 1957.

[Sutcliffe and Suttner, 2006] Geoff Sutcliffe and Christian Suttner. The State of CASC. *AI Communications*, 19(1):35–48, 2006.

[Sutcliffe, 2009] Geoff Sutcliffe. The 4th IJCAR Automated Theorem Proving Competition. *AI Communications*, 22(1):59–72, 2009.

[Tarski, 1930a] Alfred Tarski. Fundamentale Begriffe der Methodologie der deduktiven Wissenschaften I. *Monatshefte für Mathematiik und Physik*, 37:361–404, 1930. English translation "Fundamental Concepts of the Methodology of the Deductive Sciences" in [Tarski, 1956, pp.60–109].

[Tarski, 1930b] Alfred Tarski. Über einige fundamentalen Begriffe der Metamathematik. *Comptes rendus des séances de la Société des Sciences et Lettres de Varsovie (Classe III)*, 23:22–29, 1930. English translation "On Some Fundamental Concepts of Metamathematics" in [Tarski, 1956, pp.30–37].

[Tarski, 1935] Alfred Tarski. Grundzüge der Systemenkalküls. *Fundamenta Mathematicae*, 25:503–526, 1935. Part II of article in Volume 26, pp.283–301. English translation "Foundations of the Calculus of Systems" in [Tarski, 1956, pp.342–383].

[Tarski, 1956] Alfred Tarski. *Logic, Semantics, Metamathematics.* Clarendon, Oxford, 1956.

[Teller, 1989] Paul Teller. *A Modern Formal Logic Primer.* Prentice-Hall, Englewood Cliffs, NJ, 1989.

[Tennant, 1978] Neil Tennant. *Natural Logic.* Edinburgh Univ. Press, Edinburgh, 1978.

[Thom, 1981] P. Thom. *The Syllogism.* Philosophia Verlag, Munich, 1981.

[Thom, 1993] P. Thom. Spodeictic Ecthesis. *Note Dam Journal of Formal Logic*, 34: 193–208, 1993.

[Thomason, 1970] Richmond Thomason. *Symbolic Logic.* Macmillan, New York, 1970.

[Toledo, 1975] Sue Toledo. *Tableau Systems for First Order Number Theory and Certain Higher Order Theories.* Springer-Verlag, Berlin, 1975. Volume **447** of *Lecture Notes in Mathematics.*

[Urquhart, 1995] Alasdair Urquhart. The complexity of propositional proofs. *Bulletin of Symbolic Logic*, 1:425–467, 1995.

[van Dalen, 1980] Dirk van Dalen. *Logic and Structure.* Springer-Verlag, Berlin, 1980.

[von Plato, 2008a] Jan von Plato. Gentzen's logic. In D. Gabbay and J. Woods, editors, *Handbook of the History and Philosophy of Logic, Vol. 5: From Russell to Church*, pages 607–661. Elsevier, Amsterdam, 2008.

[von Plato, 2008b] Jan von Plato. Gentzen's proof of normalization for natural deduction. *Bulletin of Symbolic Logic*, 14:240–257, 2008.

[von Sydow, 2010] Björn von Sydow. ALFIE, 2010. `http://www.cs.chalmers.se/~sydow/alfie`.

[von Wright, 1951a] Georg von Wright. Deontic logic. *Mind*, 60:1–15, 1951.

[von Wright, 1951b] Georg von Wright. *An Essay in Modal Logic.* North-Holland, Amsterdam, 1951.

[Wenzel, 2002] Markus Wenzel. The ISABELLE/ISAR reference manual, 2002. `http://isabelle.in.tum.de/doc/isar-ref.pdf`.

[Wiedijk, 2007] Freek Wiedijk. The QED manifesto revisited. In R. Matuszwski and A. Zalewska, editors, *From Insight to Proof: Festschrift in Honour of Andrzej Trybulec*, pages 121–133. University of Białystok, Białystok, Poland, 2007. Volume 10(23) of the series *Studies in Logic, Grammar and Rhetoric*, available online at `http://logika.uwb.edu.pl/studies/index.html`.

[Zucker and Tragesser, 1978] Jeffery Zucker and Robert Tragesser. The adequacy problem for inferential logic. *Journal of Philosophical Logic*, 7:501–516, 1978.

ELEMENTARY LOGIC TEXTBOOKS DESCRIBED IN TABLE 1

[Anderson and Johnstone, 1962] John Anderson and Henry Johnstone. *Natural Deduction: The Logical Basis of Axiom Systems*. Wadsworth Pub. Co., Belmont, CA, 1962.

[Arthur, 2010] Richard Arthur. *Natural Deduction: An Introduction to Logic*. Wadsworth Pub. Co., Belmont, CA, 2010.

[Barwise and Etchemendy, 2002] Jon Barwise and John Etchemendy. *Language, Proof and Logic*. CSLI Press, Stanford, 2002.

[Bergmann *et al.*, 1980] M. Bergmann, J. Moor and J. Nelson. *The Logic Book*. Random House, New York, 1980.

[Bessie and Glennan, 2000] Joseph Bessie and Stuart Glennan. *Elements of Deductive Inference*. Wadsworth Pub. Co., Belmont, CA, 2000.

[Bonevac, 1987] Daniel Bonevac. *Deduction*. Mayfield Press, Mountain View, CA, 1987.

[Bostock, 1999] David Bostock. *Intermediate Logic*. Oxford UP, Oxford, 1999.

[Byerly, 1973] H. Byerly. *A Primer of Logic*. Harper & Row, New York, 1973.

[Carter, 2005] C. Carter. *A First Course in Logic*. Pearson/Longman, New York, 2005.

[Cauman, 1998] L. Cauman. *First-Order Logic: An Introduction*. Walter de Gruyter, Hawthorne, NY, 1998.

[Chellas, 1997] Brian Chellas. *Elementary Formal Logic*. Penny Lane Press, Calgary, 1997.

[Copi, 1954] Irving Copi. *Symbolic Logic*. Macmillan Co., NY, 1954.

[Curry, 1963] Haskell Curry. *Foundations of Mathematical Logic*. McGraw Hill, New York, 1963.

[DeHaven, 1996] S. DeHaven. *The Logic Course*. Broadview Press, Peterborough, Ontario, 1996.

[Fitch, 1952] Fredric Fitch. *Symbolic Logic*. Roland Press, NY, 1952.

[Forbes, 1994] Graeme Forbes. *Modern Logic*. Oxford Univ. Press, Oxford, 1994.

[Gamut, 1991] L.T.F. Gamut. *Logic, Language and Meaning, Vol. I, II*. University of Chicago Press, Chicago, 1991.

[Georgacarakos and Smith, 1979] G. Georgacarakos and R. Smith. *Elementary Formal Logic*. McGraw Hill, New York, 1979.

[Goldfarb, 2003] Warren Goldfarb. *Deductive Logic*. Hackett Pub. Co., Indianapolis, 2003.

[Goodstein, 1957] Rueben Goodstein. *Mathematical Logic*. Leicester UP, Leicester, UK, 1957.

[Gustason and Ulrich, 1973] William Gustason and Dolph Ulrich. *Elementary Symbolic Logic*. Holt, Rinehart & Winston, NY, 1973.

[Guttenplan, 1997] S. Guttenplan. *The Language of Logic*. Blackwells, Oxford, 1997

[Harrison, 1992] Frank Harrison. *Logic and Rational Thought*. West Pub. Co., St. Paul, 1992.

[Hurley, 1982] Patrick Hurley. *A Concise Introduction to Logic*. Wadsworth, Belmont, CA, 1982.

[Iseminger, 1968] Gary Iseminger. *Introduction to Deductive Logic*. Appleton-Century-Crofts, New York, 1968.

[Jacquette, 2001] D. Dacquette. *Symbolic Logic*. Wadsworth, Belmont, CA, 2001.

[Jennings and Friedrich, 2006] R. Jennings and N. Friedrich. *Truth and Consequence*. Broadview, Peterborough, Ontario, 2006.

[Kalish and Montague, 1964] Donald Kalish and Richard Montague. *Logic: Techniques of Formal Reasoning*. Harcourt, Brace, World, NY, 1964.

[Klenk, 1983] V. Klenk. *Understanding Symbolic Logic*. Prentice-Hall, Englewood Cliffs, NJ, 1983.

[Kozy, 1974] J. Kozy. *Understanding Natural Deduction*. Dickenson Publishng Company, Encino, CA, 1974.

[Leblanc and Wisdom, 1965] H. Leblanc and W. Wisdom. *Deductive Logic*. Allyn & Bacon, Boston, 1965.

[Lemmon, 1965] E. John Lemmon. *Beginning Logic.* Nelson, London, 1965.
[Machina, 1978] K. Machina. *Basic Applied Logic.* Scott-Foreman, Glenview, IL, 1978.
[Martin, 2004] R. Martin. *Symbolic Logic.* Broadview Press, Peterborough, Ontario, 2004.
[Massey, 1970] G. Masssey. *Understanding Symbolic Logic.* Harper & Row, New York, 1970.
[Mates, 1965] Benson Mates. *Elementary Logic.* Oxford UP, NY, 1965.
[Myro *et al.*, 1987] G. Myro, M. Bedau, and T. Monroe. *Rudiments of Logic.* Prentice-Hall, Englewood Cliffs, NJ, 1987.
[Pollock, 1969] J. Pollock. *Introduction to Symbolic Logic.* Holt, Rinehart & Winston, New York, 969.
[Purtill, 1971] Richard Purtill. *Logic for Philosophers.* Harper & Row, New York, 1971.
[Quine, 1950] Willard V. Quine. *Methods of Logic.* Henry Holt & Co., New York, 1950.
[Resnick, 1970] M. Resnick. *Elementary Logic.* McGraw-Hill, New York, 1970.
[Simco and James, 1976] N. Simco and G. James. *Elementary Logic.* DIckenson, Encino, CA, 1976.
[Simpson, 1987] R. Simpson. *Essentials of Symbolic Logic.* Broadview Press, Peterborough, Ontario, 1987.
[Suppes, 1957] Patrick Suppes. *Introduction to Logic.* Van Nostrand/Reinhold Press, Princeton, 1957.
[Tapscott, 1976] B. Tapscott. *Elementary Applied Symbolic Logic.* Prentice-Hall, Englewood Cliffs, NJ, 1976.
[Teller, 1989] Paul Teller. *A Modern Formal Logic Primer.* Prentice-Hall, Englewood Cliffs, NJ, 1989.
[Tennant, 1978] Neil Tennant. *Natural Logic.* Edinburgh Univ. Press, Edinburgh, 1978.
[Thomason, 1970] Richmond Thomason. *Symbolic Logic.* Macmillan, New York, 1970.
[van Dalen, 1980] Dirk van Dalen. *Logic and Structure.* Springer-Verlag, Berlin, 1980.
[Wilson, 1992] J. Wilson. *Introductory Symbolic Logic.* Wadsworth Publishing Company, Belmont, CA, 1992.

A HISTORY OF CONNEXIVITY

Storrs McCall

1 TWO THOUSAND THREE HUNDRED YEARS OF CONNEXIVE IMPLICATION.

Connexive implication is a type of implication first defined in the 4th Century B.C., a time of active debate when it was said that the very crows on the rooftops were croaking about what conditionals were true. In Sextus Empiricus' *Outlines of Pyrrhonism*, which discusses four varieties of implication, including [1] material (Philonian) and [2] strict (Diodorean) implication, we read:

> "[3] And those who introduce the notion of connection say that a conditional is sound when the contradictory of its consequent is incompatible with its antecedent." [Kneale, 1962, 129]

It follows from this definition that no conditional of the form "If p then not-p" can be true, since the contradictory of not-p, i.e. p, is never incompatible with p. Accepting this in turn requires that "compatibility" be essentially a *relational* concept, and that whether or not A is compatible with B cannot be determined by examining A and B separately. Thus even "$p\& \sim p$" is not incompatible with itself, and "If $p\& \sim p$, then not-($p\& \sim p$)" is connexively false.

In Aristotle's *Prior Analytics* we find a very interesting passage, in which Aristotle seems to be saying that it is never possible for a proposition to be implied by its own negation:

> "It is impossible that the same thing should be necessitated by the being and by the not-being of the same thing. I mean, for example, that it is impossible that B should necessarily be great if A is white, and that B should necessarily be great if A is not white. For if B is not great A cannot be white. But if, when A is not white, it is necessary that B should be great, it necessarily results that if B is not great, B itself is great. But this is impossible." (*An. pr.* 57b3-14, translation in Łukasiewicz (1957), 49-50)

What Aristotle is trying to show here is that two implications of the form "If p then q" and "If not-p then q" cannot both be true. The first yields, by contraposition, "If not-q then not-p", and this together with the second gives "If not-q then q" by transitivity. But, Aristotle says, this is impossible: a proposition cannot be implied by its own negation. His conclusion accords with Sextus' reference to a "connection" between antecedent and

Handbook of the History of Logic. Volume 11: Logic: A History of its Central Concepts.
Volume editors: Dov M. Gabbay, Francis Jeffry Pelletier and John Woods.
General editors: Dov M. Gabbay and John Woods.

consequent in true conditionals, since the contradictory of the consequent q cannot be incompatible with not-q. We shall henceforth refer to the principle that no proposition can be implied by its own negation, in symbols "$\sim (\sim p \rightarrow p)$" as *Aristotle's thesis*, the name sometimes being also used to denote "$\sim (p \rightarrow \sim p)$". The connexive principle $\sim [(p \rightarrow q) \& (\sim p \rightarrow q)]$ will be referred to as *Aristotle's second thesis*.

Nine centuries later, we find in Boethius' *De hypotheticis syllogismis*, written between 510 and 523 AD, an elaborate system of inference-schemata in the tradition of the Stoic "indemonstrables" which constitute the first systematic treatment of propositional logic [Dürr, 1951; Łukasiewicz, 1967, 66–87; Marenbon, 2003, 50–56]. Boethius divides his inference schemata into eight classes, and among them we find an appeal to an analogue of Aristotle's second thesis. The first schema of the second group of the second of these classes, which the original text gives as follows:

> '*Si est A, cum sit B, est C; ... atqui cum sit B, non est C; non est igitur A.*'
> (Prantl (1855), 706 note 157, lines 9-10; Dürr p. 37)

may be transliterated thus:

> If p, then if q then r,
> if q then not-r,
> therefore, not-p.

The reasoning that led Boethius to assert the validity of this schema was presumably this. Since the two implications "If q then r" and "If q then not-r" are incompatible, the second premiss contradicts the consequent of the first premiss. Hence, by *modus tollens*, we get the negation of the antecedent of the first premise, namely "not-p".

Is reasoning of this kind correct? Not, certainly, if we interpret the "if...then" of Boethius' schemata as material, strict or any other known variety of implication, *except* Sextus' third variety of nine centuries earlier. By contraposition and transitivity, $q \rightarrow r$ and $q \rightarrow \sim r$ yield $q \rightarrow \sim q$, which violates the condition of connexivity since $\sim\sim q$ is not incompatible with q. Hence $q \rightarrow \sim q$ is a false conditional, $\sim (q \rightarrow \sim q)$ is true, and $q \rightarrow r$ and $q \rightarrow \sim r$ cannot both be true, i.e. $q \rightarrow r$ implies $\sim (q \rightarrow \sim r)$. The corresponding conditional, "If $q \rightarrow r$ then $\sim (q \rightarrow \sim r)$" will be denoted *Boethius' thesis*, and serves with the thesis $\sim (p \rightarrow \sim p)$ as the distinguishing mark of connexive logic.

Not every logician agrees that Boethius' schemata rest on a non-standard implication operator. Jonathan Barnes takes the view that they *might*, when he asks whether there is "a coherent account of the truth-conditions of hypothetical propositions which will support Boethius' contention that (1) "If p then q", and (2) "If p then not-q" are contradictories?" Barnes does not know, though he believes the question is "worth the consideration of contemporary students of logic" [Barnes, 1981, 84]. Other philosophers, by contrast, take the view that Boethius was either simply mistaken in treating (1) and (2) above as contradictories, or, more radically, that *De hypotheticis syllogismis* is not really a work on the logic of *propositions*, but a work on the logic of *terms*. Thus John Marenbon writes that "So far from proposing an odd form of sentence logic, what Boethius seems to be doing is to work out, painstakingly, by example, a term logic of hypotheticals that mimicks the results of standard modern sentence logic" [Marenbon, 2003, 54]. Marenbon cites Christopher Martin as the source of this interpretation when he says that Boethius' logic is "not propositional at all" [Martin, 1991, 279]; that Boethius "does not think of negation as a content-forming operator on propositional contents" (p. 283); and that Boethius "cannot treat sentential connectives as propositional content-forming operators on propositional contents" (p. 288, amended).

It is true that one can read the first premise of Boethius' schema quoted above, *Si est A, cum sit B, est C,* in the way Martin does, as "If it's *A*, then when it's *B*, it's *C*" (cf. p. 287), where "it" is some common unspecified subject characterized successively by "*A*", "*B*", and "*C*". This would indeed indicate that Boethius' logic was a logic of terms. But it is also possible to read the "*est A*", "*est B*", etc. of Boethius' schemata as genuine propositional variables like the Stoics' "the first" and "the second", replaceable not by terms but by whole sentences. Boethius in fact gives examples of conditionals in which exactly such replacements are made: "If fire is hot, the heavens are spherical" [Martin, 1991, 287], and "If the Earth should stand between the sun and the moon, then an eclipse of the moon would follow" [Martin, 1987, 384]. Furthermore, the negation operator in Boethius acts not only on simple propositions such as "est *A*", but also on whole conditionals. Thus in Boethius' commentary on Cicero's *Topics* we read:

> "For because it is understood to be consequent and true, if it's day, that it's light, it is repugnant and false, if it's day, that it not be light. Which denied it is once more in this way true: 'not if it's day, then it's not light'." [Martin, 1991, 293]

This passage makes it clear that Boethius takes the *negation of the conditional* "If it's day, then it's not light" to be "Not (if it's day, then it's not light)". Which demonstrates that his logic is a true logic of propositions. Furthermore it also shows that, for Boethius, $\sim(p \rightarrow\sim q)$ follows from $p \rightarrow q$.

Moving ahead another five centuries, we find in Peter Abelard's *Dialectica* four connexive principles that Abelard makes "the centerpieces of his theory of conditionals" [Martin, 1987, 389]. These principles are the following:

1. Aristotle's second thesis $\sim[(p \rightarrow q) \,\&\, (\sim p \rightarrow q)]$
2. Aristotle's thesis $\sim(\sim p \rightarrow p)$
3. "Abelard's first principle" $\sim[(p \rightarrow q) \,\&\, (p \rightarrow\sim q)]$
4. Variant of Aristotle's thesis $\sim(p \rightarrow\sim p)$

[Martin, 1987, 381, 389; 2004, 190]

Abelard says, with reference to (2): "Suppose it be conceded that when 'animal' is denied, man may persist; yet it was formerly conceded that 'man' necessarily requires 'animal', viz in [the conditional]: if there is man there is animal. And so it [may happen] that what is not animal, be animal; for what the antecedent admits, the consequent admits. ... But this is impossible." (Abelard, quoted in [Bochenski, 1961, xi].)

Another century later we find in the commentary on the *Prior Analytics* of Robert Kilwardby a new criticism of Aristotle's second thesis, which asserts the incompatibility of "If *p* then *q*" and "If not-*p* then *q*". Kilwardby gives two examples of pairs of such propositions which are *not* incompatible:

(i) If you are seated, God exists
If you are not seated, God exists.

(ii) If you are seated, then either you are seated or you are not seated
If you are not seated, then either you are seated or you are not seated.

[Thomas, 1954, 137; Kneale, 1962, 275-6]

The first pair is true because 'God exists', being a necessary proposition, follows from anything – *quia necessarium sequitur ad quodlibet*: an early formulation of the positive paradox of strict implication. But here we must distinguish, Kilwardby says, two kinds of implication; *consequentia essentialis* or *naturalis*, and *consequentia accidentalis*. In the former the consequent must be 'understood' (*intelligitur*) in the antecedent, and such is not the case with 'If you are seated, God exists'. The latter is a *consequentia accidentalis, 'et de tali non intelligendum sermo Aristotelis'*.

The second pair, on the other hand, are both *consequentiae naturales*, and here it does seems as if the same thing could follow from two contradictory propositions. Kilwardby tries to defend Aristotle's position nonetheless, saying that the Philosopher intended only to deny that the same proposition could follow from two contradictories "in virtue of the same part of itself" (*gratia eiusdem in ipso*). But it is doubtful that Aristotle intended any such thing, and Kilwardby seems to be leaning over backwards here. It appears we must accept the fact that the type of implication for which Aristotle's thesis holds cannot consistently admit of conditionals of the form "if p, then either p or q".

Three centuries later than Kilwardby, we find Paul of Venice who, in listing no fewer than 10 interpretations of the meaning of "if ... then", reiterates Sextus Empiricus' connexive category: "Tenthly people say that for the truth of a conditional it is required that the opposite of the consequent be incompatible with the antecedent" [Bochenski, 1961, 196].

Jumping ahead to the end of the 19th century we find, in a short and amusing note in *Mind*, what has come to be known as Lewis Carroll's Barbershop Paradox.[1] Carroll's argument runs as follows:

Uncle Joe and Uncle Jim are going to a barbershop run by Allen, Brown and Carr, and Uncle Jim hopes that Carr will be in to shave him. Uncle Joe says he can prove he will be in by an argument having as premisses two hypotheticals. First, if Carr is out, then if Allen is out, Brown must be in (since otherwise there'd be nobody to mind the shop). Secondly if Allen is out Brown is out (since Allen, after a recent bout of fever, always takes Brown with him). Taking 'A' to stand for 'Allen is out', 'B' for 'Brown is out', etc., we have:

(i) If C then (if A then not-B)

(ii) If A then B,

[1] Lewis Carroll [1894], not to be confused with the set-theoretical paradox of the barber in the little town who shaves all those and only those who do not shave themselves. Carroll's paradox has been discussed in Sidgwick [1894] and [1895]; Johnson [1894] and [1895]; Russell [1903, p. 18n.]; Constance-Jones [1905a] and [1905b]; an anonymous "W" (Johnson?) in Mind [1905]; Burks and Copi [1950, pp. 219-22]; McKinsey [1950, pp. 222-3]; Burks [1951, pp. 377-8].

and these two premises, according to Uncle Joe, imply not-C, because of the incompatibility of the two hypotheticals 'If A then B' and "If A then not-B'. The result is, of course, paradoxical, because under the stated conditions Carr can perfectly well be out when the other two are in, or even when Allen alone is in. The question is, at what point is Uncle Joe's argument fallacious?

What Burks and Copi call the 'received' solution is that of Johnson and Russell. According to them, the two hypotheticals 'If A then B' and 'If A then not-B' are *not* incompatible: they may in fact both be true when A is false, as is the case in classical two-valued logic. Hence we cannot infer not-C by *modus tollens*. The thought underlying this solution is that 'If A then not-B' does not properly negate 'If A then B'.

Burks and Copi disagree, however. When interpreted as 'causal' implications rather than as material implications, the two hypotheticals above *are* in their opinion incompatible, and this is in general true of 'causal' implication. To take another example:

> "If one politician argues that if the Conservatives win the election in 1950 then Britain's economic situation will improve, and another argues that if the Conservatives win in 1950 then Britain's economic situation will not improve, there is a genuine disagreement. It would indeed be an over-zealous proponent of material implication who would expect the disputants to agree that another Labour victory at the polls would make both their statements true!". [Burks and Copi, 1950, 220]

For Burks and Copi, then, the inference from $p \rightarrow q$ to $\sim (p \rightarrow \sim q)$ — Boethius' thesis — holds for causal implication, a form of implication which, they maintain, Uncle Joe's hypotheticals exemplify. Hence the fallacy in the argument must be sought elsewhere.

The fallacy, according to them, lies in an impropriety in the statement of the first premise. Uncle Joe has

(iii) If Carr is out, then if Allen is out Brown is in,

but in fact the conditions of the problem permit only

(iv) if Carr is out and if Allen is out, Brown is in.

Nor does (iv) either mean the same as or imply (iii). This would entail that the principle of exportation "If $(p \& q) \rightarrow r$, then $p \rightarrow (q \rightarrow r)$" does not hold for causal implication. See sections 5 and 6 below, where exportation is rejected.

Another late 19th century logician, Hugh MacColl, also made notable contributions to the theory of connexive implication. Bertrand Russell credits MacColl with being the first logician to define class-inclusion using the implication operator. In Russell's words, the relation of inclusion asserted in "$a \subseteq b$" is for MacColl "derivative, being in fact the relation of implication between the statement that a thing belongs to the one class, and the statement that it belongs to the other."[2]

Making use of the concept of implication, MacColl proceeds to base Aristotelian syllogistic, at that time the standard school-book "logic", upon the logic of propositions

[2]Russell [1906, 255]. In footnote 1 of his review of MacColl [1906], Russell corrects his previous attribution of the definition of "$\subseteq$" in terms of "$\rightarrow$" to Peano.

[MacColl, 1878, 177-186; 1906, 44-70]. Today the customary translation of 'All A is B', using quantifiers, is $(x)(Ax \to Bx)$, but MacColl's approach does not involve quantifiers. Take any individual S at random, he says, out of the universe of discourse (which need not be restricted to the universe of existent things but could include Meinongian items such as round squares and unicorns). Then let the propositional variable 'a' mean 'S belongs to the class A'. Representing implication by '$\to$', the proposition 'All A is B' will be denoted by '$a \to b$', (or '$a : b$' in MacColl's notation). The syllogistic mood Barbara ('If all A is B, and all B is C, then all A is C') then becomes:

$$[(a \to b) \,\&\, (b \to c)] \to (a \to c),$$

and MacColl is on his way to reducing syllogistic to propositional logic. His square of opposition is the following:

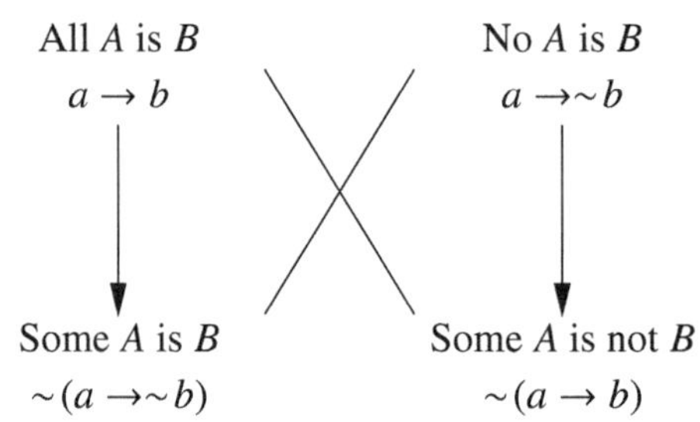

Note that the Aristotelian entailment of "Some A is B" by "All A is B" requires, in MacColl's system, Boethius' thesis $(a \to b) \to\sim (a \to\sim b)$. Using connexive implication as his basis, MacColl is able to show that all nineteen moods of the traditional syllogistic are valid in his notation. More will said about the reduction of syllogistic to propositional logic below, in Section 7.

To conclude this brief history of connexive logic, three 20th century logicians should be mentioned: F.P. Ramsey, E.J. Nelson and R.B. Angell.

In an essay entitled "General Propositions and Causality" Ramsey says: "If two people are arguing 'If p will q?' and are both in doubt as to p, they are adding p hypothetically to their stock of knowledge and arguing on that basis about q; so that in a sense 'If p, q' and 'If p, not-q' are contradictories" [Ramsey, 1931, 247]. Ramsey is confronting exactly the situation that confronted Boethius fourteen centuries earlier.

A more thorough-going espousal of connexive implication is found in a 1930 article entitled "Intensional relations" published by Everett Nelson. Nelson defines 'p intensionally implies q' as 'p is inconsistent with the contradictory of q' [Nelson, 1930, 444-5] and attacks the notion of consistency put forward by C.I. Lewis in his *Survey of Symbolic Logic* (1918). Lewis, who bases his modal system on the ideas of Hugh MacColl, defines 'p and q are consistent' as 'it is possible that p and q are both true' [Lewis, 1918, 293]. It follows, Nelson says, that an impossible proposition, such as '$2 + 2 \neq 4$', is inconsistent with every proposition, including itself. But what this shows is that Lewis has adopted a mistaken notion of consistency, since 'from the mere fact that p is false or impossible it cannot be determined that it is inconsistent with q. The meanings of both propositions are required to determine the relation' [Nelson, 1930, 443]. Similar considerations lead Nelson to reject Lewis's notion of strict implication, which has as one of its consequences that an impossible proposition implies any proposition. Like consistency, implication is 'essentially relational: it depends upon the meaning of *both* propositions' [Nelson, 1930, p. 446].

As an alternative to Lewis's notions, Nelson accordingly offers a new primitive relation of consistency, symbolized as '$p \circ q$', and a concept of intensional implication or entailment 'E', defined in such a way that p entails q if and only if p is inconsistent with the contradictory of q:

$$pEq =_{df} \sim (p \circ \sim q).$$

It is plain that this concept of implication is equivalent that of Sextus' third variety, and is to be contrasted with Lewis's definition of strict implication:

$$p \rightarrow q =_{df} \sim \Diamond(p \& \sim q).$$

Nelson's definition of entailment leads naturally to the two most distinctive laws characterizing his system: '$(pEq)\, E \sim (pE \sim q)$' and '$p \circ p$'. The former asserts that 'If p implies q, p does not imply not-q' (Boethius' thesis), and the latter says that every proposition is consistent with itself, which together with Nelson's definition of 'E' yields Aristotle's thesis '$\sim (pE \sim p)$'. To Nelson goes the credit for being the first logician to give formal equivalents of Sextus' implication operator and Aristotle's and Boethius' connexive laws.

Although Nelson formalized the notion of connexive implication, and offered a handful of axioms from which he deduced a handful of theorems, he made no attempt to incorporate his insights into a full-fledged logical system. Such an attempt was made in 1962 by R.B. Angell, whose aim was to develop a logic of 'subjunctive' or 'counterfactual' conditionals as in the following pairs:

1. If that match had been scratched, it would have lighted
   If that match had been scratched, it would not have lighted.[3]

2. If it rains, the match will be cancelled
   If it rains, the match will not be cancelled.

Not only was Angell able to construct an intuitively plausible axiomatic system incorporating a 'principle of subjunctive contrariety' $\sim [(p \rightarrow q) \& (p \rightarrow \sim q)]$ which formalizes the incompatibility of the above pairs, but in addition he accomplished the important task of demonstrating the *consistency* of his axioms. No one had done this before: no one perhaps had even thought of doing it. Angell's accomplishment will be discussed further in section 5 below.

2 CONNEXIVE CONDITIONALS: AN EMPIRICAL APPROACH

Varied opinions about the truth or falsehood of connexive theses over the ages have been discussed. But there is one voice that has not yet been heard – that of those who reason,

[3]See Nelson Goodman's classic paper [1947] and book [1955].

but are philosophically naïve. To sound their opinions, the following questionnaire was constructed. It contains concrete instances of Aristotle's and Boethius' theses, mixed in randomly with a list of (a) logical propositions which might or might not be true, and (b) arguments which might or might not be valid. To avoid difficulties about variables, each item was written in concrete form; e.g. 'If Hitler is dead, then Hitler is dead' — true or false? A "don't know" box was added to discourage guessing. The questionnaire was given to 89 students in the first lecture of the elementary logic course at McGill University, with half an hour provided in which to answer it. The students, who did not sign their names, and did not receive a mark, had previously had no logic, and, in the majority of cases, no philosophy of any kind. Though not exactly a cross-section of humanity at large, their answers give us a glimpse of how the outside world views matters. Table 1 on the next page reproduces the questionnaire, and table 2 summarizes the answers to it, defines "expert respondent", and gives the opinion of "experts" on key items. Table 3 on page 425 provides an overall "popularity ordering".

The answers to the questionnaire are exceedingly interesting, particularly from the point of view of how they reflect on material implication, and on the theses of Aristotle and Boethius. Table 3 ranks the statements held to be truest and the arguments considered most valid, together with comments. To anticipate a question sure to be asked by those reflecting on the results: how much importance should be attached to the logical opinions of admittedly total amateurs? The answer to this question depends on how much logical acumen the amateurs display. As an indicator of this, the questions on the sheet fell into three groups.

(a) Numbers 1, 3, 10, 12, 13, 14 and 16 were more or less straightforward, with right-or-wrong answers. They served as a measure of the answerer's ability. The performance of the class in answering these questions was reasonable but not outstanding. Questions 1, 3 and 12 were each answered correctly by 88% of the class or more, but those who succeeded with question 13 numbered only 35%. Students who got every one of this group right constituted only 12%; these I designated experts.

(b) The remainder of the questions were tendentious, and the answers to them correspondingly more interesting. Table 2 compares the opinion of the experts on these questions with that of the rest.

(c) Among the tendentious questions, numbers 7, 11 and 15 bear directly on the intuitive status of connexive implication. Table 3 contains the surprising result that immediately below the law of identity and the rule to infer "*p*" from "*p* or *q*" and "not-*q*", three connexive principles stand highest in the students' popularity rating. These are Aristotle's and Boethius' theses, and the rule corresponding to the latter. Of the remaining items, two are fallacies (nos 3 and 13), while the rest are principles of classical two-valued logic handed down to countless students by numberless university instructors.

Plainly, the questionnaire supports the concept of connexivity. As an empirical hypothesis, I conjecture that any time it is given to a group of college freshmen, before they

Table 1. Logical Questionnaire

A.	**Whether Hitler is dead or not, are the following statements true?**	YES	NO	DON'T KNOW
1.	If Hitler is dead then Hitler is dead.			
2.	If Hitler is dead then either Hitler is dead or there is life on Venus.			
3.	If Mussolini is dead then (Mussolini is dead and Hitler is dead).			
4.	If Hitler is dead and smoking produces cancer then Hitler is dead.			
5.	If Hitler is dead then (if the moon is made of green cheese then Hitler is dead).			
6.	If Hitler is dead then (if Hitler is not dead then shrimps whistle).			
7.	It is not the case that (if Hitler is dead then Hitler is not dead).			
8.	If Hitler is dead then (if Hitler is not dead then Hitler is dead).			
9.	If (if Hitler is not dead then Hitler is dead) then Hitler is dead.			
10.	If (if Hitler is dead then von Rümer is a liar) then (if von Rümer is not a liar then Hitler is not dead).			
11.	If (if Hitler is dead then von Rümer is a liar) it is not the case that (if Hitler is dead then von Rümer is not a liar).			
B.	**Are the following arguments valid?**	YES	NO	DON'T KNOW
12.	Either snow is white or my eyes deceive me. My eyes do not deceive me. Therefore, snow is white.			
13.	If the fires are lit the battle is over. The fires are not lit. Therefore, the battle is not over.			
14.	You can't both have your cake and eat it too. You've eaten it. Therefore, you can't have it.			
15.	If the fires are lit the battle is over. Therefore, it is false that if the fires are lit the battle is not over.			
16.	If the fires are lit the battle is over. Therefore, if the battle is not over the fires are not lit.			

Table 2. Replies

Question	Yes	No	Don't Know	Not Answered	Percentage Answering 'Yes'	Percentage of Experts answering 'Yes' to tendentious questions
1	86	2	0	1	97%	
2	36	46	7	0	40	73%
3	5	78	6	0	6	
4	69	18	1	1	78	91
5	36	46	6	1	40	73
6	11	65	11	2	12	11
7	78	6	3	2	88	100
8	19	60	10	0	21	18
9	33	39	15	2	37	55
10	54	29	5	1	61	
11	75	7	4	3	84	100
12	81	7	1	0	91	
13	55	31	3	0	62	
14	63	23	2	1	71	
15	76	12	1	0	85	100
16	68	18	2	1	77	

have studied formal logic, the results will be similar. If some readers wish to experiment, I would be glad to hear of the results at `storrs.mccall@mcgill.ca`.

3 PARADOXES OF IMPLICATION

It was W. E. Johnson who first used the expression 'paradox of implication', explaining that a paradox of this sort arises when a logician proceeds step by step, using accepted principles, until a formula is reached which conflicts with common sense [Johnson, 1921, 39]. In classical propositional logic, the two formulae which most notoriously conflict with common sense in this way, namely '$p \rightarrow (q \rightarrow p)$' (a true proposition is implied by any proposition), and '$p \rightarrow (\sim p \rightarrow q)$' (a false proposition implies any proposition), are known respectively as the positive and the negative paradoxes of material implication.

These paradoxes lead to immediate conflicts with connexive logic. A consequence of the positive paradox is $q \rightarrow (p \rightarrow p)$, which yields $\sim (p \rightarrow p) \rightarrow (p \rightarrow p)$ by substitution, and a consequence of the negative paradox is $(p\& \sim p) \rightarrow q$, whence $(p\& \sim p) \rightarrow \sim (p\& \sim p)$. Both of these violate Aristotle's thesis. Systems of strict implication, in which a necessary proposition is implied by any proposition, and an impossible proposition implies anything, fare no better. Therefore no adequate basis upon which to build a consistent system of connexive logic can be found in classical 2-valued logic, or

Table 3. Popularity

Schema No.	Subscribed to by	Name of schema	Remarks
1	97%	Law of identity	Obvious, but see Sextus no. 4
12	91%	Modus tollendo ponens	Obvious
7 15 11	88% 85% 84%	Aristotle Rule of Boethius Boethius	Surprise choice in view of the history of these schemata. The experts' judgement even more favourable.
4	78%	Law of conjunctive simplification	
16	77%	Rule of contraposition	Fundamental; introduced for comparison with 15
14	71%	Stoics' 3rd indemonstrable	Obvious
13	62%	Denying the antecedent	A fallacy unrecognized by a surprisingly large number.
10	61%	Law of contraposition	Fundamental; cf. 16
5	40%	Positive paradox of material implication	
2	40%	Law of addition	See Kilwardby.
9	37%	Consequentia mirabilis	
8	21%	Combination of both positive and negative paradoxes of material implication.	Seemingly more difficult to accept than 5.
6	12%	Negative paradox of material implication	Not as popular as 5.[4]
3	6%		Fortunately most saw through this obvious fallacy.

[4] This result confirms A.N. Prior's views as to the relative plausibility of the positive and negative paradoxes of material implication. See [Prior, 1962, p. 259].

in the Lewis systems of strict implication.

Relevance logics, which demand a relation of *relevance* between the antecedent and the consequent of true conditionals which is lacking in the above examples, offer more promise. A *locus classicus* of relevant implication is the following:

> "Imagine, if you can, a situation as follows. A mathematician writes a paper on Banach spaces, and after proving a couple of theorems he concludes with a conjecture. As a footnote to the conjecture, he writes: 'In addition to its intrinsic interest, this conjecture has connections with other parts of mathematics which might not immediately occur to the reader. For example, if the conjecture is true, then the first-order functional calculus is complete; whereas if it is false, then it implies that Fermat's last conjecture is correct.' The editor replies that the paper is obviously acceptable, but he finds the final footnote perplexing; he can see no connection whatever between the conjecture and the 'other parts of mathematics', and none is indicated in the footnote. So the mathematician replies, 'Well, I was using 'if... then...' and 'implies' in the way that logicians have claimed I was: the first-order functional calculus *is* complete, and necessarily so, so anything implies that fact — and if the conjecture is false it is presumably impossible, and hence implies anything. And if you object to this usage, it is simply because you have not understood the technical sense of 'if...then...' worked out so nicely for us by logicians.' And to this the editor counters: 'I *understand* the technical bit all right, but it is simply not correct. In spite of what most logicians say about us, the standards maintained by this journal require that the antecedent of an 'if... then...' statement must be *relevant* to the conclusion drawn. And you have given no evidence that your conjecture about Banach spaces is relevant either to the completeness theorem or to Fermat's conjecture." [Anderson and Belnap, 1962, 33; 1975, 17][5]

In the paradoxes of material and strict implication, the absence of relevance comes out plainly. In $q \rightarrow (p \rightarrow p)$ and $(p\& \sim p) \rightarrow q$ the antecedent and consequent may be statements the meanings of which are totally unconnected. This is reflected in the lack of any variable shared by the two. Consequently a way of avoiding paradox and ensuring relevance suggests itself: require that in all true conditionals antecedent and consequent share at least one variable. Relevance logics are based on *variable-sharing*, and, as will be seen in systems of natural deduction, on the requirement that an hypothesis A must be *used* in proving a conclusion B before the conditional $A \rightarrow B$ can be derived.

4 THE AVOIDANCE OF PARADOX

Part of Anderson's and Belnap's logical inventiveness shows itself in their discovery of how to use many-valued truth-matrices (a) to ensure variable-sharing, and (b) to avoid

[5] In this passage we find faint echoes of Russell: "inference will only in fact take place when the proposition 'not-p or q' (i.e. '$p \rightarrow q$') is *known* otherwise then through knowledge of not-p or knowledge of q" [Russell, 1919, 153].

what they call ‘paradoxes of necessity’. Part also lies in their method of ensuring relevance in systems of natural deduction. As will be seen, good use of all these discoveries may be made in developing systems of connexive logic.

Beginning with variable-sharing, a pure implicational logic in which $\rightarrow$ is the sole logical operator can be shown, by means of the following 4-valued matrix, to conform to variable-sharing in all formulae the matrix satisfies. (In the matrix, the ‘designated’ values are starred, and the matrix ‘satisfies’ a formula X iff X takes only designated values for all values of its variables.)

Matrix 1.

$\rightarrow$	1	2	3	4
$*1$	1	4	4	4
$*2$	1	2	3	4
3	1	4	2	4
4	1	1	1	1

To show the impossibility of any formula $X \rightarrow Y$ being satisfied when X and Y share no variable, assign 1 to all the variables in X, and 2 to all the variables in Y. Then, since $1 \rightarrow 1$ is 1 and $2 \rightarrow 2$ is 2, $X \rightarrow Y$ takes the value 4.

Anderson and Belnap define a “paradox of necessity” as generated by a provable formula of the type $X \rightarrow (Y \rightarrow Z)$, where ‘$\rightarrow$’ is ‘entails’ and X is a propositional variable replaceable by a factual proposition such as ‘Snow is white’. Paradox occurs when a proposition that is contingently true entails a proposition that depends for its truth on logical considerations. For example, the propositions ‘Snow is white’ and ‘That snow is white implies that snow is white’ are both true, but it is not the case that that first *entails* the second. The colour of snow is irrelevant to the truth of the latter, which is true by logical necessity. Matrix 2 shows that the pure implicational relevance logics which satisfy it avoid paradoxes of necessity.

Matrix 2.

$\rightarrow$	1	2	3
$*1$	2	3	3
$*2$	2	2	3
3	2	2	2

To show that no wff (well-formed formula) $X \rightarrow (Y \rightarrow Z)$ can be satisfied by matrix 2, where X is a propositional variable replaceable by a contingently true statement, take the case where X receives the value 1. Since $Y \rightarrow Z$ takes only the values 2 or 3, and since $1 \rightarrow 2$ and $1 \rightarrow 3$ are both 3, $X \rightarrow (Y \rightarrow Z)$ is rejected by the matrix. Hence Anderson’s and Belnap’s system, which satisfies Matrix 2, avoids fallacies of necessity.

Formal systems of connexive implication also avoid, in ways described below, fallacies of relevance and necessity. We turn now to the question of consistency, and the development of a consistent and complete system of connexive logic.

5 A CONSISTENT SYSTEM OF CONNEXIVE LOGIC

It was mentioned in section 1 that R.B. Angell [1962] contains a consistency proof for a ‘logic of subjunctive conditionals’ in which Aristotle’s and Boethius’ theses are theorems.

Angell's proof is based on the following matrices:

Matrix 3.

$\to$	1	2	3	4	$\sim$
$*1$	1	4	3	4	4
$*2$	4	1	4	3	3
3	1	4	1	4	2
4	4	1	4	1	1

&	1	2	3	4
1	1	2	3	4
2	2	1	4	3
3	3	4	3	4
4	4	3	4	3

The matrices satisfy Aristotle and Boethius, $\sim(p \to\sim p)$ and $(p \to q) \to\sim(p \to\sim q)$, and the rule of *modus ponens*, to infer $\vdash B$ from $\vdash A$ and $\vdash A \to B$. They define a consistent deductive system of connexive logic, named CC1. The axiomatic basis of CC1 is the following [McCall, 1966, 425]:

Primitive connectives: $\to, \sim, \&$

Definition: $A \vee B =_{df} \sim(\sim A \& \sim B)$

Rules of inference:
R1. Substitution for propositional variables.
R2. Modus ponens
R3. Adjunction: from $\vdash A$ and $\vdash B$ infer $\vdash A\&B$.

Axioms:
1. $(p \to q) \to [(q \to r) \to (p \to r)]$ Syl
2. $[(p \to p) \to q] \to q$ Wajsberg
3. $(p \to q) \to [(p\&r) \to (r\&q)]$
4. $(q\&q) \to (p \to p)$
5. $[p\&(q\&r)] \to [q\&(p\&r)]$
6. $(p\&p) \to [(p \to p) \to (p\&p)]$
7. $p \to [p\&(p\&p)]$
8. $[(p \to\sim q)\&q] \to\sim p$
9. $[p\&\sim(p\&\sim q)] \to q$
10. $\sim[p\&\sim(p\&p)]$
11. $\{\sim p \vee [(p \to p) \to p]\} \vee \{[(p \to p) \vee (p \to p)] \to p\}$
12. $(p \to p) \to\sim(p \to\sim p)$[6]

McCall's paper contains completeness proofs for the system, which demonstrate (i) that all wffs satisfied by matrix 3 are CC1 theorems, and also (ii) that CC1 is Post-complete. The latter means that if any non-theorem of CC1 is added to the system, the propositional variable q is derivable as a theorem; consequently, by substitution, any wff whatsoever. Like 2-valued logic, CC1 is complete in the strongest possible sense.

CC1 is related to a variety of other non-classical logics, as shown by the containment relations of Figure 1.

Classical logic PC and CC1 appear as Post-complete systems, and the many-valued logics of Lukasiewicz, Kleene and Bochvar are all sub-systems of PC. The remaining systems are shown in their pure implicational fragments only. Thus C is 2-valued pure implication, IC is the pure implicational fragment of intuitionist logic, $C5$–$C3$ are the pure

[6] Axiom 12, a substitution of Boethius, is the only non-classical axiom.

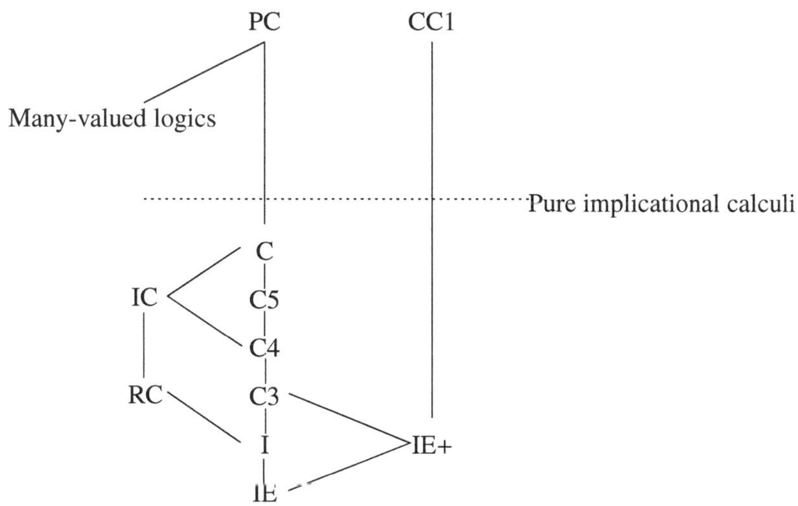

Figure 1.

strict implicational fragments of $S5$–$S3$ respectively, and RC and I are the implicational parts of Anderson's and Belnap's systems R and E respectively. IE is the part of I containing only wffs in which all variables occur an even number of times, and consequently excludes the I-axiom Frege (see below). Finally, $IE+$ is the conjectured axiomatization of Angell's pure implicational matrix 3. $IE+$ contains an additional rule of inference $R4$ which states that where A and B are pure implicational formulae, $\vdash A \rightarrow B$ is derivable from $\vdash A$ and $\vdash B$. In $IE+$, any two theorems imply one other. (The rule $R4$ does *not* hold in $CC1$ when A or B contains $\sim$ or &.) $IE+$ contains theses such as $(p \rightarrow p) \rightarrow (q \rightarrow q)$ which are not found in IE.

It is worth remarking that all the standard non-classical logics are sub-systems of 2-valued logic. Connexive logics, on the other hand, contain theorems such as Aristotle and Boethius that are 2-valuedly false, and consequently resemble non-Euclidean geometries that replace the parallel postulate by alternative axioms. When constructing connexive systems, it is important not only to *include* theses such as $\sim(p \rightarrow \sim p)$, but also to *exclude* the derivability of $B \rightarrow A$ from $A \rightarrow B$. The equivalence relation $\equiv$ is symmetrical, and also satisfies $\sim (p \equiv \sim p)$ and $(p \equiv q) \equiv \sim (p \equiv \sim q)$. But connexive implication is not equivalence, and Angell's matrix rejects $(p \rightarrow q) \rightarrow (q \rightarrow p)$, as all connexive logics must.

Although CC1 is consistent and Post-complete, it is an awkward system in many ways. Axioms 4, 6, 7, 10 and 11, for example, have little or no intuitive logical content. Nor does CC1 capture all plausible deductive theses. The absence of conjuctive simplication $(p\&q) \rightarrow p$ is understandable, since from $(p\& \sim p) \rightarrow p$ and the contrapositive of $(p \& \sim p) \rightarrow \sim p$, $(p\& \sim p) \rightarrow \sim (p\& \sim p)$ follows by transitivity. The latter contradicts Aristotle. But as is noted by Routley and Montgomery [1968], CC1 also lacks the plausible theses $(p\&p) \rightarrow p$ and $p \rightarrow (p\&p)$. (However, on this see the end of section 9 below). Since

CC1 has a finite characteristic matrix, some of its axioms do no more than reflect the fact that the matrix in question has four and only four values. Such things have nothing to do with connexive implication *per se*. Furthermore, deriving theorems from axioms in CC1 can be a tedious business. In the next section a more user-friendly subproof version of connexive logic is presented, in which relevance is based, as in Anderson and Belnap, on the requirement that premisses be *used* in deriving conclusions.

6 CONNEXIVE LOGIC IN SUBPROOF FORM

This section contains new foundations for connexive logic: an easy-to-operate Fitch-style system of natural deduction. The system incorporates Anderson's and Belnap's method of keeping track of hypotheses and ensuring relevance, using sets of "dependence numerals". The following proof of Boethius' thesis illustrates the construction of proofs using subproof rules:

(1)	1.	| $p \to q$	hyp
(2)	2.	| | $p \to\sim q$	hyp
(3)	3.	| | | p	hyp
(1)	4.	| | | $p \to q$	1 reit
(2)	5.	| | | $p \to\sim q$	2 reit
(2,3)	6.	| | | $\sim q$	3,5 MP
(1,2,3)	7.	| | | $\sim p$	4,6 MT
(1,2)	8.	| | $p \to\sim p$	3,7 CP
(1)	9.	| $\sim(p \to\sim q)$	2,8 RAC
(0)	10.	$(p \to q) \to\sim(p \to\sim q)$	1,9 CP

Features of these rules for connexive logic are (i) relevance or dependence numerals in brackets on the left, (ii) restrictions on the use of the rules MP and MT to cases in which the dependence sets of the two premises have zero intersection, (iii) the requirement that for conditional proof CP to be applied, the conclusion of a subproof must depend on its hypothesis, (iv) the restriction on reiteration stating that only implications can be reiterated, and finally (v) replacement of the rule for indirect proof (*reductio ad absurdum*) by the new connexive rule RAC. The subproof rules for the connexive system named CN*, containing only implication and negation operators, are the following, where $a, b, c, \ldots$ are sets of dependence numerals:

hyp An hypothesis may be introduced at any line of a proof. The hypothesis of the nth subproof receives dependence-numeral (n).

rep An item $(a)x$ may be repeated within the same subproof.

reit $(a)x$ may be reiterated from subproof n into subproof $n + m$, provided that x is an implication.

MP From items $(a)x$ and $(b)x \rightarrow y$ in subproof n, provided the intersection $ab = 0$, derive $(a \cup b)y$ in subproof n.

CP From the hypothesis $(n)x$ and item $(n \cup a)y$ in subproof n, derive $(a)x \rightarrow y$ in subproof $n - 1$.

MT From items $(a) \sim y$ and $(b)x \rightarrow y$ in subproof n, where $ab = 0$, derive $(a \cup b) \sim x$ in subproof n.

DNI From $(a)x$ derive $(a) \sim\sim x$ in subproof n.

DNE From $(a) \sim\sim x$ derive $(a)x$ in subproof n.

RAC From the hypothesis $(n)x$ and item $(n \cup a)y \rightarrow\sim y$ in subproof n derive $(a) \sim x$ in subproof $n - 1$.

Using Anderson's and Belnap's method of successively collapsing the subproofs of a Fitch-style proof, beginning with the innermost [1975, 24-26], we may reduce any proof in CN* to a sequence of items in the "main" proof (depending on the empty set of hypotheses). When all the items in the main proof needed to collapse subproofs employing the different rules of CN* are collected together, they are found to be derivable, using substitution and *modus ponens*, from the following set.

1. $(p \rightarrow q) \rightarrow [(q \rightarrow r) \rightarrow (p \rightarrow r)]$ Syl
2. $[(p \rightarrow p) \rightarrow q] \rightarrow q$ Wajsberg
3. $(\sim p \rightarrow q) \rightarrow (\sim q \rightarrow p)$ Trans 3
4. $p \rightarrow\sim\sim p$ Doub 1
5. $(p \rightarrow q) \rightarrow\sim(p \rightarrow\sim q)$ Boethius

We define the axiomatic system CN, containing the above axioms, as the system equivalent to the user-friendly subproof system CN*. CN is a more "intuitive" connexive system than CC1.

It is possible to introduce conjunction, and extend CN* to a system CNK* that includes introduction and elimination rules for conjunction. CNK* employs "decimalized" dependence numerals, as in the rule KE:

KE From $(a)x\&y$ in subproof n derive $(a.1)x$ and $(a.2)y$ as consecutive items in subproof n.

KI From $(a)x$ and $(b)y$ in subproof n, where $ab = 0$, derive $(a \cup b)x\&y$ in subproof n.

MPT From $(a)x$ and $(b) \sim (x\&y)$ in subproof n, where $ab = 0$, derive $(a \cup b) \sim y$ in subproof n.

RAK From the hypothesis $(n)x$ and item $(n \cup a)(y\&\sim y)$ in subproof n, derive $(a) \sim x$ in subproof $n - 1$.

Here is a sample proof in CNK^*, in which decimalized sets of relevance numerals are employed:

(1)	1.	$(p\&q) \rightarrow r$	hyp	
(2)	2.	$p\& \sim r$	hyp	
(3)	3.	$(p\&q) \rightarrow r$	1 reit	
(2.1)	4.	p	2 KE	
(2.2)	5.	$\sim r$	2 KE	
(1,2.2)	6.	$\sim (p\&q)$	3,5 MT	
(1,2)	7.	$\sim q$	4,6 MPT	(Note $(2.1 \cup 2.2) = 2$)
(1)	8.	$(p\& \sim r) \rightarrow \sim q$	2,7 CP	
(0)	9.	$[(p\&q) \rightarrow r] \rightarrow [(p\& \sim r) \rightarrow \sim q]$	1,8 CP	Antilogism

The "collapsing theorem" for CNK^* is lengthier and more complex than that for CN^*, but when carried through results in the following new axioms for conjunction. These axioms, added to the axioms for CN, form the basis for the connexive system CNK.

1K. $(p \rightarrow q) \rightarrow [(p\&r) \rightarrow (r\&q)]$

2K. $[p\&(q\&r)] \rightarrow [(p\&q)\&r]$

3K. $[(p\&q)\&r] \rightarrow [p\&(q\&r)]$

4K. $[p\&(p \rightarrow q)] \rightarrow q$

5K. $[(p \rightarrow p) \rightarrow p] \rightarrow [q \rightarrow (p\&q)]$

6K. $(p \rightarrow p) \rightarrow \sim (p\& \sim p)$

7K. $[p\& \sim (p\&q)] \rightarrow \sim q$.

As a final point, it may be asked why it is necessary to exclude from CN and CNK any wff in which a variable occurs an odd number of times. The answer is that such wffs engender contradictions in connexive logic. For example the formula

1. $[(p \rightarrow (q \rightarrow r)] \rightarrow [(p \rightarrow q) \rightarrow (p \rightarrow r)]$ Frege

has the following consequences:

2.	$[(\sim p \rightarrow p) \rightarrow (\sim p \rightarrow p)] \rightarrow \{[(\sim p \rightarrow p)] \rightarrow \sim p] \rightarrow$ $[(\sim p \rightarrow p) \rightarrow p]\}$	(1)
3.	$[(\sim p \rightarrow p) \rightarrow (\sim p \rightarrow p)]$	Law of identity
4.	$[(\sim p \rightarrow p)] \rightarrow \sim p] \rightarrow [(\sim p \rightarrow p) \rightarrow p]$	(2,3)
5.	$[(\sim p \rightarrow p) \rightarrow p] \rightarrow \sim [(\sim p \rightarrow p) \rightarrow \sim p]$	Boethius
6.	$[(\sim p \rightarrow p) \rightarrow \sim p] \rightarrow \sim [(\sim p \rightarrow p) \rightarrow \sim p]$	(4,5)

Line 6 contradicts Aristotle. Hence Frege, which is part of all standard non-classical logics, leads to inconsistency. Frege is rejected by Angell's matrix when the value 4 is assigned to all three variables, and the stipulation in the rules of CN* that excludes it is the requirement in MP and MT that $ab = 0$.

7 CONNEXIVE LOGIC AND THE SYLLOGISM

In section 1 it was noted that in the late 1800s Hugh MacColl based Aristotelian syllogistic on the logic of propositions, translating "All A is B" as $a \rightarrow b$, "No A is B" as $a \rightarrow\sim b$, "Some A is B" as $\sim (a \rightarrow\sim b)$, and "Some A is not B" as $\sim (a \rightarrow b)$. Today, using quantifiers, the usual translation of 'All A is B' is $(x)(Ax \rightarrow Bx)$, and of 'Some A is B' is $(\exists x)(Ax \& Bx)$. It is notorious that although for Aristotle 'All A is B' implies 'Some A is B', $(x)(Ax \rightarrow Bx)$ does not imply $(\exists x)(Ax \& Bx)$ when there exist no As. If "$\rightarrow$" is read as material implication, $\sim (\exists x)Ax$ implies $(x)(Ax \rightarrow Bx)$, so that if there exist no As, "All As are Bs" is trivially true. But for MacColl, the non-existence of round squares did not imply that all round squares were triangles, and it may be true that some unicorns have golden horns even though there are no unicorns. MacColl adopted an ontology that included "unrealities" such as round squares, centaurs, golden mountains, etc., all existing independently in their own right. In contrast, Russell argued that every one of these classes of object was identical with a single object: the null class [Russell, 1905; 1906]. MacColl however was unconvinced, and stoutly defended his category of unreal entities [MacColl, 1905a; 1905b].

Returning to syllogistic, one can combine MacColl's connexive arrow $\rightarrow$ with quantifiers and arrive at the following square of opposition (see [McCall, 1967a]):

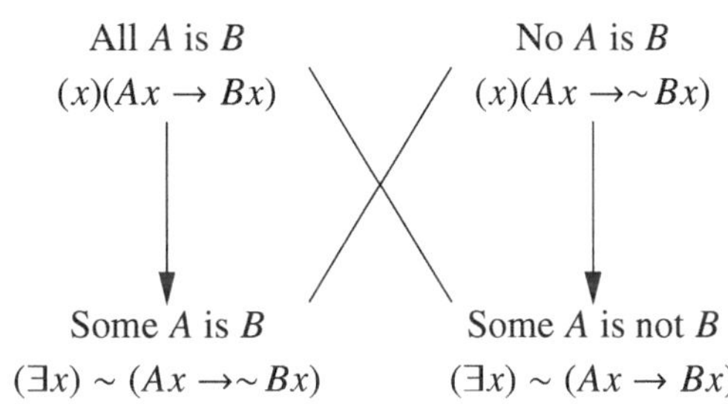

As stated in section 1, MacColl is able to demonstrate the validity of all nineteen moods of Aristotelian syllogistic. Using their traditional names ("Barbara" is the mood AAA in figure 1, "Festino" is EIO in figure 2, etc.) these moods, which include fourth-figure moods, are the following:

Figure:	1	2	3	4
	Barbara	Cesare	Darapti	Bramantip
	Celarent	Camestres	Disamis	Camenes
	Darii	Festino	Datisi	Dimaris
	Ferio	Baroco	Felapton	Fesapo
	(Barbari)	(Cesaro)	Bocardo	Fresison
	(Celaront)	(Camestrop)	Ferison	(Camenop)

In addition to the 19 traditional moods, there are 5 with weakened conclusions, e.g. Barbari which comes from Barbara. Their names are placed in brackets, yielding 24 valid moods in all. All these are derivable in MacColl's system, which faithfully reproduces traditional syllogistic logic.

8 CONNEXIVE CLASS LOGIC

It turns out that connexive propositional logic has an analogue in the algebra of classes. Where $\cap$ is intersection and $'$ is complementation, in Boolean algebra the null class 0, defined as $a \cap a'$, is a subset of every class. For all $b, 0 \subseteq b$. In connexive class logic by contrast 0 is a subset only of itself, and conversely the universal set 1, defined as $0'$, has only itself as a subset. The following is a formal axiomatization CA of connexive class logic, which stands to Boolean algebra as connexive propositional logic stands to 2-valued logic.

Primitive symbols

Variables: $a, b, c, \ldots$

Connectives: $\subseteq$, $'$, $\cap$ (algebraic); $\supset$, $\sim$ (propositional)

Definitions

1. $A\&B =_{df} \sim(A \supset \sim B)$
2. $ab =_{df} a \cap b$
3. $a = b =_{df} (a \subseteq b)\&(b \subseteq a)$
4. $a \cup b =_{df} (a'b')'$
5. $0 =_{df} aa'$
6. $1 =_{df} 0'$

Axioms

1. Axiomata-schemata for 2-valued propositional logic
2. $(a \subseteq b \;\&\; b \subseteq c) \supset a \subseteq c$
3. $a \subseteq b \supset (ac \subseteq cb)$
4. $a(b \cup c) \subseteq ab \cup ac$
5. $a(bc) \subseteq (ab)c$
6. $(ab' \subseteq 0 \;\&\; \sim(a \subseteq 0) \;\&\; \sim(b' \subseteq 0)) \supset a \subseteq b$
7. $b' \subseteq a' \supset a \subseteq b$
8. $aa' \subseteq bb'$
9. $a \subseteq aa$

10. $a \subseteq a''$

11. $a \subseteq 0 \supset 0 \subseteq a$

12. $0 \subseteq a \supset a \subseteq 0$

13. $\sim(a \subseteq a')$

Rules of inference

1. Substitution of variables and terms a' and $a \cap b$ for variables.

2. Modus ponens.

Note that of the thirteen axioms, only 12 and 13 fail in Boolean algebra. In connexive class logic, the axioms hold true when the variables are interpreted as ranging over non-empty, non-universal classes. In McCall [1967b] the formal system CA is proved complete in the sense that any addition would reduce it to an algebra containing only a finite number of non-equivalent terms, in the same way that any addition to Boolean algebra would make it finite. CA can be regarded as formally parallel to a connexive propositional system of first-degree formulae, in which no nesting of the arrow $\rightarrow$ is permitted (see next section).

9 FIRST-DEGREE CONNEXIVE FORMULAE

A large part of contemporary interest in connexive implication lies in the area of *causal* implication, e.g. "If X takes aspirin, X's fever goes down". Where "$p \rightarrow q$" is causal, the values of "p" and "q" are not themselves "if...then" formulae, at least not typically. Consequently it is worth studying the logical powers of *first-degree connexive formulae* "$p \rightarrow q$", in which no nesting of the arrow is permitted, but where, for example, transitivity of causal implication is captured by "$[(p \rightarrow q)\&(q \rightarrow r)]\supset(p \rightarrow r)$", contraposition by "$(p \rightarrow q)\supset(\sim q \rightarrow\sim p)$", Boethius' thesis by "$(p \rightarrow q)\supset \sim(p \rightarrow\sim q)$", etc. Most if not all of the interesting features of connexivity are found within the sphere of first-degree formulae.

In Anderson and Belnap [1975, pp. 434-450], the author constructs a formal system CFL of connexive first-order logic, based upon the notion of a *connexive algebra* (p. 440). Briefly, without entering into details, a connexive algebra is a septuple $\langle B, \wedge, \vee, ', D, \cup, \leq\rangle$, where B is a Boolean algebra with the logical operations $\wedge$ (meet), $\vee$ (join), and $'$ (complementation) defined on it, D is a prime filter[7] in B, U is the complement of D in B, and $\leq$ is a binary relation on B defined as follows:

$$x \leq y \text{ iff (i) } x \vee y = y \text{ and (ii) either } x, y \in D \text{ or } x, y \in U.$$

[7] D is *prime* iff for all x, y, $x \vee y \in D$ iff $x \in D$ or $y \in D$, and D is a *filter* iff for all x, y, $x \wedge y \in D$ iff $x \in D$ and $y \in D$.

The principal respect in which a connexive algebra differs from a Boolean algebra is that in it $x \leq x'$ never holds, just as in systems of connexive logic $p \rightarrow\sim p$ never holds. It is shown how a connexive first-degree logic CFL may be associated with a connexive algebra in [McCall, 1975, 440–2], and an axiomatic basis for the system CFL is given:

Axioms

1. $(q \rightarrow r) \supset [(p \rightarrow q) \supset (p \rightarrow r)]$
2. $(p \rightarrow q) \supset [(p\&r) \rightarrow (r\&q)]$
3. $[p\&(q \vee r)] \rightarrow [(p\&q) \vee (p\&r)]$
4. $[p\&(q\&r)] \rightarrow [(p\&q)\&r]$
5. $(p \rightarrow\sim q) \supset (q \rightarrow\sim p)$
6. $(p\& \sim p) \rightarrow (q\&\sim q)$
7. $p \rightarrow (p\&p)$
8. $(p\&p) \rightarrow p$
9. $\sim\sim p \rightarrow p$
10. $(p \rightarrow q) \supset \sim(p \rightarrow\sim q)$
11. $\{[(p\& \sim q) \rightarrow (p \& \sim p)]\&(p\vee \sim q)\}\supset(p \rightarrow q)$
12. $p\supset[p \rightarrow (p\supset p)]$
13. $[p \rightarrow (p\supset p)]\supset p$
14. $(p\supset q)\supset[(q\supset r)]\supset(p\supset r)]$

Rules of inference

R1. Substitution for propositional variables, with the restriction that no nesting of arrows may result.

R2. From $A, A\supset B$ infer B.

Note that in CFL the plausible theses $p \rightarrow (p\&p)$ and $(p\&p) \rightarrow p$, remarked by Routley and Montgomery as lacking in CC1, occur as axioms 7 and 8. The system CFL corresponds to the set of connexive algebras composed of any arbitrary number of elements. It is complete in the sense of possessing the *Scroggs property*, meaning that CFL has an infinite truth value matrix, but that any consistent proper extension of CFL has a finite characteristic matrix. In this it differs sharply from CC1, which has a characteristic 4-valued matrix.

10 CAUSAL IMPLICATION

Causal conditionals come in indicative, subjunctive, and counterfactual form:

(i) If that match is scratched, it will light.

(ii) If that match were to be scratched, it would light.

(iii) If that match had been scratched, it would have lighted.

Roughly and broadly, the antecedent of a causal conditional denotes a state of affairs that, in conjunction with unspecified "background conditions", is *causally sufficient* for the obtaining of the consequent. In every case, the conditional "$p \to q$" is contradicted by "$p \to\sim q$", consequently the arrow of causal implication is connexive. But constructing a formal system of causal conditionals is not easy. Consider for example axiom 2 of the system CFL in the previous section:

2. $(p \to q) \supset [(p \& r) \to (r \& q)]$.

From the conditional "If that match is scratched, it will light", it does *not* follow that "If that match is scratched and soaked in water, it will be soaked in water and light". In this section proposals for a logic of causal implication are explored in a preliminary way. But despite having been worked on for decades, causal logic is still very much an ongoing project, and no agreed-on formulation of it has yet been achieved.

In his classic paper "The problem of counterfactual conditionals" [1947], p. 113, Nelson Goodman considers:

(1) If that piece of butter had been heated to 150°F, it would have melted.

If the "if" of (1) were the two-valued truth functional "if" of classical logic, the falsehood of the antecedent "That piece of butter is heated to 150°F" would entail not only that (1) is true, but that (2) is equally true:

(2) If that piece of butter had been heated to 150°F, it would not have melted.

But intuitively, (2) contradicts (1):– (1) is true and (2) is false. Goodman points out that (1) can be contraposed into a conditional with true antecedent and consequent:

(3) Since that butter didn't melt, it wasn't heated to 150°F.

He concludes that in (1)–(3) "what is in question is a certain kind of connection between the two component sentences", and that the truth of truth of (1) and (3), and the falsehood of (2), "depends not upon the truth or falsity of the components but on whether the intended connection obtains". With that assertion, Goodman places himself (unknowingly) in the tradition of Sextus Empiricus in the fourth century B.C. But Goodman is no more successful in elucidating what *kind* of connection exists between antecedent and consequent in true counterfactuals than Sextus was, or Abelard, or Lewis Carroll. Goodman's examples are plainly based on causation, but what would a *theory* of causal implication look like? Can it be axiomatized into a formal system? Can causal implication be provided with a semantics?

Robert Stalnaker, in "A theory of conditionals" [1968] takes the problem a step further in considering the 1960's conditional:

(4) If the Chinese enter the Vietnam conflict, the United States will use nuclear weapons.

Stalnaker considers Ramsey's proposal (section 1 above) for deciding whether (4) is true: Add the antecedent p hypothetically to your stock of beliefs, making whatever adjustments are required to maintain consistency, and then consider whether the consequent q is true in the augmented belief-set [Stalnaker p. 169, page numbers from the [1975] reprint of [1968]]. Plainly if q is *already* believed to be true — if it is believed that the U.S. will use nuclear weapons whether or not the Chinese enter the conflict — then (4) passes the Ramsey test in a trivial sense. But if q's acceptance depends upon adjustments to one's belief set when p is added, then the outcome of the Ramsey test hangs on whether or not antecedent p and consequent q are believed to be linked. The Ramsey test, in Stalnaker's words, "accounts for the relevance of 'connection' when it is relevant without making it a necessary condition of the truth of a conditional" (p. 168).

Finally, there is the case where the antecedent is already taken to be false, as in counterfactual conditionals. Here you cannot add it to your belief set without introducing a contradiction. Adjustments must be made by "deleting or changing those beliefs which conflict with the antecedent", and there will in general, Stalnaker says, be more than one way in which this can be done. The limitations of the Ramsey test lie in the fact that it deals with "belief-conditions" for conditionals rather than "truth-conditions", and the time has come to make the transition to the latter.

To effect a change from belief- to truth-conditions, the ideal instrument is the concept of a *possible world*, which is the ontological equivalent of a stock of hypothetical beliefs. Stalnaker proposes the following Ramsey-like truth conditions for causal and counterfactual conditionals $A \to B$:

> "Consider a possible world in which A is true and which otherwise differs minimally from the actual world. 'If A, then B' is true (false) just in case B is true (false) in that possible world." (p.169)

A key concept here is the notion of a possible world which "differs minimally" from the actual world. How does one select a minimally different world? Stalnaker's answer is to propose a "selection function" $f(A, x) = y$, where x is the actual world (or more generally the possible world from which one departs in adding A), A is the antecedent, and y is the world which differs minimally from x. Then $A \to B$ is true in world x if and only if B is true in $f(A, x)$. Stalnaker says that this interpretation "shows conditional logic to be an extension of modal logic". His view is endorsed by Lennart Aqvist in his 1973 paper "Modal logic with subjunctive conditionals and dispositional predicates" [Aqvist, 1973]. Aqvist introduces a new logical operator "$*$" — the circumstantial operator — and reads "$*A$" as "A under the circumstances" or (David Lewis' suggestion) "*Things are the way they would be if it were the case that* A" [Aqvist, 1973, p. 2; Lewis, 1973, p. 61]. Using $*$, the Goodman conditional $A \to B$:

(1) If that piece of butter had been heated to 150°F, it would have melted

would be symbolized by $\Box(*A \supset B)$.

The philosopher who has done the most to explore Stalnaker's intuition that conditional logic is an extension of modal logic is David Lewis. In his book *Counterfactuals* [1973], Lewis constructs an elegant and elaborate possible-worlds semantics for counterfactuals, based on the concept of comparative overall similarity of worlds. Imagine a structured set of worlds, centered on a given world i and arranged in concentric three-dimensional spheres around i. Each world is a point within a 3-D sphere. The spheres are ordered so that any world w_1 which is more similar to i than w_2 finds itself located in a sphere *closer* to i than w_2, while any world w_3 which is overall less similar to i than w_2 finds itself in a sphere *farther away* from i than w_2 (page 14). This "Ptolemaic system" of concentric spheres provides an attractive modal semantics for counterfactuals and other conditionals. Thus $A \rightarrow B$ is (non-vacuously) true at a world i according to a system of spheres $\$_i$ if and only if (i) some sphere S in $\$_i$ contains at least one A-world, and (ii) every world in S is such that it is a B-world if it is an A-world (i.e. that $A \supset B$ holds at every world in S) ([Lewis, p. 16] omitting conditions for "vacuous" truth). The following diagram from Lewis, page 17, shows that (i) some sphere contains a ϕ-world, and (ii) every world in that sphere is such that if it is a ϕ-world it is also a ψ-world.

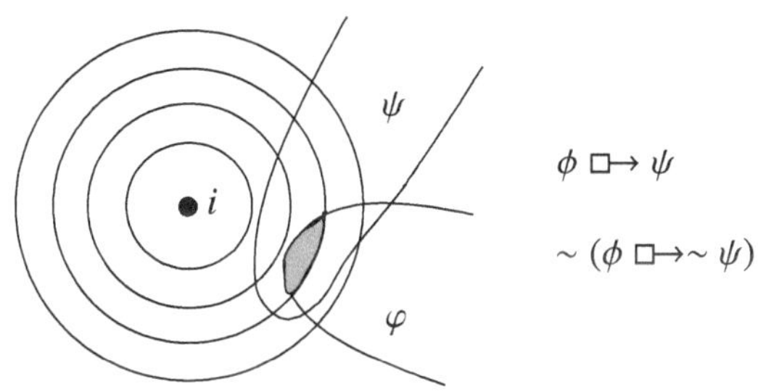

Figure 2.

Figure 2 shows that the conditional $\phi \mathrel{\Box\!\!\rightarrow} \psi$ (or as we would write it, "$\phi \rightarrow \psi$") is true in a Lewis semantic model. From the point of view of connexive implication, it is significant that in any of Lewis's models in which $\phi \mathrel{\Box\!\!\rightarrow} \psi$ is true, $\sim(\phi \mathrel{\Box\!\!\rightarrow} \sim\psi)$ is also true. Therefore Boethius' thesis $(A \rightarrow B) \supset \sim(A \rightarrow \sim B)$ holds in the Lewis semantics.

Although it would be tempting to devote a dozen pages to exploring Lewis's complex truth-conditions, there are fortunately better-known and more traditional Kripke-style modal semantics available for causal conditionals. We may, in the modal system T, define $A \rightarrow B$ as follows:

Df 1. $A \rightarrow B =_{Df} \Box(A \supset B) \,\&\, (\Diamond B \supset \Diamond A) \,\&\, (\Box B \supset \Box A)$.

This definition is found in Claudio Pizzi [1991, p. 619], where Pizzi discusses what he calls "*consequential implication*" (see also [Pizzi, 1977; 1993; 1996; Pizzi and Williamson, 1997]). Using one-sided semantic tableaux in the style of Jeffrey [1967] we can show

Aristotle's thesis $\sim(p \rightarrow\sim p)$ to be valid in the modal system T. (In the following 3-world tableau, an "x" indicates that the path above it is closed.)

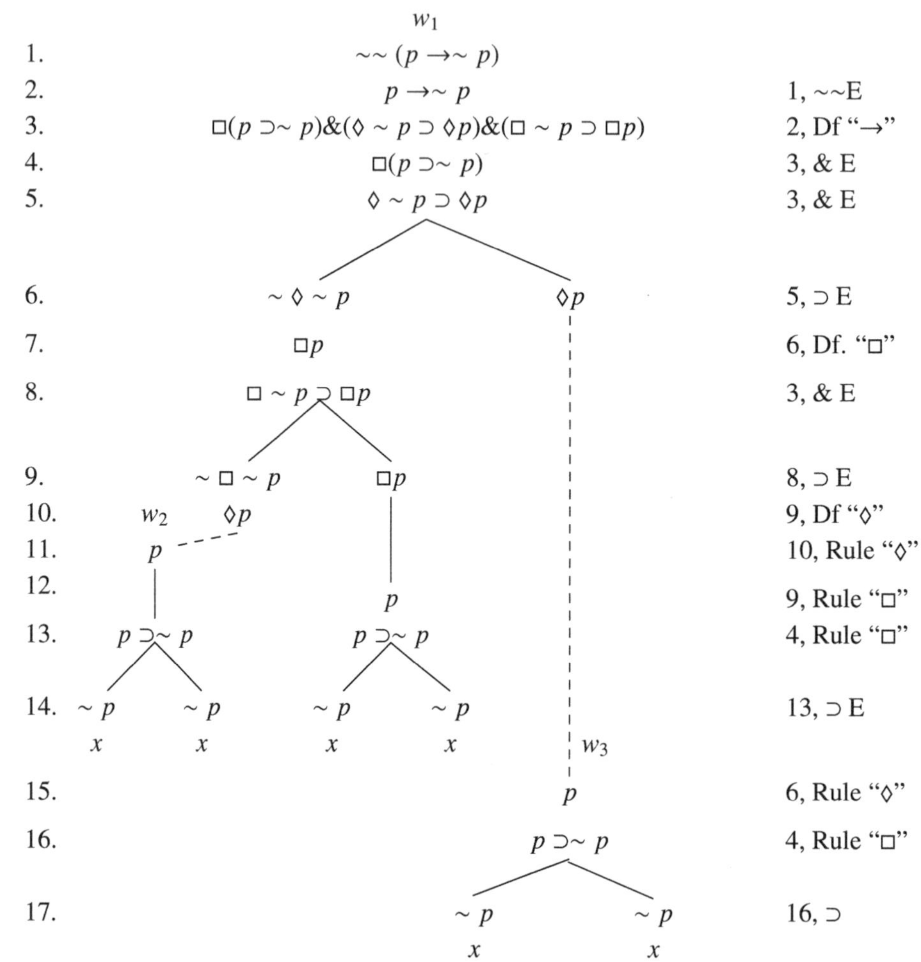

Tableau 1.

It is easily verified that if "$A \rightarrow B$" is defined as in Df 1, then transitivity $[(A \rightarrow B) \& (B \rightarrow C)] \supset (A \rightarrow C)$ and contraposition $(A \rightarrow B) \supset (\sim B \rightarrow\sim A)$ are both valid. These theses are controversial in counterfactual semantics.[8] The conditionals Lowe is interested in are connexive, as witnessed by the presence of Boethius' thesis in his list of "plausible" inference-patterns in [1995, p. 47]. Both transitivity and contraposition are invalid for Lewis (see [1973, p. 34], and the following seems plainly invalid:

[8]Transitivity has been extensively discussed by E.J. Lowe: see [Lowe, 1984; 1985; 1990; 1995], also [Wright, 1984].

(5) If the switch is turned, the light goes on.
If the light goes on, the generator is working.
Therefore, if the switch is turned, the generator is working.

Does (5) show that causal implication is not transitive after all? No. A causal conditional "If *A*, then *B*" asserts the existence of a *causal relation* between *A* and *B*, i.e. that given a certain set of background conditions, the obtaining of *A* is causally sufficient for the obtaining of *B*. Where this is so, I shall say that the background conditions *satisfy* the conditional. The time order of *A* and *B* is also important. Normally we take earlier events to be causally sufficient for later events, as when icy road conditions are sufficient for the danger of skidding. But a later event *A* can also be sufficient for an earlier event *B*, as when the death of a canary in a mine is sufficient for lack of oxygen. To reflect time order, I shall distinguish between *anterior sufficiency* of *A* for *B* (sufficiency of ice for skidding), and *retroactive sufficiency* of *A* for *B* (sufficiency of thunder for lightning).

For an inference such as (5) involving causal conditionals to be valid, three conditions must be met. (i) There must exist a single consistent set *S* of background conditions satisfying all the premisses; (ii) Every set *S* that satisfies the premisses must also satisfy the conclusion; and (iii) If the premisses form a "causal sufficiency chain" (*A* sufficient for *B*, *B* sufficient for *C*, ... etc.) then the sufficiencies in question must all be anterior, or must all be retroactive. Condition (iii) rules out any valid conclusion being drawn from the two premises in (5), which are of the form: "*A* is anteriorly sufficient for *B*" and "*B* is retroactively sufficient for *C*". From such chains no valid conclusion involving *A* and *C* can be drawn.

Similar conclusions hold for Stalnaker's and Lewis's alleged counter-examples to transitivity. Stalnaker's is the following:

(6) If J. Edgar Hoover were today a Communist, then he would be a traitor.
If J. Edgar Hoover had been born a Russian, then he would today be a Communist.
Therefore, if J. Edgar Hoover had been a Russian, he would be a traitor. (p. 173)

The premisses and conclusion of (6) are, in a broad sense, causal conditionals, in which the antecedent furnishes sufficient conditions for the truth of the consequent. At first glance, it might seem as if (6) invalidated causal transitivity, since there are background conditions that make the premisses true but do not make the conclusion true. If Hoover had *really* been born in Russia, he would probably later be a Communist though not a traitor. But the background conditions presupposed by the truth of the first premiss are inconsistent with those presupposed by the truth of the second. The first premiss rests on the supposition that J. Edgar Hoover is an American: in those circumstances (given his position as head of the FBI) he would certainly be a traitor if he were a Communist. The second premiss, on the other hand, invites us to consider the very different set of background circumstances in which Hoover is born a Russian. (Difficult to imagine certainly, but we must try.) Using Ramsey's truth-conditions, we must add "Hoover is born a Russian" hypothetically to our background beliefs, make the necessary adjustments, and see if the addition would be causally sufficient for Hoover's being a Communist. The Hoover example violates condition (i) above, that there must exist a single consistent set of background conditions satisfying all the premisses. Consequently it is not a counter-example to causal transitivity.

Similarly for Lewis's proposed counter-example:

(7) If Otto had gone to party, then Anna would have gone.
If Anna had gone, then Waldo would have gone.
Therefore, if Otto had gone, then Waldo would have gone.

As Lewis explains the background conditions, Otto is Waldo's successful rival for Anna's affections. Hence the first premiss is true. Secondly, Waldo still tags around after Anna, so the second premiss is true. Thirdly, Waldo never runs the risk of meeting Otto, so, Lewis says, the conclusion is false. But in fact Lewis's own background conditions, that Otto and Waldo are rivals for Anna's affections and that Otto's suit has prevailed, coupled with the fact that Waldo still morosely pursues Anna, constitute a consistent set (condition (i)) that also satisfies the conclusion (condition (ii)). Furthermore (7) is an example of an anterior causal sufficiency chain (condition (iii)). It follows that although (7) may be invalidated by Lewis's semantics, it is a model of validity by criteria (i)-(iii).

Next, consider an example of a *retroactive sufficiency chain*:

(8) If George is driving the car, then he holds a valid driving permit.
If George has a driving permit, then he is at least 16 years old.
Therefore, if George is driving, he is at least 16 years old.

This inference should be uncontentious, even though standard examples of causal sufficiency work forward in time rather than backwards. Given the background conditions of most jurisdictions, possessing a driving permit is a necessary condition of being able to drive, therefore driving is a sufficient condition of having a permit. Similarly, having a permit is a sufficient condition of being at least 16 years old. The conclusion follows, via a retroactive sufficiency chain.

Finally, consider a standard transitive causal argument that is valid:

(9) If the match is scratched, it lights.
If it lights, heat is generated.
Therefore, if the match is scratched, heat is generated.

Here the background conditions for the two premisses satisfy the conclusion: the match is dry, it is well-made, oxygen is present, etc. Under these conditions, the scratching of the match always results in the generation of heat, which is what the conclusion asserts. Contrast this with an invalid form of causal argumentation, the strengthening of the antecedent, which takes the form "$A \rightarrow B$, therefore $(A\&C) \rightarrow B$":

(10) If the key is turned, the car starts.
Therefore, if the key is turned and the battery is dead, the car starts.

The background conditions for (10)'s premiss assume that the engine turns over, that the carburetor supplies gasoline vapour, that the magneto produces a spark, etc. But these conditions explicitly contradict the circumstances under which the conclusion is true (if indeed such circumstances exist). Where C is an arbitrary proposition, taking values independently of the values of A and B, background conditions for the truth of $A \rightarrow B$ may commonly be found which fail to satisfy $(A \& C) \rightarrow B$, violating requirement (ii) above. Therefore, in causal implication, strengthening the antecedent is invalid.

Another contentious inference form is causal contraposition, if $A \rightarrow B$, then $\sim B \rightarrow \sim A$, which fails in Lewis's semantics. Unlike strengthening the antecedent, contraposition is valid in causal implication:

(11) If that match had been scratched, it would have lighted.
Therefore, if that match hadn't lighted, it wouldn't have been scratched.

There are background conditions (oxygen being present, etc.) that, combined with the match being scratched, are anteriorly sufficient for its lighting. But the same conditions, combined with the match's not having lighted, are retroactively sufficient for the match's not having been scratched. (Condition (iii) above does not rule out the combination of anterior sufficiency in the premiss with retroactive sufficiency in the conclusion.) Therefore causal contraposition is a valid form of inference.

Before concluding, there is a quantity of recent work on connexive implication that I survey briefly in the following section.

11 CONTEMPORARY WORK ON CONNEXIVE IMPLICATION: MEYER, ROUTLEY, MORTENSEN, PRIEST, LOWE, PIZZI, WANSING, RAHMAN AND RÜCKERT.

It was stated in the previous section that Claudio Pizzi [1991] gives a definition of the connexive arrow based on the modal notions of necessity and possibility:

Df 1. $A \rightarrow B =_{Df} \Box(A \supset B) \& (\Diamond B \supset \Diamond A) \& (\Box B \supset \Box A)$.

This definition has the immense advantage of suggesting ordinary modal logic — the Lewis systems S1-S5, plus the system T — as a conceptual and formal basis for connexive semantics. Although it is not certain that Pizzi knew of it, an earlier suggestion of this kind had been made by R.K. Meyer in a contribution to the *Relevance Logic Newsletter* entitled "S5 — The Poor Man's Connexive Implication" [Meyer, 1977]. What Meyer shows is that all the first-degree connexive theses of McCall's system CFL can be derived in S5 by defining the connexive arrow as follows:

Df. 2 $A \rightarrow B =_{Df} \Box(A \supset B) \& (A \equiv B)$,

where "$\equiv$" is material equivalence. Since S5 has a well-established, easy to operate many-worlds semantics, this semantics is available for first-order connexive logic.

A very different connexive semantics, along the lines of the Routley-Meyer semantics for relevant implication, is to be found in Routley [1978]. Formally, Routley's semantics are based on frames containing the three-term relations *Rabc* and *Sabc*, and Routley is able to derive soundness and completeness theorems for his connexive semantics. But the conditions imposed on the relations R and S for the proofs of these theorems are complex and non-intuitive, with the result that Routley-style connexive models will never achieve the elegance and convenience of the many-worlds semantics.

Routley's proposal for connexive modeling is taken up again by Chris Mortensen, who recognizes its defects [Mortensen, 1984, p. 111], and proposes in its place a truth-value semantics for connexive logic. Mortensen constructs the following three-valued truth-matrices (re-written so that the designated values 1 and 2 are reversed, and 3 replaces the sole undesignated value 0 in Mortensen's matrices):

Matrix 4.

$\to$	1	2	3	$\sim$
∗1	2	3	3	3
∗2	2	2	3	2
3	2	2	2	1

&	1	2	3
∗1	1	2	3
∗2	2	2	3
3	3	3	3

$\vee$	1	2	3
∗1	1	1	1
∗2	1	2	2
3	1	2	3

If we name the set of all wffs satisfied by matrix 4 "M3V", it is evident that M3V is inconsistent, since for example both "$p \to p$" and "$\sim (p \to p)$" are satisfied. M3V is a connexive logic, since Aristotle and Boethius are also satisfied. It is therefore an inconsistent connexive system that is not a "trivial" system, i.e. a Post-inconsistent system in which every wff is derivable. At the end of his paper Mortensen suggests that since the pure implicational fragment of M3V includes Anderson's and Belnap's pure calculus E of entailment, matrix 4 shows that the addition of Aristotle to E yields a *paraconsistent* logic, i.e. a logic which is inconsistent but not trivially so. This may be, but paraconsistency does not fall within the scope of the present study of connexivity. What we are looking for is a consistent connexive logic. Incidentally it should be remarked that the pure implicational fragment of matrix 4 is identical with matrix 2 of section 4, the matrix which shows that pure implicational relevance logics avoid paradoxes of necessity.

In Routley [1978, p. 395], the author suggests that connexivist principles in logic stem from a "cancellation" account of negation, whereby $\sim A$ "cancels out" A. To assert A, and then to follow it by $\sim A$, is to cancel the first assertion by the second, and to end up saying nothing at all. The conjunction $A \& \sim A$ has no content, and if we regard implication as requiring content inclusion (i.e. that the content of the consequent be included in that of the antecedent), $(A \& \sim A) \to A$ and $(A \& \sim A) \to \sim A$ fail. A and $\sim A$ in general have some content, and so cannot be included in the content of $A \& \sim A$, which is zero. Routley goes on to state that similar arguments show that $A \& \sim A$ cannot imply $\sim (A \& \sim A)$, and that in general $\sim (B \to \sim B)$ holds for every B when negation is understood as cancellation. Consequently the "cancellation" view of negation can serve as the basis for connexive logic.

Graham Priest in [1999] takes up the idea of cancellation negation, and argues that the cancellation concept can serve as a conceptual basis and foundation for the notion of connexivity. But Priest himself does not think that negation is correctly characterized by cancellation (p. 146), and to seek to establish the basis of connexive implication as a property of negation rather than as a connection between the antecedent and the consequent of a true implication, as Sextus Empiricus does, seems a lost cause. There is certainly interesting and important logical capital to be made out of the notion of cancellation negation, and Priest's comparison between the *total*, *partial*, and *null* (or cancellation) accounts of negation (p. 141) will interest all logicians. But the *idée maitresse* of connexive logic lies elsewhere.

In 1995 E.J. Lowe published an important paper on conditionals [Lowe, 1995] that complements and extends the analysis of causal conditionals of section 10 above. To recall, a causal conditional "If *A*, then *B*" asserts the existence of a causal relation between *A* and *B* to the effect that, given a set of background conditions, the obtaining of *A* is causally sufficient for the obtaining of *B*. Lowe's analysis extends this line of thinking to cases where the obtaining of *A* is causally sufficient neither for the obtaining of *B* nor for the non-obtaining of *B*, but where the relation between *A* and *B* is weaker than causal. For example, I am offered a lottery ticket, with one chance in a million of winning a prize. Although a rational ticket-purchaser would not accept the truth of the causal conditional

(1) If I buy that ticket, I shall win the prize,

he would not lay out the expense of the ticket purchase unless he accepted a weaker conditional implied by (1):

(2) If I buy that ticket, I may win the prize.

Lowe endorses the implication of (2) by (1), and notes that the correct logical form of (2) is:

(3) It is not the case that if I buy that ticket, I shall not win the prize.

Since (1) implies (2), and since (2) is equivalent to (3), (1) implies (3). This is where connexivity comes in, since the implication "If (1), then (3)" is a concrete instance of Boethius. Lowe does not mention connexive implication in his paper, but on p. 47, in a list of what he describes as "plausible inference-patterns", he includes

(P2) $p \rightarrow q \vdash \sim(p \rightarrow \sim q)$.

What is important about Lowe's approach is that it extends the category of causal implication, $A \rightarrow B$, from cases in which *A* is causally sufficient for *B* to cases in which, given *A*, there is only a chance of the occurrence of *B*. The lottery example is a case in point. Nevertheless, the connexive nature of the relationship between *A* and *B* still holds, as evidenced by Lowe's (P2) above.

The philosopher who has written most extensively in recent times on the notion of connexivity is Claudio Pizzi, who in definition 1 of the previous section offers a definition of the connexive arrow. This definition holds in the modal system *T* and stronger modal systems. Pizzi distinguishes two different varieties of what he calls "consequential implication", namely an "analytic" and a "synthetic" version. The "synthetic" version is analogous to, may even be identical with, the "causal implication" of section 10. Pizzi goes to some lengths to show that the law of monotonicity, i.e. the law of the factor $(A \rightarrow B) \supset [(A\&C) \rightarrow (B\&C)]$, is not derivable in either analytic or synthetic consequential implication. Pizzi shows that this law, which is a theorem of McCall's first-order system CFL of section 9, is intuitively undesirable for systems of consequential implication, as well as for causal implication in general, although he gives a weaker form of it which is acceptable [1991, p. 619]. For synthetic consequential implication he introduces the operator $*$, where as mentioned above, in section 10, $*A$ can be read as "A under the circumstances", or "Things are as they would be if it were the case that A". Using the star operator, Pizzi constructs at least two different systems of synthetic consequential implication, which correspond roughly with section 10's "causal" implication [Pizzi, 1977, pp. 292-98; 1991, p. 622 ff.; 1996, p. 649].

Heinrich Wansing makes two interesting contributions to connexive logic, the first entitled "Connexive modal logic" [2005], and the second a comprehensive review article "Connexive logic" in the *Stanford Encyclopedia of Philosophy* [2006]. In [2005] Wansing constructs what he calls a "basic" system of connexive logic, the principal feature of which is that the negation of any implication $A \rightarrow B$ is equivalent to $A \rightarrow\sim B$, i.e. $\sim(A \rightarrow B) \leftrightarrow (A \rightarrow\sim B)$. In this system, the pure implicational fragment of which is the system IC of intuitionist logic (see figure 1 of section 5), Aristotle and Boethius hold, $\sim\sim A$ is equivalent to A, and contraposition $(A \rightarrow\sim B) \rightarrow (B \rightarrow\sim A)$ fails. But despite having an interesting semantics (see below), Wansing's system cannot be regarded as a "basic" system of connexive logic. The difficulty lies in the converse of Boethius' thesis:

(4) $\sim(A \rightarrow B) \rightarrow (A \rightarrow\sim B)$.

(4) is a highly unintuitive implicational thesis, to which many counter-examples spring to mind. For example, it may be true that:

(5) It is not the case that if snow is white, grass is green,

but it is surely not the case that:

(6) If snow is white, then grass is not green.

If the price of having a "basic" system of connexive logic is accepting the implication of (6) by (5), then that price is simply too high to pay. If connexive implication is based in some way or other on a "connection" between the antecedent and consequent of any true implication, then this connection is palpably lacking in (6).

Wansing's system, named the system C, has an interesting semantics, founded on the notion of information, which rules out counter-examples like that above by establishing that (5) and (6) are neither true nor false. A C-frame is a pair $\langle W, \leq\rangle$, where $\leq$ is a reflexive and transitive binary relation on the set W. Let $\langle W, \leq\rangle^+$ be the set of all subsets X of W such that if $u \in X$ and $u \leq w$, then $w \in X$. A C-model is a structure $M = \langle W, \leq, v^+, v^-\rangle$, where $\langle W, \leq\rangle$ is a C-frame and v^+ and v^- are valuation functions from the atomic wffs of C into $\langle W, \leq\rangle^+$. Intuitively, Wansing says, [2005, p. 371], W is a set of information states, and the function v^+ sends an atomic sentence p to the states that support the truth of p, whereas v^- sends p into the states that support the falsity of p. (Wansing presumably means that v^+ is a *relation* connecting atomic sentences to the set of states that support their truth, and similarly for v^-.) $M = \langle W, \leq, v^+, v^-\rangle$ is the model based on the frame $\langle W, \leq\rangle$, and the relations $M, t \vdash^+ A$ (M supports the truth of A at t), and $M, t \vdash^- A$ (M

supports the falsity of A at t) are inductively defined as follows:

$$
\begin{array}{lll}
M,t \vdash^{+} p & \text{iff} & t \in v^{+}(p) \\
M,t \vdash^{-} p & \text{iff} & t \in v^{-}(p) \\
M,t \vdash^{+} {\sim} A & \text{iff} & M,t \vdash^{-} A \\
M,t \vdash^{-} {\sim} A & \text{iff} & M,t \vdash^{+} A \\
M,t \vdash^{+} (A \rightarrow B) & \text{iff} & \text{For all } v \geq t, (M,v \vdash^{+} A \text{ implies } M,v \vdash^{+} B) \\
M,t \vdash^{-} (A \rightarrow B) & \text{iff} & \text{For all } v \geq t, (M,v \vdash^{+} A \text{ implies } M,v \vdash^{-} B)
\end{array}
$$

Given these semantics, it can be seen why counter-examples like (5)-(6) are not admissible. The reason is, that (5) and (6) are neither true nor false according to Wansing's semantics. If we replace (6) by (7), which *is* true:

(7) It is not the case that if grass is green, grass is red,

Then (8) follows:

(8) If grass is green, then grass is not red.

Wansing's semantics avoids counter-examples such as (5)–(6) by violating the law of bivalence and allowing propositions like (5) and (6) which are neither true nor false. Intuitively, however, (5) would be considered to be true, and (6) false.

Shahid Rahman and Helge Rückert published a challenging and highly original paper entitled "Dialogical connexive logic" [2001], in which they develop an idea suggested by Paul Lorenzen in 1958. Dialogical logic was originally introduced as a "pragmatic semantics" for classical and intuitionist logic, and has been extended to connexive logic by Rahman and Rückert. In essence, dialogical logic is a kind of two-person game in which the "proponent" of a logical thesis proposes the thesis, and the "opponent" attacks it. X wins if it is Y's turn but he cannot move (whether to attack or defend).

This sounds simple, but in practice the rules of dialogical logic, in particular connexive dialogical logic, are so complicated that it is doubtful any logician would have sufficient understanding of them to be able to play a game after reading Rahman's and Rückert's paper. On pages 122–32 the authors extend their ideas to what they call "dialogical tableaux", which are semantic tableaux derived from dialogical logic. But the rules for constructing these tableaux are also exceedingly complex. Although connexive dialogical logic is an interesting idea, it seems unlikely that in the foreseeable future it will achieve the status of other forms of connexive logic.

BIBLIOGRAPHY

[Anderson and Belnap, 1962] A. R. Anderson and N. D. Belnap. The pure calculus of entailment, *The Journal of Symbolic Logic* (JSL) 27, 19-52, 1962.

[Anderson and Belnap, 1975] A. R. Anderson and N. D. Belnap. *Entailment. The Logic of Relevance and Necessity*, 1975.

[Angell, 1962] R. B. Angell. A propositional logic with subjunctive conditionals, *JSL* 27, 327-43, 1962.

[Aqvist, 1973] L. Aqvist. Modal logic with subjunctive conditionals and dispositional predicates. *Journal of Philosophical Logic*, 2, 1–76, 1973.

[Barnes, 1981] J. Barnes. Boethius and the study of logic. In M. Gibson (ed.) *Boethius*, pp. 73-89, 1981.
[Bochenski, 1961] I. M. Bochenski. *A History of Formal Logic*, 1961.
[Burks, 1951] A. W. Burks. The logic of causal propositions, *Mind* 60, 363-82, 1951.
[Burks and Copi, 1950] A. W. Burks and I. M. Copi. Lewis Carroll's Barber Shop Paradox, *Mind* 59, 219-22, 1950.
[Carroll, 1894] L. Carroll. A logical paradox, *Mind* 3, 436-8, 1894.
[Constance-Jones, 1905] E. E. Constance-Jones. *Mind* 14, 146-48 and 576-78, 1905.
[Dürr, 1951] K. Dürr. *The Propositional Logic of Boethius*, 1951.
[Goodman, 1947] N. Goodman. The problem of counterfactual conditionals, *The Journal of Philosophy* 94, 113-128. Reprinted in Goodman (1955), pp. 3-27, 1947.
[Goodman, 1955] N. Goodman. *Fact, Fiction, and Forecast*, 1955.
[Jeffrey, 1967] R. C. Jeffrey. *Formal Logic: Its Scope and Limits*, McGraw Hill, 1967.
[Johnson, 1894/1895] W. E. Johnson. *Mind* 3, 583; and *Mind* 4, 143-4, 1894/1895.
[Johnson, 1921] W. E. Johnson. *Logic*, part 1, 1921.
[Kneale and Kneale, 1962] W. Kneale and M. Kneale. *The Development of Logic*, 1962.
[Lewis, 1918] C. I. Lewis. *A Survey of Symbolic Logic*, 1918.
[Lewis, 1973] D. Lewis. *Counterfactuals*, Blackwells, 1973.
[Lowe, 1984] E. J. Lowe. Wright *versus* Lewis on the transitivity of conditionals, *Analysis* 44, 180-183, 1984.
[Lowe, 1985] E. J. Lowe. Reply to Wright on conditionals and transitivity, *Analysis* 45, 200-202, 1985.
[Lowe, 1990] E. J. Lowe. Conditionals, context and transitivity, *Analysis* 50, 80-87, 1990.
[Lowe, 1995] E. J. Lowe. The truth about counterfactuals, *The Philosophical Quarterly* 45, 41-59, 1995.
[Łukasiewicz, 1957] J. Łukasiewicz. *Aristotle's Syllogistic from the Standpoint of Modern Formal Logic*, 1957.
[Łukasiewicz, 1967] J. Łukasiewicz. On the history of the logic of propositions. In S. McCall (ed) *Polish Logic 1920-1939*, pp. 66-87, 1967.
[MacColl, 1878] H. MacColl. The calculus of equivalent statements (II), *Proceedings of the London Mathematical Society* 1877-78, pp. 177-86, 1878.
[MacColl, 1905a] H. MacColl. Symbolic reasoning (VI), *Mind* 14, pp. 74-81, 1905.
[MacColl, 1905b] H. MacColl. The existential import of propositions, *Mind* 14, pp. 401-402, 1905.
[MacColl, 1906] H. MacColl. *Symbolic Logic and its Applications*, 1906.
[McCall, 1966] S. McCall. Connexive implication, *JSL* 31, 415-433, 1966.
[McCall, 1967a] S. McCall. Connexive implication and the syllogism, *Mind* 76, 346-56, 1967.
[McCall, 1967b] S. McCall. Connexive class logic, *JSL* 32, 83-90, 1967.
[McCall, 1975] S. McCall. Connexive implication, in Anderson and Belnap (1975), pp. 434-452, 1975.
[McKinsey, 1950] J. C. C. McKinsey. Review of Burks and Copi (1950), *JSL* 15 (1950), 222-23.
[Marenbon, 2003] J. Marenbon. *Boethius*, 2003.
[Martin, 1987] C. J. Martin. Embarrassing arguments and surprising conclusions in the development of theories of the conditional in the twelfth century. In Jolivet and de Libera (eds), *Gilbert de Poitiers et ses Contemporains*, pp. 377-400, 1987.
[Martin, 1991] C. J. Martin. The logic of negation in Boethius, *Phronesis* 36, 277-304, 1991.
[Martin, 2004] C. J. Martin. Logic. In Brower and Guilfoy (eds) *The Cambridge Companion to Abelard*, pp. 158-99, 2004.
[Meyer, 1977] R. K. Meyer. S5—The Poor Man's Connexive Implication, *The Relevance Logic Newsletter*, 1977.
[Mortensen, 1984] C. Mortensen. Aristotle's thesis in consistent and inconsistent logics, *Studia Logica*, 43, 107—116, 1984.
[Nasti de Vincentis, 2006] M. Nasti de Vincentis. Conflict and connectedness: Between modern logic and history of ancient logic. In Mangione, Ballo and Franchella (eds) *Logic and Philosophy in Italy*, pp. 229–51, 2006.
[Nelson, 1930] E. J. Nelson. Intensional relations, *Mind* 39, 440-53, 1930.
[Pizzi, 1977] C. Pizzi. Boethius' Thesis and conditional logic, *Journal of Philosophical Logic* 6, 283-302, 1977.
[Pizzi, 1991] C. Pizzi. Decision procedures for logics of consequential implication, *Notre Dame Journal of Formal Logic* 32, 618-36, 1991.
[Pizzi, 1993] C. Pizzi. Consequential Implication: A Correction, *Notre Dame Jouranl of Formal Logic*, 34, 621-4, 1993.
[Pizzi, 1996] C. Pizzi. Weak vs. strong Boethius' Thesis: A problem in the analysis of consequential implication. In Ursini and Agliano (eds) *Logic and Algebra*, pp. 647-54, 1996.
[Pizzi and Williamson, 1997] C. Pizzi and T. Williamson. Strong Boethius' Thesis and consequential implication, *Journal of Philosophical Logic* 26, 569-88, 1997.

[Prantl, 1855] C. Prantl. *Geschichte der Logik im Abendlande*, vol 1, 1855.
[Priest, 1999] G. Priest. Negation as cancellation, and connexive logic, *Topoi* 18, 141-48, 1999.
[Prior, 1962] A. N. Prior. *Formal Logic*, 2nd edition, 1962.
[Rahman and Rückert, 2001] S. Rahman and H. Rückert. Dialogical connexive logic, *Synthese* 127, 105-39, 2001.
[Ramsey, 1931] F. P. Ramsey. General propositions and causality. In *The Foundations of Mathematics*, 1931.
[Routley, 1978] R. Routley. Semantics for connexive logics. I, *Studia Logica* 37, 393-412, 1978.
[Routley and Montgomery, 1968] R. Routley and H. Montgomery. On systems containing Aristotle's thesis. *Journal of Symbolic Logic*, 33, 82–96, 1968.
[Russell, 1903] B. Russell. *The Principles of Mathematics*, 1903.
[Russell, 1905] B. Russell. The existential import of propositions, *Mind* 14, 398-401, 1905.
[Russell, 1906] B. Russell. Review of MacColl (1906) in *Mind* 15, 255-260, 1906.
[Russell, 1919] B. Russell. *Introduction to Mathematical Philosophy*, George Allen and Unwin, 1919.
[Sidgwick, 1894/1895] A. Sidgwick. *Mind* 3, 582; and *Mind* 4, 144, 1894/1895.
[Stalnaker, 1968] R. C. Stalnaker. A theory of conditionals. In N. Rescher (ed.) *Studies in Logical Theory*, 1968. Reprinted in E. Sosa (ed.) *Causation and Conditionals* (1975), 165-79.
[Thomas, 1954] I. Thomas. Maxims in Kilwardby, *Dominican Studies* 7, 129-46, 1954.
[Wansing, 2005] H. Wansing. Connexive modal logic. In *Advances in Modal Logic* 5, 367-83, 2005.
[Wansing, 2006] H. Wansing. Connexive logic. In the *Stanford Encyclopedia of Philosophy*, 2006.
[Wright, 1984] C. Wright. Comment on Lowe, *Analysis* 44, 183-5, 1984.

A HISTORY OF TYPES*

Fairouz Kammareddine, Twan Laan, and Rob Nederpelt

In this article, we study the prehistory of type theory up to 1910 and its development between Russell and Whitehead's *Principia Mathematica* ([Whitehead and Russell, 1910], 1910–1912) and Church's simply typed λ-calculus of 1940. We first argue that the concept of types has always been present in mathematics, though nobody was incorporating them explicitly as such, before the end of the 19th century. Then we proceed by describing how the logical paradoxes entered the formal systems of Frege, Cantor and Peano concentrating on Frege's *Grundgesetze der Arithmetik* for which Russell applied his famous paradox[1] and this led him to introduce the first theory of types, the Ramified Type Theory (RTT). We present RTT formally using the modern notation for type theory and we discuss how Ramsey, Hilbert and Ackermann removed the orders from RTT leading to the simple theory of types STT. We present STT and Church's own simply typed λ-calculus ($\lambda\!\rightarrow_C$[2]) and we finish by comparing RTT, STT and $\lambda\!\rightarrow_C$.

1 INTRODUCTION

Nowadays, type theory has many applications and is used in many different disciplines. Even within logic and mathematics, there are many different type systems. They serve several purposes, and are formulated in various ways. But, before 1903 when Russell first introduced a type theory (see appendix of [Russell, 1903]), there were no formulations of any type theory. It is only since the second half of the twentieth century that we see explosions of type theories. In this article, we follow the evolution of type theory up to 1940. We give a historical account as to why formulations of type theory came into being and we describe the very first formulation of type theory: Russell and Whitehead's ramified theory of types (based on Russell's work [Russell, 1908] which succeeded [Russell, 1903]). In addition, we describe the simple theory of types that resulted from simplifications made by Hilbert and Ackermann and Ramsey to the ramified theory of types. Finally, we give the simply typed λ-calculus based on Church's seminal paper of 1940. It is important to

*This article originally appeared in the *Bulletin for Symbolic Logic* Volume 8, Number 2, 185–245, June 2002 under the title "Types in logic and mathematics before 1940".

[1]Russell discovered his paradox when he read Cantor's work.

[2]We write $\lambda\!\rightarrow_C$ for the original calculus of Church as presented in [Church, 1940]. Note that this is different from the calculus $\lambda\!\rightarrow$ used in frameworks like the Barendregt cube and the pure type systems found in [Barendregt, 1992].

stress that we do not give an extensive history of the subject.[3] Other developments deserve attention and we refer the reader especially to Ivor Grattan-Guinness's book [Grattan-Guinness, 2001] for an excellent historical account of many of the concepts discussed in this paper. We also refer the reader to [Cocchiarella, 1984; Cocchiarella, 1986] and [Landini, 1998].

Following the historical line from Frege (1879) and urged by the threat of the paradoxes, Russell and Whitehead developed the ramified theory of types [Russell, 1903; Russell, 1908; Whitehead and Russell, 1910] which was later simplified (or deramified) by Ramsey [Ramsey, 1926] and Hilbert and Ackermann [Hilbert and Ackerman, 1928] into the simple theory of types. The simple theory of types existed before the λ-calculus was invented by Church in 1932. Nevertheless, nowadays, when one refers to simple type theory, one usually means Church's simply typed λ-calculus of 1940. It should be noted furthermore, that Russell's type structure was different from that of Church. The former was set-based with linear sequences of types. The latter was function-based.

Our article is not only intended to introduce the prehistory of type theory (up to 1910) and its development between *Principia Mathematica* ([Whitehead and Russell, 1910]) and Church's simply typed λ-calculus of 1940, but also we will present the Russell's ramified theory of types and the simple theory of types due to to Ramsey, in a modern setting.[4] The presentation of the ramified theory of types in a modern setting was already given in [Laan and Nederpelt, 1996] but there, neither the deramification nor the connection of the ramified type theory with that of Church's simple theory of types was given. Similarly, although these theories have already been described in a modern framework, the relation between the modern description and the original system has not always been made clear. This is particularly the case because the original systems are quite far from the modern framework with respect to notation, level of formality and/or purpose. We will describe the ramified and simple theories of types within the modern framework in such a way that:

- We respect the ideas and the philosophy underlying the original system;
- We meet contemporary requirements on formality and accuracy.

The explicit and formal use of types (and thus an early form of what is presently called "type theory") was originally intended to prevent the paradoxes that occurred in logic and mathematics at the end of the 19th and the beginning of

[3]Curry, in his work on combinatory logic, introduced before 1940 an influential notion of typing that is still used nowadays when one refers to typing à la Curry as opposed to typing à la Church. Similarly, Quine in [Quine, 1937] introduced his New Foundations which retained typing axioms, but abandoned the idea of representing types formally as Russell did. Quine's NF presupposes the very simple linear type theory with types 0 for individuals, 1 for sets of individuals, 2 for sets of sets of individuals, etc.

[4]Note that there was a large unsatisfaction with ramified types and the reducibility axiom and this led to various calls for deramification. We already mentioned Hilbert and Ackermann, there is also the work of Leon Chwistek amongst others. In this paper, we concentrate on the simple theory of types as envisaged by Ramsey.

the 20th century. But it was not the only method developed for this purpose. Another tool was the fine-tuning of Cantor's Set Theory [Cantor, 1895; Cantor, 1897] by Zermelo [Zermelo, 1908], and the iterative conception of set (see [Boolos, 1971]) that resulted from the foundation axiom of Zermelo-Fraenkel's set theory ZF. Although it was clear that in ZF, the foundation axiom does not help in avoiding the paradoxes, it was added as a technical refinement. The separation axiom which replaced the unrestricted *comprehension axioms* is the one responsible for avoiding the paradoxes. This axiom goes as follows [Feferman, 1984; Bar-Hillel *et al.*, 1973]:

(*Comprehension*)
For each open well-formed formula Φ, $\exists y \forall x[(x \in y) \Leftrightarrow \Phi(x)]$ where y is not free in $\Phi(x)$

This unrestricted comprehension leads to a paradox by taking $\Phi(x)$ to be the formula $\neg(x \in x)$:
$\exists y \forall x[(x \in y) \Leftrightarrow \neg(x \in x)] \Longrightarrow \exists y[(y \in y) \Leftrightarrow \neg(y \in y)]$.
Such a comprehension axiom assumes that each open well-formed expression determines a concept whose extension exists and is the set of all those elements which satisfy the concept. *Iterative sets* were proposed to avoid the paradox and came to being by altering not the language, but the axioms of the theory. The most straightforward such theory is ZF (Zermelo-Fraenkel) where the axioms are made to fit the limitation of size doctrine. As an example, the above comprehension principle is altered to the following:

(*Separation*)
For each open well formed formula $\Phi, \exists y \forall x[(x \in y) \Leftrightarrow (x \in z) \wedge \Phi]$ where y does not occur in Φ

It is this new axiom which is responsible for the elimination of the paradox: to prove the existence of $\{x : \neg(x \in x)\}$ we need a z big enough so that $\{x : \neg(x \in x)\}$ is included in z. But we cannot show the existence of such a z. More precisely the paradox is restricted in ZF as follows:
Take $\Phi(z)$ to be $\neg(z \in z)$, and take $y = \{x : (x \in z) \wedge \neg(x \in x)\}$

- If $(y \in y) \Longrightarrow (y \in z)$ and $\neg(y \in y)$ contradiction,
- If $\neg(y \in y) \Longrightarrow$
  - if $(y \in z) \Longrightarrow (y \in y)$ contradiction,
  - if $\neg(y \in z)$ then we are fine.

Note however that we still have the syntactical ability to consider whether a set belongs to itself or not, but we are not committed to any set actually belonging to itself. For example, if $\Phi(x)$ is $x \in x$ then $\exists z \forall x[(x \in z) \Leftrightarrow (x \in \alpha \wedge x \in x)]$; in this case, although $x \in x$ is well-formed, it is likely to be false in the intended interpretation for any value of x. In the middle period of the development of

ZF, it was felt that the following foundation axiom (which is independent of and consistent with all other axioms of ZF) has to be added:[5]

(FA) $(\exists x)(x \in a) \Rightarrow (\exists x \in a)(\forall y \in x)\neg(y \in a)$

As a corollary of (FA), we have that there is no set a which has itself as its only element, for if there was then take $x = a$ in the antecedent of (FA) above and you get $(\exists x \in a)(\forall y \in x)\neg(y \in a)$, which is absurd.

It is worth pointing out that although very different conceptually, both the simple theory of types and ZF (which includes (FA)), give rise to an iterative concept of set. That is, both require the elements of a set be present before a new set can be constructed [Boolos, 1971]. We cannot however stop our discussion of set theory here. Quine's stratification in NF [Quine, 1937], and ML [Quine, 1940], (two non-iterative set theories) are sufficiently type-like to merit some discussion. Quine restricted the axiom of comprehension, to obtain the following:

(SCP) $\exists x \forall y[(y \in x) \Leftrightarrow \Phi(y)]$ where x is not free in $\Phi(y)$ and $\Phi(y)$ is stratified.[6]

Quine's NF has attracted a lot of research. Specker [Specker, 1953] refuted the axiom of Choice in NF, Jensen [Jensen, 1969] established that NF with Urelements is consistent (even when augmented with Choice, Infinity and unrestricted mathematical induction). However, consistency of NF remains an open problem. Moreover, NF is weak for mathematical induction (for propositions not expressible in type theory). Also, NF is said to lack motivation because its axiom of comprehension is justified only on technical grounds and one's mental image of set theory does not lead to such an axiom. To overcome some of the difficulties, Quine replaced (SCP) by two axioms, one for class existence and one for elementhood. The rule of class existence provides for the existence of the classes of all elements satisfying any condition Φ, stratified or not. The rule of elementhood is such as to provide the elementhood of just those classes which exist for NF. Therefore, the two axioms of comprehension of ML:

(*Comprehension by a set*)

$\exists y \forall x(x \in y \Leftrightarrow \Phi(x))$, where x and y range over sets, $\Phi(x)$ is stratified with set variables only in which y does not occur free

(*Impredicative comprehension by a class*)

$\exists y \forall x(x \in y \Leftrightarrow \Phi(x))$, where x ranges over sets, $\Phi(x)$ is any formula in which y does not occur free

ML was liked both for the manipulative convenience we regain in it and the symmetrical universe it furnishes. The earlier version of the first edition of [Quine, 1940] was subject to the Burali-Forti paradox — The well ordered set Ω of all

[5]This changed in the 1980s when Peter Aczel introduced his non-well founded set theory which relied on the Anti-Foundation axiom.

[6]Assume a first-order theory where for each primitive predicate $F(x_1, \ldots, x_n)$, we have integer constants $F_1, \ldots, F_n$. A formula Φ in the language of that theory is said to be *stratified* if there is an integer-valued function σ with domain the set of variables appearing in ϕ with the property that in each atomic formula $F(x_{i_1}, \ldots, x_{i_n})$ which appears in ϕ, and each integer $1 \leq j \leq n$, we have $\sigma(x_{i_j}) - \sigma(x_{i_1}) = F_j - F_1$.

ordinals has an ordinal which is greater than any member of Ω and hence is greater than Ω. In the second edition of [Quine, 1940], Quine corrected the axiomatization of ML (following a suggestion of Hao Wang) so that it is demonstrably consistent if NF is consistent. This latter version does not face the Burali-Forti paradox.

The approach of type theory however, is completely different from the set-theoretical approach. First, in the type theoretical approach, it is the language that is altered in order to avoid the paradox, and not the axioms[7]. Moreover, since Church's λ-calculus being extended with simple types in 1940, type theory has continued to focus on the notion of *function* in logic and mathematics. Since 1940, functions have remained one of the main objects of study for type theorists.

The historical remarks in this article have been taken from various resources. The most important ones are [Beth, 1959; Curry, 1963; van Heijenoort, 1967; Kneebone, 1963; Peremans, 1994; van Rooij, 1986; Wilder, 1965].

In Section 2 we discuss the prehistory of type theory. We first argue that the concept of types has always been present in mathematics, though nobody was incorporating them explicitly as such, before the end of the 19th century (Section 2.1). We study the way in which types implicitly occurred in logic and mathematics before there was an explicit theory of types. Then, we proceed by describing how the logical paradoxes entered the formal systems of Frege, Cantor and Peano in Section 2.2. We pay special attention to the formalisation of logic that is made in Frege's *Begriffsschrift* [Frege, 1879] and *Grundgesetze der Arithmetik* [Frege, 1892; Frege, 1903], as in this system many basic ideas are presented that are later used in type theory. Moreover, the system of *Grundgesetze der Arithmetik* is the one for which Russell derives his famous paradox, and this paradox was the reason for Russell to introduce the first theory of types.

This first type theory is the subject of Section 3. Whitehead and Russell present their theory, the Ramified Type Theory (RTT), in an informal way. Several rough descriptions of this theory have been given in the literature (see for instance [Church, 1940; Church, 1976; Hilbert and Ackerman, 1928; Ramsey, 1926]) but

---

[7]The first two accounts of avoiding the paradox by restricting the language were due to Russell and Poincaré. They both disallowed impredicative specification: only predicative specification (as will be defined below) was to be permitted. Russell's own solution (in [Russell, 1908]) was to adopt the vicious circle principle which can be roughly stated as follows: "*No entity determined by a condition that refers to a certain totality should belong to this totality*". Poincaré (in [Poincaré, 1902]) took refuge in banning "*les définitions non prédicatives*" which were taken by him to be: *Definitions by a relation between the object to be defined and all individuals of a kind of which either the object itself to be defined is supposed to be a part or other things that cannot be themselves defined except by the object to be defined.* So both Russell and Poincaré required only predicative sets to be considered, where $A = \{x : \Phi(x)\}$ is predicative iff Φ contains no variable which can take A as a value. This helps because it is otherwise very easy to get a vicious circle fallacy if we let the arguments of a certain propositional function (or the elements of a set) presuppose the function (or the set) itself. Russell's and Poincaré's solution was to use predicative comprehension, instances of which start with individuals, then generate sets, then new sets and so on as in the following example: Take 0 at level 0, $\{0\}$ at level 1, $\{0, \{0\}\}$ at level 2, and so on. Russell's ramified theory of types in Principia Mathematica applied the vicious circle principle, assuming all the elements of the set before constructing it. This theory obviously overcomes the paradox for the sentence Φ denoting $\neg(y \in y)$ is not predicative.

we present a formalisation of RTT given in [Laan and Nederpelt, 1996] that is directly based on the presentation of RTT in Whitehead and Russell's *Principia Mathematica* ([Whitehead and Russell, 1910]). The construction of this formalisation is not a simple task. Whitehead and Russell do not present a clear syntax for their so-called *propositional functions* in [Whitehead and Russell, 1910], neither do they make a clear difference between syntax and semantics. We constantly explain/defend our formalisation by using actual text from *Principia Mathematica*. We explain how the formal definition of propositional function is faithful to the original ideas exposed in *Principia Mathematica* and how the formalisation of the notion of propositional function makes it possible to express the notion of substitution of *Principia Mathematica* in terms of λ-calculus.

In 1926, Ramsey [Ramsey, 1926] proposes an important simplification of RTT, the *simple* theory of types. This simple type theory has become the basis for many modern type systems, and for the simply typed λ-calculus of Church [Church, 1940]. The simplification consisted of the removal of one of the two hierarchies from the RTT. The hierarchy of types is maintained, while the hierarchy of orders is removed. In Section 4 we discuss this process of so-called *deramification*.[8]

In Section 5 we follow this process of deramification to present Hilbert and Ackermann's [Hilbert and Ackerman, 1928] and Ramsey's [Ramsey, 1926] simple theory of types STT. We also present the well-known simple theory of types of Church $\lambda{\rightarrow}_C$ [Church, 1940]. We compare RTT, STT and $\lambda{\rightarrow}_C$.

We conclude in Section 6.

2 PREHISTORY OF TYPES

In this section, we discuss the development of type theory before it was actually baptised. This may sound like a contradiction. But types have played an important (though not very apparent) role in mathematics even before the theory of types was explicitly introduced by Russell [Russell, 1908]. Moreover, knowledge of the development of logic and mathematics before 1908, and especially of the occurrence of the logical paradoxes at the turn of the 20th century, provides insight in the way in which Russell and others formulated their theories of types.

When the first formalisations of parts of mathematics and logic appeared, the types were left implicit. Cantor's Set Theory [Cantor, 1895; Cantor, 1897], Peano's formalisation of the theory of natural numbers in [Peano, 1889], and Frege's *Begriffsschrift* [Frege, 1879] and *Grundgesetze der Arithmetik* [Frege, 1892; Frege, 1903] did not have a formal type system. The type of an object is indicated by means of natural language ("Let a be a proposition") or is taken for granted. Types were informally present in the background of these theories, but a formal

[8]Note that though the orders do not occur in the mainstream of type theories, they still provide an important intuition for logicians and play an important role in "categorising" logical theories: first-order, second-order, higher-order. For a discussion of the use of orders in modern systems, and its relation to Russell's notion of orders, the reader is referred to [Kamareddine and Laan, 2001].

representation of the types was not incorporated: one could say that they were separated from logic and mathematics.

However, even without a formalisation of the notion of types, the introduction of formal language had considerable advantages in the description of mathematical notions. The formalisation made it easier to give a precise definition of important abstract concepts, like the concept of function. The precise formulation allowed for a generalisation of the notion of function to include not only functions that take numbers as an argument, and return a number, but also functions that can take and return other sorts of arguments (like propositions, but also functions). Unfortunately, this also allowed logical paradoxes to enter the formal theory, without the (informal) type mechanism being able to prevent that.

2.1 *Paradox threats*

The most fundamental idea behind type theory is being able to distinguish between different classes of objects (*types*).[9] Until the end of the 19th century it had hardly ever been necessary to make this ability explicit. The mathematical language itself was predominantly informal, and so was the use of classes of objects.

It is, however, difficult to argue that there were no types before Russell "invented" them in 1903. Already around 325 B.C., Euclid began his *Elements* (page 153 of [Euclid]) with the following primitive definitions:

1. A *point* is that which has no part;

2. A *line* is breadthless length.

From these two basic notions of "point" and "line", Euclid defined more complex notions, like the notion of "circle":

15. A *circle* is a plane figure contained by one line such that all the straight lines falling upon it from one point among those lying within the figure are equal to one another.

At first sight, these three observations are mere definitions. But these three pieces of text do not only *define* the notions of point, line and circle, they also show that Euclid *distinguished* between points, lines and circles. Throughout the *Elements*, Euclid always mentioned to which class an object belonged (the class of points, the class of lines, etc.). In doing so, he prevented undesired results, like the intersection of two points (instead of two lines).

Undesired results? Euclid himself would probably have said: *impossible* results. When talking of an intersection, intuition implicitly forced him to think about what we would nowadays call the *type* of the objects of which he wanted to construct the intersection. As the intersection of two points is not supported by intuition, he did not even *try* to undertake such a construction.

[9]Note that it is controversial whether types should be literally taken to be classes of objects. We do not use this correspondence literally.

Euclid's attitude to, and implicit use of type theory was maintained by the mathematicians and logicians of the next twenty-one centuries. From the 19th century on, mathematical systems became less intuitive, for several reasons:

1. The system itself was complex, or abstract. An example was the theory of convergence in real analysis;

2. The system is a formal system, for example, the formalisation of logic in Frege's *Begriffsschrift*;

3. (In the second half of the 20th century:) It is not a human being working with the system, but something with less intuition, in particular: a computer.

We will call these three situations *paradox threats*. In all these cases, there is not enough intuition to activate the (implicitly present) type theory to warn against an impossible situation. One proceeds to reason within the impossible situation and then obtains a result that may be wrong or paradoxical: an *undesired* situation. We mention examples related to the three situations above:

S 1. The controversial results on convergence of series in analysis obtained in the 17th and 18th century, due to lack of knowledge on what real numbers actually are;

S 2. The logical paradoxes that arose from self-application of functions. Self-application *may lead* to intuitively impossible situations, but this is easily forgotten when working in a formal system in which such self-application can be expressed. The result is undesirable: a logical paradox;[10]

S 3. An untyped computer program may receive instructions from a not too watchful user to add the number 3 to the string `four`. The computer, unaware of the fact that `four` is not a number, starts his calculation. It is not programmed to handle the calculation of $3 + \texttt{four}$. The result of this calculation is unpredictable. The computer may: a) give an answer that is clearly wrong; b) give no answer at all; c) give an answer that is not so clearly wrong (for example, 6); or d) accidentally give the right answer. Especially situation c) is highly undesirable.

The example S 2 is the main subject of the next section (Section 2.2).

[10]Note that there are logical systems in which self-application is both consistent and useful. Consider untyped λ-calculus or second-order polymorphic λ-calculus, not to mention systems related to Quine's "New Foundations". For a modern intuition, trained in systems that arose during or after the developments up to 1940, self-application is much more obviously problematic than it was then (though certainly it was problematic for Russell!). For example, self-application appears not to have been essentially problematic for Curry, who was developing combinatory logic during that period.

2.2 *Paradox threats in formal systems*

In the 19th century, the need for a more precise style in mathematics arose. Controversial results had appeared in analysis. Many of these controversies were solved by the work of Cauchy. For instance, he introduced a precise definition of convergence in his *Cours d'Analyse* [Cauchy, 1821]. Due to the more exact definition of real numbers given by Dedekind [Dedekind, 1872], the rules for reasoning with real numbers became even more precise.

In 1879, Frege published his *Begriffsschrift* [Frege, 1879], in which he presented the first formalisation of logic that was uncommonly precise for those days. Until then, it had been possible to make mathematical and logical concepts more clear by textual refinement in the natural language in which they were described. Frege was not satisfied with this:

> "...I found the inadequacy of language to be an obstacle; no matter how unwieldy the expressions I was ready to accept, I was less and less able, as the relations became more and more complex, to attain the precision that my purpose required."
>
> (*Begriffsschrift*, Preface)

Frege therefore presented a completely formal system, whose

> "first purpose is to provide us with the most reliable test of the validity of a chain of inferences and to point out every presupposition that tries to sneak in unnoticed, so that its origin can be investigated."
>
> (*Begriffsschrift*, Preface)

FUNCTIONS AND THEIR COURSE OF VALUES

The introduction of a very general definition of function was the key to the formalisation of logic. Frege defined what we will call the *Abstraction Principle*:

Abstraction Principle 1.

> "If in an expression, [...] a simple or a compound sign has one or more occurrences and if we regard that sign as replaceable in all or some of these occurrences by something else (but everywhere by the same thing), then we call the part that remains invariant in the expression a function, and the replaceable part the argument of the function."
>
> (Begriffsschrift, *Section 9*)

Up to this section in Begriffsschrift, Frege put no restrictions on what could play the role of an argument. An argument could be a number (as was the situation in analysis), but also a proposition, or a function. Similarly, the result of applying a function to an argument did not necessarily have to be a number. (In Section 11 of his Begriffsschrift, Frege makes restrictions as we will see below.) Functions of more than one argument were constructed by a method that is very close to the method presented by Schönfinkel [Schönfinkel, 1924]:

Abstraction Principle 2.

> "If, given a function, we think of a sign[11] that was hitherto regarded as not replaceable as being replaceable at some or all of its occurrences, then by adopting this conception we obtain a function that has a new argument in addition to those it had before."
>
> (Begriffsschrift, *Section 9*)

With this definition of function, two of the three possible paradox threats mentioned on p. 458 occurred:

1. The generalisation of the concept of function made the system more abstract and less intuitive. The fact that functions could have different types of arguments is at the basis of the Russell Paradox;
2. Frege introduced a *formal* system instead of the informal systems that were used up till then. Type theory, that would be helpful in distinguishing between the different types of arguments that a function might take, was left informal.

So, Frege had to proceed with caution. And so he did, at this stage. He remarked that

> "if the [...] letter [sign] occurs as a function sign, this circumstance [should] be taken into account."
>
> (*Begriffsschrift*, Section 11)

This could be interpreted as if Frege was aware of some typing rule that does not allow to substitute functions for object variables or objects for function variables. In his paper *Function and Concept* [Frege, 1891], Frege more explicitly stated:

> "Now just as functions are fundamentally different from objects, so also functions whose arguments are and must be functions are fundamentally different from functions whose arguments are objects and cannot be anything else. I call the latter first-level, the former second-level."
>
> (*Function and Concept*, pp. 26–27)

A few pages later he proceeded:

> "In regard to second-level functions with one argument, we must make a distinction, according as the role of this argument can be played by a function of one or of two arguments."
>
> (*Function and Concept*, p. 29)

Therefore, we may safely conclude that Frege avoided the two paradox threats in the *Begriffsschrift*. In *Function and Concept* we even see that he was aware of the fact that making a difference between first-level and second-level objects is essential in preventing certain paradoxes:

[11]We can now regard a sign that previously was considered replaceable as replaceable also in those places in which up to this point it was considered fixed. [footnote by Frege]

> "The ontological proof of God's existence suffers from the fallacy of treating existence as a first-level concept."
>
> (*Function and Concept*, p. 27, footnote)

The *Begriffsschrift*, however, was only a prelude to Frege's writings. In *Grundlagen der Arithmetik* [Frege, 1884] he argued that mathematics can be seen as a branch of logic. In *Grundgesetze der Arithmetik* [Frege, 1892; Frege, 1903] he actually described the elementary parts of arithmetics within an extension of the logical framework that was presented in the *Begriffsschrift*.

Frege approached the paradox threats for a second time at the end of Section 2 of his *Grundgesetze*. There he defined the expression "the function $\Phi(x)$ has the same course-of-values as the function $\Psi(x)$" by

> "the functions $\Phi(x)$ and $\Psi(x)$ always have the same value for the same argument."
>
> (*Grundgesetze*, p. 7)

Note that functions $\Phi(x)$ and $\Psi(x)$ may have equal courses-of-values even if they have different definitions. For instance, let $\Phi(x)$ be $x \wedge \neg x$, and $\Psi(x)$ be $x \leftrightarrow \neg x$, for all propositions x. Then $\Phi(x) = \Psi(x)$ for all x. So $\Phi(x)$ and $\Psi(x)$ are different functions, but have the same course-of-values.

Frege denoted the course-of-values of a function $\Phi(x)$ by $\grave{\varepsilon}\Phi(\varepsilon)$.[12] The definition of equal courses-of-values could therefore be expressed as

(1) $\grave{\varepsilon}f(\varepsilon) = \grave{\varepsilon}g(\varepsilon) \longleftrightarrow \forall a[f(a) = g(a)]$.

In modern terminology, we could say that the functions $\Phi(x)$ and $\Psi(x)$ have the same course-of-values if they have the same *graph*.

Frege did not provide a satisfying intuition for the formal notion of course-of-values of a function. He treated courses-of-values as ordinary objects. As a consequence, a function that takes objects as arguments could have its own course-of-values as an argument. In modern terminology: a function that takes objects as arguments can have its own graph as an argument. All essential information of a function is contained in its graph. So intuitively, a system in which a function can be applied to its own graph should have similar possibilities as a system in which a function can be applied to itself. Frege excluded the paradox threats from his system by forbidding self-application, but due to his treatment of courses-of-values these threats were able to enter his system through a back door.

[12]This may be the origin of Russell's notation $\hat{x}\Phi(x)$ for the class of objects that have the property Φ. According to a paper by J. B. Rosser [Rosser, 1984], the notation $\hat{x}\Phi(x)$ has been at the basis of the current notation $\lambda x.\Phi$ in λ-calculus. Church is supposed to have written $\wedge x\Phi(x)$ for the function $x \mapsto \Phi(x)$, writing the hat in front of the x in order to distinguish this function from the class $\hat{x}\Phi(x)$. For typographical reasons, the $\wedge$ is supposed to have changed into a λ. On the other hand, J. P. Seldin informed us [Seldin, 1996] that he had asked Church about it in 1982, and that Church had answered that there was no particular reason for choosing λ, that some letter was needed and λ happened to have been chosen. Moreover, Curry had told him that Church had a manuscript in which there were many occurrences of λ already in 1929, so three years before the paper [Church, 1932] appeared.

THE RUSSELL PARADOX IN THE *Grundgesetze*

In 1902, Russell wrote a letter to Frege [Russell, 1902], in which he informed Frege that he had discovered a paradox in Frege's *Begriffsschrift*. Russell gave his well-known argument, defining the propositional function $f(x)$ by $\neg x(x)$ (in Russell's words: "to be a predicate that cannot be predicated of itself"). He assumed $f(f)$. Then by definition of f, $\neg f(f)$, a contradiction. Therefore: $\neg f(f)$ holds. But then (again by definition of f), $f(f)$ holds. Russell concluded that both $f(f)$ and $\neg f(f)$ hold, a contradiction.

Only six days later, Frege answered Russell that Russell's derivation of the paradox was incorrect [Frege, 1902]. He explained that the self-application $f(f)$ is not possible in the *Begriffsschrift*. $f(x)$ is a function, which requires an *object* as an argument, and a function cannot be an object in the *Begriffsschrift* (see above).

In the same letter, however, Frege explained that Russell's argument could be amended to a paradox in the system of his *Grundgesetze*, using the course-of-values of functions. Frege's amendment was shortly explained in that letter, but he added an appendix of eleven pages to the second volume of his *Grundgesetze* in which he provided a very detailed and correct description of the paradox.

The derivation goes as follows (using the same argument as Frege, though replacing Frege's two-dimensional notation by the nowadays more usual one-dimensional notation). First, define the function $f(x)$ by: $\neg\forall\varphi[(\grave{\alpha}\varphi(\alpha) = x) \longrightarrow \varphi(x)]$ and write $K = \grave{\varepsilon}f(\varepsilon)$. By (1) we have, for any function $g(x)$, that $\grave{\varepsilon}g(\varepsilon) = \grave{\varepsilon}f(\varepsilon) \longrightarrow g(K) = f(K)$ and this implies

(2) $f(K) \longrightarrow ((\grave{\varepsilon}g(\varepsilon) = K) \longrightarrow g(K))$.

As this holds for any function $g(x)$, we have

(3) $f(K) \longrightarrow \forall\varphi[(\grave{\varepsilon}\varphi(\varepsilon) = K) \rightarrow \varphi(K)]$.

On the other hand, for any function g,

$$\forall\varphi[(\grave{\varepsilon}\varphi(\varepsilon) = K) \rightarrow \varphi(K)] \longrightarrow ((\grave{\varepsilon}g(\varepsilon) = K) \rightarrow g(K)).$$

Substituting $f(x)$ for $g(x)$ results in:

$$\forall\varphi[(\grave{\varepsilon}\varphi(\varepsilon) = K) \rightarrow \varphi(K)] \longrightarrow ((\grave{\varepsilon}f(\varepsilon) = K) \rightarrow f(K))$$

and as $\grave{\varepsilon}f(\varepsilon) = K$ by definition of K,

$$\forall\varphi[(\grave{\varepsilon}\varphi(\varepsilon) = K) \rightarrow \varphi(K)] \longrightarrow f(K).$$

Using the definition of f, we obtain

$$\forall\varphi[(\grave{\varepsilon}\varphi(\varepsilon) = K) \rightarrow \varphi(K)] \longrightarrow \neg\forall\varphi[(\grave{\varepsilon}\varphi(\varepsilon) = K) \rightarrow \varphi(K)],$$

hence by reductio ad absurdum, $\neg\forall\varphi[(\grave{\alpha}\varphi(\alpha) = K) \rightarrow \varphi(K)]$, or shorthand:

(4) $f(K)$.

Applying (3) results in $\forall\varphi[(\grave{\alpha}\varphi(\alpha) = K) \rightarrow \varphi(K)]$,
which implies $\neg\neg\forall\varphi[(\grave{\alpha}\varphi(\alpha) = K) \rightarrow \varphi(K)]$, or shorthand:

(5) $\neg f(K)$.

(4) and (5) contradict each other.

HOW WRONG WAS FREGE?

In the history of the Russell Paradox, Frege is often depicted as the pitiful person whose system was inconsistent. This suggests that Frege's system was the only one that was inconsistent, and that Frege was very inaccurate in his writings. On these points, history does Frege an injustice.

In fact, Frege's system was much more accurate than other systems of those days. Peano's work, for instance, was less precise on several points:

- Peano hardly paid any attention to logic, especially not to quantification theory;
- Peano did not make a strict distinction between his symbolism and the objects underlying this symbolism. Frege was much more accurate on this point (see also his paper *Über Sinn und Bedeutung* [Frege, 1892a]);
- Frege made a strict distinction between a proposition (as an object of interest or discussion) and the *assertion* of a proposition. Frege denoted a proposition, in general, by $-A$, and the assertion of the proposition by $\vdash A$. The symbol $\vdash$ is still widely used in logic and type theory. Peano did not make this distinction and simply wrote A.

Nevertheless, Peano's work was very popular, for several reasons:

- Peano had able collaborators, and in general had a better eye for presentation and publicity. For instance, he bought his own press, so that he could supervise the printing of his journal *Rivista di Matematica* and *Formulaire* [Peano, 1894]
- Peano used a symbolism much more familiar to the notations that were used in those days by mathematicians (and many of his notations, like $\in$ for "is an element of", and $\supset$ for logical implication, are also used in Russell's *Principia Mathematica*, and are actually still in use).

Frege's work did not have these advantages and was hardly read before 1902[13]. In the last paragraph of [Frege, 1896], Frege concluded:

[13] When Peano published his formalisation of mathematics [Peano, 1889] he clearly did not know Frege's *Begriffsschrift*, as he did not mention the work, and was not aware of Frege's formalisation of quantification theory. Peano considered quantification theory to be "abstruse" in [Peano, 1894], on which Frege proudly reacted:

> "In this respect my conceptual notion of 1879 is superior to the Peano one. Already, at that time, I specified all the laws necessary for my designation of generality, so that nothing fundamental remains to be examined. These laws are few in number,

> ...I observe merely that the Peano notation is unquestionably more convenient for the typesetter, and in many cases takes up less room than mine, but that these advantages seem to me, due to the inferior perspicuity and logical defectiveness, to have been paid for too dearly — at any rate for the purposes I want to pursue.
> *Ueber die Begriffschrift des Herrn Peano und meine eigene*, p. 378

Frege's system was not the only paradoxical one. The Russell Paradox can be derived in Peano's system as well, by defining the class $K \stackrel{\text{def}}{=} \{x \mid x \notin x\}$ and deriving $K \in K \longleftrightarrow K \notin K$.

THE IMPORTANCE OF RUSSELL'S PARADOX

Russell's paradox was certainly not the first or only paradox in history. Paradoxes were already widely known in antiquity. The first known paradox is the Achilles paradox of Zeno of Elea. It is a purely mathematical paradox. Due to a precise formulation of mathematics and especially the concept of real numbers, the paradox can now be satisfactorily solved.

The oldest logical paradox is probably the Liar's Paradox, also known as the Paradox of Epimenides. It can be very shortly formulated by the sentence "This sentence is not true". The paradox was widely known in antiquity. For instance, it is referred to in the Bible (Titus 1:12). It is based on the confusion between language and meta-language.

The Burali-Forti paradox ([Burali-Forti, 1897], 1897) is the first of the modern paradoxes. It is a paradox within Cantor's theory on ordinal numbers.[14] Cantor's paradox on the largest cardinal number occurs in the same field. It must have been discovered by Cantor around 1895, but was not published before 1932.

The logicians considered these paradoxes to be out of the scope of logic: the paradoxes based on the Liar's Paradox could be regarded as a problem of linguistics, and the paradoxes of Cantor and Burali-Forti occurred in what was considered in those days a highly questionable part of mathematics: Cantor's Set Theory.

The Russell Paradox, however, was a paradox that could be formulated in all the systems that were presented at the end of the 19th century (except for Frege's *Begriffsschrift*). It was at the very basics of logic. It could not be disregarded, and a solution to it had to be found. Russell's solution to the paradoxes was to use type theory.

> and I do not know why they should be said to be abstruse. If it is otherwise with the Peano conceptual notation, then this is due to the unsuitable notation."
> ([Frege, 1896], p. 376)

[14]It is instructive to note that Cantor was aware of the Burali-Forti paradox and did not think that it rendered his system incoherent.

3 TYPE THEORY IN PRINCIPIA MATHEMATICA

When Russell proved Frege's *Grundgesetze* to be inconsistent, Frege was not the only person in trouble. In Russell's letter to Frege (1902), we read:

> "I am on the point of finishing a book on the principles of mathematics"
> (*Letter to Frege*, [Russell, 1902])

Therefore, Russell had to find a solution to the paradoxes, before he could finish his book.

His paper *Mathematical logic as based on the theory of types* [Russell, 1908], in which a first step is made towards the Ramified Theory of Types, started with a description of the most important contradictions that were known up till then, including Russell's own paradox. He then concluded:

> "In all the above contradictions there is a common characteristic, which we may describe as self-reference or reflexiveness. [...] In each contradiction something is said about *all* cases of some kind, and from what is said a new case seems to be generated, which both is and is not of the same kind as the cases of which *all* were concerned in what was said."
>
> (*Ibid.*)

Russell's plan was, therefore, to avoid the paradoxes by avoiding all possible self-references. He postulated the "vicious circle principle":

Vicious Circle Principle 3.

> *"Whatever involves all of a collection must not be one of the collection."*
> *([Laan, 1997], p. 465)*

Russell applies this principle very strictly. He implemented it using types, in particular the so-called *ramified* types. The theory presented in *Mathematical logic as based on the theory of types* was elaborated in Chapter II of the Introduction to the famous *Principia Mathematica* [Whitehead and Russell, 1910]. In the *Principia*, Whitehead and Russell founded mathematics on logic, as far as possible. The result was a very formal and accurate build-up of mathematics, avoiding the logical paradoxes.

The logical part of the *Principia* was based on the works of Frege. This was acknowledged by Whitehead and Russell in the preface, and can also be seen throughout the description of Type Theory. The notion of function is based on Frege's Abstraction Principles 1 and 2, and the *Principia* notation $\hat{x}f(x)$ for a class looks very similar to Frege's $\grave{\varepsilon}f(\varepsilon)$ for course-of-values.

An important difference is that Whitehead and Russell treated functions as first-class citizens. Frege used courses-of-values as a way of speaking about functions (and was confronted with a paradox); in the *Principia* a direct approach was possible. Equality, for instance, was defined for objects as well as for functions by means of Leibniz equality ($x = y$ if and only if $f(x) \leftrightarrow f(y)$ for all propositional functions f — see [Whitehead and Russell, 1910], $*13{\cdot}11$).

The description of the Ramified Theory of Types (RTT) in the *Principia* was, though extensive, still informal. It is clear that Type Theory had not yet become an independent subject. The theory

> "only recommended itself to us in the first instance by its ability to solve certain contradictions"
>
> (*Principia Mathematica*, p. 37)

And though

> "it has also a certain consonance with common sense which makes it inherently credible"
>
> (*Principia Mathematica*, p. 37)

(probably, Whitehead and Russell refer to the implicit, intuitive use of types by mathematicians as explained in Section 2.1), Type Theory was not introduced because it was interesting on its own, but because it had to serve as a tool for logic and mathematics. A formalisation of Type Theory, therefore, was not considered in those days.

Though the description of the ramified type theory in the *Principia* was still informal, it was clearly present throughout the work. It was not mentioned very often, but when necessary, Russell made a remark on the ramified type theory. This is an important difference with the earlier writings of Frege, Peano and Cantor.

If we want to compare RTT with contemporary type systems, we have to make a formalisation of RTT. Though there are many descriptions of RTT available in the literature (like [Church, 1940; Church, 1976; Hilbert and Ackerman, 1928; Ramsey, 1926] and Section 27 of [Schütte, 1960]), none of these descriptions presents a formalisation that is both accurate and as close as possible to the ideas of the *Principia*. [Laan, 1997; Laan and Nederpelt, 1996] fill up this gap in the literature. In this section, we review the formalisation of [Laan, 1997; Laan and Nederpelt, 1996] explaining how it faithfully represents the intentions of Russell and Whitehead.

Formalisation the ramified type theory is by no means easy:

- Important formal notions, especially the notion of substitution, remained completely unexplained in the *Principia*;
- The accuracy of Frege's work was not present in Russell's. This was already observed by Gödel, who said that the precision of Frege was lost in the writings of Russell, and who, due to the informality of some basic notions of the *Principia*, had to give his paper [Gödel, 1931] the title *Über formal unentscheidbare Sätze der Principia Mathematica und* verwandter *Systeme*.

In Section 2.2 we saw that Frege generalised the notion of function from analysis. For Russell's formalisation of mathematics within logic, a special kind of these functions was needed: the so-called *propositional* functions. A propositional function (pf) always returns a *proposition* when it is applied to suitable arguments. In Section 3.1, we introduce a formalised version of these pfs. This makes it possible to compare pfs with other formal systems, like λ-calculus (which we do in Section 3.2), and to give a precise definition of substitution (Section 3.4).

In Section 3.5 we give a formalisation of Russell's notion of ramified type, followed by a formal definition of the notion *the pf f is of type t*. We motivate this definition by referring to passages in the *Principia*. As the formalisation of pf is precise enough to be translated to λ-calculus, we can make a comparison between RTT and current type systems.

Thanks to our formal notation and its relation to λ-calculus, we are able to prove properties of RTT in an easy way, using properties of modern type systems. This will be done in Section 3.8. Due to the new notation it is relatively easy to see that we have proved variants of well-known theorems from Type Theory, like Strong Normalisation, Free Variable Lemma, Strengthening Lemma, Unicity of Types and Subterm Lemma.

In Section 3.9 we answer in full detail the question which pfs are typable. We also make a comparison between our notion of typable pf, and the corresponding notion in the *Principia*, and conclude that these two notions of typable pf coincide.

3.1 Principia's propositional functions

In this section we present a formalisation of the propositional functions (pfs) of the *Principia* by introducing a syntax that is as close as possible to the ideas of the *Principia*. Intuition about this syntax is provided by translating pfs into λ-terms in Section 3.2. A special attention is devoted to the notion of substitution. This notion is clearly present in the *Principia*, but not formally defined. Due to our translation of pfs to λ-calculus, we are able to give a precise definition.

DEFINITION
The definition of propositional function in the *Principia* is as follows:

> "By a "propositional function" we mean something which contains a variable x, and expresses a *proposition* as soon as a value is assigned to x."
>
> (*Principia Mathematica*, p. 38)

Pfs are, however, *constructed* from propositions with the use of the Abstraction Principles: they arise when in a proposition one or more occurrences of a sign are replaced by a variable. Therefore we have to begin our formalisation with certain basic propositions, certain basic signs, and signs that indicate a replaceable object. For this purpose we use

- A set $\mathcal{A}$ of *individual symbols* (the basic signs);
- A set $\mathcal{V}$ of *variables* (the signs that indicate replaceable objects);

- A set $\mathcal{R}$ of *relation symbols* together with a map $\mathfrak{a} : \mathcal{R} \to \mathbb{N}^+$ indicating the *arity* of each relation-symbol (these are used to form the basic propositions).

We want to have a sufficient supply of individual symbols, variables and relation symbols and therefore assume that $\mathcal{A}$ and $\mathcal{V}$ are infinite (but countable), and that $\{R \in \mathcal{R} \mid \mathfrak{a}(R) = n\}$ is infinite (but countable) for each $n \in \mathbb{N}^+$. We assume that $\{\mathtt{a}_1, \mathtt{a}_2, \ldots\} \subseteq \mathcal{A}$, $\{\mathtt{x}, \mathtt{y}, \mathtt{z}, \mathtt{x}_1, \ldots\} \subseteq \mathcal{V}$ and $\{\mathtt{R}, \mathtt{S}, \ldots\} \subseteq \mathcal{R}$. We use $a_1, a_2, \ldots$ as metavariables over $\mathcal{A}$; $x, y, z, x_1, \ldots$ as metavariables over $\mathcal{V}$ and $R, S, \ldots$ as metavariables over $\mathcal{R}$. For technical reasons we assume that there is an *order* (e.g. alphabetical) on $\mathcal{V}$. We write $x < y$ if x is ordered before y, and not equal to y (so: $<$ is *strict*). In particular, we assume that $\mathtt{x} < \mathtt{x}_1 < \ldots \mathtt{y} < \mathtt{y}_1 < \ldots \mathtt{z} < \mathtt{z}_1 \ldots$ and: for each x there is a y with $x < y$.

Definition 4. (Atomic propositions) A list of symbols of the form: $R(a_1, \ldots, a_{\mathfrak{a}(R)})$ is called an *atomic proposition*.

Other names used for these atomic propositions in the *Principia* are *elementary judgements* and *elementary propositions* (cf. [Whitehead and Russell, 1910], pp. xv, 43–45, and 91).

Propositional functions in *Principia Mathematica* are generated from atomic propositions by two means:

- The use of logical connectives and quantifiers;
- Abstraction from (earlier generated) propositional functions, using the abstraction principles.

This leads to the following formal definition of propositional function.

Definition 5. (Propositional functions) We define a collection $\mathcal{P}$ of *propositional functions* (pfs), and for each element f of $\mathcal{P}$ we simultaneously define the collection $\textsc{fv}(f)$ of *free variables* of f:

1. If $i_1, \ldots, i_{\mathfrak{a}(R)} \in \mathcal{A} \cup \mathcal{V}$ then $R(i_1, \ldots, i_{\mathfrak{a}(R)}) \in \mathcal{P}$.
   $\textsc{fv}(R(i_1, \ldots, i_{\mathfrak{a}(R)})) \stackrel{\text{def}}{=} \{i_1, \ldots, i_{\mathfrak{a}(R)}\} \cap \mathcal{V}$;
2. If $f, g \in \mathcal{P}$ then $f \vee g \in \mathcal{P}$ and $\neg f \in \mathcal{P}$.
   $\textsc{fv}(f \vee g) \stackrel{\text{def}}{=} \textsc{fv}(f) \cup \textsc{fv}(g)$; $\textsc{fv}(\neg f) \stackrel{\text{def}}{=} \textsc{fv}(f)$;
3. If $f \in \mathcal{P}$ and $x \in \textsc{fv}(f)$ then $\forall x[f] \in \mathcal{P}$.
   $\textsc{fv}(\forall x[f]) \stackrel{\text{def}}{=} \textsc{fv}(f) \setminus \{x\}$;
4. If $n \in \mathbb{N}$ and $k_1, \ldots, k_n \in \mathcal{A} \cup \mathcal{V} \cup \mathcal{P}$, then $z(k_1, \ldots, k_n) \in \mathcal{P}$.
   $\textsc{fv}(z(k_1, \ldots, k_n)) \stackrel{\text{def}}{=} \{z, k_1, \ldots, k_n\} \cap \mathcal{V}$.
   If $n = 0$ then we write $z()$ in order to distinguish the pf $z()$ from the variable z[15];

[15]It is important to note that a variable is not a pf. See for instance [Russell, 1903, Chapter VIII: "The variable", p. 94 of the 7th impression.]

5. All pfs can be constructed by using the construction-rules 1, 2, 3 and 4 above.

We use the letters f, g, h as meta-variables over $\mathcal{P}$.

Note that in clause 4. of the above definition, the variable binding in pf arguments of terms $z(k_1, \ldots, k_n)$ may be quite unexpected. We explain this feature in detail in Section 3.3 and especially in Remark 12.

Definition 6. **(Propositions)** A propositional function f is a *proposition* if $\text{FV}(f) = \varnothing$.

Example 7. We give some examples of (higher-order) pfs of the form $z(k_1, \ldots, k_n)$ in ordinary mathematics. To keep the link with mathematics clear, we use some extra logical connectives like $\leftrightarrow$ and $\rightarrow$.

1. The pfs $\mathtt{z(x)}$ and $\mathtt{z(y)}$ in the definition of equality according to Leibniz: By definition $\mathtt{x} = \mathtt{y}$ if and only if $\forall\mathtt{z}[\mathtt{z(x)} \leftrightarrow \mathtt{z(y)}]$;

2. The pfs $\mathtt{z(0)}$, $\mathtt{z(x)}$ and $\mathtt{z(y)}$ in the formulation of the principle of mathematical induction:

   $\forall\mathtt{z}[\mathtt{z(0)} \rightarrow (\forall\mathtt{x}\forall\mathtt{y}[\mathtt{z(x)} \rightarrow (\mathtt{S(x,y)} \rightarrow \mathtt{z(y)})]) \rightarrow \forall\mathtt{x}[\mathtt{z(x)}]]$.

   (we suppose that the relation symbol $\mathtt{S}$ represents the successor function: $\mathtt{S}(x, y)$ holds if and only if y is the successor of x);

3. $\mathtt{z()}$ in the formulation of the law of the excluded middle: $\forall\mathtt{z}[\mathtt{z()} \vee \neg\mathtt{z()}]$.

3.2 *Principia's propositional functions as λ-terms*

The binding structure and the notion of free variable of pfs become more clear if we translate pfs to λ-terms. Moreover, such a translation will be useful at several places in this article, for instance when we give a definition of substitution.

We first translate one of the examples of Example 7. Then we give a formal definition of the translation that we have in mind. After that we provide additional remarks and intuition on pfs.

Example 8. Consider the pf $f \equiv \forall\mathtt{z}[\mathtt{z(x)} \leftrightarrow \mathtt{z(y)}]$ of Example 7.1. Two objects x and y are Leibniz-equal if and only if they share the same properties. These objects are represented by the variables $\mathtt{x}$ and $\mathtt{y}$. The variable $\mathtt{z}$ is a variable for properties of objects, in other words: predicates over objects. Such a predicate is a function that takes the object as argument, and returns a truth value. The expression $\mathtt{z(x)}$ indicates that the predicate that is taken for $\mathtt{z}$ must be applied to the object that is taken for $\mathtt{x}$. Therefore, we translate $\mathtt{z(x)}$ by an application of $\mathtt{z}$ to $\mathtt{x}$ in λ-calculus: $\mathtt{zx}$. Similarly we translate the expression $\mathtt{z(y)}$ by $\mathtt{zy}$.

Just as in [Church, 1940], we can interpret logical connectives as functions. Therefore we can translate $\mathtt{z(x)} \leftrightarrow \mathtt{z(y)}$ by the λ-term $\leftrightarrow(\mathtt{zx})(\mathtt{zy})$. We handle the translation of universal quantification also as in [Church, 1940], hence $\forall\mathtt{z}[\ldots]$ translates to $\forall(\lambda\mathtt{z}.\ldots)$. As an effect we get a λ-term $\forall(\lambda\mathtt{z}.\leftrightarrow(\mathtt{zx})(\mathtt{zy}))$

with two free variables, $\mathtt{x}$ and $\mathtt{y}$. But we want to have a *function* taking two *arguments*. This can be solved by a double λ-abstraction. The final result is $\lambda\mathtt{x}.\lambda\mathtt{y}.\forall(\lambda\mathtt{z}.(\leftrightarrow(\mathtt{zx})(\mathtt{zy})))$.

We remark that the pf f has two free variables, $\mathtt{x}$ and $\mathtt{y}$. These two free variables correspond to the two arguments that the propositional function takes, and therefore to the two λ-abstractions that are at the front of the translation of f.

In the following definition, we translate the propositional functions to λ-terms in a similar way as we did in Example 8. Let $f \in \mathcal{P}$ and let $x_1 < \cdots < x_m$ be the free variables of f. We define a λ-term $\overline{f}$. We do this in such a way that $\overline{f} \equiv \lambda x_1.\cdots\lambda x_m.F$, where F is a λ-term that is not of the form $\lambda x.F'$. To keep notations uniform, we also give translations $\overline{a}$ for $a \in \mathcal{A}$ and $\overline{x}$ for $x \in \mathcal{V}$. To keep notations short, we use $\lambda_{i=1}^{m} x_i.F$ as shorthand for $\lambda x_1.\cdots\lambda x_m.F$.

Definition 9. (Translating propositional functions to λ-terms)

- $\overline{a} \stackrel{\text{def}}{=} a$ for $a \in \mathcal{A}$;
- $\overline{x} \stackrel{\text{def}}{=} x$ for $x \in \mathcal{V}$;
- Now assume $f \in \mathcal{P}$ has free variables $x_1 < \cdots < x_m$. Use induction on the structure of f:
  - $f \equiv R(i_1, \ldots, i_{\mathfrak{a}(R)})$. Then $\overline{f} \stackrel{\text{def}}{=} \lambda_{i=1}^{m} x_i.Ri_1\cdots i_{\mathfrak{a}(R)}$;
  - $f \equiv f_1 \vee f_2$. We can assume that for $j = 1, 2$, $\overline{f_j} \equiv \lambda_{i=1}^{m_j} y_i^j.F_j$, where $y_1^j < \cdots < y_{m_j}^j$ are the free variables of f_j. Then $\overline{f} \stackrel{\text{def}}{=} \lambda_{i=1}^{m} x_i.\vee F_1F_2$;
  - $f \equiv \neg f'$. We can assume that $\overline{f'} \equiv \lambda_{i=1}^{m} x_i.F$, because $x_1 < \cdots < x_m$ are the free variables of f'. Let $\overline{f} \stackrel{\text{def}}{=} \lambda_{i=1}^{m} x_i.\neg F$;
  - $f \equiv z(k_1, \ldots, k_n)$. Let $\overline{f} \equiv \lambda_{i=1}^{m} x_i.z\overline{k_1}\cdots\overline{k_n}$;
  - $f \equiv \forall x[f']$. We can assume that $\overline{f'} \equiv \lambda_{i=1}^{j-1} x_i.\lambda x.\lambda_{i=j}^{m} x_i.F$, because $x_1, \ldots, x_m, x$ are the free variables of f'. Define $\overline{f} \equiv \lambda_{i=1}^{m} x_i.\forall(\lambda x.F)$.

Example 10.

f	$\overline{f}$
$\mathtt{R(x)}$	$\lambda\mathtt{x.Rx}$
$\mathtt{z(R(x),S(a))}$	$\lambda\mathtt{z.z}(\lambda\mathtt{x.Rx})(\mathtt{Sa})$
$\mathtt{z_1(a)} \vee \mathtt{z_2()}$	$\lambda\mathtt{z_1}.\lambda\mathtt{z_2}.\vee(\mathtt{z_1a})\mathtt{z_2}$
$\mathtt{z(y(R(x)))}$	$\lambda\mathtt{z.z}(\lambda\mathtt{y.y}(\lambda\mathtt{x.Rx}))$
$\forall\mathtt{x}[\mathtt{R(x)}]$	$\forall(\lambda\mathtt{x.Rx})$

Lemma 11. (Properties of $^{-}$) Let $f \in \mathcal{P}$.

1. $\text{FV}(\overline{f}) = \varnothing$;

2. $\overline{f}$ is in β-normal form;

3. $\overline{f}$ is a λI-term;

4. If $x_1 < \cdots < x_m$ are the free variables of f, then $\overline{f} \equiv \lambda_{i=1}^{m} x_i.F$, where F is not of the form $\lambda x.F'$.

PROOF: By induction on the structure of f. ⊠

Observe that we use FV for indicating both the free variables of a pf and the free variables of a λ-term. We take care that it will always be clear in which meaning we use FV. In the above definition we also assume familiarity with the notion of λI-term (see [Barendregt, 1984]): in λI, terms of the form $\lambda x.F$ are only allowed if x appears as a free variable in F.

3.3 Remarks on Principia's pfs and their translation in λ-calculus

Remark 12. We show that the propositional functions of Definition 5 are indeed objects that exist in the theory of Russell.

1. In Rule 1 we describe the atomic propositions, and the atomic propositions in which one or more individuals have been replaced by variables due to one or more applications of the abstraction principles. The abstraction principles are not only present in the works of Frege, but also in the *Principia* (cf. for instance ∗9·14 and ∗9·15);

2. Rule 2 describes the use of the logical connectives $\vee$ and $\neg$. These logical connectives are also used in the *Principia*. Implication[16], conjunction[17] and logical equivalence[18] are defined in terms of negation and disjunction. In examples, we sometimes use symbols for implication, conjunction and logical equivalence as abbreviations;

3. Rule 3 describes the use of the universal quantifier. It is explicitly stated in the *Principia* (cf. pp. 14–16) that the pf $\forall x[f]$ can only be constructed if f is a pf that contains x as a variable. Existential quantification[19] is defined in terms of negation and universal quantification;[20]

[16] cf. *Principia*, ∗1·01, p. 94
[17] cf. *Principia*, ∗3·01, p. 109
[18] cf. *Principia*, ∗4·01, p. 117
[19] cf. *Principia*, ∗10·01, p. 140
[20] In the original system of *Principia*, existential quantification is not defined in terms of negation and universal quantification, but negation of universally quantified statements is defined in terms of existential quantification. But in *Principia*, ∗10·01, p. 140, we see that Russell does present the definition of the existential quantifier as an alternative approach.

4. Rule 4 is also an instantiation of the abstraction principle. The pfs that can be constructed by using the construction-rules 1–3 only are exactly the pfs of what in these days would be called first-order predicate logic. With rule 4, higher-order pfs can be constructed. This is based on the following idea. Let f be a (fixed) pf in which $k_1, \ldots, k_n$ occur. We can interpret f as an *instantiation* of a function that has taken arguments $k_1, \ldots, k_n$. We now generalise this to $z(k_1, \ldots, k_n)$, representing *any* function z taking these arguments. Such a construction is also explicitly present in the *Principia*:

> "the first matrices[21] that occur are those whose values are of the forms φx, $\psi(x, y)$, $\chi(x, y, z, \ldots)$, *i.e.* where the arguments, however many there may be, are all individuals. Such [propositional] functions we will call '*first-order* functions.' We may now introduce a notation to express 'any first-order function.' "
>
> (*Principia Mathematica*, p. 51)

Remark 13. The definition of free variable needs some special attention. We must notice that, for instance, $\text{FV}(\mathtt{z}(\mathtt{R}(\mathtt{x}), \mathtt{S}(\mathtt{a}))) = \{\mathtt{z}\}$ and not $\{\mathtt{x}, \mathtt{z}\}$. Similarly, $\text{FV}(\mathtt{z}(\mathtt{R}(\mathtt{x}), \mathtt{y})) = \{\mathtt{y}, \mathtt{z}\}$ and not $\{\mathtt{x}, \mathtt{y}, \mathtt{z}\}$. The reason for this is that the notion of free variable should harmonise with the intuitive notion of "argument place" of Frege and Russell. As was indicated in Remark 12.4, in the first example above, $\mathtt{z}$ represents an arbitrary function that takes $\mathtt{R}(\mathtt{x})$ and $\mathtt{S}(\mathtt{a})$ as arguments and returns a proposition. This means that we do not have to supply an argument for $\mathtt{x}$ "by hand". As soon as we feed a suitable[22] argument f to $\mathtt{z}$ in $\mathtt{z}(\mathtt{R}(\mathtt{x}), \mathtt{S}(\mathtt{a}))$, f will take the arguments $\mathtt{R}(\mathtt{x})$ and $\mathtt{S}(\mathtt{a})$, and return a proposition.

This idea is also clearly reflected in the translation of $\mathtt{z}(\mathtt{R}(\mathtt{x}), \mathtt{S}(\mathtt{a}))$ to the λ-term $\lambda\mathtt{z}.\mathtt{z}(\lambda\mathtt{x}.\mathtt{Rx})(\mathtt{Sa})$. The variable $\mathtt{x}$ is bound in a subterm $\lambda\mathtt{x}.\mathtt{Rx}$ that is an argument to the variable $\mathtt{z}$. The full λ-term is a function of $\mathtt{z}$ only. See example 24.

Remark 14. It appears that there is also an alternative way of constructing pfs in the *Principia*. Whitehead and Russell distinguish between quantifier-free pfs (so-called *matrices*, i.e. the pfs that can be constructed using construction-rules 1, 2 and 4). Then they form pfs by defining that

- Any matrix is a pf;
- If f is a pf and $x \in \text{FV}(f)$ then $\forall x[f]$ is a pf with free variables $\text{FV}(f) \setminus \{x\}$.

This definition is a little different from our Definition 5, as a pf of the form $z(\forall x[f])$ is not a matrix and therefore not a pf according to this alternative definition. Nevertheless we feel that Whitehead and Russell intended to give our Definition 5. In the *Principia* ([Whitehead and Russell, 1910], ∗54) they define the natural

[21] See Remark 14 [footnote of the authors].

[22] At this stage, we cannot provide a formalisation of "suitable". This can only be done after we have introduced types, and formalised the notion "the pf f is of type t".

number 0 as the propositional function $\forall \mathbf{x}[\neg \mathbf{z}(\mathbf{x})]$.[23] In defining the principle of induction on natural numbers, one needs to express the property "0 has the property y", or: $y(0)$. But $y(0)$ is not a pf according to this alternative definition, as 0 contains quantifiers.

Therefore we feel that our Definition 5, which is also based on the definition of function by Frege and on the definition of propositional function on p. 38 of the *Principia*, is the definition that was meant by Whitehead and Russell.

Remark 15. Note that pfs as such do not yet obey to the vicious circle principle 3 (on page 465). For example, $\neg \mathbf{z}(\mathbf{z})$ (the pf that is at the basis of the Russell paradox) is a pf. In Section 3.5 we will assign types to some pfs, and it will be shown (Remark 85) that no type can be assigned to the pf $\neg \mathbf{z}(\mathbf{z})$.

Remark 16. Before we make further developments of the theory based on pfs, we must decide which of the two syntaxes introduced above shall be used in the sequel. It looks attractive to use the syntax of λ-calculus:

- This syntax is well-known;
- It is used for many other type systems, so it makes the comparison of ramified type theory with modern type systems easier;
- There is a lot of meta-theory on typed and untyped λ-calculus. This can be useful when proving certain properties of the formalisation of the ramified theory of types that is to be introduced in the next sections;
- The syntax of λ-calculus gives a better look on the notion of free variable than the syntax of pfs.

Nevertheless, we shall only indirectly use λ-calculus for our further study of the ramified type theory in this article. We have several reasons for that:

- There are much more λ-terms than there are pfs. More precisely, the mapping $^{-}$ is not surjective. As we want to study the theory of *Principia Mathematica* as precise as possible, we only want to study the propositional functions, which are directly related to the syntax used by Russell and Whitehead. Not using pf-syntax may result in a system in which it is not clear which term belongs to the original ramified type theory and which term does not;
- The syntax of λ-calculus is strongly curried. This would give problems in the definition of substitution. In a pf $\mathrm{R}(\mathbf{x}, \mathbf{y})$ we may want to substitute some object a for $\mathbf{y}$ without substituting anything for $\mathbf{x}$. In λ-calculus, substitution

[23]This definition is based on Frege's definition in *Grundlagen der Arithmetik* [Frege, 1884]. See [Whitehead and Russell, 1910], vol. II, p. 4. In [Frege, 1884], the natural number n is defined as the class of predicates f for which there are exactly n objects a for which $f(a)$ holds. Hence 0 is the class of predicates f for which $f(a)$ does not hold for any object a. So 0 can be described by the pf $\forall \mathbf{x}[\neg \mathbf{z}(\mathbf{x})]$. We prefer this notation to the notation $\iota`\emptyset$ (which also lends itself), because the logical operators $\neg$ and $\forall$ are more freely available in type theory.

should be translated to application followed by β-reduction to β-normal form. If we want to substitute something for $\mathtt{y}$ in the translation $\lambda\mathtt{x}.\lambda\mathtt{y}.\mathtt{Rxy}$ of $\mathtt{R}(\mathtt{x},\mathtt{y})$, we have to substitute something for $\mathtt{x}$ first. Choosing a different representation of propositional functions does not help: the representation $\lambda\mathtt{y}.\lambda\mathtt{x}.\mathtt{Ryx}$ would have given problems if we wanted to substitute something for $\mathtt{x}$ without substituting something for $\mathtt{y}$;

- The translation of pfs to λ-calculus makes it possible to use the meta-theory and the intuition of λ-calculus when we need it without losing control over the original system.

3.4 Principia's substitution and related notions

We proceed our discussion of pfs by defining a number of related notions. If a pf $z(k_1,\ldots,k_n)$ takes an argument f for the variable z, the list $k_1,\ldots,k_n$ indicates what should be substituted for the free variables of f (cf. also Remark 12.4). We therefore call this list the list of *parameters* of $z(k_1,\ldots,k_n)$.

Another important notion is the notion of α-equality[24]. We want the pfs $\mathtt{R}(\mathtt{x})$ and $\mathtt{R}(\mathtt{y})$ to be the same. However, we want the pfs $\mathtt{S}(\mathtt{x},\mathtt{y})$ and $\mathtt{S}(\mathtt{y},\mathtt{x})$ to be different. The reason for this is the alphabetical order of the variables $\mathtt{x}$, $\mathtt{y}$. As $\mathtt{x}<\mathtt{y}$, we will consider $\mathtt{x}$ to be the "first" variable of the pfs $\mathtt{S}(\mathtt{x},\mathtt{y})$ and $\mathtt{S}(\mathtt{y},\mathtt{x})$, and $\mathtt{y}$ the "second" variable . The place of the "first" variable in $\mathtt{S}(\mathtt{x},\mathtt{y})$, however, is different from the place of the "first" variable in $\mathtt{S}(\mathtt{y},\mathtt{x})$.[25] We therefore present the following definition of α-equality:

Definition 17. **(α-equality)** Let f and g be pfs. We say that f and g are *α-equal*, notation $f =_\alpha g$, if there is a bijection $\varphi : \mathcal{V} \to \mathcal{V}$ such that

- g can be obtained from f by replacing each variable that occurs in f by its φ-image;
- $x<y$ iff $\varphi(x)<\varphi(y)$.

This definition corresponds to the definition of α-equality in λ-calculus in the following way:

Lemma 18. Let $f, g \in \mathcal{P}$. $f =_\alpha g$ if and only if $\overline{f} =_\alpha \overline{g}$.

[24]Historically, it is not correct to use this terminology when discussing Type Theory of the *Principia*, which dates from the first decade of the 20th century. The term α-equality originates from Curry and Feys' book *Combinatory Logic* [Curry and Feys, 1958], which appeared only in 1958. In that book, conversion rules for the λ-calculus are numbered with Greek letters α, β, which led to the names α- and β-conversion. In earlier papers of Church, Rosser and Kleene, these rules were numbered with Roman capitals I, II, and the terminology α-conversion, β-conversion, was not used.

[25]Compare this with their equivalents in λ-calculus $\lambda\mathtt{x}.\lambda\mathtt{y}.\mathtt{Sxy}$ and $\lambda\mathtt{x}.\lambda\mathtt{y}.\mathtt{Syx}$, which are not α-equal, either. We do not want to use the λ-notation for determining which variable is "first" and which is "second", for reasons to be explained in Remark 31. See also Remark 16.

Sometimes, we are not that precise, and want the pfs $\mathtt{S(x,y)}$ and $\mathtt{S(y,x)}$ to be α-equal. This can be a consideration especially if we are not interested in which free variable is "first" and which is "second". We call this weakened notion of α-equality: α_P-equality (α-equality modulo permutation):

Definition 19. **(α-equality modulo permutation)** Let f and g be pfs. We say that f and g are *α_P-equal*, notation $f =_{\alpha_P} g$, if there is a bijection $\varphi : \mathcal{V} \to \mathcal{V}$ such that g can be obtained from f by replacing each variable that occurs in f by its φ-image.

As function construction in *Principia Mathematica* can be compared to β-expansion plus removing an argument in λ-calculus, this suggests that instantiation in the *Principia* must be comparable to application plus β-reduction in λ-calculus. In [Laan, 1994] we showed that this is indeed the case. There, we gave a definition of instantiation using the syntax of and the intuition behind pfs. We showed that this definition is faithful to the original ideas of the *Principia* and that it can be imitated in λ-calculus using a translation similar to the one in Definition 9. This allows us to give a definition of substitution for pfs that is based on that imitation in λ-calculus, as we do below.

As was argued in Remark 16, the mapping $f \mapsto \overline{f}$ is not perfectly suited for a definition of substitution. This was due to the currying of the λ-abstractions that are at the front of the term $\overline{f}$. We therefore take a slightly different notation and remove these front abstractions from $\overline{f}$:

Definition 20. Let $f \in \mathcal{P}$ with free variables $x_1 < \cdots < x_m$. Then $\overline{f} \equiv \lambda_{i=1}^m x_i.F$ for some λ-term F by Lemma 11.4. Let $\widetilde{f} \stackrel{\text{def}}{=} F$.

Example 21.

f	$\widetilde{f}$
$\mathtt{R(x)}$	$\mathtt{Rx}$
$\mathtt{z(R(x),S(a))}$	$\mathtt{z(\lambda x.Rx)(Sa)}$
$\mathtt{z_1(a) \vee z_2()}$	$\mathtt{\vee(z_1 a)z_2}$
$\mathtt{z(y(R(x)))}$	$\mathtt{z(\lambda y.y(\lambda x.Rx))}$
$\mathtt{\forall x[R(x)]}$	$\mathtt{\forall(\lambda x.Rx)}$

The mapping $f \mapsto \widetilde{f}$ has similar properties as $f \mapsto \overline{f}$ (cf. Lemma 11):

Lemma 22. **(Properties of $\widetilde{\ }$)**

1. $\mathrm{FV}(f) = \mathrm{FV}\left(\widetilde{f}\right)$;
2. $\widetilde{f}$ is in β-normal form for all f;
3. $\widetilde{f}$ is a λI-term for all f;
4. $\overline{f}$ is a closure of $\widetilde{f}$;
5. If $\widetilde{f} =_\alpha \widetilde{g}$, then $f =_\alpha g$.

$$\begin{array}{ccc} f[x_1,\ldots,x_n{:=}g_1,\ldots,g_n] & = & h \\ \downarrow & & \uparrow \\ (\lambda x_1 \cdots x_n.\widetilde{f})\overline{g_1}\cdots\overline{g_n} & \twoheadrightarrow_\beta^{\mathrm{nf}} & \widetilde{h} \end{array}$$

Figure 1. Substitution via β-reduction

With the λ-notation we can rely on the notions of β-reduction and β-normal form to give the following definition of substitution:

Definition 23. **(Substitution)** Let $f \in \mathcal{P}$, assume $x_1,\ldots,x_n$ are distinct variables, and $g_1,\ldots,g_n \in \mathcal{A} \cup \mathcal{V} \cup \mathcal{P}$. Assume that the λ-term $(\lambda x_1 \cdots x_n.\widetilde{f})\overline{g_1}\cdots\overline{g_n}$ has a β-normal form H. Assume $h \in \mathcal{P}$ such that $\widetilde{h} \equiv H$ (If such an h exists, it is unique due to Lemma 22.5). We define the simultaneous substitution of $g_1,\ldots,g_n$ for $x_1,\ldots,x_n$ in f by: $f[x_1,\ldots,x_n{:=}g_1,\ldots,g_n] \stackrel{\mathrm{def}}{=} h$.

We sometimes abbreviate $f[x_1,\ldots,x_n{:=}g_1,\ldots,g_n]$ to $f[x_i{:=}g_i]_{i=1}^n$ or $f[\vec{x}{:=}\vec{g}]$.

So substitution in RTT can be seen as application plus β-reduction to β-normal form in λ-calculus. Definition 23 is schematically reflected in Figure 1. Notice that $f[x_1,\ldots,x_n{:=}g_1,\ldots,g_n]$ should be seen as a *simultaneous* substitution of $g_1,\ldots,g_n$ for $x_1,\ldots,x_n$. As the $\overline{g_i}$s are either closed λ-terms, or individuals, or variables, it is no problem to define this simultaneous substitution via a list of applications that results in a list of *consecutive* substitutions.

Example 24.

1. $\mathtt{S}(\mathtt{x}_1)[\mathtt{x}_1{:=}\mathtt{a}_1] \equiv \mathtt{S}(\mathtt{a}_1)$, as $(\lambda \mathtt{x}_1.\mathtt{S}\mathtt{x}_1)\mathtt{a}_1 \twoheadrightarrow_\beta^{\mathrm{nf}} \mathtt{S}\mathtt{a}_1$;

2. $\mathtt{S}(\mathtt{x}_1)[\mathtt{x}_2{:=}\mathtt{a}_2] \equiv \mathtt{S}(\mathtt{x}_1)$, as $(\lambda \mathtt{x}_2.\mathtt{S}\mathtt{x}_1)\mathtt{a}_2 \twoheadrightarrow_\beta^{\mathrm{nf}} \mathtt{S}\mathtt{x}_1$;

3. $\mathtt{z}(\mathtt{S}(\mathtt{x}_1),\mathtt{x}_2,\mathtt{a}_2)[\mathtt{x}_1{:=}\mathtt{a}_1] \equiv \mathtt{z}(\mathtt{S}(\mathtt{x}_1),\mathtt{x}_2,\mathtt{a}_2)$ as
   $(\lambda \mathtt{x}_1.\mathtt{z}(\lambda \mathtt{x}_1.\mathtt{S}\mathtt{x}_1)\mathtt{x}_2\mathtt{a}_2)\mathtt{a}_1 \twoheadrightarrow_\beta^{\mathrm{nf}} \mathtt{z}(\lambda \mathtt{x}_1.\mathtt{S}\mathtt{x}_1)\mathtt{x}_2\mathtt{a}_2$.
   This illustrates that the λ-notation is more precise and convenient with respect to free variables. In $\mathtt{z}(\mathtt{S}(\mathtt{x}_1),\mathtt{x}_2,\mathtt{a}_2)$, it is not immediately clear whether $\mathtt{x}_1$ is a free variable or not and one might tend to write:

   $$\mathtt{z}(\mathtt{S}(\mathtt{x}_1),\mathtt{x}_2,\mathtt{a}_2)[\mathtt{x}_1{:=}\mathtt{a}_1] \equiv \mathtt{z}(\mathtt{S}(\mathtt{a}_1),\mathtt{x}_2,\mathtt{a}_2).$$

   The λ-notation is more explicit in showing that $\mathtt{x}_1 \notin \mathrm{FV}(\mathtt{z}(\mathtt{S}(\mathtt{x}_1),\mathtt{x}_2,\mathtt{a}_2))$;

4. $\mathtt{z}(\mathtt{R}(\mathtt{a}),\mathtt{S}(\mathtt{a}))[\mathtt{z}{:=}\mathtt{z}_1() \vee \mathtt{z}_2()] \equiv \mathtt{R}(\mathtt{a}) \vee \mathtt{S}(\mathtt{a})$, as
   $(\lambda \mathtt{z}.\mathtt{z}(\mathtt{Ra})(\mathtt{Sa}))(\lambda \mathtt{z}_1\mathtt{z}_2.\vee\mathtt{z}_1\mathtt{z}_2) \rightarrow_\beta$
   $(\lambda \mathtt{z}_1\mathtt{z}_2.\vee\mathtt{z}_1\mathtt{z}_2)(\mathtt{Ra})(\mathtt{Sa}) \twoheadrightarrow_\beta^{\mathrm{nf}} \vee(\mathtt{Ra})(\mathtt{Sa})$;

5. $\mathtt{x}_2(\mathtt{x}_1,\mathtt{R}(\mathtt{x}_1))[\mathtt{x}_2{:=}\mathtt{x}_4(\mathtt{x}_3)] \equiv \mathtt{R}(\mathtt{x}_1)$ as
   $(\lambda \mathtt{x}_2.\mathtt{x}_2\mathtt{x}_1(\lambda \mathtt{x}_1.\mathtt{R}\mathtt{x}_1))(\lambda \mathtt{x}_3\mathtt{x}_4.\mathtt{x}_4\mathtt{x}_3) \twoheadrightarrow_\beta$

$(\lambda \mathtt{x}_3\mathtt{x}_4.\mathtt{x}_4\mathtt{x}_3)\mathtt{x}_1(\lambda \mathtt{x}_1.\mathtt{R}\mathtt{x}_1) \twoheadrightarrow_\beta$
$(\lambda \mathtt{x}_1.\mathtt{R}\mathtt{x}_1)\mathtt{x}_1 \to_\beta^{\mathrm{nf}} \mathtt{R}\mathtt{x}_1$.

Remark 25. $f[x_1,\ldots,x_n{:=}g_1,\ldots,g_n]$ is not always defined. For its existence we need:

- The existence of the normal form H in Definition 23. For instance, this normal form does not exist if we choose $n = 1$, and take $f \equiv \mathtt{x}_1(\mathtt{x}_1)$ and $g_1 \equiv \mathtt{x}_1(\mathtt{x}_1)$: then we obtain for the calculation of $f[\mathtt{x}_1{:=}g_1]$ the famous λ-term $(\lambda \mathtt{x}_1.\mathtt{x}_1\mathtt{x}_1)(\lambda \mathtt{x}_1.\mathtt{x}_1\mathtt{x}_1)$;
- The existence of a (unique) h such that $\widetilde{h} \equiv H$. For instance, if we take $n = 1$, $f \equiv \mathtt{z}(\mathtt{a})$ (with $\mathtt{z} \in \mathcal{V}$ and $\mathtt{a} \in \mathcal{A}$) and $g_1 \equiv \mathtt{a}$, then $H \equiv \mathtt{aa}$ and there is no $h \in \mathcal{P}$ such that $\widetilde{h} \equiv \mathtt{aa}$.

In Corollary 61, we will prove that, as long as we are within the type system RTT (to be introduced in Section 3.5), both H and h always exist uniquely. Until then, we implicitly assume that the substitution *exists* in the notation $f[x_1,\ldots,x_n{:=}g_1,\ldots,g_n] = h$.

Remark 26. If we compute a substitution $f[x_1,\ldots,x_n{:=}g_1,\ldots,g_n]$, we have to reduce the λ-term $(\lambda x_1\cdots x_n.\widetilde{f})\widetilde{g_1}\cdots\widetilde{g_n}$ to its β-normal form (if there is any). One might wonder whether this is too restrictive: In a reduction path to this normal form, there may be an intermediate result H that could be interpreted as the final result of the substitution $f[\vec{x}{:=}\vec{g}]$. However, this never happens, as any term that can be interpreted as such a result is always of the form $\widetilde{h}$, and is therefore always in β-normal form (Lemma 22.2).

Remark 27. The alphabetical order of the variables plays a crucial role in the substitution process, as it determines in which order the free variables of a pf f are curried in the translation $\widetilde{f}$. For example, look at the substitutions $\mathtt{z}(\mathtt{a},\mathtt{b})[\mathtt{z}{:=}\mathtt{R}(\mathtt{x},\mathtt{y})]$ and $\mathtt{z}(\mathtt{a},\mathtt{b})[\mathtt{z}{:=}\mathtt{R}(\mathtt{y},\mathtt{x})]$. The result of the first one is obtained via the normal form of $(\lambda\mathtt{z}.\mathtt{zab})(\lambda\mathtt{xy}.\mathtt{Rxy})$, which is equal to $\mathtt{Rab}$, translated: $\mathtt{R}(\mathtt{a},\mathtt{b})$. The second one is calculated via $(\lambda\mathtt{z}.\mathtt{zab})(\lambda\mathtt{xy}.\mathtt{Ryx})$, resulting in $\mathtt{Rba}$ and $\mathtt{R}(\mathtt{b},\mathtt{a})$.

Remark 28. Now that substitution has been properly defined, we could define that f is an *abstraction* of g if there are $x_1,\ldots,x_n \in \mathrm{FV}(f)$ and $h_1,\ldots,h_n \in \mathcal{A}\cup\mathcal{P}$ such that $f[\vec{x}{:=}\vec{h}] \equiv g$.
Or, in λ-calculus notation: $(\lambda x_1\cdots x_n.\widetilde{f})\overline{h_1}\cdots\overline{h_n} \twoheadrightarrow_\beta \widetilde{g}$. The set of abstractions of a pf g is therefore comparable with the set of β-expansions of the λ-term $\widetilde{g}$.

Some elementary calculation with substitutions can be done using the following lemma:

Lemma 29.

1. Assume $(f_1 \vee f_2)[\vec{x}{:=}\vec{h}]$ exists. Then $f_j[\vec{x}{:=}\vec{h}]$ exists for $j = 1, 2$, and

$$(f_1 \vee f_2)[\vec{x}{:=}\vec{h}] \equiv (f_1[\vec{x}{:=}\vec{h}]) \vee (f_2[\vec{x}{:=}\vec{h}]);$$

2. Assume $(\neg f)[\overrightarrow{x}{:=}\overrightarrow{h}]$ exists. Then $f[\overrightarrow{x}{:=}\overrightarrow{h}]$ exists, and

$$(\neg f)[\overrightarrow{x}{:=}\overrightarrow{h}] \equiv \neg(f[\overrightarrow{x}{:=}\overrightarrow{h}]);$$

3. Assume $(\forall x{:}t^a[f])[\overrightarrow{x}{:=}\overrightarrow{h}]$ exists, and $x \notin \vec{x}$. Then $f[\overrightarrow{x}{:=}\overrightarrow{h}]$ exists, and

$$(\forall x{:}t^a[f])[\overrightarrow{x}{:=}\overrightarrow{h}] \equiv \forall x{:}t^a[f[\overrightarrow{x}{:=}\overrightarrow{h}]];$$

4. Assume $z(k_1,\ldots,k_n)[z{:=}f]$ exists, and $x_1 < \cdots < x_n$ are the free variables of f. Then $f[\overrightarrow{x}{:=}\overrightarrow{k}]$ exists, and

$$z(k_1,\ldots,k_n)[z{:=}f] \equiv f[\overrightarrow{x}{:=}\overrightarrow{k}];$$

5. Assume $z(k_1,\ldots,k_n)[\overrightarrow{x}{:=}\overrightarrow{h}]$ exists, $z \equiv x_p$, and $y_1 < \cdots < y_n$ are the free variables of $k_p \in \mathcal{P}$. Define $k_i' \equiv h_j$ if $k_i \equiv x_j$, and $k_i' \equiv k_i$ otherwise. Then $k_p[\overrightarrow{y}{:=}\overrightarrow{k'}]$ exists, and

$$z(k_1,\ldots,k_n)[\overrightarrow{x}{:=}\overrightarrow{h'}] \equiv k_p[y_j{:=}\overrightarrow{k'}].$$

PROOF: Directly from the definition of substitution. ⊠

3.5 Principia's Ramified Theory of Types

In order to give a precise description of the type theory underlying the *Principia*, we need to explicitly introduce types (there is no such introduction in *Principia*), and to formalise the notion "the propositional function f has type t".

TYPES

Types in the *Principia* have a double hierarchy: one of (simple) types and one of orders. First, we introduce the first hierarchy, then we extend this hierarchy with orders, resulting in the ramified types of the *Principia*.

SIMPLE TYPES

As we saw in Section 2.2, Frege already distinguished between objects, functions that take objects as arguments, and functions that take functions as arguments. He also made a distinction between functions that take one and functions that take two arguments (see the quotations from *Function and Concept* on p. 460). In the *Principia*, Whitehead and Russell use a similar principle. Whilst Frege's argument for this distinction was only that functions are fundamentally different from objects, and that functions taking objects as arguments are fundamentally different from functions taking functions as arguments, Whitehead and Russell are more precise:

"[The difference between objects and propositional functions] arises from the fact that a [propositional] function is essentially an ambiguity, and that, if it is to occur in a definite proposition, it must occur in such a way that the ambiguity has disappeared, and a wholly unambiguous statement has resulted."

(*Principia Mathematica*, p. 47)

There is no definition of "type" in the *Principia*, only a definition of "being of the same type":

"*Definition of "being of the same type."* The following is a step-by-step definition, the definition for higher types presupposing that for lower types. We say that u and v "are of the same type" if

1. both are individuals,
2. both are elementary [propositional] functions[26] taking arguments of the same type,
3. u is a pf and v is its negation,
4. u is $\varphi\hat{x}$[27] or $\psi\hat{x}$, and v is $\varphi\hat{x}\vee\psi\hat{x}$, where $\varphi\hat{x}$ and $\psi\hat{x}$ are elementary pfs,
5. u is $(y).\varphi(\hat{x}, y)$[28] and v is $(z).\psi(\hat{x}, z)$, where $\varphi(\hat{x}, \hat{y})$, $\psi(\hat{x}, \hat{y})$ are of the same type,
6. both are elementary propositions,
7. u is a proposition and v is $\sim u$[29], or
8. u is $(x).\varphi x$ and v is $(y).\psi y$, where $\varphi\hat{x}$ and $\psi\hat{x}$ are of the same type."

(*Principia Mathematica*, $*9{\cdot}131$, p. 133)

The definition has to be seen as the definition of an equivalence relation. For instance, assume that $\varphi\hat{x}$, $\psi\hat{x}$ and $\chi\hat{x}$ are elementary pfs. By rule 4, $\varphi\hat{x}$ and $\varphi\hat{x}\vee\psi\hat{x}$ are of the same type, and so are $\varphi\hat{x}$ and $\varphi\hat{x}\vee\chi\hat{x}$. By (implicit) transitivity, $\varphi\hat{x} \vee \psi\hat{x}$ and $\varphi\hat{x} \vee \chi\hat{x}$ are of the same type.

The definition seems rather precise at first sight. But there are several remarks to be made:

- The notion "being of the same type" seems to be defined for pfs taking one argument only. On the other hand, rules 2 and 5 suggest that such a definition should be extended to pfs taking two arguments. How this should be done is not made explicit;

[26] See Definition 4 for the notion of elementary proposition. In the *Principia*, the term *elementary functions* refers to a pf that has only elementary propositions as value, when it takes suitable (well-typed) arguments. See *Principia*, p. 92.

[27] Whitehead and Russell use $\varphi\hat{x}$ to denote that φ is a pf that has, amongst others, x as a free variable. Similarly, they use $\varphi(\hat{x}, \hat{y})$ to indicate that φ has x, y amongst its free variables.

[28] Whitehead and Russell write $(x).\varphi(x)$ where we would write $\forall x[\varphi]$.

[29] $\sim u$ is *Principia* notation for $\neg u$.

- According to this definition, $\mathtt{z}_1() \vee \neg\mathtt{z}_1()$ is not of the same type as $\mathtt{z}_1()$. The only rules by which could be derived that $\mathtt{z}_1()$ and $\mathtt{z}_1() \vee \neg\mathtt{z}_1()$ are of the same type, are rules 2 and 4. But if we want to use these rules, $\mathtt{z}_1()$ must be an elementary pf, which it is not: It can take the argument $\forall\mathtt{x}[\mathtt{R}(\mathtt{x})]$, which has as result the proposition $\forall\mathtt{x}[\mathtt{R}(\mathtt{x})]$. This is not an elementary proposition and therefore $\mathtt{z}_1()$ is not an elementary pf.

So there are significant omissions in this definition. However, the intention of the definition is clear: pfs that take a different number of arguments, or that take arguments of different types, cannot be of the same type.

In order to make precise what is meant by "being of the same type", it is easier to explain what these types "are". The notion "being of the same type" can then be replaced by "having the same type". The notion of simple type as defined below is due to Ramsey [Ramsey, 1926]. Historically, it is incorrect to give Ramsey's definition of simple type before Russell's definition of ramified type, as Russell's definition is of an earlier date, and Ramsey's definition is in fact based on Russell's ideas and not the other way around. On the other hand, the ideas behind simple types were already explained by Frege (see the quotes from *Function and Concept* on page 460). Moreover, knowledge of the intuition behind simple types will make it easier to understand the ramified ones.[30] Therefore we present Ramsey's definition first.

Definition 30. (Ramsey's Simple types)

1. $\mathsf{0}$ is a simple type;

2. If $t_1, \ldots, t_n$ are simple types, then also $(t_1, \ldots, t_n)$ is a simple type. $n = 0$ is allowed: then we obtain the simple type $()$;

3. All simple types can be constructed using the rules 1 and 2.

We use $t, u, t_1, \ldots$ as metavariables over simple types.

Here, $(t_1, \ldots, t_n)$ is the type of pfs that should take n arguments (have n free variables), the ith argument having type t_i. The type $()$ stands for the type of the propositions, and the type $\mathsf{0}$ stands for the type of the individuals.

Remark 31. To formalise the notion of ith argument that a pf takes, we use the alphabetical order on variables that was introduced in Section 3.1. The ith argument taken by a pf will be substituted for the ith free variable of that pf, according to the alphabetical order.

Now it becomes clear why we considered the alphabetical order of variables in the definition of α-equality 17: we want α-equal pfs to have the same type.

[30]See [Holmes, 1993; Holmes, 1999] for a further discussion of the difference between simple and ramified type theory, especially in connection with Quine's new foundations for which there is a consistency result for its predicative version (and hence one can get models of predicative type theory in which very strong versions of "systematic ambiguity" hold). In particular, [Holmes, 1999] contains a discussion of the relationship between a predicative *linear* type scheme (with types indexed by the natural numbers) and the full ramified type scheme of *Principia*.

However, if f has type (t_1, t_2) and two free variables $x < y$, and g is the same as f except that the roles of x and y have been switched, then g will have type (t_2, t_1). Therefore we demand that the renaming of variables must maintain the alphabetical order. See also Remark 43.

Example 32. The propositional function $\mathtt{R(x)}$ should have type $(\mathtt{0})$, as it takes one individual as argument.

The propositional function $\mathtt{z(R(x), S(a))}$ (see Remark 12.4) takes one argument. This argument must be a pf that can take $\mathtt{R(x)}$ as its first argument (so this first argument must be of type $(\mathtt{0})$), and a proposition (of type $()$) as its second argument. We conclude that in $\mathtt{z(R(x), S(a))}$, we must substitute pfs of type $((\mathtt{0}), ())$ for $\mathtt{z}$. Therefore, $\mathtt{z(R(x), S(a))}$ has type $(((\mathtt{0}), ()))$.

The intuition presented in Remark 31 and Example 32 will be formalised in Definition 40. Theorem 54 shows that this formalisation follows the intuition.

Notation 33. From now on we will use a slightly different notation for quantification in pfs. Instead of $\forall x[f]$ we now explicitly mention the type (say: t) over which x is quantified: $\forall x{:}t[f]$. We do the same with the translations of pfs to λ-calculus: instead of $\lambda x.F$ we write $\lambda x{:}T(t).F$.

Ramified types

Up to now, the type of a pf only depends on the types of the arguments that it can take. In the *Principia*, a second hierarchy is introduced by regarding also the types of the variables that are bound by a quantifier (see *Principia*, pp. 51–55). Whitehead and Russell consider, for instance, the propositions $\mathtt{R(a)}$ and $\forall \mathtt{z}{:}()[\mathtt{z}() \vee \neg \mathtt{z}()]$ to be of a different level. The first is an atomic proposition, while the latter is based on the pf $\mathtt{z}() \vee \neg \mathtt{z}()$. The pf $\mathtt{z}() \vee \neg \mathtt{z}()$ involves an arbitrary proposition $\mathtt{z}$, therefore $\forall \mathtt{z}{:}()[\mathtt{z}() \vee \neg \mathtt{z}()]$ quantifies over all propositions $\mathtt{z}$. According to the vicious circle principle 3, $\forall \mathtt{z}{:}()[\mathtt{z}() \vee \neg \mathtt{z}()]$ cannot belong to this collection of propositions.

This problem is solved by dividing types into *orders* (not to be confused with the *alphabetical order* on the variables). An order is simply a natural number. Basic propositions are of order 0, and in $\forall \mathtt{z}{:}()[\mathtt{z}() \vee \neg \mathtt{z}()]$ we must mention the order of the propositions over which is quantified. The pf $\forall \mathtt{z}{:}()^n[\mathtt{z}() \vee \neg \mathtt{z}()]$ quantifies over all propositions of order n, and has order $n+1$.

The division of types into orders gives *ramified* types.

Definition 34. (Ramified types)

1. $\mathtt{0}^0$ is a ramified type;

2. If $t_1^{a_1}, \ldots, t_n^{a_n}$ are ramified types, and $a \in \mathbb{N}$, $a > \max(a_1, \ldots, a_n)$, then $(t_1^{a_1}, \ldots, t_n^{a_n})^a$ is a ramified type (if $n = 0$ then take $a \geq 0$);

3. All ramified types can be constructed using the rules 1 and 2.

If t^a is a ramified type, then a is called the *order* of t^a.

Remark 35. In $(t_1^{a_1}, \ldots, t_n^{a_n})^a$, we demand that $a > a_i$ for all i. This is because a pf of this type presupposes all the elements of type ${t_i}^{a_i}$, and therefore must be of an order that is higher than a_i.

Example 36. 0^0; $(0^0)^1$; $\left((0^0)^1, (0^0)^4\right)^5$; and $\left(0^0, ()^2, \left(0^0, (0^0)^1\right)^2\right)^7$ are all ramified types. However, $\left(0^0, \left(0^0, (0^0)^2\right)^2\right)^7$ is not a ramified type.

In the rest of Section 3 we simply speak of *types* when we mean *ramified* types, as long as no confusion arises.[31]

In the type $(0^0)^1$, all orders are "minimal", i.e., not higher than strictly necessary. This is, for instance, not the case in the type $(0^0)^2$. Types in which all orders are minimal are called *predicative* and play a special role in the Ramified Theory of Types. A formal definition:[32]

Definition 37. (Predicative types)

1. 0^0 is a predicative type;
2. If ${t_1}^{a_1}, \ldots, {t_n}^{a_n}$ are predicative types, and $a = 1 + \max(a_1, \ldots, a_n)$ (take $a = 0$ if $n = 0$), then $(t_1^{a_1}, \ldots, t_n^{a_n})^a$ is a predicative type;
3. All predicative types can be constructed using the rules 1 and 2 above.

3.6 A Formalisation RTT *of the Ramified Theory of Types*

In this section we formalise the intuition on types presented in Example 32 and (in Church's notation) in Definition 80 together with the intuition on orders. Before we can do this we must introduce some additional terminology.

In the pf $\mathtt{R}(\mathtt{x})$ we implicitly assume that $\mathtt{x}$ is a variable for which objects of type 0 must be substituted. For our formalisation we want to make the information on the type of a variable explicit. We do this by storing this information in so-called *contexts*. Contexts, common in modern type systems, are not used in the *Principia*.

Definition 38. (Contexts) Let $x_1, \ldots, x_n \in \mathcal{V}$ be distinct variables, and assume $t_1^{a_1}, \ldots, t_n^{a_n}$ are ramified types. Then $\{x_1{:}t_1^{a_1}, \ldots, x_n{:}t_n^{a_n}\}$ is a *context*. The set $\{x_1, \ldots, x_n\}$ is called the *domain* of the context or $\mathrm{dom}(\{x_1{:}t_1^{a_1}, \ldots, x_n{:}t_n^{a_n}\})$. We will use Greek capitals Γ, Δ as meta-variables over contexts.

[31]Russell seems not to like the idea that propositions make up a type (see page 48 of the *Principia*). We do however use the type (or types) of propositions because at various places in *Principia*, Russell talks as if there are types of propositions, uses quantifiers over propositions and discusses orders of propositions. For example, in 9*131 Russell refers to elementary propositions as "being of the same type", in spite of the things he says elsewhere.

[32]This definition comes straight from Whitehead and Russell. It should be noted that ramified types which are not predicative are certainly not "impredicative" in the usual sense of that word.

The pfs $\mathsf{z}_1(\mathsf{y}_1)$ and $\mathsf{z}_2(\mathsf{y}_2)$ are α-equal, according to Definition 17. But in a context $\Gamma \equiv \left\{\mathsf{y}_1{:}0^0, \mathsf{z}_1{:}(0^0)^1, \mathsf{y}_2{:}(0^0)^1, \mathsf{z}_2{:}\left((0^0)^1\right)^2\right\}$ one does not want to see $\mathsf{z}_1(\mathsf{y}_1)$ and $\mathsf{z}_2(\mathsf{y}_2)$ as equal, as the types of y_1 and y_2 differ, and the types of z_1 and z_2 differ as well. Therefore, we introduce a more restricted version of α-equality:

Definition 39. Let Γ be a context and f and g pfs. We say that f and g are α_Γ-*equal*, notation $f =_{\alpha,\Gamma} g$, if there is a bijection $\varphi : \mathcal{V} \to \mathcal{V}$ such that

- g can be obtained from f by replacing each variable that occurs in f by its φ-image;
- $x < y$ iff $\varphi(x) < \varphi(y)$;
- $x{:}t \in \Gamma$ iff $\varphi(x){:}t \in \Gamma$.

We will now define what we mean by $\Gamma \vdash f : t^a$, or, in words: f is of type t^a in the context Γ.[33] If $\Gamma \equiv \varnothing$ then we will write $\vdash f : t^a$. In this definition we will try to follow the line of the Principia as much as possible. For remarks, discussion and examples, see Section 3.7 below.

Definition 40. (Ramified Theory of Types: RTT)
The *judgement* $\Gamma \vdash f : t^a$ is inductively defined as follows:

1. **(start)** For all a we have: $\vdash a : 0^0$.

   For all atomic pfs f we have: $\vdash f : ()^0$;

2. **(connectives)** Assume $\Gamma \vdash f{:}(t_1^{a_1}, \ldots, t_n^{a_n})^a$, $\Delta \vdash g{:}(u_1^{b_1}, \ldots, u_m^{b_m})^b$, and $x < y$ for all $x \in \mathrm{dom}(\Gamma)$ and $y \in \mathrm{dom}(\Delta)$. Then

$$\Gamma \cup \Delta \vdash f \vee g : \left(t_1^{a_1}, \ldots, t_n^{a_n}, u_1^{b_1}, \ldots, u_m^{b_m}\right)^{\max(a,b)};$$

   and

$$\Gamma \vdash \neg f : (t_1^{a_1}, \ldots, t_n^{a_n})^a;$$

3. **(abstraction from parameters)** If $\Gamma \vdash f : (t_1^{a_1}, \ldots, t_m^{a_m})^a$, $t_{m+1}^{a_{m+1}}$ is a *predicative* type[34], $g \in \mathcal{A} \cup \mathcal{P}$ is a parameter of f, $\Gamma \vdash g : t_{m+1}^{a_{m+1}}$, and $x < y$ for all $x \in \mathrm{dom}(\Gamma)$, then

$$\Gamma' \vdash h : (t_1^{a_1}, \ldots, t_{m+1}^{a_{m+1}})^{\max(a, a_{m+1}+1)}.$$

   Here, h is a pf obtained by replacing all parameters g' of f which are α_Γ-equal to g by y. Moreover, Γ' is the subset of the context $\Gamma \cup \{y : t_{m+1}^{a_{m+1}}\}$ such that $\mathrm{dom}(\Gamma')$ contains all and only those variables occurring in h[35];

[33] The symbol $\vdash$ in $\Gamma \vdash f : t^a$ is the same symbol that Frege used to assert a proposition. It enters Type Theory in 1934 [Curry, 1934], via Curry's combinatory logic. Curry defines a functionality combinator F in such a way that $\mathsf{F}XYf$ holds, exactly if f is a function from X to Y. To denote the assertion of $\mathsf{F}XYf$, Curry uses Frege's symbol $\vdash$.

[34] The restriction to predicative types only is based on *Principia*, pp. 53–54.

[35] In Lemma 52 we prove that this context always exists.

4. **(abstraction from pfs)** If $(t_1^{a_1},\ldots,t_m^{a_m})^a$ is a *predicative* type, $\Gamma \vdash f : (t_1^{a_1},\ldots,t_m^{a_m})^a$, $x < z$ for all $x \in \mathrm{dom}(\Gamma)$, and $y_1 < \cdots < y_n$ are the free variables of f, then
$$\Gamma' \vdash z(y_1,\ldots,y_n) : (t_1^{a_1},\ldots,t_m^{a_m},(t_1^{a_1},\ldots,t_m^{a_m})^a)^{a+1},$$
where Γ' is the subset of the context $\Gamma \cup \{z{:}(t_1^{a_1},\ldots,t_m^{a_m})^a\}$ such that $\mathrm{dom}(\Gamma') = \{y_1,\ldots,y_n,z\}$;

5. **(weakening)** If Γ, Δ are contexts, $\Gamma \subseteq \Delta$, and $\Gamma \vdash f : t^a$, then also $\Delta \vdash f : t^a$;

6. **(substitution)** If y is the ith free variable in f (according to the order on variables), and $\Gamma \cup \{y : t_i^{a_i}\} \vdash f : (t_1^{a_1},\ldots,t_n^{a_n})^a$, and $\Gamma \vdash k : t_i^{a_i}$ then
$$\Gamma' \vdash f[y{:=}k] : (t_1^{a_1},\ldots,t_{i-1}^{a_{i-1}},t_{i+1}^{a_{i+1}},\ldots,t_n^{a_n})^b.$$
Here, $b = 1 + \max(a_1,\ldots,a_{i-1},a_{i+1},\ldots,a_n,c)$,
and $c = \max\{j \mid \forall x{:}t^j \text{ occurs in } f[y{:=}k]\}$
(if $n = 1$ and $\{j \mid \forall x{:}t^j \text{ occurs in } f[y{:=}k]\} = \varnothing$ then take $b = 0$)
and once more, Γ' is the subset of $\Gamma \cup \{y : t_i^{a_i}\}$ such that $\mathrm{dom}(\Gamma')$ contains all and only those variables occurring in $f[y{:=}k]$;

7. **(permutation)** If y is the ith free variable in f (according to the order on variables), and $\Gamma \cup \{y{:}t_i^{a_i}\} \vdash f : (t_1^{a_1},\ldots,t_n^{a_n})^a$, and $x < y'$ for all $x \in \mathrm{dom}(\Gamma)$, then
$$\Gamma' \vdash f[y{:=}y'] : (t_1^{a_1},\ldots,t_{i-1}^{a_{i-1}},t_{i+1}^{a_{i+1}},\ldots,t_n^{a_n},t_i^{a_i})^a.$$
Γ' is the subset of $\Gamma \cup \{y{:}t_i^{a_i}, y'{:}t_i^{a_i}\}$ such that $\mathrm{dom}(\Gamma')$ contains all and only those variables occurring in $f[y{:=}y']$;

8. **(quantification)** If y is the ith free variable in f (according to the order on variables), and $\Gamma \cup \{y{:}t_i^{a_i}\} \vdash f : (t_1^{a_1},\ldots,t_n^{a_n})^a$, then
$$\Gamma \vdash \forall y{:}t_i^{a_i}[f] : (t_1^{a_1},\ldots,t_{i-1}^{a_{i-1}},t_{i+1}^{a_{i+1}},\ldots,t_n^{a_n})^a.$$

Definition 41. (Legal propositional functions) A pf f is called *legal*, if there is a context Γ and a ramified type t^a such that $\Gamma \vdash f : t^a$.

Remark 42. In our attempt to faithfully implement Russell's ramified theory of types in the above definition, we face a limitation in the terms typable by our system. For example, it is not possible to type either the pf $x_1(x_2(x_1))$ or the pf $x(\forall y.(y() \vee \neg y()))$.[36] In fact, Russell intended (cf. page 165 of *Principia*) that non-predicative orders in his hierarchy are always obtained from predicative ones by quantification. Rule 8 of the above definition is the only one which creates non-predicative types but the increase of order is only at the top level of the type. This

[36] We are grateful for Randall Holmes for drawing our attention to this point.

means that we cannot type terms $z(k_1, \ldots, k_n)$ where one of the k_i's happens to be of non-predicative type. In fact, Theorem 69 will prove that terms $z(k_1 \ldots k_n)$ are typable only if the k_i's can be assigned predicative types. This may be considered as a serious restriction but our aim is to faithfully represent Russell's ramified theory of types.

A drawback to our system is that, without the ability to assign non-predicative types to variables, one cannot even *state* the axiom of reducibility. Russell himself may have noted the need for variables with non-predicative types when he introduced on page 165 of *Principia* a convention for variable functions without assigned order which he used in the formal statement of the axiom of reducibility. However, Russell did not allow quantification over such variables. In our paper we ignore the representation of the axiom of reducibility. In Section 4.2, we discuss the controversial nature of this axiom leading to the deramification and we leave the extension of our formalisation of Russell's ramified type theory to include the reducibility axiom as future work.

Finally, based on our above discussion, note that the third and fourth types given in Example 36 cannot be assigned as types to a legal pf in the sense of Definition 40. Future extensions must also address these examples.

3.7 Discussion and examples

We will make some remarks on Definition 40. First of all, we motivate the eight rules of Definition 40 by referring to passages in the *Principia*. Then we make some technical remarks, and give some examples of how the rules work. It will be made clear that the substitution rule is problematic, because substitution is not clearly defined in the *Principia*.

Remark 43. We will motivate RTT (Definition 40) by referring to the Principia:

1. Individuals and elementary judgements (atomic propositions) are, also in the *Principia*, the basic ingredients for creating legal pfs;[37]

2. We can see rule 2 "at work" in $*12$, p. 163 of the *Principia*[38]:

   > "We can build up a number of new formulas, such as [...] $\varphi!x \vee \varphi!y$, $\varphi!x \vee \psi!x$, $\varphi!x \vee \psi!y$, [...] and so on."
   > (*Principia Mathematica*, $*12$, p. 163))

   The restriction about contexts that we make in rule 2 has technical reasons and is not made in the *Principia*. It will be discussed in Remark 45;

3. Rule 3 is justified by $*9{\cdot}14$ and $*9{\cdot}15$ in the *Principia*. It is an instantiation of the abstraction principles 1 and 2 for functions that was already proposed

[37]As for individuals: see *Principia*, $*9$, p. 132, where "Individual" is presented as a primitive idea. As for elementary judgements: See *Principia*, Introduction, pp. 43-45.

[38]In the *Principia*, Whitehead and Russell write $\varphi!x$ instead of φx to indicate that φx is not only (what we would call) a pf, but even a *legal* pf.

by Frege (see Section 2.2). In Frege's definition one does not have to replace all parameters g' that are α_Γ-equal to g, but one can also take some of these parameters. In Section 3.9 we show that this is not a serious restriction.

The restriction to *predicative* types is in line with the *Principia* (cf. *Principia*, pp. 53–54);

4. Rule 4 is based on the Introduction of the *Principia*. There, pfs are constructed, and

> "the first matrices that occur are those whose values are of the forms φx, $\psi(x, y)$, $\chi(x, y, z, \dots)$, *i.e.* where the arguments, however many there may be, are all individuals. Such [propositional] functions we will call '*first-order* functions.' We may now introduce a notation to express 'any first-order function.' "
>
> (*Principia Mathematica*, p. 51)

This quote from the *Principia* is, again, an instance of Frege's abstraction principles, and so is rule 4 of our formalisation. It results in second order pfs, and the process can be iterated to obtain pfs of higher orders.

Rule 4 makes it possible to introduce variables of higher order. In fact, leaving out rule 4 would lead to first-order predicate logic, as without rule 4 it is impossible to introduce variables of types that differ from 0^0.

The use of predicative types only is inspired by the *Principia*, again;

5. The weakening rule cannot be found in the *Principia*, because no formal contexts are used there. It is implicitly present, however: the addition of an extra variable to the set of variables does not affect the well-typedness of pfs that were already constructed;

6. The rule of substitution is based on $*9{\cdot}14$ and $*9{\cdot}15$ of the *Principia*, and can be seen as an inverse of the abstraction operators in rules 3 and 4. Notice that we do not know yet whether the substitution $f[y{:=}k]$ exists or not. Therefore, we limit the use of rule 6 to the cases in which the substitution exists. In Section 3.8 we show that it always exists if the premises of rule 6 are fulfilled;

7. In the system above, the (sequential) order of the t_is is related to the alphabetic order of the free variables of the pf f that has type $(t_1, \dots, t_n)$ (see the remark before Definition 17, Remark 31, and Theorem 54). This alphabetic order plays a role in the clear presentation of results like Theorem 54, and in the definition of substitution (see Remark 27).

With rule 7 we want to express that the order of the t_is in $(t_1, \dots, t_n)$ and the alphabetic order of the variables are not characteristics of the *Principia*, but are only introduced for the technical reasons explained in this remark. This is worked out in Corollary 55;

8. Notice that in the quantification rule, both f and $\forall x{:}t_i^{a_i}.f$ have order a. The intuition is that the order of a propositional function f equals one plus the maximum of the orders of *all* the variables (either free or bound by a quantifier) in f. This is in line with the *Principia*: see [Whitehead and Russell, 1910], page 53. See also the introduction to Definition 34, and the proof of Lemma 56 below.

Remark 44. Rules 3 and 4 are a restricted version of the abstraction principles of Frege, with less power. It is, for instance, not possible to imitate all the abstractions of Remark 12 by using rules 3 and 4 only. But in combination with the other rules, rule 3 and 4 are sufficient (see Example 50 for the cases of Remark 12, and Section 3.9, especially Theorem 69).

Remark 45. In rule 2 of RTT, we make the assumption that the variables of Γ must all come before the variables of Δ. The reason for this is that we want to prevent undesired results like

$$\mathtt{x}_1{:}0^0 \vdash \mathtt{R}_1(\mathtt{x}_1) \vee \mathtt{R}_2(\mathtt{x}_1) : (0^0, 0^0)^1.$$

In fact, $\mathtt{R}_1(\mathtt{x}_1) \vee \mathtt{R}_2(\mathtt{x}_1)$ has *only one* free variable, so its type should be $(0^0)^1$ and not $(0^0, 0^0)^1$ (see Example 49, second part). For technical reasons (the order of the $t_i^{a_i}$s; see also Theorem 54) we strengthen the assumption such that for $x \in \mathrm{dom}(\Gamma)$ and $y \in \mathrm{dom}(\Delta)$, $x < y$ must hold.

As Whitehead and Russell do not have a formal notation for types, they do not forbid this kind of construction in the *Principia*.

In Lemma 67 we show that our limitation to contexts with disjoint domains as made in rule 2 is not a real limitation: all the desired judgements can still be derived for contexts with non-disjoint domains.

Remark 46. In both rules 3 and 4 we see that it is necessary to introduce at least one new variable. It is, for instance, not possible to interpret the proposition $\mathtt{R}(\mathtt{a})$ as a (constant) pf of type $(0^0)^1$. This is in line with the abstraction principles of Frege and Russell. In Frege's definition 1, for example, it is explicitly mentioned that the object that is to be replaced occurs at least once in the expression.

Translated to λ-calculus this means that the *Principia* have λI-terms, only. See also Lemma 11.3 and Lemma 22.3.

Remark 47. Contexts as used in RTT contain, in a sense, too much information: not only information on all free variables, but also information on non-free variables (cf. rules 3, 6 and 7). The set of non-free variables contains more than only the variables that are bound by a quantifier. For example, in the pf $\mathtt{z}(\mathtt{R}(\mathtt{x}))$, $\mathtt{x}$ is neither free, nor bound by a quantifier.

Remark 48. The system is based on the abstraction principles of Frege. In a context Γ, one cannot introduce a variable of a certain type t unless one has a pf (or an individual) f that has type t in Γ. This is different from modern, λ-calculus based systems, where one can introduce a variable of a type u without knowing whether or not there are terms of this type u.

We give some examples, in order to illustrate how our system works. Example 49 shows applications of the rules. Example 50 makes a link between the intuitive notion of abstraction that was explained in Remark 12 and the abstraction rules 3 and 4 of our system.

We will use a notation of the form $\dfrac{X_1 \cdots X_n}{Y} N$, indicating that from the judgements $X_1, \ldots, X_n$, we can infer the judgement Y by using the RTT-rule of Definition 40 with number N. As usual, this is called a *derivation step*. Subsequent derivation steps give a *derivation*. A derivation *of a judgement* Y is a derivation tree with Y as root (the final conclusion). The types in the examples below are all predicative (as a pf of impredicative type must have a quantifier, and the examples below are quantifier-free). To avoid too much notation, we omit the orders.

Example 49. The following type derivations are valid in RTT.

1. $\vdash \mathtt{S}(\mathtt{a}_1, \mathtt{a}_2) : ()$ by rule 1 of Definition 1;

2. $$\dfrac{\vdash \mathtt{R}_1(\mathtt{a}_1) : () \qquad \vdash \mathtt{R}_2(\mathtt{a}_1) : ()}{\vdash \mathtt{R}_1(\mathtt{a}_1) \vee \mathtt{R}_2(\mathtt{a}_1) : ()} \text{ rule 2}$$

   but *not*:

   $$\dfrac{\mathtt{x}_1 : 0 \vdash \mathtt{R}_1(\mathtt{x}_1) : (0) \qquad \mathtt{x}_1 : 0 \vdash \mathtt{R}_2(\mathtt{x}_1) : (0)}{\mathtt{x}_1 : 0 \vdash \mathtt{R}_1(\mathtt{x}_1) \vee \mathtt{R}_2(\mathtt{x}_1) : (0, 0)} \text{ rule 2}$$

   ($\mathtt{x}_1 \not< \mathtt{x}_1$ because $<$ is strict). To obtain $\mathtt{R}_1(\mathtt{x}_1) \vee \mathtt{R}_2(\mathtt{x}_1)$ we must make a different start:

   $$\dfrac{\dfrac{\vdash \mathtt{R}_1(\mathtt{a}_1) : () \qquad \vdash \mathtt{R}_2(\mathtt{a}_1) : ()}{\vdash \mathtt{R}_1(\mathtt{a}_1) \vee \mathtt{R}_2(\mathtt{a}_1) : ()} \text{ rule 2} \qquad \vdash \mathtt{a}_1 : 0}{\mathtt{x}_1 : 0 \vdash \mathtt{R}_1(\mathtt{x}_1) \vee \mathtt{R}_2(\mathtt{x}_1) : (0)} \text{ rule 3;}$$

3. $$\dfrac{\begin{array}{rcl} \mathtt{x}_1 : 0, \mathtt{x}_2 : 0, \mathtt{z}_1 : ((0), (0)) & \vdash & \mathtt{z}_1(\mathtt{R}(\mathtt{x}_1), \mathtt{R}(\mathtt{x}_2)) : (((0), (0))) \\ \mathtt{x}_1 : 0, \mathtt{x}_2 : 0, \mathtt{z}_1 : ((0), (0)) & \vdash & \mathtt{R}(\mathtt{x}_1) : (0) \end{array}}{\mathtt{z}_1 : ((0), (0)), \mathtt{z}_2 : (0) \vdash \mathtt{z}_1(\mathtt{z}_2, \mathtt{z}_2) : (((0), (0)), (0))} \text{ rule 3}$$

   As $\mathtt{R}(\mathtt{x}_1)$ is α-equal to $\mathtt{R}(\mathtt{x}_2)$ in the context, both $\mathtt{R}(\mathtt{x}_1)$ and $\mathtt{R}(\mathtt{x}_2)$ are replaced by the newly introduced variable $\mathtt{z}_2$;

4. $$\dfrac{\mathtt{x}_1 : 0, \mathtt{x}_2 : 0 \vdash \mathtt{S}(\mathtt{x}_1, \mathtt{x}_2) : (0, 0)}{\mathtt{x}_1 : 0, \mathtt{x}_2 : 0, \mathtt{z} : (0, 0) \vdash \mathtt{z}(\mathtt{x}_1, \mathtt{x}_2) : (0, 0, (0, 0))} \text{ rule 4;}$$

5. $$\dfrac{\dfrac{\mathtt{x}_1 : 0 \vdash \mathtt{R}_1(\mathtt{x}_1) \vee \mathtt{R}_2(\mathtt{x}_1) : (0) \qquad \vdash \mathtt{a}_1 : 0}{\vdash \mathtt{R}_1(\mathtt{a}_1) \vee \mathtt{R}_2(\mathtt{a}_1) : ()} \text{ rule 6;}}{\mathtt{x}_1 : 0 \vdash \mathtt{R}_1(\mathtt{a}_1) \vee \mathtt{R}_2(\mathtt{a}_1) : ()} \text{ rule 5}$$

6. $$\dfrac{\begin{array}{rcl} \mathtt{x}_1 : 0, \mathtt{x}_2 : 0, \mathtt{x}_3 : (0, 0) & \vdash & \mathtt{R}(\mathtt{x}_1) \vee \neg \mathtt{x}_3(\mathtt{x}_1, \mathtt{x}_2) : (0, 0, (0, 0)) \\ \mathtt{x}_1 : 0, \mathtt{x}_2 : 0 & \vdash & \mathtt{T}(\mathtt{x}_1, \mathtt{x}_1, \mathtt{x}_2) : (0, 0) \end{array}}{\mathtt{x}_1 : 0, \mathtt{x}_2 : 0 \vdash \mathtt{R}(\mathtt{x}_1) \vee \neg \mathtt{T}(\mathtt{x}_1, \mathtt{x}_1, \mathtt{x}_2) : (0, 0)} \text{ rule 6}$$

$\mathtt{T}(\mathtt{x}_1, \mathtt{x}_1, \mathtt{x}_2)$ is substituted for $\mathtt{x}_3$.

Example 50. We give a formal derivation of the examples of the abstraction rules that were given in Remark 12. Again, we omit the orders.

- Constructing $\mathtt{z(a)} \vee \mathtt{S(a)}$ from $\mathtt{R(a)} \vee \mathtt{S(a)}$ cannot be done with the use of rule 4 only. The following derivation is correct (to save space, we use "rx" to denote "rule x"):

$$\dfrac{\dfrac{\dfrac{\vdash \mathtt{a}{:}0}{\mathtt{z}{:}(0) \vdash \mathtt{a}{:}0}\,\text{r5} \qquad \dfrac{\dfrac{\vdash \mathtt{a}{:}0 \qquad \vdash \mathtt{R(a)}{:}()}{\mathtt{x}{:}0 \vdash \mathtt{R(x)}{:}(0)}\,\text{r3}}{\mathtt{x}{:}0, \mathtt{z}{:}(0) \vdash \mathtt{z(x)}{:}(0,(0))}\,\text{r4}}{\mathtt{z}{:}(0) \vdash \mathtt{z(a)}{:}((0))}\,\text{r6} \qquad \vdash \mathtt{S(a)}{:}()}{\mathtt{z}{:}(0) \vdash \mathtt{z(a)} \vee \mathtt{S(a)} : ((0))}\,\text{r2.}$$

  To obtain $\mathtt{z(a)}$ instead of $\mathtt{z()}$, we must transform $\mathtt{R(a)}$ into a pf $\mathtt{R(x)}$ by abstracting from $\mathtt{a}$. Then we can construct $\mathtt{z(x)}$ by abstraction from pfs (rule 4). In this way, the "frame" for $\mathtt{z(a)}$ is of the right form. Substituting $\mathtt{a}$ for $\mathtt{x}$ gives $\mathtt{z(a)}$ (and "neutralises" the application of rule 3 at the top of the derivation). Simply applying rule 4 on the judgement $\vdash \mathtt{R(a)} : ()$ does not work: it results in $\mathtt{z()} \vdash \mathtt{z()} : (())$;

- Constructing $\mathtt{z}_1() \vee \mathtt{z}_2()$ is easier: $\mathtt{z}_1()$ can be obtained by abstracting from $\mathtt{R(a)}$, and $\mathtt{z}_2()$ similarly from $\mathtt{S(a)}$. Result:

$$\dfrac{\dfrac{\vdash \mathtt{R(a)}{:}()}{\mathtt{z}_1{:}() \vdash \mathtt{z}_1(){:}(())}\,\text{rule 4} \qquad \dfrac{\vdash \mathtt{S(a)}{:}()}{\mathtt{z}_2{:}() \vdash \mathtt{z}_2(){:}(())}\,\text{rule 4}}{\mathtt{z}_1{:}(), \mathtt{z}_2{:}() \vdash \mathtt{z}_1() \vee \mathtt{z}_2() : ((),())}\,\text{rule 2.}$$

  We see that in fact two abstractions are needed to construct this pf: we must abstract from $\mathtt{R(a)}$ as an instance of the pf $\mathtt{z}_1()$, and from $\mathtt{S(a)}$ as an instance of the pf $\mathtt{z}_2()$. As rule 4 does not work on parts of pfs, these abstractions have to be made before we use rule 2. Applying rule 4 on $\vdash \mathtt{R(a)} \vee \mathtt{S(a)} : ()^0$ would result in $\mathtt{z} : () \vdash \mathtt{z()} : (())$;

- We can extend the derivation of $\mathtt{z}_1{:}(), \mathtt{z}_2{:}() \vdash \mathtt{z}_1() \vee \mathtt{z}_2() : ((),())$ to obtain a type for $\mathtt{z(R(a), S(a))}$:

$$\dfrac{\dfrac{\dfrac{\mathtt{x}_1{:}(), \mathtt{x}_2{:}() \vdash \mathtt{x}_1() \vee \mathtt{x}_2() : ((),())}{\mathtt{x}_1{:}(), \mathtt{x}_2{:}(), \mathtt{z}{:}((),()) \vdash \mathtt{z}(\mathtt{x}_1,\mathtt{x}_2){:}((),(),((),()))}\,\text{rule 4}}{\mathtt{x}_2{:}(), \mathtt{z}{:}((),()) \vdash \mathtt{z}(\mathtt{R(a)},\mathtt{x}_2) : ((),((),()))}\,\text{rule 6}}{\mathtt{z}{:}((),()) \vdash \mathtt{z(R(a),S(a))} : (((),()))}\,\text{rule 6}$$

  (for reasons of space, we omitted the premises $\mathtt{z}{:}((),()), \mathtt{x}_2{:}() \vdash \mathtt{R(a)}{:}()$ and $\mathtt{z}{:}((),()) \vdash \mathtt{S(a)}{:}()$ of the first and second application of the substitution rule);

- For the derivation of the type of $\mathtt{z}(\mathtt{R}(\mathtt{x}),\mathtt{S}(\mathtt{a}))$ we first make a derivation of the "frame" $\mathtt{z}(\mathtt{y}_1,\mathtt{y}_2)$ of this pf (we use "rx" to denote "rule x"):

$$\dfrac{\dfrac{\dfrac{\dfrac{\vdash \mathtt{a}{:}0}{\mathtt{y}_1{:}(0)\vdash \mathtt{a}{:}0}\,\mathrm{r5} \qquad \dfrac{\dfrac{\vdash \mathtt{a}{:}0 \qquad \vdash \mathtt{R}(\mathtt{a}){:}()}{\mathtt{x}{:}0\vdash \mathtt{R}(\mathtt{x}){:}(0)}\,\mathrm{r3}}{\mathtt{x}{:}0,\mathtt{y}_1{:}(0)\vdash \mathtt{y}_1(\mathtt{x}){:}(0,(0))}\,\mathrm{r4}}{\mathtt{y}_1{:}(0)\vdash \mathtt{y}_1(\mathtt{a}){:}((0))}\,\mathrm{r6} \qquad \dfrac{\vdash \mathtt{R}(\mathtt{a}){:}()}{\mathtt{y}_2{:}()\vdash \mathtt{y}_2(){:}(())}\,\mathrm{r4}}{\mathtt{y}_1{:}(0),\mathtt{y}_2{:}()\vdash \mathtt{y}_1(\mathtt{a})\vee \mathtt{y}_2() : ((0),())}\,\mathrm{r2}}{\mathtt{y}_1{:}(0),\mathtt{y}_2{:}(),\mathtt{z}{:}((0),())\vdash \mathtt{z}(\mathtt{y}_1,\mathtt{y}_2) : ((0),(),((0),()))}\,\mathrm{r4}.$$

  Then we derive $\mathtt{x}{:}0\vdash \mathtt{R}(\mathtt{x}){:}(0)$ and $\vdash \mathtt{S}(\mathtt{a}){:}()$, and after applying the weakening rule, we can substitute $\mathtt{R}(\mathtt{x})$ for $\mathtt{y}_1$ and $\mathtt{S}(\mathtt{a})$ for $\mathtt{y}_2$. As a result, we get

$$\mathtt{z}{:}((0),()),\mathtt{x}{:}0\vdash \mathtt{z}(\mathtt{R}(\mathtt{x}),\mathtt{S}(\mathtt{a})) : (((0),())).$$

Example 51. In the example below, the orders are important:

$$\dfrac{\dfrac{\dfrac{\vdash \mathtt{R}(\mathtt{a}) : ()^0 \qquad \dfrac{\vdash \mathtt{R}(\mathtt{a}) : ()^0}{\vdash \neg\mathtt{R}(\mathtt{a}) : ()^0}\,\text{rule 2}}{\vdash \mathtt{R}(\mathtt{a})\vee\neg\mathtt{R}(\mathtt{a}) : ()^0}\,\text{rule 2}}{\mathtt{z}{:}()^0\vdash \mathtt{z}()\vee\neg\mathtt{z} : (()^0)^1}\,\text{rule 4}}{\vdash \forall\mathtt{z}{:}()^0[\mathtt{z}()\vee\neg\mathtt{z}()] : ()^1}\,\text{rule 8}.$$

We see that $\forall\mathtt{z}{:}()^0[\mathtt{z}()^0\vee\neg\mathtt{z}()]$ does not have a predicative type. This is the case because this pf has a bound variable $\mathtt{z}$ that is of a higher order than the order of any free variable (as there are no free variables here). Therefore, the order of this pf is determined by the order of the bound variable $\mathtt{z}$.

We still need to prove that the contexts in the conclusions of rules 3, 4 and 6 exist. This follows from the following Lemma:

Lemma 52. Assume $\Gamma\vdash f : t^a$. Then

1. **(Free variable lemma)** *All variables of f that are not bound by a quantifier are in* $\mathrm{dom}(\Gamma)$*;*

2. **(Strengthening lemma)** *If Δ is the (unique) subset of Γ such that* $\mathrm{dom}(\Delta)$ *contains all and only those variables of f that are not bound by a quantifier, then $\Delta\vdash f : t^a$.*

PROOF: An easy induction on the definition of $\Gamma\vdash f : t^a$. ⊠

3.8 Properties of RTT

TYPES AND FREE VARIABLES

In this section we treat some meta-properties of RTT. Using the λ-notation for pfs, we can often refer to known results in typed λ-calculus[39]. For proofs and further details, see [Laan and Nederpelt, 1996; Laan, 1997].

Theorem 53. (First Free Variable Theorem)
Let $f \in \mathcal{P}$; $k_1, \ldots, k_n \in \mathcal{A} \cup \mathcal{V} \cup \mathcal{P}$.

$$\mathrm{FV}(f[x_1, \ldots, x_n{:=}k_1, \ldots, k_n]) = (\mathrm{FV}(f) \setminus \{x_1, \ldots, x_n\}) \cup \{k_i \in \mathcal{V} \mid x_i \in \mathrm{FV}(f)\}.$$

Theorem 54. (Second Free Variable Theorem) Assume that we can derive $\Gamma \vdash f : (t_1^{a_1}, \ldots, t_n^{a_n})^a$, and $x_1 < \cdots < x_m$ are the free variables of f. Then $m = n$ and $x_i : t_i^{a_i} \in \Gamma$ for all $i \leq n$.

PROOF: An easy induction on $\Gamma \vdash f : (t_1^{a_1}, \ldots, t_n^{a_n})^a$. For rules 6 and 7, use Theorem 53. ⊠

We can now prove a corollary that we promised in Remark 43.7:

Corollary 55. If $\Gamma \vdash f : (t_1^{a_1}, \ldots, t_n^{a_n})^a$ and φ is a bijection $\{1, \ldots, n\} \to \{1, \ldots, n\}$ then there is a context Γ' and a pf f' which is α_P-equal to f such that $\Gamma' \vdash f' : \left(t_{\varphi(1)}^{a_{\varphi(1)}}, \ldots, t_{\varphi(n)}^{a_{\varphi(n)}}\right)^a$.

We can also prove unicity of types and unicity of orders. Orders are unique in the following sense:

Lemma 56. Assume $\Gamma \vdash f : t^a$. If x occurs in f and $x : u^b \in \Gamma$, then u^b is predicative. Moreover, if also $\Gamma \vdash f : t'^{a'}$, then $a = a'$.

PROOF: By induction on the derivation of $\Gamma \vdash f : t^a$ one shows that a variable x that occurs in f always has a predicative type in Γ, and that both a and a' equal one plus the maximum of the orders of all the (free and non-free) variables that occur in f. ⊠

Corollary 57. (Unicity of types for pfs) Assume Γ is a context, f is a pf, $\Gamma \vdash f : t^a$ and $\Gamma \vdash f : u^b$. Then $t^a \equiv u^b$.

PROOF: $t \equiv u$ follows from Theorem 54; $a = b$ from Lemma 56. ⊠

Remark 58. We cannot omit the context Γ in Corollary 57. For example, the pf $\mathtt{z(x)}$ can have different types in different contexts, as is illustrated by the following derivations (we have omitted the orders as they can be calculated via Lemma 56):

$$\dfrac{\dfrac{\vdash \mathtt{R(a_1)} : () \qquad \vdash \mathtt{a_1} : \mathtt{0}}{\mathtt{x} : \mathtt{0} \vdash \mathtt{R(x)} : (\mathtt{0})}\ \text{rule 3}}{\mathtt{x} : \mathtt{0}, \mathtt{z} : (\mathtt{0}) \vdash \mathtt{z(x)} : (\mathtt{0}, (\mathtt{0}))}\ \text{rule 4}$$

versus

$$\dfrac{\dfrac{\vdash \mathtt{R(a_1)} : ()}{\mathtt{x} : () \vdash \mathtt{x()} : (())}\ \text{rule 4}}{\mathtt{x} : (), \mathtt{z} : (()) \vdash \mathtt{z(x)} : ((), (()))}\ \text{rule 4.}$$

[39]The meta-properties can also be proved directly, without λ-calculus: see [Laan, 1994].

Theorem 54 and Corollary 57 show that our system RTT makes sense, in a certain way: The type of a pf only depends on the context and does not depend on the way in which we derived the type of that pf.

As a corollary of Corollary 57 we find:

Corollary 59. If $\Gamma \vdash f : t^a$, $\Gamma \vdash k : u^b$, $x{:}u^b \in \Gamma$ and $\Gamma \vdash f[x{:=}k] : t'^{a'}$ then $a \geq a'$.

PROOF: If $x \notin \text{FV}(f)$ then $f \equiv f[x{:=}k]$ and the corollary follows from Unicity of Types (Corollary 57). If $x \in \text{FV}(f)$ then the variables that occur in $f[x{:=}k]$, occur either in f or in k, and as the order of k is smaller than the order of f ($x \in \text{FV}(f)$, so $b < a$), the corollary follows from the proof of Lemma 56. ⊠

OTHER PROPERTIES

We conclude this section with mending the two loose ends discussed in Remark 25 which play a role in RTT-Definition 40:

First, using the strong normalisation of Church's $\lambda\!\to_C$, it is easy to see that:

Theorem 60. (Existence of normal forms) Take $i \leq n$. Assume the following: $\Gamma \cup \{y{:}t_i^{a_i}\} \vdash f : (t_1^{a_1}, \ldots, t_n^{a_n})^a$ and $\Gamma \vdash k : t_i^{a_i}$ (so: the preconditions of rule 6 of RTT are fulfilled). Then $(\lambda y{:}T(t_i^{a_i}).\widetilde{f})\overline{k}$ is strongly normalising.

Substitution always exists in the case of RTT-rule 6 of Definition 40.

Theorem 61. (Existence of substitution) If $f \in \mathcal{P}$, y is the ith free variable in f, $\Gamma \cup \{y : t_i^{a_i}\} \vdash f : (t_1^{a_1}, \ldots, t_n^{a_n})^a$, and $\Gamma \vdash k : t_i^{a_i}$, then $f[y{:=}k]$ exists.

3.9 Legal propositional functions in RTT

We recall Definition 41: a pf f is called *legal* if $\Gamma \vdash f : t^a$ for some Γ and t^a. We will check whether this definition of legal pf coincides with the definition of formula that was given in the *Principia*. For this purpose we prove a number of lemmas concerning the relation between legal pfs and predicative types.

We do not distinguish between pfs that are α_P-equal, nor do we distinguish between types $(t_1, \ldots, t_n)$ and $(t_{\varphi(1)}, \ldots, t_{\varphi(n)})$ for a bijection φ. This is justified by Corollary 55 and by the fact, that pfs that are α_P-equal are supposed to be the same in the *Principia* too.

We define the notion "up to α_P-equality" formally:

Definition 62. Let $f \in \mathcal{P}$, Γ a context, t^a a type. *f is of type t^a in the context Γ up to α_P-equality*, notation $\Gamma \vdash f : t^a (\text{mod } \alpha_P)$, if there is $f' \in \mathcal{P}$, a context Γ' and a bijection $\varphi : \mathcal{V} \to \mathcal{V}$ such that

- $\Gamma' \vdash f' : t^a$;
- f' and f are α_P-equal via the bijection φ;
- $\Gamma' = \{\varphi(x){:}u^b \mid x{:}u^b \in \Gamma\}$.

We say that f is *legal* in the context Γ *up to* α_P*-equality* if there is a type u^b such that $\Gamma \vdash f : u^b (\text{mod } \alpha_P)$. We say that f is *legal up to* α_P*-equality* if there is a context Γ such that f is legal in Γ up to α_P-equality.

The following lemma states that all predicative types are "inhabited":

Lemma 63. If t^a is predicative then there are f, Γ such that $\Gamma \vdash f : t^a$.

PROOF: We use induction on predicative types. ⊠

Remark 64. From a modern point of view, this is a remarkable lemma. Many modern type systems are based on the principle of propositions-as-types. In such systems types represent propositions, and terms inhabiting such a type represent proofs of that proposition. In a propositions-as-types based system in which all types are inhabited, all propositions are provable. Such a system would be (logically) inconsistent. RTT is not based on propositions-as-types, and there is nothing paradoxical or inconsistent in the fact that all RTT-types are inhabited.

This lemma can be generalised to some non-predicative types:

Corollary 65. If $(t_1^{a_1}, \ldots, t_m^{a_m})^a$ is a type such that the $t_i^{a_i}$ are all predicative, then there are f and Γ such that $\Gamma \vdash f : (t_1^{a_1}, \ldots, t_m^{a_m})^a$.

We can also show that $z(k_1, \ldots, k_m)$ is legal if $k_1, \ldots, k_m$ are either legal pfs or variables, and z is "fresh".

Lemma 66. If $k_1, \ldots, k_n \in \mathcal{A} \cup \mathcal{V} \cup \mathcal{P}$, $t^a = (t_1^{a_1}, \ldots, t_n^{a_n})^a$ is a predicative type, $\Gamma \vdash k_i : t_i^{a_i}$ for all $k_i \in \mathcal{A} \cup \mathcal{P}$ and $k_i : t_i^{a_i} \in \Gamma$ for all $k_i \in \mathcal{V}$, and $z \in \mathcal{V} \setminus \text{dom}(\Gamma)$, then $z(k_1, \ldots, k_n)$ is legal in the context $\Gamma \cup \{z : t^a\}$ (up to α_P-equality).

It is also not hard to show that $f \vee g$ is legal if f and g are (see also Remark 45):

Lemma 67. If f and g are legal in contexts Γ_1 and Γ_2, respectively, and $\Gamma_1 \cup \Gamma_2$ is a context, then $f \vee g$ is legal in the context $\Gamma_1 \cup \Gamma_2$ (up to α_P-equality).

The following lemma is easy to prove and will be used in the proof of the main result of this section.

Lemma 68. If $R(i_1, \ldots, i_{\mathfrak{a}(R)})$ is a pf with free variables $x_1 < \cdots < x_m$, then it is legal in the context $\{x_j{:}0 \mid 1 \leq j \leq m\}$.

PROOF: Write $f = R(i_1, \ldots, i_{\mathfrak{a}(R)})$. Let $a_1, \ldots, a_m \in \mathcal{A}$ be m different individuals that do not occur in f, and replace each variable x_j in f by a_j, calling the result f'. By the first rule of RTT, f' is legal in the empty context. Re-introducing the variables $x_1, \ldots, x_m$ (by applying rule 3 of RTT m times) for the individuals $a_1, \ldots, a_m$, respectively, we obtain that f is legal in the context $\{x_j{:}0 \mid 1 \leq j \leq m\}$. ⊠

Finally, we can give a characterisation of the legal pfs:

Theorem 69. (Legal pfs in RTT) Let $f \in \mathcal{P}$. f is legal (mod α_P) if and only if:

- $f \equiv R(i_1, \ldots, i_{\mathfrak{a}(R)})$, or
- $f \equiv z(k_1, \ldots, k_n)$, $z \neq k_j$ for all $k_j \in \mathcal{V}$ and z does not occur in any $k_j \in \mathcal{P}$, and there is Γ with $\text{FV}(f) \subseteq \text{DOM}(\Gamma)$ and for all $k_j \in \mathcal{P}$, $\Gamma \vdash k_j{:}t_j^{a_j}$ for some predicative type $t_j^{a_j}$, or
- $f \equiv \neg f'$ and f' is legal (mod α_P) or
- $f \equiv f_1 \vee f_2$ and there are Γ_i and $t_i^{a_i}$ such that $\Gamma_i \vdash f_i{:}t_i^{a_i}$ (mod α_P) for $i = 1, 2$ and $\Gamma_1 \cup \Gamma_2$ is a context, or
- $f \equiv \forall x{:}t^a.f'$ and f' is legal.

PROOF: Use induction on the structure of f. ⊠

We can now answer the question whether our legal pfs (as given in Definition 41) are the same as the formulas of the *Principia*.

First of all, we must notice that all the legal pfs from Definition 41 are also formulas of the *Principia*: This was motivated in Remark 43.

Moreover, we proved (in Theorem 69) that if f is a pf, then the only reasons why f cannot be legal (according to Definition 41) are:

- There is a constituent $z(k_1, \ldots, k_m)$ of f in which z occurs in one of the k_i's;
- There is a constituent $z(k_1, \ldots, k_m)$ of f and a $j \in \{1, \ldots, m\}$ such that k_j is a pf, but not a legal pf;
- f contains two non-overlapping constituents f_1, f_2 that cannot be typed in one and the same context;
- There is a legal constituent $z(k_1, \ldots, k_m)$ of f which is not of predicative type.

Pfs of the first type cannot be legal in the *Principia*, because of the vicious circle principle. The same holds for pfs of the second type, because also in the *Principia*, parameters cannot be untyped. The third problem is a non-issue in the *Principia*. Formal contexts are not present in the *Principia*, but have been introduced in this article to make a precise analysis of RTT possible. Propositional functions of the *Principia* are always constructed in one, implicitly defined, context.[40] A formula,

[40] It is worth remarking that it is possible to formalize *Principia* without resorting to explicit contexts at all. For example, Randall Holmes has an implementation which constructs contexts from the structures of the terms analysed. Following Randall Holmes, the price of this is that the types deduced for terms by his checker are polymorphic: for STT this isn't a problem at all (it's an advantage); Holmes expresses that in RTT, the handling of polymorphic types was quite difficult – he had to allow orders defined in terms of the unknown orders of polymorphic types. Further, in RTT, the type checker had to be much smarter than the STT checker, because it had to be able to deduce identity between polymorphic types in order to successfully infer types for quite simple terms (such as the "definition of equality" $(\forall x.(x(y) \leftrightarrow x(z)))$, where the two

therefore, cannot contain two non-overlapping constituents that cannot be typed in the same context. This excludes pfs of the third type. As to the fourth type, it represents Russell's assumption that non-predicative orders in his hierarchy are always obtained from predicative ones by generalization (i.e., by quantification). Of course Russell's assumption is not true of terms $z(k_1, \ldots, k_n)$ where one of the k_i's happens to be of non-predicative type. This means that both our system and Russell's intended system are not able to type such terms.

We conclude that we have described the legal pfs of the *Principia Mathematica* with the formal system RTT.

We present some refinements of Theorem 69:

Theorem 70. Assume $\Gamma \vdash f : t^a$.

- If $f \equiv R(i_1, \ldots, i_{\mathfrak{a}(R)})$ and $x \in \text{FV}(f)$ then $x{:}0^0 \in \Gamma$;
- If $f \equiv z(k_1, \ldots, k_m)$ then there are $u_1^{b_1}, \ldots, u_m^{b_m}, b$ such that
  - $z{:}\left(u_1^{b_1}, \ldots, u_m^{b_m}\right)^b \in \Gamma$;
  - $\Gamma \vdash k_i{:}u_i^{b_i}$ for $k_i \in \mathcal{A} \cup \mathcal{P}$;
  - $k_i{:}u_i^{b_i} \in \Gamma$ for $k_i \in \mathcal{V}$.

In this section we gave a formalisation of the Ramified Theory of Types. Some of the main ideas underlying this theory were already present in Frege's Abstraction Principles 1 and 2.

RTT not only prevents the paradoxes of Frege's *Grundgesetze der Arithmetik*, but also guarantees the well-definedness of substitution (Theorem 61). This second problem was not realized in the *Principia*, where substitution did not even have a proper definition.

There is a close relation between substitution in *Principia* and β-reduction in λ-calculus (Definition 23). RTT has characteristics that are also the basic properties of modern type systems for λ-calculus.

As there is no real reduction in RTT, we don't have an equivalent of the Subject Reduction theorem. However, the fact that the Free Variable property (Theorem 54) is maintained under substitution can be seen as a (very weak) form of Subject Reduction.

Expressing Russell's propositional functions in λ-calculus has made it possible to compare these pfs with λ-terms. We found that pfs can be seen as λ-terms, but in a rather simple way:

- A pf is always a λI-term, i.e. if $\lambda x{:}A.B$ is a subterm of the translation $\widetilde{f}$ of a pf f, then $x \in \text{FV}(B)$;

variables y and z are both polymorphic, and one has to be careful to determine that they have the same type (because they are in the same argument of the same unknown pf x) before attempting the final type-checking of the term: if one is not careful about the order in which things are done, two incompatible types for x will be deduced depending on unknown and possibly different orders for y and z).

- Substitution in the *Principia* can be seen as application plus β-reduction to normal form.

Although the description of the Ramified Theory of Types in the *Principia* is very informal, it is remarkable that an accurate formalisation of this system can be made (see Theorem 69 and the discussion that follows it). The formalisation shows that Russell and Whitehead's ideas on the notion of types, though very informal to modern standards, must have been very thorough and to the point.

A characteristic of RTT that is maintained in many modern type systems is the syntactic nature of the system: type and order of a pf are determined on purely syntactical grounds. No attention is paid to the interpretation of such a pf. This is remarkable, as the propositions $\forall \mathtt{x}{:}0^0[\mathtt{R}(\mathtt{x})]$ and $\forall \mathtt{x}{:}0^0[\mathtt{R}(\mathtt{x})] \vee \forall \mathtt{z}{:}()^9[\mathtt{z}() \wedge \neg\mathtt{z}()]$ are logically equivalent in most logics[41], though they are of different type (the former pf has type $()^1$ and the latter has type $()^{10}$). In [Kamareddine and Laan, 1996], it is shown that other viewpoints are possible besides this concentration on syntax.

4 HISTORY OF THE DERAMIFICATION

4.1 *The problematic character of* RTT

The main part of the *Principia* is devoted to the development of logic and mathematics using the legal pfs of the ramified type theory. It appears that RTT is not easy to use. The main reason for this is the implementation of the so-called *ramification*: the division of simple types into orders. We illustrate this with two examples:

Example 71. (Equality) One tends to define the notion of *equality* in the style of Leibniz ([Gerhardt, 1890]):

$$\mathtt{x} =_{\mathtt{L}} \mathtt{y} \stackrel{\text{def}}{\leftrightarrow} \forall \mathtt{z}[\mathtt{z}(\mathtt{x}) \leftrightarrow \mathtt{z}(\mathtt{y})],$$

or in words: Two individuals are equal if and only if they have exactly the same properties.

Unfortunately, in order to express this general notion in our formal system, we have to incorporate *all* pfs $\forall \mathtt{z} : (0^0)^n[\mathtt{z}(\mathtt{x}) \leftrightarrow \mathtt{z}(\mathtt{y})]$ for $n > 1$, and this cannot be expressed in one pf.

The ramification does not only influence definitions in logic. Some important mathematical concepts cannot be defined any more:

Example 72. (Real numbers and least upper bounds) Dedekind constructed the real numbers from the rationals using so-called Dedekind cuts. In this construction, a real number is a set r of rationals such that

- $r \neq \emptyset$;

[41] At least in all the logical systems that Russell had in mind when he wrote the *Principia*.

- $r \neq \mathbb{Q}$;
- If $x \in r$ and $y < x$ then $y \in r$;
- If $x \in r$ then there is $y \in r$ with $x < y$.

For instance, the real number $\frac{1}{2}$ is represented by the set $\{x \in \mathbb{Q} \mid 2x < 1\}$, and the real number $\sqrt{2}$ is represented by the set $\{x \in \mathbb{Q} \mid x < 0 \text{ or } x^2 < 2\}$.

If we take $\mathbb{Q}$ as the set of individuals $\mathcal{A}$, and assume that the binary relation $<$ on $\mathbb{Q}$ is an element of $\mathcal{R}$, the set of relations, we can see real numbers as unary predicates f over $\mathbb{Q}$ such that

$$\begin{array}{ll} & \exists \mathtt{x}{:}0^0[\mathtt{z}(\mathtt{x})] \wedge \exists \mathtt{x}{:}0^0[\neg \mathtt{z}(\mathtt{x})] \wedge \\ (6) & \forall \mathtt{x}{:}0^0[\forall \mathtt{y}{:}0^0[\mathtt{z}(\mathtt{x}) \to \mathtt{y} < \mathtt{x} \to \mathtt{z}(\mathtt{y})]] \wedge \\ & \forall \mathtt{x}{:}0^0[\mathtt{z}(\mathtt{x}) \to \exists \mathtt{y}{:}0^0[\mathtt{z}(\mathtt{y}) \wedge \mathtt{x} < \mathtt{y}]] \end{array}$$

holds if we substitute f for $\mathtt{z}$. We will abbreviate the predicate (6) (with the free variable $\mathtt{z}$) as $\mathbb{R}$. It has type $\left((0^0)^1\right)^2$, and real numbers can be seen as pfs of type $(0^0)^1$. We will, for shortness of notation, write $\mathbb{R}(f)$ for $\mathbb{R}[\mathtt{z}{:=}f]$, so $\mathbb{R} \equiv \mathbb{R}(\mathtt{z})$. A real number r is smaller than or equal to another real number r' if for all x with $r(x)$, also $r'(x)$ holds. We write, shorthand, $r \leq r'$ if r is smaller than or equal to r'.

In traditional mathematics, the above would define a system that obeys the traditional axioms for real numbers. In particular, the theorem of the least upper bound holds for this system. This theorem states that each non-empty subset of $\mathbb{R}$ with an upper bound has a least upper bound. In our formalism:

$$\forall \mathtt{v}{\subseteq}\mathbb{R}\left[\begin{array}{c} \left(\begin{array}{c} \exists \mathtt{z}_1{\in}\mathbb{R}\,[\mathtt{v}(\mathtt{z}_1)] \wedge \\ \exists \mathtt{z}_2{\in}\mathbb{R}\forall \mathtt{z}_3{\in}\mathbb{R}\,[\mathtt{v}(\mathtt{z}_3) \to \mathtt{z}_3 \leq \mathtt{z}_2] \end{array}\right) \\ \to \\ \exists \mathtt{z}_1 \in \mathbb{R}\left[\begin{array}{c} \forall \mathtt{z}_2{\in}\mathbb{R}\,[\mathtt{v}(\mathtt{z}_2) \to \mathtt{z}_2 \leq \mathtt{z}_1] \wedge \\ \forall \mathtt{z}_3{\in}\mathbb{R}\left[\begin{array}{c} \forall \mathtt{z}_4{\in}\mathbb{R}\,[\mathtt{v}(\mathtt{z}_4) \to \mathtt{z}_4 \leq \mathtt{z}_3] \\ \to \mathtt{z}_1 \leq \mathtt{z}_3 \end{array}\right] \end{array}\right] \end{array}\right]$$

(We write, shorthand, $\forall v{\subseteq}\mathbb{R}[g]$ to denote $\forall v{:}\left((0^0)^1\right)^2[\forall \mathtt{u}{:}(0^0)^1[v(\mathtt{u}) \to \mathbb{R}(\mathtt{u})] \to g]$, and $\forall z{\in}\mathbb{R}[g]$ to denote $\forall z{:}(0^0)^1[\mathbb{R}(z) \to g]$).

If we try to prove this theorem within the system of Dedekind as formulated in the *Principia*-language RTT, we have to specify a type t^a for the variable $\mathtt{z}_1$. As $\mathtt{z}_1$ must be a real number, its type must be $(0^0)^1$. If we give a proof of the theorem, and construct some object f that should be the least upper bound of a set of real numbers V, f will depend on V. Therefore, a general description of f will have a variable v for V in it. As v is of order 2, f must be of order 3 or more. Therefore, f cannot be a real number, since real numbers have order 1.

This makes it impossible to give a constructive proof of the theorem of the least upper bound within a ramified type theory.

This is a consequence of the fact that it is not possible in RTT to give a definition of an object that refers to the class to which this object belongs (because of the Vicious Circle Principle). Such a definition is called an *impredicative* definition. The relation with the notion of impredicative *type* is immediate:[42] an object defined by an impredicative definition is of a higher order than the order of the elements of the class to which this object should belong. This means that the defined object f has an *impredicative* type.

Nowadays we would consider the use of the Vicious Circle Principle too strict. We consider the impredicative definition of f as a matter of *syntax*, whilst the existence of the object f has to do with *semantics*.[43] The fact that we are not able to give a predicative definition of f does not imply that such an object does not exist. Here we must remark that Russell and Whitehead did not make a distinction between syntax and semantics in the *Principia*.[44] Therefore they had to interpret the Vicious Circle Principle in the strict way above.

4.2 The Axiom of Reducibility

Russell and Whitehead tried to solve these problems explained in Section 4.1 with the so-called *axiom of reducibility*.

Axiom 73. (Axiom of Reducibility) For each formula f, there is a formula g with a *predicative* type such that f and g are (logically) equivalent.

Accepting this axiom, one may define equality on formulas of order 1 only:

$$\mathtt{x} =_1 \mathtt{y} \stackrel{\text{def}}{=} \forall \mathtt{z} : (0^0)^1[\mathtt{z}(\mathtt{x}) \leftrightarrow \mathtt{z}(\mathtt{y})].$$

If f is a function of type $(0^0)^n$ for some $n > 1$, and $\mathtt{a}$ and $\mathtt{b}$ are individuals for which the Leibniz equality $\mathtt{a} =_{\mathtt{L}} \mathtt{b}$ holds then $f(\mathtt{a}) \leftrightarrow f(\mathtt{b})$ holds: With the

[42]This terminology is again the one assumed by *Principia* and not everyone agrees with it. There is actually no problem with the formulation of "impredicative" types from a predicative standpoint: objects of these types are predicatively respectable. An object of truly impredicative type would be defined using quantifiers over *its own* type or even higher types (as is allowed in simple type theory), and would not be typable in the ramified theory at all.

[43]This is obviously a point on which one might disagree. For example, Randall Holmes is unconvinced by the remarks about "syntax" and "semantics". He believes that the syntactical criteria of ramified type theory are a correct implementation of the Vicious Circle Principle and that the Vicious Circle Principle is best understood as a criterion appropriate for *definitions*. According to Randall Holmes, if instances of abstraction or comprehension principles are to be thought of as definitions, then impredicative abstraction or comprehension is indeed questionable and hence the conclusion to be drawn is that abstraction or comprehension axioms are not definitions, but assertions of matters of fact (so "semantic" rather than "syntactic"), and so are not subject to the Vicious Circle Principle (it is not that it should be applied in a more lenient way, but that it does not apply at all). But as long as the Vicious Circle Principle is to be applied, syntactical criteria are appropriate: what a correct definition is should be a matter of syntax.

[44]Though the basic ideas for this were already present in the works of Frege. See for instance *Über Sinn und Bedeutung* [Frege, 1892a].

Axiom of Reducibility we can determine a predicative function g (so of type $(0^0)^1$), equivalent to f. As g has order 1, $g(\mathtt{a}) \leftrightarrow g(\mathtt{b})$ holds. And because f and g are equivalent, also $f(\mathtt{a}) \leftrightarrow f(\mathtt{b})$ holds. This solves the problem of Example 71. A similar solution gives, in Example 72, the proof of the theorem of the least upper bound.

The validity of the Axiom of Reducibility has been questioned from the moment it was introduced. In the introduction to the 2nd edition of the *Principia*, Whitehead and Russell admit:

> "This axiom has a purely pragmatic justification: it leads to the desired results, and to no others. But clearly it is not the sort of axiom with which we can rest content."
>
> (*Principia Mathematica*, p. xiv)

Though Weyl [Weyl, 1918] made an effort to develop analysis within the Ramified Theory of Types (but without the Axiom of Reducibility), and various parts of mathematics can be developed within RTT and without the Axiom[45], the general attitude towards RTT (without the axiom) was that the system was too restrictive, and that a better solution had to be found.

4.3 *Deramification*

The first impulse to such a solution was given by Ramsey [Ramsey, 1926]. He recalls that the Vicious Circle Principle 3 was postulated in order to prevent the paradoxes. Though all the paradoxes were prevented by this Principle, Ramsey considers it essential to divide them into two parts:

1. One group of paradoxes is removed

   > "by pointing out that a propositional function cannot significantly take itself as argument, and by dividing functions and classes into a hierarchy of types according to their possible arguments."
   >
   > (The Foundations of Mathematics, p. 356)

   This means that a class can never be a member of itself. The paradoxes solved by introducing the hierarchy of types (but not orders), like the Russell paradox, and the Burali-Forti paradox, are *logical* or *syntactical* paradoxes;

2. The second group of paradoxes is excluded by the hierarchy of orders. These paradoxes (like the Liar's paradox, and the Richard Paradox) are based on the confusion of language and meta-language. These paradoxes are, therefore, not of a purely mathematical or logical nature. When a proper distinction between object language (the pfs of the system RTT, for example)

[45] See [Jackson, 1995], where many algebraic notions are developed within the Nuprl Proof Development System, a proof checker based on the hierarchy of types and orders of RTT without the Axiom of Reducibility.

and meta-language is made, these so-called *semantical* paradoxes disappear immediately.

Ramsey agrees with the part of the theory that eliminates the syntactic paradoxes. This part is in fact RTT without the orders of the types. The second part, the hierarchy of orders, does not gain Ramsey's support, for the reasons described above. Moreover, by accepting the hierarchy in its full extent one either has to accept the Axiom of Reducibility or reject ordinary real analysis. Ramsey is supported in his view by Hilbert and Ackermann [Hilbert and Ackerman, 1928]. They all suggest a *deramification* of the theory, i.e. leaving out the orders of the types. When making a proper distinction between language and meta-language, the deramification will not lead to a re-introduction of the (semantic) paradoxes.

The solution proposed by Ramsey, and Hilbert and Ackermann, looks better than the Axiom of Reducibility. Nevertheless, both deramification and the Axiom of Reducibility are violations of the Vicious Circle Principle, and reasons (of a more fundamental character than "they do not lead to a re-introduction of the semantic paradoxes" and "it leads to the desired results, and to no others") why these violations can be harmlessly made must be given. Gödel [Gödel, 1944] fills in this gap. He points out that whether one accepts this second principle or not, depends on the philosophical point of view that one has with respect to logical and mathematical objects:

> "it seems that the vicious circle principle [...] applies only if the entities involved are constructed by ourselves. In this case there must clearly exist a definition (namely the description of the construction) which does not refer to a totality to which the object defined belongs, because the construction of a thing can certainly not be based on a totality of things to which the thing to be constructed itself belongs. If, however, it is a question of objects that exist independently of our constructions, there is nothing in the least absurd in the existence of totalities containing members, which can be described only by reference to this totality."
>
> (Russell's mathematical logic)

The remark puts the Vicious Circle Principle back from a proposition (a statement that is either true or false, without any doubt) to a philosophical principle that will be easily accepted by, for instance, intuitionists (for whom mathematics is a pure mental construction) or constructivists, but that will be rejected, at least in its full strength, by mathematicians with a more platonic point of view. It should be noted that intuitionistic mathematics is quite often impredicative although different "constructivist" mathematicians have different opinions about this.

Gödel is supported in his ideas by Quine [Quine, 1963], sections 34 and 35. Quine's criticism on impredicative definitions (for instance, the definition of the least upper bound of a nonempty subset of the real numbers with an upper bound) is not on the definition of a special symbol, but rather on the very assumption of the *existence* of such an object at all. Quine continues by stating that even for

Poincaré, who was an opponent of impredicative definitions and deramification, one of the doctrines of classes is that they are there "from the beginning". So, even for Poincaré there should be no evident fallacy in impredicative definitions.

The deramification has played an important role in the development of type theory. In 1932 and 1933, Church presented his (untyped) λ-calculus [Church, 1932; Church, 1933]. In 1940 he combined this theory with a deramified version of Russell's theory of types to the system that is known as the *simply typed λ-calculus*[46].

5 THE SIMPLE THEORY OF TYPES

5.1 *Constructing the Simple Theory of Types* STT *from* RTT

So far, we have seen the development of type theory since the appearance of *Principia Mathematica* (1910-1912) went through a process of deramification where Ramsey [Ramsey, 1926], and Hilbert and Ackermann [Hilbert and Ackerman, 1928], simplified the Ramified Theory of Types by removing the orders. The result is known as the *Simple Theory of Types* (STT).

Nowadays, STT is known via Church's formalisation in λ-calculus. However, STT already existed (1926) before λ-calculus did (1932), and is therefore not inextricably bound up with λ-calculus. In this section we show how we can obtain a formalisation of STT directly from the formalisation of RTT that was presented in Section 3 by simply removing the orders. Most of the properties that were proved for RTT hold for STT as well, including Unicity of Types and Strong Normalisation. The proofs are all similar to the proofs that were given for RTT. We also make a comparison between Church's formalisation in λ-calculus and the formalisation of STT that is obtained from RTT. It appears that Church's system is much more than only a formalisation. Because of the λ-calculus, Church's system is more expressive.[47]

It is straightforward to carry out the deramification as it was originally proposed by Ramsey, Hilbert and Ackermann: We take the formalisation of RTT that was

[46]Thus, the adjective *simple* is used to distinguish the theory from the more complicated — both in its construction with a double hierarchy and in its use *ramified* theory. The classification "simple", therefore, has nothing to do with the fact that STT, formulated with λ-calculus as described in [Church, 1940], is the simplest system of the Barendregt Cube (see [Barendregt, 1992]).

[47]The removal of orders from type theory may suggest that orders are to be blamed for the restrictiveness of RTT, and that the concept of order is problematic. [Kamareddine and Laan, 1996] shows that this is not necessarily the case by introducing a system KTT, based on Kripke's Hierarchy of Truths [Kripke, 1975], that has an approach completely opposite to STT. Whilst STT is *order*-free, and *types* play the main role, Kripke's Hierarchy of Truths is *type*-free, and *orders* play an important, though not a restrictive, role. The main difference between Kripke's and Russell's notion of order is that Russell's classification is purely syntactical, whilst Kripke's is essentially semantical. [Kamareddine and Laan, 1996] shows that RTT can be embedded in KTT and that there is a straightforward relation between the orders in RTT and the hierarchy of truths of KTT.

presented in Section 3, and leave out all the orders and the references to orders (including the notions of predicative and impredicative types). The system we obtain in this way will be denoted STT. The types used in the system are the simple types of Definition 30.

The following definitions, lemmas, theorems and corollaries, including their proofs, can be adapted to STT without any problems: Definitions 38, 39, 40, 41, Lemma 52, Theorems 53 (first free variable theorem), 54 (second free variable theorem), Corollaries 55, 57 (unicity of types), and Theorem 61 (existence of substitution).

The description of legal pfs for STT follows the same line as in Section 3.9, with straightforward adaptions of Definition 62, and Lemmas 63 (now, all simple types are inhabited), 66, 67, 68, and finally Theorem 69 (characterisation of legal pfs):

Theorem 74. (Legal pfs in STT) Let $f \in \mathcal{P}$. f is legal (mod α) if and only if:

- $f \equiv R(i_1, \ldots, i_{\mathfrak{a}(R)})$, or
- $f \equiv z(k_1, \ldots, k_n)$, $z \neq k_j$ for all $k_j \in \mathcal{V}$ and z does not occur in any $k_j \in \mathcal{P}$, and there is Γ with $\mathrm{FV}(f) \subseteq \mathrm{DOM}(\Gamma)$ and for all $k_j \in \mathcal{P}$, $\Gamma \vdash k_j{:}t_j$, or
- $f \equiv \neg f'$ and f' is legal (mod α) or $f \equiv f_1 \vee f_2$, there are Γ_i and t_i such that $\Gamma_i \vdash f_i{:}t_i$ (mod α) and $\Gamma_1 \cup \Gamma_2$ is a context, or
- $f \equiv \forall x{:}t.f'$ and f' is legal.

A comparison between the formalisations of STT and RTT can easily be made using Theorems 74 and 69. We find that

- All RTT-legal pfs are (when the ramified types behind the quantifiers are replaced by their corresponding simple types) STT-legal;
- A STT-legal pf f is RTT-legal, except when f contains a subformula of the form $z(k_1, \ldots, k_n)$, where one or more of the k_js are not RTT-legal or can only be typed in RTT by an impredicative type.

5.2 *Church's Simply typed λ-calculus $\lambda{\rightarrow}_C$*

We give a definition of the simply typed λ-calculus as introduced by Church [Church, 1940].

The types and terms in the original presentation of $\lambda{\rightarrow}_C$ are a bit different from the presentation in [Barendregt, 1992]. We give some explanation after repeating the original definition:

Definition 75. (Types of $\lambda{\rightarrow}_C$) The types of $\lambda{\rightarrow}_C$ are defined as follows:

- ι and o are types;
- If α and β are types, then so is $\alpha \rightarrow \beta$.

We denote the set of simple types by $\mathbb{T}$.

ι represents the type of individuals; o is the type of propositions. $\alpha \to \beta$ is the type of functions with domain α and range β. We use $\alpha, \beta, \ldots$ as meta-variables over types. $\to$ associates to the right: $\alpha \to \beta \to \gamma$ denotes $\alpha \to (\beta \to \gamma)$.

Definition 76. (Terms of $\lambda{\to}_C$) The terms of $\lambda{\to}_C$ are the following:

- $\neg$, $\wedge$, $\forall_\alpha$ for each type α, and $\imath_\alpha$ for each type α, are terms;
- A variable is a term;
- If A, B are terms, then so is AB;
- If A is a term, and x a variable, then $\lambda x{:}\alpha.A$ is a term.

Remark 77. We see that the constants $\neg$, $\wedge$, $\forall_\alpha$ and $\imath_\alpha$ are terms. This may need some explanation for the modern reader.

- Church considers $\neg$ and $\wedge$ to be functions. The function $\neg$ takes a proposition as argument, and returns a proposition; similarly $\wedge$ takes two propositions as arguments, and returns a proposition. In Definition 79, we see that $\neg$ and $\wedge$ are assigned the corresponding types $o \to o$ and $o \to o \to o$;
- More remarkable: $\forall_\alpha$ and $\imath_\alpha$ are just terms, and do not act as binding operators. The usual variable binding of $\forall_\alpha$ and $\imath_\alpha$ is obtained via λ-abstraction: instead of $\forall x{:}\alpha.f$, Church writes $\forall_\alpha(\lambda x{:}\alpha.f)$. In this way, $\forall_\alpha$ is a function that takes a propositional function of type $\alpha \to o$ as argument, and returns a proposition (a term of type o). In Definition 79, $\forall_\alpha$ obtains the corresponding type $(\alpha \to o) \to o$. Similarly, the unique choice operator $\imath_\alpha$ takes a propositional function of type $\alpha \to o$ as argument, and returns a term of type α. The term $\imath x{:}\alpha.f$, or in Church's notation: $\imath_\alpha(\lambda x{:}\alpha.f)$, has as interpretation: the (unique) object t of type α for which $f[x{:=}t]$ holds. Correspondingly, the type of $\imath_\alpha$ is $(\alpha \to o) \to \alpha$.

Definition 78. (Contexts of $\lambda{\to}_C$) A *context* in $\lambda{\to}_C$ is a set $\{x_1{:}\alpha_1, \ldots x_n{:}\alpha_n\}$ where the x_i are distinct variables and the α_i are types.

Some terms are typable (legal) in $\lambda{\to}_C$, according to the following derivation rules:

Definition 79. (Typing rules of $\lambda{\to}_C$) The judgement $\Gamma \vdash A : \alpha$ holds if it can be derived using the following rules:

- $\Gamma \vdash \neg : o \to o$;

  $\Gamma \vdash \wedge : o \to o \to o$;

  $\Gamma \vdash \forall_\alpha : (\alpha \to o) \to o$;

  $\Gamma \vdash \imath_\alpha : (\alpha \to o) \to \alpha$;

- $\Gamma \vdash x : \alpha$ if $x{:}\alpha \in \Gamma$;
- If $\Gamma, x{:}\alpha \vdash A : \beta$ then $\Gamma \vdash (\lambda x{:}\alpha.A) : \alpha \to \beta$;
- If $\Gamma \vdash A : \alpha \to \beta$ and $\Gamma \vdash B : \alpha$ then $\Gamma \vdash (AB) : \beta$.

We use $\vdash_{\lambda\to_C}$ if we need to distinguish derivability in $\lambda\to_C$ from derivability in other type systems.

The simply typed λ-calculus can be seen as a pure type system, and therefore has the properties of pure type systems [Barendregt, 1992]. To adapt the simply typed λ-calculus to a pure type system, some amendments are made:

- The two basic types ι, o are replaced by an infinite set of *type variables*;
- The constants $\neg$, $\wedge$, $\forall_\alpha$ and $\imath_\alpha$ are not introduced in the PTS-presentation.

These adaptions do not seriously affect the system and are only used to make $\lambda\to_C$ fit in the PTS-framework.

5.3 Comparing RTT *and* $\lambda\to_C$

Apart from the orders, RTT is a subsystem of $\lambda\to_C$ via the embeddings $\bar{\ }$ of Section 3.2 and a mapping T that we define below. There are, however, important differences between the way in which the type of a pf is determined in RTT, and the way in which the type of a λ-term is determined in λ-Church. The rules of RTT, and the method of deriving the types of pfs that was presented in Section 3.9, have a bottom-up character: one can only introduce a variable of a certain type in a context Γ, if there is a pf that has that type in Γ. In $\lambda\to_C$, one can introduce variables of any type without wondering whether such a type is inhabited or not.

Church's $\lambda\to_C$ is more general than RTT in the sense that Church does not only describe (typable) *propositional* functions. In $\lambda\to_C$, also functions of type $\tau \to \iota$ (where ι is the type of individuals) can be described, and functions that take such functions as arguments, etc..

Just as propositional functions can be translated to λ-terms, simple types (see Definition 30) can be translated to types of the simply typed λ-calculus of Church.

Definition 80. (Translating simple types to $\lambda\to_C$-types) We define a type $T(t)$ for each simple type t by induction:

1. $T(0) \stackrel{\text{def}}{=} \iota$;
2. $T((t_1,\ldots,t_n)) \stackrel{\text{def}}{=} T(t_1) \to \cdots \to T(t_n) \to o$.

A simple type t of Definition 30 has the same interpretation as its translation $T(t)$. Moreover, T is injective:

Lemma 81. If t and u are simple types (Definition 30), then $T(t) = T(u)$ if and only if $t = u$.

PROOF: Induction on the definition of simple type. ⊠

The mapping T is injective when restricted to predicative types:

Lemma 82. If t^a and u^b are predicative types, then $T(t^a) = T(u^b)$ if and only if $t^a = u^b$.

PROOF: Induction on the definition of predicative type. ⊠

Ramified types can also be translated to types of the simply typed λ-calculus. However, we lose the orders if we do so.

Definition 83. (Translating ramified types to $\lambda{\rightarrow}_C$-types) We define a type $T(t)$ for each ramified type t by induction:

1. $T(0^0) \stackrel{\text{def}}{=} \iota$;
2. $T\left((t_1^{a_1}, \ldots, t_n^{a_n})^a\right) \stackrel{\text{def}}{=} T(t_1) \to \cdots \to T(t_n) \to o$.

Now we can relate typing in RTT to that of Church's $\lambda{\rightarrow}_C$:

Theorem 84. (Typability in RTT implies typability in $\lambda{\rightarrow}_C$) If $\Gamma \vdash f : t^a$ in RTT then

1. $T(\Gamma) \vdash_{\lambda \rightarrow_C} \widetilde{f} : o$;
2. $T(\varnothing) \vdash_{\lambda \rightarrow_C} \overline{f} : T(t^a)$.

PROOF: A straightforward induction on $\Gamma \vdash f : t^a$ with the use of Theorem 54 and the Subject Reduction property for λ-Church. ⊠

Remark 85. Observe that the above theorem immediately excludes the pf that leads to the Russell Paradox from the well-typed pfs: If $\neg\mathbf{z}(\mathbf{z})$ were legal then the λ-term $\neg(\mathbf{zz})$ would be typable in λ-Church, which is not the case (see [Barendregt, 1992]).

5.4 Comparison of STT *with Church's* $\lambda{\rightarrow}_C$

The mappings T for types and $\bar{}$ for terms (see Definitions 83 and 9), adapted for STT, make it possible to compare STT with $\lambda{\rightarrow}_C$.

Regarding the types, we find that T gives an injective correspondence between types of STT and $\lambda{\rightarrow}_C$. T is clearly not surjective, as $T(t)$ is never of the form $\alpha \to \iota$ (this follows directly from Definition 80). This indicates an important difference between STT and $\lambda{\rightarrow}_C$. In RTT and STT, functions (other than propositional functions) have to be defined via relations (and this is the way it is done in *Principia Mathematica*). The value of such a function f, described via the relation R, for a certain value a is described using the $\imath$-operator: $\imath y.R(a,y)$ (to be interpreted as: the unique y for which $R(a, y)$ holds). Things get even more complicated if one realizes that the $\imath$-operator is not a part of the syntax used in *Principia Mathematica*, but an abbreviation with a not so straightforward translation (see [Whitehead and Russell, 1910], pp. 66–71). In $\lambda{\to}_C$, as everywhere in λ-calculus, functions (both propositional functions and other ones) are first-class citizens, which means that the construction with the $\imath$-operator is not the first tool to be used when constructing a function. If one has an algorithm (a λ-term) that describes the function f, the value of f for the argument a can be easily described via the term fa. And even if such an algorithm is not at hand, one can use the $\imath$-operator, which is part of the syntax of $\lambda{\to}_C$. This makes $\lambda{\to}_C$ much easier to use for the formalisation of logic and mathematics than RTT and STT.

Regarding the terms, $\bar{}$ provides an injective correspondence between terms of STT and $\lambda{\to}_C$. Again, this mapping is not surjective, for several reasons:

- T is not surjective. As there is no t with $T(t) = \iota \to \iota$, there cannot be a legal pf f such that $\overline{f} \equiv \lambda \mathtt{x}{:}\iota.\mathtt{x}$
- We already observed that $\overline{f}$ is a λI-term for all $f \in \mathcal{P}$. $\lambda{\to}_C$ also allows terms like $\lambda \mathtt{x}{:}\alpha.\mathtt{y}$;
- If $\overline{f} \equiv zH_1 \cdots H_n$ for some $z \in \mathcal{V}$ and some terms $H_1, \ldots, H_n$, the H_i's must be either *closed* λ-terms, or variables, or individuals. This means that there is no $f \in \mathcal{P}$ such that $\overline{f} \equiv \lambda \mathtt{z}{:}o{\to}o.\lambda \mathtt{x}{:}\iota.\mathtt{z}(\mathtt{Rx})$, since $\mathtt{Rx}$ contains the free variable $\mathtt{x}$ and is neither a variable nor an individual;
- We remark that $\overline{f}$ is always a closed λ-term, so there is no $f \in \mathcal{P}$ such that $\overline{f} \equiv \mathtt{x}$;
- It has already been remarked that the $\imath$-operator is part of the syntax of $\lambda{\to}_C$, and this is not the case in STT and RTT.

The discussion above makes clear that $\lambda{\to}_C$ is a far more expressive system than RTT and STT. Type-theoretically, it generalises the idea of function types of Frege and Russell from propositional functions to more general functions.

Philosophically, there is another important difference between STT and $\lambda{\to}_C$. The systems STT and RTT have a strong bottom-up approach: To type a higher-order pf one has to start with propositions of order 0. Only by applying the abstraction principles, it is possible to obtain higher-order pfs. In $\lambda{\to}_C$, one can introduce a variable of a higher-order type at once, without having to refer to terms of lower order.

6 CONCLUSION

In this article, we gave a history of type theory up to 1910 and presented in detail the first type theory RTT due to Russell which he used to prevent the paradoxes of Frege's *Grundgesetze der Arithmetik*. Then we discussed the deramification of

RTT (i.e. the removal of orders) leading to the simple theory of types STT. We also presented Church's simply typed λ-calculus $\lambda{\rightarrow}_C$ and compared the three type systems RTT, STT and $\lambda{\rightarrow}_C$.

Some of the main ideas underlying RTT were already present in Frege's Abstraction Principles 1 and 2.

RTT not only prevents the paradoxes of Frege's *Grundgesetze der Arithmetik*, but also guarantees the well-definedness of substitution, as we have shown in Corollary 61. This second problem was not realized in the *Principia*, where substitution did not even have a proper definition.

There is a close relation between substitution in *Principia* and β-reduction in λ-calculus (Definition 23). RTT has characteristics that are also the basic properties of modern type systems for λ-calculus. As there is no real reduction in RTT, we don't have an equivalent of the Subject Reduction theorem. However, the fact that the Free Variable property 54 is maintained under substitution can be seen as a (very weak) form of Subject Reduction.

Although the description of the Ramified Theory of Types in the *Principia* is very informal, it is remarkable that an accurate formalisation of this system can be made (see Theorem 69 and the discussion that follows it). The formalisation shows that Russell and Whitehead's ideas on the notion of types, though very informal to modern standards, must have been very thorough and to the point.

Apart from the orders, RTT is a subsystem of Church's $\lambda{\rightarrow}_C$ of [Church, 1940] via the embeddings $^{-}$ of Section 3.2 and T of Section 5.3. There are, however, important differences between the way in which the type of a pf is determined in RTT, and the way in which the type of a λ-term is determined in λ-Church. The rules of RTT, and the method of deriving the types of pfs (propositional functions) that was presented in Section 3.9, have a bottom-up character: one can only introduce a variable of a certain type in a context Γ, if there is a pf that has that type in Γ. In $\lambda{\rightarrow}_C$, one can introduce variables of any type without wondering whether such a type is inhabited or not.

Church's $\lambda{\rightarrow}_C$ is more general than RTT in the sense that Church does not only describe (typable) *propositional* functions. In $\lambda{\rightarrow}_C$, also functions of type $\tau \rightarrow \iota$ (where ι is the type of individuals) can be described, and functions that take such functions as arguments, etc..

A characteristic of RTT that is maintained in many modern type systems is the syntactic nature of the system: type and order of a pf are determined on purely syntactical grounds. No attention is paid to the interpretation of such a pf. This is remarkable, as the propositions $\forall \mathtt{x}{:}0^0[\mathtt{R}(\mathtt{x})]$ and $\forall \mathtt{x}{:}0^0[\mathtt{R}(\mathtt{x})] \vee \forall \mathtt{z}{:}()^9[\mathtt{z}() \wedge \neg \mathtt{z}()]$ are logically equivalent in most logics[48], though they are of different type (the former pf has type $()^1$ and the latter has type $()^{10}$).

We saw in Section 4.1 that the Ramified Theory of Types is very restrictive for the description of mathematics within logic, because it is not possible to formulate impredicative definitions in RTT.

[48]At least in all the logical systems that Russell had in mind when he wrote the *Principia*.

This was already realised by Russell and Whitehead, who tried to solve this by postulating the Axiom of Reducibility (Axiom 73). This axiom has been criticised from the moment it was written down, both by Russell and Whitehead themselves and by others. Ramsey, Hilbert and Ackermann were not satisfied by RTT's orders and therefore deramifed RTT: They removed the orders. They observed that this does not lead to known paradoxes as long as a proper distinction between language and metalanguage is made.

Gödel and Quine observed that the deramification does not violate the Vicious Circle Principle, as long as one accepts that objects and pfs exist independently of our constructions.

The main line in the history continues with non-ramified theories. For example, Church's combination of λ-calculus with simple type theory, the basis for most modern type systems, has no orders. Similarly to RTT however, $\lambda{\rightarrow}_C$ is very restrictive. The hierarchy of the simple theory of types used by $\lambda{\rightarrow}_C$ leads to a duplication of work. For example, numbers, booleans, the identity function have to be defined at every level. This led to the development of type theories that are *polymorphic* and hence avoid this unsatisfactory and inefficient duplication of work.[49] In [Barendregt, 1992], the reader may find a review of some of these type theories.

ACKNOWLEDGEMENTS

This article appears in [Kamareddine *et al.*, 2002]. We thank the Association for Symbolic Logic for their permission to reprint it in this Volume.

We are grateful for the useful feedback from and discussions with Henk Barendregt, Andreas Blass, Ivor Grattan-Guinness, Roger Hindley and Joe Wells. Randall Holmes read our work in thorough details, provided extremely useful comments and implemented an impressive proof checker for the system of Principia Mathematica which helps one to see the extent to which Russell and Whitehead were successful in their early attempt at the logical formalization of mathematics.

BIBLIOGRAPHY

[Abramsky *et al.*, 1992] S. Abramsky, Dov M. Gabbay, and T.S.E. Maibaum, editors. *Handbook of Logic in Computer Science, Volume 2: Background: Computational Structures.* Oxford University Press, 1992.

[Barendregt, 1984] H.P. Barendregt. *The Lambda Calculus: its Syntax and Semantics.* Studies in Logic and the Foundations of Mathematics 103. North-Holland, Amsterdam, revised edition, 1984.

[Barendregt, 1992] H.P. Barendregt. Lambda calculi with types. In [Abramsky *et al.*, 1992], pages 117–309. Oxford University Press, 1992.

[Bar-Hillel *et al.*, 1973] Bar-Hillel Y., Fraenkel A. and Levy A. *Foundations of set theory.* North-Holland, 1973.

[49]Note that polymorphism was already recognized by Russell as *typical ambiguity* (cf. pages 161 and 162 of *Principia*). Moreover, Quine's NF and ML are polymorphic systems.

[Benacerraf and Putnam, 1983] P. Benacerraf and H. Putnam, editors. *Philosophy of Mathematics*. Cambridge University Press, second edition, 1983.

[Beth, 1959] E.W. Beth. *The Foundations of Mathematics*. Studies in Logic and the Foundations of Mathematics. North-Holland, Amsterdam, 1959.

[Boolos, 1971] G. Boolos. The iterative conception of set. *Philosophy*, LXVIII:215–231, 1971.

[Burali-Forti, 1897] C. Burali-Forti. Una questione sui numeri transfiniti. *Rendiconti del Circolo Matematico di Palermo*, 11:154–164, 1897. English translation in [van Heijenoort, 1967], pages 104–112.

[Cantor, 1895] G. Cantor. Beiträge zur Begründung der transfiniten Mengenlehre (Erster Artikel). *Mathematische Annalen*, 46:481–512, 1895.

[Cantor, 1897] G. Cantor. Beiträge zur Begründung der transfiniten Mengenlehre (Zweiter Artikel). *Mathematische Annalen*, 49:207–246, 1897.

[Cauchy, 1821] A.-L. Cauchy. *Cours d'Analyse de l'Ecole Royale Polytechnique*. Debure, Paris, 1821. Also as *Œuvres Complètes* (2), volume III, Gauthier-Villars, Paris, 1897.

[Church, 1932] A. Church. A set of postulates for the foundation of logic (1). *Annals of Mathematics*, 33:346–366, 1932.

[Church, 1933] A. Church. A set of postulates for the foundation of logic (2). *Annals of Mathematics*, 34:839–864, 1933.

[Church, 1940] A. Church. A formulation of the simple theory of types. *The Journal of Symbolic Logic*, 5:56–68, 1940.

[Church, 1976] A. Church. Comparison of Russell's resolution of the semantic antinomies with that of Tarski. *The Journal of Symbolic Logic*, 41:747–760, 1976.

[Cocchiarella, 1984] N. B. Cocchiarella. Frege's double correlation thesis and Quine's set theories NF and ML. *Philosophical Logic*, 13, 1984.

[Cocchiarella, 1986] N. B. Cocchiarella. Philosophical perspectives on formal theories of predication. *Handbook of Philosophical Logic*, 4, 1986.

[Curry, 1934] H.B. Curry. Functionality in combinatory logic. *Proceedings of the National Academy of Science of the USA*, 20:584–590, 1934.

[Curry, 1963] H.B. Curry. *Foundations of Mathematical Logic*. McGraw-Hill Series in Higher Mathematics. McGraw-Hill Book Company, Inc., 1963.

[Curry and Feys, 1958] H.B. Curry and R. Feys. *Combinatory Logic I*. Studies in Logic and the Foundations of Mathematics. North-Holland, Amsterdam, 1958.

[Dedekind, 1872] R. Dedekind. *Stetigkeit und irrationale Zahlen*. Vieweg & Sohn, Braunschweig, 1872.

[Euclid] Euclid. *The Elements*. 325 B.C.. English translation in [Heath, 1956].

[Feferman, 1984] S. Feferman. Towards useful type-free theories I. *Symbolic Logic*, 49:75–111, 1984.

[Frege, 1879] G. Frege. *Begriffsschrift, eine der arithmetischen nachgebildete Formelsprache des reinen Denkens*. Nebert, Halle, 1879. Also in [van Heijenoort, 1967], pages 1–82.

[Frege, 1884] G. Frege. *Grundlagen der Arithmetik, eine logisch-mathematische Untersuchung über den Begriff der Zahl*. Breslau, 1884.

[Frege, 1891] G. Frege. *Funktion und Begriff, Vortrag gehalten in der Sitzung vom 9. Januar der Jenaischen Gesellschaft für Medicin und Naturwissenschaft*. Hermann Pohle, Jena, 1891. English translation in [McGuinness, 1984], pages 137–156.

[Frege, 1892] G. Frege. *Grundgesetze der Arithmetik, begriffsschriftlich abgeleitet*, volume I. Pohle, Jena, 1892. Reprinted 1962 (Olms, Hildesheim).

[Frege, 1892a] G. Frege. Über Sinn und Bedeutung. *Zeitschrift für Philosophie und philosophische Kritik*, new series, 100:25–50, 1892. English translation in [McGuinness, 1984], pages 157–177.

[Frege, 1896] G. Frege. Ueber die Begriffschrift des Herrn Peano und meine eigene. *Berichte über die Verhandlungen der Königlich Sächsischen Gesellschaft der Wissenschaften zu Leipzig, Mathematisch-physikalische Klasse 48*, pages 361–378, 1896. English translation in [McGuinness, 1984], pages 234–248.

[Frege, 1902] G. Frege. Letter to Russell. English translation in [van Heijenoort, 1967], pages 127–128, 1902.

[Frege, 1903] G. Frege. *Grundgesetze der Arithmetik, begriffsschriftlich abgeleitet*, volume II. Pohle, Jena, 1903. Reprinted 1962 (Olms, Hildesheim).

[Gerhardt, 1890] C.I. Gerhardt, editor. *Die Philosophischen Schriften von Gottfried Wilhelm Leibniz.* Berlin, 1890.

[Gödel, 1931] K. Gödel. Über formal unentscheidbare Sätze der Principia Mathematica und verwandter Systeme I. *Monatshefte für Mathematik und Physik*, 38:173–198, 1931. English translation in [van Heijenoort, 1967], pages 592–618.

[Gödel, 1944] K. Gödel. Russell's mathematical logic. In P.A. Schlipp, editor, *The Philosophy of Bertrand Russell.* Evanston & Chicago, Northwestern University, 1944. Also in [Benacerraf and Putnam, 1983], pages 447–469.

[Grattan-Guinness, 2001] I. Grattan-Guinness. *The Search for Mathematical Roots, 1870-1930.* Princeton University Press, 2001.

[Heath, 1956] T.L. Heath. *The Thirteen Books of Euclid's Elements.* Dover Publications, Inc., New York, 1956.

[van Heijenoort, 1967] J. van Heijenoort, editor. *From Frege to Gödel: A Source Book in Mathematical Logic, 1879–1931.* Harvard University Press, Cambridge, Massachusetts, 1967.

[Hilbert and Ackerman, 1928] D. Hilbert and W. Ackermann. *Grundzüge der Theoretischen Logik.* Die Grundlehren der Mathematischen Wissenschaften in Einzeldarstellungen, Band XXVII. Springer Verlag, Berlin, first edition, 1928.

[Holmes, 1993] R. Holmes. Systems of combinatory logic related to predicative and "mildly impredicative" fragments of Quine's "New Foundations". *Annals of Pure and Applied Logic*, 59:45-53, 1993.

[Holmes, 1999] R. Holmes. Subsystems of Quine's "New Foundations" with Predicativity Restrictions. *Notre Dame Journal of Formal Logic*, 40(2):183-196, 1999.

[Jackson, 1995] P.B. Jackson. *Enhancing the Nuprl Proof Development System and Applying it to Computational Abstract Algebra.* PhD thesis, Cornell University, Ithaca, New York, 1995.

[Jensen, 1969] R.B. Jensen. On the consistency of a slight modification of Quine's NF. *Synthese*, 19:250–263, 1969.

[Kamareddine and Laan, 1996] F. Kamareddine and T. Laan. A reflection on Russell's ramified types and Kripke's hierarchy of truths. *Journal of the Interest Group in Pure and Applied Logic*, 4(2):195–213, 1996.

[Kamareddine and Laan, 2001] F. Kamareddine and T. Laan. A correspondence between Martin-Löf type theory, the ramified theory of types and pure type systems. *Logic, Language and Information*, 10(3):375–402, 2001.

[Kamareddine *et al.*, 2002] F. Kamareddine, R. Nederpelt and T. Laan. Types in Logic and Mathematics before 1940. *Bulletin of Symbolic Logic* 8(2): 185-245, 2002.

[Kneebone, 1963] G.T. Kneebone. *Mathematical Logic and the Foundations of Mathematics.* D. Van Nostrand Comp., London, New York, Toronto, 1963.

[Kripke, 1975] S. Kripke. Outline of a theory of truth. *Journal of Philosophy*, 72:690–716, 1975.

[Laan, 1994] T. Laan. A formalization of the Ramified Type Theory. Technical Report 94-33, TUE Computing Science Reports, Eindhoven University of Technology, 1994.

[Laan, 1997] T. Laan. *The Evolution of Type Theory in Logic and Mathematics.* PhD thesis, Eindhoven University of Technology, 1997.

[Laan and Nederpelt, 1996] T. Laan and R.P. Nederpelt. A modern elaboration of the Ramified Theory of Types. *Studia Logica*, 57(2/3):243–278, 1996.

[Landini, 1998] G. Landini. *Russell's hidden substitutional theory.* Oxford University Press, 1998.

[McGuinness, 1984] B. McGuinness, editor. *Gottlob Frege: Collected Papers on Mathematics, Logic, and Philosophy.* Basil Blackwell, Oxford, 1984.

[Peano, 1889] G. Peano. *Arithmetices principia, nova methodo exposita.* Bocca, Turin, 1889. English translation in [van Heijenoort, 1967], pages 83–97.

[Peano, 1894] G. Peano. *Formulaire de Mathématique.* Bocca, Turin, 1894–1908. 5 successive versions; the final edition issued as *Formulario Mathematico.*

[Peremans, 1994] W. Peremans. Ups and downs of type theory. Technical Report 94-14, TUE Computing Science Notes, Eindhoven University of Technology, 1994.

[Poincaré, 1902] H. Poincaré. Du rôle de l'intuition et de la logique en mathematiques. *C.R. du IIme Cong. Intern. des Math., Paris 1900*, pages 200–202, 1902.

[Quine, 1937] W. Van Orman Quine. New foundations for mathematical logic. *American Mathematical Monthly*, 44:70–80, 1937. Also in [Quine, 1961], pages 80–101.

[Quine, 1940] W. Van Orman Quine. *Mathematical Logic.* Norton, New York, 1940. Revised edition Harvard University Press, Cambridge, 1951.

[Quine, 1961] W. Van Orman Quine. *From a Logical Point of View: 9 Logico-Philosophical Essays.* Harvard University Press, Cambridge, Massachusetts, second edition, 1961.

[Quine, 1963] W. Van Orman Quine. *Set Theory and its Logic.* Harvard University Press, Cambridge, Massachusetts, 1963.

[Ramsey, 1926] F.P. Ramsey. The foundations of mathematics. *Proceedings of the London Mathematical Society,* 2nd series, 25:338–384, 1926.

[van Rooij, 1986] A.C.M. van Rooij. *Analyse voor Beginners.* Epsilon Uitgaven, Utrecht, 1986.

[Rosser, 1984] J.B. Rosser. Highlights of the history of the lambda-calculus. *Annals of the History of Computing,* 6(4):337–349, 1984.

[Russell, 1902] B. Russell. Letter to Frege. English translation in [van Heijenoort, 1967], pages 124–125, 1902.

[Russell, 1903] B. Russell. *The Principles of Mathematics.* Allen & Unwin, London, 1903.

[Russell, 1908] B. Russell. Mathematical logic as based on the theory of types. *American Journal of Mathematics,* 30:222–262, 1908. Also in [van Heijenoort, 1967], pages 150–182.

[Schönfinkel, 1924] M. Schönfinkel. Über die Bausteine der mathematischen Logik. *Mathematische Annalen,* 92:305–316, 1924. Also in [van Heijenoort, 1967], pages 355–366.

[Schütte, 1960] K. Schütte. *Beweistheorie.* Die Grundlehren der Mathematischen Wissenschaften in Einzeldarstellungen, Band 103. Springer Verlag, Berlin, 1960.

[Seldin, 1996] J.P. Seldin. Personal communication, 1996.

[Specker, 1953] E.P. Specker. The axiom of choice in Quine's New Foundations for Mathematical Logic. *Proc. Nat. Acad. Sci. USA.,* 39:972–975, 1953.

[Weyl, 1918] H. Weyl. *Das Kontinuum.* Veit, Leipzig, 1918. Also in: Das Kontinuum und andere Monographien, Chelsea Pub.Comp., New York, 1960.

[Whitehead and Russell, 1910] A.N. Whitehead and B. Russell. *Principia Mathematica,* volume I, II, III. Cambridge University Press, first edition: 1910, 1912, 1913, second edition: 1925, 1927, 1927. All references are to the first volume, unless otherwise stated.

[Wilder, 1965] R.L. Wilder. *The Foundations of Mathematics.* Robert E. Krieger Publishing Company, Inc., New York, second edition, 1965.

[Zermelo, 1908] E. Zermelo. Untersuchungen über die Grundlagen der Mengenlehre. *Math. Annalen,* 65:261–281, 1908.

A HISTORY OF THE FALLACIES IN WESTERN LOGIC[1]

John Woods

1 INTRODUCTORY REMARKS

The concept of fallacy predates the founding of logic and the bestowal of its name by Aristotle. It is implicit in the contrast between good arguments and good-looking arguments, which in turn instantiates the more generic distinction between appearance and reality. Aristotle's predecessors knew well that bad argument can exhibit the false appearance of goodness. In many circles, not excluding their own, the Sophists were a scandal. Not only did they pedal bad arguments that looked good, but they had the nerve to expect payment for their trouble.

Plato (*c*.421-347 BC) was an implacable enemy of false appearances but a great respecter of argument. Plato knew, especially about matters inaccessible to direct observation, that one of the functions of argument was to establish the truth of propositions antecedently believed to be false or even absurd. Plato was also aware, though not quite in these words, that sometimes a valid argument is a *reductio* of its own premisses and sometimes is a sound demonstration of a shocking truth. So while he was a stalwart in the fight against false appearances and bad arguments alike, he was never so naïve as to suppose an argument to be defective just because its conclusions appeared to be false or absurd.

It takes little reflection to see that any valid argument is subject to this dualism. What one person sees as a *reductio,* another may see as the sound demonstration of an utterly counterintuitive truth. This raises a question of fundamental importance. Is there a principled way of adjudicating this kind of disagreement? Is there a principled way of regulating the appearance-reality distinct in its application to arguments?[2] The answer to these questions depends in no small part on whether

[1]A good review of the early development of Indian logic is Jonardon Ganeri "Indian Logic", this *Handbook,* volume 1 [2004]. See also [Powers, 2012] on Dignanian syllogisms. Also of note is chapter 5 of [Hamblin, 1970] on the Indian tradition. The same is true for Arabaic contributions in Tony Street's "Arabic Logic", also in volume 1. For a discussion of early developments in Chinese logic, see [Chad Hansen, 1983]. A valuable discussion of fallacies in the Mohist tradition is [Zhai, 2011]. Zhai [2011] appears in a special number of *Studies in Logic,* devoted to the history of logic in China.

[2]Plato's disdain for the sophistries of his time was mitigated by the importance he attached to the, as he thought, mistaken arguments of some of his Presocratic rivals, especially Parmenides (b. *c*.515 BC) and Protagoras (*c*.490-420 BC).

Handbook of the History of Logic. Volume 11: Logic: A History of its Central Concepts.
Volume editors: Dov M. Gabbay, Francis Jeffry Pelletier and John Woods.
General editors: Dov M. Gabbay and John Woods.

it is possible to give an accurate and nontrivial characterization of the property of being a bad argument that appears to be good one. This is where Aristotle enters the picture, and in so doing he assigned the fallacies project to logic.

Much of what I shall have to say in the chapter is centred around a *concept* and a *list*. The concept is the traditional concept of fallacy, an idea that has come to us over the centuries from its early definition in antiquity. The list is the traditional list, or what I will for light relief call the "gang of eighteen". The gang of eighteen is not a mathematically well-defined set. Its number varies somewhat from writer to writer, and some lists will contain items which other lists omit. But the gang of eighteen is a representative sample of the fallacies that we think of today as exemplars of the traditional concept of fallacy. The gang of eighteen is constituted as follows: *ad baculum, ad populum, ad verecundiam, ad hominem, ad ignorantiam, ad misericordiam,* affirming the consequent and denying the antecedent, begging the question, many questions, hasty generalization, equivocation, biased statistics, gambler's, *post hoc, ergo, propter hoc,* composition and division, faulty analogy, and *ignoratio elenchi* (of which straw man is a special case).

The earliest predecessor list was Aristotle's "gang of thirteen" in *On Sophistical Refutations.* It is made up of ambiguity (cf. the modern fallacy of equivocation), amphiboly (cf. equivocation), combination of words (cf. composition), division of words (cf. division), *secundum quid* (cf. hasty generalization), *ignoratio elenchi* (cf. straw man), begging the question (cf. begging the question), many questions (cf. many questions), consequent, non-cause as cause, accident, accent and form of expression.[3]

Roughly speaking, the traditional concept of fallacy is that of a mistake of reasoning which people in general tend to commit with a notable frequency and which, even after successful diagnosis, are subject to this same inclination to commit. The traditional list that has come down to us during this same period may be regarded as giving paradigmatic expression to this form of error.

To avoid excessive length, I shall proceed as follows, although here too there will be some occasional exceptions. I shall concentrate my exposition on individuals rather than periods. I shall start the chapter at the subject's beginning, that is, with Aristotle, and I shall close it in 1970, which is the year in which Charles Hamblin's landmark book, *Fallacies,* first appeared. Thanks largely to Hamblin there has been a flurry of activity in the fallacies project these past nearly forty-five years. But the period between 1970 and the present falls under the heading of contemporary affairs, not history. Still, certain features of its handling of fallacies are sufficiently established to permit some "pre-historical" reflection.

1.1 The No-Theory problem

In an important development, Hamblin chastises logicians for something — both historically and presently – they have neglected to do. He writes:

[3]Not on Aristotle's list are *ad hominem* argument, babbling and solecism. I will come back to these below.

> We have no *theory* of fallacy at all, in the sense in which we have theories of correct reasoning or inference. (p. 11)

Hamblin continues:

> ... we are in the position of the medieval logicians of the twelfth century [with respect to logic itself]: we have lost the doctrine of fallacy, and need to recover it. (p. 11)

This is an arresting observation, never mind that it understates the problem. The problem is not that we once had a robust theory of fallacious reasoning, from which somehow we have managed to lose contact. We never did have such a theory. There was nothing to lose contact *with*.

"Well", it might be asked, "Why don't we just get down to it and produce such a theory?" Hamblin responds:

> But it is all the more complicated than that because, these days, we set ourselves higher standards of theoretical rigour and will not be satisfied for long with a theory less ramified and systematic than we are used to in other Departments: and one of the things we may find is that the kind of theory we need cannot be constructed in isolation from them. (pp. 11-12)

Readers of *Fallacies* may think that I am making too much of the No-Theory problem. After all, isn't *Fallacies* itself a contribution to a positive solution? It is true that in chapter 8, entitled "Formal Dialectic", Hamblin makes the following claim, in answer to the question, "What sort of thing would a theory be?'

> ... we need to extend the bounds of Formal Logic; to include features of dialectical contexts with which arguments are put forward. ... Let us start, then, with the concept of a *dialectical system*. (pp. 254-255).

The rest of the chapter is devoted to expounding a dialectical system which, in all essentials, is the one that underlies the Obligation games of the mediaeval dialecticians, beginning with William of Sherwood (1200/1210-1266/1271).[4] But Hamblin discusses the mediaeval developments in chapter 3, "The Aristotelian Tradition", and gives due recognition to Obligation and like enterprises. Yet his judgement in 1970 remains that fallacies writers have got nowhere close to providing the theory that he demands. So we can only wonder at the status of Hamblin's formal dialectic by Hamblin's own lights. (I will recur to the No-Theory question at chapter's end.)

2 ARISTOTLE (384-322 BC)

We begin with the founder of systematic logic. Son of the court physician to the king of Macedonia, Aristotle was born in the city of Stagira. Following the early

[4] William of Sherwood [1996].

death of his father, Aristotle was sent to Athens by his guardian and, around the age of seventeen, he entered Plato's Academy. Twenty years later, Plato died and was succeeded by his nephew Speusippus. It is believed that Aristotle thought badly enough of this choice to leave Athens.

Following a fruitful period of research and writing in biology, Aristotle was appointed tutor to the thirteen-year-old Alexander the Great, as he would soon become. Three years later Aristotle returned to Stagira and in or about 355 BC he established in Athens his own school, the Lyceum. In 323 BC the school was in jeopardy from Aristotle's political enemies. Leaving the Lyceum in the care of Theophrastus, Aristotle sought safety on the island of Euboea. He died there in the following year, aged sixty-two.

As with any of his contemporaries, Aristotle was moved by the fact that in matters both political and scientific, decisions were often taken on the basis of arguments that looked good but were not in fact good. This same failure to suppress bad but good-looking arguments was also the occasion of philosophical doctrines heavy with flummery and error, especially among the Sophists.[5] In his early monographs, *Topics* and *On Sophistical Refutations,* Aristotle seeks for a general theory of argument with which to discipline the distinction between the merely good-looking and the actually good. He emphasized that the theoretical core of this theory of argument would be a theory of what he called "syllogisms". A syllogism is made up of two premisses and a single conclusion, in fulfillment of a number of further conditions. One is that the premisses together deductively imply (or entail) the conclusion. Another is that the conclusion not repeat a premiss. A third is that the argument not contain a redundant premiss. This alone makes syllogistic consequence a nonmonotonic relation and at the same time imposes on syllogisms a relevance condition.[6] Yet another is that the premisses be mutually consistent, thus rendering syllogisms highly paraconsistentist.[7] A fifth is that the statements making up a syllogism be categorical propositions, that is statements of the form: All A are B, Some A are B, Some A are not-B, and No A are B. (This requirement has the particular effect of banning compound statements from syllogisms.) Satisfaction of these conditions produces a particular type of syllogistic implication or syllogistic consequence. In contrast to classical consequence, it is possible to show that for any pair of syllogistically admissible premisses the number, if any, of its syllogistic consequences is exceedingly low.

Some arguments are obviously syllogisms, *Barbara* for example: (1) All A are B; (2) All B are C; (3) Therefore, All A are C. But some syllogisms don't look like syllogisms at all. This matters. The very problem for which they were invented in the first place also inflicts itself on syllogisms. Aristotle has a special name for

[5]But here, too, Aristotle had respect enough for Heraclitus (d. after 480 BC) to give his views a determined (and inconclusive) airing in Book Gamma of the *Metaphysics*.

[6]A more detailed discussion by Woods and Irvine can be found in volume 1 [2004] of this *Handbook,* chapter 2.

[7]A relation of deductive implication is paraconsistent if and only if there is at least one sentence that an inconsistent sentence doesn't deductively imply. Similarly for inconsistent sets of sentences.

this. An argument that looks like a syllogism but isn't a syllogism in fact is a *fallacy*. (*Soph. Ref.* 169^a 14-16)[8]

Implicit in Aristotle's writings is a distinction between arguments in the broad sense and arguments in the narrow sense. Arguments in the narrow sense are typified by syllogisms. They are finite sequences of context-insensitive and agent-independent statements of which the terminal member is the conclusion and the others are premisses, provided that they also fulfill the defining conditions mentioned just above. Arguments in the broad sense are social events. They are dialogues between two or more parties. Aristotle discusses four types of argument in the broad sense: refutation arguments, instruction-arguments, examination arguments and scientific demonstrations from first principles. Syllogisms can be exhaustively described by way of their syntactic and semantic properties. But refutations, indeed all arguments in the broad sense, also incorporate pragmatic factors having to do with the roles of speakers and the forms and order of their utterances, whence the importance of *context* and *agency* to arguments in the broad sense. In addition to the pragmaticization of logic in the broad sense, we owe to Aristotle the introduction into logic (also in the broad sense) of expressly competitive considerations, thus making his broad logic the ancestral home of dialogue logics that would be developed in the ensuing centuries.

The definition of fallacy arises in the discussion of refutations, but it is clear that Aristotle does not think that fallacies are limited to elenchic contexts. Fallacies can also crop up in the other three types of argument mentioned. Aristotle's dialectic is a more complex notion than the one of the same name that has come down to us over the years. For Aristotle, a dialectical argument is a quite particular way of conducting a contentious argument. An argument is not dialectical on account of its contentiousness, but rather because of two features that are definitive of it. The first defining feature has to do with questions. Dialectical arguments pivot on two kinds of dialectical question. Questions of the first kind are in the form "Is A B or is A not-B?" Questions of the second kind are in the form "Is it the case that A is B?" Questions of the second sort are answerable by a "yes" or a "no". Questions of the first sort are answered by affirming one or other of the disjuncts. Refutations originate in an answer to a question of the first sort, which now becomes the answerer's *thesis*, and which he must defend against the attackers a second party. It falls to the second party to put to the thesis-holder a series of questions of the *second* sort. The questioner's further role to use the answerer's yes-no answers as premisses of a syllogism. If the conclusion of the

[8]It may be that in later writings Aristotle came to think that he had solved the appearance-reality problem for syllogisms. Aristotle takes it as given that some syllogisms, *Barbara* for example, are *perfect*. A perfect syllogism is one whose syllogisity is obvious. In its application *here*, the appearance-reality problem for syllogisms is that a syllogism need not appear to be one; that is, there are imperfect syllogisms. However, at *Prior Analytics* I 23, Aristotle launches a perfectability proof, according to which, if an argument is an imperfect or unobvious syllogism, it is possible even so, using principles of reasoning which themselves are obviously valid, that to demonstrate the argument in question is a syllogism. Aristotle's proof is defective but reparable. [Corcoran, 1972].

syllogism is the contradictory of the answerer's thesis, then the questioner has produced a refutation of his opponent.

The second defining feature of dialectical arguments has to do with the subject-matters of dialectical questions. Aristotle requires that yes-no questions be about topics on which there is a more or less settled general agreement. It is the other way round with dialectical questions. These must be about matters concerning which general agreement might be lacking. This second feature gives to dialectical arguments a very peculiar cast. The thesis must be the object of some doubt — a contentious matter, as we might say. But the challenges must admit of ready answers, that is, answers which both the contending parties would generally agree about.

In the writings in which he first introduces the concept of fallacy, Aristotle advances, although not in these words, a pair of theses that pivot on the distinction between arguments in the narrow sense and arguments in the broad sense. The first thesis says that it is an error of logic in the narrow sense to mistake a non-syllogism for a syllogism. The second thesis says there is an interesting class of cases for which the following holds true: If you make a mistake of logic in the narrow sense, you will also make a mistake of logic in the broad sense. In particular, if you mistake a non-syllogism for a syllogism you will wreck the refutation in which it is embedded. As we saw, mistakes of the first kind are called fallacies. Mistakes of the second kind are called *sophistical refutations.*[9] Mistakes of the first kind are a general kind of error, commitable with scant regard, if any, to context. Mistakes of the second kind are mistakes of a specific kind, made so by their essential tie to contexts of refutation. Aristotle's thesis is that the sophistry that attaches to a sophistical refutation is occasioned — *caused* — by the commission of a fallacy, by the mistaking of a non-syllogism for a syllogism. This is a point routinely lost sight of, not only in the foundational writings but in most of those that came after. This, we may think, gives part of an answer to the No-Theory problem raised by Hamblin: Fallacy theorists fail to observe the difference between sophistries and fallacies.

Aristotle thinks that confusing a non-syllogism with a syllogism makes a would-be refutation fail in one or other of the thirteen ways captured by his classification of the sophistical refutations. Aristotle never expressly says that the thirteen are all and only the refutational errors committed by syllogistic mismanagement. But he leaves little doubt that these are not casual errors, more or less randomly committed. On the contrary, these are errors that people in general have a natural tendency to commit, hence are both widespread among humans and widely attractive to them. Aristotle might have added that even after diagnosis, there is a disposition to re-offend. Fallacy is easy to fall into and difficult to climb out of.

[9]The view that the sophisticality of a sophistical refutation is always occasioned by a misbehaving syllogism is not always honoured in Aristotle's writings, even in *On Sophistical Refutations.* Still, there is reason to believe that the intrinsic link is Aristotle's considered view — his official opinion so to speak. See here Hitchcock's excellent [2000]. Nor does Aristotle consistently stick to the requirement that syllogisms not be single-premissed. These, too, are slips.

We see, then, that Aristotle's *concept* of fallacy is clearly discernible in the notion that tradition has preserved to the present day. (The question is, does Aristotle's *definition* of fallacy accord with this *concept* of it?)

Aristotle catalogues and briefly discusses thirteen ways in which refutations can go wrong, that is, ways in which they can be rendered sophistical. Again, these are equivocation, amphiboly, combination and division of words, accent, forms of expression, accident, *secundum quid,* consequent, non-cause as cause, begging the question, many questions and *ignoratio elenchi.* Some of these, e.g. accent, are no longer seriously discussed. Others have a decidedly familiar ring to them, although in a number of cases it is the name that has survived, not the nominatum. Two examples are *secundum quid* and many questions. In modern treatments, *sequndum quid* is the fallacy of hasty generalization, whereas for Aristotle it is the fallacy of omitting a qualification, as in the inference of "Ali is a white man" from "Ali is a white-toothed-man". Similarly, to modern ears, many questions is the fallacy of unconceded presumption, as in "Are you still drinking two bottles a day?" For Aristotle it was the strictly technical mistake of asking a question whose answer is a compound statement rather than a categorical proposition.

At one level *On Sophistical Refutations* is a practical manual in which manoeuvres that support unsatisfactory resolutions of contentious argument are identified, and methods for spotting and blocking them are suggested. But at another level Aristotle is less interested in the practical question of how to train people to win argumentative contests than he is in developing a theory of objectively good reasoning.

At *Soph Ref* 16 175^a5-17, Aristotle explains the importance of a theory of contentious argument:

> The use of ... [contentious arguments] ..., then, is for philosophy, two-fold. For in the first place, since for the most part they depend upon the expression, they put us in a better condition for seeing in how many senses any term is used, and what kind of resemblances and what kind of differences occur between things and their names. In the second place they are useful for one's own personal researches; for the man who is easily committed to a fallacy by someone else, and does not perceive it, is likely to incur this fate of himself also on many occasions. Thirdly [sic] and lastly, they further contribute to one's reputation, viz., the reputation of being well-trained in anything: for that a party to arguments should find fault with them, if he cannot definitely point out their weakness, creates a suspicion, making it seem as though it were not the truth of the matter but merely inexperience that put him out of temper.

Aristotle classifies failed refutations into those that depend on language and those that depend on factors external to language, although it may be closer to his intentions to understand the word 'language' as 'speech'. Aristotle is aware that some fallacies arise because a given word is used ambiguously. However, when the argument in question is *spoken*, the offending word is often given a different pronounciation at different occurrences. Thus speaking the argument rather than reading it sometimes flags the ambiguous term and makes it easy for the arguer to avoid the ambiguity fallacy or at least repair the argument. For such fallacies the mediaevals used the term *in dictione*, and it would appear that what is meant are fallacies whose commission is evident by speaking the argument. Not all fallacies can be identified just in speaking the arguments in which they occur. The mediaevals translated Aristotle's classification of these as *extra dictionem,* that is, as not being identifiable by speaking them. On the other hand, Aristotle also says that there are exactly six ways of producing a 'false illusion in connection with *language*' (165^b26, emphasis added), and his list includes precisely six cases. Further, Aristotle occasionally notices that some of his *extra dictionem* fallacies also qualify for consideration as language-dependent, for example, *ignoratio elenchi* (167^a 35) and many questions (175^b 39). So the modern day practice of taking the *in dictione* fallacies to be language-dependent and the *extra dictionem* fallacies to be language-independent finds a certain justification in Aristotle's text. To this extent, interpreting the distinction in the 'discernible/indiscernible'-in-speech way is doubtful as a general explication. Here is Aristotle's classification.

Sophistical refutations

In dictione	*Extra dictionem*
equivocation	accident
amphiboly	*secundum quid*
combination of words	consequent
division of words	non-cause as cause
accent	begging the question
forms of expression	*ignoratio elenchi*
	many questions

- *Equivocation.* It is immediately evident that Aristotle's placement of these errors does not fit well with the 'discernible in speech versus not discernible in speech' distinction. For example, *equivocation* involves the exploitation of a term's ambiguity, and can be illustrated by the following argument:

  The end of life is death
  Happiness is the end of life
  Therefore, happiness is death.

  But his is a mistake that is not made evident just by speaking the argument.

  Suppose now that we schematize the structure of this argument as

  T is D
  H is E
  Therefore, H is D

in which H stands for 'happiness', E for 'end' in the sense of *goal* or *purpose*, T for 'end' in the sense of *termination*, and D for 'death'. The form is certainly invalid; it commits what later would call the 'fallacy of four terms'. It does not, however, commit the fallacy of *ambiguity*, since in it the term 'end' is fully disambiguated. Some fallacies, such as the fallacy of denying the antecedent, i.e., those having the form,

> If P then Q
> Not-P
> Therefore, not-Q

are thought to be fallacious just in virtue of their having this logical form. Having it, the very fallacy that afflicts the form is said to be inherited by any natural language argument whose form it is. Thus the natural language argument commits the fallacy of denying the antecedent just because its logical form also commits this same fallacy.

It is different with ambiguity errors. In the present example, the natural language argument commits the fallacy of ambiguity, but not because its logical form also commits it. Its logical form commits the different fallacy of four terms, and is entirely free of ambiguity. That this is so requires us to recognize two different concepts of *formal fallacy*. In the second example, denying the consequent, the natural language argument contains a formal fallacy in the sense that it has a logical form guilty of the same error. In the first example, the natural language argument commits the different fallacy of four terms.

- *Amphiboly.* Amphiboly arises from what today is called syntactic (as opposed to lexical) ambiguity. It is typified by sentences such as 'Visiting relatives can be boring', which is ambiguous as between (1) 'Relatives who visit can be bores' and (2) 'It can be boring to visit relatives'. To see how amphiboly can wreck an argument, consider,

  > All visiting relatives can be boring
  > Oscar Wilde is a visiting relative
  > Therefore, Oscar Wilde can be boring.

  If the first premiss is taken with the meaning of (1), the argument is a syllogism. If with the meaning of (2), the argument is not a syllogism but a *paralogismos*, a piece of 'false reasoning'. Even so, our case seems to collapse into an ambiguity fallacy, with 'visiting relatives' the offending term — ambiguous as between 'the visiting of relatives' and 'relatives who visit'. Here, too, an amphibolous argument seems not to be one that an arguer would be alerted to just by speaking it.

- *Combination and division of words.* The next two types of sophistical refutation, combination and division of words, can be illustrated with the example of Socrates walking while sitting. Depending on whether the words 'can sit while walking' are taken in their combined or their divided sense, the following is true or not:

> Socrates can walk while sitting.

Taken as combined, the claim is false, since it means that

> Socrates has the power to walk-and-sit at the same time.

However, in their divided sense, these words express the true proposition that

> Socrates, who is now sitting, has the power to stop sitting and to start walking.

This is a better example of an *in dictione* fallacy. 'Socrates, while sitting, CAN WALK' sounds significantly different from 'Socrates can WALK WHILE SITTING'. Composition and division fallacies of the present day are not treated as fallacies *in dictione* [Copi and Cohen, 1990, pp. 17-20]. Rather they are fallacies that result from mismanaging the part-whole relationship. Thus the modern fallacy of composition is exemplified by

> All the members of the Boston Bruins are excellent players
> Therefore, the Bruins are an excellent team.

Division fallacies make the same mistake, but in the reverse direction, so to speak;

> The Bruins are a top-ranked team
> So, all the Bruins players are top-ranked players.

We see, then, that combination and division of words is an *in dictione* fallacy for Aristotle, whereas composition and division is an *extra dictionem* fallacy for later writers.

- *Accent.* Accent is perhaps rather difficult for the reader of English to understand, since English is not qualified in the way, for example, that French is, with its accents, *aigu, grave* and *circumflex.* It is troublesome that the Greek of Aristotle's time was not accented either; that is, there no syntactic markers of accent such as 'é' (acute); 'è' (grave); and 'ô' (circumflex). Even so, Aristotle *introduces* accents in his discussion of Homer's poetry. Conceding that 'an argument depending upon accent is not easy to construct in unwritten discussion; in written discussion and in poetry it is easier' (*Soph Ref* 166^b 1-2), Aristotle notes that

  Some people emend Homer against those who criticize as absurd his expression to $\mu\varepsilon\upsilon\sigma\upsilon \quad \kappa\alpha\tau\alpha\pi\upsilon\theta\varepsilon\tau\alpha\iota \quad \sigma\mu\beta\rho\omega$. For they solve the difficulty by a change of accent, pronouncing the $\sigma\upsilon$ with an acute accent. (166^b 2-3)

The emendation changes the passage from 'Part of which decays in the rain' to 'It does not decay in the rain', a significant alteration to say the least.

- *Form of expression.* Form of expression is meant in the sense of '*shape* of expression' and involves a kind of ambiguity.

> Thus (e.g.) 'flourishing' is a word which in the form of its expression is like 'cutting' or 'building'; yet the one denotes a certain quality — i.e. a certain condition — while the others denote a certain action (166^b 16-19).

Hamblin observes (over-hastily in my view) that "it was given to J.S. Mill to make the greatest of modern contributions to this Fallacy by perpetrating a serious example of it himself ...." Mill said (*Utilitarianism,* ch. 4, p. 32):

> The only proof capable of being given that an object is visible, is that people actually see it. The only proof that a sound is audible, is that people hear it; and so of the other sources of our experience. In like manner, I apprehend, the sole evidence it is possible to produce that anything is desirable, is that people do actually desire it.

"... Mill is misled by the termination 'able'." [Hamblin, 1970, p. 26], although here, too, we seem not to have a convincing example of a fallacy discernible in speech.

- *Accident.* Turning now to *extra dictionem* fallacies, accident also presents today's reader with a certain difficulty. The basic idea is that what can be predicated of a given subject may not be predicable of its attributes. Aristotle points out that although the individual named Coriscus is different from Socrates, and although Socrates is a man, it would be an error to conclude that Coriscus is different from a man. This hardly seems so, at least when in the conclusion 'is different from a man' is taken to mean 'is not a man'. The clue to the example is given by the name of the fallacy, 'accident'. Part of Aristotle wants to say is that when individual X is different from individual Y, and where Y has the accidental or non-essential property P (e.g., being six feet tall), it doesn't follow that X is not six feet tall, too. But this insight is obscured by two details of Aristotle's example.

The first is that 'is a man' would seem to be an essential property of man. Here, however, Aristotle restricts the notion of an essential property to a *synonymous* property, such as 'is a rational animal'. The other obscuring feature of the case is that Aristotle also wants to emphasize that what is predicable of an individual is not necessarily predicable of its properties. If we take 'Coriscus is different from' as predicable of Socrates, it does not follow that it is predicable of the property *man,* which Socrates has. But why should this be so if no individual (Coriscus included) is identical to any property (including the property of being a man)? Perhaps it is possible to clarify the case by differentiating two meanings of 'is different from a man'. In the one meaning 'different from a man' is the one-place negative predicate 'is not a man'; and in its second meaning it is the negative relational predicate 'is not identical to the property of being a man'. So from the fact that Coriscus and Socrates are different men, it doesn't follow that Coriscus is not a man. But it does follow that Coriscus is not identical to the property of being any man.

Little of this treatment survives in present day accounts. For example, in Carney and Scheer the fallacy of accident is just a matter of misapplying a general principle, that is, of applying it to cases 'to which they are not meant to apply' [1980, p. 72].

- *Secundum quid* is easier to make out. 'Secundum quid' means 'in a certain respect'. The error that Aristotle is trying to identify involves confusing the sense of a term in a qualified sense with its use in its absolute, or unqualified, sense. Thus from the fact that this black man is a white-haired man, it does not follow that he is a white man. Similarly, from the fact that something exists in thought it doesn't follow that it exists in reality (Santa Claus, for example).

  It is not difficult to see something of the link between the present-day fallacy of *secundum quid* (which is a variant of hasty generalization) and Aristotle's original treatment of it, especially in examples such as the white-toothed man versus the white headed man. Perhaps this is more easily seen if we slightly change to the example to one in which, for some man who is presently out of sight, the task is to determine what features he posseses based on information at hand. If the sole piece of information were "He is white-toothed", and it were concluded from this that he is white, the inference is faulty in something like the way that overgeneralizing from an unrepresentative sample is faulty. White teeth don't make for white men. It is notable in this regard that in the *Rhetoric,* Aristotle briefly treats *secundum quid* as if it were indeed the same as hasty generalization.[10]

- *Ignoratio elenchi,* or 'ignorance of what makes for a refutation' results from Violation of any of the conditions that define a proper refutation. As we saw, a refutation is genuine when one party, the questioner, is able to fashion from the other party's (the answerer's) answers a syllogism whose conclusion is the contradictory, not-T, of the respondent's original thesis T. However, Aristotle says that there are two ways in which the questioner might be guilty of *ignoratio elenchi.* It might be the case that the conclusion is not syllogistically implied by those premisses, in which case a non-syllogism is confused

[10]In [Woods, 2004b] it is argued that *secundum quid* has an even larger, though undiscussed, provenance in Aristotle's scheme. See chapter 18.

with a syllogism. Or it might be the case that, although syllogistically implied by them, the conclusion is not the contradictory of the opponent's thesis T, in which case a non-syllogism is not confused with a syllogism. If the first type of error can be called a syllogistic error, the second can be called a *contradiction error* [Hansen, 1996, p. 321]. This is an important development. It requires that we expand the Aristotelian definition of fallacy. Fallacies are either syllogistic errors or contradiction errors, each an error of logic in the narrow sense. Aristotle even goes so far as to suggest a precise coincidence between the *in dictione-extra dictionem* distinction and the distinction between contradiction errors and syllogistic error:

> All types of fallacy, then, fall under ignorance of what a refutation is, those dependent on language because the *contradiction*, which is the proper mark of a refutation, is merely apparent, and the rest because of the definition of *syllogism*. (*Soph Ref* 169^a 19-21; emphasis added. Cf. [Hansen, 1996, p. 321]).

In present day treatments (e.g., [Copi and Cohen, 1990, pp. 105-107]), the *ignoratio elenchi* is the fallacy committed by an argument which appears to establish a certain conclusion, when in fact it is an argument for a different conclusion. There is some resemblance here to Aristotle's contradiction error, which can be considered a special case.

- *Consequent.* Aristotle says at *Soph. Ref.* 168^a 27 and 169^b 6 that the fallacy of consequent is an instance of the fallacy of accident. Bearing in mind that Aristotle thinks that consequent involves a conversion error, perhaps we can get a clearer picture of accident. As noted above, accident is exemplified by a confusing argument about Coriscus and Socrates. We might now represent that argument as follows:

  1. Socrates is a man
  2. Coriscus non-identical to Socrates
  3. Therefore, Coriscus is not a man.

  In line (1), the word 'is' occurs as the 'is' of predication. Suppose that line (1) were in fact *convertible*, that is, that (1) itself implied

  1. A man is Socrates.

  In that case, the 'is' of (1) would be the 'is' of identity, not the 'is' of predication, and the argument in question would have the valid form

  1′. $S = M$

  2′. $C \neq S$

  3′. Therefore $C \neq M$

Thus the idea that (1) is convertible and the idea that its 'is' is the 'is' of identity come to the same thing and is the source of the error. For the only interpretation under which

Socrates is a man

Is true, is when 'is' is taken non-convertibly, i.e., as the 'is' is not of identity but of predication.

Consequent is an early version of what has come to be known as the fallacy of affirming the consequent [Copi and Cohen, 1990, pp. 211, 282]. In present day treatments it is the mistake of concluding that P on the basis of the two premisses, 'If P then Q' and 'Q'. Where P is the antecedent and Q the consequent of the first premiss, the fallacy is that of affirming P on the basis of having affirmed the consequent Q. Aristotle seems to have this kind of case firmly in mind. But he also thinks of consequent as a conversion fallacy, that is, as the mistake of inferring 'All P are S' ('All mortal things are men') from 'All S are P' ('All men are mortal').

- *Non-cause as cause.* Non-cause as cause also appears to have been given two different analyses. In the *Rhetoric* it is the error that later writers call the fallacy of *post hoc, ergo propter hoc,* the error of inferring that event e is the efficient cause of event e' just because the occurrence of e' followed upon (temporarily speaking) the occurrence of e. In the *Sophistical Refutations,* however, it is clear that Aristotle means by 'cause' something like 'reason for' (but see just below). The non-cause as cause error is exemplified by the following type of case. Suppose that

  P
  Q
  Therefore, not-T,

  is a refutation of the thesis T. Then the argument in question is a syllogism, hence a valid argument. As any reader of modern logic knows, if validity is a monotonic property and the argument at hand is valid, so too is the second argument,

  R
  P
  Q
  Therefore, not-T,

  no matter what premiss R expresses. But it is not a syllogism since a proper subset of its premisses, namely, $\{P, Q\}$ also entails its conclusion. Hence our second argument cannot be a refutation. Being a syllogism is a *non-monotonic* property. This matters in the following way. Aristotle thinks of the premisses of a refutation as reasons for ('cause of') its conclusion. But since our second argument is *not* a refutation of T, R cannot be a reason for not-T.

In the course of real-life contentions, an answerer will often supply the questioner with many more answers than the questioner can use as premisses of his would-be refutation. Aristotle requires that syllogisms have no idle premisses. Thus the questioner is obliged to select from the set of his opponents' answers just those propositions, no more and no fewer, that non-circularly and self-consistently necessitate the required conclusion. The fallacy of non-cause as cause is clearly the mistake of using an idle premiss, but it may not be clear as to why Aristotle speaks of this as the error in which a non-cause masquerades as a cause. Something of Aristotle's intention may be inferred from a passage in the *Physics* (195^a 15), in which it is suggested that in syllogisms premisses are the *material causes* (the stuff) of their conclusions, i.e., that premisses stand to conclusions as parts to wholes, and hence are causes of the whole. Idle premisses fail to qualify as material causes; they can be removed from an argument without damaging the residual subargument. Real premisses are different. Take any syllogism and remove from it any (real) premisses and the whole (i.e., the syllogism itself) is destroyed. In other places, Aristotle suggests a less technical interpretation of the fallacy, in which the trouble with R would simply be its falsity, and the trouble with the argument accordingly would be the derivation of not-T from a falsehood, a false cause (167^b 21).[11]

- *Petitio principii.* Aristotle provides several different treatments of *begging the question,* or *petitio principii.* In the *Sophistical Refutations* it is a flat out violation of the definition of 'syllogism' (hence of 'refutation'). If what is to be proved is also assumed as a premises, then that premises is repeated as the conclusion, and the argument in question fails to be a syllogism, hence cannot be a refutation. On the other hand, in the *Posterior Analytics* (86^a21), begging the question is a *demonstration* error. Demonstrations are deductions from first principles. First principles are themselves indemonstrable, and in any demonstration every succeeding step is less certain than preceding steps. But if one inserts the proposition to be demonstrated among the premisses, it cannot be the case that all premisses are more certain than the conclusion. Hence the argument in question is a failed demonstration.

  Aristotle in fact recognizes five ways in which a question can be begged (*Topics* 162^b34ff):

  > People appear to beg their original question in five ways: the first and most obvious being if anyone begs the actual point requiring to be shown: this is easily detected when put in so many words; but it is more apt to escape detection in the case of different terms, or a term and an expression, that mean the same thing. A second

[11]For a more detailed analysis of non-cause as cause see Woods and Hansen [2004c].

> way occurs whenever any one begs universally something which he has to demonstrate in a particular case ....

Aristotle rightly notices that such cases are unlikely to fool actual reasoners, but he reminds us that if a synonym of not-T were used as a premiss then the premiss would *look* different from the conclusion, and the circularity might go undetected. In fact, this seems also to be what Aristotle has against the fifth way, i.e., deducing a proposition from one equivalent to it.

In the second way of begging the question, Aristotle has in mind a certain form of what came to be called 'immediate inference'. It is exemplified by the subalternation argument

> Some A are B
> Since All A are B.

Aside from the fact that the single-premiss arguments seem not to qualify as syllogisms, it is difficult to make out a logical fault here. Bearing in mind that the fault, whatever it details, is the questioner's (i.e., premiss-selector's) fault, not the answerer's, one might wonder what is wrong with a questioner's asking a question which if answered affirmatively would give him a desired conclusion in just one step. Evidently Aristotle thinks that a refutation is worth having only if every premiss (individually) is consistent with the answerer's thesis. Such is W.D. Ross' view of the matter:

> And syllogism is distinguished from *petitio principii* in this, that while in the former both premises together imply the conclusion in the latter one premise alone does so. [Ross, 1953, p. 38]

Thus in a good refutation the thesis is refuted, never mind that the thesis is consistent with each separate answer given.

The third way of begging seems quite straightforward. It is illustrated by the plainly invalid form of argument

> Since some A are B, all A are B.

Even so, as the example makes clear, begging the question is fundamentally an error of *premiss-selection*, given that the answerer's job is to elicit premisses that syllogistically imply its contradictory, 'All A are B'. Again, begging the question is selecting a premiss, and in the present case, the questioner has begged the *wrong* question, i.e., he has selected the wrong premiss. It is a premiss which does the answerer's thesis no damage; and in any extension of this continuing argument in which damage were to be done, this premises, 'Some A are B', would prove idle.

Genuinely perplexing is the fourth way of begging. By the requirement that syllogisms be constructed from propositions, it would appear that no competent syllogizer would ever wish to conclude his argument with a conjunctive statement (since these aren't propositions). However, consider the argument

P
Q
Therefore, P and Q.

It is clearly valid. What Aristotle appears to have in mind is that it is useless for the questioner first to beg for P, then for Q, if his intention is to conclude 'P and Q'. The manifest validity of the argument might deceive someone into thinking he had produced a syllogism. But the fault lies less with his premiss selection than with his choice of target conclusion. Once begged, those questions will assuredly 'get' him that conclusion, but it is a statement of a type that guarantees that his argument nevertheless is not a syllogism.

The problem of would-be refutations that derive their targets in one step from a single premiss is according to Aristotle the problem that the argument,

All A are B.
Therefore, Some A are B,

begs the question. The contradictory of its conclusion is inconsistent with each of the premisses (of which there happens to be only one). If it is correct to say that the contradictory of a syllogism's conclusion must be consistent with each premiss, then our argument is not a syllogism. On the other hand, Aristotle in places appears to accept subalternation arguments (e.g., at *Topics* 199^a32-36). Some writers interpret Aristotle in a different way, as claiming an epistemic fallacy: One could not know the premiss to be true without knowing the conclusion to be true.

This raises a further matter as to whether, e.g.,

All A are B
All C are A
Therefore, all C are B

doesn't also beg the question. Mill is said to have held that this is precisely the case with all syllogisms. It is arguable that this was not, in fact, Mill's view (see section 9 below), but Aristotle seems to consider (and reject) the possibility (*Posterior Analytics* 72^b5-73^a20). If the first interpretation is correct, then were syllogisms fallaciously question-begging as such, it could not be for the reason that affects subalternation arguments. For

All A are B
All C are A
Therefore, all C are B

cannot be a syllogism unless its premisses fail individually to derive its conclusion; and it can't commit a *petitio* of the subalternation variety unless one of its premisses does indeed yield the conclusion on its own.

As we said in the above discussion of *ignoratio elenchi*, Aristotle's fallacy of *many questions* is a very different thing from that presented in present day logic textbooks [Copi and Cohen, 1990, pp. 96-97]. In such treatments, the fallacy is typified by such questions as, 'Have you stopped beating your servant?', in which there is an unconceded presupposition, namely, 'The addressee has been a beater of his servant in the past'. Aristotle intended something quite different by the many questions fallacy. It is the error of admitting to the premiss-set of a would-be syllogism a statement that is not a proposition, in Aristotle's technical sense of 'one thing predicated of one thing'. It was mentioned above that Aristotle had technical reasons for requiring syllogisms to be made up of propositions. This can be explained as follows. In *Topics* (100^a 18-21) and *On Sophistical Refutations* (183^a 37-36), Aristotle declares that his aim is

> To discover a method from which we shall be able to reason [syllogistically] about every issue from *endoxa*, i.e., reputable premisses, and when compelled to defend a position, say nothing to contradict ourselves.

In other places, his aims are forwarded more ambitiously. At *Soph Ref* 170^a38 and171^b 6-7, Aristotle says that the strategies he has worked out will enable a person to reason correctly about anything *whatever* independently of a knowledge of its subject matter. This is precisely what the Sophists also claimed to be able to do. Aristotle scorns their claim, not because it is unrealizable, but because the Sophists lack the theoretical wherewithal to bring it off. The requisite theoretical capacity Aristotle took to be the *logic* of syllogisms.

In present day treatments, the many questions fallacy is committed by *asking a question*, e.g., 'Have you stopped beating your servant?' In Aristotle's view, the fallacy is that of using an answer to the question as a *premiss*. Suppose that the answer is 'No'. This is equivalent to

> Not-(I have beaten my servant in the past and I do so at present)

or

> Either I have not beaten my servant in the past or I do not do so at present.

In each case, the answer contains a connective, 'not' — 'and' in the first instance, and 'or' in the second. In neither case, then, is the answer a proposition in which 'one thing is said of one thing', so it is inadmissible as a premiss of a syllogism. Where the modern theorist sees the fallacy as an interrogative fallacy [Hintikka, 1987], for Aristotle sees it is a technically syllogistic flaw.

Although *On Sophistical Refutations* is the primary source of what people have come to call Aristotle's fallacies, Aristotle gives them a somewhat different characterization in his other writings. In the *Topics*, we read that an argument (not necessarily a *refutation*) is fallacious in four different ways: (1) when it appears to be a syllogism but is not a syllogism in fact; (2) when it is a syllogism but reaches 'the wrong conclusion'; (3) when it is a syllogism but the conclusion is derived from 'inappropriate' premisses; and (4) when, although valid, the conclusion is reached from false premisses. Case (1) might well be exemplified by the fallacy of affirming the consequent. In modern terms, case (2) might be thought of in this way: let the premisses all be drawn from the discipline of economics, and let the conclusion be the logical truth, 'Either it will rain today or it will not'. Although that conclusion does follow validly from those premisses — at least by modern lights — it might be objected that it is the 'wrong thing' to conclude from those premisses. Case (3) is similar. Aristotle's own example is one in which it is concluded that walking after a meal is not good for the health (a conclusion from the art or discipline of medicine) from the premiss, proposed by Zeno, that motion (hence walking after dinner) is impossible. Even if Zeno's paradoxical proposition were true, Aristotle would see it as an inappropriate premiss for a medical argument, since it is not a *medical* premiss. Case (4) is obvious: Although true conclusions are often compatible with false premisses, no true conclusion can be established by false premisses.

The following is a list, taken from Hansen and Pinto [1995, p. 9], of where in Aristotle's writings the individual fallacies are discussed:

- Equivocation: *Soph Ref* 4 (165^b 31-16^a 7); 6 (168^a 24); 7 (169^a 22-25); 17 (175^a 36-^{b}8); 19; 23 (179^a 15-19); *Rhet* II, 24 (1401^a 13-23).
- Amphiboly: *Soph Ref* 4 (166^a 7-22); 7 (169^a 22-25); 17 (175^a 36-^{b} 8); 19; 23 (179^a 19). Combination of words: *Soph Ref* 4 (166^a 23-32); 6 (168^a 22-25); 7 (169^a 25-27); 23 (179^a 12-13); *Rhet* II, 24 (1401^a 24-^{b} 3).
- Division of words: *Soph Ref* 4 (166^a 33-39); 6 (166^a 27); 7 (169^a 25-27); 20; 23 (179^a 12-13); *Rhet* II 24 (1401^a 24-^{b} 3).
- Accent: *Soph Ref* 4 (166^b 1-9); 6 (168^a 27); 7 (169^a 27-29), 21, 23 (179^a 13-14).
- Form of expression: *Soph Ref* 4 (166^b 10-19); 6 (168^a 25); 7 (169^a 30-169^a 3); 23 (179^a 20-25).
- Accident: *Soph Ref* 5 (166^b 28-37); 6 (168^a 34-168^b 10, 168^b 26-169^a5); 7 (169^b 3-6); *Rhet* II, 24 (1401^b 5-19).
- *Secundum Quid*: *Soph Ref* 5 (166^b 38-167^a 20); 6 (168^b 11-16); 7 (169^b 9-13); 25; *Rhet* II, 24 (1401^b 35-1402^a 28).
- Ignorance of refutation: *Soph Ref* 5 (167^a 21-36); 6 (168^b 17-21); 7 (169^b 9-13); 26.

- Begging the Question: *Soph Ref* 5 (167^a 37-40); 6 (168^b 25-27); 7 (169^b 13-17); 17 (176^b 27-32); *Topics*, 8 (161^b 1-18), (162^b 34-163^a 13), 13 (162^b 34-163^a 28) *Pr Anal* 24 (41^b 9); *Pr Anal B*, 16 (64^b 28-65^a 37).
- Consequent: *Soph Ref* 5 (167^b 1-20); 6 (168^b 26-169^a 5); 7 (169^b 3-9); *Pr Anal B*, 16 (64^b 33); *Rhet* II, 24 (1401^b 10-14, 20-29).
- Non-Cause: *Soph Ref* 5 (167^b 21-37); 6 (168^b 22-26); 7 (169^b 13-17); 29; *Pr Anal B* II, 17; *Rhet* II, 24 (1401^b 30-34).
- Many Questions: *Soph Ref* 5 (167^b 38-168^a 17); 6 (169^a 6-18); 7 (169^b 13-17); 17 (175^b 39-176^a 19); 30.

We come now to

- *solecism and babbling.* At *Soph Ref* 165^b 12 Aristotle writes:

> First we must grasp the number of aims entertained by those who argue as competitors and rivals to the death. These are five in number, refutation, fallacy, paradox, solecism, and fifthly to reduce the opponent in the discussion to babbling — i.e. to constrain him to repeat himself a number of times ....

Solecism is the art of inducing an opponent to use an ungrammatical expression. Some commentators see this as a variation of the form of expression sophism. My own view is that it is a separate wrong-doing, got by asking one's opponent an inadmissible question. If, for example, the question is "Is virtue teachable or is it a gift of the gods?", and the opponent's answer is in the yes-no form demanded by the rules of refutation, then the answer will be "ungrammatical" in relation to the question asked. "Yes" doesn't answer the question, and "no" doesn't answer it either.

Babbling is not much discussed by Aristotle. He may have had in mind the not uncommon situation in which having exhausted everything he has had to say in defence of his thesis, a beleagured respondent is faced with two options. He can yield, or he can stick to his guns. But how could he stick to his guns if all his supporting premisses have failed to hit the mark? Aristotle appears to think that his only recourse is simply to assert his claim over and over. This would be tantamount to begging the question.

A better interpretation takes account of Aristotle's suggestion that babbling is an iteration, of a kind that occasions infinite regression. This is an important observation, carrying the suggestion that babbling is not a fallacy in Aristotle's usual meaning of the term; that is, not the mistaking of a non-syllogism for a syllogism or a proposition for the contradictory of a given thesis. The factor of iteration is present in the much later puzzle of the tortoise and Achilles. Here is a simplified version of Carroll's paradox:

> *First party*: P, therefore Q.
> *Second party*: Even if I grant P, what obliges me to accept Q?

> *First party*: Because P, and if P then Q. So Q.
> *Second party*: Yes, I grant P and if P then Q. But where does Q come in?
> *First party*: Well, it's pretty straightforward. Consider these three premisses: P, if P then Q, and if P and if P then Q, then Q.
> *Second*: I get the premisses. I don't get the conclusion.
> *First*: Oh, dear. Here we go again![12]

The second party is a babbler. At each stage beyond the first, he gives a defence of a case which carries a premiss implying that very case. This is the kind of thing that Aristotle thinks of as iteration, and is the kind of iteration (reiteration) that leads to an infinite regress, which though a logical error, is not a fallacy in either of Aristotle's official senses.

- *Ad hominem.* Neither solecism nor babbling has made its way into the literature of the present day. *Ad hominem* is quite a different story, notwithstanding the modern theorist's disposition to link it to Locke rather than Aristotle. In fact, the *ad hominem,* in both example and name, is Aristotle's. It arises in his discussion of refutations. In several places he seems to think that refutations are proofs, but in a looser sense of "proof" than one would find in mathematical proof. In other places, refutations appear to be proofs in no sense of the word. For

> I mean, 'proving by way of refutation' to differ from 'proving' in that, in proving, one might seem to beg the question, but where someone else is responsible for this, there will be a refutation, not proof. (*Metaphysics* 1006^a 15-18)

Accordingly,

> In such matters there is no proof simply, but against a particular person, there is. (*Metaphysics* 1062^a 2-3)

This is Ross. In Barnes' version we have:

> About such matters there is no proof in the full sense, though there is proof *ad hominem*.[13]

What is it about an Aristotelian refutation that makes it a proof *ad hominem*? It turns on the fact that the premisses of the questioner's refuting syllogism are required to be selected from the *respondent*'s concessions and only them. For this condition to be met, it is not required that those premisses be true (only that the respondent thinks that they are) or that the questioner believe them to be true (indeed he may believe them to be false). So when a refutation succeeds, questioner and answerer agree that from the answerer's

[12] Shades of Lewis Carroll's "What the tortoise said to Achilles, *Mind,* 14 (1895), 278-280.
[13] Cf. *Soph Ref* 177^b 33-34, 178^b 17, 170^a 13, 183^a22, 24, and *Topics* 161^a21.

own concessions there follows a proposition that contradicts the answerer's own thesis. What this shows is not that the thesis is false, but rather that it is inconsistent for the answerer to hold it in the light of his subsequent concessions. The upshot of the proof is not the falsity of that thesis, but the inconsistency of the *person* who holds it.

2.1 Aristotle's importance

Aristotle's discussion of the *ad hominem* gives us occasion to recur to the twofold contribution to logic that is to be found in the early part of the *Organon.* No one could seriously suppose that Aristotle is the founder of logic in the broad sense, especially in those forms of it that investigate the logic of dialogue. But Aristotle is the inventor of logic in the narrow sense, with the early appearance of the syllogism. This we may take as the singular achievement of the early logic. But not far behind is Aristotle's attempt to show that there is a class of logics in the broad sense — of refutation, examination — arguments, instruction arguments, and scientific demonstrations — that cannot stand unless they embody at their cores a logic in the narrow sense. Next in order of importance are Aristotle's contributions to the logic of dialogue, especially to that part of it that deals with contentious argument. Embedded in Aristotle's discussion — and especially evident in the account of proofs of *ad hominem* - is a network of concepts that support the working vocabulary of present day dialogue logic: Question, answer, concession, commitment, retraction, and so on. Thus, a yes-answer to a question is a *concession* of its propositional content and a *commitment* to its truth. Answers may not be *retracted*, but theses must be if subsequent commitments support their contradictories. If an arguer is committed to some propositions he is also committed to any syllogistic consequence of them (though not necessarily to every non-syllogistic consequence of them). A person whose concessions necessitate the retraction of his thesis has not thereby falsified the thesis. So self-refutions are collisions of concessions, not falsifiers of refuted propositions. Aristotle may not have developed a technical lexion for "concession" "commitment", "retraction", and the rest. But there is in these writings no mistaking their conceptual presence.

There is a critical point at which Aristotle's theory is at its most vulnerable. We might grant that the presence of any of Aristotle's thirteen elements in a would-be refutation would destroy it. We might agree that anyone introducing these elements by way of a merely would-be syllogism would have committed a fallacy, since the would-be syllogism wouldn't be a syllogism in fact. But we needn't agree that there is no other way of wrecking a refutation than by fallacious syllogism-making. Consider a case. Suppose that your thesis is not-Q and that I have got you to concede that P and nothing else. Suppose that I now make the following move: "Since you hold that P, and moreover given that P implies Q, it follows that Q, which contradicts your thesis." The mistake, of course, is that in constructing this argument, I begged the question against you. I did this by helping myself to a premiss ("P implies Q") that you hadn't conceded. So while my refutation is no good, my logic is perfect. But Aristotle can't accept this. My logic is *not* perfect. It is not a *syllogism*. Its conditional premiss is not a categorical proposition. And it is this (he insists) that makes the refutation sophistical.

Although hailed as the father of logic and the originator of fallacy theory, almost no one accepts Aristotle's thesis that the sins of bad refutations are necessarily and exclusively the product of bad syllogizing. It is true that until the mid-point of the 19th century Aristotle's was the mainstream logic. But in the matter of the fallacies — apart from those few that are expressly syllogistic in character, e.g., the fallacy of undistributed middle — the distinction between fallacy and sophistical argument has long since collapsed, and the faults involved in sophistical argument are analyzed without reference to the particularities of the syllogism.

We have here the seeds of a problem which may well spell disaster for Aristotle's fallacies project. As we have said, Aristotle's doctrine is made up of the following elements.

1. The property of being a syllogism is definable for arguments in the narrow sense and only them. An argument in the narrow sense is a fallacy if and only if it is a non-syllogism mistaken for a syllogism.

2. The property of being a sophistical refutation is definable for arguments in the broad sense. A sophistical refutation is a would-be refutation embodying one or other of the items mentioned in Aristotle's list of thirteen — begging the question, many questions, equivocation, and so on.

3. These argumentative shortcomings are "sophistries". Then a would-be refutation is a sophistical refutation if and only if it reflects a sophistry.

4. In contexts of would-be refutation it is impossible to commit a sophistry except by committing a fallacy.

But as our discussion shows, there is little reason to accept thesis (4). Indeed it is easy to see that any of the thirteen could strike (and injure) an argument without a chance of its being (or embedding) a fallacy in Aristotle's technical sense of carrying the false appearance of syllogisity. Accordingly, we have what might be called

> *The Concept-List Misalignment Thesis.* Aristotle's own list of the thirteen sophistical refutations fails to instantiate Aristotle's own concept of fallacy.

If this is right, it can only be expected that, left undetected, it is a misalignment that would throw fallacy theory, both then and to come, into disarray. It did and has done.[14] Two stands of evidence stand out as especially important. One

[14]In my *Seductions and Shortcuts: Error in the Cognitive Economy*, scheduled to appear in 2013, I attempt to show that the concept-list misalignment problem, once suitably adjusted, bedevils contemporary fallacy theory. Perhaps this helps answer Hamblin's implied question, "Why haven't logicians taken up the fallacies programme?"

is the persistent and persistently unsettled rivalry between those who think that fallacies are inherently dialectical in character and those others who see them as intrinsically logical mistakes. The other is that virtually no one of genuine note in the creative mainstream of logic from Frege onwards pays the slightest heed to the fallacies programme[15] as witness again, Hamblin's *cri de coeur*: "we have no *theory* of fallacies at all ...". Hamblin thought this an embarrassing omission, for which he was more than prepared to blame logicians. But the collective and largely tacit judgement of logic's present day orthodoxies suggests a different judgement, that the blame falls on fallacies. That is, interesting and important as they may be in their own right, fallacies just aren't *logic-worthy*. That, if true, would be a fairly crunching answer to the No-Theory problem.[16]

On the face of it a closed issue — for such is the extent of the indifference of mainstream logicians — this is actually a matter on which the jury is still out, occasioned in part by developments in *informal logic* over the past forty years. I shall say something further about this at the end of the present chapter.

This would be a good place to remind ourselves that the syllogistic is not Aristotle's sole contribution to logical theory. In addition to attempts (up to five in number, according to some commentators) to extend the logic of syllogisms to modal contexts,[17] Aristotle also wrote about what later editors would mis-name the logic of "immediate" inference. The better name would be "immediate consequence", exemplified by what we find on the Square of Opposition. The square is a diagramatic device for setting out relations which hold between pairs of single categorical propositions — contradictoriness, contrariety, subcontrariety and subalternation. Consider the latter: The subalternation relation provides that "Some A are B" is a one-step consequence of "All A are B". This stands in marked contrast to syllogisms, whose conclusions are never one-step consequences from either of their premisses. Recall, syllogisms are required to have more premisses than one, whereas immediate consequences are always from single premisses. Is this a principled distinction? It is. Aristotle is clear in *On Sophistical Refutations* that if someone were to use this example of a subalternation in an argument in the broad sense, he would beg the question against his opponent. Used as arguments in the broad sense, subalternation arguments are logicically defective. Used simply as statement of what follows from what, the very same arguments are logically impeccable.

We see embedded in this distinction a distinction between *consequence-having* and *consequence-drawing*. No one doubts that "Some A are B" is a consequence that "All A are B" has. No one doubts that "Socrates is mortal" is a consequence that "All men are mortal" and "Socrates is a man" jointly have. But Aristotle is

[15] An exception is Hintikka (e.g. [1987]). Even so, Hintikka's work expressly on the fallacies is scant.

[16] Let us also note that in general the *sophistries* aren't inherently dialectical either. Someone who reasons solo that Ali is a white man from the fact that he is a white-toothed man commits the *secundum quid* error, but there need be no one around with whom, in so doing, he is contending, or wrangling, or even talking.

[17] See, for example, [Corcoran, 1974; McKirahan, 1992; Patterson, 1995].

clear. In argumentative contexts only one of those consequences can, on pain of the fallacy of question-begging, be *drawn*. This is the ancient source of the view, which was given vigorous advancement in 1970 by Gilbert Harman, to the effect that the laws of logic are not rules of inference.[18] In other words: the laws of consequence-having are not the rules of consequence-drawing. Everyone knows that consequence is a logician's principal focus. All logicians have had something to say about consequence-having. Comparatively speaking, consequence-drawing has had few mainstream takers. The qualification "comparatively speaking" is important. Any logician who has interested himself in fallacies has had a stake, whether he knows it or not, in achieving a firm grip of the having-drawing distinction. For whatever else they are, fallacies are mistakes of a kind that in some fashion or other hook up with (and spoil) consequence-drawing. Logicians who are in the descendent class of the Aristotelian approach to the fallacies might be said to have heeded this distinction by locating fallacious reasoning in argumentative contexts, contexts in which consequences are not only had, but also, when the time is ripe, drawn. Here is a simplified example.

> First party: Thesis T is the case.
>
> Second party: Do you accept that P?
>
> First party: Yes, I do.
>
> Second party: But isn't-T a consequence of P?
>
> First: Damn! So it is!
>
> Second: So *doesn't it follow* that your defence of T is inconsistent with it?
>
> First: Okay, okay. Let's go for a beer.

It is interesting to note that the consequence that Second invites First draw, and First does draw at the last line, is not a *syllogistic* consequence of preceding lines.

There is a family of distinctions in Aristotle that have been an enduring inheritance for logic ever since, and — like literal inheritances — not an always well-managed one. Members of the family aren't pairwise equivalent, but there are similarities enough to justify their membership. Here they are in two columns of connected items:

[18] Harman [1970]. See also Harman [1986, chapter 2].

Column I

- Argument in the broad sense
- Logic in the broad sense: dialogue logic
- Consequence-drawing
- Inference
- Context-sensitivity
- "Dialectic"

Column II

- Argument in the narrow sense
- Logic in the narrow sense: syllogisitc logic and the logic of immediate "inference"
- Consequence-having
- Entailment
- Context-freedom
- "Logic"

Pages ago I suggested that, as they make their way through these pages, it would be helpful for readers to bear in mind the questions raised by Hamblin's No-Theory complaint. If I am not mistaken, much the same could be said for the present septet of contrasts.

3 THE HELLENISTIC AND MEDIAEVAL PERIODS

We owe to Diogenes Laertius' *Lives of Eminent Philosophers*[19] five hundred or so years after Aristotle's death, some slight indication of what Milesian logicians made of the fallacies, although it seems quite clear that commentators such as Ebulides take a fallacy to an inherently dialectical impropriety of interrogative argument. The fallacies are not especially well-described by Diogenes and, in any event, they are for the most part warmed over presentations of what is already in Aristotle. Notable, however, are early discussion of the logical and semantic paradoxes, the Liar and the Sorites.

Lucian of Samosota (*c.* 130-180) also had a version of the Liar and Sorites paradoxes, and two others known as the Hooded Man and the Horned Man. Concerning the first: What you have not lost you still have. You have not lost horns. Therefore, you still have horns. Concerning the second; You say you know your brother. But that man who came in just now with his head covered is your brother, and you did not know him. (*Vitarum Auctio*, 221)[20]

To Diodorus (d. late 1st cent. BC) we owe the Master Argument in support of the thesis that nothing is possible which neither is nor will be. For

Everything past is necessary

The impossible doesn't follow from the possible

[19]Diogenes Laertius [1925]. See especially book II.

[20]Lucian [1915].

> What neither is nor will be impossible.
>
> Propositions (i), (ii) and (iii) are jointly inconsistent.
>
> Propositions (i) and (ii) are true.
>
> Therefore nothing is indeed possible which neither is nor will be true.[21]

We have better information about Stoic thinking, especially in Stextus Empiricus's *Outlines of Pyrrhonism* and *Against the Logicians*.[22] It would not be far off the mark to say that with the Stoics the logic of fallacious thinking, indeed logic generally, takes a step towards disciplinary independence, that is, a step towards the narrow. In Aristotle's hands, logic in the narrow sense was a service industry, and had a supporting role — albeit a major one — in the theory of argument. (Even the mature syllogistic of the *Prior Analytics* and *Posterior Analytics* subserves the ends of scientific demonstration.) A significant part of the Stoic separation from Aristotle was Chrysippus' development of a dialectically de-contextualized calculus of propositions, and with it the idea that a fallacy is an argument or claim that contravenes the rules or theorems of such systems. It is true that in moments of self-attribution, Stoic logicians made free use of the word "dialectic". But by now "dialectic" was pretty much the received word for logics of this de-contextualized kind. This is not to overlook that arguments that follow within their purview can easily occur in dialectical and interrogative settings. But when a fallacy is committed, it is a logical mistake, having no essential connection to the setting in which it arises.

Sextus (*fl. c.* 200) makes two contributions of particular note to the history of logic, one of quite general import, and the other bearing more directly on the fallacies. The general contribution was a massive (and, was he thought provable) nihilism about reason, proof and truth. But making good on the general contribution depends upon making good on the particular contribution. So we should begin with it. The particular contribution was an argument supporting the claim that every valid proof commits the fallacy of begging the question. [Sextus, 1933-1949, II, § 236-259]

> By what means, then, can we establish that the apparent thing is really such as it appears? Either, certainly, by means of a non-evident factor or by means of an apparent one. But to do so by means of a non-evident fact is absurd; for the non-evident is so far from being able to reveal anything that, on the contrary, is itself in need of something to establish it. And to do so by means of an apparent fact is much more absurd; for it is the thing in question, and nothing that is in question is capable of confirming itself.

This same difficulty infests the very enterprise of fallacy theory, and provides a hard-edged answer to the modal form of the No-Theory question:

[21] Diodorus [1933-1967].

[22] Sextus Empiricus [1933-1949], volumes one and two.

> As regards all the sophisms which dialectic seems peculiarly able to expose, their exposure is useless; whereas in all cases where the exposure is useful, it is not the dialectician who will expose them but the experts in each particular art who grasp the connection of the facts.

Of course, just as they stand, Sextus's arguments are simply asking for trouble. They purport to be good proofs that good proofs are impossible to arrive at. But Sextus sees this coming. He is content to be hoist upon his own *pétard.* (II, § 481)

> And again, just as it is not impossible for the man who has ascended to a high place by a ladder with his foot after his ascent, so also it is not unlikely that the Sceptic after he has arrived at the demonstration of his thesis by means of the argument proving the non-existence of proof, as it were by a step-ladder, should abolish this very argument.[23]

There is a further attack on the logic of fallacies. Sextus agrees that ambiguity is the natural enemy of good reasoning. It also appears to be his view that ambiguity is a permanent feature of language and that no word is wholly ambiguous or wholly unambiguous.

> And in the ordinary affairs of life we see already how people — yes, even the slave-boys — distinguish such distortions of use. Certainly if a master who had servants named alike were to bid a boy called, say, 'Manes' (supposing this to be the name common to the servants) to be summoned, the slave-boy will ask 'Which one?' And if a man who had several different wines were to say to this boy 'Pour me out a draft of wine', then too the boy will ask 'Which one?' Thus it is the experience of what is useful in each affair that brings about the distinguishing of ambiguities. (II, § 236-259)

Suppose now that we have a valid proof from premisses we take for true of what we take to be false. There is always the possibility that the appearance of falsehood arises from an a yet to be discovered ambiguity in one or more of the terms. So we might set out to discover this ambiguity. Sextus' point is that if the sole indication of this ambiguity is the apparent falsity of the conclusion of a formally valid argument from apparently sound premisses, then proofs lose all practical utility. If no term is inoculated against ambiguity, it is impermissible to be moved by any proof whose conclusion is not already secured on grounds independent of the proof itself. Theories, too, are a kind of proof. They are case-makers for the propositions they advance. Fallacy theory is also like this. It undertakes to uncover by strict methods the essential truths about fallacies. But Sextus thinks he has disabled such methods for useful work in intellectual life quite generally. He can hardly propose otherwise for the logic of the fallacies.

---

[23]See here Wittgenstein [1958, 6.54].

We owe the word "logic" in its modern sense to Alexander of Aphrodisias, a contemporary of Diogenes and Sextus. Now long lost, Alexander's is the only known commentary on *On Sophistical Refutations* before the 1200s, though some reflections of it can be found in Peter of Spain's *Treatise on the Major Fallacies,* which itself has yet to find a publisher.[24] We also owe to Alexander the doctrine of *multiplex* terms, terms possessing a double meaning, which is an attempt to systematize Aristotle's refutations dependent on language. Alexander's paraphrase of Aristotle's list is complete is paraphrased in Peter's *Summulae Logicales,*[25] the seeds of which can be found at *Soph Ref* 168a 24:

> For of the fallacies that consist in language, some depend upon a double meaning [= *multiplex*], e.g. ambiguity of words and phrases, and the fallacy of like verbal forms (for we habitually speak of everything as though it were a particular substance) — while fallacies of combination and division and accent arise because the phrase in question on the term as altered is not the same as was intended.

Aristotle continues:

> Accordingly an expression that depends upon division is not an ambiguous one. (177^b 8)

Aristotle's theory of the fallacies was not an instant hit (if the colloquialism may be forgiven) but, unlike Hume's *Treatise,* there is no reason to suppose that it fell still-born from the press. Any fine-tuned appreciation of Aristotle's influence in this regard is significant compromised by the massive disappearance of the ancient manuscripts between the sixth and twelfth centuries. Even so, not always reliable echoes of Aristotle's earlier teachings could be heard by twelfth century scholars even before the lost works were re-discovered. In good measure, the losses of this period were attenuated by translations into Arabic of *On Sophistical Refutations,* among other works, as well as commentaries by philosophers, such as Averroes (1126-1198) in *The Incoherence of the Incoherence.*[26]

The interpretations of fallacies which filled the gap between the sixth and twelfth centuries and which, to some serious extent survived the re-discovery of the old texts are called by Hamblin the "spurious doctrine" [Hamblin, 1970, pp. 104ff.]. The spurious doctrine spawned some of the most important contributions to the logics of the mediaeval period, not least of which is the doctrine of *suppositio.* Initially a theory of reference, supposition theory adumbrates the most sophisticated treatment of quantifiers before the breakthrough engineered by Peirce, Frege, Whitehead and Russell in the late 19th and early 20th centuries of the present era.[27] An important creator of the spurious doctrine was Boethius (*c.*

[24] It exists in manuscript form in the Bayerische Staatsbibliothek in Munich.

[25] Peter of Spain [1947].

[26] Averroes [1954]. For more on the Arabic influence in logic, see [Street, 2004] in addition to [Rescher, 1964].

[27] An excellent appreciation of this contribution can be found in [Powers, 2012].

445-*c.* 526), whose commentaries on Aristotle's *Categories* and *On Interpretation,* and Porphyry's *Introduction* and Cicero's *Topics* formed the basis of what in the twelfth century would be called the "old logic". Still, there is almost nothing in the old logic about fallacies, except possibly for Boethius.

In his remarks on *On Interpretation* 17^a 34, Boethius discusses the mere appearance of contradictoriness between pairs of statements under the following six headings: *Equivocation* "Cato killed himself at Utica" and "Cato did not kill himself at Utica" may both be true, since there are two different historical Catos. *Univocation* "Man walks" and "Man doesn't walk" can both be true, since "man" can refer to this man and to the species Man. *Different Part* "The eye is white" and "The eye is not white" can both be true since the eyeball is white and the pupil is not. *Different Relatum* "Ten is double" needn't be the contradictory of "Ten isn't double" depending on what ten is said to be the double *of. Different Time* "Socrates is sitting" and "Socrates is not sitting" aren't contradictories since they can be true or false at different times. *Different Modality* "The kitten can see" needn't contradict "The kitten can't see" since the first "can" might apply to a just-born kitty and the second to him at age one. Thus a Boethian fallacy is the mistake of taking the syntactic form of contradictoriness as giving the semantic reality of it. Boethius' classification of the fallacies is taken up and examined by Peter Abelard (1079-1142), arguably the major logician of his time.

In the period from Boethius to Abelard, we see a growing (or revived) interest in the study of grammar for its own sake, and a corresponding blending of the *de dictione* fallacies into this more general interest in the complexities of language. In Aristotle's hands, grammatical considerations were in the service of logic. But now we see something of a reversal of that relation in, for example, *On Sophistical Refutations* by Robert Grosseteste (*c.* 1170-1253) and St. Albert the Great (1193-1280), as well as St. Thomas Aquinas' *opusculum* on the fallacies (*c.*1244-1245). *The Introduction to Logic* by William of Sherwood (*c.*1200-c. 66) reserves a chapter for fallacies, which shows the influence of Peter of Spain's *Summulae Logicales.* Also significant is the *Sumule Dialectices* of Roger Bacon (*c.*1245).

It may well be that William's *Introduction to Logic* is the paradigm logic text of the thirteenth century. It is made up of five books corresponding roughly to five of the six books of Aritiotle's *Organon.*[28] It also contains serious discussions of *insolubilia* (paradoxes, sophismata), obligation games and *syncategoremata.* Two things stand out about this work. One is its fidelity to Aristotle. The other is the openness with which William addresses the tension between logic, in the manner of the *Prior Analytics,* and dialectic, in the manner of *On Sophistical Refutations.*

> I maintain that the substance of disputation is *nothing but* syllogism. Considered as an entity, therefore, disputation and syllogism are one and the same thing. It [= disputation] is called 'syllogism', however, in virtue of the fact that a person can organize thought by means of

[28]The omission is *Posterior Analytics.*

it.[29]

Aside from discussions of paradox, babbling and solecism, the chapter on sophisms is pretty much a reworking, albeit in greater detail, of Aristotle's original material. But here, too, the emphasis is clearly on the various phenomena of linguistic ambiguity, concerning which the mediaeval doctrine of supposition is the most sustained and systematic treatment since Aristotle's *Categories.*

Another development is the anticipation by William and others of what has come to be known in the present day as game theoretic logic, under their name for it, "the game of Obligation". Obligation is an operations manual for disputatious arguments, adumbrated in St. Thomas' *Disputed Questions on Truth.*[30] Obligation is patterned closely on Aristotle's own theory of contentious argument. Here, too, there is a thesis — or *positum* — which is the point of disagreement, together with its defender, the respondent, and its attacker, the opponent. Attacks take the form of questions by attackers designed to elicit the propositions (*proposita*) embedded in the respondent's answers. Obligation can be played in both an engaged manner and a disengaged one. Played engagingly, a respondent must sincerely believe his *positum* and the *proposita* he gives in answer to his opponent's questions. In the disengaged mode, the *positum* need not be believed by the respondent, nor need his *proposita*, while they shouldn't be obviously evasive must not be incompatible with the original *positum.* The game terminates in the manner of Aristotle, with the opponent's utterance of "*Cedat tempus*", when the opponent has been led into inconsistency with his *positum* or the opponent concedes that the respondent has successfully avoided the contradiction.

Hamblin has opined (p. 129, n. 1) that Obligation is "isomphorphic" to the deontic system presented in chapter five of von Wright's classic *An Essay on Modal Logic* (1951). I myself doubt the isomorphism, but the similarity is unmistakable. Deontic logics have rules that greatly outnumber the rules of unmodalized deductive logic, and it is with respect to these that the majority of William's sophisms arise, some fifty in all. Many of these involve the paradoxes of self-reference, and few of them are from Aristotle's original list of thirteen.

A nominalist version of Obligation is John Buridan's *Sophismata* (c. 1330). Although it contains a chapter on fallacies, *Sophismata* is dominantly a metaphysical work, whose chief objective is the promotion of nominalism in philosophy and what we would now call philosophy of science. Given this emphasis, it is hardly surprising that the *posita* of disputations inherited from Obligation are transformed by *Sophismata* into philosophical hypotheses.

An interesting treatment of composition and division is advanced by Walter Burleigh:

> Every animal is rational or irrational. Not every animal is rational. Therefore, every animal is irrational.[31]

[29]William of Sherwood [1966, p. 132].
[30]Thomas Aquinas [1953].
[31]Quoted from Bochenski [1970, p. 176ff].

This is fallacy, says Burleigh, since the conclusion is false. But the minor is not false, so the major must be the culprit. But this is absurd; the major is true. Burleigh's solution is that the major "is multiple, according to composition." In modern terms, as Bochenski notes [1970, p. 187], the difference is simply that between

$$\forall x\ (x \text{ is an animal} \supset (x \text{ is rational or } x \text{ is irrational})$$

$$\forall x\ (x \text{ is animal} \supset x \text{ is rational}) \vee \forall y\ (y \text{ is an animal} \supset y \text{ is irrational}).$$

3.1 Mediaeval supposition theory

The period covered in this section extends from the latter part of the 12th to the 14th century. In this interval logic is dominantly a fusion of Aristotle's *On Sophistical Refutations* and the *Analytics* with the mediaeval doctrine of *suppositio,* which is a semantic theory of relations of standing for, defined for ordinary Latin, but regimented in various ways to facilitate broader theoretical engagement. Supposition theory — rather in the way that present-day philosophy of language is — generated a large and technically sophisticated literature, covering different schools and, at times, conflicting theoretical frameworks. Its coverage is impressive: Terms, nouns, verbs, propositions, equipollence, conversion, hypotheticals, predicables, exponibles, categories, syllogisms, consequences, obligations, insolubles ...; the list goes on. Among supposition theory's leading exponents were William of Sherwood, Peter of Spain, Walter Burley, William of Ockham and John Buridan. Since our topic here is the place of fallacies in mediaeval logic, and mediaeval logic is a daunting thing to describe with technical accuracy, and also a lot to learn if one is starting from scratch, I shall try to give an informal exposition of the treatment of fallacies in this large literature, as free as possible from its technical arcane.

Here is an example, the fallacy of univocation [de Rijk, 1967a, p. 492], in which a proposition, whose terms occur unambiguously, may nevertheless vary in meaning depending on the proposition's adjoining expressions. Thus "'Man' is a name" and "Man is a species" *man* was thought to occur univocally, but that certain "confusions" might nevertheless ensue: *Man* is a term, and *man* is animal. Another example comes from William of Sherwood and concerns the quantifiers confusions that arise from negative and affirmative quantifiers. He allows (rightly) that "No man is an ass; no man is this ass" is deductively correct, but "Every man is an animal".

3.1.1 Modes of supposition: Fallacies thereof.

Under William of Sherwood's rules for confusion and distribution, we have "No man is an ass, therefore no man is an ass", but we are not allowed "Every man is an animal; therefore every man is this animal." (*IL*, V. 13.1 (117)) Under his rules of inference other fallacies arise. Given that a man sees only himself, it is fallacious to infer "Every man does not see a man." (*IL*, V. 13.2 (118)) Also unacceptable is "A man is not seen by Socrates; therefore Socrates does not see a man", as is "Every man sees only himself; therefore a man is seen by every man." (119) However, William's constraints allow for "Every donkey is an animal; consequently some donkey is every animal." William also wrote about composition and division, sometimes in ways that capture the spirit of *On Sophistical Refutations*. He offers this interesting example:

> Whatever is possible will be true. That a white thinks is black is possible. Therefore, that a white thing is black will be true.

Other examples, which reflect difficulties presented by tense can be found in [Kretzmann, 1966, pp. 142ff].

Concerning the later 14th century supposition theorists, in the further interests of space I'll confine my comments to opinions of the fallacies that appear to have been commonly acknowledged. For example, under the rules of descent and ascent, it is a fallacy to infer from "Every donkey is a mammal", either "Every donkey is this mammal and every donkey is that mammal and [so on for all the donkeys]" or "Every donkey is this mammal or every donkey is that mammal or [so on for all the donkeys]." Similarly the rules of ascent forbid "Some stone is not this donkey; therefore some stone is not a donkey."

Consider now "Every donkey which is not a donkey is running", which is logically false. The mediaevals thought that everything followed from a logical falsehood.[32] So presumably, "This donkey which is not a donkey is running" follows from it too. The question is whether it follows by descent, or as we would say, instantiation? It would appear not. Since the instantiation is also logically false, it ascends to the universally quantified original proposition. But that is an invalid mode of inference.

Additional fallacies arise from violations of rules for immobile distribution. From "Every man besides Socrates runs", we mustn't infer "Plato besides Socrates runs." Similarly for the rules of distributive supposition. From "Every horse is an animal", it is fallacious to descend to any conjunction or disjunction of instances of the form "Every horse is *this* animal". It is permissible to infer "Every horse is this animal or that animal or ... [for all the animals]." However "Not some donkey every animal isn't; therefore not some donkey this animal or that animal ..." Is fallacious. Equally, from "No animal is every man" it is illegitimate to infer "No animal is this man or that man or ...." (Buridan, *SD* 4.3.8.2 (277). However from "No animal is every man", it is permissible to conclude "No animal is this man and that man and ...."

3.2 The importance of this period

This was a fruitful period for logic, especially if we allow into logic's broader province dialogue systems in the manner of Obligation. The list is impressive:

[32] Buridan *TC* I. 8.3 (196) and Venice *LP* III.1 (167).

propositional logic, modal logic, temporal logic, deontic logic, none of which, to be sure, achieving the finished state of syllogisms in the *Prior Analytics*, but robust anticipations of the things yet to come. The paradoxes, too, attracted the attention of talented thinkers, and so did the fallacies. On the whole, the fallacies whose treatments in this period were largely reworkings of *On Sophistical Refutations,* made no advance in the direction of the theory whose absence Hamblin laments.It is quite true that the mediaeval supposition theorists uncovered a complex tissue of misinferences that arise in growingly complex contexts of quantification and term-ambiguation of various stripes. But it is not in the general case true that these logical errors are either natural seeming or committed with sufficient recurrent frequency to capture the notion of fallacy which preoccupied Hamblin so. The concept of fallacy in this period is a blend of fallacies in Aristotle's sense and the sophismata, that is, propositions and inferences whose logical forms are obscure. Some of Aristotle's fallacies are caught in this net, but so too are the paradoxes of self-reference as well as Buridan Ass and the like.[33]

What, then, of the centuries immediately following? Here is one view of the matter:

> From the 400 years between the middle of the fifteenth and the nineteenth century we have ... scores of textbooks but very few works that are at one new and good. [Kneale and Kneale, 1984, p. 298]

This is harsh in at least three respects. Leibniz (1646-1716) was a great logician, and a respector of traditional logic. Even so, for all his originality, Leibniz had nothing original to say about the fallacies.[34] Descartes, too, had interesting and original things to say about logic, but, here too, nothing of note about the fallacies. The same is true of Kant.[35] Bacon (1561-1626) played a pivotal role in the move from demonstrative to the experimental sciences, and in *Novum Organum* (1620) and other writing's spiritual father of the new logic of science. Bacon too has virtually nothing to say of the standard list of fallacies, but his remarks on the idols of the mind are something that would re-pay the acquiaintance of any fallacy theorist. Arnauld (1611-1694), together with his colleague Nicole (1625-1695), is another story. Not only is the Port Royal *Logique* (1662) an early source of inductive logic, it brims with fresh things to say about the fallacies.

---

[33]The mediaeval paradoxes are discussed by Mikko Yrjönsuuri in chapter 10 of volume 2 of this *Handbook* [2008].

[34]Except for his addition, in *New Essays*, Bk. 4, ch. 17, of a fifth *argumentum* to Locke's four. This is the argumentum *ad vertiginem* (giddiness). This is an argument in the form: if this proof of P is not sound then there would be no means of ascertaining that P is the case. But this is absurd; so the proof is sound. Leibniz allows that such a line of reasoning is valid in those cases in which it expresses a "primitive" or "immediate" truth, such as that "we ourselves exist."

[35]Leibniz's logic is discussed by Wofgang Lenzen in chapter 1 of volume 3 [2004] of this *Handbook*.

4 FRANCIS BACON (1561-1626)

Bacon was perhaps logic's most accomplished man of affairs, in which regard only the example of Leibniz serves as a rough comparison. Bacon was a lawyer, politician, and statesman. He was Baron Verulam and Viscount St. Albans. Member of parliament, Bacon held high office in government. He was attorney general, keeper of the great seal and lord chancellor. His political career was ended by accusations of bribery and other forms of corruption, in a series of actions brought against him in 1621. Bacon admitted wrongdoing, was fined and left parliament, never to return. Some historians have suggested that Bacon's political distinction may have been occasioned by the chicanery of his enemies rather than his own malfeasance in office.

High among Bacon's intellectual achievements is the role he played in the reconceptualization of science not as the demonstrative closure of first principles but rather as an experimental enquiry. Central to this transformation was Bacon's insistence that the data for experimental science – the observable events that motivate the enquiry in the first place, as well as the observable events that ultimately serve as the theory's experimental confirmation — are not "raw facts", but rather are facts or observations which will already have been conceptualized in a certain way by the observer's mind. Here is perhaps the first clear expression of the thesis that the data for science are inescapably "theory-laden", that the thing-in-itself, as Kant would put it, is not accessible to an observer's direct apprehension. It is an important claim, assigning to enquiring minds a psychological role in the construction of our knowledge of the world. Bacon was alert to the idealist flavour of this view, but was more open about the possible corruptions that attended psychological participation, notably the possibility — indeed the likelihood — of bias in both science and everyday affairs.indexRamus, P.

4.1 Background remarks

In matters of logic, it is necessary to say a little something about Bacon's intellectual connection to the teachings of Peter Ramus (1515-1572), who entitled his dissertation of 1537 *Everything Aristotle Said Was False*. Concerning the fallacies, Hamblin [1970, p. 156] sums up the Renaissance's resistance to the learning of the Schoolmen, as well as the subsequent traces of it:

> There have been three groups either actively opposed to Fallacies or uninterested in them: the first, Agricola and Ramus; the second, Locke and the empiricists, matched by Leibniz and the rationalists; the third, Boole, Frege and Russell. Answering revivals have been instituted by, first, Fraunce, Buscher, Bacon and Arnauld; secondly, Whately, J.S. Mill and DeMorgan; thirdly, the modern mainly American logicians [whose work on fallacies is to be found in elementary logic textbooks in general circulation in 1970].

In *Aristotelicae Animadversiones* (1543), Ramus bans fallacies from logic. Who seeking enlightenment "expect light [on the fallacies] from the author of darkness?" (pp. 70-71) It is amusing that Ramus' followers were hardly able to keep themselves from writing about the fallacies, and in due course a Ramist theory of fallacies could not be prevented from appearing, as witness Herzio Buscher's *The Theory of the Solution of Fallacies ... Deduced and Explained From the Logic of P. Ramus* (1594), a book, says Hamblin, that "does not possess any merits that would warrant our discussion." (p. 143).

4.2 Idols of the mind

Bacon is the greatest of Ramists, but is no enemy of fallacy theory in logic. In *Novum Organum,* Bacon emphasizes the psychological element in the construction of our knowledge of the objects of nature. Knowledge arises from our contact with these objects, which in our understanding of them are "creations of the mind and hand" (p. 12). In the *Advancement of Learning* (1543) he writes,

> 1X. The sole cause and root of almost every defect in the sciences is this, that while we falsely admire and extol the powers of the human mind, we do not search for the real helps.
>
> X. The subtlety of nature is far beyond that of sense or of the understanding: so that the specious meditations, speculations, and theories of mankind are but of a kind of insanity, only there is no one to stand by and observe it. (p. 12).

Bacon allows that the standard treatment of the fallacies is in perfectly good order, but he goes on to say that

> ... there is a much more important and profound kind of fallacies in the mind of man ...

He continues,

That instance which is the root of all superstition, namely, *That to the nature of the mind of all men it is consonant for the affirmative or active to affect more than the negative or privative*: so that a few times hitting as presence, countervails oft-times failing or absence; .... (1605, p. 395)

Such errors Bacon calls *idols of the tribe.* An idol is a false appearance by which we represent to ourselves "the real stamp and impression of created objects, as they are found in nature". They are "powerful in producing unanimity", they are of "familiar occurrence" and they "immediately hit the understanding and satisfy the imagination." (p. 17) Idols of the tribe are biases that are common to human experience as such. "They have their foundation in human nature itself .... [T]he human understanding is like a false mirror, which, receiving says irregularly, distorts and discloses the "nature of things by mingling its own nature with it." (p. 54). "Bias" here is a kind of representational distortion, easy to fall into and difficult to snap out of. We see in these biases the unmistakable stamp of

Aristotle's own concept of fallacy. We see in the insetted passage just above the fallacy of hasty generalization or false cause and, correspondingly the failure to notice the presence or weight of countervailing consideration.

It is well to emphasize that idols of the tribe are reflections of a philosophical problem of central importance. Knowledge of the world is effected by the right representation of it. But human beings are so constituted that even when they are functioning properly their representations are distortions. Sometimes these distortions are benign, sometimes not. The problem is that any means we may employ to discipline this distinction is itself subject to the distortive character of representation. A faulty representation can only be corrected by a correct representation. But any representation that appears correct may later, on the basis of what now appears correct, appear to be incorrect. And so on, recursively.

This is a notable development in logic. Idols of the tribe are neither unique to nor distinctive of argumentative contexts. It is true that Aristotle invented logic to serve as the theoretical core of a general theory of argument, whereupon logic's historic difficulty in sorting out logic's dialectical character (or, for that matter, dialectic's logical nature). In our section on Aristotle, we noted the distinction between consequence-having and consequence-drawing. Consequence-having is a propositional relation instantiable without the involvement of human agency. Consequence-having occurs in logical space. Consequence-*drawing* is different. Consequences drawn are consequences drawn by agents, by beings like us. Consequence-drawing occurs in the human mind. In the earlier remarks, we noted a rough concurrence between the distinction between having and drawing and the different logics that Aristotle built for them. For the first is the misnamed logic of immediate inference, better called immediate consequence. Immediate consequences are consequences that can't be drawn, that is, can't be drawn non-fallaciously. For drawing the immediate consequence of something is question-begging. To some extent, syllogistic is the logic for consequence-drawing. A syllogistic consequence can be drawn without fear of circularity. It can be drawn without fear of premissory irrelevance. Not that we have a free hand of course. There are conditions and contexts in which a syllogistic consequence shouldn't be drawn. The premisses might be false or unconceded. The consequence might fail to be the contradictory of the thesis under attack; and so on. But one thing is clear — in dialectical contexts immediate consequence must never be drawn. For if validly drawn they beg the question.

It is easy to see that, in as much as arguing is a matter of drawing consequences from the concessions of opponents, syllogistic is a suitable logic for argument, or more plainly, for *arguers*. But what Aristotle doesn't emphasize, and Bacon does, is that human arguers are *cognitive* beings whose quest for knowledge — including the knowledge got by successful consequence-drawing — is subject to the riches and limitations of how nature and we are built and interact. Thus the idols of the tribe: They are errors — including errors of consequence-drawing — that flow from the epistemic constitution of the human agent. This is the modern era's first invocation by a major thinker of *psychologism* in logic.

> *Idols of the cave* are the false appearances imposed on us by every man's own individual nature and custom, ... Which minister unto us infinite errors and vain opinions, if they be not recalled to examination. (*Advancement*, p. 396)

Bacon's name calls to mind Plato's allegory of the cave in Book nine of the *Republic.* The comparison is not apt. On Plato's telling the impediments to understanding occasioned by the cave-dwellers' alienation from the light are the quite general condition of human kind. But Bacon's cave-idols arise case-by-case from impediments to knowledge that strike human beings variously. Thus one person may be unschooled, another may be stupid, yet another might be gullible, and others still might be hot-headed and impatient. The ensuing errors are idols, made so not merely by the fact that they are errors, but also by the fact that they seem to their committors not to be errors. That is to say, they are false appearances. It is also Bacon's view that, while they vary in kind and intensity with individual reasoners, idols of the cave are the inescapable lot of mankind in general. That is, the *commission* of those errors that are disguised as idols of the cave is the lot of mankind in general. But Bacon adds to this grim diagnosis the reassuring qualification, "if they not be recalled to examination". Error might be unavoidable for beings like us, but error can be corrected by judicious examination.

Here is another juncture in Bacon's thought at which it is clear that the problem of fallacies — of their making and their subsequent unmaking - is epistemological rather than dialectical. Again, the epistemological problem is to divine means for the correction of error which are themselves immune from its recurrence. Bacon's response to this challenge and the methods which bear his name, inductive procedures that anticipate Mill's Methods. Central to Bacon's project is that in the domain of observables error will out, idols will lose their veils of respectability. To the objection that yesterday's error which has been corrected by today's observation is corrected by an observation no more protected from error than it, Bacon's advice, in effect, was to keep records. If today's correction of yesterday's error is not itself met with new observations inconsistent with it, this is some reason to suppose that our experimental methods are philosophically reliable.

Next come *the idols of the marketplace.* They are the

> ... false appearances that are imposed upon us by words, which are framed and applied according to the conceit and capacities of the vulgar sort; and although we think we govern our words, and prescribe it well, ..., yet certain it is that words ... do shoot back upon the understanding of the wisest, and mightily entangle and pervert the judgement; .... (p. 396)

In earlier writings Bacon called these the idols of *the palace,* and therein lies a useful suggestion as to meaning. Bacon is reminding us that perfectly literate speech, grammatically well-constructed and lexically legitimate (its words are *bona fide* words of English) can be the conveyance of sophistry and flummery. Such is the speech with which sly ministers might entreat the king, or the language of a present-day press release from a political action group, predecessors all of 1984's Newspeak. Bacon thinks that false doctrines and noxious lies do less harm if expressed in technical or high-faluting language that no one understands. But real trouble is afoot if the language is understandable but its content is obscure. People can speak utter nonsense in the plainest and most accessible of words.

The problem posed by idols of the marketplace is not exhausted by political Newspeak. The problem on which Bacon fixes his gaze became in due course a philosophical commonplace. It is that the grammatical form of a sentence can often obscure its logical form. It may be perfectly true that there exists a possibility that Harry will be late, but Bacon would be quick to condemn the apparently correct inference that there exists an x such that the possibility that Harry will be late $= x$.

Here, too, Bacon is an optimist. If we exercise caution and engage in critical reflection, we may not always avoid marketplace idolatry, but there is a decent chance that, upon weighing its consequences, such error can be diagnosed and steps taken to repair it.

In his *idols of the theatre,* Bacon's contempt for First Philosophy is expressed with an uncharacteristic crudeness. In an early work, Bacon named Aristotle as the worst of the sophists, stupefied by his engagement with jargon of his own devisement. In *Novum Organum,* idols of the theatre

> are Idols which have immigrated into men's minds from the various dogmas of philosophies, and also from wrong laws of demonstration. (p. 55)

They are theatrical fallacies in as much as they are doctrines and methodologies that have so little objective readily that might have been the mere artifice and make-believe of the playwright and stage director. From our perch in the early years of the 21st century, perhaps Bacon's intolerance is rather small beer, for it is an intolerance bred by the revolutionary overthrow of the old science, brokered by the harsh scepticisms that repose in the coils of any aggressive form of empiricism. But Bacon had another target. These were the curricula of universities, replete with the set-pieces that make for irreflective and close-minded canonicity — the natural enemies of intellectual freedom and open inquiry. What, then, would a student of the day derive from his courses? In matters of knowledge and understanding, he would have done as well, or better, to spend his hours at the theatre.

For logicians, the most important fixture of idols of the theatre are that they may arise "from wrong laws of demonstration", as they apply, for example, to the establishment of the first principles of science, and with it the classical tussle between Aristotle's *noûs* and Bacon's induction.

4.3 Bacon's importance

Logicians before Bacon made room for human agency. In its broad sense, logic was about argument, and arguments require arguers. The human agent is several things at once. He is an object of nature. He is a dialectical being. He is a cognitive being, that is, an object who makes his way in life by knowing things — sometimes succeeding, sometimes not. It is Bacon's view that if logic is to take account of the human agent, it must take him as he comes, warts and all. Logic must therefore take on an epistemological texture with which to acknowledge the human agent's cognitive nature. Bacon epistemologizes logic, and psycholgizes epistemology, each a serious turn toward naturalization.

The epistemologization of logic carries negative consequences for the view that logic is wholly or at least dominantly the regulation of the distinction between consequence-having and consequence-drawing in argumentative contexts. This is not an emphasis that Bacon shares. Argument may be a kind of reasoning, but Bacon is more interested in the genus that the species. *Inference* is his principle focus. Some of today's argumentation theorists try to downplay the difference between inferring and arguing, insisting that arguing is just reasoning out loud, and that solo reasoning is just arguing with oneself. But this is not a pitch which Bacon would have been prepared to catch. His epistemologization of logic was matched by a corresponding inclination to demote its dialectical features.

Bacon's further importance as a philosopher of science is widely known and appreciated, and need not detain us here. Nor need his determination to place logic in the service not of arm-chair disputation, but rather of the revolution in science which had toppled syllogistic demonstration and was on course to replace it with a new logic for experimental enquiry. There are logicians today who see the present state of inductive logic as an as yet unfinished chapter of logic's long history. Some are less sanguine than others about future prospects. Right or wrong, inductive logic is a large enterprise, and Bacon is its founder.[36]

5 ANTOINE ARNAULD (1612-1694) AND PIERRE NICOLE (1625-1695)

The twentieth child of his father, Arnauld was born in Paris. His father, also Antoine was a prominent lawyer who succeeded to his own father's post as *Procureur Général* to Queen Catherine d' è Medici, and was an outspoken critic of the Jesuits, calling in 1594 for their expulsion from France. In 1641 the "third" Arnauld was ordained and admitted to the degree of doctor of theology. On the death of Cardinal Richeleu who had opposed it, Arnauld entered the Sorbonne in 1643.

Arnauld was a leading exponent of the views of Cornelius Jansen (1585-1638), the Dutch bishop of Ypres. Jansenism is an interesting motif in the fortunes of the Arnauld family. Rejecting what it took to be scholastic excesses in Church doctrine

[36]The modern history and current state of inductive logic is examined in volume 10 of this *Handbook* [2011], of which Stephan Hartmann is third co-editor.

and practice, Jansenism proposed a return to St. Augustine's teachings about grace, and espoused a strong form of predestination. In this there is an echo of the intermittent Calvinism of Arnauld's own grandfather, although Arnauld himself was no Calvinist, since, despite this point of doctrinal overlap, Calvinists were hostile to the Jansenists for their insistence that salvation required the mediation of the official Church. Arnauld's sister was abeyess of the convent of Port Royal des Champs, a leading centre of Jansenist thought. In 1656 Arnauld lost his post at the Sorbonne and, together with other Jansenists, endured Jesuit persecution for several years following. He died in exile in Brussels in 1694.

Arnauld's first major work, *De la fréquente communion*, appeared in 1643, ironically the year of his entry to the Sorbonne, and was a significant expression of Jansenist opinion. The proximate cause of his censure and removal from the Sorbonne was the appearance of a subsequent work, *Letters to a Duke and Peer*. Since the religious members of Port Royal subscribed to the Jansenist doctrines of this work, in 1661 a royal edict effected the closure of the convent and the removal of its members, and in 1709 Port Royal des Champs was burned to the ground by order of Louis XIV. For all these difficulties, Arnauld, in collaboration with another Port Royalist, Pierre Nicole (1625-1695), anonymously brought forth the *Port-Royal Logique, ou l'Art de Penser* in 1662.

Pierre Nicole was born in Chartres, and in 1642 was sent to Paris by his barrister-father to study theology. In time he was admitted to a minor order but hesitated and eventually declined to be ordained to the priesthood. Shortly after his arrival in Paris he joined the Jansenist community at Port Royal, where for many years he taught in a school for boys. But his primary function, which he shared with Arnauld, was to serve as general editor of the massive and growing writings of the Jansenists and as co-editorship of Blaise Pascal's *Provincial Letters* (1656). Apart from his co-authorship, which he and Arnauld took to be a Cartesian "reconstruction" of Aristotle's logic, Nicole was much engaged with theological matters. A major work on transubstantiation, also in collaboration with Arnauld, was *La Perpétuité de la foi d'Eglise Catholoque touchant l'eucharistie (1669)*. This was followed in 1671 by Nicole's *Essais de morale*.

At times Arnauld and Nicole fell out of Jansenist favour and suffered some persecution for it. In 1679, they made a hasty retreat to Belgium, where their collaboration came to a stressful end. By 1633 Nicole had patched things up with his tormentors, returning to Paris, where two years later he died.

5.1 Logic

Translated as the *Art of Thinking*, the book is divided into four parts, which are preceded by a forward, two "discourses" of prefatory essays, and an introduction. The parts concern Conception, Judgement, Reasoning, and Ordering (or the methodology of knowledge and true opinion). In two of the chapters of part three are to be found Arnauld and Nicole's account of fallacies, which they call "sophisms". Chapter nineteen, "Sophisms: the different ways of reasoning badly", discusses a class of fallacies associated with mistakes in scientific method. Of the ten fallacies considered here, eight re taken over from Aristotle's original list of thirteen and, for the most part, retain a strong Aristotelian flavour, despite occasional deviations. Chapter twenty is novel in a number of ways. Entitled "Fallacies Committed in Everyday Life and in Ordinary Discourse" it treats of two categories of sophism, (1) Sophisms of Self-love, of Interest, and Passions, and (2) "Fallacious Arguments Arising from the Objects Themselves". Fallacies of the first sort, of which ten are discussed, arise from "internal" factors, that is, from the reasoner's state of mind. Fallacies of the second category, of which seven are discussed, have to do with "external" matters, that is, the propensity that situations external to the mind have to deceive us. No less important a theorist of probability than John Maynard Keynes was of the view that "the authors of the Port Royal *Logic* ... were the first to deal with the logic of probability in the modern manner" [1921, 80], and it may be that this material was contributed by the mathematician Pascal (1623-1662), famous to this day for the Wager that bears his name. Keynes also points out that "Locke follows the Port Royal Logicians very closely." (idem.) This is evident in three notable respects. Arnauld and Nicole charge that

> For it seems that ordinary philosophers hardly ever apply themselves to logic except to give rules of good and bad reasoning. Now we cannot say that these rules are useless. ... We should not, however, believe that this usefulness extends [very far]. (*Discourse* 1, p. 9)

Furthermore,

> ... experience shows that of a thousand young persons who learn logic, there are not ten who know anything about it six months after they have finished ... so because students from never seen it put into practice, they do not use it themselves are quite happy to dismiss it as trivial and worthless learning. (*Discourse* 2, p. 16)

And

> Hence there are two kinds of methods, one for discovering the truth which is known as *analysis*, or the *method of resolution*, and which can also be called the *method of discovery*. The other is for making truth understood by others once it is found. This is known as *synthesis*, or the *method of composition* and also can be called the *method of instruction*. (Part IV, Chapter 2, p. 232)

Arnauld and Nicole were of the view that analysis always precedes synthesis. In the first instance, he is making what would become a rather common complaint against Aristotelian syllogistic logic, namely, that it is of little efficacy in analyzing the operations of the mind in the process of competent inference. In the second instance, Arnauld shows himself to be a spiritual forear of what has in our own time come to be called "informal" or practical logic. Since the deductive formalities of syllogistic logic shed little light on the operations of inference, it must follow that since inference is the process by which a great many real life problems are handled, deductive logic will have little occasion to be put to "real use". In the third instance, the Port Royalists tie together these prior two objections against syllogistic logic and, in so doing, articulate a distinction that proved to be of great importance for Locke. They assert that the challenges of real life are much more to discover what things are true or probable than to demonstrate them, it follows that the rules of syllogistic, which are rules of demonstration, can have only a restricted utility, and that they soon will be forgotten by young student-logicians because of their general inapplicability to the main problems of life.

We see here a considerable debt to Descartes. Not only do Arnauld and Nicole accept the distinction between the logic of discovery and the logic of demonstration, they are shrewd to notice (where others do not) that the logic of discovery resists detailed articulation:

> That is what may be said in a general way about analysis which consists more in *judgement* and *mental skills* than in particular rules. (Part IV, Chapter 11, p. 238; emphasis added)

Or, as we might say in contemporary terms, there is little prospect for an explicit *theory* of informal logic. In this same way, the Port Royalists think that, strictly speaking, the study of sophisms is auxiliary to the study of logic, since "what is to be avoided is often more striking than what is to be imitated". That is, even if we are able to discern the presence of a sophism in a piece of reasoning, it is not true in general the case that we will be able to formulate the "particular procedures" or the positive rules which govern the discovery of truth or probable knowledge.

5.2 Sophisms

Chapters nineteen and twenty of part three of the *Art of Thinking* give the following classification of the fallacies.

I. *The different ways of reasoning badly (pp. 189-203)*

(1) *ignoratio elenchi*
(2) begging the question
(3) non-cause as cause
(4) overlooking an alternative
(5) accident
(6) composition
(7) division
(8) *secundum quid*
(9) ambiguity
(10) incomplete induction

II. *Fallacies committed in everyday life and in ordinary discourse*

(i) *Sophisms of self-love, interest and passion* (pp. 204–214)

(11) taking our own interest as reason to believe something

(12) of the heart

(13) believing oneself infallible

(14) [another case of (13)]

(15) accusing your opponent of obstinacy

(16) envy of another's achievement

(17) the spirit of contention

(18) complaisance

(19) aiming for conviction rather than truth

(20) considering an opinion for a reason other than to determine its truth

(ii) *Fallacious arguments arising from the objects themselves* (pp. 214-225)

(21) failure to realize the admixture of truth and falsity

(22) tailoring truth by long oration

(23) jumping to a conclusion, rash judgment

(24) hasty generalization

(25) rationalization

(26) authority

(27) manner

Arnauld and Nicole never claimed completeness for their classification. Concerning the sophisms of chapter nineteen (the different ways of *reasoning* badly), these are "the main sources of bad reasoning" (p. 189) and "we will limit them to only seven or eight [sic] kinds, since some are so obvious that they are not worth mentioning" (idem.) The fallacies of chapter twenty, having to do with public life and everyday affairs, are presented in recognition of the fact that "reason does not find its principal use in science; to err in science is not a grave matter, for science has but little bearing on the conduct of life." Of the sophisms here noted, "we give only a general indication of some of the causes of false judgements so common among men."

Nor are the sub-classifications perfectly disjoint. For example, fallacy number (10), incomplete induction, recurs as fallacy number (24), hasty generalization. In fact, it must be said that the entire classificatory scheme is shrouded in obscurity. It is not the Port Royalists' view that the "scientific sophisms" are committable only when reasoning about physics or the other branches of natural philosophy. There is nothing to preclude an arguer's illicitly exploiting an ambiguity in a public dispute about, e.g., whether taxes should be raised. Nor is there any reason why a prejudice for established Aristotelian science couldn't taint an argument about the natural propensity of unsupported bodies to fall and, when they do, at velocities proportional to their weight. Similarly, a scientist can consider himself infallible as readily as anyone who advances arguments about everyday affairs. Neither should we think that the scientific fallacies are fallacies of syllogistic argument only (for one thing, item (10) is not a syllogistic error). Scientific sophisms are, thus, not fallacies of demonstration only — they are not even required to be deductive fallacies. Although it may be more difficult to articulate the rules of discovery than the rules of demonstration, it is clear that the scientific fallacies can afflict analysis (or inferences aimed at the discovery of truth) as readily as synthesis (or arguments aiming at the demonstration of a truth already thought to have been discovered).

Then, too, in the sophisms of passion sub-category, we see Arnauld and Nicole anticipating Mill's distinction between moral and intellectual errors. Moral errors are errors caused or induced by psychological factors such as conceit, prejudice, animosity, self-deception, and the like. But, for Mill, a fallacy is always an intellectual error, never mind what its cause might be. On a fair reading, this is also Arnauld's position. Wanting badly enough to convince someone of a given claim may well cause one to commit the fallacy of non-cause as cause or to surrender to a vitiating ambiguity. In fact, Arnauld and Nicole offer no reason for thinking that there is any scientific fallacy that cannot be induced by at least some of the passionate causes of sub-category II (i). If there is any hard and fast distinction to be seen between category I and subcategory II (i), it is surely Mill's distinction between the "moral" cause of an error and the error itself. In particular, out of envy for another's achievement someone might construct a hostile denunciation of his adversary's opinion; but the denouncing argument might be entirely cogent. So we do have a difference. Any instance of category I is fallacious as such; that is, the mistake inheres in its nature. No instance of II (i) is fallacious as such, though it may predispose a reasoner or arguer to fallacy, even when it doesn't actually cause it. Sub-category II (ii) is harder to make out. It seems to be a mélange of intrinsic errors, as in hasty generalization and (perhaps) manner, and of causes of or predispositions to fallacious conduct, as in the failure to realize the admixture of truth and falsity.

It is possible that some people will see a clear differentiation between scientific fallacies, on the one hand, and, on the other, public and everyday fallacies, the Port Royalists, in the fact that, for, each of the former is analyzable without reference to dialectical factors, that is, to factors pertaining to what Aristotle called "contentious arguments"; and that the latter require for their analysis some consideration of such factors. In brief, then, it might be proposed that the chief distinction here is that between non-dialectical and dialectical sophisms. There is reason to doubt this idea. It is true that most of the scientific sophisms require no dialectical analysis, but not all; e.g., *ignoratio elenchi*, taken in the way that Arnauld and Nicole themselves do (See below). It is also true that some of the public and everyday sophisms do require such a treatment, e.g., tailoring truth by lying, or oration, or accusing your opponent of obstinacy. But not all are so, e.g., believing oneself infallible, or rash judgement, or taking our own interest zs reason to believe something. So, again, it would seem that, overall, the Port Royalists' classificatory scheme lacks a satisfactory motivation. If this is so, we must look for greater clarity to the details of the individual analyses, one by one.

5.3 Scientific sophisms

Ignoratio elenchi is the fallacy of "proving something other than what is at issue" (p. 189). There are commonly two ways of committing it: (1) by imputing to an adversary an opinion he does not hold and then refuting it; and (2) by challenging an opinion he does hold with consequences he refuses to accept. Concerning (1), we see that Arnauld and Nicole are giving early recognition to what has come to be known as the "straw man fallacy". Concerning (2), it is hard to make it out as a fallacy; for that one's opponent denies that his thesis carries a certain implication far from demonstrates that it does not carry it. On the other hand, there is something strategically maladroit in pressing an opponent with consequences that he will not admit to. But if fallacy there is, it is that of the disputant who will not see that his view does carry those consequences (if it does) or of his opponent who claims that it carries those consequences (if it doesn't). If, in the latte case, this is simply the "fallacy" of *non sequitur*, in the former case it is the fallacy of seeing a *sequitur* as a *non sequitur*. Although Arnauld and Nicole invoke the name of Aristotle in their discussion of *ignoratio enenchi*, it is clear that their own conception of it is broader than Aristotle's. Aristotle holds that the fallacy is the failure to honour of the several conditions which make a *syllogism* a good refutation. And nothing in Port Royalist accounts restricts the fallacy to syllogistic faults. However, on a core idea, Arnauld and Aristotle are at one. An argument commits the *ignoratio elenchi* when it correctly deduces the wrong conclusion.

Begging the question is "assuming as true what is at issue" (p. 190), and it may be that Arnauld and Nicole have in mind Aristotle's conception of it as a demonstration-error. For Aristotle, the premisses of a demonstration must be, as Arnauld and Nicole say, "clearer and better known" than its conclusion. If the conclusion repeats a premiss, this condition is clearly failed. Also failed is Aristotle's quite general condition on syllogisms, that the conclusion must be "other than" each premiss. Departing from Aristotle, the Port Royalists also recognize two further forms of begging the question, each of which has found its way into common usage. One is to attribute to an opponent an opinion he would not accept (note, here, the kinship with the straw-man fallacy); and the other is to prove something unknown on the basis of something else equally or more unknown.

Non-cause as cause or "Taking for a cause what is not a cause" (p. 192), also on Aristotle's original list, is committed in "several ways", none of which coheres with Aristotle's diagnosis in *On Sophistical Refutations*. One way is simply to mistake one event as the cause of another event, as with the claim that handling frogs is a cause of warts. Another way is to confuse a remote cause with a proximate cause, as with the assertion that swimming caused the swimmer to drown. Yet another case is one in which a cause is "invented" merely to disguise our ignorance of the true cause. Arnauld and Nicole cite *Post hoc, ego propter hoc* as a special case, as with the claim that since that heat of August follows the appearance in the heavens of the Dog Star (i.e., Sirius), the latter causes the former. In this, Arnauld and Nicole side with Aristotle's treatment of this sophism in the *Rhetoric.*

Overlooking an alternative or "imperfect enumeration" (p. 196), is not a fallacy on Aristotle's list, although it seems to have an Aristotelian flavour to it. Aristotle imposed on syllogisms the requirement that premisses be "appropriate to" conclusions, that is, relevant to them. One commits the Royalists' version of the fallacy by failing to take into account all relevant alternatives which might bear on the matter at issue. However, Aristotle's notion of relevance is that premisses and conclusion must be drawn from some common disciplines such as physics or medicine. Arnauld and Nicole's conception of relevance is different. It is omitting from consideration possible premisses which, even if drawn from the appropriate discipline, will have had a bearing on the truth of the issue in question.

The fifth kind of sophism is Judging Something by What Applies to it Only Accidentally. (p. 198) "In the Schools this fallacy is called *fallacia accidentia*" (idem.).

> Man is composed of body and soul
>
> Therefore, body and soul think

Thus, the fallacy of ambiguity is at least sometimes a special case of the fallacy of division (and of composition, too).

The last scientific sophism, not on Aristotle's original list, is that of "Drawing a general conclusion from a faulty induction." (p. 202) It is a case handled with impressive clumsiness, by asserting that it is the drawing of any general conclusion from any class of instances which is not complete. To take a common present-day example, the induction that all ravens are black is fallacious unless drawn from an examination of all ravens there ever were or will be. On the other hand, no such induction is construable as a *syllogism*, that is, as an argument whose premisses necessitate its conclusion. And it may be that what Arnauld and Nicole were really wanting to say about such cases is not that inductions are fallacious when "incomplete", but rather that the fallacy attaches to the claim that an incomplete induction is a syllogism. Saying so would be a mistake, right enough, but is hardly a mistake that would "deceive even clever men".

5.4 Public and everyday sophisms

The first group of these fallacies arise from self-love, interest and passions. First considered is the rhetorical question "What could be less reasonable, however, than taking our interest as a motive for believing something?" (p. 204) It is exemplified by the situation in which the (say) upper-classes, noting that certain privileges are in their interests, conclude that they are in fact entitled to those privileges. In the same way, "sophisms and illusions of the heart" (p. 205) "[are] those errors which consist in transferring our passions to the objects of our passions, in judging those objects to be what we desire them to be," as when a loving mother says of a wastrel son, "He is so clever and industrious; it is just that his enemies won't give him a chance." Self-love and self-regard give rise to feelings of infallibility (p. 205): "I am right, so you are wrong" and "If your opinion (which opposes mine) were correct, then I would be mistaken; but I don't make mistakes, so your opinion is false." Then, too, there are "rhetorical tricks" such as accusing one's opponent of obstinacy or chicanery, which is a destructive force and casts truth and error, justice and injustice into so profound an obscurity that the common man is unable to distinguish them. Clearly a kind of *ad hominem* manoeuvre, it also resembles the pragma-dialectical treatment of the *ad baculum fallacy*, for it has the effect of inhibiting the advancement of an opponent's point of view. Envy is also a destructive force as in, "am not the author of *Principia Mathematica*, si it is a worthless book", Contentiousness, though a milder vice than envy or self-love, is even so "no less injurious to the mind". But the contrary disposition — complaisance — is also injurious to judgement, as when people take as true anything told them (pp. 205 ff.).

In a celebrated passage, met with earlier, Hamblin excoriates logicians for what he calls the "Standard Treatment" of the fallacies, in which

> A writer throws away all logic and keeps the readers' attention, if at all, only by retailing the traditional puns, anecdotes, and *witless examples* of his forbears, 'Everything that runs has feet; the river runs; therefore the river has feet' — this is a medieval example, *but the modern ones are no better.* (1970, p. 12; emphases added)

Such treatments, says Hamblin, are useless and they leave us in a situation in which '[w]e have no *theory* of fallacies at all ...." (ibid., p. 11) It may be thought that the passionate fallacies are grist for Hamblin's mill, for the examples do seem "witless" and, in any event, are no preclude to anything that could be called a theory. In fairness, Arnauld and Nicole admit the logic of discovery is especially resistant to theory, and evidently this is also their opinion about the fallacies, considered as impediments to the discovery (and teaching) of truth. In this they may be right, but being right makes his complaint against the logic of the syllogism ring hollow, ironically so. For recall, this was the complaint that the rules of syllogistic are useless for the conduct of life and destined quickly to be forgotten by students forced to learn them. One can only think the same of their own treatment of the passionate sophisms.

Though much the same is true of the discussion of the final sub-category of fallacies, those which arise from "objects themselves", there is to be found here points of occasional historical interest. It is correct but rather dreary to be told that "while censuring a man for his errors, we must not reject the truth he advances". Nor — anticipating Locke's disdaining of figurative speech — should he allow fancy oratory to obscure the principal business of getting at the truth of things, and still less should we be content with rash judgements and faulty inductive generalizations. Also to be condemned is the habit of finding a course of action to have been wise just because it chanced to have a desirable outcome. There is little to be learned from such injunctions that one would not already have known, and one can only wonder whether a charge of "useless" might not also here apply.

Concerning sophisms of authority and manner, somewhat more interesting things can be said, but not by the Royalists. Sophisms of authority they conceive of as accepting "something as true on an authority insufficient to assure us of this truth". In this, Arnauld and Nicole anticipate the present day notion of the *argumentum ad verecundiam*, as in the characterization of Copi and Cohen:

> The fallacy of *ad verecundiam* arises when the appeal is made to parties having no legitimate claim to authority in the matter at hand .... Wherever the truth of some proposition is asserted on the basis of the authority of one who has no special competence in that sphere, the appeal to misplaced authority is ... [a] fallacy .... [1990, 95]

Arnauld and Nicole distinguishes two subcases of this sophism, each in turn anticipating conceptions that survive in the textbooks of the day. One way to commit the fallacy of authority is be acceding to "doctrines spread by sword and bloodshed" — the modern *ad baculum* nearly enough. Another way is to yield to the argument: "The majority hold this opinion; therefore, it is the truest", an error which adumbrates the modern *argumentum ad populum.*

Sophisms of manner are even less pardonable than certain cases of the sophism of authority. Arnauld and Nicole concede,

> It is true that if there are pardonable errors, they ar those that lead people to defer more than they should to the opinions of those deemed to be good people. (Part III, Ch. XX, p. 221).

But, they continue,

> But there is an illusion more absurd in itself, although quite common, which is to believe that people speak the truth because they are of noble birth or wealthy or in high office. (pp. 221-222)

This is the fallacy of manner and it greatly resembles Locke's own *ad verecundiam*, which is not, the *ad verecundiam* of the logic textbooks of the present day. In his *Essay Concerning Human Understanding* 1690, Locke wrote of

> *Four sorts of Arguments* that men, in their reasonings with others do ordinarily make use of, to prevail on their assent, or at least so to awe them, as to silence their opposition. (Bk IV, ch. XVII)

One of these

> Is to allege the opinions of men, whose parts, learning, eminency, power or some other cause has gained a name, and settled their reputation in the common esteem with some kind of authority. When men are established in any kind of dignity, [it is] thought a breach of *modesty* for others to derogate any way from it, and question the authority of men, who are in possession of it .... Whoever backs his tenets with such authorities, thinks he ought thereby to carry the cause, and is ready to style it impudence in anyone who shall stand out against them. This, I think, may be called *argumentum ad verecundiam* (ibid.; Emphasis added in the first instance).

Although Locke makes it clear that he does not think that the *ad verecundiam* is a proof "drawn from the foundations of knowledge or probability" or that it "brings true instruction with it and advances us in our way to knowledge", it is interesting that he does not condemn it or call it a fallacy. The Royslists's *manner*, on the other hand, is held to be a sophism. What explains this difference? Two possibilities come to mind. (1) What Arnauld and Nicole condemn is accepting a proposition as true on the basis of the station of him who affirms it. What Locke does not condemn is the use of such a fact in trying to win a debate with an opponent, either by getting him to assent, or at least to think twice about his own contrary opinion. What Arnauld condemns and Locke does not are not the same thing; and we know independently that Locke would condemn a person inferring the truth of a claim just because it was pronounced by a man "of parts". In that respect, the Royalists' fallacy of manner and Locke's *ad verecundiam* come to the same thing.

They are also alike in stressing the factor of modesty. A modest person will defer to his betters and so he should, depending on who is deferred to and in what his betters' betterness consists. Both the Royalists and Locke caution against both over-deference and "irrelevant" betterness. For Arnauld and Nicole, it is always a mistake to believe true anything said by someone richer than oneself and just because he is so. For Locke, it is falsely modest to defer to the opinion of "learned doctors", in as much as one of Locke's principal purposes in the *Essay* is to establish that the learned doctors or "Schoolman" ought not to be deferred to. Even so, there are differences of emphasis between the two writers as concerns the interpretation of the idea of one's betters. Arnauld and Nicole's "betters" are people of superior birth, wealth or rank. Locke's "betters" are those whose eminency and reputation have been earned (with the possible exception of some learned doctors) and whose betterness consists of superior expertise and knowledge, rather than superior wealth or social rank. Perhaps, in the end, it is this difference which explains the Royalists' comparative harshness toward deference to manner and Locke's comparative serenity towards arguments *ad verecundiam*.

Descartes is important for having emphasized the distinction between methods of proof and methods of discovery, although it was not a distinction of his own invention. In this, he sets himself against the Euclidean geometrical paradigm (although even Euclid was not indifferent to this distinction). Descartes is sometimes careless in pressing his criticism against syllogistic reasoning. In some ways, his criticisms are "old hat" though not without merit. For example, in Rules 10 and 13 of the *Regulae*, he echoes a theme that has been sounded since antiquity, namely, that syllogisms are always instances of *petition principii*, a complaint that had been forcefully advanced by Sextus Empiricus in *Outlines of Pyrrhonism*, book two, sections 134-244 (especially section 163). Aristotle himself considered this same objection and rejected it (*Posterior Analytics*, 72^b 5 — 73^a 20). On the other hand, Aristotle never thought of the theory of syllogisms as a "logic" of discovery. This or something like it was reserved for the *Topics*, in which it is proposed that the job of discovery is the ascertainment of middle terms out of which syllogisms are subsequently constructible. Descartes was also critical of Aristotle's account of topical reasoning, but in as much as most of his complaints against topical reasoning are indistinguishable from those that he levels against the syllogism, it may fairly be said that Descartes misconceives Aristotle's objectives in the *Topics*.

Where Descartes differs from Aristotle, in ways harder to dismiss as merely misconceived, is in his rejection of the idea that the way of discovery involves the ascertainment of how things are in their objective natures, as revealed by their essences. Aristotle held that discovery involved the making of explanatory deductions grounded in the "natural order". Descartes agreed with this, but in so doing he displaced Aristotle's notion of the natural order with a different conception of it. The Cartesian natural order was not the realm of the essences of things as they are objectively and independently of the enquiring mind. Rather it is the order of human understanding, of how the mind operates in uniting *ideas* of things in ways that facilitate intelligibility, clarity and explanatoriness.

Given this new conception of achieving explanations according to the natural order, a further distinction announces itself. A *rationalist* is one who holds that the procedures of discovery are essentially "geometric", whereas an *empiricist* is one who holds that discovery is rooted in the particularities of sensory experience. Rationalists and empiricists are agreed, however (and contrary to Aristotle), on what constitutes the natural order. It is, to repeat, the order which is natural for the enquiring mind, not the order of independently realized Aristotelian essences. Both the Royalists and Locke accept this rejection of the Aristotelian natural order in favour of Descartes' own conception of it. In this Locke was much influenced by the Port Royal logicians, notwithstanding that Locke was an empiricist and Arnauld and Nicole (like Descartes himself) a rationalist.

5.5 *The Port Royalists' importance*

In a great many respects Arnaud's and Nicole's work can be seen as parasitic upon Descartes'. Arnauld, though respected by Descartes, was not the original thinker that Descartes was. Nevertheless, Arnauld and Nicole matter for fallacy theory, concerning which Descartes had nothing specifically to say. It may be seen that Arnauld and Nicole are early advocates of what has come to be known as informal logic. They greatly distrusted the scientific paradigm of correct reasoning in general. Their scepticism is rooted in three related convictions. First, since the reasonings of ordinary life do not aspire to them standards of scientific rigour, they cannot be faulted for their failure to model them. Second, when ordinary reason goes wrong it will typically be for reasons different from those that infect scientific reasoning. In particular, the Royalists saw non-scientific errors as arising from emotional excess and deeply engrained prejudices. This is an important development in the evolution of the concept of fallacy. The main idea is not new. It was long since taken for granted, in Athens and Jerusalem, that emotional excess is the enemy of virtue, both moral and intellectual. What is new in Arnauld and Nicole is the belief that the sins of the passion interfere with reason in *classifiably significant variations.*

A third factor, whereas scientific reasoning is the orderly demonstrative presentation of what is already known, the reasoning of ordinary life is more an attempt to discover truth. This carries consequences for the prospects of theory. It is clear that there exists a theory of the former, viz., the logic of demonstration. If there a theory of the latter, it would be a logic of discovery. Arnauld and Nicole can be seen as denying the possibility of fallacy *theory*, except for those deductive errors that fall within the narrow ambit of syllogistic reasoning.

In this regard, it may be thought that the Royalists are somewhat behind the times, at least behind the times that would come. Work by N.R. Hanson[37] and others reflect an optimism about logics of discovery, an optimism that is also reflected, albeit circumspectly, in recent work on the logic of abduction.[38]

6 ISAAC WATTS: AN INTERLUDE

I open this section with some further words from Hamblin:

> The eighteenth century, despite giants like Hume and Kant, was another Dark Age for Logic; with only a few immature stirrings behind the scenes, from writers like Saccheri and Ploucquet, to give promise for the future. (p. 163)

Hamblin continues:[39]

> Almost the only logic book written in English during the entire century was one in 1725 by Isaac Watts, better known as the author of 'O God, our help in ages past.'

The book in question, to give it its full title, is *Logick: or the Right Use of Reason in the Enquiry after Truth, with a Variety of Rules to Guard Against Error, in the*

[37]Hanson (1958). For the contrary view see Reichenbach [1938].

[38]See for example, Gabbay and Woods [2005], and Aliseda [2006], Magnani [2001; 2009] and Woods [2012].

[39]Hamblin seems to overlook William Duncan's *The Elements of Logick,* which appeared in 1748.

Affairs of Religion and Human Life, as well as in the Sciences. Three features of *Logick* merit mention, if only briefly. One is the book's overall indebtedness to the Port Royal *Logic,* both as it pertains to the fallacies of prejudice, but also in relation to the more traditional sophisms. A second feature is Logick's absorption of some of Locke's classification in "Of Reason". The third is the originating use of the name "false cause" for the fallacy discussed by Aristotle in the *Rhetoric* under the name "non-cause as cause". In the interests of space, I shall present Watts' classification of argument *verbatim* (*Logick,* pp. 465-466):

1. If an Argument be taken from the Nature or Existence of Things, and addrest to the *Reason* of *Mankind,* 'tis called *Argumentum ad judicium.*

2. When 'tis borrowed from some convincing Testimony, 'tis *Argumentum ad Fidem,* an Address to our *Faith.*

3. When 'tis drawn from any insufficient *Medium* whatever, where the Opposer has not Skill to refute of answer it, this is *Argumentum ad Ignorantiam,* an Address to our *Ignorance.*

4. When 'tis built upon the profest Principles or Opinions of the Person with whom we argue, whether these Opinions be true or false, 'tis named *Argumentum ad Hominem,* an address to our *profest Principles. St. Paul* often uses this Argument when he reasons with the *Jews,* and when he says, *I Speak as a Man.*

5. When the Argument is fetch'd from the Sentiments of some wise, great, or good Men, whose Authority we reverence, and hardly dare oppose, 'tis called *Argumentum ad Verecundiam,* an address to our *Modesty.*

6. I add finally, when an Argument is borrowed from any Topics which are suited to engage the Inclinations and Passions of the Hearers on the Side of the Speaker, rather than to convince the Judgement, this is *Argumentum ad Passiones,* an Address to the *Passions*; or if it be made publickly, 'tis called an *Appeal to the People.*

Of course, *ad passione* and *ad populum* are not on Aristotle's list; nor is *ad fidem. Ad ignorantiam, ad hominem* and *ad verecundiam* are Locke's in name but not in substance; and *ad judicum* is too obscurely rendered for conception one way or the other as something genuinely Lockean.

7 JOHN LOCKE

Son of a county lawyer, Locke was born near Bristol, England in 1632, the same year as Spinoza. He took his early education at home and later entered Westminister School. Upon graduation, Locke went up to Oxford in 1652. In 1660, he was appointed lecturer in Greek and subsequently, Reader in Rhetoric and Censor of

Moral Philosophy. While at Oxford, he developed a keen interest in the emerging new sciences, especially chemistry and physics, influenced by his friendship with Robert Boyle. Locke also studied medicine, taking his degree late (1674), owing to the interruption of his academic career nine years earlier when he was appointed secretary to a diplomatic mission to the Elector of Brandenburg. Locke practiced medicine sparingly after 1674, having in 1667 caught the eye of the soon-to-be earl of Shaftesbury, thanks to whom he received a series of administrative appointments over many years. Shaftesbury was periodically in the political wilderness and, during these episodes, Locke lived in Oxford, France and Holland. In France he was befriended by Gassendi and came to respect the philosophical innovations of Descartes and the Cartesians, and these proved an enduring influence. With the succession of William of Orange to the English throne in 1688, Locke returned to his native country and three years later retired. He died in 1704.

7.1 Locke's Essay

Locke's two major works are the *Essay Concerning Human Understanding* and *Two Treatises of Civil Government,* both published in 1690 (although the *Essay* appeared in late 1689). The *Essay* is an especially important contribution to the empiricist tradition in philosophy. It shares with Bacon's and Arnauld's writings a dissatisfaction with scholastic influences and anticipates some of Mill's later criticisms of them. Locke's complaints were threefold. Aristotelian philosophy had become dogmatic and authoritarian; it could not satisfactorily account for modern science, and it gave an unrealistic picture of the actual workings of the human intellect.

The *Essay* contains no explicit theory of fallacies, but in t here places it deals with issues in ways that encourage conjectures as to Locke's views on these matters might have been. In Chapter Twenty-Two of Book Four, Locke examines "wrong assent, or error". Chapter Seventeen of the same book is given over to a critique of Aristotelian logic and ends with a brief discussion of arguments *ad hominem, ad ignorantiam, ad verecundiam* and *ad judicum.* Book Three, Chapters Nine to Eleven deals with the "imperfection and abuse of words."

It may be said that if Locke did have a core conception of fallacy in the *Essay,* it is to be found at the end of Book Three, Chapter Three, "Of the Abuse of Words", where he shows great hostility to "figurative speeches", the stock-in-trade of the ancient liberal art of oration. He writes that

> all the art of rhetoric, besides order and clearness, all the artificial and figurative application of words eloquence hath invented, are for nothing but to insinuate wrong *ideas,* move the passions, and thereby mislead the judgment, and so indeed are perfect cheat and ... they are certainly, in all discourses that pretend to inform or instruct, wholly to be avoided ....

The arts of rhetoric, then, are "the arts of fallacy", a "powerful instrument of error and deceit". Locke disdains to enumerate the various ways in figurative speech gives rise to fallacies — doing so would be superfluous, he says. But there can be little doubt that he thinks that emotional language "harangues" and "popular addresses" are among the main culprits.

Locke seems to have a two-part thesis about fallacies, a part dealing with the nature of fallacy and a part dealing with its origin. As to what a fallacy *is*, it is a defect of reasoning that impairs the attainment of truth and knowledge. As to how fallacies *arise*, they are dominantly, though not exclusively, the result of the inappropriate use of language, in particular the deceptive use of language. Jointly, a fallacy is a piece of incorrect reasoning made to appear correct by deceptive features of language. For all his hostility to Aristotelian thought, it is interesting that Locke's core concept resembles Aristotle's notion of language-dependent fallacies.

7.2 Wrong assent, or error

The idea that fallacies are of linguistic origin is carried over into Book Four. It becomes clear that it is not Locke's view that only rhetorical language is at fault. Rather it is language as such, since "arguments being [as for the most part they are] brought in words, *there may be a fallacy latent in them* ...". That is, fallacies are latencies of language. The more rhetorical the language, the more active the latencies, no doubt, but Locke does not hold that fallacies are unique to contexts of oratorical abuse.

Not all fallacies are solely the result of language. Errors, says Locke are mistakes of "our judgement giving assent to that which is not true." Such judgements will surely include those in which we reason to a conclusion from what we take to be evidence, and so fallacies are a species of probative error. Among these he cites mistakes we commit when we lack access to proofs to the contrary, a condition that all of us are in all of the time with respect to at least a good many of our reasonings. Further, people will often have access to considerations which, if appreciated, would steer them clear of error, but are unable to understand them or to see their relevance. And even where an error-saving proof is available and understood, it may be resisted, as when we are caught in the grip of prejudice, since "what suits our wishes is forwardly [i.e., preconceivedly] believed."

To these Locke adds three kinds of cases, not entirely disjoint from those already listed or disjoint from one another. He speaks of mistakes occasioned by our uncritical acceptance of "received hypothesis" or what we might call "common knowledge", mistakes induced by "predominant passions or inclinations", and mistakes that arise from authority. Concerning the first of these, there is ample room to commit the fallacy of *petitio principii,* for if asked why he is justified in holding that the earth is flat, a person replies that his is common knowledge, he may with some justification be thought to have begged the question. On the other hand, a person not wanting to surrender his opinion to a good argument to the contrary may dig in his heels and say, "Though I cannot answer, I will not yield," since "I know not yet all that may be said on the contrary side," a move which many a later writer would regard as the fallacy of evading the burden of proof.

7.3 Critique of logic

If Locke's idea is that a fallacy is an error of reasoning, it is fair to ask whether he has a theory of good and bad reasoning, or, a we might loosely say, a *logic*. In the Twenty-first Chapter of Book Four, Locke ventures an interesting idea. Logic, he says, is the "doctrine of signs" or anyhow, of those signs that are words. This logic is

> the business whereof to consider the nature of signs the mind makes use of for the understanding of things, or conveying knowledge to others.

Since we have direct access neither to the things of which we have ideas nor to other people's ideas of things, "the consideration, then of *ideas* and *words* ... if they were distinctly weighted and duly considered ... would afford us another sort of logic and critique than what we have hitherto been acquainted with." To modern ears, this is a conception of logic that sounds rather like semantics or psycholinguistics, and it is unfortunate that Locke never gave to it a positive and detailed formulation. But he does have interesting things to say about the logic we have "hitherto been acquainted with." It is the Aristotelian theory of syllogisms.

Known for his hostility toward syllogistic logic, Locke in Chapter Seventeen, Book Four, makes a striking concession:

> And I readily own that all right reasoning may be reduced to his [i.e., Aristotle's] forms of the syllogism.

Even so, says Locke, "they are not the only, nor the best ways of reasoning, for the leading of those into truth who are willing to find it."

Here we meet with a distinction between the *logic of discovery* and the *logic of justification* (or *demonstration*), an echo of Descartes' *Discourse on Method* and Arnauld's *The Art of Thinking,* in which the distinction is more adequately drawn. A logic of discovery would be a logic of hypothesis formation and of legitimate inferences therefrom, or, as Locke says, of "illation." A logic of justification would concern the most certain and economical *presentation* of knowledge already obtained (as the output, as it were, of the logic of discovery). We find in this a clever anticipation of Mill's celebrated doctrine that, strictly speaking, there are no deductive *inferences,* in Book One of his *A System of Logic.* If we consider the valid syllogism.

1. All men are mortal.

2. Socrates is a man.

3. Therefore, Socrates is mortal

we see, says Mill, that premiss (1) records no actual operation of the mind, in the course of inferring Socrates' mortality from his humanity. On the other hand, our syllogism is an efficient, economical and perfectly correct *presentation* of what is already known of the connection between being human and being mortal. In like fashion, the proofs of Euclid's geometry are fine for what they are. But they are no faithful record of how the work-a-day geometer goes about the business of discovering the truths of geometry, a view in which Locke concurred. Locke held that how the truths of mathematics are actually arrived at will not, in the general case, constitute a demonstration of their truth or a justification of the claim that they are true. And so,

> Tell a country gentlewoman that the wind is south-west and the weather louring and like to rain, and she will easily understand that it is not safe for her to go abroad thin-clad on such a day, after a fever; she clearly sees the probable connection of all these, viz., south-west wind, and clouds, rain wetting, taking cold, relapse, and danger of death, without trying them together in those artificial and cumbersome fetters of several syllogisms, that clog and hinder the mind, which proceeds from one part to another quicker and clearer without them. And I think that everyone will perceive in mathematical demonstrations, that the knowledge gained thereby comes shortest and clearest without syllogism.

Further, people "in their own enquiries after truth, never use syllogisms to convince themselves, because, before they put them into a syllogism, they must see the connection that is between [the evidence and the conclusion drawn from it] ... and when they see that, they see whether the inference be good or no, and so syllogism comes too late to settle it."

Locke can be seen as claiming that, in general, the recognition that our inferences are correct must precede any syllogistic justification of them. The old logic, thus, "fails our reason in the part which, if not its highest perfection, is yet certainly its hardest task, and that which we need most help in, and that is, *the finding out of proofs, and making new discoveries*." Bacon (and Arnauld, too) made the same complaint against the old logic and proposed a new one, an *ars inveniendi,* or a theory of the advancement of learning. Locke was less ambitious. He had no *ars inveniendi* to propose. Human reasoners rely essentially and inescapably upon their "sagacity," a thought which Locke shares with Arnauld. Beyond that, Locke had little to say about the logic of discovery.

Locke thought that he had a good reason for asserting that syllogisms never constitute discoveries. It is a reason which exploits a technicality in Aristotle's theory of the formal syllogism, as developed in the *Prior Analytics.* Locke's argument turns upon two factors. One is that central to his own theory of knowledge is the claim that "rightly considered, the immediate object of all our reasoning and knowledge is the claim that "rightly considered, the immediate object of all our reasoning and knowledge is nothing but particulars." The other is that, central to Aristotle's logic is the claim that all syllogisms must contain at least one general premiss. And so we must

> take note of one manifest mistake in the rules of syllogism, viz., 'that no syllogistical reasoning can be right and conclusive, but what has, at

> least, one general proposition in it.'

Locke's positive view was that a logic of discovery would characterize inferences drawn "according to reason":

> *According to reason* are such propositions whose truth we can discover by examining and tracing those ideas we have from sensation and reflection, and by *natural deduction* find to be true or probable. (Emphasis added in the second instance).

There is one respect in which Locke was entirely faithful to his Aristotelian heritage. Like Aristotle, he recognized the existence of first principles, the objects of "our highest degree of knowledge." Such knowledge is (intuitive). It "is certain, beyond all doubt, and needs no probation, nor can have any, this being the highest of all human certainty." These "maxims" are propositions everyone knows to be true "as soon as ever they are proposed to his understanding." And they are unprovable. Where Locke parted company with Aristotle was not over the existence of such maxims or first principles, but rather over where they might be found. Whereas Aristotle thought that each science possesses first principles peculiar to it and on which all its truths depend deductively, it can be questioned whether Locke would have agreed that natural sciences such as chemistry contain first principles in Aristotle's sense. In this he seems to share Bacon's suspicion of self-evident principles in science, if not Mill's outright hostility to them.

We must now ask whether Locke's views on the syllogism carry implications for a theory of fallacies. It would seem that they do. Insamuch as all right reasoning reduces to syllogisms, any fallacy that can be made out to be a deficiency of a syllogism can also be made out to constitute the "non-rightness" of the reasoning which reduces to it. To the extent that Aristotle's list of thirteen can be seen as syllogistic errors, they can be seen as fallacies that Locke himself would recognize. It seems, thus, that Locke is committed to recognizing the fallacies of ambiguity, amphiboly, combination of words, division of words, wrong accent, the forms of expression used; accident, *secundum quid, ignoratio elenchi,* begging the question, consequent, non-cause as cause, and many questions. It is, as we saw, a point of contemporary contention as to whether Aristotle's thirteen are construable as syllogistic errors, but it is the present writer's view that this is precisely how Aristotle thought of them.

7.4 The imperfection and abuse of words

In Chapter Nine, Book Four, Locke draws our attention to imperfections in language which constitute an impediment to knowledge. The meaning of some words can be given by ostension, but most words are not ostensibly explicable, especially those words in which we attempt to transact the business of science and philosophy, the very enquiries which we are inclined to look upon as "giving the most perfect knowledge." Locke admits that where ostension won't serve, common use is sometimes a helpful guide to the meaning of words. But

> Common use *regulates the meaning of words* pretty well for common conversation, but nobody having an authority to establish the precise signification of words, nor determine to what *ideas* anyone shall annex them, common use is not sufficient to adjust them to philosophical [and other technical] discourses.

This is a state of affairs which impedes knowledge, and if

> we consider, in the *fallacies* men put upon themselves as well as others and the mistakes in men's disputes and notions, how great a part is owing to words and their uncertain or mistaken signification, we shall have reason to think this no small obstacle to knowledge ... [emphasis added]

Locke admits that by rather considerable effort to correct the significations of words, some of these fallacies can be averted. But because these imperfections are "naturally in language," it would be naïve to expect a perfect outcome. In contrast with the "obscurity and confusion that is so hard to be avoided in the use of words," Locke is quicker to condemn and to be optimistic about the avoidance of "*willful faults and neglects*, which men are guilty of in this way of communication ...," and which are the subject of Chapter Ten. Such faults and neglects include the use of obscure words or words having no distinct meaning (and in some cases, no meaning at all). Particularly odious is the habit of using a word now with one meaning and later with another, which, when it occurs in the same contexts of communication, is a clear case of the illicit exploitation of a term's ambiguity. Also disapproved of is the unannounced redefinition of a familiar word. Locke has no time for "skill in *disputing*" as an indication of a person's "parts and learning," since it is part and parcel of disputes to win at all costs. "Disputes" therefore are a standing invitation to manipulate language even at the cost of truth. In addition to the faults already noted, Locke also cites the over-clever use of words, learned gibberish, jargon, each the abettor of "logical niceties" and "empty speculations." Further to this is the mistake of supposing that all terms denote real things or states of affairs. Overall, it is evident that Locke would find little fault with Aristotle's fallacies that depend on language, though it must be said that Locke shows no inclination to pursue their analyses in any detail. Concerning their avoidance, Locke has little to say except to be on guard against the imperfections of language and those who would abuse us with them.

7.5 Arguments ad

Chapter Seventeen concludes with the one passage cited by all writers who look to Locke for instruction on the fallacies. Given its celebrity, I will quote the passage in full, including the parts already discussed in the section on the Port Royalists:

> Before we quite this subject, it may be worth our while a little to reflect on *four sorts of arguments* that men, in their reasonings with others,

> do ordinarily make use of to prevail on their assert, or at least so to awe them as to silence their opposition.
>
> *First,* The first is to allege the opinions of men whose parts, learning, eminency, power, or some other cause has gained a name and settled their reputation in the common esteem with some kind of authority. When men are established in any kind of dignity, it is thought a breach of modesty for others to derogate any way from it, and question the authority of men who are in possession of it. This apt to be censured as carrying with it too much pride, when a man does not readily yield to the determination of approved authors which is wont to be received with respect and submission by others; and it is looked upon as insolence for a man to set up and adhere to his own opinion against that of some learned doctor or otherwise approved writer. Whoever backs his tenets with such authorities thinks he ought thereby to carry the cause and is ready to style it impudence in anyone who shall stand out against them. This I think may be called *argumentum ad verecundiam.*
>
> *Secondly,* Another way that men ordinarily is to drive others and force them to submit their judgments and receive the opinion in debate is to require the adversary to admit what they allege as a proof, or to assign a better. And this I call *argumentum ad ignorantiam.*
>
> *Thirdly,* A third way is to press a man with consequences drawn from his own principles or concessions. This is already known under the name of *argumentum ad hominem.*
>
> *Fourthly,* The fourth is the using of proofs drawn from any of the foundations of knowledge or probability. This I call *argumentum ad judicium.* This alone of all the four brings true instruction with it and advances us in our way to knowledge. For: (1) It argues not another man's opinion to be right because I, out of respect or any other consideration but that of conviction, will not contradict him. (2) It proves not another man to be in the right way, nor that I ought to take the same with him, because I know not a better. (3) Nor does it follow that another man is in the right way because he has shown me that I am in the wrong. I may be modest and therefore not oppose another man's persuasion; I may be ignorant and not be able to produce a better; I may be in an error, and another may show me that I am so. This may dispose me, perhaps, for the reception of truth but helps me not to it; that must come from proofs and arguments and light arising from the nature of things themselves, and not from my shamefacedness, ignorance, or error.

It is a matter of considerable interest that Locke does not think that *ad verecundiam, ad ignorantiam* and *ad hominem* arguments are fallacies. The word "fallacy" doesn't occur in this passage. Even so, it is clear that Locke would readily accede to the idea that arguments of these three types are fallacious when they are offered as arguments of the fourth type, viz., as arguments *ad judicium*. *Ad judicium* arguments satisfy a particularly strong condition. They cannot be bad arguments, for they are *proofs* which succeed in giving instruction concerning matters of knowledge and probability. In this, they bear some resemblance to Aristotle's own conception of syllogisms, particularly those known as didactic or demonstrative. Not only is there no such thing as an invalid syllogism, but didactic and demonstrative syllogisms also provide true instruction about things.

Concerning the *ad hominem*, it is instructive to return to our earlier comparisons of Locke with Aristotle. In a number of places, Aristotle recognizes a distinction between making an argument against an opponent's thesis and making an argument against one's opponent himself (*Soph Ref*, 177^b33, 178^b, 17, and 183^a 21). At 178^b 17, Aristotle draws the distinction in such a way that a Latin translation of the original Greek would be well-justified in employing the expression of "*ad hominem*". "[S]uch persons direct their solutions *against the man*, not against his argument." (Emphasis added) In all three passages, Aristotle's adjacent discussion is difficult and leaves obscure how the distinction applies to the cases considered. This being the case, it is difficult to determine the extent to which Aristotle has anticipated Locke's own conception of the *ad hominem* argument. Even so, it is clear that Aristotle thinks that sometimes, and only sometimes, an argument succeeds against "the man" and against "his argument" alike. This suggests, with Locke, that it is a mistake to suppose in general that an argument that succeeds against an opponent also succeeds against his or her thesis. An exception would seem to be the *reductio ad absurdum* argument, though Aristotle doesn't discuss it in these passages. If a critic manages to derive an absurdity from an opponent's thesis, then the thesis implies a falsehood and is itself false. The critic has done two things in making a successful *reductio*. (1) He has shown that his opponent holds a thesis inconsistent with its own consequences, and (2) he has shown that the negation of his opponent's thesis is true. But, as we saw at the beginning of this chapter, it is sometimes the case that one person's *reductio* is for another the sound demonstration of a surprising or like shocking truth.

We may see Locke as going half-way with Aristotle on this matter. A Lockean *ad hominem* is one which derives a consequence from an opponent's own principles or concessions and which is at variance with them. But it does not follow for Locke that the derivation establishes that the negation of the opponent's principle or concession is true. If it did, the derivation would qualify as an *ad judicum* argument, which is something that Locke is at pains to deny. Whatever else they are, Lockean *ad hominems* are not *reductio ad absurdum* arguments.

It also becomes apparent that Locke conceives of the *ad hominem* more narrowly than many subsequent writers, even those who invoke Locke's name with approval. Two examples will make the point clear. Consider an argument in which a proponent defends the thesis that a certain piece of environmental legislation should be resisted. If a critic objects that the proponent holds this view because she owns shares in an oil and gas exploration company, a contemporary writer would classify this remark as a "circumstantial" *ad hominem.* It advances the suggestion that the proponent holds her thesis for self-interested reasons, not because her position is objectively defensible. Whatever is to be made of such a complaint, it is not an *ad hominem* in Locke's sense, since the critic is not deriving the objection as a consequence of the proponent's thesis. The same is true of the so-called *tu quoque* complaint. If someone holds the view that cigarette smoking is a bad thing, one is said to make an *ad hominem* objection of the *tu quoque* sort in pointing out that the anti-smoker is himself a smoker, that his behaviour is inconsistent with his own principles. Though it might well be true that his behaviour is inconsistent with what he himself espouses, in drawing attention to it, the critic is not deriving this behaviour from his opponent's thesis.

It must be said, then, that post-Lockean analyses of the *ad hominem* do not, in general, conform to Locke's conception of it.

Locke's *ad verecundiam* is often called the fallacy of the appeal to authority. It is true that Arnauld in *Art of Thinking* (1662), recognized the "sophism" of authority, but it is not Locke's *ad verecundiam.* Locke's *ad verecundiam* more nearly resembles Arnauld's "sophism of manner." "Verecundiam" means modesty. A person making use of such an argument urges that his opponent would be immodest not to accede to the opinions of reputable and recognized experts.

The principal difference between Arnauld and Locke concerning modest deferral to the opinions of one's betters is this: Arnauld thinks of one's "betters" almost exclusively as those who are one's superiors in the social hierarchy. Though he is ironic in places about the learnedness of scholastic professors, Locke's conception of one's betters is, by and large, that of persons who have superior knowledge or expertise. Thus, for Arnauld, not only is not immodest not to defer to one's betters in his sense, it is a cringing and stupid thing to do. But for Locke, a modest deferral to one's betters, in his sense of the term "better" is often perfectly reasonable. Locke and Arnauld agree that the opinion of one's betters never suffices to establish the truth of a disputed proposition. For if it did suffice, then, as Locke would say, that would qualify an *ad verecundiam* argument as an *ad judicium* argument.

Similarly for the *ad ignorantiam.* It is frequently recognized that if Jones and Smith are arguing about P, with Jones *pro* and Smith *contra,* and if Jones cannot give a better argument for his side of the issue than Smith has given for his, then Smith wins the argument. Whether he does or not depends upon where the greater burden of proof lies. But, even if he does, in some sense, win the argument, this is no guarantee that Smith's position is true.

It can hardly be doubted that just because a critic is unable to make a better case for his position than his opponent has made for the contrary position, that the opponent's position cannot, in general be considered as having been verified. But it is probably too much to say that this can never happen. In certain systems of deductive logic, results by Jacques Herbrand establish that if, with regard to some thesis, a search for supporting considerations fails, then this very fact constitutes a counterexample to the thesis. This is not quite an *ad ignorantiam* sort of situation, but it is a near thing. If such "Herbrand-cases" exist, and even if they have the look of *ad ignorantiam* situations, then in Lockean terms they are *ad judicum* arguments and, hence, cannot be considered as bona fide *ad ignorantiam* arguments.

Some people will not be much impressed by the idea that it is a fallacy to confuse an *ad judicum* argument with any of the three types we are here examining. They will think that it is perfectly obvious that *ad verecundiam, ad ignorantiam* and *ad hominem* arguments cannot be *ad judicum* arguments in Locke's strong sense of that term. To the extent that this is so, they can't be fallacies either, for in offering, e.g., an *ad verecundiam* argument as a knock-down *ad judicum* argument, it is obvious that it is no such thing. And if fallacies are conceived of as errors that fail to appear to be errors, these will not be fallacies. More important, perhaps, is that saying of the three that they are not the fourth is hardly the most interesting or instructive thing that can be said of them. If that were all there were to it, it would be inexplicable that these three *ad* arguments should have excited the enormous interest that has surrounded them. More interesting, some would say, is the specification of the exact conditions under which *ad hominem, ad verecundiam* and *ad ignorantiam* arguments succeed as *non-judicum* arguments. For example (and leaving aside *reductio ad absurdum* arguments), when one disputant presses another with consequences of his principles or concessions, when do those consequences constitute a dispute-winning hand, and when not? A much harder question, but also a more interesting one.

7.6 Locke's Importance

Locke's importance to logic is slight. He can, of course, be commended for picking up on the distinction between the logic of discovery and the logic of demonstration, and for thinking, with Arnauld, that discovery is prior to demonstration. Important as the distinction may be, it is a dubious basis on which to attack Aristotle's logic, since Aristotle certainly did not think of the formal theory of the syllogism developed in the *Prior Analytics* as a logic of discovery. That Locke himself did not press forward with a positive theory of discovery, as Bacon and Descartes attempted to do may strike us as regrettable. Even if we believe that a logic of discovery is theoretically intractable, as it might well be, at least Arnauld had the wit to recognize the point and to urge it with plausible effect. Much the same reservations attach to Locke's entirely justified complaint that both language and a person's willfulness are occasions of error, but there is little detailed attention given to the ways in which such predispositions might be resisted. All the same, Locke enjoys a large reputation in regard to the *ad* fallacies. There is in this an irony as large as the reputation, in as much as Locke did not consider the *ad* arguments to *be* fallacies.

8 RICHARD WHATELY (1787-1863)

Whately was the son of a London clergyman and educated privately and at Oriel College, Oxford, from which he emerged with a double-second. He was elected Fellow of the College and in 1814 received holy orders. Whately wrote copiously, and with effect, about religious matters, and played a large reformist role at Oxford, especially in his role as principal of St. Alban Hall.

Elements of Logic appeared to considerable acclaim in 1826, and *Elements of Rhetoric* in 1828. *Elements of Logic* was widely used in the colleges and universities of Britain and North America, moving Charles Peirce to say that it was this book which, at age 12, endeared him to logic, then and ever more.

In 1831 Whately was appointed Archbishop of Dublin, but not before an unsuccessful challenge in the House of Lords. In the years following, he was much occupied with ecclesiastical and educational reforms and the improvement of the Irish poor, and wrote several further works of contemporary importance. Whately remained active until the illness that occasioned his death in 1863.

8.1 Logic

Whately published *Elements of Logic* in no small part out of a sense of dissatisfaction with the condition of logic at Oxford. Contemporary texts were of little value, notably, Henry Aldrich's *Compendium* of 1691, which was too elementary, and Jeremy Bentham's *The Book of Fallacies,* published in 1824, which — aside from not having actually been written by Bentham — was too slight and silly a book for anyone with a real interest in logic.[40] Whately's purpose was to help get logic back on its feet.

> It is on Logical principles therefore that I propose to discuss the subject of Fallacies; and it may, indeed, seem to have been unnecessary to make my apology for so doing, after what has been formerly said, generally, in defence of Logic; but that the generality of Logic writers have usually followed so opposite a plan. Whenever they have to treat of any thing that is beyond the mere elements of Logic, they totally lay aside all reference to the principles they have been occupied in establishing and explaining, and have recourse to a loose, vague and popular kind of language; ...." (*Elements of Logic,* Bk. III, *Intro.,* pp. 101-102).

Here, a hundred and forty-four years earlier, is Hamblin's complaint of 1970. When they discuss them at all, logicians fail to bring to the fallacies anything close to the disciplined theoretical rigour that characterizes their treatment of the principles of logic. In challenging logicians to correct this omission, Hamblin was well aware of Whately's own ambitions. So we must suppose on Hamblin's part a not wholly

[40]Note the book's full title: *The Book of Fallacies: From Unfinished Papers of Jeremy Bentham.* By a Friend, London: Hunt, 1824. The "friend" was a young man named Peregrine Bingham.

approving estimate of Whately's success in fulfilling them. Even so, Hamblin finds value in various aspects of Whately's contributions (*Fallacies,* pp. 168-175, and 195-199), and rightly so.

Perhaps Whately's most notable contribution to the study of fallacies is to be found not in *Elements of Logic,* but in *Elements of Rhetoric,* published two years later in 1826, where he stresses the importance of presumption and burden of proof. Notwithstanding the title of this work, Whately's views on these matters are less concerned with their connection to persuasiveness than with the dialectical requirements of case-making. Also of interest is that Whately seeks instruction about these matters from English jurisprudence:

> According to the most correct use of a term, a 'Presumption' in favour of any supposition, means, not (as has been sometimes erroneously imagined) a preponderance of probability in its favour, but, such a pre-occupation at the ground, as implies that it must stand good till some sufficient reason is adduced against it; in short, that the *Burden of proof* lies on the side of him who disputes it. (*Rhetoric,* Part I, ch. III, § 2)

He continues:

> Thus, it is a well-known principle of the Law, that every man (including a prisoner brought up for trial) is to be *presumed* innocent till his guilt be established. This does not, of course, mean that we are to *take it for granted* he is innocent; for if that were the case he would be entitled to immediate liberation: nor does it mean that it is antecedently *more likely than not* that he is innocent; or, that the majority of those brought to trial are so. It means only that the 'burden of proof' lies with the accuser; ....

There is one element in this passage about which Whately is wrong in law. Jurors, he says, must not "take it for granted" that the accused is innocent, since it would otherwise follow that the accused is "entitled to immediate liberation." In fact, that is precisely what the jury's position should be, and that the accused *is* entitled to liberation unless and until the prosecutor proves otherwise beyond any reasonable doubt.

Whately's particular contribution lies in the claim that the manner in which the factors of presumption and burden are managed in law is transposable to the logic of case-making quite generally. If, in any non-legal context in which an argument has arisen,

> You have the 'Presumption' on your side, and can but *refute* [= rebut] all the arguments brought against you, you have, for the present at least, gained a victory: but if you abandon this position, by suffering the Presumption to be forgotten, which is in fact *leaving out one of, perhaps, your strongest arguments,* you may appear to be making a feeble attack, instead of a triumphant defense.

Consider a case. You take it as given ("take for granted", presume) that the next tiger your neighbour will see will be four-legged. Your neighbour challenges this. "What's your evidence?", he demands. Whately's advice would be to tell your neighbour to stop being so silly; and he would advise you not to take the bait. For suppose you did. Suppose you said, "Evidence? You want evidence? *All* tigers are four-legged. That's my evidence!" Game to your opponent. He will have nabbed you for question-begging.

There now arises a question of central importance. Does a proposition's presumptive status have evidentiary or probative force, or is it merely a procedural advantage? In the lines just quoted, Whately's words are ambiguous. If you forget that presumption is on your side, he says, you "leave out" what might be "one of ... your strongest arguments." If "argument" means "argument in support of the truth of", then, contrary to what Whately thinks, presumptions have probative force. But if "argument" here means "manoeuvre to keep one's opponent at bay", presumption is more a procedural advantage. Textual evidence leaves this question unsettled. However, taken in the procedural way, it is easy to see that presumption favours received or established positions. Whately's critics were bothered by this conservatism, especially its use in supporting the Christian religion. The criticism is not without interest. How, it might be asked can it in any intellectually honest way be a support of it that Christianity is the default position in early nineteenth century England? If the presumption of its truth lends Christian doctrine no probative support, and offers instead only the lack of any dialectical obligation to prove its teachings, there is in England no more support for Christianity than that rendered to Zoroastrianism in Persia.

On thinking it over, there is a world of difference between the presumption that the next-observed tiger will be four-legged, or that Paris didn't burn to the ground last week, and the presumption that Spike is innocent until his guilt has been proved beyond a reasonable doubt. A good part of this difference is that the presumptions about four-leggedness and Paris do in fact have evidential force, albeit defeasibly, whereas the presumption of innocence is entirely a matter of social policy — designed as a discouragement of unjust convictions even at the cost of erroneous acquittals. Indeed, the very sense in which the expectations about the next tiger and about Paris are presumptions is precisely the sense in which, in well-managed jurisdictions of criminal law, the very fact that Spike has been charged carries a presumption of his actually having committed the crime. In non-corrupt judicial systems, prosecutors don't bother with doubtful cases.

There is in this a moral of some general significance. Roughly speaking it is this: when a concept of interest to logicians has a central place in legal thinking and practice, do not suppose until you learn otherwise any guidance for the logical concept from how its counterpart concept works in the law. This is certainly true of the concept of presumption. But it is little or no less true of concepts such as proof, relevance, evidence, and probability.[41]

---

[41]See here [Woods, 2007a; 2010; Gabbay and Woods, 2010].

Similar reservations apply to Whately's "legalization" of the logician's (or dialectician's) idea of where the burden of proof lies. In English criminal law and its descendents in other Commonwealth countries and the United States, the burden of proof shifts between prosecution and defence as the trial moves from element to element. However it is certainly true that on the matter of conviction the burden rests with the prosecution. But here, too, there is a difference (and sometimes a conflict) between legal and extralegal provisions. In criminal law, the legal burden is levied even when, in a non-legal sense, the burden falls on the defendant. Consider a case: Spike has widely and loudly announced his intention to murder the mayor. Now that the mayor has indeed been murdered, it emerges that Spike was spotted at the scene of the crime, and his fingerprints were on the murder weapon. It is just common sense that *dialectically* the burden in those circumstances would be Spike's. But if Spike goes to trial, the burden will have shifted from him to the state as a matter of legal policy.

In *Logic*, Bk III, Whately presents a classification of the types of fallacy, whose complexity outreaches even the Port Royalist list. Here is a non-tabular paraphrase of it. The fallacies divide into logical (conclusions don't follow) and non-logical or material (conclusion does follow). The logical subdivide into the purely logical (invalid by form of expression) and semi-logical (ambiguity of the middle term). Purely logical fallacies include undistributed middle and illogical process. A fallacy may be semi-logical "in itself", either "accidently" or from "some connexion between the different senses", or from context. Accident-fallacies include fallacies of resemblance, analogy and cause and effect; and in-context fallacies include composition and diversion, as well as "fallacia accidentis".

Non-logical fallacies divide into those in which premisses are unduly assumed and those in which the conclusion is irrelevant. Premiss-fallacies include *petitio principii,* of which circularity is one instance, as well as assuming a proposition "unfairly implying" the question at issue. Conclusional irrelevancies include the fallacies of "objection" (to something wholly irrelevant) and "shifting ground" (from premiss to premiss alternately). Other conclusional irrelevancies are the use of (over)-complex and (over)-general terms, and such appeals to the passions — see here the influence of the Port Royalists — as the *ad hominem* and *ad verecundiam.*

Whately follows Locke in distinguishing *ad hominem* and *ad verecundiam* from *ad judicium* (or *ad rem*) and adds to this list *ad populum*. *Ad populum* is an appeal to "the prejudices, passions , etc. of the multitude (Bk. III, § 15); and the *ad hominem* (while retaining something of its Lockean character, picks up the added feature of shifting the burden of proof, not always "unjustly". *Ad ignoratiam* retains no Lockean trace, and is now the entire category of fallacies of conclusional irrelevance — much in accordance with its present-day meaning.

Whately is best known for his discussion of the *tu quoque* example of burden of proof. How, he asks, does the critic justify his hostility to blood sports when he himself is content to "feed on the flesh of ... harmless sheep and ox?" Whately continues in a partly Lockean and partly Arnauldian fashion:

> Such a conclusion it is often both allowable and necessary to establish,

> *in order to silence those who will not yield to fair general argument;* or *to convince those whose weakness and prejudices would not allow them to assign to it its due weight* ... provided it is done plainly, and *avidely*; ... The fallaciousness depends upon the *deceit*, or attempt to deceive. (Emphasis added in the first two instances)

The importance of deceptive intent is also apparent in Whately's understanding of the other *ad* fallacies. They are fallacies only when "unfairly used".

8.2 Whately's importance

Although Aristotelian and Lockean influences are clearly at work in them, Whately's logical writings reinforce the Port Royalist emphasis on the prejudices and passions in human thinking and discourse, including the intention of adversaries to trick and deceive. Whately also has original, though perhaps mistaken, things to say about the influence on logic of the legal notions of presumption and burden of proof. In all these respects, Whately is recognizably present in the writings of modern fallacy theorists, both before Hamblin and afterwards. Like many of them, most of Whately's contributions to logic were contained in a popular undergraduate textbook. In some ways, Whately was the Irving Copi of a good part of nineteenth century undergraduate instruction.[42]

9 JOHN STUART MILL (1806-1873)

Born in London, Mill was educated by his father, James Mill, who was Bentham's friend and collaborator, and who, seeing no need to send his son to school, didn't. James was an officer of the East Indian Company and was joined by his son in 1823 as a clerk. Rising to senior levels in the Company, Mill *fils* wrote a defence of the Company's application for the renewal of its licence in 1857. When the application failed, he left the Company and in 1865 stood for Parliament as member for Westminster and was elected. Defeated in the election of 1868, he divided his time between London and Avignon until his death in the latter place five years later.

Mill began his intellectual life as a philosopher radical in the manner of Bentham and his father. Following an emotional crisis at the age of twenty, he came to regard his former radicalism as too skeptical and extreme. There followed a fertile period of thinking, and Mill's maturing views were given expression in a large number of essays on economics, politics, sociology and philosophy. Dismissed by some as a "mere" pamphleteer incapable of systematic thought, Mill answered his critics with the publication in 1843 of *A System of Logic, Ratiocinative and Inductive: Being A Connected View of the Principles of Evidence and the Methods of Scientific Investigation.* Here he argues that ethics, politics and the social sciences are

---

[42]For an insightful discussion of Whately's importance as a logician, see James Van Evra's contribution to *British Logic in the Nineteenth Century,* volume 4 of this *Handbook,* [2008].

indeed systematically intelligible once developed by the methods of the natural sciences. There is not much evidence of this connection in the parts of his book in which he discusses the fallacies. Even so, the *Logic* enjoyed a large success and was adopted as a text at Oxford and Cambridge, in turn.

9.1 Deduction and inference

It is Mill's view that the proper subject of logic is proof. Proof he understands to encompass *deduction, generalization* and *observation.* Mill thus inherits from Bacon the idea that logic, properly conceived of, is a theory of scientific method. But, in book three of the *Logic*, Mill adds that "a complete logic of the sciences would also be a complete logic of practical business and common life" (Bk III, Ch. I, Sec. 1).

Mill is a radical about logic. He shows himself the anticipator of developments for which others have won credit. Anyone who knows Lewis Carroll's celebrated essay "What the tortoise said to Achilles" (*Mind* 1894), will be interested to know that there is nothing in this piece that isn't in Mill's *Logic* book two, chapter six, section five. Then, too, Mill greatly distrusted the idea that the formal sciences are analytic and was led to the view that arithmetic is an empirical science a full century and more before it became a received idea in some quarters of respectable philosophy. Even so, it may be said that Mill was, until Frege decades later, the strongest nineteenth century voice against psychologicism in logic.

Mill is also notorious for having held that syllogisms as such commit the fallacy of "begging the question", *petitio principii*, or circularity. Although this view has been held by various people since antiquity — it was, for example, considered and rejected by Aristotle himself (*Posterior Analytics* 72^b 5 and 73^a 20), it is still debatable as to whether Mill was one of them. In fact, it is an opinion he declared to be "fundamentally erroneous" since it misrepresents "the true character of the syllogism". In its "true character", there isn't the slightest doubt of his admiration of syllogistic logic. As he says in his *Autobiography* of 1873 at chapter one, section 12:

> My own consciousness and experience ultimately led me to appreciate ... the value of an early practical familiarity with school logic [i.e., syllogistic logic]. I know nothing, in my own education, to which I think myself more indebted for whatever capacity of thinking I have attained.

Even so, Mill disagreed with Whately, whose own book, *Elements of Logic*, appeared in 1825 and was approvingly reviewed by Mill. In a sentence that might have been penned by Locke, a hundred and thirty-five years earlier, Whately wrote (in book four, chapter one, section one) that

> ... *all* reasoning, on whatever subject, is one and the same process, which may be clearly exhibited in the form of syllogisms.

Certainly by 1843, Mill no longer accepted this position, if ever he did. But what he is rejecting is deductivism, not deductive reasoning as such. Deductivism is the view that all correct reasoning is syllogistic, hence is Whately's own position.

On the question of the circularity of syllogistic reasoning, it is difficult to make Mill out. On some readings (e.g. Wilson, volume 4, this *Handbook*), Mill, in effect, allies himself with Sextus's ancient claim that syllogisms are circular as such, but without accepting Sextus' all-purpose skepticism about logic. Seen Mill's way, the principal contribution of syllogistic logic to human thinking is as a check against inconsistency. Take *Barbara* as an example:

i. All A are B

ii. All B are C

iii. Therefore, all A are C.

Given that (iii) follows deductively from these premisses, it is a check on future premiss-selections that the negation of (iii) is inadmissible for selection as long as (1) and (ii) are retained.

On the alternative interpretation, Mill's position was that (1) when a syllogism is forwarded as a proof and (2) when its general premiss is analyzed in conformity with a common view, then the syllogism would indeed be circular. But it is possible that Mill took this to be a *reductio ad absurdum* of that view of generality. It can also be taken as discouragement of the idea that inference should always aspire to the status of a proof.

It was commonly supported that a general proposition such as "All humans are mortal" is strictly equivalent to the conjunction of all and only its positive instances: "a is a human and is mortal *and* b is a human and is mortal *and* c is a human and is mortal *and* ..." and so on. Consider the syllogism

1. All humans are mortal

2. Socrates is a human

3. Therefore, Socrates is mortal

In as much as premiss (1) is just a reformulation of its own exhaustive conjunction of positive instances, then the conclusion (3) is already asserted by premiss (1) at the conjunct "Socrates is a human and Socrates is mortal". There is reason to think that Mill agrees with Whately that the *petitio* is the fallacy

> In which the premise either appears manifestly to be the same as the conclusion or is actually proved from the conclusion, or is such sa would naturally and properly be so proved. (*Elements of Logic*, Bk III, Sec. 13)

Mill himself goes on to say of this passage:

> By this last clause I presume is meant that it is not susceptible of any other proof, for otherwise there would be no fallacy. (Bk V, Ch. vii, Sec. 2)

It becomes evident that Mill is here endorsing a twofold conception of the *petitio* which some theorists of the present day call the "equivalence" and the "dependency" conceptions. (See Woods and Walton, 1989/2007b, chapter three). In Mill's diagnosis, the received analysis of general propositions is the locus of the difficulty each tiem. For let us suppose that our sample syllogism is not forwarded as proof. It is still circular in the sense of clause one of Whately's definition, i.e., in the equivalency sense, for part of its first "premise ... appears manifestly to be the same as the conclusion". But if taken as a proof, things worsen, By "proof", Mill means what Aristle means by a *demonstration*. A demonstration is a syllogism in which each succeeding line is less certain than its predecessor. But in as much as the conclusion (3) is *part* of premiss (1), then there can be no proof of (1) which isn't also a proof of (3). Thus, when considered as proofs, syllogisms on the received view of generality commit the *petitio* in Whately's second sense, i.e., in the dependency sense.

Mill is of the opinion that, correctly interpreted, syllogisms do not commit the *petitio* fallacy on either conception. To appreciate what Mill thinks is the correct interpretation of syllogisms (or, as we may now say, of deductive reasoning more broadly conceived), it is necessary to expose the essentials of Mill's own theory of inference. First, and contrary to Whately, Mill holds that there are three kinds of inference: deductive, inductive and particular. Deductive reasoning is reasoning in which the conclusion is less general than its least general premiss. Particular reasoning is reasoning in which the conclusion and all premisses are particular propositions. Moreover, reasoning from particulars to particulars is "not only valid but ... the foundation of both induction and deduction" (Bk II, Ch. I, Sec. 3).

A second pillar of Mill's theory of inference is his distinction between *verbal inferences* and *real inferences*. A verbal inference is one which is deductively valid. Verbal inferences may be correct in their way, but they do not advance knowledge. Real inferences do indeed advance knowledge, and for this to be so, Mill thinks that all real inferences must be

> from particulars to particulars: General propositions are merely registers of such inferences already made, and short formulae for making more. The major [i.e. general] premise of a syllogism, consequently, is a formula of this description: and the conclusion is not an inference drawn *from* the formula, but an inference drawn *according to* the formula: the real logical antecedent, or premise, being the particular facts from which the general proposition was collected by induction. (Bk II, Ch. iii, Sec. 4)

Here, then, we meet with the idea that anticipates Lewis Carroll's article in *Mind*, namely, that general propositions are "registers" of inference *rules* and are not, as such, eligible to be premisses of real inferences.

The third basic component of Mill's theory of inference is absolutely original. It is that inductive generalizations may be all right for "big science". After all, institutional science has immense resources and is not subject to the same pressures of time as is the lowly individual. It takes little reflection to appreciate the immensity of the task of constructing a "clean" induction even to the homely generalization "All ravens are black". In fact, given the age of that species and its reproductive zeal, there is no practical possibility of examining all ravens with respect to correlations with blackness. Compare this with the task that confronts the youngster playing with matches, who learns in one encounter with a burnt finger not to do *that* again. As Mill has said, generalizations are a kind of record-keeping of real inferences previously transacted.

Mill is not a tidy writer and, given the originality of his views, it is not surprising that his exposition is so often misunderstood. Linking together the three basic components of his account of inference is a fourth. It makes for an obscure connection, well worth persisting with. To see how it works, consider a syllogism just like the one we considered previously except that the conclusion is affirmed of a living person, say the neighbour's ten-year old niece, Sarah. So we have the argument

(1′) All humans are mortal.

(2′) Sarah is a human.

(3′) Therefore, Sarah is mortal.

Though not itself a real inference (since it is deductively valid and hence a verbal or "book-keeping" inference), it is Mill's view that it is *underlain* by a real inference. If this is right, then, on Mill's own account of the matter, this real inference (a) must be an inference of a particular from particulars: (b) the induction implicit in premise (1′) must be reconcilable to that fact; and (c), the deduction which overlies the real inference must likewise be reconcilable to fact (a). Mill's contention is that the real inference in question is as follows:

(1*) Sarah resembles in a relevant way those things that are positively correlated with having died and concerning which there is no known negative instance.

(2*) Therefore, Sarah too will die [i.e., is mortal].

We see, then, that for Mill, when reasoning from particulars to particulars, the reasoner makes a book-keeping entry in the form of a general proposition, such as (1′), then his reasoning is *analogical reasoning.* What is more, in as much as most inductions can't be generated in a timely way, if at all, most inductions are disguised analogical inferences. Further still, the book-keeping function of general propositions accommodates what Mill has to say about the "true character" and role of deduction:

> An induction from particulars to generals, followed by a syllogistic process from those generals to other particulars, is a form in which we may always state our reasonings if we please. It is not a form in which we *must* reason, but it is a form in which we *may* reason, and into which it is indispensable to throw our reasoning, when there is any doubt of its validity ... (BK II, Ch. iii, Sec. 5).

To see what Mill is getting at, consider three arguments, one from particulars to particular (PAR), another from particular to general (IND, and a third from general to particular (DED):

PAR	IND
p_1	p_1
$\vdots$	$\vdots$
p_n	p_n
$\therefore$ Bentham is mortal	$\therefore$ All humans are mortal

DED
Al humans are mortal
Fact F
$\therefore p_n + m$

Concerning IND, it is valid if PAR is, and if PAR is valid, so is IND. Thus for Mill PAR arguments and their corresponding IND arguments are validated by exactly the same evidence. Concerning DED, given that F is a fact, then should the new particular $\mathrm{p}_n + m$ be false, it would follow that "All humans are mortal" is false. This would mean, in turn that IND is a defective argument and PAR too. So the role of deductive arguments is to test the adequacy of the non-deductive inferences that underlie them.

With these things said, it is clear that no real inference associated with a deduction is guilty of the *petitio* fallacy. For consider again

(1*) Sarah resembles in a relevant way those things that are positively correlated with having died and concerning which there is no known negative instance.

(2*) Therefore, Sarah too will die.

It is obvious on inspection that (2*) is not affirmed in (1*). It is equally clear that (1*) is provable, if at all, well short of being a proof of (2*). The inferences embedded in deductions are circular in neither the equivalency nor the dependency sense of that term.

Although he clearly escapes the claim that inferences of this kind are fallaciously circular, the doctrine that all real inference is non-deductive commits Mill to the view that mathematical proofs are either inductive arguments or that they are deductive registers of the real thinking from particulars to particulars that underlies them. This may prove a troublesome and eventually an unsatisfactory position, but there can be little doubt that Mill is struggling to mark a valuable and important distinction between deductive arguments (which Mill calls "ratiocinations") and inferences. If this is so, Mill can be said to be attempting a clarification of another claim he inherited from Bacon. Bacon saw logic as a branch of rational psychology. Thinking so drew accusations of psychologism, the scorned view that the laws of logic are to some extent dependent upon or constrained by psychological factors.

But Mill also inherits the notion that an argument (when good) is a structure of propositions which is truth-preserving or probability-enhancing, and is so independently of any psychological fact. There is plenty of evidence that Mill thought of syllogisms in precisely this way. Yet it is also clear that, in the emphasis he gives to operations of the mind which, as it were, underlie good arguments, Mill adopts a mentalistic conception of inference. Inference, then, is the revision of beliefs in the light of the interplay of new information upon old. Seen this way, inference, unlike argument, *is* constrained by psychological factors.

It follows that the rules of good argument may not always be rules of good inference. This is certainly Mill's view in book two. As we have seen, deductions do not describe their "underlying" inferences. So rules of deductive argument are, at least sometimes, not rules of inference. Implicit in this is the idea that the term "logic" is importantly ambiguous. It is one thing when considered as a theory of argument in our present sense of that term. It is quite another thing when taken as a theory of inference. In this Mill anticipates a much later development to the same effect associated with the work of such contemporary writers as Gilbert Harman. (See [Harman, 1986, pp. 4-6]).

Bacon's own harshness towards syllogistic structures is now explicable (op. cit., chapter XII). It seems to be his view that syllogistic is a branch of the theory of argument, and the theory of argument is in certain important respects a bad theory of inference. Therefore syllogistic is bad *logic.* So it is in one sense of the term only, the sense in which logic is a theory of inference. What Bacon seems to have overlooked is that in the sense in which logic is a theory of argument, the syllogistic does much better than he gives it credit for. Mill's own position, implicitly at least, is that logic has done rather better as a theory of argument, that is, in its first sense, than it has done in its second sense, that is, as a theory of inference. One of Mill's objectives in the *Logic* is to repair that deficiency. It is regrettable that Mill himself was often forgetful of the ambiguity of "logic", and that he rather routinely failed to acknowledge the distinction between argument and inference. Even so, it is quite clear that in one of these two senses of the term "logic", Mill is a strong opponent of psychologism.

9.2 Fallacy Theory

Book five of the *Logic,* entitled "On Fallacies" runs to seven chapters and ninety-six pages in the definitive edition of the University of Toronto Press. It is evident throughout that, concerning the nature of induction, Mill owes a considerable debt to Bacon's *The New Organon*, and, concerning the fallacies, that he was much influenced by Whately's *Element of Logic.* In chapter one, "Of Fallacies in General", Mill writes,

> In the conduct of life — in the practical business of mankind — wrong inferences, incorrect interpretations of experience, unless after much culture of the thinking faculty, are absolutely inevitable; and with most people, after the highest degree of culture they ever attain, such erroneous inferences, producing corresponding errors in conduct, are lamentably frequent. (Bk V, Ch. I, Sec. 1)

A fallacy for Mill is the mistaking of apparent evidence for real evidence. Concerning the principal objective of book five, Mill proposes

> To examine, then, the various kinds of apparent evidence which are not evidence at all, and of apparently conclusive evidence which do not really amount to conclusiveness. ... (Bk V, Ch. I, Sec. 3)

Here is a conception of fallacy that resonates even today. It is a conception according to which fallacies are errors that are widely committed, easy and natural to commit, and difficult to correct. They are important not just because they lead to false opinion but also to "lamentable" errors in conduct. Further, although not every mistake is a fallacy (causal errors of inattention, for example, are not), it is possible to catalogue the most common patterns of attractive error, and doing so is part of the proper job of logic.

Not only are fallacies to be distinguished from causal errors, they must also be separated from "moral errors", which in turn contrast with "intellectual" errors. Fallacies are always intellectual errors. Moral errors sub-divide into *indifference to truth* and *bias*. Errors of bias involve the drawing of conclusions on the basis of one's psychological states, especially one's emotional states. It appears to be Mill's view that although there are moral causes of error, and of fallacy too, no *analysis* of a fallacy involves reference to its moral causes. A fallacy is an intellectual mistake even if it is caused by moral weakness. Thus someone biased against women could be induced to accept the inference that all women are bad drivers just because some are. The bias might cause the inference to be drawn, but that would not make it fallacious. The inference would be bad *irrespective* of what caused it to be drawn.

9.3 *Fallacies Classified*

In Chapter two, "Classification of Fallacies", Mill attempts to disarm those critics who hold that in as much as there are "infinite ways to err", any finite list of fallacies will be arbitrary and unmotivated. Mill is of the view that any time an error of inference is made it is made on the basis of some fact or putative fact which appears to be "evidentiary" but is not. Whenever this happens there must be a property or a relation, either in the fact or in our way of considering it, which has "an invariable relation to a general formula" and which leads to the error of thinking that the "evidentiary" fact is constantly conjoined with the "concluded" fact. Evidently Mill thinks that every mode of mistaking non-constant for constant conjunction is discernible in principle, and that such mistakes must be of low finite number. Whatever we might think of Mill's reasoning on this point, the fallacies are classified as follows:

FALLACIES	
Of Simple Inspection	*Of Inference*
1. Fallacies *à priori*	

Fallacies of inference, in turn, subclassify:

FROM EVIDENCE DISTINCTLY CONCEIVED		FROM EVIDENCE INDISTINCTLY CONCEIVED
Inductive Fallacies	*Deductive Fallacies*	5. Fallacies of Confusion
2. Fallacies of Observation	4. Fallacies of Ratiocination	
3. Fallacies of Generalization		

In a concession that anticipates a modern development known as the Asymmetry Thesis [Massey, 1975], Mill allows that his classification may be *somewhat* arbitrary. This he explains by the fact that erroneous inferences "do not admit of such a sharply cut division as valid arguments do" (BK V, Ch. ii, Sec. 3). Even so, if fully and unambiguously expressed, any erroneous argument "must ... be so in some one of these five modes unequivocally." (*op. cit.*) The third confusion he catches. Arguments fully expressed and free of ambiguity cannot commit a fallacy of confusion. But given that real life arguments are rarely complete and entirely untouched by ambiguous or vague language, "[a]lmost all fallacies, therefore, might in strictness be brought under our fifth class, Fallacies of Confusion". (op. cit.)

9.3.1 Fallacies of simple inspection

Chapter three deals with fallacies of simple inspection, or *à priori* fallacies. These are a special case, since strictly speaking they are not errors of inference. They are propositions taken as sufficiently obvious as neither to require nor admit of argument, and so they simulate the fallacy of false assumption. These include superstitions and common misconceptions (themselves fallacies in a widely recognized sense of that word). Mill does not here invoke the term "*argumentum ad populum*", but it is clear what his own analysis of it would be. We might imagine someone holding that an *ad populum* fallacy is an argument in the following form:

1. Proposition P is a popular belief

2. Therefore, P is true

Mill would demur from this analysis. For one thing, it is a transparently silly argument, hence not a seductive one, hence not a fallacy in Mill's sense. The fallacy, rather, is that of affirming P without argument; and what makes it so is that popular beliefs often have the property of seeming to be correct even when they are not, and these are well-exemplified by biased opinion, a particularly "vulgar error". It may seem that Mill is again forgetting his own distinction between moral and intellectual errors, concerning the second of which only are fallacies attributable. The appearance is mistaken, Mill allows that bias may cause an error, but bias is not what the error consists in. Biased beliefs can be true as well as false. Even when true, a fallacy occurs when the biased reasoner mistakes non-evidence for evidence or inconclusive evidence for conclusive evidence. All the same, Mill does deviate from his own account in another respect. A fallacy of simple inspection is not the mistake of over-rating the evidential backing of a proposition; it is the mistake of supposing that the proposition in question requires no evidence. It is not the distinction between moral and intellectual errors that Mill is forgetting. It is rather his own core concept of fallacy.

Another example is one against which Mill argues at length in book two, the idea that whatever "is inconceivable must be false" — a principle that was falsely directed against the Copernican idea of a vast empty space which was thought inconceivable, hence non-existent. Further examples include the natural inclination to ascribe objective existence to abstractions, and the philosophical principle of sufficient reason. This principle is sometimes invoked in connection with the law of inertia:

> A body at rest cannot, it is affirmed, begin to move unless acted upon by some external force; because, if it did, it must move up or down, forward or backward, and so forth; but if no outward force acts upon it, there can be *no reason* for its moving up rather than down, or down rather than up. Etc., *ergo,* it will not move at all. (Bk V, Ch. iii, Sec. 5)

"This reasoning" Mill conceives "to be entirely fallacious", and so it may be. But it is not an example faithful to Mill's own category of simple inspection. Such fallacies are not fallacies of reasoning or inference, and Mill would have been better served had he cited the principle itself as fallacious, that is, the "self-evident" principle that if there is no reason for something to happen, then it cannot happen.

Other examples involve the fallacious principle that distinctions in language invariably mirror differences in nature, an echo of Locke; that an event cannot have more than one cause; and that the causes of something must resemble — have the same properties as — the thing caused.

Perhaps the best summary statement of Mill's position on simple inspection fallacies can be found in chapter seven, paragraph four, of his *Autobiography*:

> The notion that truths external to the mind may be known by intuition or consciousness, independently of observation and experience, is, I am persuaded, in these times, the great intellectual support of

> false doctrines and bad institutions. By the aid of this theory every inveterate belief and every intense feeling, of which the origin is not remembered, is unable to dispense with the obligation of justifying itself by reason, and is erected into its own all-sufficient voucher and justification. There never was such an instrument devised for consecrating all deep-seated prejudices.

9.3.2 Fallacies of observation

Chapter four takes us to fallacies of observation which, unlike those we have been examining, namely prejudices "superceding proof", are instead "those which lie in the incorrect performance of the proving process". "Proof" here is intended "in its widest extent ... [and] embraces one or more, or all, of three processes, Observation, Generalization, and Deduction ... "

Mill cites the belief that "a fortune-teller was a true prophet". This involves two fallacies, only one of which is a fallacy of observation. The observation fallacy is the mistake of attending to observations which exclude negative instances and further observations which qualify the others. Thus if one's observations don't include those instances in which the fortune-teller's predictions turn out false, or those cases in which the accuracy of the observed predictions is the result of some kind of trick, then it is Mill's view that the very having of those original observations is a fallacy. It is also a fallacy of a different kind to infer from those tainted observations, e.g., that the fortune-teller is a prophet. This is a generalization fallacy.

Mill reveals that he has learned well a lesson from Bacon, whom he approvingly mentions. In its most general and lethal form, an observation fallacy is the fallacy of collecting evidence without any principle of organization. No set of observations collected willy-nilly is ever the justified basis for a generalization from them. And so Mill expressly forbids that "weak" form of induction, the collecting of evidence merely by simply enumeration of observations chanced upon. More generally still, an observation fallacy is the mistake of overlooking relevant considerations. These considerations may themselves be directly observable or they may be inferable from what is observed. Even so, Mill decides to include the latter in the category of observation fallacy.

9.3.3 Fallacy of Generalization

Fallacies of generalization are the business of chapter five, and Mill says that the "class of Fallacies of which we are now to speak, is the most extensive of all; embracing a greater number and variety of unfounded inferences than any of the other classes, and which it is even more difficult to reduce to sub-classes or species". (Bk V, Ch. v, Sec. 1)

This difficulty aside, one kind of fallacy is the groundless generalization, such as that of inferring that the whole universe must have the same character as our solar system. The example carries an important presupposition. It is that whatever we currently know of the universe it is knowledge of only a part of it — of our solar system, to make a simplifying and outdates assumption. So the universe can be subdivided conceptually into the known KU and the unknown UU. Now it is a fundamental principle of induction, says Mill, that any generalization about UU on the basis of KU must involve observed or correctly inferred constant conjunctions of the form "KU and UU". But in the nature of the case, we have no data of the UU-kind, and so we lack the wherewithal for any induction to properties of UU. Mill's point is not well-formulated. An ambiguity lurks in the term "unknown". If by the "unknown" is meant those reaches of reality to which present inductive resources are inapplicable, then it is an empty truism to go on to say that it is an inductive error to reason from the known to the unknown. On the other hand, if by the "unknown" is meant only those aspects of reality that have not yet been experienced — for example, future settings of the sun — then it is quite untrue to say that present inductive resources are inapplicable to the unknown in this sense.

All the same, Mill is onto something important. His principle seems to be that a generalization from the known to the unknown is legitimate if it is sustained by constant conjunctions that are governed by certain implicit limitations on the generalization's range. Although we don't normally say so explicitly, when we make the generalization that all ravens are black, we are not intending to say that all ravens on Mars are black or that all ravens in the genetic aftermath of a nuclear holocaust are black. Certainly conditions of locale and of normalcy are presupposed, difficult as they may be to state satisfactorily, and Mill is saying that without them generalizations are at grave risk. More particularly, if a would-be generalizer proposes to exceed such constraints, that is, to generalize beyond them, he must have principled reasons for doing so, and knowing that his generalization holds *within* those constraints (e.g. earthbound ravens are black) is not reason enough to hold that it holds beyond them (e.g., Martian ravens are black). In fact, Mill surely knew that there weren't any ravens on Mars, but this nicely makes his point. For if Mars is raven-free, the generalization about the ravens being black there fails, except in the trivial sense that there are no non-black ones there either. In this, Mill can be seen as making the point that induction is always relativized to blocks of background information. Since we may assume that we don't have, or that he didn't in 1843 have appropriate kinds of background information about the outer galaxies, there will be certain generalizations that hold in our galaxy which might well fail in those beyond. The same is true, as we now see, even of inductions of a more local character.

Mill also does well with his next type of example, namely, reductionist theories such as those which claim that heat is just motion of a certain kind or that consciousness is but a state of the nervous system. We commit a fallacy of reductive generalization, as we ourselves might call it, when we confuse the true claim that phenomena of type K *supervene* upon phenomena of type L with the false or unproved claim that K-phenomena are identical to L-phenomena when and only when there is no change among the K-phenomena without some corresponding change in the L-phenomena. In this sense, it might well be true that the phenomena of consciousness supervene upon the phenomena of neural activity, that there is no change in one's consciousness without some specific change in one's neural states, but his would not show that the conscious mind just is the nervous system. Thinking so would be a species of generalization fallacy in Mill's sense.

A further type of generalization error is that of inferring from observations collected in such a way as to involve an observation fallacy, lately discussed. Here too the principal culprit is the "law" of simple enumeration, and Mill again applauds "Bacon's emphatic denunciation of it ..." In its most general form it is the inference.

> This, that, and the other are A and B, I cannot think of any A which is not B, therefore every A is B. (BK V, CH V, Pt. 4)

The premisses Mill is prepared to concede may well reflect "empirical" laws or regularities between As and Bs, but these may represent no causal connection. So we may take it that the core mistake is that of investing mere correlations with a causal significance that they do not lack, or concerning which there is no independent evidence.

Mill also treats of the *post hoc, ergo propter hoc*, which he is careful to distinguish *hoc* from the former case. This might surprise today's readers, in as much as many a contemporary writer analyses the *post hoc* as precisely the error of confusing accidental correlations with causal connections. Mill, at any rate, sees them as different. In the previous example, the fallacy involves a failure to appreciate how a causal enquiry should be conducted. In the present case, that failure doesn't occur. The error is committed by someone who has an adequate general appreciation of how casual inferences are go be drawn but who is misled by the *appearance* that a prior event is causal with regard to a succeeding event into thinking that it is in fact causal with respect to it.

There follows a discussion of false analogies.

> An argument from analogy, is an inference [sic] that what is true in a certain case, is true in a case known to be somewhat similar, but not known to be exactly parallel, that is, to be similar in all the material circumstances. (BK V, Ch. v, Sec. 6)

This constitutes the first of two kinds of analogical argument. Let X and Y have property P and let X not be known to possess property Q which Y does possess, and let Q "not be connected with" P. Then the "conclusion to which the analogy points" is that X also has property Q. Concluding this is an analogical inference, and indispensable to its being so is that we have "not the slightest reason to suppose any real connexion between the two properties". Consider the planets. The earth is inhabited and the earth bears very many similarities to the other planets, none of which is known to be inhabited. Further there is no known connection between any of the properties shared by the planets and the property of the earth's being inhabited. Nonetheless, such a connection there *may* be, and Mill is ready to say that if it is even slightly less probable that an alien planet would be inhabited if it did not resemble the earth, then the analogical inference is a defensible one. In the light of this, it would be well to revisit Mill's point about (in our example) inducing that ravens on Mars are black. Mill now seems to be saying that such a claim might have something to be said for it, however slightly, when considered as an analogy. If so, this cannot be an *induction*, which is precisely what Mill says it is not. Mill is here contradicting, or revising, his earlier claim that all inferences are inductive. Analogical reasoning of the kind we have just discussed is a kind of inference, but it is not inductive according to Mill.

A second kind of analogical argument is quite different. Analogical arguments of the first kind are in no sense inductive, as we have just seen. Of the second kind, they are arguments having an inductive character but which fail to be "real inductions". Mill has in mind the case in which conditions $C_1, \ldots, C_n$ are (as it happens) parts of the cause of an event-type E. Inferring that whenever $C_1, \ldots, C_n$ occur so does E is not, therefore, a real induction, but rather a flawed or incomplete induction. Even so, if the probability of E's occurrence *given* the occurrence of $C_1, \ldots, C_n$, is greater than the probability alone of E's occurrence, then Mill is of the view that the inference in question is analogical in our second sense and that it may be allowed as a modestly justified inference.

It is evident that Mill recognizes what has come to be known as inference by conditional probabilities. If the probability of an event A's occurring *conditional* upon an event B's occurring is high enough, say, 0.8 or higher, it might be said that the inference

1. The probability of A, given B is greater than 0.8

2. B occurs

3. Therefore, A occurs

registers an acceptable analogical conjecture, until such time as it may be overriden by additional evidence.

Now, Mill certainly had the concept of conditional probability from his reading of Laplace's *Essais philosophiques sur les probabilities*, but he was critical of Laplace, especially in relation to Principle VI of the *Essais*, said by some to be an anticipation of Bayes' Theorem. What is interesting is Mill's desire to assimilate the idioms of probability to a theory of analogy never mind that the project was largely uncompleted.

There are two ways of committing the *fallacy* of analogy, according to Mill. One way is simply "overrating the probative force" of a correct analogical argument of either kind. The other way, "more deserving of the name of fallacy", involves ignoring independent evidence that properties concerning which a connection is proposed on analogical grounds are *not* in fact connected. "This is properly the Fallacy of False Analogies".

It appears that here too Mill has run foul of his own core concept of fallacy. A fallacy is a kind of misinference, the finding of evidence when there is none, or the over-rating of such evidence as there may be. But Mill also thinks that the conclusions of analogical reasoning are always singular propositions. This matters in two ways. One is that analogical mistakes can hardly be fallacies of *generalization*, as Mill claims. The other is that, in as much as all inference is inductive and fallacies are errors of inductive inference, it can hardly be true that there are any fallacies of analogy, contrary to fact and to what Mill himself claims.

9.3.4 Fallacies of Ratiocination

Up to this point Mill would acknowledge that what he has been calling fallacies are not generally called so. Chapter six returns us to a more common usage. For we "have now, in our progress through the class of Fallacies, arrived at those to which, in the common books of logic, the appellation is in general exclusively appropriated; those which have their seat in the ratiocinative or deductive part of the investigation of truth". The deductive fallacies don't hold much interest for Mill, for two reasons. One is that he thinks that Whately's *Logic*, published seventeen years earlier, handles this topic "most satisfactorily". The other is "the rules of the syllogism are a complete protection"; all we need do is to show a deduction's syllogistic form, and "we are sure to discover if it be bad, or at least if it contains any fallacy of this class". (BK V, Ch. vi, Sec. 1)

It may seem that by book five, chapter six, Mill has entirely forgotten his position of books one and two: deductions are not inferences, but only apparent inferences, and all inferences are inductive. This is so if we hold Mill strictly to the view that a fallacy is always an offence against evidence and that a syllogism is not a reasoning from evidence. On the other hand, it is more realistic to attribute to Mill the view that there is a conception of fallacy appropriate to (inductive) inference and a different conception of fallacy, definable for syllogisms or radiocinations. In any event, the discussion of deductive fallacies is somewhat *pro forma*. Fallacies of conversion are noted, as in the inference from "All A are B" to "All B are A". They are the fallacies that have come to be known as affirming the consequent and denying the antecedent. Also recognized is the error of confusing the contrary of a proposition with its contradictory; and then, too, the syllogistic fallacies of four terms and undistributed middle.

Special attention is reserved for what are "certainly the most dangerous fallacies of this class", namely *secundum quid* mistakes, and these it seems can infect inferences and radiocinations alike. They are fallacies which result from reasoning from premisses, on which an essential qualification has been ignored, as when from the premiss, "This black man is white-haired", it is concluded "Some black men are white". In so saying, Mill gives contemporary recognition to a type of fallacy, of the same time, first noted by Aristotle in *On Sophistical Refutations*, and the analysis he gives to it is scarcely different from Aristotle's own.

9.3.5 *Fallacies of Confusion*

Finally, in Chapter seven, come the fallacies of confusion. Under this heading fallacies of ambiguity are discussed at some length, and there is much quotation from Whately's *Logic*.

Mill accepts Whately's suggestion that "[o]ne not unusual form of the Fallacy of Ambiguous Terms, is known technically as the "Fallacy of Composition and Division". (BK V, Ch. vii, Sec. 1) This fallacy occurs "when the middle term is collective in the premises, or *vice versa*; or when the middle term is collective in one premiss, distributive in the other". (*op. cit.*) Thus the fallacy of division is committed when it is true collectively that Italians are charming and yet concluded that Guido is charming; for it would not be true that even Italians are charming in a distributive sense; i.e., it is not true of each and every Italian that he or she is charming. Composition is the reverse of this. It may be true, distributively, of every citizen of a country that he or she is thrifty, yet not true, collectively, that the citizenry as a whole is thrifty, for it may have a spendthrift government.

Next discussed is the fallacy of *petitio principii*, a distinct kind of confusion fallacy. Mill writes that

> Every case where a conclusion which can only be proved from certain premises is used for the proof of those premises, is a case of *petitio principii*, (Bk V, ch. vii, Sec. 2)

a fallacy which "includes a very great proportion of all incorrect reasoning". An astonishing claim and, on the face of it, beyond believability, it seems clear that Mill has in mind the following kind of example. Let "P" be any term for which there is a synonym, "Q". Suppose that someone asserts "S is P" (say, "Jones is a bachelor"). A critic challenges: "You can't prove that". The other replies, "Oh yes, I can, since S is Q" ("Jones is a man who has never married"). An argument is constructible out of this exchange:

1. Jones is a man who has never married.

2. Therefore, Jones is a bachelor.

Mill thinks that, although (2) follows from (1), (1) is not provable except by (2); and this is a *petitio* fallacy of as great a commonality as there are synonymous terms "P" and "Q".

Mill has committed a great howler, of course. It is certainly not true that the only way to establish that Jones is a man who has never married is to marshall the proposition that he is a bachelor. One could have asked "Have you ever married?" and have been told, "No"; or one could have checked the record of marriages for all jurisdictions in which Jones has lived during the period of his nubility. Mill confuses two things. It may be that synonymous terms "P" and "Q" are interdefinable and that neither is definable in any other way. But it does not follow from this that "S is P" and "S is Q" are interprovable or that neither is probable in any other way.

The third and final subdivision of the fallacies of confusion gives us the fallacy of *ignoratio elenchi.* Whereas the fallacy of ambiguity involves "misconceiving the import of the premises", and the *petitio principii* is a matter of "forgetting what the premises are", the present fallacy consists "in mistaking the conclusion which is to be proved", and this is the fallacy of *ignoratio elenchi* "in the widest sense of the phrase". As Mill observes, the fallacy is called by Whately the "Fallacy of Irrelevant Conclusion". Mill remarks that the works of controversial writers "are seldom free from this fallacy". He means that they have a way not of committing it themselves but in provoking the commission of it by their critics. Thus the best-known argument against Berkeley's proofs of the non-existence of matter is Dr. Johnson's "I refute him thus", said as he kicked a stone. But this is "a palpable fallacy", an argument, so to speak, in which the conclusion "I, Samuel Johnson, have just kicked a stone" is, though quite true, irrelevant as a refutation of the claim that things like stones possess a propertyless material substratum.

9.3.6 Mill's Importance

One of Mill's objects in writing *A System of Logic* was to produce a general theory of inference which would set the methodological standard for ethics and the social sciences. There is not much evidence of this connection in his treatment of the fallacies, although elsewhere in the treatise the connection is developed in greater detail. Even so, Mill is of the view that considered as general and recurring mistakes of inference, fallacies stand as much chance of infecting politics or sociology as astronomy or chemistry, and in this he is surely correct.

With the exception of the ratiocinative and confusion fallacies, Mill's treatment can be seen as an extension and refinement of Bacon's earlier effort. From Bacon he learned the folly of trying to generalize from disorganized and undisciplined observations. But in various ways, Mill exceeded Bacon's reach. He seems to have recognized that good inductive generalizations are always embedded in contexts of background information, and he caught the importance of inference by conditional probabilities. He was also alert to the critical difference between supervenience and identity, and he recognized the importance of distinguishing between inferences and argument.

As for the fallacies of ratiocination and of confusion, there is less that is novel in Mill. For the most part, they are fallacies that were first noted and as capably handled by Aristotle, and Mill's discussion of them often is little more than a recapitulation of Whately. But of all that Mill is known for in his writings on logic, it is supremely ironic that his most celebrated claim, that valid deductive inferences commit the fallacy of *petitio principii,* turns out to be a claim that he may well have denied. [43]

[43]An excellent overview of Mill's contribution to logic may be found in Fred Wilson's contribution to this *Handbook,* volume 4 [2008]

10 AUGUSTUS DEMORGAN (1806-1871)

Almost an exact contemporary of Mill (1806-1873), DeMorgan was born in Madura, India, to a military family. Following his early education, he entered Trinity College, Cambridge, from which in 1827 he graduated fourth in his class in mathematics (in Cantabridgean parlance, he was "fourth wrangler", a considerable distinction). Religious scepticism denied him an appointment at Cambridge, but it was no bar to his being founding holder of the chair of mathematics at the fledgling University of London.

One of the leading mathematicians of his day, DeMorgan did important work in the foundations of algebra and in mathematical methodology. He was also a pioneer of modern logic, anticipating the algebraic approach to propositional logic of George Boole (1815-1864) and "the logic of relatives" of Charles Peirce (1839-1914). In the present day, every beginning student of logic knows DeMorgan for the equivalences bearing his name that establish the duality of conjunction and disjunction. He is perhaps second best-known for his *Formal Logic*, which was published in 1847. This is a comparatively early work. DeMorgan's mature writings were contributed to the *Cambridge Philosophical Transactions* in the period of 1846 to 1862. A brief account of these works was published in 1860 as the *Syllabus of a Proposed System of Logic.*

10.1 On Fallacies

Chapter thirteen of *Formal Logic*, some fifty pages in all, is given over to the fallacies, DeMorgan is famously of the view that

> There is no such thing as a classification of the way in which men may arrive at an error: it is much to be doubted whether there ever *can be.*

It appears that such a sentiment makes DeMorgan hostile to the fallacy theories of his predecessors. However, this is an illusion. None of the great classifiers, whether Aristotle, Ramus, Arnauld, or Whately ever supposed that his classification was complete.

In any event, DeMorgan did think that all *deductive* errors could be identified and classified by a complete method.

> As to mere inference, the main object of this work, it is reducible to rules; these rules being all obeyed, an inference, as an inference, is good; consequently a bad inference is a breach of one or more of these rules.

This has consequences for fallacy theory. If all that is wanted is a theory of deductive fallacies, there is no need to produce a separate chapter; for a deductive fallacy just is a breach of a rule of deductive inference. Thus DeMorgan's view is not that a special theory of deductive fallacies is illegitimate or impossible to produce; rather it is superfluous to a general logic of deductive inference.

If such a claim sounds odd to modern ears, it should be emphasized that what DeMorgan means by "inference" many others understand as a syllogism in something resembling Aristotle's sense of the term, that is, as a finite sequence of propositions the terminal member of which (the conclusion) is necessitated by, and is different from, the preceding members (the premisses). Not only did he not, like Mill, take pains to distinguish syllogisms from inferences, DeMorgan regards them as coming to the same thing. Thus, any mentalistic or epistemic interpretation of "inference", whereby inference is an operation of the mind, is excluded.

If a special theory of deductive (or syllogistic) fallacies is unnecessary, there is another kind of fallacy whose analysis must take us beyond general deductive logic.

> There are many points connected with the matter of premisses, to which it is very desirable to draw a reader's attention.

DeMorgan is evidently trying to draw a distinction between errors that are reflected in the deductive structure of an argument, that is, in the formal relationship between premisses and conclusions, and errors that afflict an argument by way of pecularities of the premisses themselves. A case in point is ambiguity. Depending on how the premiss "John is a bachelor" is taken, whether as saying of John that he has yet to marry or that he is the holder of a first-level university degree, an argument to the conclusion "Therefore, John has no wife" will be made correct or not.

DeMorgan recognizes that fallacies of this type are rather whimsically exemplified, as "they have been handed down from book to book". These "jests, puns, &c, are for the most part only fallacies so obvious that they excite laughter", and calling them fallacies "has itself the taste of the ludicrous". Here, in words that C.L. Hamblin would have relished is the common complaint not that fallacy theory in this second sense is impossible, but rather that so often it is done in a silly way. What Hamblin calls the "Standard Treatment" of fallacies was anticipated and regretted by DeMorgan well over a century before the publication of Hamblin's influential criticisms.

DeMorgan appropriates the term "paralogism" and gives it a twofold sense. A paralogism is either a syllogistic or deductive error, or it is a fallacy in the sense of "defective or erroneous statement or premiss". Fallacies, do not "belong to logic", since "the middle writers", i.e., those who followed Aristotle until the end of the mediaeval period, "abandon technicalities almost entirely" (a point also made by Hamblin). Whether redundant or not, DeMorgan's own contribution to fallacy theory is to fallacies in the first sense, or, as he himself puts it, to "logical fallacies". To understand this choice it is useful to point out that DeMorgan's motivation way by and large born of reaction to the inductive logic of Bacon and that he further anticipated and rejected the contemporary view that each discipline has its own peculiar or "material" logic (concerning which, see [Toulmin, 1958]).

DeMorgan's fallacies are Aristotle's own, mainly the "sophisms" of the *Sophistical Refutations*. They are thus the fallacies of: equivocation, amphiboly, accent, forms of expression, accident, *secundum quid, petitio principii, ignoratio elenchi, ad hominem* (recognized by Aristotle, but not on his original list of thirteen), consequent, non-cause as cause, and many questions. With the exception of the treatment of *petitio principii*, there is little in DeMorgan's account that is novel and not in the spirit of Aristotle's own discussion, although it must be said that DeMorgan's expositions are typically more comprehensive and clearer than Aristotle's.

What DeMorgan has to say about the *petitio* stands alone and is genuinely interesting.

10.2 Does the syllogism beg the question?

It is this question that serves as the focus for seven of the most interesting pages of the chapter "On Fallacies". Sextus Empiricus had long since laid the charge that all syllogisms committed the fallacy of begging the question or *petitio principii* (*Outlines of Phyrrhonism*, chapter 17). It is commonly thought that Mill, too, leveled the same complaint, and that it is principally against Mill that DeMorgan is speaking. This, as we saw, might not be right. In chapter one of *A System of Logic*, Mill expressly can be read as denying that syllogisms are inherently question-begging, and nowhere in the seven pages of DeMorgan where this idea is reviewed is the name of Mill mentioned.

Though he wavers on the point, sometimes indicating that a *petitio* fallacy is a dialectical or pragmatic error, DeMorgan persists in the view that question-begging is a logical or syllogistic error. Thus, if the charge of Sextus Empiricus were true — that all syllogisms commit this fallacy — DeMorgan shrewdly realized that it ought to be discernible in the formal structure of any syllogism. DeMorgan considers a standard example:

1. All men are mortal
2. Plato is a man
3. Therefore, Plato is mortal.

According to Sextus Empiricus, the argument exemplifies what Woods and Walton [1975] call the dependency conception of circularity; that is, in order to know premiss one we need first know the conclusion — thus the premiss depends on the conclusion. In answer to this, DeMorgan insists that question-begging is definable only for pairs of single propositions, one a premiss and the other a conclusion. In this he is faithful to Aristotle's claim in the *Sophistical Refutations* that no argument is a syllogism if its conclusion repeats a premiss. By these lights,

1. Plato is a bachelor
2. Therefore, Plato has yet to marry

would be a valid yet question-begging argument, since its single premiss is repeated in the conclusion.

It is easy to see that DeMorgan is now well-placed to reply to Sextus. He can say that it is not true that the only way to know premiss (1) is by prior knowledge of the conclusion, (3) — for lots of people know that all men are mortal who have never heard of Plato. So (3) is not circular in relation to (1). In fact, says DeMorgan, "the whole objection tacitly assumes the superfluity of the minor [premiss]; that is, tacitly assumes we know Plato to be man, as soon as we know him to be Plato". Furthermore, since circularity is definable only for pairs of single propositions, one a premiss and the other the conclusion, it cannot be true that (3) is circular in relation to (1) and (2) together. Together, these two replies to Sextus constitute what Woods and Walton have called "DeMorgan's Deadly Retort" [1975]

It is possible that DeMorgan's reservations about one-premiss arguments derives from Aristotle. Aristotle required that *negation* of the conclusion of a syllogism be consistent with each premiss singly and inconsistent with them only jointly. Aristotle appears to believe that if it is true of no premiss that it "says the same thing" as the conclusion, it cannot be true of the premisses conjoined that they "say the same thing" as the conclusion. Why this line of reasoning doesn't commit the fallacy of composition is something Aristotle doesn't consider (nor DeMorgan either). It is a significant omission in as much as Aristotle (and apparently DeMorgan too) is aware that the property being of consistency with the negation of a conclusion of a syllogism is true of premisses singly but not collectively.

It is perhaps too much to think that the Deadly Retort is conclusive against Sextus, indeed that it is deadly in any wholly convincing way. But it does put Sextus is a difficult position. If Sextus were to demand to know *why* circularity couldn't be defined between conclusions and multiple premisses, a principled answer lies in wait for him. It is this. Everyone knows *intuitively* that valid arguments aren't fallacies as such. But if circularity were definable for conclusions in relation to multiple premisses, every valid argument with more than one premiss *would* be fallacious. And that is a good reason to exclude circularity of the multiple premisses type.

10.3 DeMorgan's importance

DeMorgan and his contemporary Boole (1815-1864), mark beginning of the most fateful transition in logic since Aristotle's invention of the syllogism. Struck by the similarity of the connectives of propositional logic and the operator symbols of algebra, DeMorgan was drawn to the idea that logic - that is, logic in what I've been calling the narrow sense - might in some way be reducible to mathematics. If so, it would be possible without relevant loss to re-express the truths of logic as statements of mathematics. Forty-three years later saw the publication of the *Begriffschrift* of Frege (1848-1925), and with it a shift to the converse reduction. According to Frege and Russell, it ought to be possible to reduce mathematics to logic. This is *logicism*, a thesis in the epistemology of mathematics. It was a remarkable idea: For each true sentence of arithmetic there was a truth-preserving, but *not* content-proving, map to a theorem of logic. The theorems of number theory could be true even in the absence of numbers from the furniture of the world.

It was more than a remarkable idea. It was a manifestly impossible one. No narrow logic at mid-19th century had the capacity to express arithmetic. Repairs would be needed. Indeed logic would have to be radically restructured; whereupon in due course Frege's *Grundgesetze* (1893, 1903) and Whitehead and Russell's *Principia* (1910-1913).

The old logic was designed to serve as the theoretical core of a general theory of argument in the broad sense. The new logic was designed to serve the reductive ambitions of a philosophical thesis about mathematics. Something of the complexity and power of the new logic is given in a number of the volumes of this *Handbook*, notably volumes 3, 5, 6, 7, 8 and 9. Also evident is the transformative extent to which the new logic would absorb the energies of the coming breed of logicians, then and ever since.

It would not be wrong to locate the rise of mathematical logic in the work and ideas of DeMorgan, whether, as DeMorgan himself thought, as the mathematicization of logic or, as Frege and Russell supposed, the logicization of mathematics. Either way, the new logic dealt with uninterpreted symbolic languages and technically recondite examinations of the properties of such languages. Uninterpreted languages are languages in name only. There is in them little capacity for communication and none at all for real life-argument making. Except for that scant sample of the traditional formal fallacies, the new logic is neither designed for nor adept at the exposure of the logical structure of the gang of eighteen. Most of the gang of eighteen are context-sensitive and agent-oriented, but the new logic lacks both context and agency. The new logic downed tools on the fallacies project, not as a matter of policy, but as a consequence of its lack of wherewithal to provide for it. DeMorgan had something worthwhile to say about *petitio,* but his further importance for fallacy theory lies in the largeness of his role in killing it.[44]

11 THE GREAT DEPRESSION: 1848-1970

Like great empires, it is rare for paradigms simply to crash. They tend to wither away (old soldiers, too). The old ways of logic, including such attention as it gave to fallacies, endured. Undergraduates had to be taught, and textbooks were needed to fill that need. Occasionally substantial books were published, not least H.W.B. Joseph's *Introduction to Logic* (1906; ref. ed. 1916) and the three volumes of John Cook-Wilson's *Statement and Inference* (1926). But, as the old century closed and the new one got well into its own rhythms, the fallacies steadily became,

[44]De Morgan's place in logic is well described in Michael Hobart's and Joan Richards' contribution to volume 4 of this *Handbook,* [2008].

in words borrowed from Imke Lakatos, a degenerating research programme. In due course, the press of the new logic would be felt by undergraduates, and textbooks would appear, some rather good and many rather awful, in which chapters of the new logic would mingle with a chapter or so on the "standard treatment" of the fallacies. But rarely in this period did a logician of note publish any paper of note on fallacies.

Then, in 1970, Hamblin blew the whistle. Logicians had let the fallacies programme degenerate to the level of witless puns and worn-out examples, as useless as they were threadbare.

A possible exception to this gloom is Alfred Sidgwick (1850-1943) who, apart from having England's greatest moral philosopher as a cousin, wrote six books on logic.[45] Hamblin is equivocal about Sidgwick's accomplishments. On the one hand, Sidgwick "is so far as I can discover, the only person ever to have tried to develop a complete theory of Logic around the fallacies." (p. 176) But Hamblin is dissatisfied with its execution, and suggests that Sidgwick's having been "passed over' might not have been wholly undeserved – notwithstanding the latter's attention to "neglected topics like Presumption and Burden of Proof ...." (Hamblin means "neglected in 1970.")

In the late 1990s, and briefly after, there was something by of renewal of Sidgwick, occasioned by Flemming Steen Nielsen's doctoral thesis of 1997 and a presentation two years later at a conference at Brock University, in St. Catharines, Ontario.[46] The thesis is in Danish and the talk, although in English, was published on the CD-Rom of the conference's proceedings. Present day enthusiasts are right in saying that there are aspects of Sidgwick's handling of argument (or what, these days, seems destined to be called "argumentation"),[47] which prefigure subsequent developments. In some respects, contemporary approval of Sidgwick is occasioned by the latter's adumbration of some of its own views.[48] Hamblin is well-acquainted with Sidgwick's opinions. His doubts about their value for the analysis of the fallacies is not an unreflective one. To the extent that Sidgwick's work does indeed anticipate later developments, it would not be misconceived to think that Hamblin's negative judgement of Sidgwick might also transpose to it. Perhaps this is going too far. Hamblin's untimely death in 1985 denied him both access to and participation in the subsequent development of argumentation theory. As we have it now, argumentation theory is a sprawling enterprise, in which a good many disciplinary influences are at work. I said at the beginning that the post-1970 events are current affairs, not history. This is a chapter on history. The

[45]Here are two: *Fallacies: A View From the Practical Side* (1984), the earliest, and *Elementary Logic* (1914), the last.

[46]Nielsen [1997; 1999].

[47]Not to be peevish, "argument" in English is both a count noun and a mass term. In this latter role, it functions as an abstract noun for the practice of arguing, i.e. making arguments (via that word's first function). All purveyors of "argumentation" acknowledge its use as a mass term. Some also give it employment as a count noun. The cost of the first habit is redundancy. The cost of the second is the loss of grammar.

[48]See here [Walton, 2000].

argumentation movement is history in the making.

12 NOW

The past forty-one years (I write this in the fall of 2011) has seen the convergence of five scholarly and/or pedagogical developments, some directly provoked by Hamblin, and the others less directly implicated.[49] One is the revival of the fallacies programme in the philosophy of logic.[50] Another is the emergence of the informal logic movement.[51] A third is the development of college and university courses which emphasize critical thinking. Yet another is the growth of argumentation theory in departments of speech communication and discourse analysis.[52] A fifth influence was the production of elementary college textbooks reflecting these new developments.[53] In due course — indeed a good deal later — there would be expressions of interest from computer science[54] and psychology.[55,56]

In virtually all these developments there has been some attempt to take the fallacies project to levels of sophistication that would release it from Hamblin's indictment. Even so, there are two points to take particular note of. One is that over the four decades here under review, the emphasis on fallacies has noticeably waned, notwithstanding an unbroken absorption in the structures and functions of arguments in the broad sense. The other is that not even in the best and most ambitious of these treatments is there to be found anything like a theory of the kind called for by Hamblin in the opening pages of *Fallacies.* That is to say, there is nothing in these writings that approaches the metamathematical sophistication of a sound and complete treatment of deductive consequence or the categoricity of an axiomatic theory. There is in these writings no promise of a *Principia* for fallacious reasoning.

In a way, this is hardly surprising. Metamathematically sophisticated theories of natural language behaviour have proved notoriously unforthcoming in linguistics. Why would we expect philosophers to succeed where linguists have not? Perhaps, in the end, we should concede that Hamblin's No-Theory problem is incapable of resolution by fallacy theorists. But should we not add to this concession that this is failure was guaranteed by the very impossibility of presuppositions of Hamblin's demand? After all, we can't get blood from a turnip.

However, in bringing this chapter to a close, let us take brief note of an exception to this line of thinking. There have been notable formal successes in dialogue and game-theoretic logics, notwithstanding that fallacies are rarely their target

[49]See Gabbay *et al.*, [2002].

[50]See [Woods and Walton, 1989/2007b].

[51]See [Johnson, 1996] and [Hitchcock, 2007].

[52]See [van Eemeren and Grootendorst, 1984].

[53]Johnson and Blair [1977] (informal logic), Woods and Walton [1982] (fallacy theory), Fisher [1988], (critical thinking) and Scriven [1976] (critical thinking).

[54][Ashley, 1990; Atkinson *et al.*, 2006; Reed and Norman, 2003].

[55]Adler and Rips [2008].

[56]These are research communities of considerable size. Some estimate of the sweep of their work and of the identities of whose who do it, can be got by inspecting the archives of the two principal journals, *Informal Logic* and *Argumentation.*

or motivation.[57] More recent developments from computer scientists, information technologists and mathematically oriented logicians[58] suggest the possibility of another answer to the No-Theory issue: The reason that the fallacies project has not achieved the status of a highly sophisticated metamathematical theory is not the principled impossibility of providing it, but rather that the people working on fallacies post-1970 haven't known how to produce one.

If so, the fallacies project remains an open question for logical theory, and, on the No-Theory question, the jury is still out.

Acknowledgements: Boundless appreciation to Charles Hamblin for getting it all started; to Douglas Walton who helped so much with the Woods-Walton Approach; to David Hitchcock, Else Barth, and Erik Krabbe for their scholarly as well as logical acumen; to critical stimulation from Frans van Eemeren and the rest of the Dutch group, and for their generosity and hospitality over many years in both Amsterdam and Gronigen; to Ralph Johnson, Tony Blair and Bob Pinto for the journal *Informal Logic* and those vital conferences and symposia in Windsor; and for friendship and intellectual example: David Hitchcock, Jonathan Cohen, Maurice Finocchiaro, Hans Hansen, Harvey Siegel, Mark Weinstein; Jonathan Adler, Lorenzo Magnani, Jim Freeman, Fabio Paglieri, and Larry Powers. Thanks, too to the co-editors of this volume, Dov Gabbay and Jeff Pelletier, and to the two most indispensable people on earth, Jane Spurr and Carol Woods.

BIBLIOGRAPHY

[Abelard, 1919–1927] Peter Abelard, *Logica Ingredientibus.* In *Peter Abaerlards Philosophische Schriften,* B. Meyer, editor, Münster: Wien, 1919-1927.

[Adler and Rips, 2008] Jonathan Adler and Lance J. Rips, editors, *Reasoning: Studies of Human Inference and its Foundations,* New York: Cambridge University press, 2008.

[Aliseda, 2006] Atocha Aliseda, *Abductive Reasoning: Logical Investigation into the Processes of Discovery and Evaluation.* Amsterdam: Springer, 2006.

[Aquinas, 1953] Thomas Aquinas, "*De veritate*". In Thomas Aquinas, *Quaestiones Disputatae,* 9th edition, volume 1, edited by R.M. Siazzi, Torino and Rome: Marietti, 1953.

[Aristotle, 1984] Aristotle, *The Complete Works of Aristotle: The Revised Oxford Translation,* Two volumes. Jonathan Barnes, editor, Princeton: Princeton University Press, 1984.

[Ashley, 1990] K. D. Ashley, *Modeling Legal Argument: Reasoning with Cases and Hypotheticals,* Cambridge, MA: MIT Press, 1990.

[Atkinson *et al.*, 2006] K. Atkinson, T.J.M. Bench-Capon and Peter McBurney, "Computational representation of practical argument", *Synthese,* 152 (2006), 157-206.

[Averroes, 1954] Averroes, Ibn Rush, Muhammae Ibn, Ahmad, *The Incoherence of the Incoherence,* transl. S. van den Bergh, two volumes. London: Luzac, 1954.

[d'Avila Garcez *et al.*, forthcoming] Artur d'Avila Garcez, Howard Barringer, Dov M. Gabbay and John Woods, *Neuro-Fuzzy Argumentation Networks,* forthcoming from Springer.

[Bacon, 1858–1874] Francis Bacon, *Works,* six volumes, London: Spedding, Ellis and Heath, 1858 1874. Volume 3 is *The Advancement of Learning.* Volume 4 contains *Novum Organum* (1620).

[57]Early works of note are Lorenzen and Lorenz (1978) and Barth and Krabbe (1982). See also Walton and Krabbe (1995), Rahman and Rückert (2001), Pauly and Parikh (2003), Barringer *et al.* (2005, 2008), Besnard and Hunter (2008), Garcez *et al.,* forthcoming.

[58]See, for example, Vreeswijk (1997).

[Bentham, 1824] Jeremy Bentham, *The Book of Fallacies: From Unfinished Papers of Jeremy Bentham.* By a Friend, London: Hunt, 1824.

[Barringer *et al.*, 2005] Howard Barringer, Dov Gabbay and John Woods, "Temporal dynamics of support and attack networks: From argumentation to zoology". In Dieter Hutter and Werner Stephan, editors, *Mechanizing Mathematical Reasoning: Essays in Honor of Jörg H. Siekmann on the Occasion of His 60th Birthday,* Lecture Notes on Artificial Intelligence2605, pp. 59-98. Berlin: Springer, 2005.

[Barringer *et al.*, 2008] Howard Barringer, Dov M. Gabbay and John Woods, "Network modalities". In G. Gross and K.U. Schulz, editors, *Linguistics, Computer Science and Language Processing: Festschrift for Franz Guenthner on the Occasion of his 60th Birthday,* pp. 70-102, Tribute Series 6, London: College Publications, 2008.

[Barth and Krabbe, 1982] E.M. Barth and E.C.W. Krabbe, editors, *From Axiom to Dialogue: A Philosophical Study of Logics and Argumentation.* Berlin: De Gruyter, 1982.

[Besnard and Hunter, 2008] Philippe Besnard and Anthony Hunter, *Elements of Argumentation,* Cambridge, MA: MIT Press, 2008.

[Bochenski, 1970] I.B. Bochenski, *A History of Formal Logic,* edited and translated by Ivo Thomas, Notre Dame: University of Notre Dame Press, 1970.

[Buridan, 1966] John Buridan, *Sophismata,* edited as *John Buridan's Sophisms on Meaning and Truth,* transl. with an introduction by Theodore Kermit Scott. New York: Appleton-Century-Crofts, 1966.

[Buridan, 1976] John Buridan, *Tractatus de Cionsequentiis,* 14th century. In Hubert Hubien, editor, *Johannis Buridani tractatus de consequentics: Edition critique,* volume XVI of Philosophers médiévaus, Université de Louvain, 1976. Translated in Klima (2001).

[Buridan, 2001] John Buridan, *Summulae de Dialectica,* fourteenth century. Translated in Klima (2001).

[Burley, 1955] Walter Burley, Philotheus Boehner, editor, *Walter Burleigh: Depuritate artis logicae tractatus longior, with a Revised Edition of the Tractatus brevior,* The Franciscan Institute, St. Bonaventure, NY, 1955. Translated in Spade (2000).

[Carroll, 1895] Lewis Carroll, "What the tortoise said to Achilles, *Mind,* 14 (1895), 378-380.

[Cook Wilson, 1926] John Cook Wilson, *Statement and Inference.* Oxford: Clarendon Press, 1926.

[Copi and Cohen, 1990] Irving M. Copi and Carl Cohen, *Introduction to Logic,* 8th edition. New York: MacMillan, 1990.

[Corcoran, 1972] John Corcoran, "Completeness of an ancient logic", *Journal of Symbolic Logic,* 37 (1972), 696-702.

[Corcoran, 1974] John Corcoran, "A panel discusson on future research in ancient logical theory". In John Corcoran, editor, *Ancient Logic and its Modern Interpretation,* pages 189-208. Dordrecht: Reidel, 1974.

[DeMorgan, 1847] Augustus DeMorgan, *Formal Logic or The Calculus of Inference.* London: Taylor and Walton, 1847.

[DeMorgan, 1860] Augustus DeMorgan, *Syllabus of a Proposed System of Logic,* London: Walton and Malbery, 1860.

[De Rijk, 1962–1967] L.M. De Rijk, editor, Abelard's *Logica Modernorum: A Contribution to Early Terminist Logic,* volume one, *Glosses on the Sophistical Refutions, Parvipontanus Fallacies, Summa of Sophistical Refutations and Viennese Fallacies.* Assen: Van Gorcum, 1962-1967.

[Diogenes, 1925] Diogenes Laertius, *Lives of Eminent Philosophers,* volumes one and two, transl. R.D. Hicks. London: Loeb, 1925/ Originally published c. 230 AD.

[Diodorus, 1933–1967] Diodorus Siculus, *The Library of History of Didodorus of Sicily,* transl. C.H. Oldfather, 12 volumes, London: Loeb (1933-1967).

[Duncan, 1948] William Duncan, *The Elements of Logick,* in Robert Dodsley, *The Preceptor: Containing a general cause of instruction, wherein the first principles of polite learning are laid down in a way most suitable for trying the genius and advancing the instruction of youth.* London: J. Dodsley, 1948.

[Fisher, 1988] Alec Fisher, *The Logic of Real Arguments,* Cambridge: Cambridge University Press, 1988.

[Frege, 1879] Gottlob Frege, *Begriffsschrift: eine der authuretischen nachgebeldete Forurelsprache des servien Denkins,* Halle, 1879.

[Frege, 1964] Gottlob Frege, *The Basic Laws of Arithmetic,* a translation by Montgomery Furth of selections from *Grudgesetze der Arithmetik* (1893). Berkeley: University of California Press, 1964.

[Gabbay *et al.*, 2002] Dov M. Gabbay, R.H. Johnson, H.J. Ohlbach and John Woods, editors, *Handbook of the Logic of Argument and Inference,* Amsterdam: North-Holland, 2002.

[Gabbay and Woods, 2004a] Dov M. Gabbay and John Woods, editors, *Handbook of the History of Logic,* volume 1, *Greek, Indian and Arabic Logic.* Amsterdam: North-Holland, 2004a.

[Gabbay and Woods, 2004b] Dov M. Gabbay and John Woods, editors, *Handbook of the History of Logic,* volume 3, *The Rise of Modern Logic: From Leibniz to Frege.* Amsterdam: North-Holland, 2004b.

[Gabbay and Woods, 2006] Dov M. Gabbay and John Woods, editors, *Handbook of the History of Logic.* volume 7, *Logic and the Modalities in the Twentieth Century.* Amsterdam: North-Holland, 2006.

[Gabbay and Woods, 2005] Dov M. Gabbay and John Woods, *The Reach of Abduction: Insight and Trial,* volume 2 of *A Practical Logic of Cognitive Systems,* Amsterdam: Elsevier, 2005.

[Gabbay and Woods, 2007] Dov M. Gabbay and John Woods, editors, *Handbook of the History of Logic.* volume 8, *The Many Valued and Nonmonotonic Turn in Logic.* Amsterdam: North-Holland, 2007.

[Gabbay and Woods, 2009] Dov M. Gabbay and John Woods, editors, *Handbook of the History of Logic.* Volume 5, *Logic from Russell to Church.* Amsterdam: North-Holland, 2009.

[Gabbay *et al.*, 2011] Dov M. Gabbay, Stephan Hartmann and John Woods, editors, *Handbook of the History of Logic,* volume 10, *Inductive Logic.* Amsterdam: North-Holland, 2011.

[Gabbay *et al.*, 2012] Dov M. Gabbay, Akihiro Kanamori and John Woods, editors, *Handbook of the History of Logic,* volume 6, *Sets and Extensions in The Twentieth Century.* Amsterdam: North-Holland. To appear in 2012.

[Gabbay *et al.*, to appear] Dov M. Gabbay, Jörg Siekmann and John Woods, editors, *Handbook of the History of Logic,* volume 9, *Computational Logic,* to appear.

[Ganeri, 2004] Jonardon Ganeri, "Indian Logic". In Dov M. Gabbay and John Woods, *Handbook of the History of Logic.* Volume 1, *Greek, Indian and Arabic Logic,* pp. 309-395. Amsterdam: North-Holland, 2004.

[Hamblin, 1970] Charles L. Hamblin, *Fallacies,* London: Methuen, 1970.

[Hansen, 1983] Chad Hansen, *Language and Logic in Ancient China.* Ann Arbor: Michigan Studies on China, 1983.

[Hanson, 1958] N.R. Hanson, *Patterns of Discovery.* Cambridge: Cambridge University Press, 1958.

[Harman, 1970] Gilbert Harman, "Introduction: A discussion of the relevance of the theory of knowledge to the theory of induction". In Marshall Swain, editor, *Induction, Acceptance and Rational Belief,* Dordrecht: Reidel, 1970.

[Harman, 1986] Gilbert Harman, *Change in View,* Cambridge, MA: MIT Press, 1986.

[Hintikka, 1987] Jaakko Hintikka "The fallacy of fallacies", *Argumentation,* 1 (1987) 211-238.

[Hintikka, 1997] Jaakko Hintikka, "What was Aristotle doing in his early logic anyway? A reply to Woods and Hansen", *Synthese* 113 (1997) 241-249.

[Hitchcock, 2000] David Hitchcock, "Fallacies and formal logic in Aristotle", *History and Philosophy of Logic,* 21 (2000), 207-221.

[Hitchcock, 2007] David Hitchcock, "Informal logic and the concept of argument". In Dale Jacquette, editor, *Philosophy of Logic,* pp. 101-129. A volume in Dov M. Gabbay and John Woods, editors, *The Handbook of the Philosophy of Science,* Amsterdam: North-Holland, 2007.

[Hobart and Richards, 2008] Michael E. Hobart and Joan L. Richards, "De Morgan's Logic". In Dov M. Gabbay and John Woods, editors, *British Logic in the Nineteenth Century,* pp. 283-329. Volume 4 of the *Handbook of the History of Logic.* Amsterdam: North-Holland, 2008.

[Johnson and Blair, 1977] Ralph H. Johnson and J. Anthony Blair, *Logical Self Defense,* Toronto: McGraw Hill-Ryerson, 1977.

[Johnson, 1996] Ralph H. Johnson, *The Rise of Informal Logic,* Newport News, VA: Vale Press, 1996.

[Joseph, 1916] H.W. B. Joseph, *An Introduction to Logic,* Oxford: Clarendon Press, 1916.

[Klima, 2001] G. Klima, editor and translator with an introduction, John Buridan: *Summulae de Dialectica,* New Haven: Yale University Press, 2001.

[Kneale and Kneale, 1984] W. Kneale and M. Kneale, *The Development of Logic,* Oxford: Clarendon Press, 1984.

[Kretzmann, 1966] Norman Kretzmann, editor and translator, *William of Sherwood's Introduction to Logic*, Minneapolis: University of Minnesota Press, 1966.

[Leibniz, 1949] Gottfried Wilhelm Leibniz, *New Essays Concerning Human Understanding,* transl. with notes by A.G. Langley, La Salle, IL: Open Court, 1949. Completed in 1703-05.

[Lenzen, 2004] Wolfgang Lenzen, "Leibniz's logic". In Dov M. Gabbay and John Woods, editors, *The Rise of Modern Logic: From Leibniz to Frege,* pp. 1-83. Volume 3 of the *Handbook of the History and Philosophy of Logic,* Amsterdam: North-Holland 2004.

[Lorenzen and Lorenz, 1978] Paul Lorenzen and Kuno Lorenz, *Dialogische Logik,* Darmstadt: Wissenschaftlich Buchgesellschaft, 1978.

[Lucian of Samosata, 1915] Lucian of Samosata, *Works,* transl. A.M. Harman, London: Loeb, 1915.

[McKierahan, 1992] Richard D. McKierahan, *Principles and Proofs,* Princeton: Princeton University Press, 1992.

[Magnani, 2001] Lorenzo Magnani, *Abduction, Reason and Science: Processes of Discovery and Explanation,* New York: Kluwer, Plenum 2001.

[Magnani, 2009] Lorenzo Magnani, *Abductive Cognition: The Epistemological and Eco-Cognitive Dimensions of Hypothetical Reasoning,* Berlin: Springer Verlag, 2009.

[Mill, 1843] J.S. Mill, *A System of Logic, Ratiocinative and Inductive,* London: Longmans, 1843.

[Nielsen, 1999] Flemming Steen Nielsen, *Alfred Sidgwicks Argumentationsteori,* Copenhagen: Museum Tusculanums Forlag, 1999.

[Nielsen, 1999a] Flemming Steen Nielsen, "Alfred Sidgwick's rogative approach to argumentation". In Christopher W. Tindale, Hans V. Hansen and Sacha Raposo, editors, *Argumentation at Century's Turn,* CD-ROM, St. Catharines, Ontario: Society for the Study of Argumentation, 1999.

[Parsons, 2008] Terence Parsons, "The development of supposition theory in the later 12th through 14th centuries". In Dov M. Gabbay and John Woods, editors *Handbook of the History of Logic,* volume 2, *Mediaeval and Renaissance Logic,* pages 157-280. Amsterdam: North-Holland, 2008.

[Patterson, 1995] Richard Patterson, *Aristotle's Modal Logic,* Cambridge: Cambridge University Press, 1995.

[Paul of Venice, 1984] Paul of Venice, *Logica Parva,* 14th century. Reprinted by Georg Olms Verlag in Hildesheim, 1970. Translated in Perreiah (1984).

[Perreiah, 1984] A. Perreiah, editor and translator, *Logica Parva: Translation of the 1472 Edition.* Munich: Philosophia Verlag, 1984.

[Peter of Spain, 1947] Peter of Spain, *Summulae Logicales,* J.M. Bocheski, editor, Torino and Rome: Marietti, 1947.

[Pauly and Parikh, 2003] Marc Pauly and Rohit Parikh, editors, *Game Logic,* a special issue of *Studia Logica,* 72 (2003), 163-256.

[Powers, 2012] Lawrence Powers, *Non-Contradiction,* London: College Publications, 2012.

[Ramus, 1964] Petrus Ramus, *Aristotelicae Animadversiones,* together with *Dialectica Institutiones.* Facsimile of the first editions, Paris, 1543, with an introduction by W. Risse. Stuttgart-Bad Cannstatt: Frommann, 1964.

[Prakken, 2005] Henry Prakken, "Coherence and flexibility in dialogue games for argumentation", *Journal of Logic and Computation,* 15 (2005), 1009-1040.

[Rahman and Rückert, 2001] Shahid Rahman and H. Rückert, editors, *New Perspectives in Dialogical Logic,* a special issue of *Synthese,* 78 (2001).

[Reed and Norman, 2003] Chris Reed and T.J. Norman, editors, *Argumentation Machines: New Frontiers in Argument and Computation,* Dordrecht: Kluwer, 2003.

[Reichenbach, 1958] Hans Reichenbach, *Experience and Prediction.* Chicago: University of Chicago Press, 1958

[Rescher, 1964] Nicholas Rescher, *The Development of Arabic Logic,* Pittsburgh: University of Pittsburgh Press, 1964.

[Scriven, 1976] Michael Scriven, *Reasoning,* New York: McGraw-Hill, 1976.

[Spade, 2000] Paul Spade, editor and translator, *Walter Burley's On the Purity and Art of Logic,* New Haven: Yale University Press, 2000.

[Empiricus, 1933–1949] Sextus Empiricus, *Outlines of Pyrrhonesm and Against the Logicians,* vomumes one and two respectivelu, of *Works,* transl. R.G. Bury, London: Loeb, 1933-1949. First published 300-4000 AD.

[Sidgwick, 1884] Alfred Sidgwick, *Fallacies: A View From the Practical Side.* New York: D. Appleton, 1884.

[Sidgwick, 1914] Alfred Sidgwick, *Elementary Logic.* Cambridge: Cambridge University Press, 1914.

[Street, 2004] Tony Street, "Arabic Logic". In Dov M. Gabbay and John Woods, editors, *Handbook of the History of Logic,* volume 1, *Greek, Indian and Arabic Logic,* pages 523-596. Amsterdam: North-Holland, 2004.

[Toulmin, 1958] Stephen Toulmin, *The Uses of Argument,* Cambridge: Cambridge University Press, 1958.

[van Eemeren and Grootendorst, 1984] Frans van Eemeren and Rob Grootendorst, *Speech Acts in Argumentative Discussions: A Theoretical Model for the Anslysis of Discussions Towards Solving Conflicts of Opinion.* Dordrecht: Floris, 1984.

[Van Evra, 2008] James Van Evra, "Richard Whately and Logical Theory". In Dov M. Gabbay and John Woods, editors, *British Logic in the Nineteenth Century,* pp. 75-91. Volume 4 of the *Handbook of the History of Logic.* Amsterdam: North-Holland, 2008.

[Vreeswij, 1997] G. Vreeswij, "Abstract argumentation systems", *Artificial Intelligence, 90 (1997), 225-279.*

[von Wright, 1951] Georg H. von Wright, *An Essay in Modal Logic,* Amsterdam: North-Holland, 1951.

[Walton and Krabbe, 1995] Douglas Walton and E.C.W. Krabbe, *Commitment in Dialoque: Basic Concepts of Interpersonal Reasoning.* Albany: State University of New York Press, 1995.

[Walton, 2000] Douglas Walton, "Alfred Sidgwick: A little-known precursor of informal logic and argumentation", *Argumentation,* 14 (2000), 175-179.

[Watts, 1725] Isaac Watts, *Logick, or the Right Use of Reason in the Enquiry after Truth with a Variety of Rules to Guard Against Error, in the Affairs of Religion and Human Life,* London: John Clark and Richard Hett, 1725

[Whately, 1963] Richard Whately, *Elements of Rhetoric, Comprising an Analysis of the Laws of Moral Evidence and of Persuasion, with Rules for Argumentative Composition and Elocution,* 1828. Edited by Douglas Ehninjer, Carbondale: Southern Illinois University Press, 1963.

[Whitehead and Russell, 1910] Alfred North Whitehead and Bertrand Russell, *Principia Mathematica,* 3 volumes. Cambridge: Cambridge University press, 1910, 1912 and 1913.

[William of Sherwood, 1966] William of Sherwood, *Introductes in logicam,* 13th century. Graham Martin, editor, Sitzungsberichle der Bayerischen Akademic der Wissenschaften, Philosophische-historische Abteilung, Jahrgang 1937, H.O. Munich, 1937. Trnslated in Kretzmann (1966).

[Wilson, 2008] Fred Wilson, "The logic of John Stuart Mill". In Dov M. Gabbay and John Woods, editors, *British Logic in the Nineteenth Century,* pp 229-281. Volume 4 of the *Handbook of the History of Logic.* Amsterdam: North-Holland, 2008.

[Wittgenstein, 1961] Ludwig Wittgenstein, *Tractatus Logico-Philosophicus,* London: Routledge and Kegan Paul, 1961. First published in 1922.

[Woods, 2004b] John Woods, *The Death of Argument: Fallacies in Agent-Based Reasoning.* Dordrecht and Boston: Kluwer, 2004b.

[Woods, 2007a] John Woods, "Should we legalize Bayes' theorem?". In Hans V. Hansen and Robert C. Pinto, editors, *Reason Reclaimed: Essays in Honor of J. Anthony Blair and Ralph H. Johnson,* pp. 257-267. Newport News, VA: Vale Press, 2007a.

[Woods, 2009] John Woods, "SE 176a 10-12: Many questions for Julius Moravcsik". In Dagfinn Follesdall and John Woods, editors, *Logos and Language: Essays in Honour of Julius Moravcsik,* pp. 211-220. London: College Publications 2009.

[Woods, 2010] John Woods, "Abduction and proof: A criminal paradox". In Gabbay *et al.*, editors, *Approaches to Legal Rationality,* pp. 217-238. Dordrecht: Springer, 2010.

[Woods, 2012] John Woods, "Cognitive economics and the logic of abduction", *Review of Symbolic Logic,* 592012), 148–161.

[Woods, 2013] John Woods, *Seductions and Shortcuts: Error in the Cognitive Economy,* to appear in 2013.

[Woods and Walton, 1975] John Woods and Douglas Walton, "Petitio principii", *Synthese,* 31 (1975) 107-128.

[Woods and Hansen, 1997] John Woods and Hans Hansen, "Hintikka on Aristotle's fallacies", *Synthese,* 113 (1997) 217-239.

[Woods and Hansen, 2004] John Woods and Hans Hansen, "The subtleties of Aristotle on non-cause", *Logique et Analyse,* 176 (2004) 395-415.

[Woods and Irivine, 2004] John Woods and Andrew Irvine, "Aristotle's Early Logic". In Dov M. Gabbay and John Woods, editors, *Handbook of the History of Logic,* volume 1, *Greek, Indian and Arabic Logic,* pages 27-99. Amsterdam: North-Holland, 2004.

[Woods and Walton, 2007] John Woods and Douglas Walton, *Fallacies: Selected Papers 1972-1982,* 2nd edition with a Foreword by Dale Jacquette. London: College Publications, 2007. First edition Dordrecht: Foris, 1989.

[Woods and Gabbay, 2010] John Woods and Dov Gabbay, "Relevance in the law: A logical perspective". In Gabbay *et al.,* editors, *Approaches to Legal Rationality,* pp. 261-289. Dordrecht: Springer, 2010.

[Yrjönsuuri, 2008] Mikko Yrjönsuuri, "Treatments of the paradoxes of self-reference". In Dov M. Gabbay and John Woods, editors, *Handbook of the History of Logic,* volume 2, *Mediaeval and Renaissance Logic,* pages 579-608. Amsterdam: North-Holland, 2008.

[Zhai, 2011] Jincheng Zhai, "A new interpretation of reasoning patterns in Mohist logic", *Studies in Logic,* 4 (2011), 126-143.

A HISTORY OF LOGIC DIAGRAMS

Amirouche Moktefi and Sun-Joo Shin

1 INTRODUCTION

Diagrams, however we may define them, are one of the most widely used tools by humans — from ordinary life communication to brainstorming for complicated problems or outlining the overall structure of talks, papers, etc. Some philosophers and psychologists extend the territory of diagrams by embracing mental images under the same category. Diagrams, external or internal, cover such a vast area of human activities that different disciplines, not surprisingly, have approached the topic from different angles.

This chapter carves up a certain part of the diagram-territory, logic diagrams, and provides the reader with both historical data about and theoretical implications of logic diagrams. The topic — logic diagrams — as the name implies, is about diagrams and has something to do with logic. The reader might wonder how diagrams, say circles, lines, etc., could be related to a discipline where we have encountered rules and symbols only. We, either as a student or a teacher, have drawn diagrams in mathematics and logic classes, but only as an aid to understanding complicate logical steps which are written in symbols. By 'logic diagrams' we do not mean to refer diagrams drawn as a reasoning aid, a heuristic tool, or a brainstorming tool, but diagrams which carry out logical reasoning independently.

Are there logic diagrams? Can diagrams serve as logical languages in a strict sense? To our surprise, we can go back through centuries and find evidence to show that logic diagrams have existed just as symbolic logic does. Our goal is not only to examine historical logic diagrams but also to explore the theoretical justification and advantage of logic diagrams. Hence, before we present various kinds of logic diagrams in an (almost) chronological order, we invite the reader to explore the relation between diagrams and logic at a purely theoretical level.

A main goal of the discipline logic is to study and enhance valid reasoning. We may call reasoning, say from P to Q, valid if the information Q is extractable from information P. Hence, talk about valid reasoning, in almost all situations, presupposes representation, and logic mainly investigates logical systems which represent information and allow us to extract information from given information. The semantics and the rules of inference of a logical system carry out these two important tasks — representing and manipulating information. Furthermore, a logical system equipped with syntactic manipulation rules and formal semantics can be checked as to whether the system is sound and complete.

Handbook of the History of Logic. Volume 11: Logic: A History of its Central Concepts.
Volume editors: Dov M. Gabbay, Francis Jeffry Pelletier and John Woods.
General editors: Dov M. Gabbay and John Woods.

The reader must agree that we just laid out a straightforward and uncontroversial view of logic. One question is whether the basic mission statement demands that a system should be symbolic. That is, could we have a logical system (i) which has a list of transformation rules, (ii) which has a formal semantics, but (iii) whose vocabulary is diagrammatic? For now, we would like to answer at the level of principle, not by example. There is no intrinsic reason why syntax and semantics are tied up with a form of representation. All syntax does is to stipulate vocabulary, well-formed units, and a manual of permissible transformations. Semantics defines a function-style match from syntactic elements with non-syntactic entities so that ambiguity is avoided. Therefore, we believe diagrammatic logical systems are possible in principle. At the same time, in reality we have many diagrammatic logical systems throughout history. These are the logic diagrams we intend to explore both historically and theoretically in this chapter. We present them in a mixture of chronological order and by the type of diagrammatic vocabulary a system adopts.

In the next section (2), we will discuss the most familiar tradition of logic diagrams, known as spatial diagrams, popularised by Leonhard Euler. We will see how Euler's method leads to some ambiguities that will be later removed by John Venn and Charles S. Peirce. This section is aimed at familiarising the reader with spatial diagrams that will be used, in the two subsequent sections, to discuss more broadly the representation and manipulation of information with diagrams. In section (3), we will discuss the principles of representation on which are founded logic diagrams. Other traditions of diagrams will be discussed here, notably linear and tabular diagrams. Section (4) will be devoted to the use of diagrams to solve logic problems, namely syllogistic and elimination problems. As far, sections (2), (3) and (4) were concerned with term/class logic.

When we move to full-blown predicate logic, the presence of logic diagrams is more striking than ever. Section (5) examines two different systems of logic diagrams invented by two founders of modern logic — Gottlob Frege's Conceptual Notation and Peirce's Existential Graphs. The reader might be somewhat surprised to realize that modern quantified logic appeared first in graphical notation, since (modern) formal logic has been strongly identified with symbolic logic. The section on what we call the Frege-Peirce Affair will, we hope, raise legitimate skepticism toward the equation between formalization and symbolization, by studying Frege's and Peirce's graphs and analyzing their original intentions and defenses against their contemporaries.

At this point, we anticipate the following questions: "Why has the topic — logic diagrams — been neglected until recently?", "Why have symbolic systems been dominant in logic?". The last section suggests some answers to these questions in the course of covering recent research on logic diagrams. A different age has a different demand, its focus being shifted: At the turn of the 20th century, formalization captured mathematicians' full attention not only because predicate logic made formalization more powerful but also because surprises in the world of mathematics put mathematicians and logicians on alert. Accurate formal systems were sought after. In the age of the computer, however, we are in search of not only accurate but also more efficient logical systems. Targeting both accuracy and efficiency, we have realized advantages of combining different modes of representation, as our natural reasoning often does. Hopefully, the historical and theoretical study of logic diagrams in this chapter will make a small contribution to this newly energized effort for multi-modal formalization.

Preliminaries

Before discussing the history of logic diagrams in the following sections, we would like here to insist on few preliminary issues of some importance to understand the aim and scope of this chapter. Firstly, we did not intend here to offer a systematic survey of diagrams used by logicians and logic students, past and present, though the schemes discussed in this chapter give a nice overview of the richness and variety of diagrammatic methods used in logic.[1] Our aim is rather to discuss the main issues that faced the inventors and users of such diagrams and how their needs and expectations evolved and accompanied the development of logic itself. In a way, our purpose is not as much to introduce and assess particular diagrammatic schemes as it is to stress the existence and evolution of a diagrams-related scholarship in logic. As such, we plead here for the study of diagrams within the context in which they have been conceived, the practices that led to their introduction and thus the uses those diagrams were invented for. From this perspective, the history of logic diagrams certainly says something significant on the development of modern logic.

Secondly, as explained above, we are interested here in diagrams that "carry out logical reasoning independently". As such, we focused in this chapter on diagrams that analyses inferences and could be used to solve logic problems, though the definition of what stands for a logic problem may vary depending on the historical period, as we will see in the subsequent sections. Thus, it should be kept in mind that we are concerned here with "analytical diagrams", an expression seemingly coined by Venn [Venn, 1894, p. 506]. This restriction does not minimise the importance of other types of diagrams, widely used in logic for illustrative, mnemonic or heuristic purposes, notably in educational settings. For instance, diagram [Fig. 1], reproduced from [Holman, 1892, p. 65], shows the position of the middle term in the four traditional figures of syllogisms.

[1]Such surveys will be found in [Hamilton, 1871; Venn, 1880b; Gardner, 1958] and [Mac Queen, 1967].

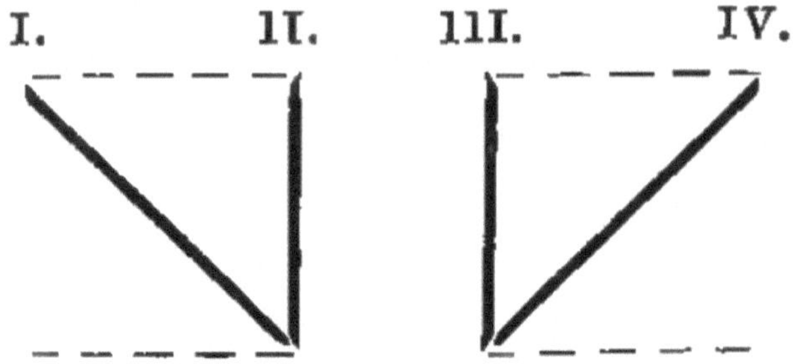

Fig. 1

Fig. 2

Such visual devices need not to be necessarily about formal rules in logic however. Indeed, some authors appealed to illustrations in order to make their subject more comprehensible to the reader. As an example, the drawing [Fig. 2], titled "Logic pays a visit", is included in Alfred J. Swinburne's *Picture Logic* to illustrate the view that language stands as a medium between logic and thought. The scene is described as follows:

> Scene, a man's mouth. Thought peeping from the throat. Language telling logic (who has come to visit Thought) that his orders are that no one can see Thought, and that all communications must be made through him (Language). 'So you'd better by 'alf tell me what you want 'owever.'
>
> And Logic mutters, 'What a coarse medium! But there is no help for it. Alas what mistakes and confusions will arise!' [Swinburne, 1887, p. 49-50]

To say that we didn't deal with such diagrams in this chapter doesn't mean that we exclude entirely the possibility of using illustrative diagrams as visual aids in order to make some immediate inferences. Such is particularly the case of diagrams representing structures where concepts or propositions are interconnected with lines defining specific relations. Diagrams of this kind, such as structures of oppositions and logic trees, have long been used in logic. The square of oppositions was the most popular diagram of the first type and was included in most logic textbooks until the nineteenth century. Several extensions have been since proposed, such as the octagon of opposition [Fig. 3] published by John Neville Keynes [Keynes, 1894, p. 113].[2] A well-known example of logic trees is the porphyry tree used by logicians to show the hierarchy of different substances and to explain the concepts of genus, species and differentia, as shown by [Fig. 4] included in a nineteenth century textbook [Jevons, 1872, p. 104].[3]

[2]Keynes acknowledges the help of William E. Johnson in the making of this octagon. On the square of oppositions, its history and its extensions see [Peckhaus, 2005; Parsons, 2006; Moretti, 2009; Béziau, 2012] and [Béziau and Jacquette, 2012].

[3]On the history of logic trees, see [Hacking, 2007]. Note that Lewis Carroll designed in 1894 a

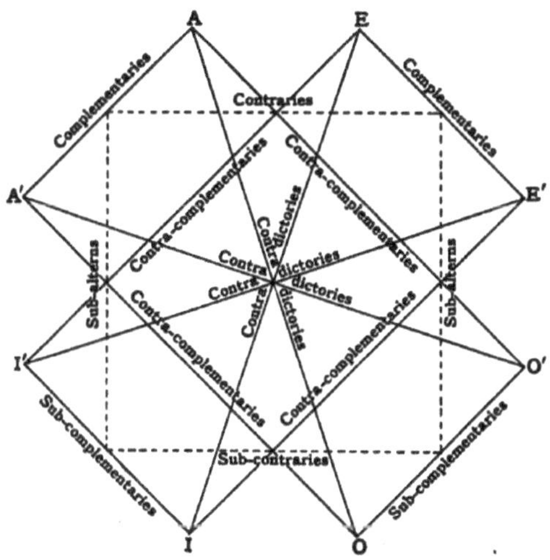

Fig. 3

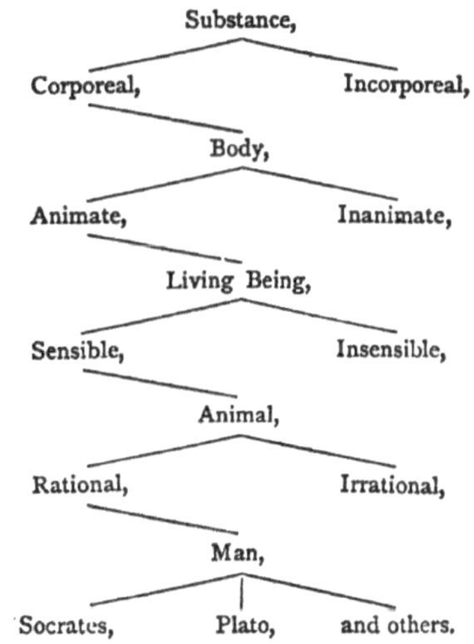

Fig. 4

Finally, a last remark that one has to make before discussing the history of logic diagrams in the following sections is that we did not attempt here to provide a precise definition on what ought to be called a logic diagram. Martin Gardner argued that a "logic diagram is a two-dimensional geometric figure with spatial relations that are isomorphic with the structure of a logical statement" [Gardner, 1958, p. 28]. This gives a good starting point to which it is possible to adhere in a soft way. However, one has to be prepared to depart occasionally from Gardner's definition when needed. Also, one should keep in mind that most schemes and notations, on varied levels, combine both diagrammatic and symbolic methods. For instance, [Fig. 5] represents the proposition "Any C is not any D" in Sir William Hamilton's notation.

Fig. 5

In this scheme, the colon ':' stands for '*all*', the horizontal line represents the affirmative copula, while the vertical line through the copula represents its negation [Hamilton, 1871, p. 530]. It is uneasy to tell whether this mode of representation is diagrammatic or symbolic. This example shows how a strict separation between these modes is difficult and unsuitable.

"method of trees" to solve complex logic problems. On this method which combines diagrammatic and symbolic modes, see [Bartley, 1986, p. 279-319; Abeles, 1990; 2005] and [Moktefi, 2008]. On the use of trees in twentieth-century logic, see [Anellis, 1990].

2 THE GOLDEN AGE OF LOGIC DIAGRAMS

In this section, we will discuss the development of one particular tradition of logic diagrams, commonly known as spatial diagrams because they appeal to relations between closed curves (mostly circles) to represent logical relations.[4] Other diagrammatic methods will be discussed in subsequent sections. The focus on spatial diagrams is mostly motivated by their popularity and the fact that their evolution is highly representative of the crucial developments in logic at the time. These diagrams were known before the eighteenth-century[5] and were commonly used by several early eighteenth-century logicians, including Gottfried Leibniz. However, one has to wait Euler's systematic treatment for a wide use among logicians and logic students, all along the nineteenth-century. This period (18th — 19th centuries) might thus fairly be recognised as the golden age of logic diagrams, and it will not be surprising that the analytic diagrams that we will discuss in this chapter belong mostly to this period.

2.1 Euler's circles

Euler introduced his logic diagrams in the second volume of his *Letters to a German Princess* [Euler, 1768].[6] The diagrams appear for the first time in Letter *CII*, dated 14 February 1761, where he expresses how to represent the four classical propositions with the use of his scheme:

> These four species of propositions may likewise be represented by figures, so as to exhibit their nature to the eye. This must be a great assistance towards comprehending more distinctly wherein the accuracy of a chain of reasoning consists.
>
> As a general notion contains an infinite number of individual objects, we may consider it as a space in which they are all contained. Thus, for the notion of *man* we form a space [...]
>
> 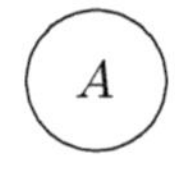
>
>
> in which we conceive all men to be comprehended. [Euler, 1833, p. 339]

This approach makes it easy to represent logic propositions with two circles that include, intersect or exclude each other. For instance, to represent the universal

[4]Other accounts of the history of spatial diagrams will be found in [Gardner, 1958; Baron, 1969; Coumet, 1977; Shin, 1994], and [Greaves, 2002].

[5]A. W. F. Edwards reproduced an eleventh-century spatial diagram in [Edwards, 2006].

[6]Euler wrote these 234 letters between 1760 and 1762. The two first volumes appeared in 1768 and a third appeared in 1772. The diagrams appear in the second volume, in the letters devoted to logic (letters *CII* to *CVIII*, dated from 14 February 1761 to 7 March 1761).

affirmative proposition "All A is B", Euler draws a circle A that is completely included within a circle B [Fig. 6]. Similarly, Euler draws two disjoint circles to represent the universal negative proposition "No A is B" [Fig. 7]. The representation of particular propositions requires two intersecting circles. If the proposition is affirmative ("Some A is B"), the letter 'A'' (corresponding to the subject) is inserted within the overlapping space [Fig. 8]. Else, the proposition is negative ("Some A is not B") and the letter 'A'' is inserted within the space of the circle A which is outside the circle B [Fig. 9].

Fig. 6

Fig. 7

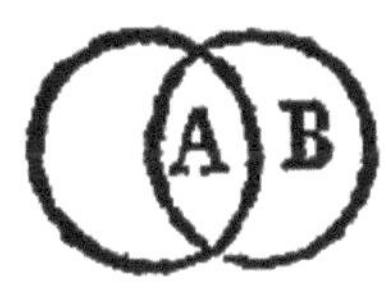

Fig. 8

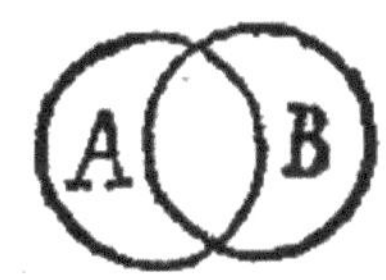

Fig. 9

Euler circles represent the actual relation of the classes. As such, this scheme has one important limit because of the ambiguity of the correspondence between the propositions and the diagrams. Indeed, one given proposition might need more than one figure to express adequately its "potential" information. Though traditional logicians used to handle just 4 types of categorical proposition (A, E, I, O), Joseph D. Gergonne has shown in 1817 that there are 5 possible relations between two given classes A and B: (a) A is strictly included in B; (b) A strictly includes B; (c) A coincides completely with B; (d) A and B are completely disjoint; (e) A and B partly overlap [Gergonne 1817]. These relations, today known as the Gergonne relations [Faris, 1955; Grattan-Guinness, 1977], are represented with circles in [Fig. 10].

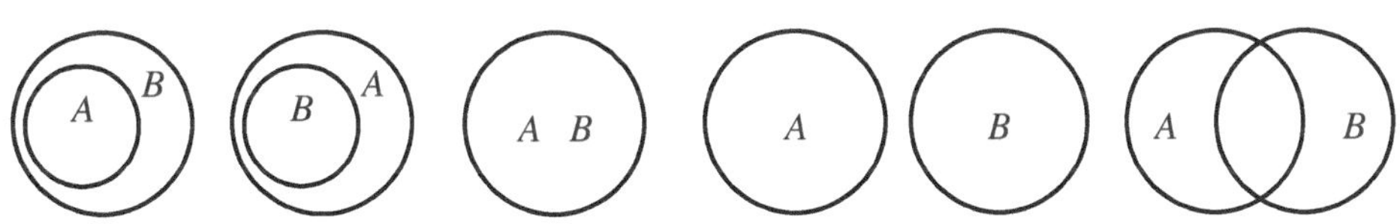

(a) (b) (c) (d) (c)

Fig. 10

It is easy to point the imprecise correspondence between these relations and the different propositions. Two disconnected circles A and B [Fig. 10-d] suffice to represent satisfactorily the proposition "No A is B". However, one needs 3 figures to make the topological relation of the circles represent adequately the proposition "Some A is not B", depending on whether B is inside A [Fig. 10-b], B is outside A [Fig. 10-d], or B overlaps with A [Fig. 10-e]. Each of these 3 figures would represent just one possible relation between A and B, permitted by the proposition

"Some A is not B". As we do not know what the actual relation between A and B is, the three figures are legitimate candidates.

In order to remove this ambiguity, some nineteenth-century logicians appealed to a graphical convention that allows combining all potential cases in one single diagram, by indicating the uncertainty with discontinuous lines [Thomson, 1849, 271; Ueberweg, 1871, p. 217-218]. Let's take for instance the proposition "All A is B" which needs two figures to represent fully its meaning depending on whether A is strictly included in B, or A is identical to B. One figure would show the circle A inside the circle B [Fig. 11] and the other figure would depict one single circle which is both A and B [Fig. 12]. On can mix these two figures and obtain just one, as shown in [Fig. 13].

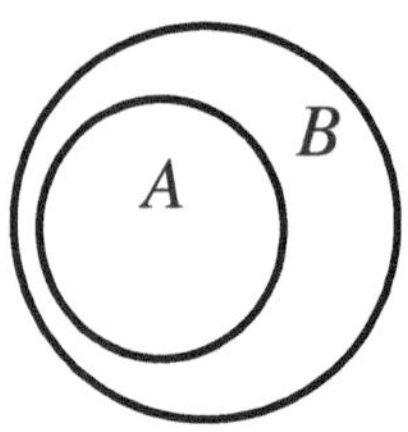

Fig. 11

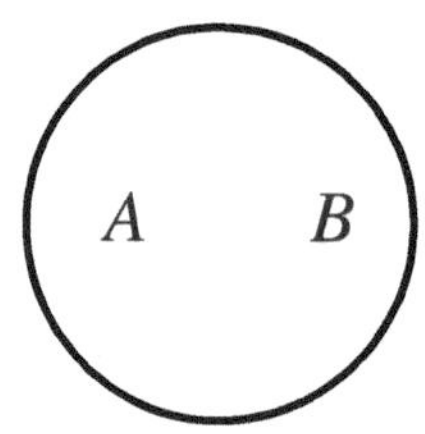

Fig. 12

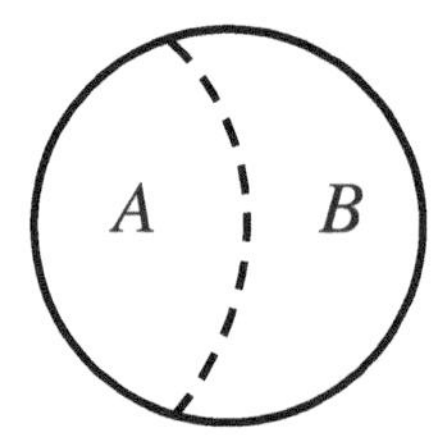

Fig. 13

The discontinuous line indicates that we do not know for sure whether that line exists or not. Either the line exists (i.e. the discontinuous line becomes continuous) and [Fig. 13] will then be equivalent to [Fig. 11], or the line doesn't exist (i.e. the discontinuous line disappears) and [Fig. 13] will then look similar to [Fig. 12]. This solution avoids the ambiguity of Euler's scheme, but it is obvious that it looses the simplicity and intuitiveness of the Eulerian method where one has just to "look" in order to "see" the proposition. In this new method, one has additionally to "read" the diagrams. Still, this method of representing uncertainty with specific devices is the one that will be pursued and carried out by Venn when he later developed his own diagrams.

2.2 Venn's diagrams

Venn introduces his diagrams in an 1880 paper which he published in *The Philosophical Magazine* [Venn, 1880a]. He describes and uses them more lengthily in his logical treatise: *Symbolic Logic*, first published in 1881, enlarged in 1894. He explains that it was his dissatisfaction with Euler's scheme that led him to invent his own diagrams [Venn, 1894, p. 114]. Venn uses circles, as Euler did. However, he makes a very different use of them. Indeed, instead of representing directly the actual relation of the classes, Venn draws first a primary diagram representing all possible sub-classes that one obtains by intersecting the terms involved in the argument. Then, he adds distinctive marks to represent propositions according to the emptiness or occupation of the sub-classes. This two-step representation (classes first, then propositions) is the main innovation of Venn's scheme in comparison with Euler's use. Let's look at [Fig. 14]:

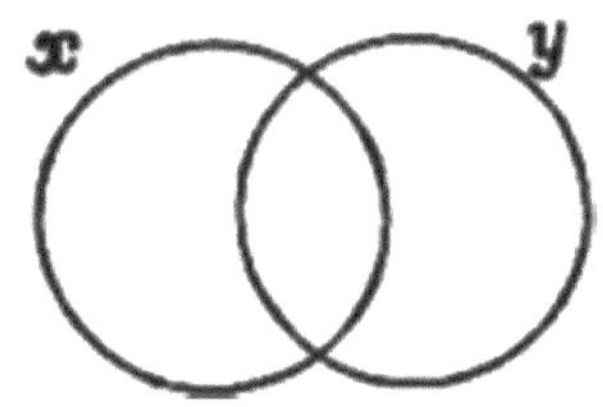

Fig. 14

In Euler's scheme, this figure would represent the actual intersection of the classes x and y. In Venn's approach however, this diagrams represents merely the four compartments produced by the terms x and y, which are: $x \quad y$ (the space common to the circles x and y), x *not-y* (the space of the circle x which is not part of the circle y), *not-x y* (the space of the circle y which is not part of the circle x), and finally *not-x not-y* (the space which is outside the two circles x and y). As far, this diagram tells nothing on the actual relation of the classes x and y. To represent propositions, one has to add some marks to indicate the condition of the compartments in this primary diagram. For instance, Venn shades a compartment to tell that it is empty. Consequently, if one want to represent the universal affirmative proposition "All x are y", which is equivalent in Venn's logic to the universal negative proposition "No x is *not-y*", it suffices to shade the compartment x *not-y*, as shown in [Fig. 15].

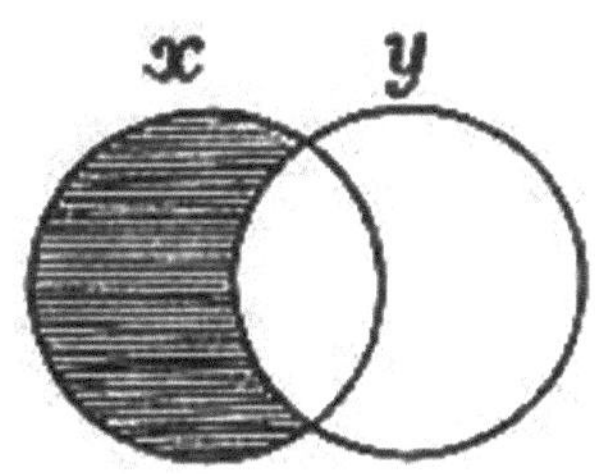

Fig. 15

Contrary to universal propositions which he represents easily and adequately, Venn was ambiguous and hesitating in dealing with particular propositions. He even remained completely silent on this issue in his 1880 paper and in the first edition of his *Symbolic Logic* [Venn, 1881]. It is interesting to note that Euler did also feel uneasy with his method of representing particular propositions. This led him in his late diagrams (letter *CV*) to mark the non-emptiness of a region with a star '*' [Fig. 16 & Fig. 17].[7] This convention anticipates the method of representation used later by Venn when his added graphical marks to indicate the non-emptiness of a compartment.

[7]On the use of this convention in Euler, see [Hammer and Shin, 1998, p. 10-11]. Note that in [Fig. 17], it is not visually clear what the circle A encloses.

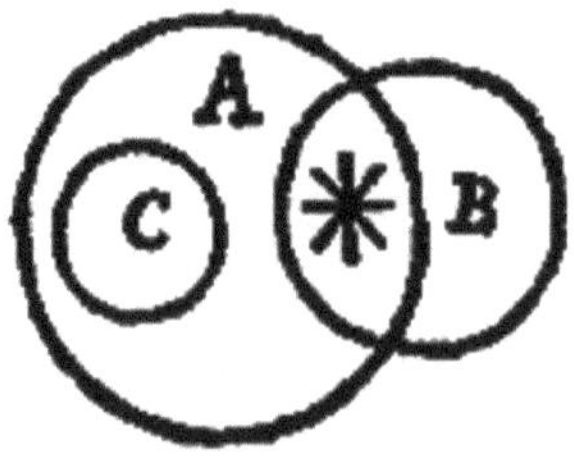

Fig. 16

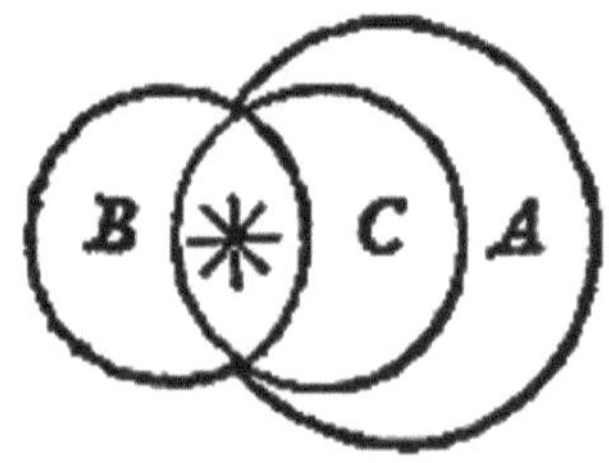

Fig. 17

Venn seems to discuss the representation of particular propositions for the first time in his 1883 review of Peirce's (and his students') *Studies in Logic* [Peirce, 1883]. He suggests using bars as distinctive marks to express the occupation of the compartments:

> If we introduce particular propositions also, we must of course employ some additional form of diagrammatical notation [...] We might, for example, just draw a bar across the compartments declared to be saved ; remembering of course that, whereas destruction is distributive, [...] the salvation is only alternative or partial... [Venn, 1883, p. 599-600]

Venn didn't provide a graphical illustration of his method. Consequently, we do not know precisely what kind of bars he had in mind, and especially how the bars had to be inserted within the compartments (vertically, horizontally, etc.). It is not difficult however to understand how this convention could be applied to represent particular propositions. For instance, one can just draw a bar across the compartment xy to mean that it is not empty and thus to represent the proposition "Some x are y". In the following years, Venn defended this method of representing particular propositions [Venn, 1887],[8] and several of his contemporaries and successors used it in their logical treatises [Keynes, 1894, p. 137; Johnson, 1921, p. 151-154]. Venn was not happy with his invention however, as he didn't include it in the second edition of his *Symbolic Logic* (1894) where he suggests the use of some other methods to represent existential statements [Venn, 1894, p. 130-132].

It is needless here to discuss in detail all the methods used by Venn to represent particular propositions, as they didn't ultimately satisfy him. A brief survey shows however how he was uncomfortable with the problem. Instead of the bar-convention, Venn suggests first the use of different shadings to express the emptiness or occupation of the compartments. For instance, vertical shadings would be used to tell that a compartment is empty (as is the case with universal propositions) while horizontal shadings would tell that a compartment is occupied (as is the case with particular propositions). Then, Venn dismisses immediately this

[8] In this short note [Venn, 1887], Venn replied to Alfred Sidgwick who praised Lewis Carroll's diagrammatic representation of particular propositions [Sidgwick, 1887]. Carroll published his diagrams in a booklet, *The Game of Logic*, first published in 1886 [Carroll, 1886].

solution and suggests another one which is to enumerate particular propositions, and then mark the compartments which are (or might be) saved by the numeral(s) of the proposition(s) that save(s) them. It is obvious that when only one particular proposition is represented, one could use a distinct mark (a star or a cross) instead of a numeral. That's what Venn did when, in rare instances, he used a cross to indicate the non-emptiness of a compartment on a Marquand diagram (that we will describe later) [Venn, 1894, p. 208 & 376]. He even once used a cross on one of his own diagrams in a letter to Lewis Carroll, published by the latter [Carroll, 1897, p. 182].

Venn's hesitations show the difficulties that are raised when it comes to representing diagrammatically particular propositions which have already proved to be difficult to represent symbolically too. Contrary to universal propositions which erase compartments, particular propositions save compartments. Consequently, universals are much easier to represent because erasing a class implies erasing all its sub-divisions. For particulars however, saving a class doesn't imply saving all its sub-divisions. It rather means that at least one of its sub-divisions is not empty and, thus, should be saved. Particular propositions need the expression of a disjunction which is not easy to represent diagrammatically. Venn's failure, or at least lack of satisfaction, led Peirce to suggest some improvements in order to deal adequately with both particular and disjunctive propositions.

2.3 *Peirce's improvements*

Peirce developed his own graphs as we will see later in this chapter. In this section we will focus merely on his improvements of Venn diagrams, which he continued to name as the Eulerian scheme, though he accepted Venn's innovations. Peirce identified some deficiencies in Euler-Venn diagrams and worked on their improvement. For this purpose, he introduced new graphical conventions: one needs just to put a "0" on a compartment (rather than shading it) to tell that it is empty, and similarly one has just to put a cross "X" on a compartment to tell that it is not empty. For instance, to represent the proposition "Some x are y", one has just to put a cross 'X' on the compartment xy [Fig. 18]. Similarly, one has just to put a '0' on the compartment xy to represent the proposition "No x is y" [Fig. 19].

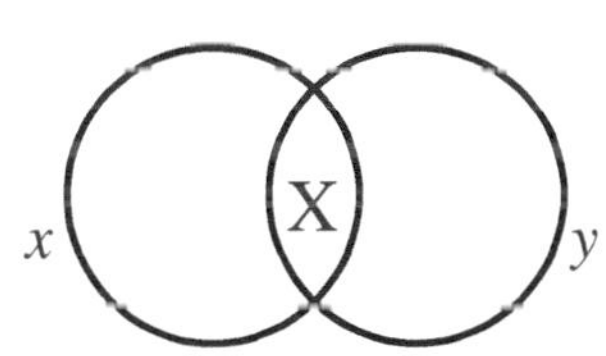

Fig. 18

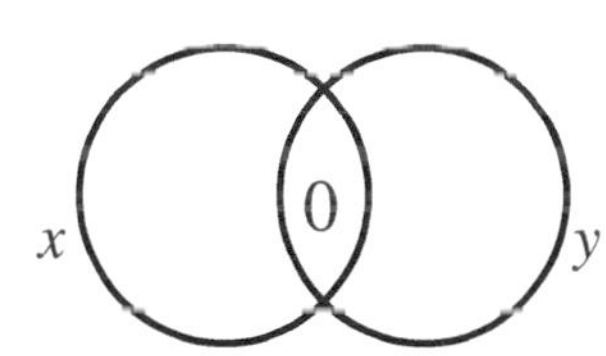

Fig. 19

The use of a cross (or any other distinctive mark) permits an easy representation of existential statements, and the use of a '0' prevents from shading compartments as Venn did. In order to represent disjunctive propositions, Peirce introduces one more convention: after representing separately propositions involved in the disjunction, one has just to connect the marks with a line to indicate the disjunction. For instance, in order to represent the proposition ' "Some x are y" or "No *not-x* is *not-y*" ', one has first to represent independently the propositions "Some x are y" (by putting a cross 'X' in the compartment xy) and the proposition "No *not-x* is *not-y*" (by putting a '0' on the region *not-x not-y*). Then, one must connect the two symbols 'X' and '0' with a line as shown in [Fig. 20].

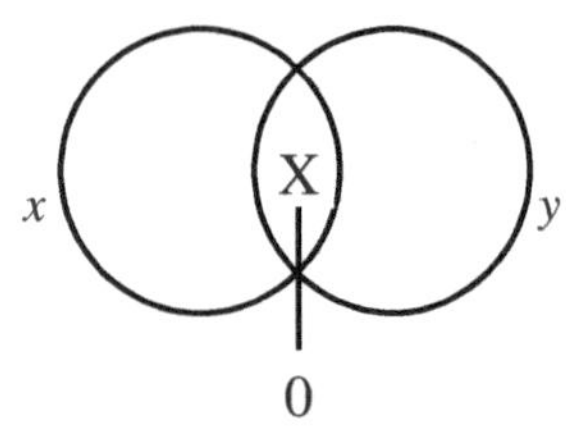

Fig. 20

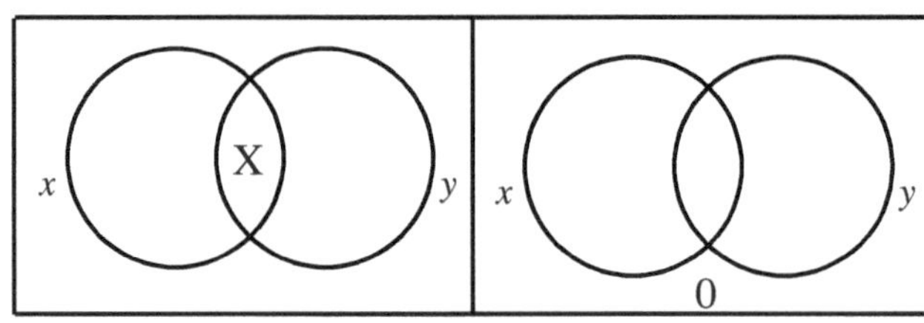

Fig. 21

Thanks to his new conventions, Peirce represents adequately particular and disjunctive propositions. It is clear however, that his diagrams loose the visual naturalness of Euler's original scheme as one need to interpret the different symbols in order to understand the information that is represented, especially when the number of propositions involved increases. Peirce was conscious of this visual limitation and, thus, suggested occasionally other methods to represent disjunctives. For instance, he suggests putting the cross 'X' on the boundary between two compartments rather putting a cross 'X' in each compartment and connecting them with a line, as he used to do to express disjunction [Peirce, 1933, §4.363]. Anther method would be to put side-to-side diagrams that represent the different possibilities permitted by the disjunction:

> It is merely that there is a greater complexity in the expression than is essential to the meaning, There is, however, a very easy and very useful way of avoiding this, It is to draw an Euler's Diagram of Euler's Diagrams each surrounded by a circle to represent its Universe of Hypothesis, There will be no need of connecting lines in the enclosing diagram, it being understood that its compartments contain the several possible cases. [Peirce, 1933, §4.365]

For instance, in order to represent the disjunctive proposition alluded to earlier ' "Some x are y" or "No *not-x* is *not-y*" ', one has to put side-to-side diagrams representing respectively the propositions "Some x are y" and "No *not-x* is *not-y*" as shown in [Fig. 21], with the implicit understanding that they are the terms of the disjunction.

3 REPRESENTING INFORMATION WITH DIAGRAMS

In the previous sections we discussed the development of spatial diagrams from Euler, through Venn, to Peirce. A look at how these diagrams evolved gives an idea on the (new) needs and requirements of the logicians who invented those diagrams. [Fig. 22] gives an overview of the diagrams used by Euler, Venn[9] and Peirce to represent the four canonical propositions A, E, I, and O. It is easy to see how the diagrammatic schemes get more and more expressive power but all along the way lose more and more visibility. Euler's principles of representation are quite simple and intuitive, but his diagrams lack rigor (in the form that he originally used them) and are unable to represent partial information. Venn handled the matter but at the price of adding new devices which makes the diagrams less accessible at first glance. Finally, Peirce added more devices and ultimately got diagrams that are more powerful but offer much less visual aid than Euler's scheme.

This dilemma was already pointed out by Eric Hammer and Sun-Joo Shin:

> It is interesting that Venn and Peirce adopted the same kind of solution in order to achieve these improvements. The solution was to introduce new syntactic objects. Venn invented shadings, and Peirce the symbols X and 0 and lines connecting these symbols. These syntactic objects brought mixed results to their systems. As a positive result, the systems became much more general and flexible in being able to represent many kinds of information. However, on the negative side, these revised systems suffer from a loss of visual naturalness. [Hammer and Shin, 1998, p. 14]

It is obvious that Euler and Venn appeal to different methods of representation and that Peirce followed Venn's method after accepting his improvements. Euler represents the actual relations between the classes while Venn shows first a general structure and then marks compartments to represent information. Hereafter, we will refer to the first method as Euler-type method, and to the second as Venn-type. Similarly, diagrams appealing to the Euler-type method will be called Euler-type diagrams, while diagrams using the Venn-type method will be designated as Venn-type diagrams. This distinction will prove to be useful in the following sections, as we will see other styles of diagrammatic representation, and we will show that they still appeal to either method: Euler-type or Venn-type. However, let us keep in mind that these names are purely conventional and that it is not meant that these methods have been originated by Euler or Venn. Also, the conceptual shift from the Euler-type to the Venn-type method of representation, as we observed it in the development of spatial diagrams, should not be interpreted as a necessary development. Neither is one method more fundamental than the other, nor

[9]As Venn didn't specify how the bars, used to represent existential statements, must be, we represented them horizontally for convenience.

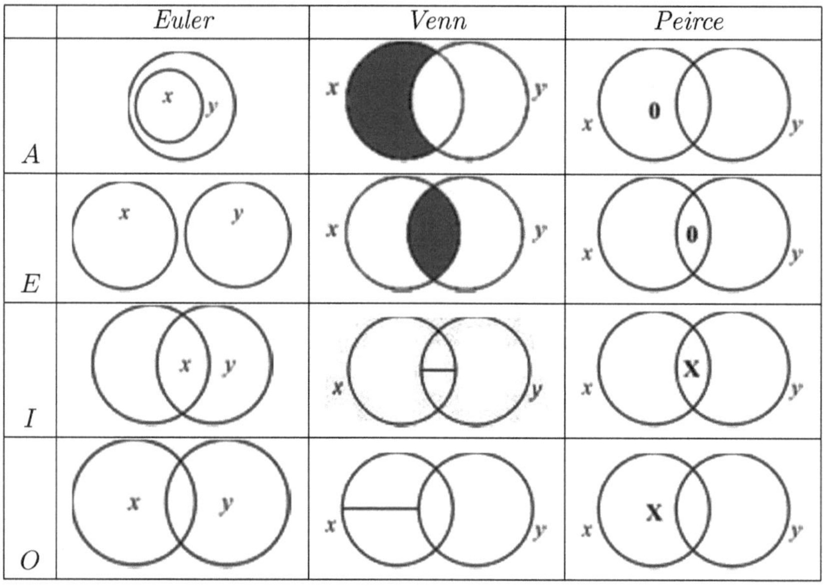

Fig. 22

did any method supersede the other. All we mean is to point the existence of these two methods, to identify the crucial differences between them, and to remind that each method has its advantages and its inconveniences. In a way, the history of logic diagrams can be accounted for as a continuous search for the best balance between visual aid and expressive power.

3.1 Division and dichotomy

For a better understanding of the difference between these two methods of representation, we will briefly examine the motivations of Venn's departure from the Euler-type method, and how this move does reflect the logical line in which he was working. Venn clearly says that he first worked with Euler's circles and that it was his dissatisfaction that led him to invent his own diagrams [Venn, 1980a, p. 4]. Venn argued that it was not possible with Euler's method to proceed gradually with the solution of complex problems, as one finds in Boolean logic, because Euler's scheme doesn't allow one to represent partial information:

> The weak point in [Euler's scheme], and in all similar schemes, consists in the fact that they only illustrate in strictness the actual relation of classes to each other, rather than the imperfect knowledge of these relations which may possess, or may wish to convey by means of the proposition. [Venn, 1894, p. 510]

In a way, Euler's diagrams, because they represent actual knowledge, require that one knows what the relation is before one represents it. Venn's approach is entirely different as it first represents imperfect knowledge, and allows for additional information to be added, without drawing more than one diagram. We have already seen how Venn used his idea with benefit, when he represented propositions involving 2 terms. The situation naturally becomes more complex when more terms are involved in the argument, as is the case with syllogisms where two propositions and 3 terms are to be represented in a single diagram. Indeed, even if one sticks to the original Eulerian scheme, this requires the discussion of all the combinations permitted by these propositions. For instance, in order to represent the two propositions "Some A are B" and "All C are A", Euler lists in his letter *CIV*, three possible cases depending on the relationship of C and B. Indeed C might be completely outside B [Fig. 23], completely inside B [Fig. 24], or intersect partially with B [Fig. 25].[10]

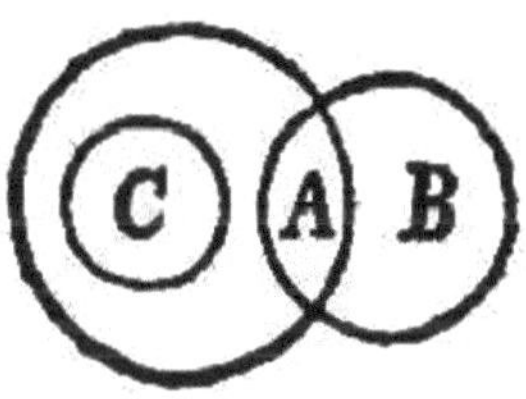

Fig. 23

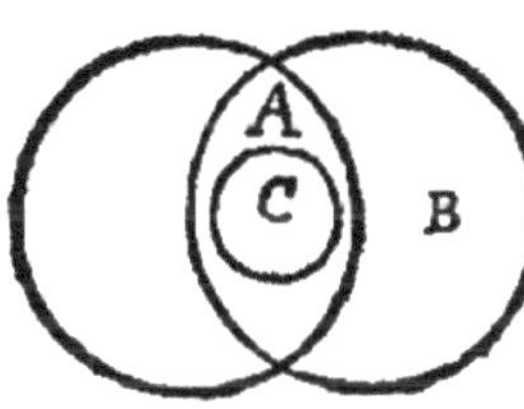

Fig. 24

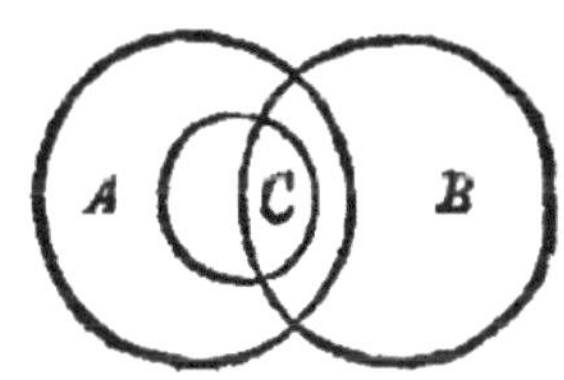

Fig. 25

This example shows how the representation of actual relation between classes obliges one to use more than one diagram and thus makes it difficult to grasp at once what that actual relation is. As we have seen previously, some nineteenth-century logicians appealed to dotted lines to remove this difficulty and combine all diagrams in just one. Hence, in the case of Euler diagrams for three classes, dotted lines would indicate the various relations that the third class might have with the two first ones. For instance, [Fig. 26] reproduced from Julius Bergmann's *Allgemeine Logic* (1879), shows two disjoint classes M and P, while the three dotted circles indicates that the class S, which contains class M, is either completely outside P, or overlaps partly with P or contains completely class P [Bergmann, 1879, p. 372]. Using the same technique, [Fig. 23], [Fig. 24] and [Fig. 25] could be combined in just one diagram as shown in [Fig. 27].

[10]We observe that in [Fig. 25], Euler puts erroneously the letter 'A' in the region of A which is outside B, while it should be in the intersecting region of A and B as is shown in [Fig. 23] and [Fig. 24]. Also, if we follow Euler's typology of propositions, one could have a fourth case where "some C is not B", which gives a diagram with the circle C inside the circle A and intersecting with the circle B, like [Fig. 25], where the letter 'C' would be in the region of circle C which is outside the circle B however.

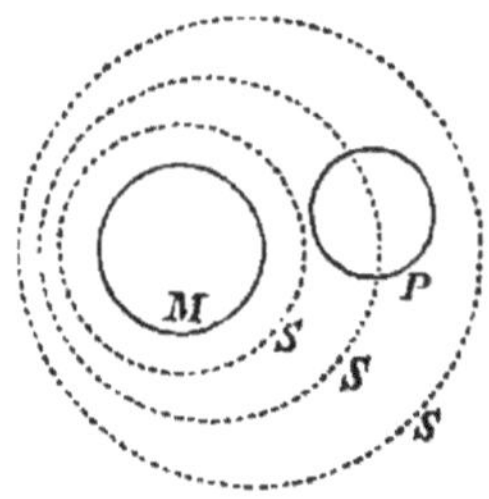

Fig. 26

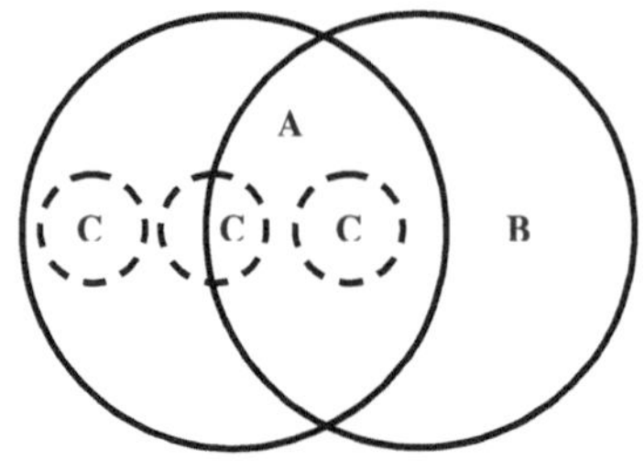

Fig. 27

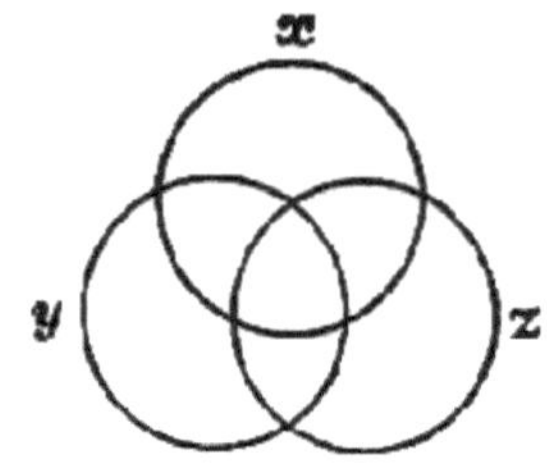

Fig. 28

This state of affairs never occur using Venn diagrams. Indeed, Venn uses from the beginning a single primary diagram which could be used to represent all possible relations between the classes in the argument. In order represent sets of propositions involving 3 terms (x, y, z), Venn uses 3 circles which interest in such a way as to divide the universe into eight compartments, corresponding to the 8 combinations: *xyz*, *x y not-z*, *x not-y z*, *x not-y not-z*, *not-x y z*, *not-x y not-z*, *not-x not-y z*, and *not-x not-y not-z* [Fig. 28]. Note that the class *not-x not-y not-z* is represented by the space outside the 3 circles. Then, one has to add marks, as we explained, to indicate occupation or emptiness of the compartments.

It is important to insist on the difference in the principles of representation between Euler and Venn to understand the change that occurred here. Euler uses the circles to divide the space into subdivisions that are assumed to exist and which are topologically related in the same way as the classes they represent do. As such, Euler's circles (or more strictly speaking, the space within the circles) represent directly the classes (understood as extensions of the terms). In Venn diagrams however, none of the subdivisions is assumed to exist. Strictly speaking, Venn does not represent the classes at all, but rather compartments, which when marked, tells whether the corresponding class is empty or occupied, as he clearly explains here:

> The best way of introducing this question will be to enquire a little more strictly whether it is really classes that we thus represent, or merely compartments into which classes may be put? [...] The most accurate answer is that our diagrammatic subdivisions, or for that matter our symbols generally, stand for compartments and not for classes. We may doubtless regard them as representing the latter, but if we do so we should never fail to keep in mind the proviso, "if there be such things in existence." And when this condition is insisted upon, it seems as if we expressed our meaning best by saying that what our symbols stand for are compartments which may or may not happen to be occupied. [Venn, 1894, p. 119-120]

To appraise the significance of this move made by Venn, it is important to understand how his method of constructing logic diagrams, by dividing *dichotomically* the universe into compartments, reflects the development of symbolic logic from the mid-nineteenth century onwards. Of course, Division and dichotomy were long well-known in logic (as shown by the porphyry tree), but they got particular attention from nineteenth-century logicians. Division was one of the two formal methods used by logicians to form classes, the other being classification. In the former method, on divides an existing class into two or more (new) classes, while in the latter, one puts things together in a group such as to form a class.[11] Keynes provides an interesting survey on the "doctrine of division", as he called it, in an appendix to his logical treatise [Keynes, 1906, pp. 441-449]. He identifies two main features that are commonly attributed to division: first, the sub-classes obtained by division should be exclusive each other (that means that they should not contain common individuals); and second the sub-classes issued from a division should be exhaustive (that means that all the individuals of the divided class should be in either sub-class).

As such, dichotomy is the simplest case of logical division, as one divides an existing class X into just two sub-classes XA and X *not-A*. Nineteenth-century symbolist logicians gave a prominent role to dichotomy after George Boole expressed it algebraically as: "$x^2 = x$", and recognized it as "the fundamental equation of thought" [Boole, 1854, p. 50-51]. Venn clearly says that dichotomy is at the foundation of his compartmental view of logic:

> At the basis of our Symbolic Logic, however represented, whether by words by letters or by diagrams, we shall always find the same state of things. What we ultimately have to do is to break up the entire field before us into a definite number of classes or compartments which are mutually exclusive and collectively exhaustive. The nature of this process of subdivision will have to be more fully explained in a future chapter, so that it will suffice to remark here that nothing more is demanded than a generalization of a very familiar logical process, viz. that of dichotomy. [Venn, 1894, p. 111]

One great advantage of dichotomy, which made William S. Jevons consider that dichotomy-based system was the "inevitable and only system which is logically perfect" [Jevons, 1883, p. 694], is that all other divisions can be reduced to successive dichotomy divisions where some of the sub-classes are empty. Take for instance the following division: a class X is divided into 3 sub-classes A, B, and C which are mutually exclusive and which together exhaust the class X, as shown in [Fig. 29].

[11] E. C. Constance Jones writes that: "It may be said that Division and Classification are the same thing looked at from different points of view; any table presenting a Division presents also a Classification. A Division starts with unity, and differentiates it, a Classification starts with multiplicity, and reduce it to unity, or, at least, to order" [Jones, 1905, p. 101]. There are however few essential differences. For instance, the logical Universe can be formed only by classification, as the universe forms no part of a class other than itself, and from which it would have been formed by division.

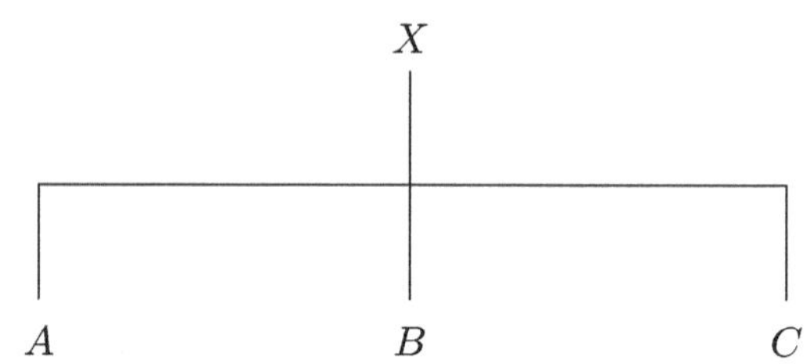

Fig. 29

This same division can be reached dichotomically by dividing first X into A and *not-A*, then dividing each of these sub-classes into two parts: one where B is affirmed and the other where B is denied, and so on, until one obtains the complete division shown on [Fig. 30]. This figure shows all the possible subdivisions obtained by combining the classes A, B, and C. Now, in order to get the desired division, one has to mark the non-existing branches (i. e. branches 1, 2, 3, 5, and 8), by equating them to 0.

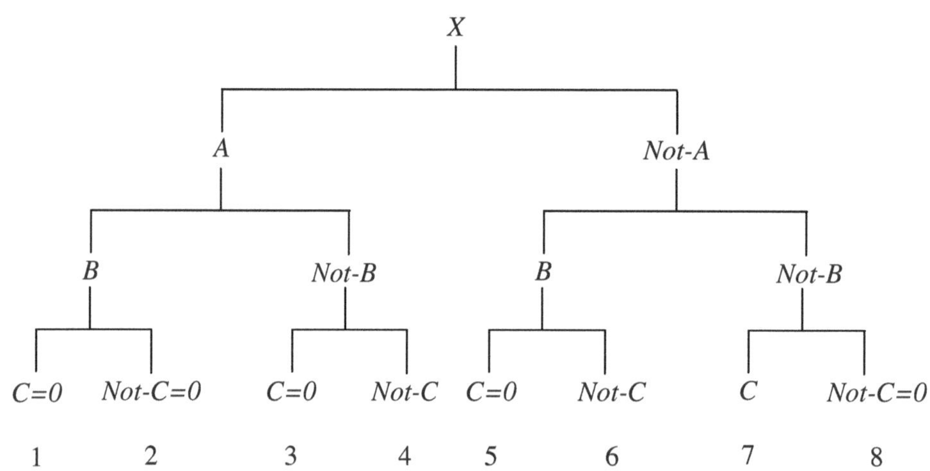

Fig. 30

It is easy to see the difference between these two methods of division and how they perfectly reflect the difference in the methods of representation between Euler and Venn. The first method, in [Fig. 29], shows the actual division of class X, and as such represents, as Euler does, the actual relations between the logical classes A, B, and C. The second method, as Venn does, lists rather all possible combinations and then indicates the status of the classes. Just like Venn's primary diagram, [Fig. 30] would represent no information if the non-existing branches were not indicated. Note that Boole proceeds the same way with his symbolic notation. For instance, to represent the proposition "No x is y", Boole makes first the classes x and y intersect and then indicates that the intersecting class (xy) is empty by equating it to 0. Thus, he obtains the notation: "$xy = 0$". Venn was a great admirer of Boole, as he adopted his equational notation. The move made by Venn in the way he represents information with his diagrams reflects the line carried out by symbolic logic in the nineteenth century by working on formal processes without paying attention to actual existence.

3.2 Linear diagrams

As far, we discussed the development and the principles of representation used in spatial diagrams, that is schemes that use spaces to represent logical propositions. We have seen how one distinguishes two major different methods of representation, one being used by Euler and the other by Venn. Of course, this distinction should not be understood strictly as intermediary approaches combining both methods might exist, as is the case of the dotted-line method. Still, it is important to understand the difference between the two methods and how the move from one to another changes substantially the status and the use of the diagrams. As explained previously, our appeal to spatial diagrams to illustrate our purpose was made for pure convenience, given that those diagrams are the most popular among logicians and in logic textbooks. It is however important to keep in mind that other diagrammatic schemes, such as linear and tabular diagrams, have been and are still used in logic. In the following section, we will introduce some of those schemes and discuss their specific features. Note that our aim is not as much to give a historical account of their development as it is to show how the same modes of representation (Euler-type *vs* Venn-type) are also to be found in these schemes.

Linear diagrams seem to have been as old as spatial diagrams. Both methods were known and used by Leibniz, though he didn't publish any in his lifetime. Couturat gave in 1901 an account of these diagrams [Couturat, 1901], based on Leibniz's manuscripts which he reprinted shortly later [Couturat, 1903, p. 311-312]. In these papers, Leibniz made different uses of the diagrams and it is difficult to give one single and regular reading. However, the principles of representation are similar to those used later by Euler: a line represents a class, and the spatial relations between the lines represent the logical relations between the classes. This is not surprising as Leibniz already knew and used spatial diagrams in the same way Euler was going to do few decades later. For instance, in order to represent the proposition "All S are P" with lines, one has just to draw two horizontal straight lines corresponding to S and P respectively in such a way as line S is strictly above line P, as shown by the vertical dotted lines[12] which delimit the segment under consideration [Fig. 31].

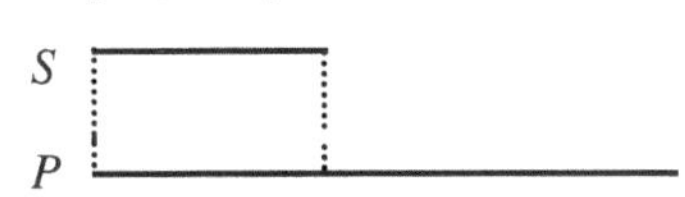

Fig. 31

[12]These vertical dotted lines should not be confused with the dotted lines which have been discussed in the section on spatial diagrams and which will be mentioned again later on in this section.

Similarly, one represents E and I propositions, using Leibniz linear diagrams, as [Fig. 32] and [Fig. 33] respectively.

Fig. 32

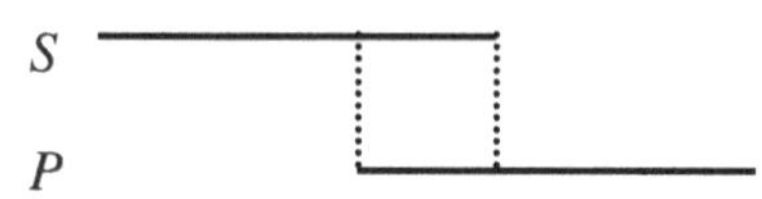

Fig. 33

As is the case with Euler's circles, Leibniz's lines are used to represent the actual relation between the classes. As such, they are subject to the same criticism: one often needs more than one diagram to represent all the information in one proposition. For instance, to represent satisfactorily the proposition "All S are P", one needs two diagrams depending on whether S coincides with P [Fig. 34] or S is strictly included in P [Fig. 35].[13]

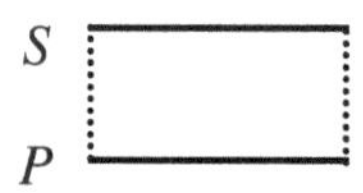

Fig. 34

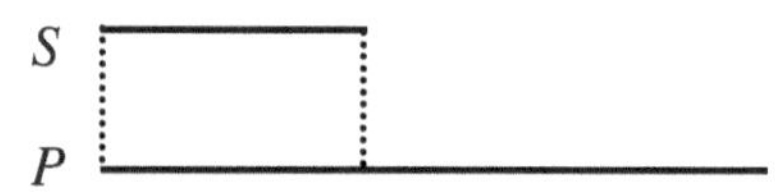

Fig. 35

One way to avoid this limit is to combine the diagrams with dotted lines in the same way as we did with spatial diagrams in order to represent uncertainty. Leibniz himself appealed (at times) to this technique [Couturat, 1901, p. 30-31]. Leibniz's work being unpublished at the time however, it seems that it is through Johann H. Lambert, in his *Neues Organon*, that this device happened to be known and commonly used by nineteenth-century logicians who resorted to linear diagrams [Lambert 1764]. Using Lambert's dotted-line method, one represents the A proposition above as shown in [Fig. 36].

Fig. 36

The reader, who is familiar with the dotted method used in spatial diagrams as we explained it earlier, will easily understand here that the continuous segments stands for the case where the two classes S and P are identical, while the dotted-lines save the possibility of having S strictly included in P. It is obvious that there are many different ways of introducing the dotted lines, even for the very same proposition, depending on whether the continuous segments are in the centre, on the right or on the left of the diagram. Even when the continuous segment is

[13]Other examples, concerning proposition A, I and O, will be found in [Keynes 1906, p. 164].

fixed (say on the left), Keynes rightly points that it is still possible to exhibit the uncertainty in two different ways, depending on whether the dotted-line is drawn with the predicate as shown in [Fig. 37] or with the subject as shown in [Fig. 38]. Hence, both figures represent the proposition "All S are P" [Keynes, 1906, p. 165].[14]

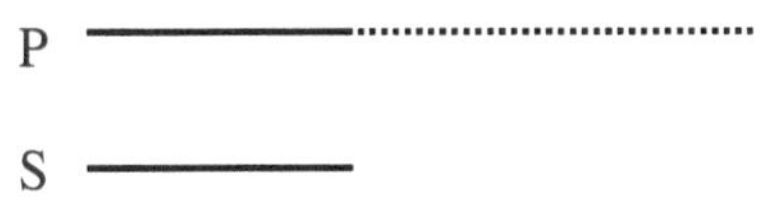

Fig. 37

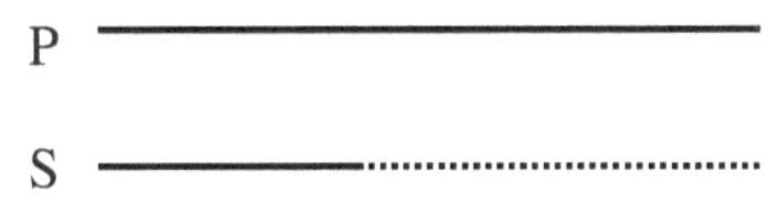

Fig. 38

Both Leibniz and Lambert appealed to an Euler-type method of representation, in the sense that they aimed at representing the actual relations between the terms. Though most logicians who used linear diagrams worked on this line, we must mention here one notable exception. Indeed, the logician James Welton introduced (and thoroughly used) in 1891 a new kind of linear diagrams worked out thanks to the Venn-type method of representation [Welton 1891, p. 252-255]. For this purpose, Welton proceeds the same way as Venn did with his spatial diagrams, using a line instead of space and segments instead of compartments. In Welton's method, one has to divide a line into 4 segments corresponding to the combinations of the terms S and P. Then, in order to represent propositions, one has to mark the segments to indicate the emptiness or occupation of the corresponding sub-classes. Welton provides no illustration of his primary diagram — that is the diagram before any proposition is represented on. However, from his discussion, we suggest that such a primary diagram must have been as shown in [Fig. 39]. Indeed, Welton considered that doubtful classes (that is which we do not know whether they exist or not) should be represented by dotted segments. As this must always be the case before we are given a proposition to represent, it follows that this must be the status of all segments in the primary diagram.

Fig. 39

From Welton's use of dotted lines to indicate uncertainty, it is easy to guess what other conventions he used: one has simply to make the segment continuous

[14]Interestingly, in a way that reminds us the difficulty that Euler and Venn faced in order to represent particular propositions with spatial diagrams, those propositions caused much trouble to the logicians who appealed to linear diagrams too. We will not discuss the issue in this chapter. However, the reader is invited to consult the writings of Venn [1894, p. 518-519], Keynes [1906, p. 133-136], and Peirce [1933, p. 297] to see how these authors failed to make sense of Lambert's diagram for particular propositions. A convincing solution is provided in [Shin, 1994, p. 36-38].

to signify occupation and to erase entirely the segment to signify emptiness. Hence, in order to represent the proposition "All S are P", one has just to erase the S not-P segment and to make continuous the segment S P (in accordance with Welton's logical theory). The other segments remain dotted, as shown in [Fig. 40].

Fig. 40

Similarly, [Fig. 41] represents the proposition "No S is P" (the segment S *not-P* is considered occupied, because it's emptiness would involve that the class S is empty, a situation that was not admitted in Welton's logical theory).

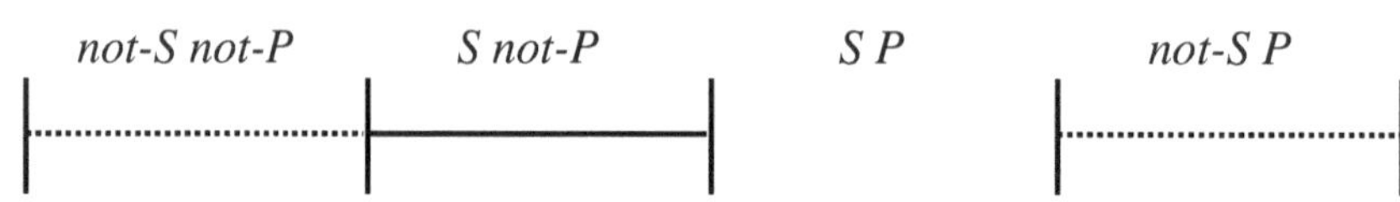

Fig. 41

In this section, we have seen how linear diagrams have been constructed and used by several logicians since the eighteenth-century. Though less popular than spatial diagrams, linear diagrams were certainly well-known, and some logicians did even prefer them to spatial diagrams [Keynes, 1906, p. 165]. Also, linear diagrams[15] are quite distinguishable from spatial diagrams, even if they appeal mostly to the same principles of representation. Tabular diagrams are different in the sense that they are also spatial. Remember that the use of circles was mostly for convenience and that one might very well have used squares, triangles, rectangles, or any other closed curve, as long as the representation principles are respected. What really makes tabular diagrams interesting is that their shape almost naturally raises several issues (such as the representation of the universe of discourse, the status of negative terms, etc.) of importance to the logician. We will discuss these issues in the next section.

3.3 *Tabular diagrams*

The notion of 'universe of discourse' has been introduced in logic by Augustus De Morgan in 1846. Rejecting the indefinite character of negative terms, De Morgan introduced this notion to define the extension of *not-x* as simply the complement of x to fulfil the logical universe:

[15]In the forms discussed as far. We will see later how Alexander Macfarlane's "logical spectrum" offers an intermediary diagram between linear and tabular-spatial methods.

> Writers on logic, it is true, do not find elbow-room enough in anything less than the whole universe of possible conceptions: but the universe of a particular assertion or argument may be limited in any matter expressed or understood. And this without limitation or alteration of any one rule of logic [...] By not dwelling upon this power of making what we may properly (inventing a new technical name) call the universe of a proposition, or of a name, matter of express definition, all rules remaining the same, writers on logic deprive themselves of much useful illustration. [De Morgan, 1966, p. 2]

This notion has not been used by Boole in his 1847 treatise on *The Mathematical Analysis of Logic* [Boole, 1847], although he made use of the notion of "Universe" which he represented by the symbol "1". In his *Laws of Thought* (1854) however, Boole used De Morgan's notion of a limited universe and coined the expression 'Universe of discourse' [Boole, 1854, p. 42]. Most of De Morgan's and Boole's followers made a thorough use of this notion. Venn is an exception however as, although he recognized the existence of a limited universe within which every argument should be understood, he considered the issue as "extra-logical" and argued that it was "entirely a question of the application of our formulae, not of their symbolic statement" [Venn, 1894, p. 250]. Thus, in accordance with his theory of logic, Venn chose not to represent the universe of discourse in his diagrams. One graphical inconvenience of this choice is the difficulty to shade the outside region of the diagram in order to express its emptiness. Carroll criticized Venn on this ground:

> It will be seen that, of the four Classes, whose peculiar Sets of Attributes are $xy, xy', x'y$, and $x'y'$, only three are here provided with closed Compartments, while the fourth is allowed the rest of the Infinite Plane to range about in!
>
> This arrangement would involve us in very serious trouble, if we ever attempted to represent No x' are y'. [Carroll, 1897, p. 175][16]

Willard Van Orman Quine considered Carroll's criticism unfair as "he knew full well that we easily shade enough of it to get on with" [Quine, 1977, p. 1019]. However, that was not Venn's approach. Indeed, when this situation occurred in one of his examples, Venn says simply that he did not "trouble" to shade the outside of the diagram [Venn, 1881, p. 281; 1894, p. 352]. In another similar case, Venn avoids the difficulty by explaining that: "The outside [...] should also be shaded, to make it complete" [Venn, 1881, p. 271]. It is obvious here that Venn felt uneasy with representing the emptiness of the outer region in his diagram. In fact, Venn's difficulty results from a pure "technical" problem which could be easily avoided by using a different convention to indicate emptiness, as the one used by Peirce where the problem disappears immediately, without delimitating the

[16] x' stands for *not-x*, and y' stands for *not-y*.

universe. So, even if related each other, the issue of representing the emptiness of the outer region should not be confused with the issue of representing the universe.

Venn's decision not to represent the universe of discourse might first be explained by his logical theory where he already expressed his dissatisfaction with that notion which he considered as extra-logical.[17] It seems however that what bothered Venn was not as much the delimitation of the universe in his diagrams as the extent of those limits:

> I draw a circle to represent X; then what is outside of that circle represents not-X, but the limits of that outside are whatever I choose to consider them. They may cover the whole sheet of paper, or they may be contracted definitely by drawing another circle to stand as the limit of the Universe; or, better still, we may merely say that the limits of the Universe are somewhere outside the figure but that there is not the slightest ground of principle or convenience to induce us to indicate them. [Venn, 1894, p. 252-253]

As we have seen, Venn chose the last option and preferred not to indicate in his diagrams the limits of the Universe. It is difficult not to dispute Venn's argument. Indeed, according to him, quantitative considerations as to the size of the compartments and extent of the terms should be avoided [Venn 1894, p. 525-527]. Consequently, if one considers the delimitation of the universe suitable, any limit would have fitted, and it would have been much more convenient to indicate it to avoid the difficulties with shading the outer region, as we explained above.[18]

Although frequently attributed to Carroll (whose diagrams will be discussed later on in this chapter), in his *Game of Logic* [Carroll, 1886], the graphical representation of the universe is in fact prior to the publication of Venn diagrams themselves. Indeed, in 1879, Alexander Macfarlane included in his *Algebra of Logic* Euler diagrams with a closed square around to represent the universe as shown in [Fig. 42], where x is the intersection of y and z [Macfarlane, 1879, p. 23]. Similar diagrams were included in Macfarlane's review of the first edition of Venn's *Symbolic Logic*. Macfarlane reproaches there explicitly Venn for not having represented the Universe:

> A difficulty arises in the application of this process of shading-out in the case of the contrary class [...] owing to the fact that Mr. Venn does not represent the whole class of things considered by an enclosure such as the square [...]. Not only does the shading-out process require this, but the logical-diagram machine insists upon it. [Macfarlane, 1881, p. 62]

---

[17]This has been noticed by Alexander Macfarlane in his review of Venn's *Symbolic Logic* [Macfarlane, 1881, p. 62-63].

[18]There is one instance where Venn reproduces diagrams with the outer region represented by a limited "small irregular compartment". However, those figures do not actually represent logic diagrams, but rather the plans of a "logical-diagram machine" [Venn, 1894, p. 135-137].

Finally, few years later, Macfarlane reproduces Venn diagrams (not Euler diagrams anymore) with a rectangle around enclosing the Universe as shown in [Fig. 43] [Macfarlane 1885, p. 286].

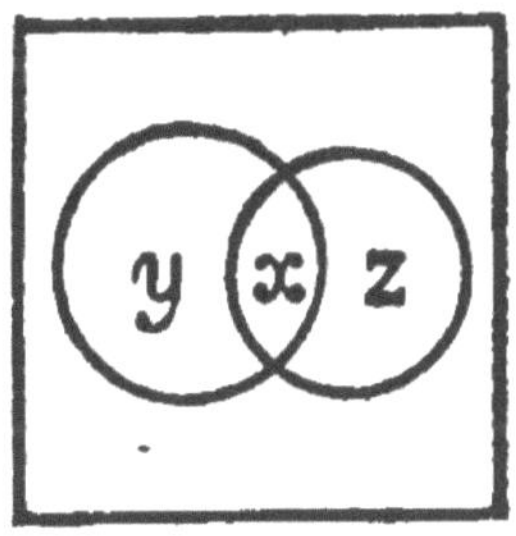

Fig. 42

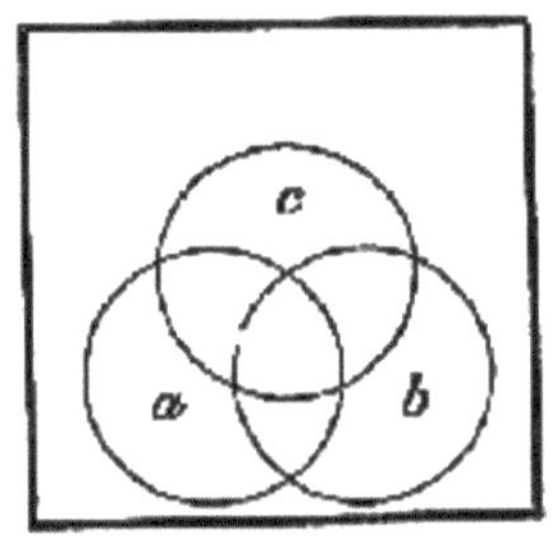

Fig. 43

We saw above instances where the universe encloses, as if it has been added around, a Venn diagram as is commonly done in modern logical manuals. However, another possible approach would be to represent the universe first and then divide it in such a way as to engender the different compartments for the terms involved in the argument. This is precisely the principles on which are based tabular diagrams. Such diagrams existed prior to Venn's diagrams, and can also be found in Macfarlane's 1879 *Algebra of Logic*, as shown in [Fig. 44]. This diagram depicts a class x within the Universe, and shows a class y included in the class x (as it occupies the lower part of the space devoted to x) [Macfarlane, 1879, p. 42]. This diagram shows the actual relation between the classes x and y, and should thus be considered as an Euler-type diagram.

However, many tabular diagrams will be published later in reaction to the introduction of Venn diagrams. For instance, Allan Marquand, one of Peirce's students, introduced in 1881 a new kind of diagrams where the "logical universe" is represented (for convenience) by a square which is then divided dichotomically depending on the number of terms involved in the argument. Thus, for 2 terms A and B, one has to divide the square into four compartments AB, A *not-B*, *not-A* B, and *not-A not-B* as shown in [Fig. 45] (where a stands for *not-A*, and b for *not-B*) [Marquand 1881]. This figure does not represent the actual relations of the classes A and B, but rather a framework showing the four combinations of the terms, on which one needs to add distinctive marks to indicate emptiness and occupation, and so to represent propositions. Thus, this diagram is a Venn-type diagram. From the above, in the same way as it has been shown with spatial and linear diagrams, we note that tabular diagrams do also appeal to the two modes of representation, known as the Euler-type and Venn-type methods. The former method represents actual information while the latter depicts uncertain knowledge, keeping the possibility to represent actual information with the aid of additional marks however.

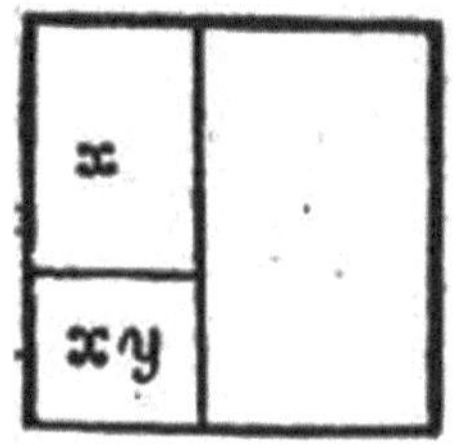

Fig. 44

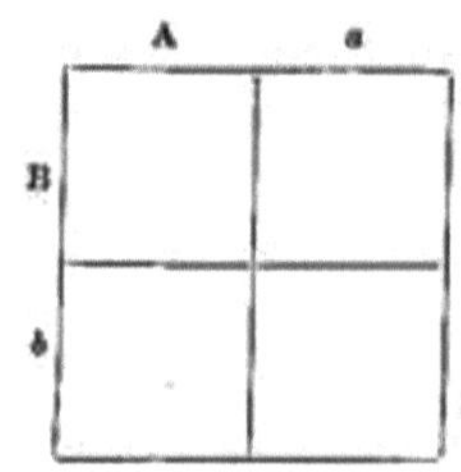

Fig. 45

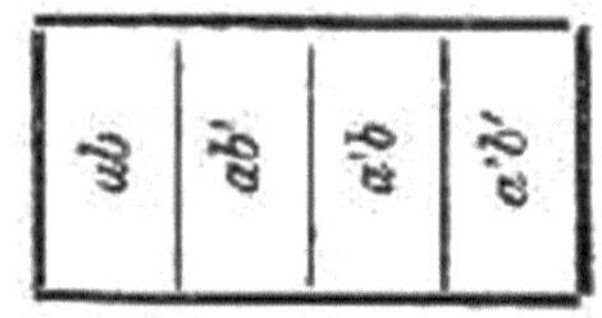

Fig. 46

Interestingly, Marquand made the same reproach to Venn (as Macfarlane did and as Carroll will do), concerning the inappropriateness of his diagrams when the outer region is empty:

> Attention may be drawn to the fact that these diagrams differ from those suggested by Mr. Venn in having a compartment for the *absence of all* the characters or objects [...] This compartment may need to be shaded out, and hence should be indicated on a complete logical diagram. [Marquand, 1881, p. 269]

Another scheme based on the same principle of representation as Marquand's, but with a different method of division, has been published in 1885 by Macfarlane, again. He represented the universe by a rectangular strip which he divides always on the same side as shown in [Fig. 46] [Macfarlane, 1885]. This scheme, which Macfarlane named "the logical spectrum", is in a sense an intermediary representation, half-way between linear and tabular diagrams.

Both Marquand's square and Macfarlane's spectrum are Venn-type diagrams, because they require the addition of marks to represent propositions on those primary diagrams. However, they differ from Venn's diagrams in many aspects: they represent a limited universe, they use rectangular lines, and they are more easily constructed when the number of terms involved in the argument becomes higher (this issue will be discussed in the next section). There is one additional instance where tabular diagrams are more suitable. Indeed such constructions allow more easily the introduction of quantitative considerations (if wanted or needed),[19] as has been admitted by Venn himself [Venn, 1894, p. 526]. For instance, tabular diagrams allow a symmetric division of the universe by devoting equal space to the opposite sub-classes A and *not-A*, which thus would be regarded on the same footing. This issue is important when one works within post-Boolean logic. Indeed, as dichotomy is a purely formal process, the distinction between positive and negative terms is purely conventional. Carroll used a syllogism with negative terms, where the outer region was to be found occupied in the conclusion, in order to show the superiority of his diagrams over Euler and Venn schemes [Carroll, 1897, p. 179-183].

[19] Although Venn insisted on excluding matters of size and extent, it is obvious that he tried himself to construct symmetric diagrams with like-size classes in order to get diagrams that "look" better.

We already saw how Venn's neglect of the outer region (which corresponds to the intersection of all negative terms) led to him to difficulties in representing its emptiness. However, Carroll's criticism was unfair against Euler's diagrams which were designed for syllogistic reasoning where negative terms were dismissed and the outer region was assumed to exist anyway. Keynes attempted to adapt Euler's scheme in order to handle negative terms [Keynes, 1906, p. 170-174]. Hence, he identified and represented diagrammatically 7 possible relations between classes *S*, *not-S*, *P*, and *not-P*:

> In Euler's diagrams, as ordinarily given, there is no explicit recognition of *not-S* and *not-P* ; but it is of course understood that whatever part of the universe lies outside *S* is *not-S*, and similarly for *P*, and it may be thought that no further account of negative terms need be taken. Further consideration, however, will shew that this is not the case; and, assuming that *S*, *not-S*, *P*, *not-P* all represent existing classes, we shall find that *seven*, not five, determinate class relations between them are possible [Keynes, 1906, p. 170].

indexGergonne relation

Keynes relations (as one might call them) have been obtained simply by substituting for each Gergonne relation, two sub-cases: one in which the outer region is empty and one where it is not.[20] For instance, the Gergonne case where classes *S* and *P* are completely disjoint engender two sub-cases, depending on whether there is "something" outside [Fig. 47] or nothing [Fig. 48].

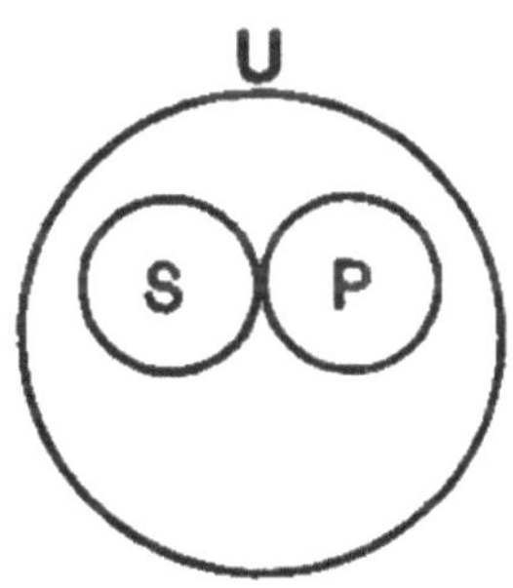

Fig. 47

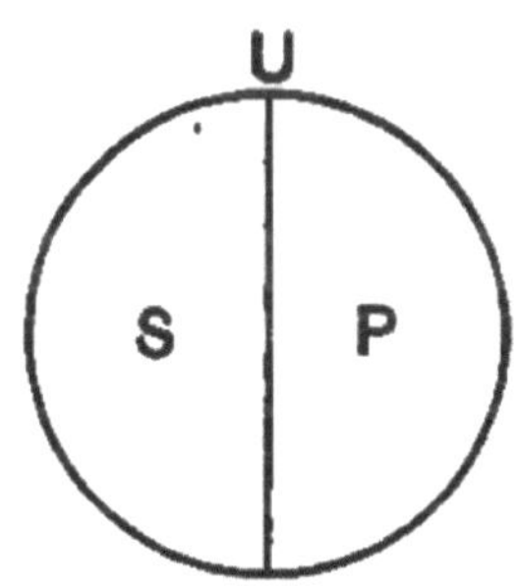

Fig. 48

In both diagrams, *U* represents the Universe. Interestingly, Keynes represented his relations also using linear and tabular diagrams with a limited Universe [Keynes, 1906, p. 173-176]. As curious as Keynes diagrams might look, they are simply necessary if one wants to work rigorously logical problems involving negative terms with the use of Euler-type diagrams.

[20]Among the 10 relations obtained by distinguishing two sub-cases for each of the 5 Gergonne relations, Keynes excluded 3 cases where either *not-S* or *not-P* (or both) was empty, and so kept just 7 relations. Keynes exposed first these relations in the third edition of his *Formal Logic* [Keynes, 1894, p. 140-146].

4 MANIPULATING INFORMATION WITH DIAGRAMS

In the previous sections, we discussed lengthily how various logic diagrams have been invented and introduced in the realm of logic, and how they were used to represent information. However, as we explained in the introduction of this chapter, our aim was to discuss analytical diagrams. We meant diagrams that were used not merely for illustration, but were rather needed in order to solve logical problems faced by logicians and logic students. Until the nineteenth-century, syllogisms were the classical form of inferences as one finds them in traditional logic textbooks and treatises. The following passage from Euler's *Letters to a German Princess* gives a fair idea about the importance devoted to syllogisms in traditional logic:

> Every syllogism, then, consists of three propositions; the two first of which are called the *premises* and the third the *conclusion*. Now, the advantage of the all these [valid] forms to direct our reasoning is this, that if the premises are both true, the conclusion infallibly is so.
>
> This is likewise the only method of discovering unknown truths. Every truth must always be the conclusion of a syllogism, whose premises are indubitably true. [Euler, 1833, p. 350]

It must not be inferred however that syllogisms remained as they were first introduced by Aristotle. Indeed, syllogistic certainly new several improvements, mostly extensions in order to handle new kinds of arguments. Still, the syllogism remained at the centre of logical theory. Even when its status declined with the growth of symbolic logic [Van Evra, 2000], it remained an essential passage for every logic student. Hence, quite naturally, most of the diagrammatic systems discussed as far were mainly conceived to solve syllogistic problems, of the kind that we will discuss in the next section.

4.1 Working syllogisms

In syllogistic, the logic problem that one has commonly to face, is to be offered a syllogistic form (two premises and a conclusion) and to decide whether it is valid or not (whether the conclusion follows from the premises). The usual method is to compare the given form with a list of valid forms that one has already learned by rote, and check whether it coincides with one of them. The number of such valid forms depends on some issues logicians disagree about (such as the existential import of propositions), but most traditional logicians listed nineteen valid forms, each having a specific name such as *Barbara*, *Festino*, *Datisi*, etc.

Euler lists those nineteenth valid forms, as most logicians do. However, his use of diagrams enables him to check the validity of any given form, without appealing to the list. Indeed, the diagrams can be worked independently and effectively for solving syllogisms, as they are in accordance with the two general principles which Euler considered to be the foundation of syllogistic reasoning:

> The foundation of all these [valid] forms is reduced to two principles, respecting the nature of *containing* and *contained.*
>
> I. *Whatever is in the thing contained, must likewise be in the thing containing.*
>
> II. *Whatever is out of the containing, must likewise be out of the contained.* [Euler, 1833, p. 350].

In order to check whether a given syllogistic form (two premises and a conclusion) is valid, one has just to represent diagrammatically the information contained in the two premises, and then to check whether the information contained in the given conclusion does also appear in the diagram. Importantly, it is not necessary that the given conclusion contains all the information represented by the diagram. Indeed, the conclusion could just be incomplete, which does not mean that it is incorrect.

Suppose that we are given the two propositions "All S are M" and "All M are P" as premises of a syllogistic form whose conclusion is "All S are P". Using Euler's method, it is simple to represent the premises on a single diagram, as shown in [Fig. 49]. It is also easy to observe that the conclusion is represented by the diagram because the circle S is inside the circle P. Hence, the given syllogistic form is valid.

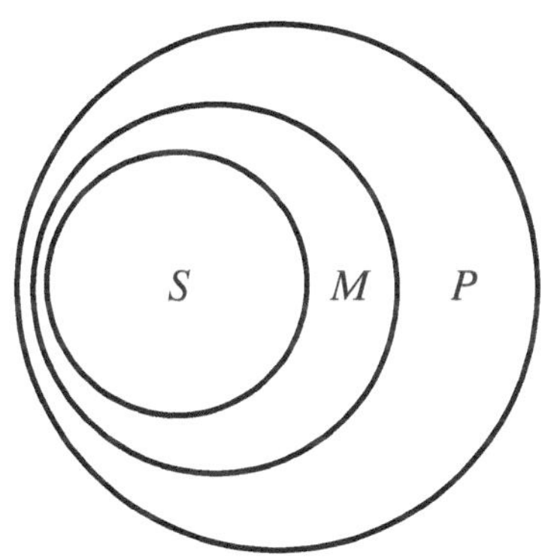

Fig. 49

Euler's method looks simple and intuitive. Of course, the example would have been more complex if it involved particular propositions. Still, the method gives much visual aid as it shows the actual relationship between the two terms of the conclusion, and so, it allows an easy checking of whether the given conclusion is compatible with the diagram. However, as we saw previously, a rigorous use of Euler's circles would involve us in much trouble. Indeed, a close look shows that there are in fact three possible cases, other than the one shown in [Fig. 49], that should be considered in order to handle accurately the problem [Fig. 50].[21]

[21] For a detailed development, see [Moktefi, 2010].

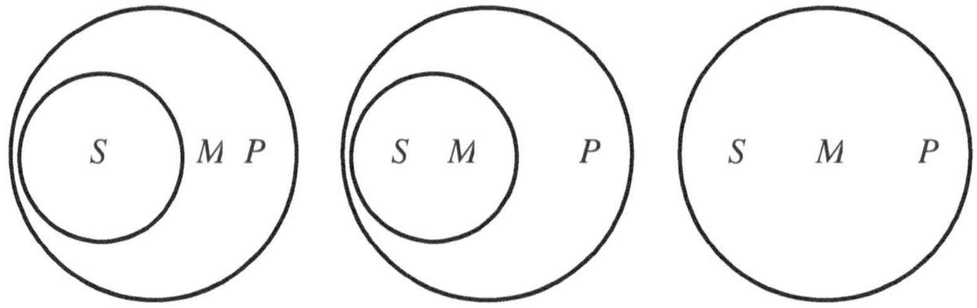

(a) (b) (c)

Fig. 50

So, a rigorous working of Euler's diagram would require listing all the possible combinations of the terms involved in the given premises, and checking whether the given conclusion is in accordance with all those combinations. If just one of the diagrams obtained contradicts the given conclusion, that would mean that the given syllogistic form is not valid. In our previous example, it is simple to ensure that in all four cases, we always have "All S are P", which means that the reasoning is valid. The task would have been however much more complex if we were given particular propositions, as the number of combinations quickly increases, and finding the combinations itself could be much of a challenge.[22]

This criticism would not affect Venn, whose method requires one (and only one) diagram for a given number of terms. It must be said however that the logical problems that Venn designed his diagrams for are slightly different from the syllogistic problems described above. Of course Venn was also able to use his diagrams in order to check whether a give conclusion can be deduced from a set of premises. However, Venn mostly used his diagrams in order to find what conclusion would follow from a given set of premises.[23] In the following we will explain how Venn used his diagrams for problems involving 3 terms (as syllogisms do). Problems involving more terms will be discussed later. Just like Euler, Venn first represents diagrammatically the two given premises, and looks for the conclusion in the diagram he gets. The main difference is that, unlike Euler, Venn requires only one diagram for each problem. This advantage is however due to the difference in the way of representing the propositions, not to any difference in the method of solving problems itself.

Suppose we have been offered the two premises discussed in the previous example ("All S are M" and "All M are P") and were asked what conclusion they produce as to the relation between the terms S and P. In order to solve such a

[22] Though most Euler diagrams' users appealed (and continue to do so) to the direct method, several logicians used the rigorous method, even by admitting that it was "very complex" [Keynes, 1906, p. 344].

[23] This small change is interesting and tells how symbolic logic was influenced by mathematics. Indeed, in the "new logic", propositions were represented as equations (at least in Boole, Jevons and Venn systems) and solving logical problems requires solving systems of equations. The solution is not given, it is to be found.

problem with Venn diagrams, one has to use the 3-term diagram with 8 compartments formed by the combination of the circles S, M, and P. In order to represent the first premise "All S are M", one has to shade all the S compartments which are outside M[24] to express their emptiness. Similarly, in order to represent the second premise "All M are P", one has to shade all the M compartments which are outside P to express their emptiness. Hence, we obtain the diagram shown in [Fig. 51].[25] Now, one has to eliminate M, and tell what the relation between the terms S and P is. From the diagram one observes that the compartment "S *not-P*" is empty, i.e. there is no S that is *not-P*. Hence, the conclusion is "All S are P".

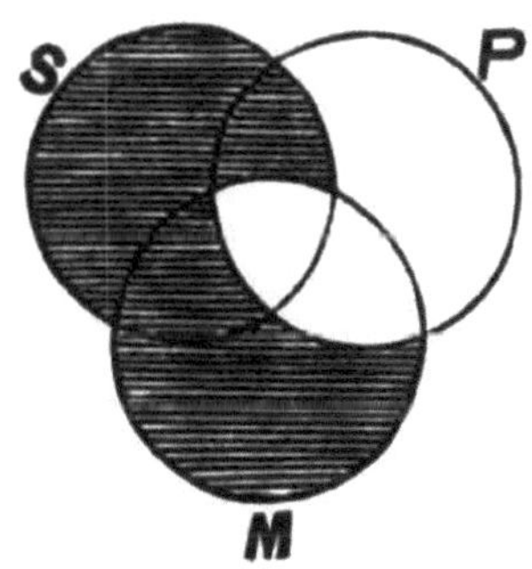

Fig. 51

Venn's advantage over Euler is the brevity and precision of his method, though part of the visual suggestiveness of Euler diagrams is lost here. However, one difficulty with Venn's scheme is that the extraction of the conclusion from the diagram is a practice that requires some training in order to be accomplished safely, especially if we were given premises involving particular propositions. This criticism has been expressed by Couturat as follows:

> This diagrammatic method has, however, serious inconveniences as a method for solving logical problems. It does not show how the data are exhibited by cancelling certain constituents, nor does it show how to combine the remaining constituents so as to obtain the consequences sought. In short, it serves only to exhibit one single step in the argument, namely the equation of the problem; it dispenses neither with the previous steps, *i.e.*, "throwing of the problem into an equation" and the transformation of the premises, nor with the subsequent steps, *i.e.*, the combinations that lead to the various consequences. Hence it is of very little use, inasmuch as the constituents can be represented by algebraic symbols quite as well as by plane regions, and are much easier to deal with in this form. [Couturat, 1914, p. 75]

[24]In Venn's logical theory, the proposition "All S are M" is equivalent to the proposition "No S is *not-M*".

[25]This figure is reproduced from [Keynes, 1887, p. 242].

Couturat is severe here, at least concerning the so-called "previous steps". Indeed, when one starts from concrete propositions, those steps occur in both symbolic and diagrammatic methods, and thus, there is no reason to blame diagrammatic methods apart. On the so-called "subsequent steps", Couturat's criticism is in way a variation of what was known to nineteenth century logicians, after Jevons, as the inverse problem [Keynes, 1906, p. 525-535]. In this case, the issue is that when one represents the given data (involving 3 terms) on a diagram, he has to find out what compartments are occupied or empty. However, when one looks for the conclusion, he proceeds the inverse way: he needs to establish from the emptiness or occupation of the compartments what data is available as to the relation between the 2 remaining terms (other than the eliminated term).

A solution to this problem can be found in Carroll's diagrammatic method which we are going to present in the following. Carroll's diagrams, first published in 1986, are Venn-type diagrams where the universe is represented with a square [Carroll, 1886]. It is not clear whether Carroll worked his diagrams as an improvement or modification of Venn's, or independently of him. Still, Carroll's scheme looks like a "mature" method summing up several improvements that have been introduced by his predecessors and contemporaries (Venn-type method of representation, limitation of the universe, symmetrical division, efficient representation of existential statements, etc.).[26] For 2 terms x and y, Carroll divides the square into 4 compartments, and obtains the so-called biliteral diagram [Fig. 52] (where x' stands for *not-x* and y' for *not-y*).[27] For 3 terms x, y, and m, Carroll adds a smaller square in order to get 8 compartments as shown by his triliteral diagram [Fig. 53].

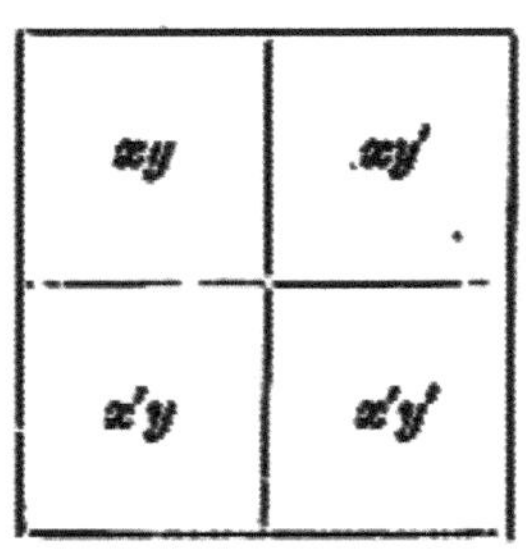

Fig. 52

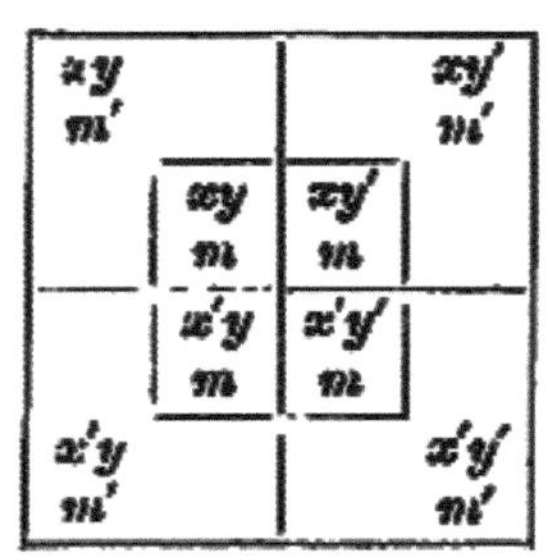

Fig. 53

In order to represent propositions, one has to add marks. A compartment is empty if it is marked with a '0' and is occupied if it is marked with a 'I'. For

[26]For a detailed discussion of Carroll's diagrams, see [Abeles, 2007] and [Moktefi, 2008].

[27]We substituted here the letters x, y, m for the letters P, M, Q, used previously, in order to fit better with Carroll's symbolism. In addition, Carroll and Venn have different conceptions on the existential import of propositions, which makes their diagrammatic methods of solving logic problems difficult to compare in the way we present them here. Thus, the aim of this section is simply to show how Carroll proceeds with his diagrams and how his method is (at least partly) not subject to the criticism made by Couturat.

instance, in order to represent the proposition "All x are y", one has to put a '0" on the x *not-y* compartment and a 'I' on the xy compartment [Fig. 54].[28] Finally, Carroll introduces one more device that will prove to be useful in his scheme. Suppose that one wants to represent the proposition "Some x are m " on a triliteral diagram. This means that either xym or $xy'm$ is occupied ("or" is understood here inclusively). To represent this uncertainty, Carroll puts the symbol 'I' (for occupation) on the boundary between those two compartments, as shown in [Fig. 55] [Carroll, 1897, p. 26].

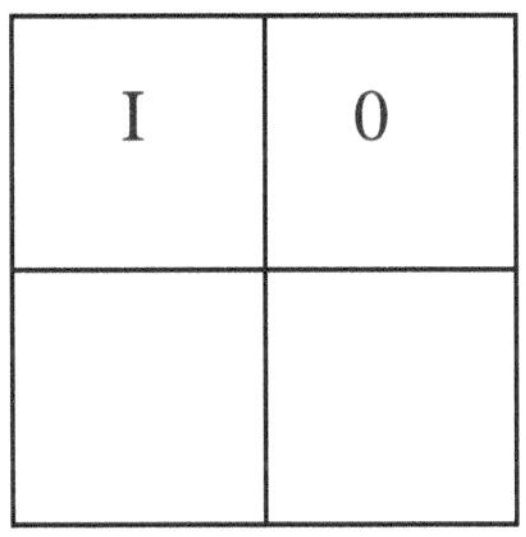

Fig. 54

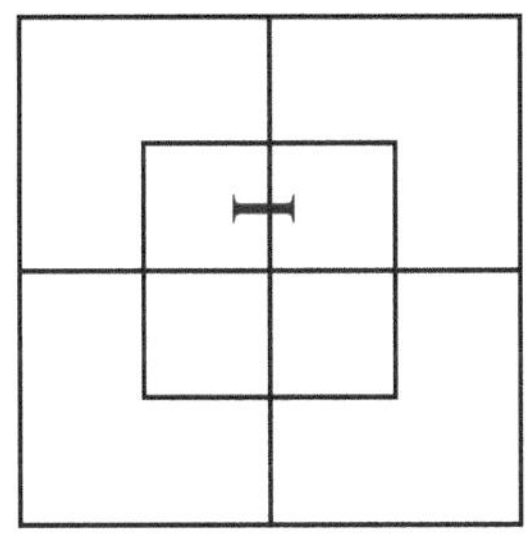

Fig. 55

In order to find the conclusion of a syllogism, Carroll first proceeds in the same way as Venn did. He represents the data expressed by the two premises on a triliteral diagram. Let the premises be: "All x are m" and "All m are y". Their representation is shown in [Fig. 56]. Given that the compartment $xy'm$ is empty and that either xym or $xy'm$ is occupied, the diagram can be arranged as to obtain a simplified triliteral diagram [Fig. 57]. Contrary to Venn who extracts the conclusion directly from the 3-term diagram, Carroll transfers the data shown by the triliteral diagram into a biliteral diagram involving only the 2 terms that should appear in the conclusion (and thus eliminating the middle term).

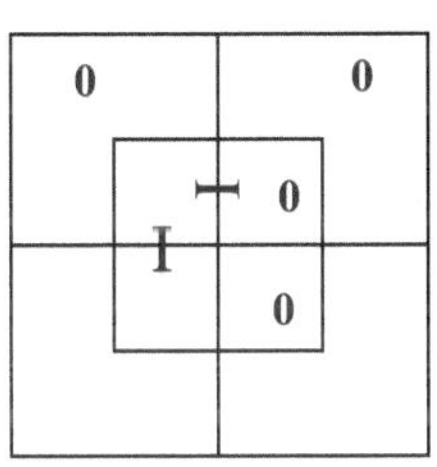

Fig. 56

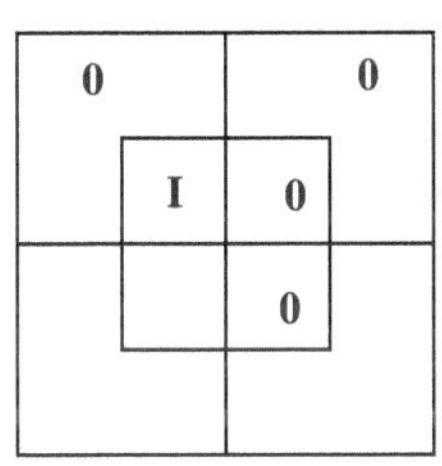

Fig. 57

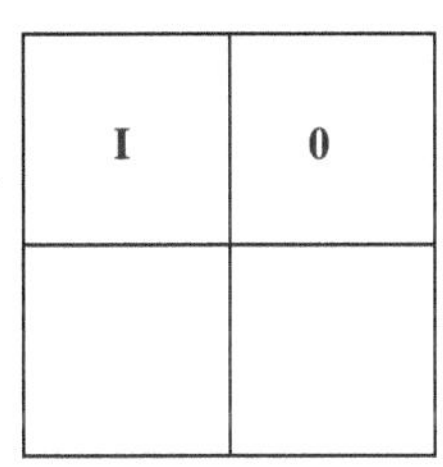

Fig. 58

This transfer is made following two rules defined by Carroll who applies them on the 4 quarters of the triliteral and biliteral diagrams [Carroll, 1897, p. 53]:

[28]Carroll considered that A propositions assert the existence of their subject and predicate. See [Abeles 2005] and [Moktefi 2008].

- Rule A: If the quarter of the triliteral diagram contains a 'I' in *either* Cell, then it is certainly occupied, and one may mark the corresponding quarter of the biliteral diagram with a "I" to indicate that it is occupied.
- Rule B: If the quarter of the triliteral diagram contains two "0"s, one in *each* cell, then it is certainly empty, and one may mark the corresponding quarter of the biliteral diagram with a "0" to indicate that it is empty.

Let's apply these rules to our previous syllogism who led us to [Fig. 57]. Now, we need to transfer the information from this triliteral diagram into a biliteral diagram. We observe that the North-West quarter of the triliteral diagram contains a 'I' in one of its cells, and thus we mark the N.-W. quarter of the biliteral diagram with a "I" (according to rule A). Also, we note that the North-East quarter of the triliteral diagram contains two "0"s, one in *each* cell, and so we mark the N. – E. quarter of the biliteral diagram with a "0" (according to rule B). Nothing can be said about the two other quarters of both diagrams. Consequently, we obtain the bilateral diagram [Fig. 58] which holds the conclusion: "All x are y".[29]

Carroll's method of transfer performs Couturat's requirement that the diagrammatic method should "show how to combine the remaining constituents so as to obtain the consequences sought". As such it provides an artefact that simplifies the working of syllogistic problems. This is particularly true when one uses the board and counters that Carroll joined to his logic books: the board contains the triliteral and the biliteral diagrams and the counters (red or grey) should be used to indicate occupation or emptiness. This way, the reader who wants to work the examples provided in Carroll's books does not need to draw diagrams for each problem, and has just to mark the compartments by putting or slipping the counters on the board.

4.2 Beyond syllogisms

Only few logicians before Venn have attempted to represent logic diagrams where more than three terms were involved. Firstly, most logicians at the time were working within the syllogistic tradition where there was hardly a need for such diagrams; and secondly Euler-type diagrams were anyway not designed for complex logical problems. The diagram [Fig. 59], from Francis Garden's *Outline of Logic* (1867), shows one instance where an (incomplete) Euler-type scheme has been used to represent an argument involving 5 terms, one of them (A) being overlapped by the four others (B, C, D, E) [Garden, 1867, p. 39].

[29]Note that the finding of the conclusion would have been much easier if one omits the existential import of A propositions, as is the case in Venn's use.

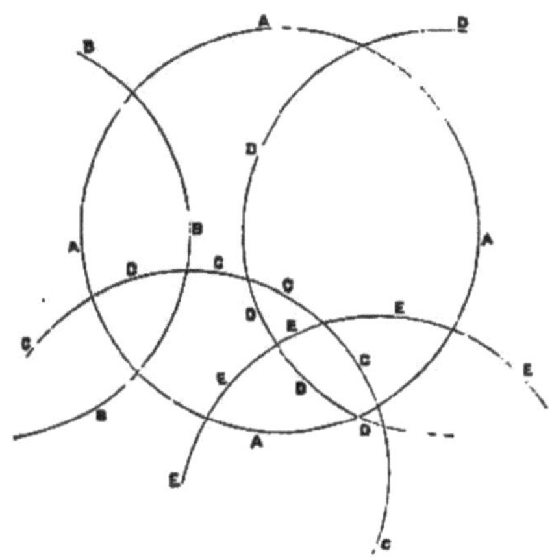

Fig. 59

However, with the development of symbolic logic, which Venn championed, diagrams for more than 3 terms were much desired. Indeed, Boole and his followers aimed at constructing a logical theory which would be more general than syllogistic logic, and as such would deal with problems containing any number of premises with any number of terms. The problem is then to find the logical relation between any two (or more) given terms that one wishes to keep in the conclusion. This is how Boole describes this problem, known to nineteenth-century logicians as the elimination problem:

> As the conclusion must express a relation among the whole or among a part of the elements involved in the premises, it is requisite that we should possess the means of eliminating those elements which we desire not to appear in the conclusion, and of determining the whole amount of relation implied by the premises among the elements which we wish to retain. Those elements which do not present themselves in the conclusion are, in the language of the common Logic, called middle terms; and the species of elimination exemplified in treatises on Logic consists in deducing from two propositions, containing a common element or middle term, a conclusion connecting the two remaining terms. But the problem of elimination, as contemplated in this work, possesses a much wider scope. It proposes not merely the elimination of one middle term from two propositions, but the elimination generally of middle terms from propositions, without regard to the number of either of them, or to the nature of their connexion [Boole, 1854, p. 8]

It is obvious that this new approach transforms the syllogism into a particular (the simplest) case of elimination. As such, it does not deny the interest or validity of syllogistic. However, it expands the scope of logic in such a way as to deal with new problems which could not be reduced into a series of syllogisms. This "generalization" of the common logic, as it has been called by Venn [1894, p. xxvii], asks for the invention of new methods (symbolic, diagrammatic and sometimes mechanical) to solve these new problems, and that is precisely what most Boolean logicians were doing in the second half of the nineteenth-century [Green, 1991].

Venn, who was one of Boole's most fervent followers, paid a careful attention to the problem of elimination and its solution using diagrammatic methods. Hence, he proposed new diagrams to deal with problems beyond syllogisms. For 4 terms, Venn abandoned the use of circles because it was not possible to make them intersect appropriately so as to produce the 16 required compartments. Consequently, Venn used ellipses and proposed the diagram shown in [Fig. 60] where the star indicates the compartment *not-x y z w* [Venn, 1894, p. 116]. For 5 terms however, Venn failed to combine 5 ellipses in the desired manner.[30] Thus, he decided to add the fifth term z with an annulus [Fig. 61], so that the inside and outside of the annulus forms the (discontinuous) class *not-z* [Venn, 1894, p. 117].

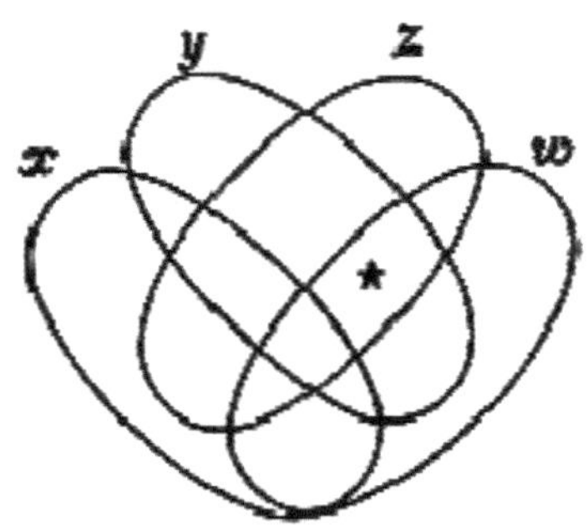

Fig. 60

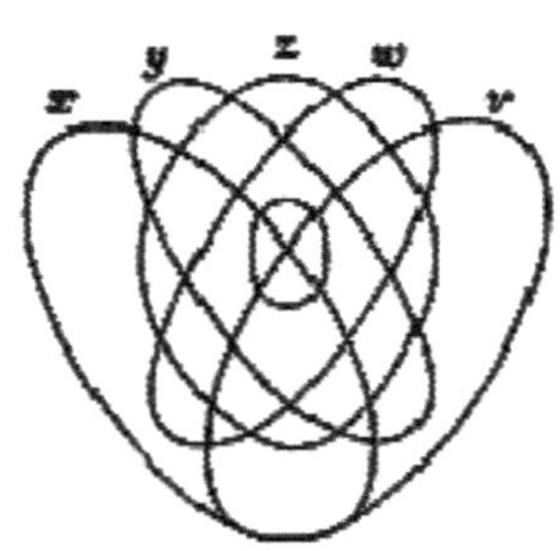

Fig. 61

For more than 5 terms, Venn did not believe in the worthiness of his diagrams, as they get more complex and as it was rare to face problems that require that number of terms. Still, he unconvincingly gives some indications on how to draw such diagrams without providing visual illustrations. For instance, in order to represent a 6-term diagram, Venn suggests putting side to side two 5-term diagrams, one corresponding to the affirmation of the fifth term and the other to its negation [Venn, 1894, p. 117-118].

Marquand and Macfarlane shared Venn's interest in symbolic logic and in constructing diagrams for n terms. Both made high claims about their methods. Marquand explains that the object of his paper was to construct diagrams which "may be indefinitely extended to any number of terms, without losing so rapidly their special function, viz. that of affording visual aid in the solution of problems" [Marquand, 1881, p. 266]. Similarly, Macfarlane says that his method "is capable of representing quite generally the universe subdivided by any number of marks" [Macfarlane, 1885, p. 287].

[30]It has been since proved that it is possible to draw a 5-term diagram with ellipses. See: [Grünbaum, 1975], [Schwenk, 1984] and [Hamburger and Pippert, 2000].

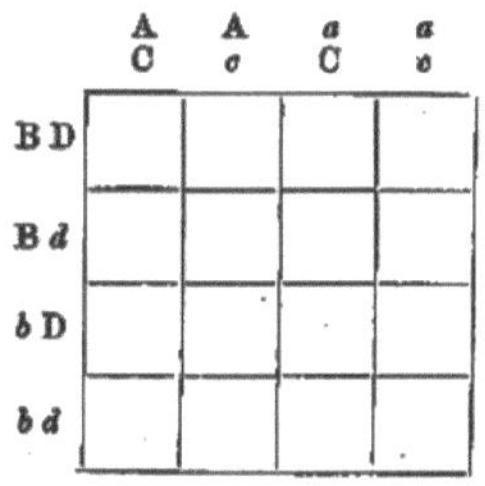

Fig. 62

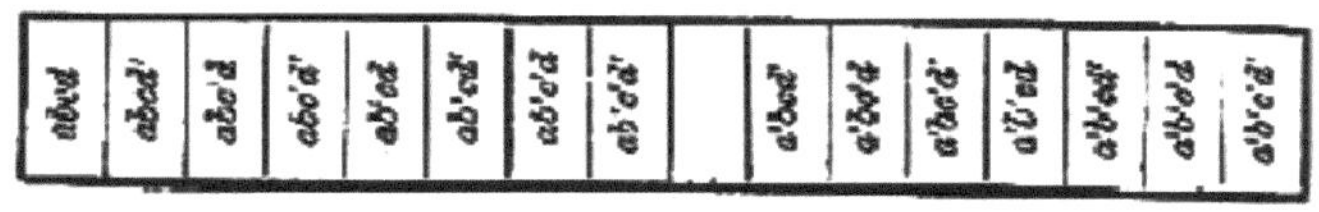

Fig. 63

However, in order to get the target, Marquand and Macfarlane made no attempt to keep their diagrams continuous, as is shown by [Fig. 62] and [Fig. 63], which represent 4-term diagrams using Marquand and Macfarlane's methods respectively. In [Fig. 62] (where *a* stands for *not-A*, *b* for *not-B*, etc.), classes *A*, *not-A*, *B*, and *not-B* are continuous but classes *C*, *not-C*, *D*, and *not-D* are not. In [Fig. 63] (where a' stands for *not-a*, b' for *not-b*, etc.), only classes *a* and *not-a* are continuous. Hence, the approach is different from that of Venn, who kept the continuity of the figures up to four terms, and abandoned it afterwards unhappily. Marquand and Macfarlane seem to care more about the regularity (rather than the continuity) of the figures, as Macfarlane explains:

> This method allows all the *a* part to be contiguous; but the *b* part is broken up into two portions, the *c* part into four portions, the *d* part into eight portions. However the regularity of the spectrum enables us easily to find all the portions belonging to any one mark. [Macfarlane, 1885, p. 287]

Carroll, though he used tabular diagrams, seems closer to Venn than to Marquand and Macfarlane. Indeed, he also maintained his diagrams continuous up to 4 terms [Fig. 64] but failed to do so for 5 terms [Carroll, 1897, p. 177]. Indeed, the fifth class is partitioned into 16 sub-divisions located above diagonal lines that divide each compartment of the 4-term diagram [Fig. 65].

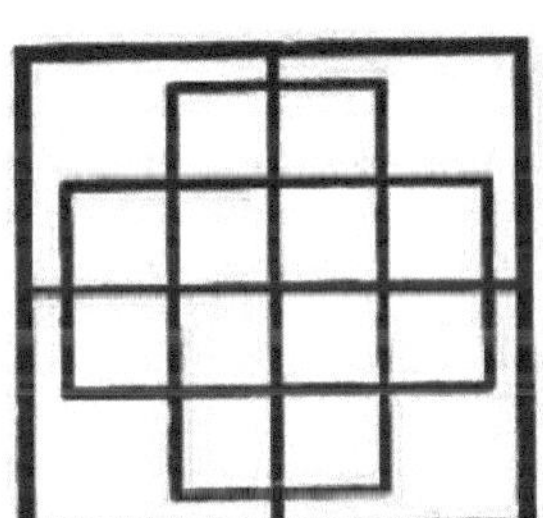

Fig. 64

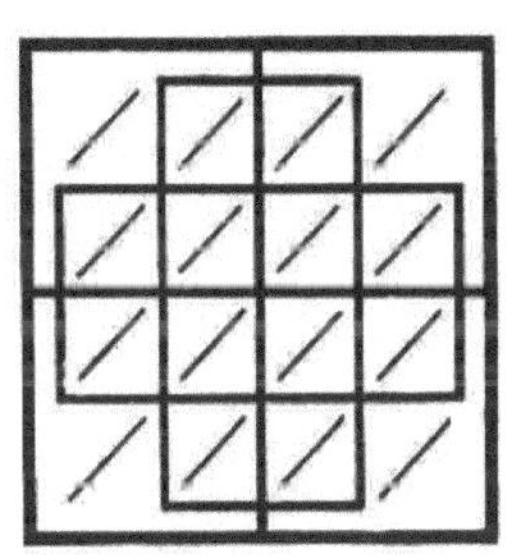

Fig. 65

Contrary to Venn, Carroll didn't stop here, and continued to provide illustrations of diagrams up to 8 terms and to explain how to construct them up to 10

[Carroll, 1897, p. 177-179]. Carroll is however severe when he writes that: "Beyond six letters Mr. Venn does not go" [Carroll, 1897, p. 176]. Indeed, Venn stopped at 6 terms because he didn't believe it worth to go further, not because he didn't know how to do so. In fact, Venn has even invented an inductive method for constructing continuous diagrams for any number of terms, but didn't make use of it because he was not happy with the asymmetrical diagrams he got [Venn, 1880a, p. 8; Venn, 1894, p. 118].[31]

Carroll made high claims about his diagrams and their superiority over Venn's, when it comes to constructing diagrams for more than 3 terms [Carroll, 1897, p. 174-179].[32] A close look shows however that Carroll's and Venn's methods are similar. What makes Carroll's diagrams (and more generally tabular diagrams) look better and easier is not the extension method itself (as Venn's method is quite the same) but rather the unity and regularity of the scheme thanks to the representation of the universe and its symmetric division. Interestingly, Venn recognised this superiority as he himself appealed to Marquand diagrams (rather than his own) when he worked examples requiring too many terms [Venn, 1894, p. 139-140 & 373-376].[33]

We will not illustrate here how such diagrams for more than 3 terms are used to solve logic problems. Many such examples could be found in Venn's *Symbolic Logic* [Venn 1894].[34] The principle is exactly the same as for syllogisms: one represents first the information contained in the premises on the appropriate diagram (depending on the number of terms), then one has to look for the specific relation (or relations) between the term (or terms) that might interest him, simply by eliminating the undesired terms. From the diagrams we discussed above, it is clear that they get progressively more difficult to use when the number of terms increases and the figures get more complex. For problems involving more than 5 or 6 terms, on can hardly recommend using diagrams as they do not give anymore

[31]On Venn's inductive method, see: [Bowles, 1971]. A. W. F. Edwards reproduces a (corrected) diagram for 7 terms, constructed by Peirce using this method [Edwards, 2004, p. 31]. Mathematicians have since developed many methods for drawing continuous Venn diagrams for any number of terms. One often obtains interesting mathematical figures with unexpected artistic qualities. However, such figures could hardly be used in logic as they were not designed for such a purpose and thus are not easy to manipulate and do not provide the visual aid one would expect from such representations. See [Hocking, 1909; More, 1959; Henderson 1963; Anderson and Cleaver, 1965] and [Edwards, 1989]. For a general discussion on the history of constructing logic diagrams for n terms, see [Edwards, 2004] and [Moktefi and Edwards, 2011].

[32]This claim has been often repeated since. See for instance [Lewis, 1918, p. 180; Davenport, 1952, p. 153] and [Macula, 1995, p. 269].

[33]The superiority of tabular methods for constructing diagrams for more than, say, 5 terms motivated the invention of many such methods, after Venn, at the end of the nineteenth and beginning of the twentieth centuries. In addition to Marquand, Macfarlane and Carroll, see for instance [Hawley, 1896] and [Newlin, 1906]. Though theses diagrams have not survived as such, several tabular schemes using the same principles are used in modern logic (truth tables) and computer science (switching theory charts). On the former, see [Shosky, 1997; Anellis, 2004] and [Grattan-Guinness, 2005]. On the latter, see [Veitch, 1952; Karnaugh, 1953] and [Nadler, 1962].

[34]See also [Macfarlane 1890]. Curiously, Carroll didn't publish any example of diagrammatic solutions to logic problems requiring more than three terms. A rare example may be seen in a manuscript reproduced in [Abeles, 2010, p. 59].

the visual aid one would expect.[35] However, it must be kept in mind that this limit is purely an issue of ease and convenience. Otherwise, there are no 'theoretical' objections for the use of diagrammatic methods for solving such logic problems, whatever is the number of terms, provided that one accepts to face some practical difficulties.

The construction of diagrams for many terms shows again how logicians had to search for a balance between visual aid and logical power. That doesn't mean that the logicians who invented new diagrams failed in their attempts to make use of them. The diagrams discussed so far were designed mostly as visual tools to solve very specific logic problems, not to construct languages or logical systems as modern logicians do.[36] Thus, one has always to keep in mind that these schemes have been introduced to deal with specific purposes within particular contexts. As such, many of the inconveniences that one might reproach to those diagrams are in fact the limits of the logical theory (for which they have been designed) itself, not a limitation of the diagrammatic approach proper. The diagrams we discussed in sections II, III and IV, were mainly concerned with term/class logic, either in the old syllogistic form or in the post-Boolean form. In the next section, we will deal with another style of diagrams that accompanied the advances of logic in the late nineteenth-century, notably the development of propositional logic and quantification theory to which Peirce and Frege contributed greatly.

5 THE FREGE-PEIRCE AFFAIR

We suspect that not many Frege scholars are reading Peirce's work and vice versa. Encouraging them to do so is not our goal, either. If one works on Frege's philosophy of language or one's interest is Peirce's pragmatism or epistemology, these two philosophers' relevant work might be disjoint from each other. On the other hand, logicians, especially those who are interested in the history of modern logic, must have often found both names on the same page. As we will explain below, the current topic, logic diagrams, also puts Frege and Peirce on the same page but with several twists. We call the entire matter the Frege-Peirce affair.

Let us start with the simplest aspect of the Frege-Peirce affair. Until recently, Frege received the entire credit for being the founder of modern logic, Peirce being neglected. Publication dates have mattered: Frege's *Begriffsschrift* (where the concept and the notation of quantificational logic appeared) was published in 1879 [Frege, 1879], and Peirce's paper "On the algebra of logic: A contribution to the philosophy of notation" (where the final form of quantified logic was shaped) came

[35] Though unhappy with his 5-term diagram, Venn maintained that for such a number of terms, he still preferred to use his diagrams rather than symbolic methods [Venn, 1880a, p. 7-8]. This claim made Hugh MacColl react immediately to defend the superiority of (at least his own) symbolic methods [MacColl, 1880].

[36] Still, these diagrams can be adapted and used for the purpose of constructing logic systems, as Shin did with spatial diagrams [Shin, 1994] or George Englebretsen with linear diagrams [Englebretsen, 1998].

out in 1885 [Peirce, 1885]. However, it turned out that these two years, 1879 versus 1885, cannot be the sole factor for deciding who established quantificational logic. Since Peirce developed his system independently without any knowledge of Frege's work, the consensus has been that a six-year difference should not matter. At the same time if we were obsessed with the chronological order, Peirce's 1870 paper, "Description of a notation for the logic of relatives, resulting from an amplification of the conceptions of Boole's calculus" [Peirce, 1870], precedes Frege's work as far as the discussion of relational logic is concerned. Now that Peirce's long journey toward modern logic has been explored, we name both Frege and Peirce as the founders of modern predicate logic.

If we stop here, the reader might end up having the following story of the Frege-Peirce affair:

- Frege and Peirce invented modern predicate logic independently of each other.
- Since Frege's publication preceded Peirce's, it was Frege who was acknowledged as the founder of modern logic.
- But, the record got straightened and both logicians have been getting the credit they deserve.

It is not an incorrect story, but this story might incorrectly implicate the following thought: Peirce's work on modern logic had to wait until the late 20th century to be recognized while Frege's work on logic has well been received and recognized from its publication in 1879 until now. Nothing could be further from truth than this.

There is a missing link between the invention of modern logic and current evaluations of the two logicians' work: It is Peirce's notation, not Frege's, which was accepted and adopted by their contemporaries. This missing part of the story is directly related to our current topic, logic diagrams, and at the same time is the heart of the Frege-Peirce affair we would like to explore in this section. There are two places where an ironical twist takes place in the affair: The first irony is that in spite of its earlier publication Frege's 1879 system was ignored, while Peirce's 1885 system was adopted by contemporary logicians like Schröder and Peano. Let us call this the "Peirce-first" twist. The other irony is that it was Frege alone who got credit for founding modern logic until the late 20th century. This we may call the "Frege-first" twist.

Explanations are demanded for both the Peirce-first and the Frege-first phenomena. Why was Frege's system rejected at the beginning? What caused Frege's work to be revived and at the same time Peirce's contribution to be underevaluated for several decades? We are interested in the first part of the Frege-Peirce affair, that is, the Peirce-first phenomenon.[37] Here is the much cited passage from Putnam:

[37] The Frege-first phenomenon has been explained and speculated in literature. Russell's and Whitehead's joint work *Principia Mathematica* (1910), many believe, drew our attention to

Schröder does mention Frege's discovery, though just barely; but he does not explain Frege's notation at all. The notation he both explains and adopts (with credit to Peirce and his students, O. H. Mitchell and Christine Ladd-Franklin) is Peirce's. And this is no accident: Frege's notation (like one of Peirce's schemes, the system of existential graphs) repelled everyone (although Whitehead and Russell were to study it with consequential results). Peirce's notation, in contrast, was a typographical variant of the notation we use today. Like modern notation, it lends itself to writing formulas on a line (Frege's notation is two-dimensional) and to a simple analysis of normal-form formulas into a prefix (which Peirce calls the Quantifier) and a matrix (which Peirce calls the "Boolean part" of the formula). [Putnam, 1982, p. 290-291]

According to Putnam, the Peirce-first phenomenon should be attributed to different styles of Frege's and Peirce's systems. Frege's 1879 notation is two-dimensional, that is, diagrammatic, while Peirce's 1885 is linearly ordered as is now usual. For example, our formula, say $\forall x(Px \rightarrow Qx)$, would correspond to the following representations:[38]

- Frege's Conceptual notation (henceforth, 'CN')

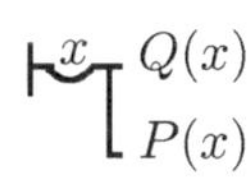

- Peirce's notation

$$\prod_x (\bar{P}x + Qx)$$

We would like to mention a couple of misconceptions that will be relevant to our further discussion. As we pointed out, after a certain point the Frege-first twist was made and, hence, some of us have not even realized for a long time that our current first-order logic notation is much closer to Peirce's, not to Frege's. But, as shown in the above comparison, Peirce's notation is much closer to what we have now. On the other hand, when it comes down to graphic representation, we find almost the opposite direction of a mistake: After Peirce's graphical systems became well publicized, many of us have regarded Peirce (not Frege) as a pioneer

Frege's work. Hintikka explains the Frege-first phenomenon in a larger context, that is, the competition between the tradition of universality of logic versus the model-theoretic tradition. According to Hintikka, the Frege-first phenomenon is a symptom of the predominance of the universality of logic [Hintikka, 1990 & 1997]. For these two different traditions, see [van Heijenoort, 1967] and [Goldfarb, 1979]. For further discussion, see [Shin, 2002, §2.1.1]. Some scholars would not forget to point out Peirce's personal problems and misfortunes as one of the reasons why Peirce's projects, including his invention of modern logic, were not appreciated for a long time in the States.

[38] To focus on our main point, we made a slight modification about the choice of symbols. For example, neither notation adopts x as a variable, but a lower-case Gothic for Frege and alphabet letter i for Peirce.

for diagrammatic reasoning. It is not our intention to let Frege and Peirce compete in the history of modern logic, but to frame the Frege-Peirce affair and fill in significant parts of the affair to get the record straight. The process, as we will see, requires deeper understanding of Frege's and Peirce's motivations for each of their systems. In this section we not only present and interpret their own justifications but also pinpoint specific aspects of each system to illustrate their views on logic diagrams.

In the first subsection, from the outset we fully accept Putnam's explanation of the Peirce-first phenomenon: Frege's pioneering work *Begriffsschrift* was shunned by his contemporaries mainly because of its graphical aspect. Schröder's well-known criticisms, which we will examine shortly, address this aspect. However, we find Frege's own defense not as well-known as the criticisms. Frege presented forceful arguments for his calculated and well-reasoned choice for two-dimensional representation. To be faithful to Frege's spirit, we approach Frege's system as graphically as possible. At the end, we support Frege's motivation and reasoning by presenting a case study.

Peirce's graphical representation was invented as an alternative and superior system to his own symbolic notation. Until recently, the history, however, was not kind to Peirce's innovation of diagrammatic representation. After removing the source of the misconception of his iconic system, in the second subsection we examine a more comprehensive and intuitive approach to Existential Graphs. We also revisit Peirce's passages on visual features of Existential Graphs and discuss a way to implement Peirce's principle into our understanding the systems so that they may become more efficient.

The Frege-Peirce affair is not limited to the Frege-first versus Peirce-first controversies. It is not even an interesting part of the story at all. Their relation is much more intricate than being captured in terms of competition. Both founders invented graphical systems at the dawn of modern logic and a strong prejudice against non-familiar two-dimensionality treated both systems equally — in a harsh way. Applying to each system our view that realizing fundamental differences between symbolic and diagrammatic representation provides us with a new way to appreciate logic diagrams, we find ourselves being ready for a Frege-Peirce rendezvous. That will be our third subsection.

5.1 *Frege's two-dimensional notation*

The main goal of Frege's lifelong project was to secure the foundations of arithmetic, and he believed that if it could be shown that arithmetic is founded on logical truth, the goal would be accomplished. Hence, his subgoal was to reduce arithmetical truth to logical truth by re-writing the arithmetical concepts and the laws of arithmetic purely in terms of logical concepts and logical laws. Even though the mission was unsuccessfully executed, his project required a more powerful logic than Boole's monadic logic, which he did provide in the *Begriffsschrift* (1879). Accordingly, the *Begriffsschrift* became the birth place and 1879 the birth date for modern logic.

The first logical notation Frege put out, however, is quite different from any notation we are familiar with. For example, none of the following strings is found in his *Begriffsschrift*: $A \to B, A \wedge \neg B, \forall x(Px \to Qx), \exists x(Px \wedge \neg Qx)$. A difference between Frege's notation (henceforth, 'CN') and modern notations is not a matter of different choices of symbols,[39] but a matter of how symbols are arranged: Frege's CN requires two-dimensional arrangement while modern symbolic notations are linearly ordered. As the following brief summary shows,[40] the first impression of Frege's CN is that it is graphic or diagrammatic, as opposed to symbolic.

Conceptual Notation for sentential logic (CNSL) [Frege, 1879, §§1-7]

<u>Vocabulary</u>

- Sentence symbols: $A_1, A_2, \ldots$
- Horizontal stroke: ——
- Vertical stroke: |
- Negation stroke: |
- Conditional stroke:[41] 

<u>Syntax and semantics</u>[42]

1. If A_i is a sentence symbol, then the following notation is *CNSL*.

   We translate it into 'A_i'

2. If D is *CNSL* and we attach the negation stroke to the underside of a horizontal stroke of D, then the result is also *CNSL*. For example the following notations are all *CNSL*

   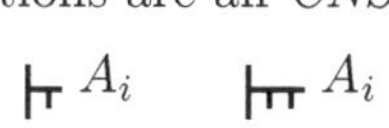

   We translate them into '$\neg A_i$' and '$\neg\neg A_i$', respectively.

3. If D is *CNSL* and we attach the conditional strike to the underside of a horizontal stroke of D with a sentence symbol at the end of the conditional stroke, then the result is also *CNSL*. For example, the following notations are all *CNSL*.

[39] For example, $\to$ versus $\supset$, $\neg$ versus $\sim$, $\forall$ versus $\prod$, $\exists$ versus Σ.

[40] Our presentation has a slight modification of Frege's original presentation without changing its essence.

[41] Note that the conditional stroke consists of the vertical and horizontal components, and below we may call the horizontal component the horizontal stroke.

[42] We carry out its semantics by translation of CNSL into modern logical notation.

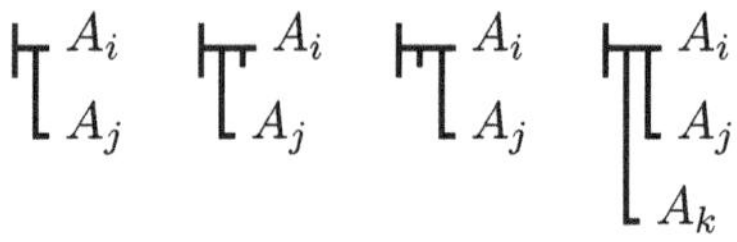

We translate them into '$A_j \rightarrow A_i$', '$A_j \rightarrow \neg A_i$', '$\neg(A_j \rightarrow A_i)$', '$A_k \rightarrow (A_j \rightarrow A_i)$', respectively.

Hence, the system is truth-functionally complete.

Conceptual Notation for predicate logic (CNPL) [Frege, 1879, §§9–12]

Vocabulary

- Predicate symbols: $P_1, P_2, \ldots$
- Constants: $a_1, a_2, \ldots$
- Variables[43]: $x_1, x_2, \ldots$
- Horizontal stroke: ————
- Vertical stroke: |
- Negation stroke: |
- Conditional stroke: 
- Horizontal stroke with a concavity inserted: 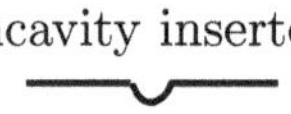

Syntax and semantics

Definition: α is an *atomic formula* iff $\alpha = P(t_1, \ldots, t_n)$, where P is an n-ary predicate and each t_i (where $1 \leq i \leq n$) is either a constant or a variable.

1. An atomic formula is a formula.
2. If Φ is a formula, the following notation is *CNPL*

   We translate it into 'Φ'.
3. If $\Phi(x)$ is a formula and x is a variable in $\Phi(x)$, then the following notation is *CNPL*

   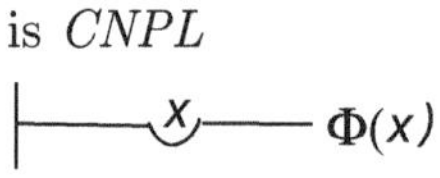

   We translate it into '$\forall x \Phi(x)$'.

[43] Frege adopts lower-case Gothic letters for variables.

4. If D is *CNPL* and we attach the negation stroke to the underside of a horizontal stroke of D, then the result is also *CNPL*. Hence, the following notations are *CNPL*

We translate them into '$\neg\forall x\Phi(x)$' and '$\forall x\neg\Phi(x)$', respectively.

5. If D is *CNPL* and we attach the conditional strike to the underside of a horizontal stroke of D with a sentence symbol at the end of the conditional stroke, then the result is also *CNPL*. For example, the following notations are all *CNPL*.

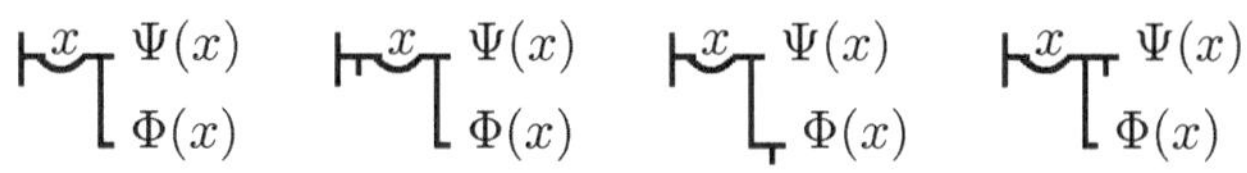

We translate them into '$\forall x(\Phi(x) \rightarrow \Psi(x))$', '$\neg\forall x(\Phi(x) \rightarrow \Psi(x))$', '$\forall x(\neg\Phi(x) \rightarrow \Psi(x))$', and '$\forall x(\Phi(x) \rightarrow \neg\Psi(x))$', respectively

Even though Frege has received full credit as a founder of modern logic, his notation itself was not welcomed and subsequently was not adopted and appreciated by logicians of the time.[44] Now we are about to explore how Frege explained, justified, and defended his new notation.

Frege's contemporaries were familiar with symbolization, for example, Boole's and Jevons' works. Hence, symbolization itself was not new to them, but Frege's various kinds of symbols and his way of arranging those symbols were so new that CN almost alienated his contemporaries. The following passage illustrates that feeling:

> a "conceptual notation" ... which makes a strange and chilling impression with its long and short, vertical and horizontal strokes; with its concavities and snaky lines; its double colon and function symbols; its German, Greek, and italic large and small letters. [Michaëlis, 180; Bynum, 1972, p. 212]

Ordinary language is too ambiguous, too complex, and too vague to be adopted in a rigorous logical project. Hence, a move from ordinary language to artificial symbolization, e.g. A, B, etc., was granted for logical purposes. But, Frege's notation demands two additional steps away from ordinary language. One is to introduce various forms of symbols, i.e. italicized capital Greek letters, lower-case Gothic letters, and italicized Latin letters. Different grammatical categories are visually exhibited in this new notation, unlike ordinary language.[45] The other move Frege's CN demands is to replace the linear order of ordinary language with

---

[44] Terrell Ward Bynum collected six reviews of *Begriffsschrift* by Frege's contemporaries. See Appendix I of [Bynum, 1972, p. 209-235].

[45] For our presentation of CN we were not strict to Frege's original work, but followed the modern notation. We recommend [Macbeth, 2005, §2.3] for an interesting discussion of the significance of Frege's various choices of letters.

two-dimensional spatial arrangements of new symbols. The second deviation is much more radical and makes, we believe, CN graphic.

Not surprisingly one of the main criticisms of CN was raised against the second aspect of CN, that is, the two dimensional aspect of the notation. Schröder expressed his uneasiness at CN's unconventional aspect in a blatantly harsh way:

> In fact, the author's formula language not only indulges in the Japanese practice of writing vertically, but also restricts him to only one row per page, or at most, if we count the column added as explanation, two rows! This monstrous waste of space which, from a typological point of view (as is evident here), is inherent in the Fregean "conceptual notation", should definitely decide the issue in favour of the Boolean school — if, indeed, there is still a question of choice. [Schröder, 1880; Bynum, 1972, p. 229]

CN's spatial arrangement was considered to be a waste of space without any special merit. What's worse, according to Schröder's view, Frege's notation is even inferior to Boole's, and the following case was presented:

> [I]n order to represent, for example, the disjunctive "or" — namely, to state that *a holds or b holds, but not both* — the author [Frege] has to use the schema
>
> 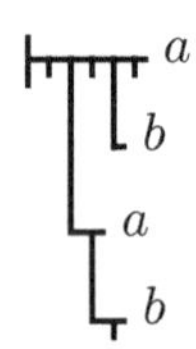
>
>
> which definitely appears clumsy compared to the Boolean mode of writing:
>
> $$ab_1 + a_1b = 1$$
>
> or also
>
> $$ab + a_1b_1 = 0.$$
>
> [Schröder, 1880; Bynum, 1972, p. 227][46]

This is a slightly different criticism from CN being two-dimensional, but a lack of number of connectives in CN, compared with the Boolean notation. CN, as Schröder understood, has only two connectives, $\neg$ and $\rightarrow$.[47] Hence, there is no good reason to adopt CN over the existing Boolean expression, Schröder's crowd must have concluded.

[46] $a_1$stands for *not-a*, andb_1 stands for *not-b*.

[47] A similar criticism will be raised when we discuss Peirce's graphical system in the next subsection.

Is Frege's spatial notation as cumbersome, inefficient, and wasteful as Schröder claims? We believe there are two related underlying reasons for the criticism: One is the unfamiliarity of two-dimensional notation and the other is the lack of understanding of the necessity of two-dimensional notation. Our ordinary language is linearly ordered, hence, one-dimensional and non-spatial. On the other hand, Frege's CN is spatial and two-dimensional.

Throughout his responses to criticisms against CN, Frege made it clear that it was his intention to make his CN two-dimensional, as opposed to the linear order of ordinary language. Below we focus on Frege's own explanations found in his two papers, "On the scientific justification of a conceptual notation," [Frege, 1882a] and "On the aim of 'conceptual notation' " [Frege, 1882b].

First, it should be noted that Frege's sense of 'symbols' is much broader than in our ordinary sense:

> [W]ithout symbols we could scarcely lift ourselves to conceptual thinking. Thus, in applying the same symbol to different but similar things, we actually no longer symbolize the individual thing, but rather what [the similars] have in common: the concept. This concept is first gained by symbolizing it; for since it is, in itself, imperceptible, it requires a perceptible representative in order to appear to us. [Frege, 1882a; Bynum, 1972, p. 84]

The main (and possibly sole) function of symbols is to represent invisible concepts in terms of a visible medium. Symbols do not have to be limited to letters, like a, b, etc., but could be a circle or a line. For example, Boole's system is symbolic, not because it has alphabetical letters displayed in a linear order, but because it represents concepts, e.g. identity, quantity, and so on. Hence, we claim that 'symbolic system' means 'representation system' according to Frege.

What do we need for representation? Frege's suggestion is:

> We need a system of symbols from which every ambiguity is banned, which has a strict logical form from which the content cannot escape. [Frege, 1882a; Bynum, 1972, p. 86]

It is interesting to note that Frege assumes that a logical system is supposed to reveal logical form. In order to reveal a logical form, Frege states, a representation system should be free from ambiguity. Without raising questions as to the exact meaning of 'logical form,' Frege moves on to the comparison among different kinds of representation systems.

A comparison between audible[48] versus visible symbols is highlighted as a contrast between one-dimensional time versus two-dimensional space, and Frege opts for written symbols based on their two-dimensional aspect. Why?

[48]Talking about audible symbols is another piece of evidence for our claim that Frege's use of 'symbol' is much broader than its ordinary sense.

> The spatial relations of written symbols on a two-dimensional writing surface can be employed in far more diverse ways to express inner relationships than the mere following and preceding in one-dimensional time ... In fact, simple sequential ordering in no way corresponds to the diversity of logical relations through which thoughts are interconnected. [Frege, 1882a; Bynum, 1972, p. 87]

Frege makes a subtle, but crucial move here — from the comparison between audible and written symbols to the (potential) comparison among written symbols themselves. Frege prefers written symbols to audible symbols based on the two-dimensionality of written symbols. Audible symbols can be nothing but one-dimensional, while written symbols can be spatial. Strictly speaking, Frege's comparison, then, is about how symbols are arranged, i.e., sequential versus non-sequential arrangement of symbols, *not* about kinds of symbols, e.g. audible or written symbols. As a logical system, both audible and sequential written systems are in the same boat: Both are inferior to non-sequential spatial written systems, according to Frege.

Not all written symbols are spatially arranged, but almost every natural language is linearly ordered. This is why we can read aloud a text written in a natural language to make it audible.[49] It is important to note that Frege does not argue for a non-linear form for every written representation, but only for a system which aims to represent logical relations. An artificial logical system has its own goal: We do not have to be able to read it aloud[50] but it aims to represent logical relations. Frege's firm belief is that logical relations come in different varieties and, hence, a simple sequential ordering would not be an efficient way to represent them. It is no wonder that Frege himself invented a non-linear representational system by fully facilitating the aspect of two-dimensionality of written symbols.

Frege's response to Schröder's review — that CN is a waste of space and follows a Japanese vertical writing style — reveals how simple-minded these criticisms are. Instead of asking the motivation behind CN's two-dimensional representation, the critics revealed their prejudice against unfamiliar forms of representation. According to Frege, this is a small price to pay for the merit of the system, that is, 'perspicuity.'

> I cannot deny that my expression takes up more room than Schröder's, ... The disadvantage of the waste of space of the "conceptual notation" is converted into the advantage of perspicuity; the advantage of terseness for Boole is transformed into the disadvantage of unintelligibility. The "conceptual notation" makes the most of the two-dimensionality of

[49]In this sense, natural language texts are similar to audible symbols. It is an interesting question to ask why a natural language text is linear. It might be related to the way our ordinary thoughts are formed.

[50]Frege explicitly says that his CN is not meant to be converted into audible symbols: "The arithmetic language of formulas is a conceptual notation since it directly expresses the facts without the intervention of speech." [Frege, 1882a; Bynum 1972, p. 88]

> the writing surface by allowing the assertible contents to follow one below the other while each of these extends [separately] from left to right. Thus, the separate contents are clearly separated from each other, and yet their logical relations are easily visible at a glance. For Boole's, a single line, often excessively long, would result. [Frege, 1882b; Bynum, 1972, p. 97]

By arranging statements spatially, the logical relation among them is visibly presented, while maintaining their contents separately, according to Frege. In what sense does two-dimensionality facilitate clear and precise representation of logical relations? We take up Frege's graphical representation of conditional relations as a case study to explore the inquiry — how much perspicuity is added by two-dimensional representation in CN.

Frege presents the following diagrammatic scheme as the representation of the conditional relation between two statements:

Fig. 66

A sentence attached to the underside of a horizontal stroke by the conditional stroke, in this case B, is a sufficient condition for A to occur. Let us call it the Conditional Stroke Convention. That is, it denies the possibility that B is affirmed but A is denied. Then, suppose another proposition, say C, is added as a sufficient condition for A. Then, all we need to do is to attach sentence C by a conditional stroke in the following way:

Fig. 67

The occurrence of B and C is a sufficient condition for A to occur.

$$(B \wedge C) \rightarrow A \tag{1}$$

Hence, we cannot have A to be affirmed while B or C is denied.

$$\neg[(B \wedge C) \wedge \neg A] \tag{2}$$

Now the story gets somewhat interesting. We said any sentence hanging under the horizontal line by a conditional stroke is a sufficient condition for the content of its right part. For example, in the case of [Fig. 66], B is hanging as a sufficient condition for its right part, that is, the content of A; hence we obtain $(B \rightarrow A)$. Applying the same principle, one may read off [Fig. 67] as C being a sufficient condition of the right part of the diagram, i.e. the following diagram:

Fig. 68

Again, this subpart [Fig. 68] of the diagram says $(B \rightarrow A)$. Then, we may translate [Fig. 67] into

$$C \rightarrow (B \rightarrow A) \tag{3}$$

Please note that we obtained the sentences (1), (2), and (3) from one and the same diagram [Fig. 67]. Depending on which aspect of a given diagram is focused on, we produced three different looking sentences.

We could only imagine what Frege would make out of this case: There is a logical relation among the three statements A, B, and C which [Fig. 67] represents, and when we have a language in a linear form, there is more than one way to express this logical relation, e.g. sentences (1), (2), and (3). However, when we become free from the linear constraint and adopt two-dimensional notation, we draw one diagram, [Fig. 67], and the logical relation is once and for all clearly represented. We believe this is one example of the perspicuity Frege had in mind. But we can push the issue further.

Let's go back to [Fig. 66]. Expressing the same proposition as [Fig. 67], what if one attaches the sentence C between A and B as follows?

Fig. 69

Both sentences, B and C, are hanging under the horizontal line attached to the sentence A. According to the Conditional Stroke Convention, this diagram should be read off as the occurrence of C and B being a sufficient condition for A to occur. We obtain the following translations, corresponding to (1) and (2) in a trivial sense:

$$(C \wedge B) \rightarrow A \tag{4}$$

$$\neg[(C \wedge B) \wedge \neg A] \tag{5}$$

However, when we get to the way we read [Fig. 67] off as sentence (3), things are not trivial any more. [Fig. 69] says B is a sufficient condition for the following content:

Fig. 70

[Fig. 70] says ($C \rightarrow A$). Hence, we get the following translation of [Fig. 69]:

$$B \rightarrow (C \rightarrow A) \tag{6}$$

One may point out that [Fig. 67] and [Fig. 69] are different diagrams. However, according to the Conditional Stroke Convention, there is no substantial difference[51] so that we may easily see the same logical form displayed in these two diagrams. On the other hand, it is far from being easy to see the logical equivalence among the six sentences (1)–(6). Nonetheless, the logical equivalence among these six sentences shows that all of them represent the same logical relation, but only at a meta-level by using inference rules may we make the claim like that. This is a contrast with the case of CN: The sameness of logical relation is exhibited in CN since it is one and the same diagram. This is, we believe, what Frege had in mind when he claimed that logical relations are better represented in a non-linear form.

5.2 *Peirce's chef-d'oeuvre*

Peirce's contribution to modern logic has been recently re-evaluated in two major aspects: One is to recognize him as a founder of modern quantified logic, along with Frege. The other is to re-discover the novelty of Peirce's Existential Graphs (henceforth, 'EG'). Many important questions could be raised about the re-evaluations of Peirce as logician, but we limit ourselves to the issues related to our current topic, logic diagrams.

Interestingly enough, *both* Peirce as a founder of modern logic and Peirce as an inventor of graphical logic are intricately related to Frege. As we discussed at the beginning of the section, the Frege-Peirce affair outlines Frege's and Peirce's positions in the history of modern logic, and hence, the acknowledgment of Peirce's contribution to predicate logic could not be made without mentioning Frege's work. Then, how does the re-discovery of Peirce's EG bring in Frege again?

Citing Putnam above, we attributed the Peirce-first phenomenon – the adoption of Peirce's logical notation – to the non-familiar two-dimensionality of Frege's notation. In that quoted passage, Putnam, saying "Frege's notation (like one of Peirce's schemes, the system of existential graphs)," put Frege's CN and Peirce's EG in the same group. Considering why the Peirce-first phenomenon took place, we should not be surprised that Peirce's EG was not welcomed, either. One may fill in more in this part of the story, but we believe it is the correct framework to situate the fate of Peirce's EG.

However, there is one major difference between Frege's CN and Peirce's EG with respect to its role as a system in each logician's work. The two-dimensional notation presented in CN in 1879 is Frege's only logical system, but Peirce presented two different, but logically equivalent, systems — first a linear symbolic system (1885) and later a spatial graphical system (1897). Why did Peirce investigate another form of representation, in spite of the acceptance of his own linear notation

[51] We may make a quite trivial inference rule between these two types of diagrams.

and the unwelcoming reception of Frege's two-dimensional notation? Furthermore, Peirce himself firmly believed his EG to be superior to his previous linear system. There have been various suggestions on this question, and many of Peirce scholars have related Peirce's semiotics to EG. We do not doubt the birth of EG was deeply rooted in Peirce's grand philosophy. But our goal is to focus on the main diagrammatic/visual features of EG, rather than to get involved in the theoretical background of the invention of EG.

A brief introduction of EG is in order. EG consists of three parts: Alpha, Beta, and Gamma, and they correspond to sentential, first-order, and modal logic, respectively. In order to make our main points clear, we focus on the simplest EG, i.e. Alpha graphs, and discuss relevant aspects of Beta diagrams. We strongly suspect and hope that more scholars will work on Peirce's modal system, that is, Gamma, from a similar point of view as we present here.[52]

Alpha Graphs

Vocabulary

1. Sentence symbols: A_1, A_2, ...
2. Cut

Syntax and Semantics[53]

1. An empty space is an Alpha diagram. We translate it into $\top$.
2. A sentence symbol is an Alpha diagram. It translates into A_i.

$$A_i$$

3. If D is an Alpha diagram, then so is a single cut of D (we write '$[D]$'), and a cut represents negation. For example,

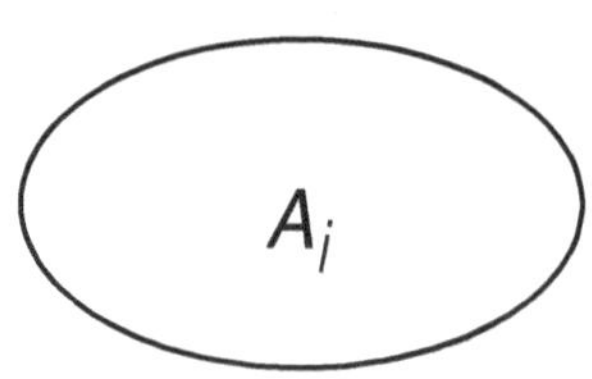

is translated into '$\neg A_i$.'

[52]See [Roberts 1973, Ch. 5].

[53]We combine syntax and (intuitive) semantics to save space as well as to give equal treatment with Frege's system in the previous subsection. For more lengthy and rigorous presentations, see [Shin, 2002, Ch. 4].

4. If D_1 and D_2 are well-formed diagrams, then so is the juxtaposition of D_1 and D_2 (write '$D_1 D_2$'). Juxtaposition represents conjunction. Hence,

$$A_i \quad A_j$$

is translated into '$(A_i \wedge A_j)$.'

5. Nothing else is a well-formed diagram.

The semantics presented above is the *endoporeutic* reading, which Peirce presented in the following passage and which subsequent Peirce scholars and logicians followed without much reflection:

> The interpretation of existential graphs is *endoporeutic*, that is proceeds inwardly; so that a nest sucks the meaning from without inwards unto its centre, as a sponge absorbs water. [Peirce, Ms 650], quoted in [Roberts, 1973, p. 39 - n. 13]

The order, i.e. from outside inward, is crucial. Hence, Roberts' warning is appropriate:

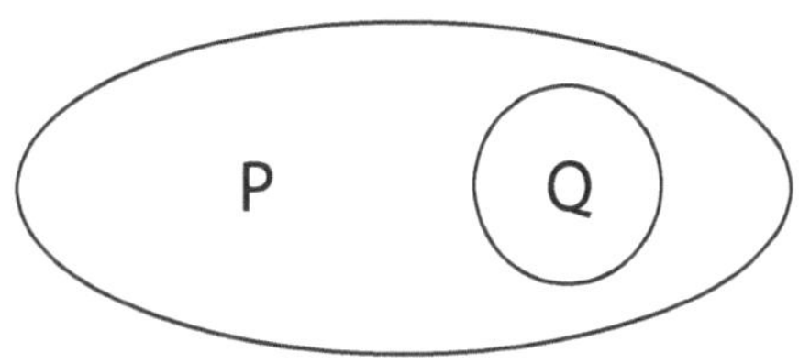

> Notice that we do not read [the above diagram]: 'Q is true and P is false', even though Q is evenly enclosed and P is oddly enclosed.... we read the graph from the outside (or at least enclosed part) and we proceed inwardly ... [Roberts, 1973, p. 39]

An analogy to symbolic representation would be the difference between the formulas '$(\neg P \wedge Q)$' and '$\neg(P \wedge \neg Q)$'.

The endoporeutic reading gets the result correct, but that does not mean it is the best way to understand the Alpha system. First of all, when we have nested cuts, we have a complicated-looking sentence. For example, the following diagram is translated into the following sentence, according to the endoporeutic reading:

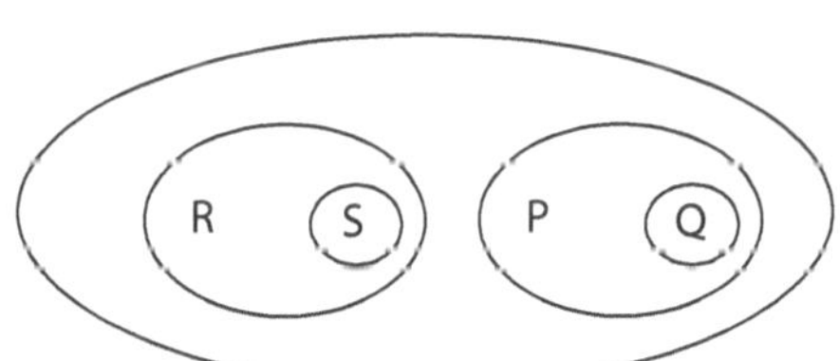

Fig. 71

$$\neg(\neg(R \wedge \neg S) \wedge \neg(P \wedge \neg Q)) \tag{7}$$

That is, Peirce's Alpha diagrams were considered to be quite similar to a sentential language which has only two kinds of connectives, i.e. $\neg$ and $\wedge$. This misconception yielded two undesirable results — one is unfortunate and the other is plainly wrong. Just as we prefer a sentential language with $\neg$, $\wedge$, $\vee$, $\rightarrow$, and $\leftrightarrow$ to a language with $\neg$ and $\wedge$, which is hardly even used, Alpha diagrams have likewise not been put to use. After all, De Morgan's laws are for symbolic sentences, not for Alpha diagrams. Hence, the endoporeutic method prevented us from exploring any special features of Peirce's EG, by treating the system as a two-connective logical system. However, this treatment is false, and below we present Shin's multiple reading method, which highlights a fundamental difference between Alpha diagrams and symbolic sentences.

Multiple Readings[54]

1. If D is an empty space, then it is translated into $\top$.
2. If D is a sentence letter, say A_i, then it is translated into A_i.
3. Suppose the translation of D is α. Then, $[D]$ is translated into $(\neg\alpha)$.
4. Suppose the translation of D_1 is α_1 and the translation of D_2 is α_2.
   (a) a translation of $D_1 D_2$ is $(\alpha_1 \wedge \alpha_2)$,
   (b) a translation of $[D_1 D_2]$ is $(\neg\alpha_1 \vee \neg\alpha_2)$,
   (c) a translation of $[D_1[D_2]]$ is $(\alpha_1 \rightarrow \alpha_2)$, and
   (d) a translation of $[[D_1][D_2]]$ is $(\alpha_1 \vee \alpha_2)$.

Please note two major differences between the traditional and the multiple readings:

(i) While the traditional endoporeutic reading gets us a sentence with only two connectives (i.e. $\neg$ and $\wedge$), the algorithm of multiple readings directly yields a sentence with $\neg$, $\wedge$, $\vee$, and $\rightarrow$.

(ii) For a given diagram, the new method might produce more than one sentence as its translation, while the endoporeutic method yields one and only one sentence.

The first point might prompt the reader to ask the following question: Does this mean that Alpha diagrams are similar to a sentential language with four connectives as opposed to one with two connectives? We firmly answer "no" for the following important reasons: No additional syntactic device was introduced in the Alpha system to get a sentence with more connectives. In the case of symbolic

[54]For more details, see [Shin, 2002, §4.3.2].

systems, more connectives are introduced and so are more inference rules. That is, to make our life more convenient the syntax of a system needs to be more elaborate. But in the case of the Alpha system, without modifying any part of the syntax we are able to read off negative, conjunctive, disjunctive, and conditional information directly from a given diagram. Where does this difference come from?

Let us start with the clause 4(b) above where the new algorithm allows us to bring in '$\vee$.' According to the old endoporeutic algorithm, the diagram $[D_1 D_2]$ is translated into $\neg(\alpha_1 \wedge \alpha_2)$. Then, how does the new method direct us to a logically equivalent disjunctive sentence, $(\neg\alpha_1 \vee \neg\alpha_2)$? We may observe different features from the diagram $[D_1 D_2]$, and Peirce's original reading urges us to focus on the following two features in the following order: First, the subdiagram $D_1 D_2$ is enclosed by a cut. Second, $D_1$and D_2 are juxtaposed with each other. However, there are other visual features the reader might notice. For example, D_1 is enclosed by a cut, D_2 is enclosed by a cut, and the juxtaposition of D_1 and D_2 takes place in an area which is enclosed by a cut, etc. Shin's new algorithm aims to reflect all of these visual features in the reading without a constraint on the order in which visual features are observed. Clause 4(b) tells us how to read off these extra visual features to get disjunctive information directly from the diagram $[D_1 D_2]$. Similarly, clauses 4(c) and 4(d) add further visual features to the reading list. Let us go back to [Fig. 71] to illustrate this point:

As seen above, the endoporeutic reading gets the following translation:

$$\neg(\neg(R \wedge \neg S) \wedge \neg(P \wedge \neg Q)) \tag{7}$$

The multiple readings method also yields the same translation if one applies clauses 4(a) and 3 in that order. But, suppose the reader happens to notice the pattern [[ ][ ]] first in [Fig. 71]. Then, clause 4(d) is there to be applied. So, the following translation is obtained:

$$(R \wedge \neg S) \vee (P \wedge \neg Q) \tag{8}$$

Suppose that the pattern [[ ]] caught the reader's attention. Then, clause 4(c) will be applied. At the same time, depending on how the subpart $[R[S]]$ is perceived, we may get different translations as follows:

$$\neg(R \wedge \neg S) \rightarrow (P \wedge \neg Q) \tag{9}$$

$$(\neg R \vee S) \rightarrow (P \wedge \neg Q) \tag{10}$$

$$(R \rightarrow S) \rightarrow (P \wedge \neg Q) \tag{11}$$

Hence, for a given diagram we may obtain more than one sentence as its translation, which we pointed out above as the second difference between the two reading algorithms. The main reason why multiplicity has been ignored is, we suspect, that diagrams have been considered as a form of symbols, and hence, no attention was paid to fundamental differences between diagrammatic and symbolic representation.

Shin's algorithm of multiple readings focuses on one prominent feature of diagrammatic representation: Diagrams, being spatial, may be perceived in more than one way, as the classic Gestalt phenomenon illustrates. Then, it is natural and intuitive to let an algorithm be flexible so that it may fully facilitate various kinds of perception. At the same time we claim it is more efficient than the traditional method which results in only one reading. Sentences (7)–(11) are logically equivalent to one another. In the case of a symbolic system, we need a proof using inference rules to prove the logical equivalence among them. But, the existence of one and the same Alpha diagram is a proof for the equivalence. When one utilizes the spatiality of diagrams and does not limit oneself to a linearly ordered perception, a more natural, more intuitive, and more efficient reading is born.

At this point, we should remind the reader that Peirce himself did not come up with the algorithm of multiple readings for his own system. This is a good piece of evidence that Peirce did not catch the Gestalt phenomenon as a feature which distinguishes his EG from his own linear notation presented in 1885. Instead Peirce's passion for EG is to represent relations in a diagrammatic way. Peirce's effort as logician was concentrated on the representation of relations, since Peirce correctly believed that the logic of relations requires a significant leap from Boole's logic.[55] After inventing linear representation, Peirce kept searching for better representations of relations, which led him to EG. Hence, one may say that the Beta system, being the logic of relations, is where Peirce's goal was accomplished. In this subsection we do not plan to introduce the syntax and the semantics of the Beta system as we did for Alpha,[56] but will examine important visual features of EG which Peirce himself had in mind. We will also connect Peirce's spirit to Shin's idea for multiple readings.

Beta Graphs

How are relations represented in the Beta system? How about quantifiers and variables? The answers for these two crucial questions, we believe, will convince us why Peirce's EG are diagrammatic and non-symbolic. Peirce was inspired by the representation of chemical compounds based on the doctrine of valency. We can see easily how Peirce's EG on the right side was modeled after a representation of the water molecule on the left:

[55] For more details and Peirce's citations, see [Shin and Hammer, 2011].

[56] Zeman and Roberts produced the pioneering work on Peirce's Beta diagrams [Zeman 1964; Roberts, 1973]. See [Shin, 2002, Ch. 5] for the summary of literature on the Beta system. Shin also presents an alternative understanding of Beta in the same chapter.

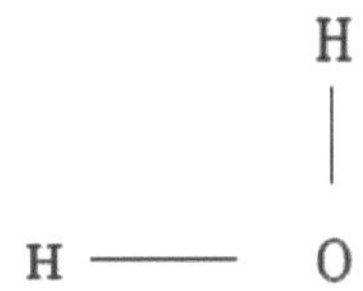

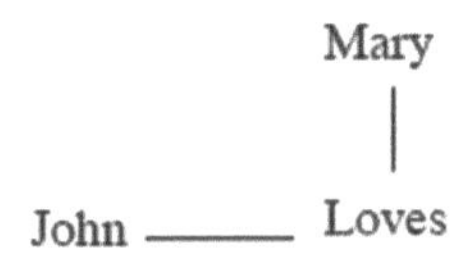

Relations are represented by Peirce's syntactic device, a line (called a line of identity) [Peirce, 1933, §§ 4.423 & 4.442], and the number of arguments for a given predicate is represented by the number of lines which branch out from the predicate.

In Beta diagrams we do not find syntactic vocabulary which corresponds to universal and existential quantifiers,[57] but here are Peirce's explanations as to how these concepts are expressed in his system:

> *[A]ny line of identity whose outermost part is evenly enclosed refers to* something, *and any one whose outermost part is oddly enclosed refers to* anything *there may be.* [Peirce ,1933, §4.458][58]

Applying this passage to the following Beta diagrams,

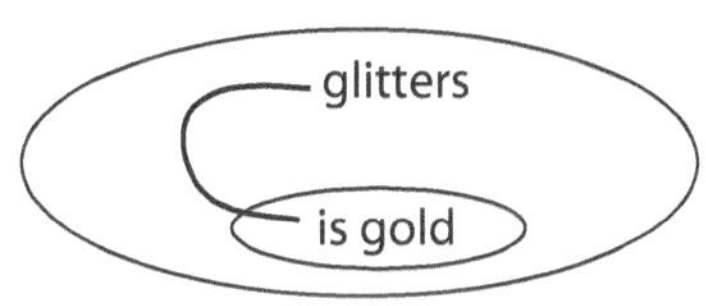

the diagram on the left side has an identity line whose outermost part is enclosed by zero cuts while the outermost part of the identity line is enclosed by one cut in the diagram on the right. Hence, the former says "Something that glitters is gold," and the latter "Everything that glitters is gold."

Two points we would like to make: First, if Peirce had only the endoporeutic reading in mind, there would be no need to make a visual distinction between the outermost part of a line being evenly versus oddly enclosed. The diagram on the right side could be read as "It is not the case something that glitters is not gold."[59] For some reason, Peirce did not push the endoporeutic reading here but suggested to read off a universal statement directly, not in terms of the negation of an existential statement. Second, both existential and universal statements are read off without adding any new syntactic device. Let's recall how the method of multiple readings for Alpha diagrams let us read off disjunctive and conditional information without adding any new syntactic object, but only by paying attention

[57] In Peirce's own linear symbolic notations, Π is adopted for a universal quantifier and Σ for an existential quantifier [Peirce, 1885].

[58] X is evenly (oddly) enclosed if and only X is enclosed by an even (odd) number of cuts.

[59] Zeman pushed the endoporeutic reading in the Beta system as well [Zeman Ch. 2]. For its merits and drawbacks, see [Shin 2002, §5.1.1].

to various visual features present in diagrams. We witness the same idea spelled out by Peirce for the expression of 'everything' and 'something' in Beta diagrams.

When a system does not depend on the linearity of syntactic objects and it is capable of expressing both universal and existential statements directly, the next question is how to solve the scope problem among multiple quantifiers. In the case of a symbolic system, linearity takes care of the scope issue. Hence, a difference between '$\forall x \exists y x < y$' and '$\exists y \forall x x < y$' is formalized in the semantics. Beta diagrams being non-linear, Peirce realized that there should be a different way to solve the problem, and again, he counted on a visual feature as follows: Compare the outermost parts of identity lines. The less enclosed it is, the larger the scope it gets.

We would like to borrow Roberts' following well-cited example to illustrate the scope matter nicely [Roberts, 1973, p. 52]:[60]

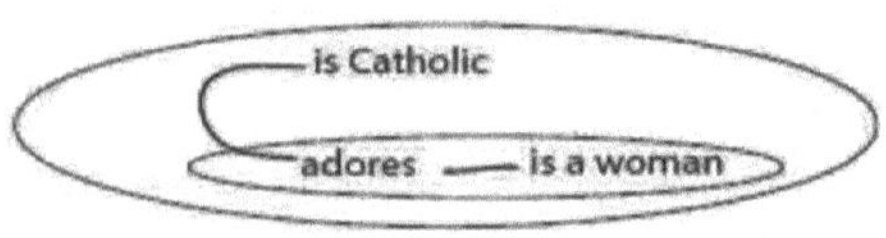

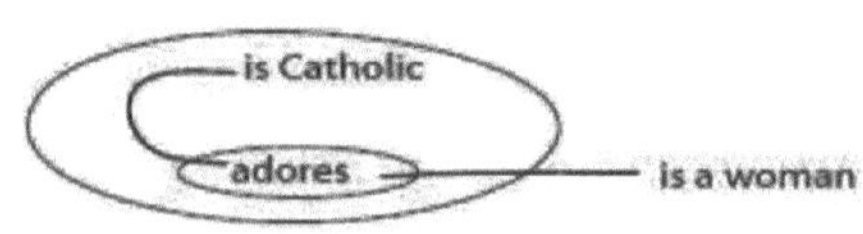

In the graph on the left side, one identity line has its outermost part enclosed by one cut (hence, it is read off as a universal statement) and the other has its outermost part enclosed by two cuts (hence, it is read an existential statement). Therefore, the universal quantifier gets larger scope than the existential quantifier. It says "Every Catholic admires some woman." The following first-order sentence makes things totally non-ambiguous:

$$\forall x\ (Catholic\ (x) \rightarrow \exists y\ [Adores\ (x,y) \wedge\ Woman\ (y)])$$

The order of two quantifiers in the diagram on the right side is opposite. Hence, it says "There is a woman every Catholic admires." That is,

$$\exists y\ (Woman\ (y) \wedge \forall x[Catholic\ (x) \rightarrow Adores\ (x,y)])$$

As our short discussions show, Peirce himself explained much more about how visual features were used in the Beta system than in the Alpha system. After coming up with cuts (as negation) and juxtaposition (as conjunction),[61] Peirce did not dwell on Alpha diagrams much since the system is truth-functionally complete.[62] After all, Peirce's heart was always with the logic of relations. Applying Peirce's main idea about iconicity or visuality to Alpha system, we could see more clearly why Peirce considered EG as his *chef-d'oeuvre*. Let us pause to prepare for the *rendez-vous* which the Frege-Peirce affair has anticipated.

[60]The two graphs are found in [Peirce, 1933, §4.452].

[61]In his first graphical system, Entitative Graph, Peirce took juxtaposition as disjunctive. For a transition from Entitative Graph to EG, see [Shin, 2002, §3.2].

[62]Peirce discussed the pattern [[ ]] (called 'scroll' by Peirce), which expresses conditional information. See [Peirce, 1933, §4.437] and [Shin, 2002, §4.3.1].

5.3 The Frege-Peirce Rendez-vous

After some years' serious discussions and research, the logic community reached the consensus that not Frege alone, but both Frege and Peirce are the founders of modern predicate logic, even though Frege's publication (1879) preceded Peirce's (1885) and even though Peirce's linear symbolic notation, not Frege's two-dimensional notation, was adopted by contemporary logicians. Many scholars raised the question "Why was the Frege-first phenomenon prevalent for a while even though Peirce's notation has been widely accepted and used?" At the beginning of the section, we explored the other side of the story — "Why was Frege's notation ignored even though his CN came out six years earlier than Peirce's paper on predicate logic?" Reviews of CN suggest that the harsh reception had something to do with the two-dimensional aspect of the system. We examined Frege's vehement responses to critics, where Frege made it clear how important and crucial it is for a logical system to be two-dimensional.

Here is an interesting irony from our perspective: When we started the section, the Frege-Peirce affair was mainly about an intricate relation between Frege's CN and Peirce's 1885 symbolic notation in the history of modern logic. However, the critical reception of CN and Frege's defense of his own system have shed a new light on the Frege-Peirce affair. Now we realize that it is Peirce's 1897 EG (not Peirce's 1885 linear representation) which is much closer to Frege's CN in several ways. Both CN and EG are two-dimensional. Hence, both suffered an unkind reception. Our discussions added one item to the list: Both notations may allow the reader to perceive a given diagram in more than one way. Let's compare the sentences (1)–(6) obtained from Frege's diagram and the sentences (7)–(11) from Peirce's diagram. Each group consists of logically equivalent sentences and all of the sentences in each group are the results of reading off one and the same diagram. We would like to call this efficiency. Did Frege or Peirce have efficiency in mind?

According to Frege, logical relations are too diverse to be represented in a linear form. In order to increase the perspicuity of logical relations, Frege chose a form of two-dimensional representation. Peirce's motive for graphic representation is too complicated for us to sum up in this short space, but Peirce firmly believed that iconic representation of relations helps us carry out "fruitful reasoning."[63] We leave this section with a strong speculation that efficiency is one way to cash out Frege's perspicuity and Peirce's fruitfulness. When logicians and mathematicians have become interested in efficiency partly due to the computer age, we should not be surprised to witness the increasing interest in Peirce's EG and the re-evaluation of Frege's CN. We would very much consider this short section as a small step toward the big project to come.

[63]See [Peirce, 1933, §3.457] and [Shin and Hammer, 2011, §3].

6 REVIVAL IN A NEW AGE

The events of the last section took place in the late 19th century, and we confirmed that neither Frege's CN nor Peirce's EG held the spotlight when they first appeared in the world. It is not our sole intention to discuss both systems at length from a historical point of view, but there is also a new demand which urges us to do so. There is no doubt that we witness trends for diagrammatic representation in various disciplines — philosophy, logic, mathematics, computer science, psychology, cognitive science, and so on. Recalling what Frege and Peirce presented more than a century ago, we would like to draw attention to the aspect of revival in current diagrammatic research. Like every revival, it has new features the old achievements did not have, but by focusing on the angle of what has been resurrected, we can shed new light not only on the past but also on what we are currently doing. Then, an immediate questions is "What happened between the Frege-Peirce era and now?", i.e., "What buried their diagrams then and what resurrected them now?"

6.1 Accuracy pursued

How revolutionary was the revolution of logic Frege and Peirce initiated during the third quarter of the 19th century? There would have been no Hilbert's program, no Gödel's theorems, no Löwenhein-Skolem theorem, no non-standard models, etc. An upshot is that mathematical logic, as we know it, would not have existed.[64] All of the interesting theorems and theses in mathematical logic were born after modern logic was founded.

What is the essence of the Frege-Peirce revolution? Many might think it is the formalization that elevated the new logic to a different dimension, compared with Aristotelian or medieval logic. However, the crux of the Frege-Peirce revolution lay in the extension of the territory of logic, from being monadic to relational, which corresponds to an extension from sentential to quantified logic. As we all know, Frege's main goal was to reduce arithmetical truth to logical truth. And the prerequisite for the project was to develop a logical system expressive enough to represent arithmetical definitions and axioms. Hence, quantificational logic was a necessary stepping stone to get to Frege's logicism. On the other hand, Peirce's goal was simply (!) to formalize relations, since early on he realized that the logic of relations is fundamentally different from the logic of non-relatives and Boole's logic is limited to non-relatives.[65]

Not many would object to the way we have identified the novelty of Frege-Peirce's modern logic in the above; that is, *what* to formalize is extended by these two logicians. However, not many have paid attention to another kind of extension the Frege-Peirce revolution showed us, that is, *how* to formalize is extended.

[64] We do not mean to say the revolution would not have taken place without Frege or Peirce or there would be no mathematical logic without either of them, but we are exploring the novelty of their achievements, whoever would have made that possible.

[65] For more details, see [Shin and Hammer, 2011].

Both Frege's CN and Peirce's EG, as seen in the last section, are two-dimensional graphical systems. Hence, a clear departure of modern logic from Boole's logic lies not only in the content of formalization but also in the manner of formalization.

It is quite interesting to realize that the world of logicians and mathematicians immediately appreciated and embraced an expansion of the scope of logic,[66] but not a new mode of representation. Analyzing some of the non-friendly reviews of Frege's *Begriffsschrift*, in the last section we attributed unfamiliarity as a main source of the harsh reception of CN. Reviewers did not see any advantage in CN while it seemed to waste more space and looked more cumbersome. Understanding CN's reception, it should hardly come as a surprise to learn that Peirce's EG was not well welcomed either. However, in the case of Peirce's EG, there is another level of irony. Peirce invented EG as an alternative to his own symbolic linear system, which is similar to Boole's notation.[67] Moreover, Peirce himself strongly preferred his EG over his linear representation, but the rest of the world did not agree with Peirce's evaluation of his own systems. The same reaction as in the case of evaluating Frege's CN, that is, seeing it as a waste of space and cumbersome notation, might have played a role in this case as well. Is that all?

Some of us pointed out there has been a long standing prejudice for symbolic systems and against diagrammatic systems. The prejudice is real, we believe, and it is consistent with the reception of CN and EG we discussed above. But that cannot be the end, but only the beginning, of the story, since the existence of the prejudice is not an explanation but a description of the phenomenon in which we are interested. For almost every prejudice in life, there are some bases, rational or irrational, why and how a given bias has been formed. When we get to understand the rationales behind the birth of a prejudice, we will be in a better position to evaluate the situation so that we may fight against the preconception, if necessary. Below we will unfold a quite involved story of the early 20th century and propose a more comprehensive account of the prevalent preference for symbolic systems which has prevailed since the dawn of modern logic.

There are many ways the preference for symbolic systems has been manifest. As discussed in the previous section, Peirce's linear notation was widely adopted over Frege's two-dimensional notation in spite of Frege's publication being earlier than Peirce's. Between Peirce's two kinds of systems, the symbolic notation was accepted much better than his EG in spite of Peirce's own preference for EG over the other kind of representation. Even though contemporary logicians and mathematicians followed a new larger territory of logic to which both Frege and Peirce led them, they more or less stuck to the existing Boolean style of logical notation. Inertia might have played a role in the initial skepticism toward a new form of notation, but cannot explain several decades' neglect. However, the initial

[66]The remarkable advance made in mathematical logic in the early 20th century is a solid piece of evidence for an enthusiastic reception for modern logic.

[67]Please note the title of Peirce's paper where his first serious attempt for relational symbolic logic appeared — "Description of a notation for the logic of relatives, resulting from an amplification of the conceptions of Boole's calculus." [Peirce, 1870]

response contributed to forming an incorrect equation between formalization and symbolization. This unfortunate identification combined with the following history to create a long standing preference for symbolic systems, we will argue.

To say the least, the turn of the 20th century was quite eventful in the world of mathematics and logic. Thanks to the Frege-Peirce revolution, much energy, talent, and enthusiasm were pouring into this newly charged discipline and remarkable progress was made. Lots of triumphs were celebrated, but not all surprises were good ones, it turned out. Non-Euclidean geometries shook our confidence in the certainty of mathematical truth. Cantor's simple elegant theory contained a contradiction. Frege's logicism was found to be built on this paradox. Hilbert's ambitious program was proven to be a mission not to be completed. Were Gödel's earth-breaking theorems good news or bad news? Either way, they were news which provoked serious discussions of the nature of mathematics and logic. Hence, it is not a coincidence that the early 20th century is marked as an era to explore the philosophical and foundational issues of mathematics.

At the core of the turmoil, certainty as we know it was in danger and at least was seriously examined. In the past it was taken for granted that mathematics is the bedrock of accuracy, but the lesson at the turn of the 20th century was that some serious work is required to guarantee accuracy even in mathematics. Hence, all of those involved sought to secure and justify accuracy for mathematics, even though they differed in how to do so. We believe this special environment pushed the entire discipline to be more obsessed with formalization than ever. A more standardized and mechanical menu would prevent us from committing fallacies in reasoning. So far so good.

Contemporary concept of formalization, however, as we said earlier, was narrow in that it was identified with symbolization. Nothing but formalization, hence, nothing but symbolization, could be taken as an accurate system of representation. The more formalization has been demanded, the more prevalent has been symbolization. At the same time, we would like to point out a rationale behind the identification between symbolization and formalization in the context of the early 20th century.[68] When the pursuit of accuracy was almost the sole agenda, the awareness of possible misuses of diagrams in rigorous reasoning must have driven mathematicians and logicians to avoid using them in a system. They did not pause to question whether it is possible to formalize graphs or diagrams, but ruled them out from the beginning. We all had to wait until the era of frenzy to pursue accuracy was over to realize that there are other desiderata for a formal system and we may obtain some of them without sacrificing accuracy.

[68]We understand some might still believe that symbolization is the only method to formalize a system, and we will get to this issue more deeply later in the section where we discuss more recent works on diagrammatic reasoning.

6.2 *Efficiency revisited*

We claimed that accuracy became the primary concern for the early 20th century research in mathematics and logic. Also, we pointed out the extension of formalization was pretty much limited to symbolic systems. Both the accuracy and the formalization story are consistent with the inventions of Frege's CN and Peirce's EG. First, these two diagrammatic systems were presented before the accuracy issue became intense. Please note that Frege's *Begriffsscrift* came out in 1879 and Peirce's EG appeared in 1897.[69] More importantly, neither Frege nor Peirce limited formalized systems to symbolic ones. That does not mean that their systems do not care about accuracy. Let us ask what else was at stake in their systems in addition to accuracy so that we may connect their insights to the revival of diagrammatic representation in recent years.

In the last section, we identified motivations behind both Frege's and Peirce's non-linear representation of predicate logic. Frege himself emphasized 'perspicuity' and Peirce used the phrase 'fruitful reasoning' to justify and advocate their graphical systems. In other words, Frege aimed to invent an accurate and perspicuous system and Peirce wanted an accurate system that leads us to more productive reasoning. That is, more than one accurate system being given, we may make a comparison among them, Frege and Peirce both seemed to think. A more perspicuous one is better, Frege would say, and a system which leads us to reasoning in a more productive way is a superior one, according to Peirce. These phrases — 'perspicuity' and 'fruitful reasoning' — are not free from ambiguity, and we believe they have too many implications to be exhausted in our short paper, but one obvious point is that both sought a system which achieves something beyond accuracy and both believed that two-dimensionality would help achieve that goal. We claim that this something else both Frege and Peirce wanted to obtain is the *efficiency* of a logical system.

If we need to choose between accuracy and efficiency, of course accuracy should get priority. An accurate system without efficiency is usable, while we cannot imagine an efficient non-accurate system (?). But, do we need to choose one over the other as far as accuracy and efficiency go? Why not both? In the present computer age, efficiency is not just a familiar topic, but has been a key issue. Hence, the point is not to explain why efficiency is important, but to find out how it is achieved. For better or for worse, it is difficult, albeit important, task to define what efficiency is or how to measure efficiency. Our immediate goal is, instead of discussing efficiency in a broader context, to connect half of the Frege-Peirce revolution with recent research on diagrammatic reasoning.

We pointed out the Frege-Peirce revolution consisted of two kinds of expansion of logical systems: One is about the content of formalization, from being monadic to relational, and the other is about the mode of formalization, from being symbolic to diagrammatic. The extension of the mode of formalization, we believe, is

---

[69]Soon Frege himself was found in the middle of the intense debate of the philosophy of mathematics, but Peirce was quite sheltered from that hysteria.

directly related to the efficiency issue, which did not receive attention from contemporaries, and not until recently. Then, in what sense are both Frege's CN and Peirce's EG (which are diagrammatic) efficient? First, efficiency is a relative concept such that we would like to say system A is more efficient than system B, rather than saying system A is efficient. Second, efficiency is a context-sensitive concept such that we would like to say system A is more efficient than system B in doing X. Of course, that does not mean that we cannot come up with an overall definition (or something close to it) of efficiency, but again, that is not the purpose of our current discussion. Rather we aim to relate our discussions of Frege's CN and Peirce's EG in the previous section with this relative context-sensitive concept, efficiency.

Frege's 'perspicuity' of his CN is taken in the following way: When we adopt two-dimensional representation of logical form, we see more clearly, i.e. more perspicuously, logical relations. We illustrated his main idea by presenting a case study through [Fig. 67]. Spatial representation gave us freedom in how to read off one and the same diagram so that we may obtain different but logically equivalent sentences. This is clearly more efficient than dealing with multiple sentences as far as logical consequence or equivalence goes. Peirce's EG also has several aspects of efficiency. By adopting Shin's multiple reading algorithm, we concluded that logical equivalence among sentences is much more efficiently obtained than in symbolic systems. Peirce himself facilitated visual features of graphs so that EG does not have as many as syntactic devices as the corresponding symbolic systems. For example, Peirce noted that there is no need to introduce diagrammatic objects which represent universal and existential quantifiers, but all that is needed is to direct the reader to different visual features, that is, whether a line started (or ended) in an area enclosed by an even or odd number of cuts. If a system allows us to directly read off existential and universal statements, but without necessarily having two kinds of syntactic vocabulary, that could only add to the efficiency of the system. As far as the matter of scope goes, Peirce appealed to our intuitive way to regard scope: The less the outermost part is enclosed the larger the scope it gets. When we expand linear to spatial representation, we find that more natural and more intuitive readings become available. We believe all of these elements are boosting the efficiency of the system.

6.3 Multi-modal reasoning and logic diagrams

For several reasons we do not even attempt to cover the ground of various recent diagrammatic reasoning projects. First, as we will briefly mention below, diagrammatic reasoning projects have been approached from various disciplines with different purposes, and many of their agendas are beyond the scope of our topic, logic diagrams. Second, like other on-going projects, it is a difficult and delicate task to give an objective view in a small space. Third, and most importantly, a simple survey of the project could be too superficial and too shallow to be of any use, especially when the reader could obtain scattered, but more plentiful online information on the topic. Instead, we decided to explore a major branch of the diagrammatic reasoning project called "Heterogeneous Logic," which is about logic diagrams and at the same time shares an underlying theme with the tradition of logic and representation systems — especially in light of our views about the Frege-Peirce revolution.

The discipline 'Artificial Intelligence' was baptized in the middle of the 20th century by Turing Award recipient and computer scientist John McCarthy. As the phrase suggests, it is a research field to engineer an intelligent machine. Depending on which aspects of intelligence we would like to simulate, we have different machines, but how to present knowledge and how to implement reasoning constitute important parts of the task. Hence, logic is an essential component of building intelligent devices. Facing the limits of standard first-order logic, it did not take much time for researchers to go beyond it; hence, we have seen a sudden burst of interest in non-monotonic, modal, tense logic, etc. since Artificial Intelligence took off. These extensions are, however, limited to the homogenous mode, that is, symbolic representation. Then, how about bringing in different modes of representation so that we may allow multi-modal, heterogeneous, reasoning, as natural intelligence does all the time?[70]

Toward the end of the 1980's two Stanford logicians, Jon Barwise and John Etchemendy, launched the project "Heterogeneous Reasoning."[71] Their enterprise started with a simple observation and a simple idea: We human beings use various kinds of media to carry out reasoning, that is, sentences, maps, pictures, diagrams, charts, sounds, smells, etc., while the discipline logic was exclusively occupied with symbolic systems. What explains this discrepancy between our ordinary reasoning and formal logic as we know it? Both the theoretical implications and the practical applications of the project are immense and quite complicated. At a theoretical level, the heterogeneous reasoning project was challenging the traditional concept and territory of logic, by providing a new mathematical framework[72] and by opening up a philosophical debate. At its practical side, Barwise and Etchemendy have been developing computer software to facilitate our multi-modal reasoning and some of their programs have become the most popular tools for teaching logic at universities throughout the world. In the following two paragraphs, we will respectively outline how each aspect of the research has been developed.

Can we develop logic out of non-symbolic expressions? Are non-symbolic systems possible? If not, what in that case would be special about symbolic representation? Shin took up one of the simplest and the most popular uses of diagrams, Venn diagrams, as a case study to explore these questions [Shin 1994]. She first set up the syntax and the semantics of Venn diagrams. The syntax takes care of the definition of well-formed diagrams and the set of transformation rules, and

[70]Whether artificial intelligence could be better off by imitating natural intelligence has been one of the points of debate. However, the heterogeneous reasoning research does not have to be limited to the context of artificial intelligence, but it is one way to cast the project in a bigger context.

[71]A brief summary of the project is found in [Shin, 2004].

[72]See [Barwise and Etchemendy, 1991; 1995] and [Barwise and Seligman, 1997].

the semantics tells us what a given diagram means. Then, we can check whether each transformation rule is valid (so that we obtain only logical consequences), and whether we have a sufficient number of rules to obtain all of the logical consequences. That is, Venn diagrams constitute a logical system which can be proven to be sound and complete. When the Venn system was acknowledged, the extension of the concept of formalization was not limited to symbolic systems only, but got expanded to diagrammatic systems. Hence, this case study not only confirms the initial speculation of the heterogeneous reasoning project that in principle we may formalize reasoning close to our ordinary natural reasoning (let us call it 'heterogeneous logic') but also directs us to the neglected part of the Frege-Peirce revolution, that is, a new way of formalization (let us call it 'diagrammatic formalization'). Interesting and important works on various diagrammatic representations have been produced along these lines [Sowa, 1984; Harel, 1988; Glasgow *et al.*, 1995; Hammer, 1995; Allwein and Barwise, 1996; Shin and Lemon, 2008]. Both heterogeneous logic and diagrammatic formalization will help us revisit time-honored topics like logical consequences, logical form, logical concepts, etc., since traditional treatments of the topics were tied up with one mode of representation — symbolic representation [Etchemendy, 1999].

As for practical visualization, the team led by Barwise and Etchemendy applied the spirit of heterogeneous logic to produce remarkable visual software for teaching logic. After more than a decade's trial, their logic text book *Language, Proof, and Logic* uses the software *Tarski's World* which presents a block world so that sentential information and visual information could be combined in a natural way as we reason in ordinary life [Barwise and Etchemendy, 2002]. Many logic teachers expressed their high opinion about the crucial role of this visual computer program: Students are comprehending major logical concepts more easily mainly because multi-modal reasoning is much closer to home than the traditional way of teaching those concepts — using only symbolic representation. This welcoming reception supports our view that accuracy and efficiency do not have to be an exclusive-or choice, but compatible with each other, and at the same time it shows how important it is to embrace various modes of representation.

Independently of the heterogeneous reasoning project, as an important revival of logic diagrams, we would like to draw attention to John Sowa's conceptual graphs [Sowa, 1984]. Sowa, a computer scientist, was on the frontier of reviving Peirce's EG, which supports our view about the demand for efficiency and the revival of graphical systems. When diagrams become legitimate logical vocabulary, it did not escape computer scientists' attention. The Visual Modeling Group led by John Howse is an excellent research center which pursues visual logic from computer scientists' point of view and the group has produced much interesting and important work [Howse *et al.*, 2001].[73]

If we broaden our scope to embrace visual reasoning in general, we find more disciplines involved in the topic. The role of diagrams and images in our cognition and mental representation had been debated in cognitive science and philosophy

[73]See the center's website at: http://www.cmis.brighton.ac.uk/research/vmg

even before logic diagrams were on the research map. The time-honored imagery debate in psychology is a prime example [Kosslyn, 1980; Pylyshyn, 1981; Block, 1983; Tye, 1991], and the philosophy of mind has an even longer history in exploring the status of visual imagery in mental discourse [Cummins, 1996]. While both psychology and philosophy of mind more or less focus on diagrams/pictures as mental images [Shepard and Metzler, 1971; Simon 1978; Johnson-Laird, 1983], some philosophers have examined a role of diagrams (as external representation) in the practice of mathematics and have found interesting data throughout the history of mathematics [Mumma, 2006; Miller, 2007; Giaquinto, 2007; Mancosu, 2008].

Diagrams, being widely used in almost every corner of life, have been not only revived in a new era which has new demands from new technology, but they also connect us with a forgotten part of history. The subject, interestingly, has led many of us to an interdisciplinary arena where mathematics, psychology, philosophy, literature, and art find something in common. We do not claim all of these diagram-related research projects are about logic diagrams, but like any other interdisciplinary research, we do not doubt that different perspectives could shed a new and creative light on the topic. We only wish we had knowledge and space to accommodate a bigger picture (!) of the current story.

ACKNOWLEDGMENTS

This paper benefitted from enjoyable and instructive discussions that we shared with several friends and colleagues to whom we would like here to express our thanks and dept: Francine Abeles, Catherine Allamel-Raffin, Jean-Yves Béziau, Anthony Edwards, Valeria Giardino, Sybille Krämer, Jørgen Fischer Nilsson, Fabien Schang, Frederik Stjernfelt and Jan Wöpking. We would like to thank particularly Prof. John Woods and Jane Spurr for their patience and support all along the preparation of this chapter.

BIBLIOGRAPHY

[Abeles, 1990] F. F. Abeles. Lewis Carroll's method of trees: its origin in *Studies in logic*. *Modern Logic*, 1 (1), 25-35, 1990.

[Abeles, 2005] F. F. Abeles. Lewis Carroll's formal logic. *History and Philosophy of Logic*, 26 (1), 33-46, 2005.

[Abeles, 2007] F. F. Abeles. Lewis Carroll's visual logic. *History and Philosophy of Logic*, 28 (1), 1-17, 2007.

[Abeles, 2010] F. F. Abeles, ed. *The Logic Pamphlets of Charles Lutwidge Dodgson and Related Pieces*, New York: LCSNA – Charlottesville and London: University Press of Virginia, 2010.

[Allwein and Barwise, 2006] G. Allwein and J. Barwise, eds. *Logical Reasoning with Diagrams*, New York - Oxford: Oxford University Press, 1996.

[Anderson and Cleaver, 1965] D. E. Anderson and F. L. Cleaver. Venn-type diagrams for arguments of N terms. *Journal of Symbolic Logic*, 30(2), 113-118, 1965.

[Anellis, 1990] I. H. Anellis. From semantic tableaux to Smullyan trees: a history of the development of the falsifiability tree method. *Modern Logic*, 1 (1), 36-69, 1990.

[Anellis, 2004] I. H. Anellis. The genesis of truth-table device. *Russell: the Journal of Bertrand Russell Studies*, 24 (1), 55-70, 2004.

[Baron, 1969] M. E. Baron. A note on the historical development of logic diagrams: Leibniz, Euler and Venn. *Mathematical Gazette*, 53 (384), 113-125, 1969.

[Bartley, 1986] W. W. Bartley III, ed. *Lewis Carroll's Symbolic Logic,* New York: Clarkson N. Potter, 1986 (1st ed., 1977).

[Barwise and Etchemendy, 1991] J. Barwise and J. Etchemendy. Visual information and valid reasoning. In W. Zimmerman and S. Cunningham, eds., *Visualization in Teaching and Learning Mathematics*, pp. 9–24. Washington, DC: Mathematical Association of America, 1991.

[Barwise and Etchemendy, 1995] J. Barwise and J. Etchemendy. Heterogeneous reasoning. In [Glasgow *et al.*, 1995, pp. 211-234].

[Barwise and Etchemendy, 2002] J. Barwise and J. Etchemendy. *Language, Proof, and Logic*, Stanford, CA.: CSLI Publications, 2002.

[Barwise and Seligman, 1997] J. Barwise and J. Seligman. *Information Flow: the Logic of Distributed Systems*, Cambridge – New York: Cambridge University Press, 1997.

[Bergmann, 1879] J. Bergmann. *Allgemeine Logic*, Berlin: Mittler, 1879.

[Béziau, 2012] J.-Y. Béziau. The power of the hexagon. *Logica Universalis*, 6 (1-2), 1–43, 2012.

[Béziau and Jacquette, 202] J.-Y. Béziau and D. Jacquette. *Around and Beyond the Square of Opposition*, Basel: Birkhäuser (Springer), 2012.

[Block, 1983] N. Block.'Mental pictures and cognitive science. *Philosophical Review*, 92, 499-541, 1983.

[Boole, 1847] G. Boole. *The Mathematical Analysis of Logic.* Cambridge: Macmillan, Barclay, and Macmillan - London: George Bell, 1847.

[Boole, 1854] G. Boole. *An Investigation of the Laws of Thought.* London: Walton and Maberly - Cambridge: Macmillan, 1854.

[Bowles, 1971] L. J. Bowles. Logic diagrams for up to n classes. *Mathematical Gazette*, 55(394), 370-373, 1971.

[Bynum, 1972] T. W. Bynum, ed. *Frege: Conceptual Notation and Related Articles.* Oxford: Clarendon Press, 1972.

[Carroll, 1886] L. Carroll. *The Game of Logic*,. London: Macmillan, 1886 (new ed., 1887).

[Carroll, 1897] L. Carroll. *Symbolic Logic. Part I: Elementary.* 4th ed., London: Macmillan, 1897 (1st ed., 1896)

[Coumet, 1977] E. Coumet. Sur l'histoire des diagrammes logiques, 'figures géométriques'. *Mathématiques et Sciences Humaines,* 15 (60), 31-62, 1977.

[Couturat, 1901] L. Couturat. *La Logique de Leibniz.* Paris : Felix Alcan, 1901.

[Couturat, 1903] L. Couturat. *Opuscules et Fragments Inédits de Leibniz.* Paris : Felix Alcan, 1903.

[Couturat, 1914] L. Couturat. *The Algebra of Logic.* Chicago - London: Open Court, 1914.

[Cummins, 1996] R. Cummins.*Representation, Targets, and Attitudes.* Cambridge, Mass.: MIT Press, 1996.

[Davenport, 1952] C. K. Davenport. The role of graphical methods in the history of logic. *Methodos*, 4, 145-162, 1952.

[De Morgan, 1966] A. De Morgan. *On the Syllogism: and Other Logical Writings*, edited with an introduction by Peter Heath. London: Routledge and Kegan Paul, 1966.

[Edwards, 1989] A. W. F. Edwards. Venn diagrams for many sets. *New Scientist*, 121 (1646), 51-56, 1989.

[Edwards, 2004] A. W. F. Edwards. *Cogwheels of the Mind: the Story of Venn Diagrams.* Baltimore, Maryland: Johns Hopkins university press, 2004.

[Edwards, 2006] A. W. F. Edwards. An eleventh-century Venn diagram. *BSHM Bulletin*, 21 (2), 119-121, 2006.

[Englebretsen, 1998] G. Englebretsen. *Line Diagrams for Logic.* New York: Edwin Mellen Press, 1998.

[Etchemendy, 1999] J. Etchemendy. Reflection on consequence. Manuscript, 1999.

[Euler, 1768] L. Euler. *Lettres à une Princesse d'Allemagne*, vol. 2. Saint Petersburg : Imprimerie de Académie Impériale des Sciences, 1768.

[Euler, 1833] L. Euler. *Letters of Euler on Different Subjects in Natural Philosophy Addressed to a German Princess*, vol. 1. New York: J. and J. Harper, 1833.

[Faris, 1955] J. A. Faris. The Gergonne relations. *Journal of Symbolic Logic*, 20 (3), 207-231, 1955.

[Frege, 1879] G. Frege. *Begriffsschrift.* Halle: Louis Nebert, 1879; translated in [Bynum, 1972, pp. 109–203].

[Frege, 1882a] G. Frege. Über die wissenschaftliche berechtigung einer begriffsschrift. *Zeitschrift für Philosophie und philosophische Kritik*, 81, 48-56, 1882; translated in [Bynum, 1972, pp. 83–89].

[Frege, 1882b] G. Frege. Über den Zweck der Begriffsschrift. *Jenaische Zeitschrift für Naturwissenschaft*, 16, Supplement, 1-10, 1882-1883. (Lecture at the January 27, 1882 meeting of Jena's Society for Medicine and Natural Science); translated in [Bynum, 1972, pp. 90–100].

[Garden, 1867] F. Garden. *An Outline of Logic.* London: Rivingtons, 1867.

[Gardner, 1958] M. Gardner. *Logic Machines and Diagrams.* New York – Toronto – London: McGraw-Hill Book Company, 1958. (2nd ed., 1983).

[Gergonne, 1817] J. D. Gergonne. Essai de dialectique rationnelle. *Annales de Mathématiques Pures et Appliquées*, 7 (7), 189-228, 1817.

[Giaquinto, 2007] M. Giaquinto. *Visual Thinking in Mathematics.* Oxford: Oxford University Press, 2007

[Glasgow *et al.*, 1995] J. Glasgow, N. H. Narayanan and B. Chandrasekaran, eds. *Diagrammatic Reasoning: Cognitive and Computational Perspectives.* Menlo Park, Calif.: AAAI Press - Cambridge, Mass.: MIT Press, 1995.

[Goldfarb, 1979] W. Goldfarb. Logic in the twenties: The nature of the quantifier. *Journal of Symbolic Logic*, 44 (3), 351-368, 1979.

[Grattan-Guinness, 1977] I. Grattan-Guinness. The Gergonne relations and the intuitive use of Euler and Venn diagrams. *International Journal of Mathematical Education in Science and Technology*, 8 (1), 23-30, 1977.

[Grattan-Guinness, 2005] I. Grattan-Guinness. Comments on Stevens' review of *The Cambridge Compagnon* and Anellis on truth tables. *Russell: the Journal of Bertrand Russell Studies*, 24 (2), 185-188, 2004–2005.

[Greaves, 2002] M. Greaves. *The Philosophical Status of Diagrams.* Stanford, California: Center for the Study of Language and Information (CSLI) publications, 2002.

[Green, 1991] J. Green. The problem of elimination in the algebra of logic. In T. Drucker, ed., *Perspectives in the History of Mathematical Logic*, pp. 1–9. Boston–Basel–Berlin: Birkhäuser, 1991.

[Grünbaum, 1975] B. Grünbaum. Venn diagrams and independant families of sets. *Mathematics Magazine*, 48 (1), 12-23, 1975.

[Hacking, 2007] I. Hacking. Trees of logic, trees of porphyry. In J. L. Heilbron, ed., *Advancements of Learning: Essays in honour of Paolo Rossi*, pp. 221–163. Firenze: L. S. Olschki, 2007.

[Hamburger and Pippert, 2000] P. Hamburger and R. E. Pippert. Venn said it couldn't be done. *Mathematics Magazine*, 73 (2), 105-110, 2000.

[Hamilton, 1871] Sir W. Hamilton. *Lectures on Logic.* Boston: Gould and Lincoln, 1871.

[Hammer, 1995] E. Hammer. *Logic and Visual Information.* Stanford, CA: Center for the Study of Language and Information, 1995.

[Hammer and Shin, 1998] E. Hammer and S.-J. Shin. Euler's visual logic *History and Philosophy of Logic*, 19 (1), 1-29, 1998.

[Harel, 1988] D. Harel. On visual formalisms. *Communications of the ACM*, 13(5), 514 530, 1988.

[Hawley, 1896] T. D. Hawley. *Infallible Logic.* Lansing, Michigan: Robert Smith, 1896.

[Henderson, 1963] D. W. Henderson. Venn diagrams for more than four classes. *American Mathematical Monthly*, 70 (4), 424-426, 1963.

[Hintikka, 1990] J. Hintikka. Quine as a member of the tradition of the universality of language. In R. Barret and R. Gibson, eda., *Perspective on Quine*, pp. 159–174. Oxford: Blackwell, 1990.

[Hintikka, 1997] J. Hintikka. The place of C.S. Peirce in the history of logical theory. In J. Brunning and P. Forster, eds., *The Rule of Reason*, pp. 13–33. Toronto: University of Toronto Press, 1997.

[Hocking, 1909] W. E. Hocking. Two extensions of the use of graphs in elementary logic. *University of California Publications in Philosophy*, 2 (2), 31-44, 1909.

[Holman, 1892] H. Holman. *Questions on Logic: Part I.* London: W. B. Clive and Co., 1892.

[Howse *et al.*, 2001] J. Howse, F. Molina, J. Taylor, S. Kent and J. Gil. Spider diagrams: a diagrammatic reasoning system. *Journal of Visual Languages and Computing*, 12 (3), 299-324, 2001.

[Jevons, 1872] W. S. Jevons. *Elementary Lessons in Logic*, 3rd ed. London: Macmillan, 1872.

[Jevons, 1883] W. S. Jevons. *The Principles of Science*, London: Macmillan, 1883 (1st ed., 1874).

[Johnshon-Laird, 1983] P. Johnshon-Laird. *Mental Models: Towards a Cognitive Science of Language, Inference and Consciousness*. Cambridge, Mass.: Harvard University Press, 1983.

[Johnson, 1921] W. E. Johnson. *Logic: Part I*. Cambridge: University Press, 1921.

[Jones, 1905] E. E. C. Jones. *A Primer of Logic*. London: John Murray, 1905.

[Karnaugh, 1953] M. Karnaugh. The map method for synthesis of combinational logic circuits. *Transactions of the American Institute of Electrical Engineers, Part 1*, 72, 593-599, 1953.

[Keynes, 1887] J. N. Keynes. *Studies and Exercises in Formal Logic*, 2nd ed. London: Macmillan, 1887 (1st ed., 1884).

[Keynes, 1894] J. N. Keynes. *Studies and Exercises in Formal Logic*, 3rd ed. London: Macmillan, 1894

[Keynes, 1906] J. N. Keynes. *Studies and Exercises in Formal Logic*, 4th ed. London: Macmillan, 1906

[Kosslyn, 1980] S. Kosslyn. *Image and Mind*. Cambridge, MA: Harvard University Press, 1980.

[Lambert, 1764] J. H. Lambert. *Neues Organon*. Leibzig: Johann Wendler, 1764.

[Lewis, 1918] C. I. Lewis. *A Survey of Symbolic Logic*. Berkeley: University of California Press, 1918.

[Macbeth, 2005] D. Macbeth. *Frege's Logic*. Cambridge, MA: Harvard University Press, 2005.

[MacColl, 1880] H. MacColl. On the diagrammatic and mechanical representation of propositions and reasonings. *Philosophical Magazine*, 10, 168-171, 1880.

[Macfarlane, 1879] A. Macfarlane. *Principles of the Algebra of Logic*. Edinburgh: David Douglas, 1879.

[Macfarlane, 1881] A. Macfarlane. Review of J. Venn's *Symbolic Logic* (London: Macmillan, 1881). *Philosophical Magazine*, 12 (72), 61-64, 1881.

[Macfarlane, 1885] A. Macfarlane. The logical spectrum. *Philosophical Magazine,* 19, 286-290, 1885.

[Macfarlane, 1890] A. Macfarlane. Application of the method of the logical spectrum to Boole's problem. *Proceedings of the American Association for Advancement of Science*, 39, 57-60, 1890.

[Mac Queen, 1967] G. Mac Queen. *The Logic Diagram*. MA Thesis, McMaster University, 1967.

[Macula, 1995] A. J. Macula. Lewis Carroll and the enumeration of minimal covers. *Mathematics Magazine*, 68(4), 269-274, 1995.

[Mancosu, 2008] P. Mancosu. *The Philosophy of Mathematical Practice*. Oxford – New York: Oxford University Press, 2008.

[Marquand, 1881] A. Marquand. Logical diagrams for n terms. *Philosophical Magazine*, 12, 266-270, 1881.

[Michaëlis, 1880] C. T. Michaëlis. Review of G. Frege's *Begriffsschrift* (Halle: Louis Nebert, 1879). *Zeitschrfit für Völkerpsychlogie und Sprachwissenscharft*, 12, 232-240, 1880; translated in [Bynum, 1972, pp. 212–218].

[Miller, 2007] N. Miller. *Euclid and his Twentieth Century Rivals: Diagrams in the Logic of Euclidean Geometry*. Stanford, CA.: Center for the Study of Language and Information, 2007.

[Moktefi, 2008] A. Moktefi. Lewis Carroll's logic. In D. M. Gabbay and J. Woods, eds., *The Handbook of the History of Logic*, vol. 4: *British Logic in the Nineteenth-Century*, pp. 457–505. Amsterdam: North-Holland (Elsevier), 2008.

[Moktefi, 2010] A. Moktefi. La face cachée des diagrammes d'Euler. *Visible*, 7, 149-157, 2010 (erratum in : *Visible*, 8, 233, 2011).

[Moktefi and Edwards, 2011] A. Moktefi and A. W. F. Edwards. One more class: Martin Gardner and logic diagrams. In M. Burstein, ed., *A Bouquet for the Gardener*, pp. 160-174. New-York: LCSNA, 2011.

[More, 1959] T. More Jr. On the construction of Venn diagrams. *Journal of Symbolic Logic*, 24 (4), 303-304, 1959.

[Moretti, 2009] A. Moretti. The geometry of oppositions and the opposition of logic to it. In: U. Savardi, ed., *The Perception and Cognition of Contraries*, pp. 29–60. Milano: McGraw-Hill, 2009.

[Mumma, 2006] J. Mumma. *Intuition Formalized: Ancient and Modern Methods of Proof in Elementary Geometry.* PhD dissertation, Carnegie Mellon University, 2006.

[Nadler, 1962] M. Nadler. *Topics in Engineering Logic.* Oxford: Pergamon Press, 1962.

[Newlin, 1906] W. J. Newlin. A new logical diagram. *Journal of Philosophy, Psychology and Scientific Methods*, 3 (20), 539-545, 1906.

[Parsons, 2006] T. Parsons. The traditional square of opposition. *Stanford Encyclopedia of Philosophy*, 2006. `http://plato.stanford.edu/entries/square/`

[Peckhaus, 2005] V. Peckhaus. Algèbre de la logique, théorie de la quantification et carré des oppositions. In P. Joray, ed., *La Quantification dans la Logique Moderne*, pp. 53–71. Paris : L'Harmattan, 2005.

[Peirce, 1870] C. S. Peirce. Description of a notation for the logic of relatives, resulting from an amplification of the conceptions of Boole's calculus. *Memoirs of the American Academy*, 9, 317-378, 1870.

[Peirce, 1883] C. S. Peirce, ed. *Studies in Logic.* Boston: Little, Brown, and Company, 1883.

[Peirce, 1885] C. S. Peirce. On the algebra of logic: A contribution to the philosophy of notation. *American Journal of Mathematics*, 7 (2), 180-202, 1885.

[Peirce, 1933] C. S. Peirce. *Collected Papers*, vols. 3 and 4. Cambridge, MA: Harvard University Press, 1933.

[Putnam, 1982] H. Putnam. Peirce as logician. *Historia Mathematica*, 9, 290-301, 1982; (reprinted in H. Putnam, *Realism with a Human Face*, pp. 252–260. Cambridge, MA.: Harvard University Press, 1990).

[Pylyshyn, 1981] Z. W. Pylyshyn. The imagery debate. analogue media versus tacit knowledge. *Psychological Review*, 88 (1), 16-45, 1981.

[Quine, 1977] Willard Van Orman Quine. The algebra of attributes. *The Times Literary Supplement*, 3937, pp. 1018–1019, 1977.

[Roberts, 1973] D. Roberts. *The Existential Graphs of Charles S. Peirce.* The Hague: Mouton, 1973.

[Schwenk, 1984] A. J. Schwenk. Venn diagrams for five sets", *Mathematics Magazine*, 57(5), 297, 1984.

[Shepard and Mezler, 1971] R. N. Shepard and J. Metzler. Mental rotation of three-dimensional objects. *Science*, 171 (3972), 701-703, 1971.

[Shin, 1994] S.-J. Shin. *The Logical Status of Diagrams.* New York: Cambridge University Press, 1994.

[Shin, 2002] S.-J. Shin. *The Iconic Logic of Peirce's Graphs.* Cambridge, MA.: MIT Press, 2002.

[Shin, 2004] S.-J. Shin. Heterogeneous reasoning and its logic. *Bulletin of Symbolic Logic*, 10 (1), 86-106, 2004.

[Shin and Lemon, 2008] S.-J. Shin and O. Lemon. Diagrams. *Stanford Encyclopedia of Philosophy*, 2008, `http://plato.stanford.edu/entries/diagrams/`

[Shin and Hammer, 2011] S.-J. Shin and E. Hammer. Pierce's Logic. *Stanford Encyclopedia of Philosophy*, 2011, `http://plato.stanford.edu/entries/peirce-logic`

[Shosky, 1997] J. Shosky. Russell's use of truth tables. *Russell: the Journal of Bertrand Russell Studies*, 17 (1), 11-26, 1997.

[Schröder, 1880] E. Schröder. Review of G. Frege's *Begriffsschrift* (Halle: Louis Nebert, 1879). *Zeitschrift für Mathematik und Physik*, 25, 81-94, 1880; translated in [Bynum, 1972, pp. 218–232].

[Sidgwick, 1887] A. Sidgwick. Review of L. Carroll's *Game of Logic* (London: Macmillan, 1887). *Nature*, 36 (914), 3-4, 1887.

[Simon, 1978] H. A. Simon. On the forms of mental representation. In C. W. Savage, ed., *Perception and Cognitive: Issues in the Foundation of Psychology*, pp. 3–18. Minneapolis: University of Minnesota Press, 1978. (Series *Minnesota Studies in the Philosophy of Science*, 9)

[Swinburne, 1887] A. J. Swinburne. *Picture Logic*, 5th ed. London: Longmans, Green, and Co., 1887.

[Thomson, 1849] W. Thomson. *An Outline of the Necessary Laws of Thought*, 2nd ed. London: William Pickering – Oxford: W. Graham, 1849.

[Tye, 1991] M. Tye. *The Imagery Debate.* Cambridge, MA.: The MIT Press, 1991

[Ueberweg, 1871] F. Ueberweg. *System of Logic and History of Logical Doctrines.* London: Longmans, Green and Co., 1871

[Van Evra, 2000] J. Van Evra. The development of logic as reflected in the fate of the syllogism 1600-1900. *History and Philosophy of Logic*, 21, 115-134, 2000.

[Van Heijenoort, 1967] J. van Heijenoort. Logic as calculus and logic as language. *Synthese*, 17, 324-330, 1967.

[Veitch, 1952] E. W. Veitch. A chart method for simplifying truth functions. *Proceedings of the Association for Computing Machinery*, pp. 127–133, 1952.

[Venn, 1880a] J. Venn. On the diagrammatic and mechanical representation of propositions and reasonings. *Philosophical Magazine,* 10 (59), 1-18, 1880.

[Venn, 1880b] J. Venn. On the employment of geometrical diagrams for the sensible representation of logical propositions. *Proceedings of the Cambridge Philosophical Society*, 4, pp. 47–59, 1880.

[Venn, 1881] J. Venn. *Symbolic Logic*, 1st ed. London: Macmillan, 1881.

[Venn, 1883] J. Venn. Review of C. S. Peirce's *Studies in Logic* (Boston: Little and Brown, 1883). *Mind*, 8 (32), 594-603, 1883.

[Venn, 1887] J. Venn. The game of logic. *Nature*, 36 (916), 53-54, 1887.

[Venn, 1894] J. Venn. *Symbolic Logic*, 2nd ed. London: Macmillan, 1894.

[Welton, 1891] J. Welton. *A Manual of Logic*, vol. 1. London: W. B. Clive and Co., 1891.

[Zeman, 1964] J. Zeman. *The Graphical Logic of C. S. Peirce.* PhD dissertation, University of Chicago, 1964.

INDEX

Kolibri 360 GmbH
Ferdinand-Jühlke-Straße 7
99095 Erfurt, Deutschland
produktsicherheit@kolibri360.de